Olefin Metathesis and Metathesis Polymerization

To the soldiers of metathesis

Olefin Metathesis and Metathesis Polymerization

K. J. IVIN
Emeritus Professor of Chemistry,
The Queen's University of Belfast

J.C. MOL
Faculty of Chemistry,
University of Amsterdam

ACADEMIC PRESS
San Diego London Boston
New York Sydney Tokyo Toronto

This book is printed on acid-free paper

Academic Press, Inc.
525 B Street, Suite 1900, San Diego, California 92101-4495, USA
http://www.apnet.com/

Academic Press Limited
24–28 Oval Road, London NW1 7DX, UK
http://www.hbuk.co.uk/ap/

ISBN 0-12-377045-9

A catalogue record for this book is available from the British Library

Typeset by Tradespools Ltd, Graphic House, Frome, Somerset
Printed and bound in the United Kingdom
Transfered to Digital Printing, 2011

Contents

Preface

It is 13 years since the publication of *Olefin Metathesis* by one of us. Since then the literature on the subject has more than doubled and considerable advances have been made. In particular, the preparation of well-defined metal carbene complexes of tungsten, molybdenum, and more recently ruthenium, has led first, to a much better understanding of the mechanism; second, to the preparation of a new family of block copolymers, including semi-conducting polymers, through the use of 'living' systems; and third, to an explosion of applications to the synthesis of biologically active compounds, particularly by the ring-closing metathesis of dienes, enynes, and dienynes. Furthermore, highly active catalysts of good stability are now available that can tolerate all kinds of functional groups and give good yields of products with high selectivity.

Numerous friends and colleagues in the field of metathesis (the 'soldiers' to whom we dedicate this book) have encouraged us to believe that a new book incorporating these recent developments would be both timely and welcome. We felt, however, that the book should still outline the historical development of the subject and not just be a supplement to the original book. This has necessarily meant some compression of earlier material and some restriction of discussion. The title has been expanded to include the words 'Metathesis Polymerization', which embraces not only ring-opening metathesis polymerization (ROMP), but also the metathesis condensation reactions of acyclic dienes (ADMET) and the addition reactions of acetylenes. The division of the material and the subjects of the chapters follow the same pattern as before. The literature has been covered up to mid-1996.

We thank all those who have helped us in the preparation of this book, in particular our friends in the Universities of Durham and Belfast; also Marco de Wit for assistance in the gathering of the literature, Leslie Taylor for the preparation of part of the manuscript, and Christopher Ivin for guiding one of us into the world of computer technology. Thanks are also due to Professor Lajos Bencze for permission to reproduce his cartoon of the 'Olefin Chopper', first seen at the 2nd International Symposium on Olefin Metathesis (ISOM 2) held near Amsterdam in 1977.

Olefin metathesis and metathesis polymerization covers a broad field, of interest to chemists of all kinds, and to material physicists. We hope that readers of all backgrounds will find something of value in this book, and that it may serve both as a link to the past and a guide to the future.

Hans Mol, Amsterdam

Ken Ivin, South Petherton
September 1996

Abbreviations

ADMET	acyclic diene metathesis
9-BBN	9-borabicyclononane
Boc	*t*-butoxycarbonyl
COD	cycloocta-1,5-diene
COSY	correlated spectroscopy
Cp	cyclopentadienyl
Cp*	pentamethylcyclopentadienyl
CTA	chain transfer agent
Cy	cyclohexyl
DCPD	dicyclopentadiene
DEPT	distortionless enhancement by polarization transfer
DME (dme)	1,2-dimethoxyethane
dmpe	1,2-bis(dimethylphosphinoethane)
DMSO	dimethyl sulfoxide
DP	average degree of polymerization $= M_n/M_1$
DRS	diffuse reflectance spectroscopy
DSC	differential scanning calorimetry
ESR	electron spin resonance
EXAFS	extended X-ray absorption fine structure
FABMS	fast-atom bombardment – mass spectrometry
FTIR	Fourier-transform infrared
GC	gas chromatography
GPC	gel permeation chromatography (size exclusion chromatography)
HH, HT, TT	head–head, head–tail, tail–tail
HPLC	high-performance liquid chromatography
I	initiator
INEPT	insensitive nuclei enhanced by polarization transfer
IR	infrared
M	mol dm^{-3} (mol $litre^{-1}$)
M	monomer
MAS	magic-angle spinning
M_1	molecular weight of monomer
mes	mesitylene
M_n	number-average molecular weight
MO	molecular orbital
MS	mass spectroscopy
Mt	metal
MW	molecular weight
M_w	weight-average molecular weight

MWD	molecular weight distribution
NBE	norbornene
NMR	nuclear magnetic resonance
NOE	nuclear Overhauser enhancement
Np	neopentyl
OTs	*p*-toluene sulfonate
p-xyl	*p*-xylene
PDI	polydispersity index $= M_w/M_n$
py	pyridine
RCM	ring-closing metathesis
RI	refractive index
ROMP	ring-opening metathesis polymerization
SAXS	small-angle X-ray scattering
SEM	scanning electron microscopy
TBDMS	*t*-butyldimethylsilyl
TBDPS	*t*-butyldiphenylsilyl
TEM	transmission electron microscopy
TFA	trifluoroacetic acid
T_g	glass transition temperature
THF (thf)	tetrahydrofuran
TOF	turnover frequency
TON	turnover number
TPD	temperature-programmed desorption
TPR	temperature-programmed reduction
UV/vis	ultraviolet/visible
XANES	X-ray absorption near-edge spectroscopy
XPS	X-ray photoelectron spectroscopy
XRD	X-ray diffraction
ZNP	Ziegler–Natta polymerization

Ligand order in the formulae of metal complexes

There is no universally adopted system in the literature. Within this book we have generally given the metal first, followed by the multiply bonded ligands, followed by the other ligands in accordance with IUPAC rules (as quoted in the *Handbook of Chemistry and Physics*), namely, anionic ligands first, in the specified order, followed by neutral ligands in the specified order. For cocatalysts such as tetramethyltin (Me_4Sn) or ethyl aluminium dichloride ($EtAlCl_2$) we have used the same order of ligands in the formula as normally used in the words.

1

Introduction

1.1 The olefin metathesis reaction

The word mĕtăth′ĕsĭs, with the accent on the second syllable, is derived from the Greek *meta* (change) and *tithemi* (place). As a grammatical term it means the transposition of sounds or letters in a word. In chemistry it refers to the interchange of atoms between two molecules. In olefin chemistry it describes the (apparent) interchange of carbon atoms between a pair of double bonds.

Exchange

$$\begin{matrix} PhCH{=}CHPh \\ + \\ CH_2{=}CH_2 \end{matrix} \rightleftharpoons \begin{matrix} PhCH \\ \| \\ CH_2 \end{matrix} + \begin{matrix} CHPh \\ \| \\ CH_2 \end{matrix} \qquad (1)$$

Ring-opening metathesis polymerization

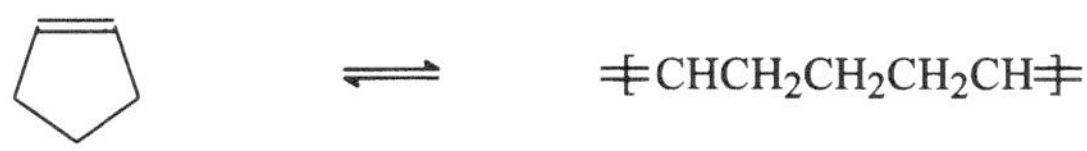

(2)

Ring-closing metathesis

(3)

Olefin metathesis reactions fall into three broad groups in which the *overall* chemistry (but not the mechanism) is represented by eqns. (1)–(3). The reactions are generally reversible and, with the right catalyst system, equilibrium can be attained in a matter of seconds, even with substrate/catalyst ratios of 10^4; a truly remarkable reaction.

Catalyst systems for olefin metathesis are described in detail in Ch. 2 and almost invariably contain a transition metal compound. These are sometimes effective by themselves but often require the presence of a second compound (cocatalyst), and sometimes a third (promoter). The systems most commonly used are based on the chlorides, oxides, or other easily accessible compounds of Mo, Ru, W, or Re. Os or Ir compounds are sometimes used, and occasionally Ti, V, Cr, Co, Nb, Rh, or Ta compounds. Typical cocatalysts are $EtAlCl_2$, R_3Al and R_4Sn (R = Ph, Me, Et, Bu), while promoters often contain oxygen, e.g. O_2, EtOH, PhOH. There is now much evidence, detailed in Ch. 3, to show that such ‘first-generation’ catalyst systems act through the initial formation of metal carbene complexes which initiate and then propagate the reaction.

The most important advance over the past 15 years has been the preparation of numerous well-defined metal carbene complexes which can act directly as initiators of all types of olefin metathesis reaction. These 'second-generation' catalysts allow much closer control and better understanding of the mechanism of the olefin metathesis reaction. The initiating and propagating species can be closely monitored and in some cases the intermediate metallacyclobutane complexes can also be observed. Well-defined metallacyclobutane complexes also can sometimes be used as initiators.

The forward reaction (1) is an example of *cross-metathesis* between two different olefins and provides a route to styrene. The reverse of (1) is a *self-metathesis* reaction; such reactions may be either *productive* as in (4), or *non-productive* (also called *degenerate*) as in (5).

Productive metathesis

$$\begin{array}{c} CH_3CH{=}CH_2 \\ + \\ CH_3CH{=}CH_2 \end{array} \longrightarrow \begin{array}{c} CH_3CH \\ \| \\ CH_3CH \end{array} + \begin{array}{c} CH_2 \\ \| \\ CH_2 \end{array} \qquad (4)$$

Non-productive metathesis

$$\begin{array}{c} CH_3CH{=}CH_2 \\ + \\ CH_2{=}CHCH_3 \end{array} \longrightarrow \begin{array}{c} CH_3CH \\ \| \\ CH_2 \end{array} + \begin{array}{c} CH_2 \\ \| \\ CHCH_3 \end{array} \qquad (5)$$

Isotopic labelling studies (see Ch. 5) have shown that with terminal olefins non-productive metathesis is generally much faster than productive metathesis. Reaction (4) at one time found practical application in the petrochemical industry as a means of converting surplus propene into useful products. Reactions of the higher alkenes provide routes to sex pheremones of insects.

The forward reaction (2) is an example of *ring-opening metathesis polymerization* (ROMP). On paper the first stage of polymerization can be written as eqn. (6) and the ring-expansion process then repeated until high polymer is formed, but in fact the reaction is propagated by metal carbene complexes, resulting in linear high polymer as well as cyclic oligomers formed either concurrently or subsequently by intramolecular metathesis (backbiting) reactions. Low concentration of monomer favours the intramolecular reactions, which is an advantage in the preparation of the cyclic dimer of cyclooctene, an intermediate in the perfume industry. When ROMP is initiated by metal carbene complexes, the reaction shows all the well-known features of a 'living' polymerization, and block copolymers can be made by sequential addition of different monomers.

The forward reaction (3) is an example of *ring-closing metathesis* (RCM). Such reactions have become an important tool in the hands of the synthetic chemist and

(6)

containing functional groups or hetero-atoms (Si, O, N); see Ch. 8. Dienes can undergo intermolecular as well as intramolecular olefin metathesis. *Intramolecular* metathesis (RCM) reactions tend to be favoured when the product is a 5-, 6-, 7-, or sometimes an 8- or higher-membered ring compound, as in reaction (7), especially at low substrate concentration.

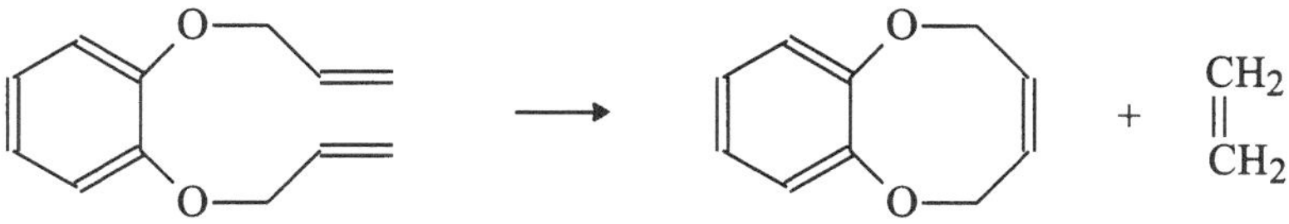

(7)

Intermolecular metathesis reactions, such as eqn. (8), lead eventually to high polymers and proceed very cleanly when initiated by metal carbene complexes; these are known as ADMET (*acyclic diene metathesis*) polymerizations; see Ch. 8.

$$\begin{array}{c} CH_2{=}CH(CH_2)_2CH{=}CH_2 \\ + \\ CH_2{=}CH(CH_2)_2CH{=}CH_2 \end{array} \longrightarrow \begin{array}{c} CH_2{=}CH(CH_2)_2CH \\ \| \\ CH_2{=}CH(CH_2)_2CH \end{array} + \begin{array}{c} CH_2 \\ \| \\ CH_2 \end{array} \qquad (8)$$

Catalysts are now known that facilitate the metathesis of olefins containing all types of functional groups. Such reactions are described for acyclic olefins in Ch. 7, and for cyclic olefins in Ch. 12 and 13. Not only can allyl compounds be metathesized (Mol 1979), reaction (9), but even acrylonitrile will undergo cross-metathesis (Crowe 1995).

$$\begin{array}{c} CH_2{=}CHCH_2OEt \\ + \\ CH_2{=}CHCH_2OEt \end{array} \xrightarrow{Re_2O_7/Al_2O_3/Me_4Sn} \begin{array}{c} CH_2 \\ \| \\ CH_2 \end{array} + \begin{array}{c} CHCH_2OEt \\ \| \\ CHCH_2OEt \end{array} \qquad (9)$$

In olefin metathesis reactions, a *cis* reactant may yield all-*cis* products, or all-*trans* products, or a mixture of *cis* and *trans*, depending on the catalyst and substrate. For example, in the ROMP of 5,5-dimethylnorbornene, reaction (10), $ReCl_5$-initiation gives an all-*cis* polymer, while $RuCl_3$-initiation gives an all-*trans* polymer.

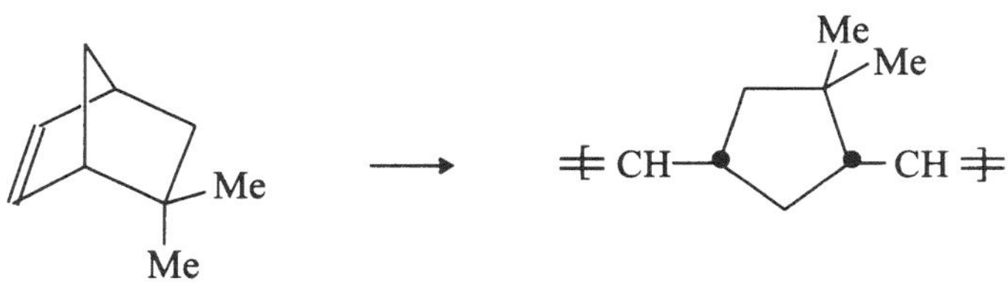

(10)

Much interest therefore centres on the stereospecificity of the reaction since this provides clues to the mechanism. Additional information can be derived from the structural sequences in such polymers: the double bonds may be present in *cc*, *ct*, or *tt* dyads; successive rings may have an isotactic (*m*) or syndiotactic (*r*) relationship; and the substituents may be oriented randomly to give head–head (HH), head–tail (HT), and tail–tail (TT) structures, or may align always in the same direction (HT); (see Ch. 11, 12 and 13).

At one time it seemed likely that the high-*trans* polymer of cyclopentene formed in reaction (2) would find use as an all-purpose elastomer. Raw materials were cheap and the product had properties akin to those of natural rubber. However, its early promise has not been fulfilled. Other high-*trans* polymers, such as those formed from norbornene, eqn. (11), and cyclooctene, eqn. (12), have found their way on to the market as components of elastomeric products, and the ROMP of *endo*-dicyclopentadiene, eqn. (13), is being used to produce large objects by reaction injection moulding (RIM); see Ch. 17.

The principal side reactions that can appear during olefin metathesis using first-generation catalysts are alkylation, isomerization (especially double-bond shift reactions), cyclization, and addition across the double bond. These can generally be prevented by taking one or more of the following precautions: (i) the proper choice of solvent – alkylation reactions are less likely to occur in chlorobenzene or dichloromethane than in benzene; (ii) the use of bases to suppress cationic side reactions – alkali metal hydroxides can be added to supported catalysts, and tertiary amines or other polar additives to catalyst systems in solution (Nishiguchi 1981); (iii) careful selection of the order of mixing of catalyst, cocatalyst, and substrate – this applies especially when oxygen is the promoter (Warwel 1977a); and (iv) by

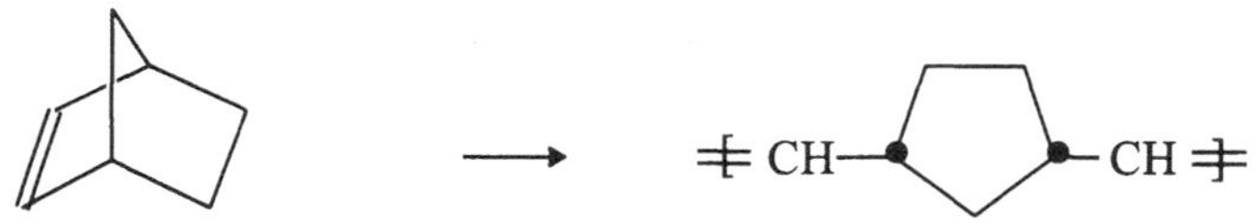

(11)

(12)

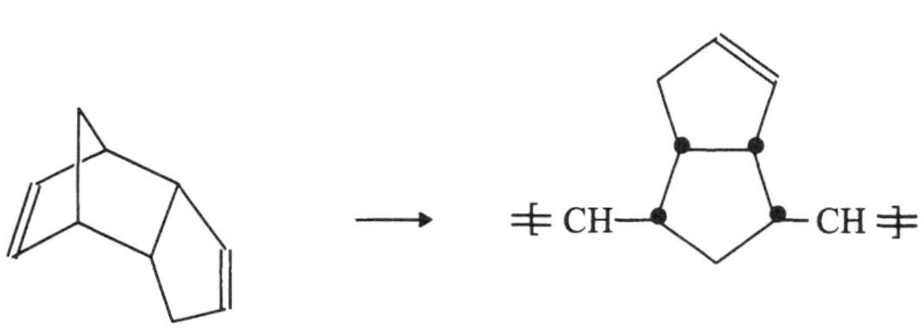

(13)

working at as low a temperature as possible. With second-generation catalysts, side reactions are seldom a problem. Molybdenum and tungsten carbene complexes must be handled in a dry box, but ruthenium carbene complexes can be used in the presence of both air and water. With all types of catalyst it is not uncommon to achieve selectivities of more than 99%.

1.2 Brief history

The expression 'olefin metathesis' was first used by Calderon (1967a). Until then, the chemistry of exchange reactions like (1) and (4), and of polymerization reactions like (2), (6), and (10)–(13), had developed independently. Both types of reaction were remarkably late starters on the modern scene, the first open publications being those of Truett (1960) on reaction (11) and Banks (1964) on reaction (4). These were foreshadowed by the patent disclosures of Eleuterio (1957) on reactions (2), (11), etc. and of Peters (1960) on reaction (4). The full story of these discoveries has been published by Eleuterio (1991) and Banks (1986a).

The connection between the two types of reaction was not immediately apparent mainly because different catalysts and conditions were generally involved. On the one hand disproportionation (or dismutation) of acyclic olefins, as such reactions were then called, was usually carried out on supported oxide catalysts such as MoO_3/Al_2O_3 at elevated temperature (160°C), whereas ring-opening polymerization of cycloalkenes was mostly initiated with catalysts of the Ziegler–Natta type such as $MoCl_5/Et_3Al$ at room temperature or below. Supported oxide catalysts were in fact used by Eleuterio but gave only poor yields of polymer.

The discovery that $WCl_6/EtAlCl_2/EtOH$ (1/4/1) would bring about not only the very rapid polymerization of cyclooctene and cycloocta-1,5-diene (Calderon 1967c) but also the disproportionation of pent-2-ene (Calderon 1967a,b, 1968) at room temperature provided the bridge that led to the realization that these were examples of one and the same chemical reaction. Furthermore, the reaction between but-2-ene and but-2-ene-d_8 led only to but-2-ene-d_4, demonstrating with great elegance the remarkable fact that the double bonds are completely broken in the reaction, leading to an exchange of alkylidene moieties; reaction (14) (Calderon 1967b, 1968). Henceforward the reaction became known as 'olefin metathesis', a term that applies to the reactions of both cyclic and acyclic olefins.

$$\begin{matrix} CH_3CH{=}CHCH_3 \\ + \\ CD_3CD{=}CDCD_3 \end{matrix} \xrightarrow{WCl_6/EtAlCl_2/EtOH} \begin{matrix} CH_3CH \\ \| \\ CD_3CD \end{matrix} + \begin{matrix} CHCH_3 \\ \| \\ CDCD_3 \end{matrix} \qquad (14)$$

Independent evidence was provided by Mol (1968) using [2-^{14}C]propene; reaction (15). No radioactivity is found in the ethene, whereas the but-2-ene has twice the specific radioactivity of the reactant propene.

$$\begin{matrix} CH_3{}^{14}CH{=}CH_2 \\ + \\ CH_3{}^{14}CH{=}CH_2 \end{matrix} \xrightarrow{Re_2O_7/Al_2O_3} \begin{matrix} CH_3{}^{14}CH \\ \| \\ CH_3{}^{14}CH \end{matrix} + \begin{matrix} CH_2 \\ \| \\ CH_2 \end{matrix} \qquad (15)$$

A somewhat different approach is required to prove that the double bond is completely broken during ROMP of cycloalkenes. Dall'Asta (1971b) prepared copolymers of [1-^{14}C]cyclopentene/cyclooctene using $WOCl_4/Et_2AlCl/(PhCOO)_2$ as catalyst. After complete degradation of the copolymer by ozonolysis and conversion of the products into the α,ω-diol acetates, all of the ^{14}C is essentially localized in $AcO(CH_2)_5OAc$. Had the ring-opening occurred at a single bond, a different result would have been obtained. Similar results were obtained using copolymers of [1-^{14}C]cyclobutene/3-methylcyclobutene prepared with Mo-, Ti-, and Ru-based catalysts. In this case, the ^{14}C is found essentially only in $AcO(CH_2)_4OAc$ (Dall'Asta 1972b).

The quite extraordinary nature of the olefin metathesis reaction took chemists by surprise. No one would have predicted in the 1950s or early 1960s that a reaction in which the double bond was apparently cleaved and the pieces put back together again was even remotely possible. Yet not only is it possible but in some cases it can proceed to equilibrium within seconds.

The metathesis reactions of alkynes provide some surprises. They fall into two distinct groups: (i) metathesis involving complete cleavage of the triple bond; and (ii) polymerization involving cleavage of two of the three bonds.

The metathesis of pent-2-yne was first reported by Pennella (1968), using WO_3/SiO_2 at 350°C. Cleavage of the triple bond was proved by ^{14}C labelling; eqn. (16) (Mortreux 1972).

$$\begin{matrix} BuC{\equiv}{}^{14}CH \\ + \\ BuC{\equiv}{}^{14}CH \end{matrix} \xrightarrow{MoO_3/SiO_2/350°C} \begin{matrix} BuC \\ \||| \\ BuC \end{matrix} + \begin{matrix} {}^{14}CH \\ \||| \\ {}^{14}CH \end{matrix} \qquad (16)$$

Since then catalysts have been found which allow the reaction to proceed at 25°C (Bencheick 1982). Especially significant is the fact that the reaction can be initiated by metal carbyne complexes (McCullough 1984). Details are given in Ch. 10.

Polymerization of acetylenes by metathesis-type catalysts such as $MoCl_5$ and WCl_6/Ph_4Sn was first observed in the 1970s (Woon 1974; Masuda 1974, 1976), but the nature of the chain carrier was then in some doubt. However, it was soon found: (i) that metal carbene complexes would initiate the polymerization of $Me_3CC{\equiv}CH$ at 60°C (Katz 1980a); (ii) that end-groups derived from such initiators could be detected in polymers of $PhC{\equiv}CH$ (Kunzler 1988b); and (iii) that triblock copolymers could be made by successive addition of norbornene, acetylene and norbornene to such initiators (Schlund 1989). All types of acetylene can be polymerized in this way and the reactions proceed by a ROMP-type mechanism; see Ch. 10.

(i) that metal carbene complexes would initiate the polymerization of $Me_3CC{\equiv}CH$ at 60°C (Katz 1980a); (ii) that end-groups derived from such initiators could be detected in polymers of PhC≡CH (Kunzler 1988b); and (iii) that triblock copolymers could be made by successive addition of norbornene, acetylene and norbornene to such initiators (Schlund 1989). All types of acetylene can be polymerized in this way and the reactions proceed by a ROMP-type mechanism; see Ch. 10.

1.3 The metal carbene mechanism

Evidence for the mechanism is discussed in detail in Ch. 3. Here is a brief outline. The basic question concerns the role of the catalyst and especially the transition metal in the olefin metathesis mechanism. How does it facilitate the exchange of alkylidene moieties? At first it was thought that two double bonds came together in the vicinity of the transition metal site and that the orbitals of the transition metal overlapped with those of the double bonds in such a way as to allow exchange to

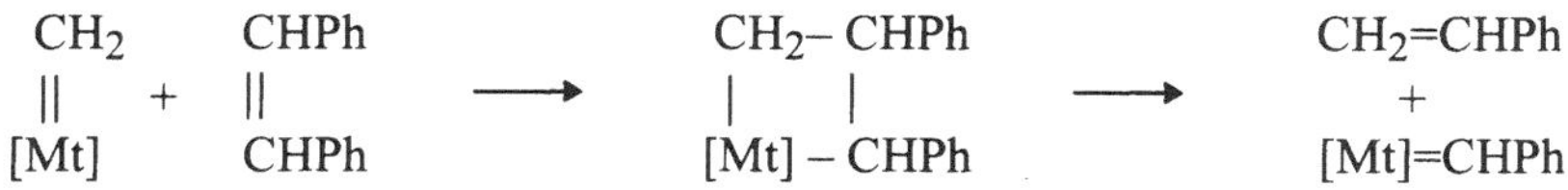

(17a)

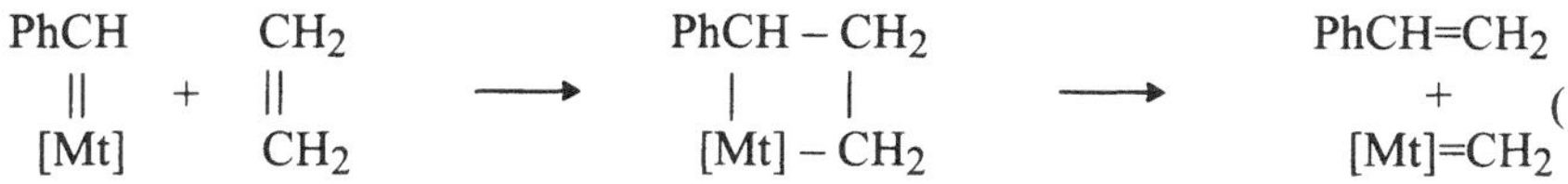

(17b)

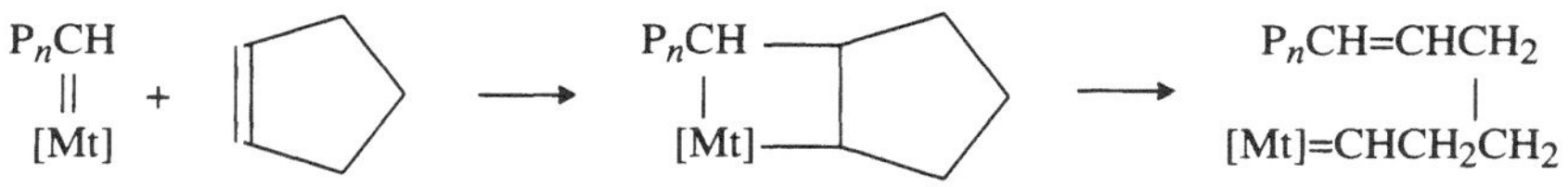

(18)

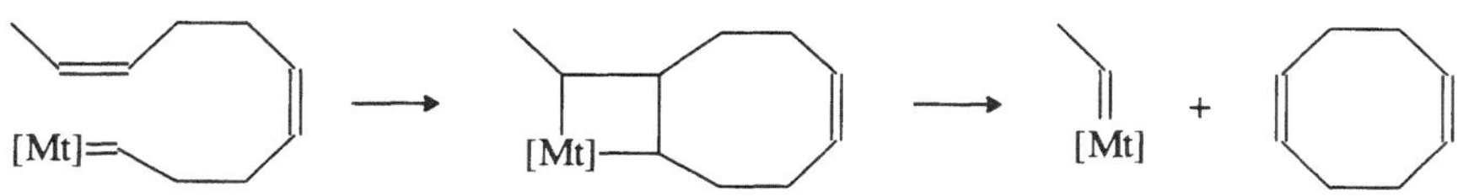

(19a)

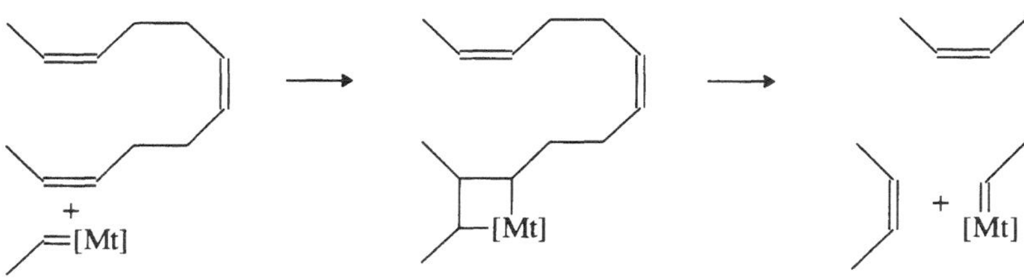

(19b)

Stable metal carbene complexes, such as $W[=C(OMe)Me](CO)_5$, were first prepared by Fischer, E.O. (1964) and many hundreds are now known. Some, like **1**, react with other compounds in such a way as to indicate that the carbene ligand is electrophilic; others, such as **2**, react in the opposite way, indicating that the carbene ligand is nucleophilic (Grubbs 1978; Parshall 1980).

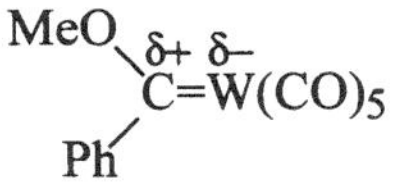

(1)

$\overset{\delta-}{C}H_2=\overset{\delta+}{Ta}Me(\eta^5\text{-}C_5H_5)_2$

(2)

The idea that a metal carbene species might be involved in olefin metathesis was first proposed by Hérisson and Chauvin (1971) on the basis of the initial products of cross-metathesis of cyclic and acyclic olefins. The fact that cycloocta-1,5-diene gives a series of cyclic oligomers $(C_4H_6)_n$, $n = 4–13$ (Scott 1969), also finds a ready explanation in terms of backbiting reactions, of the kind shown in eqn. (19a), whereas the pairwise mechanism would be expected to give initially only the even-numbered members of this series. The observation that metal carbene complexes such as $W(=CPh_2)(CO)_5$ could be used as single-component initiators of olefin metathesis provided early reassurance of the validity of the metal carbene mechanism (Katz 1976c). Since then the identification of both initiating and propagating metal carbenes, also the intermediate metallacyclobutanes, by ^1H and ^{13}C NMR spectroscopy, has completely confirmed this mechanism, at least for certain types of catalyst system (Schrock 1980; Wengrovius 1980; Kress 1982, 1985, 1987b, 1988). In the case of certain Ti-based catalysts, stable titanacyclobutanes can be identified as intermediates in the non-productive metathesis of alk-1-enes; see Ch. 2.

Information about the nature of the *initiating* metal carbene has been obtained in many cases by examining the first-formed products in the reactions of acyclic olefins in situations where these are distinguishable from the main metathesis products; see Ch. 3. The end groups in polymers formed by ROMP also carry information about the initiating carbene, provided they can be distinguished from those formed in termination or transfer reactions. This is relatively easy if a labelled initiator is used, for example, $^{13}CH_2=$ can be detected as an end group in the ROMP of norbornene catalyzed by $WCl_6/(^{13}CH_3)_4Sn$ (Ivin 1987). In living ROMP the initiating group derived from the metal carbene initiator is easily identified and, if the living chains are terminated by reaction with aldehydes, the end group derived from the aldehyde is also well defined, but with the first-generation catalysts the nature of the end groups formed in the termination reaction is not usually well understood.

Catalyzed olefin metathesis reactions are chain reactions with high turnover numbers. In some cases a metal carbene complex can exchange its alkylidene ligand in a single step reaction with an unsaturated compound. These are known as *stoichiometric metathesis* reactions.

Of particular interest is the fact that many olefin metathesis catalyst systems are of the Ziegler–Natta type. This raises the question of the relationship between the mechanism of olefin metathesis and that of Ziegler–Natta polymerization; this aspect is discussed in Ch. 4.

1.4 Equilibria and stereoselectivity

The olefin metathesis reaction clearly provides a route for the establishment of equilibrium between all species formed by exchange of alkylidene moieties. Thermodynamic data may therefore be used to predict the ultimate equilibrium position; alternatively, experimental equilibrium concentrations may be used to provide new thermodynamic data and to check existing data.

In the simplest case, such as eqns (20) or (21), the system contains a single equilibrium and the heat of reaction is very small.

$$2\,CH_2{=}CD_2 \rightleftharpoons CH_2{=}CH_2 + CD_2{=}CD_2 \qquad (20)$$

$$cis\text{-}MeCH{=}CHMe \rightleftharpoons trans\text{-}MeCH{=}CHMe \qquad (21)$$

In the more general case there will be a number of interlocking equilibria, as in system (22), and the route by which equilibrium is ultimately reached will depend on the relative rate constants of the individual components of the mechanism.

$$\begin{array}{ccccc} 2\,cis\text{-}MeCH{=}CHEt & \rightleftharpoons & cis\text{-}MeCH{=}CHMe & + & cis\text{-}EtCH{=}CHEt \\ \updownarrow & & \updownarrow & & \updownarrow \\ 2\,trans\text{-}MeCH{=}CHEt & \rightleftharpoons & trans\text{-}MeCH{=}CHMe & + & trans\text{-}EtCH{=}CHEt \end{array} \qquad (22)$$

With some catalysts the ‘horizontal’ equilibria of (22) are established much more quickly than the ‘vertical’ *cis/trans* equilibria and, in the initial stages, such metatheses may be very stereoselective. With other catalysts or with other substrates, the reverse is often true and there is little stereoselectivity. More complex still are the reactions involving more than two types of alkylidene moiety, derived from two or more reactant alkenes. Reactions of acyclic olefins are discussed in detail in Ch. 5–9.

Cycloalkenes offer a wider range of possible equilibria, since the products may consist of rings or chains in a variety of sizes, and with different proportions and distributions of *cis* and *trans* double bonds. Consideration of such equilibria may be simplified by regarding the monomer as the first member of the cyclic oligomeric series; see (23).

Cycloalkene

$\swarrow\!\!\nearrow$ $\searrow\!\!\nwarrow$

Cyclic oligomer $\rightleftharpoons$ Linear polymer (23)
(distribution of ring sizes; and of *cis* and *trans* double bonds) (distribution of chain lengths; and of *cis* and *trans* double bonds)

In accordance with the theory of aggregation processes (Jacobson 1950), linear polymer is formed only if the total oligomer concentration (including monomer) exceeds a certain critical value (the equilibrium concentration). With cyclohexene, this value is never exceeded, no high polymer is formed and only minor quantities of cyclic oligomer are produced at low temperature. With cyclopentene, the strain energy in the monomer is low (of the order of 17 kJ mol^{-1}) and at equilibrium there is an appreciable concentration of monomer as well as cyclic oligomer and high polymer. On the other hand, with cyclooctene and cycloocta-1,5-diene the strain energy in the monomer is higher and the equilibrium concentration of monomer, relative to that of cyclic oligomer, is very low indeed.

With a high initial cycloalkene concentration the relative rates of the individual steps of the mechanism may be such that, in the early stages of the reaction, an excess of linear polymer is formed which is then slowly converted in part or in whole into cyclic oligomers. In other cases only cyclic oligomers are formed in the initial stages, but once the critical concentration is exceeded linear polymer begins to form. At equilibrium the proportion of double bonds having the *cis* configuration in the linear polymer or in the macrocyclic oligomer is generally 15–20%. However, with certain catalysts very high *cis* stereoselectivity (95–100%) is sometimes observed; high initial *trans* stereoselectivity is also possible. In some cases there is evidence for an initial non-random distribution of *cis* and *trans* double bonds along the chain. If the catalyst is not too stereoselective, the *cis* content of the polymer formed in a given reaction mixture changes with time and ultimately approaches the equilibrium value, and a random *cis/trans* distribution. Such changes result from secondary metathesis reactions of the double bonds in the polymer and may also be observed by placing a polymer with high *cis* content in contact with a catalyst of low stereoselectivity.

For the monomers cyclopropene through to cycloheptene, only the *cis* isomers are known, but for the higher members of the series the *trans* isomers exist and, with increasing ring size, become progressively more stable relative to the *cis* isomer. The isomers of the cyclic oligomers of the cycloalkenes have different free energies according to the number and sequence of *cis* and *trans* double bonds, but in backbiting reactions the initial proportions of isomers are kinetically controlled and only reach their equilibrium proportions after prolonged exposure to metathesis catalyst.

The proportions of the cyclic oligomers at equilibrium provide an excellent means of testing the Jacobson–Stockmayer theory of ring-chain equilibria. This predicts that, so long as the cyclic oligomers are unstrained, the equilibrium concentration of oligomer should decrease with increasing ring size in proportion to $x^{-2.5}$, where x is the number of monomer units in the cyclic oligomer. This prediction has been confirmed for a number of systems. When refined, the theory also predicts correctly the absolute values of the equilibrium concentrations; see Ch. 11.

The metathesis reactions of cycloalkenes are discussed in detail in Ch. 11–14 and their cross-metathesis reactions with acyclic olefins in Ch. 15. Degradation reactions of unsaturated polymers by olefin metathesis are covered in Ch. 16. Industrial applications are described in Ch. 17.

The following is a selection of recent reviews on various aspects of the subject: Amass (1989), Breslow (1993), Dräxler (1988), Feast (1989a,b), Feldman (1991), Finkel'shtein (1992a), Gibson, V.C. (1994), Grubbs (1989, 1994a, 1995), Grünert (1992b), Imamoglu (1990), Makovetsky (1994), Mol (1987), Novak (1992), Schrock (1986a, 1990a, 1993, 1994a).

2

Survey of Catalyst Systems

2.1 Introduction

The number of catalyst systems that initiate olefin metathesis is very large indeed. A single compound is sometimes effective but more commonly the catalyst system contains two, three, or more components. The systems may be heterogeneous or homogeneous, though it is sometimes difficult to be sure whether a system is truly homogeneous and to know whether the catalytic activity resides in the homogeneous or the heterogeneous part.

The most important catalyst systems are derived from compounds of the nine transition elements shown in upright type in the table below; those shown in bold type are generally the most effective. Catalysis by compounds of non-transition elements is very much the exception. Rare examples appear to be $EtAlCl_2$ (Ivin 1978c), Me_4Sn/Al_2O_3 (Ahn, H-G. 1992) and $MgCl_2$ (Buchacher 1996).

IVA	VA	VIA	VIIA	VIII	
Ti	*V*	*Cr*			*Co*
Zr	Nb	**Mo**	*Tc*	Ru	*Rh*
	Ta	**W**	**Re**	Os	Ir

The range of effective catalysts for a given metathesis reaction is governed to some extent by the exothermicity of the reaction. Thus norbornene and cyclobutene contain rather strained rings, and their ROMP can be readily initiated by a wide range of catalyst systems. Less strained rings such as cyclopentene generally require more active catalyst systems, used under more carefully controlled conditions. Metathesis of acyclic olefins is usually close to thermoneutral and, if equilibrium is to be attained quickly, only the most active systems can be used. The range of effective transition metal compounds is continually being extended by manipulation of the ligands. Cross-metathesis experiments indicate that it is the initiation rather than the propagation step that is generally more sensitive to the catalyst and substrate olefin.

Half-lives for olefin metathesis can vary from a few seconds to several days. The activity of a particular catalyst system is dependent on a number of factors including: (i) the proportions of the components (there is frequently a maximum at some particular composition); (ii) pretreatment procedures, especially for catalysts supported on Al_2O_3 or SiO_2; (iii) the order in which the components are mixed; and (iv) in the case where the olefin is added last, the period of incubation before adding the olefin. Some catalyst systems, such as $WCl_6/EtAlCl_2/EtOH$, reach maximum activity very quickly after mixing, and it is then usually best to add the olefin before the cocatalyst ($EtAlCl_2$); others, such as $MoCl_2(NO)_2(py)_2/Et_3Al_2Cl_3$ (Hughes 1970a), require an hour or so to develop their full power, and it is then better to premix the catalyst components and wait before adding the olefin. Optimization of a given catalyst system by adjustment of the various parameters in turn can be quite a lengthy procedure but is less essential for the exothermic ROMP reactions than for the thermoneutral reactions of acyclic olefins.

The activity of a given catalyst system will be determined by both the nature and the concentration of the active species. In Ch. 3 we present the evidence that the active species is a metal carbene complex, formed in some way from the transition metal compound. It is useful to keep this in mind when reviewing the different types of catalyst system.

Catalyst systems may be divided into three types: (i) those consisting of an actual metal carbene, such as $W(=CPh_2)(CO)_5$; (ii) those containing an alkyl or allyl group in one of the components, e.g. $EtAlCl_2$, from which a carbene ligand can readily be generated; and (iii) those having neither a preformed carbene nor an alkyl group in any component. In the last case a metal carbene can only be formed by interaction of the substrate olefin itself with the transition metal centre. Some of the suggested mechanisms of formation of the active metal carbenes are shown in Scheme 2.1. Individual cases will be discussed elsewhere.

Supported catalysts are frequently used for gas-phase olefin metathesis in flow systems at 0–400°C. The convenience of such an arrangement for industrial operation will be self-evident. Such catalysts can be made in several ways: (i) by dry mixing of the individual components; (ii) by coprecipitation of the catalyst and support; (iii) by impregnation of the support with substances that decompose at high temperature to leave the catalyst deposited on the support, for example, Al_2O_3 can be impregnated with an aqueous solution of NH_4ReO_4 and then heated; (iv) by sublimation of the catalyst, e.g. Re_2O_7, on to the support; and (v) by reaction of a compound such as $(\pi\text{-}C_3H_5)_4Mo$ with the surface hydroxyl groups of Al_2O_3 to produce a surface compound (see Section 2.4.2).

Supported catalysts are generally activated by pretreatment, usually at high temperature (550°C), with gases such as H_2, O_2, CO or HCl. Various refractory materials have been tested as supports, in particular the oxides of Si, Al, Th, Zr, and Ti, of which SiO_2 and Al_2O_3 are the most commonly employed. Phosphates of Al, Zr, Ti, Mg, and Ca have also been used (Heckelsberg 1969a), as well as partially oxidized activated charcoal (Rooney 1977). The role of the support is seldom passive. In many cases there are indications of chemical interaction between the support and transition metal compounds, so it is always necessary to quote the support in defining the overall catalyst system. The surface area of the support is

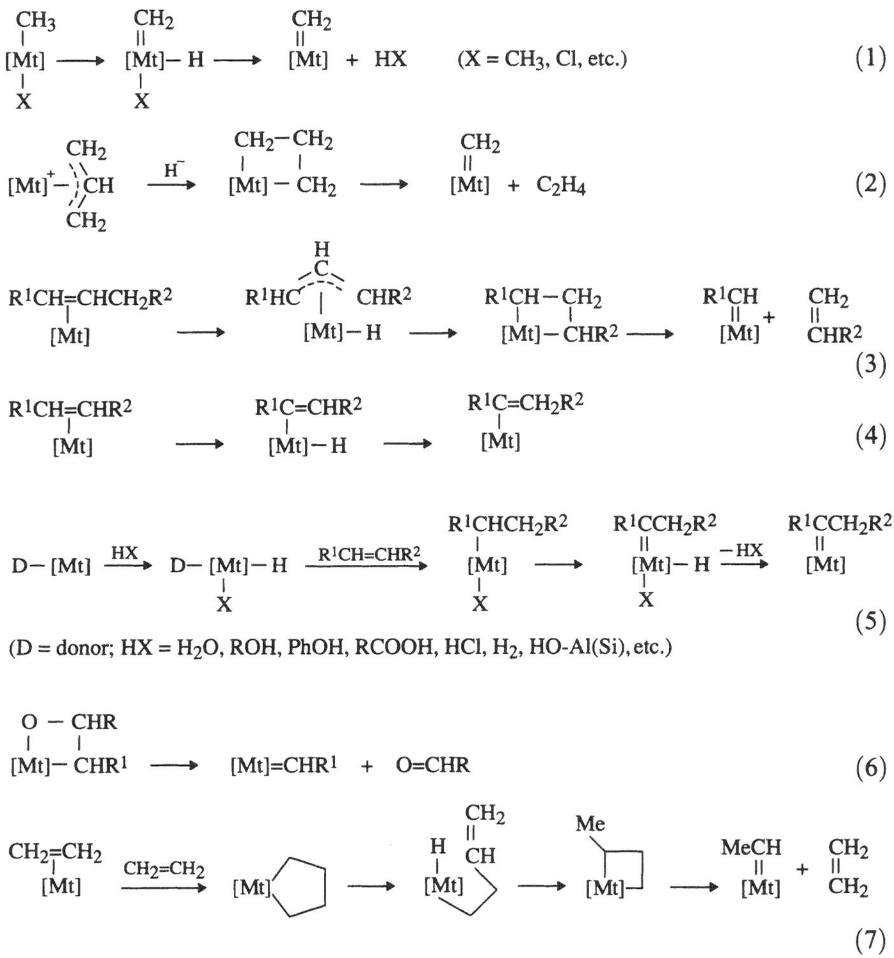

Scheme 2.1 Possible routes to metal carbene initiating species in olefin metathesis.

typically more than 200 $m^2\ g^{-1}$ while the catalyst content is generally 1–20% by weight.

Chemically bound polymer-supported complexes have been tested for use with solutions of olefins. With cross-linked polystyrene having the functional group $-CH_2W(CO)_3(\eta^5\text{-}C_5H_5)$ attached to a phenyl group in the *para* position, after activation with *i*-$BuAlCl_2/O_2$ in chlorobenzene, good conversion of *trans*-hept-3-ene can be achieved in 60 min and the activity is maintained even after treatment with several batches of reactant solution (Warwel 1978). $W(CO_3)R(\eta^5\text{-}C_5H_5)$-, $(R = H)$, when bonded via the cyclopentadienyl ligand to cross-linked polystyrene, gives a highly active heterogeneous catalyst component of high stability when treated with *i*-$BuAlCl_2$. It brings about the metathesis of *trans*-hept-3-ene, accompanied by double-bond shift reactions; the catalyst can be reused up to 20 times (Heveling 1987). When R = Cl the activity is comparable to that for R = H. When R = CH_3 the activity is lower but can be enhanced by treatment with O_2. Infrared (IR) spectroscopy indicates that for R = H or CH_3 the reusable catalyst component is the chloro compound (pol)$-CH_2(\eta^5$-$C_5H_4)W(CO_3)Cl$, formed by interaction with the cocatalyst (Heveling 1990). The functional group $-CH_2PPh_2W(CO)_5$ attached to the polymer resin, again in conjunction with *i*-$BuAlCl_2/O_2$, provides an active but less stable catalyst (Warwel 1978), while (pol)$-CH_2PPh_2Mo(CO)_5$ activated by $EtAlCl_2$ gives a rather feeble catalyst for the metathesis of *cis*-pent-2-ene (Basset 1975). The catalyst $MoCl_2(PPh_3)_2(NO)_2$ can be attached to the polymer resin through a phosphine link and activated with $Me_3Al_2Cl_3$ (Grubbs 1977/78). $MoCl_2(NO)_2(py)_2$ immobilized on a copolymer of hydroxyethyl methacrylate and ethene dimethacrylate, and activated with $EtAlCl_2$, gives a catalyst for the metathesis of *cis*-pent-2-ene with an activity and stereoselectivity almost identical with those observed for homogeneous analogues (Berglund 1986a); see Ch. 6. A resin with $-SnMe_3$ functional groups reacts with WCl_6 to give a catalyst of moderate activity, which can be used in a flow system (Grubbs 1977/78). Polymer-anchored $-AlEtCl$ or $-AlEt_2$ treated with WCl_6 gives an active catalyst for *cis*-pent-2-ene and *cis*-oct-2-ene metathesis, but only when extra $EtAlCl_2$ is added ($Al/W > 2$) (Uchida, A. 1985). The functional group $-(bpy)W(CO)_4$ activated by $EtAlCl_2/O_2$ is very effective for metathesis of pent-2-ene (Tamagaki 1978). The first polymer-supported metal carbene catalysts to be reported were those based on $Ru(=CHCH=CPh_2)(Cl)_2(PR_3)_2$ attached to a 2% cross-linked polystyrene through the phosphine ligands. When R = Cy such a catalyst is active not only for the ROMP of norbornene, but also for the ROMP of cyclooctene and the metathesis of *cis*-pent-2-ene (Nguyen 1995).

Active catalysts can also be made by reacting dry silica gel with a solution of WCl_6 or $WOCl_4$ in benzene. The solution goes colourless leaving a coloured solid, which, after washing and treatment with Me_4Sn, is active for metathesis of pent-2-ene or methyl oleate (van Roosmalen 1980b). Treatment of WCl_6 or $WOCl_4$ bonded to silica–alumina with Et_3Al as cocatalyst is active for pentene metathesis, while WBr_5/SiO_2 activated with Et_3Al gives a catalyst for propene metathesis (Maksimowski 1991). Treatment of SiO_2 or Al_2O_3 with $(\pi$-$C_3H_5)_4Mo$ or $(\pi$-$C_3H_5)_4W$ also gives an active catalyst; see Sections 2.4.2, 5.3.5 and 5.6.

Cocatalysts normally consist of an organometallic compound of a non-transition metal in groups I–IV. The most commonly used are BuLi, $EtAlCl_2$, R_3Al, and R_4Sn (R = Me, Bu, Ph, etc.). Their function may be several-fold. First, they may provide an alkyl ligand at the transition metal which can be subsequently converted into an alkylidene ligand by reductive elimination; Scheme 2.1, mechanism (1). Secondly, reduction of the transition metal opens up sites for the coordination of the reactant olefin. Thirdly, the cocatalyst itself may act as a ligand, attaching itself to the catalytic complex in some way, thereby modifying the electron density at the transition metal atom. In this connection it is significant that the activating ability of the commonly used aluminium cocatalysts lies in the order

$EtAlCl_2 > Et_2AlCl > Et_3Al$ (Khodzhemirov 1976a), which is the order of their electron-accepting ability. $AlBr_3$ is extremely effective in activating $W(=CHCMe_3)(Br)_2(OCH_2CMe_3)_2$ for the metathesis of pent-2-ene. There is no significant change in $J(C_\alpha\text{–}H)$ in the W(=CHR) moiety nor any significant displacement of the OCH_2R resonances in the ^1H or ^{13}C NMR spectra, which shows that activation occurs by attachment of $AlBr_3$ to the Br ligands (Kress 1982).

Third components, especially hydroxylic compounds, hydridic compounds, and hydrogen itself, may well assist the formation of Mt-H bonds, for which there is IR and NMR evidence in some systems (Laverty 1976b). These in turn can provide a route to the formation of the initial metal carbene; Scheme 2.1, mechanism (5). Certain single-component catalyst systems, such as WCl_6 and $ReCl_5$, sometimes work better when 'aged', suggesting that in reality they are two-component systems, with adventitious moisture serving the role of second component. The reaction of WCl_6 with alcohols results in displacement of chlorine by alkoxy groups (Khodzhemirov 1976a) and compounds containing aryloxy ligands have frequently been used as metathesis catalysts in conjunction with cocatalysts such as $EtAlCl_2$ or Et_4Pb (Quignard 1985a; Dodd 1988; Bell, A. 1992; Nugent 1995). The importance of OR ligands in stabilizing the propagating metal carbene complexes has been highlighted by Schrock (1995a).

The presence of a cocatalyst or additive appears not always to be essential for activity, and the only possible route to a metal carbene is then by direct interaction of the olefin with the transition metal site; Scheme 2.1, mechanisms (3) and (4).

Actual metal carbene complex catalysts can be divided into two broad classes, 'Fischer-type' and 'Schrock-type'. The Fischer-type carbene complexes are low-valent and generally characterized by the presence of one or two heteroatoms (O, N, or S) bonded to the carbene carbon. Such complexes do not normally initiate the chain metathesis of olefins, since they are both coordinatively and electronically (18e) saturated. However, they can sometimes be activated for metathesis by heating, or by reaction with a cocatalyst, or photochemically. Some examples are listed in Table 2.1.

The Schrock-type carbene complexes do not have a heteroatom (other than Si) bonded to the α-carbon and are of the general formula $Mt(=CR^1R^2)(L)_n$, where R^1, $R^2 = H$, alkyl, aryl, and $(L)_n = n$ ligands, not all the same. Examples of their use as metathesis initiators will be found throughout this book; a small selection is given in Table 2.2. The effectiveness of these initiators sometimes depends on the solvent. Ethers, such as THF and DME, may be strongly bound and then have a moderate or strong retarding effect on metathesis, as for **3**, **12**, and **16**; also see Fox (1994a). In other cases they are less strongly bound, exchanging readily with substrate olefins, and then have no serious effect on the rate of metathesis, as for **8** and **9**.

For the d^0 complexes of Mo(VI), W(VI) and Re(VII) (counting the alkylidene ligands as ionic) the reactivity is greatly enhanced when the alkoxy ligands are more electron-attracting: $OCMe(CF_3)_2 > OCMe_2CF_3 \gg OCMe_3$. Compare, for example, **5**, which can bring about the metathesis of both alk-1-enes and alk-2-

enes, with **4**, which can do neither; also see Feldman (1989a,b) and Schrock (1995a). The same enhancement of reactivity can sometimes be achieved through the use of a Lewis-acid cocatalyst, which can coordinate with, or remove, one of the ligands, thereby reducing the electron density at the metal centre. The use of $GaBr_3$ as a cocatalyst for **7** is a case in point. However, for the d^4 complexes of Ru(IV) in Table 2.2 (again counting the alkylidene ligands as ionic), the reactivity surprisingly *increases* with the electron richness of the PR_3 ligands (Nguyen 1993).

The ruthenium carbene complexes are notable in being stable to air and water. The first tungsten(VI) carbene complex stable to air and moisture was synthesized by Blosch (1991, 1992b), viz. $W(=CHCMe_3)(=O)(Cl)(Tp')$ where Tp' = hydrido-tris(3,5-dimethyl-1-pyrazolyl)borate, but it requires the use of $AlCl_3$ as cocatalyst to initiate the ROMP of cyclooctene or norbornene; also see Blosch (1992a) and Gamble (1993). A tris(pyrazolyl)borate–molybdenum–imido–alkylidene complex is also effective for the ROMP of norbornene when activated by $AlCl_3$ (Vaughan 1995).

Table 2.1 Low-valent, mainly Fischer-type complexes as sources of initiating species for olefin metathesis

Complex	Activator	Substrate	Reference
$(CO)_5Mo{=}C(OPh)NBu_4$[a]	$MeAlCl_2$	$PrCH{=}CH_2$	Kroll 1971
$(CO)_5W{=}C(OMe)Et$	Bu_4NCl	$PrCH{=}CH_2$	Kroll 1971
$(CO)_5W{=}C(OEt)Bu$	$TiCl_4$	Cyclopentene	Chauvin 1976
$(CO)_5W{=}CPh_2$[b]	Heat (38°C)	*cis*-Cycloalkenes	Katz 1976c
	21–25°C	$EtCH{=}CHMe$	Katz 1977b
$(CO)_5W{=}C(OMe)Ph$	Heat (50°C)	Cyclobutene, norbornene	Katz 1976d
	Heat (40°C)	$RC{\equiv}CR$	Katz 1980a
	$PhC{\equiv}CH$	Cyclopentene	Katz 1980b
	hν	Cycloocta-1,5-diene	Kolesnikov 1981
	Halides	$PrCH{=}CH_2$	Banks 1982
$(CO)_5Re{-}Re[{=}C(OMe)Me](CO)_4$	*i*-$BuAlCl_2$	$PrCH{=}CHEt$	Warwel 1983b
$(CO)_4W{=}C(OMe)$	None (20°C)	Hept-1-yne	Liaw 1985
↗ |	None (60°C)	Norbornene	Doherty 1986
$CH_2{=}CH(CH_2)_2$	Et_3Al (20°C)	Hept-1-yne	Soum 1986
	Et_3Al (60°C)	Norbornene	Doherty 1986
	$AlCl_3$ (25°C)	Norbornene	Leymet 1989
	BCl_3 (30°C)	$CH{\equiv}CCMe_3$	Liaw 1993

[a] The first stable carbene complex to be used as part of a metathesis catalyst system. [b] Highly stereospecific to give *cis* products.

Some very active molybdenum carbene initiators are listed in Table 2.2. A very convenient way of generating the highly active complex $Mo(=CHCMe_3)(=NAr)(OR)_2(Me_3CNH_2)$ *in situ* is to react the very robust complex $Mo(=NAr)(=NCMe_3)(CH_2CMe_3)_2$ with ROH in the presence of the olefinic substrate (Bell, A. 1994a).

Table 2.2 Examples of metal carbene complexes with a count of less than 18 electrons; effectiveness as initiators of olefin metathesis/ROMP[a]

	Complex[b]	Number of ligands	Electron count[c]	Metathesis reactivity: Pent-2-ene	Metathesis reactivity: Norbornene	Reference
1	Nb(=CHCMe$_3$)(Cl)(OCMe$_3$)$_2$(PMe$_3$)	5	12	Yes		Rocklage 1981
2	Ta(=CHCMe$_3$)(OAr)$_3$(THF)	5	12	Yes[d]	Yes[e]	Wallace 1987, 1988
3	Ta(=CHCMe$_3$)(TIPT)$_3$	4	10	No	Yes[f]	Wallace 1987, 1988
4	Mo(=CHCMe$_3$)(=NAr)(OCMe$_3$)$_2$	4	12	No	Yes	Schrock 1990a
5	Mo(=CHCMe$_2$R)(=NAr)[OCMe(CF$_3$)$_2$]$_2$ (R = Me, Ph)	4	12	Yes[g]	Yes	Schrock 1990b; Fox 1993, 1994a
6	W(=CHCMe$_3$)(O)(Cl)$_2$(PEt$_3$)	5	14	Yes		Wengrovius 1980
7	W(=CHCMe$_3$)(Br)$_2$(ONp)$_2$	5	12	No[h]	Yes[i]	Kress 1982, 1983, 1987c
8	W(=CHCMe$_3$)(Cl)(Np)(OAr$'$)$_2$(O-*i*-Pr$_2$)	6	14	Yes	Yes	Quignard 1985b
9	[W]=CHCMe$_3$; see formula **1** of Ch. 3	6	14	Yes	Yes	Couturier 1992
10	W[=CHC$_6$H$_4$(OMe)-2](=NAr$''$)[OCMe(CF$_3$)$_2$]$_2$	4	12	Yes	Yes	Johnson 1993
11	W(=CHSiMe$_3$)(=NPh)(CH$_2$SiMe$_3$)L[j]	5	14	No	Yes	van der Schaaf 1993, 1994
12	Re(=CHCMe$_3$)(≡CCMe$_3$)[OCMe(CF$_3$)$_2$]$_2$	4	14	Yes[k]	Yes	Toreki 1993
13	Re(=CHCMe$_3$)(≡CCMe$_3$)(CH$_2$CMe$_3$)(OTf)L[l]	5	16	Yes		LaPointe 1995
14	Re(=CHCH=CPh$_2$)(O)[OCMe(CF$_3$)$_2$]$_3$(THF)	6	16	Yes[m]		Flatt 1994
15	Ru(=CHCH=CPh$_2$)(Cl)$_2$(PPh$_3$)$_2$[n]	5	16	No	Yes	Nguyen 1992; Wu 1993
16	Ru(=CHCH=CPh$_2$)(Cl)$_2$(PCy$_3$)$_2$[n]	5	16	Yes[p]	Yes	Nguyen 1993; Grubbs 1994b; Lynn 1996
16a	Ru(=CHR)(Cl)$_2$(PR$'_3$)$_2$; R$'$ = PPh$_3$, PCy$_3$; R = H, Me, Et, Ph, *p*-ClC$_6$H$_4$	5	16		Yes	Schwab 1995, 1996; Lynn 1996

[a] For synthetic routes to metal carbene complexes see Quignard 1985b, Murdzek 1987a, Aguero 1988, Schoettel 1989, Feldman 1989a, Schrock 1990b, Johnson 1990, Novak 1992, Oskam 1993b, de la Mata 1996. [b] Ar, C$_6$H$_3$-*i*-Pr$_2$-2,6; Ar$'$, C$_6$H$_3$-Ph$_2$-2,6; Ar$''$, C$_6$H$_3$-Me$_2$-2,6; TIPT, S-C$_6$H$_2$-*i*-Pr$_3$-2,4,6; Np, CH$_2$CMe$_3$; Cy, cyclohexyl; OTf, triflate. [c] Nb, Ta, 5; Mo, W, 6; Re, 7; Ru, 8; Cl, Br, OAr, OCR$_3$, TIPT, 1; =CHR, =NAr, PR$_3$, OR$_2$, 2; ≡CR, 3. [d] Short-lived. [e] At 50°C, via isolable metallacyclobutane complex. [f] Reaction inhibited in THF. [g] Also active for terminal olefins. [h] Becomes very active in the presence of a Lewis acid such as AlBr$_3$ or GaBr$_3$, which removes a bromide ion from the complex. [i] Becomes much more active in the presence of GaBr$_3$; the intermediate metallacyclobutane complex can then be detected at low temperature. [j] L is a ligand such as 8-quinolinolate containing a nitrogen chelated to the tungsten. [k] Much retarded in THF. [l] L = MeCN; short-lived (< 1 h). [m] When activated by GaBr$_3$. [n] Mixture of isomers in which the phosphine ligands are either *cis* (20%) or *trans* (80%). [p] Rate dependent on solvent: CD$_2$Cl$_2$ > C$_6$D$_6$ > THF-d_8 (relative rates 103:26:11, respectively). [q] R = Me, Ph.

There follows a summary, group by group, of the effective catalysts based on transition metal compounds and a final section on photochemically stimulated catalyst systems.

2.2 Group IVA

2.2.1 Titanium

There are only a few reports of metathesis of *acyclic* olefins initiated by Ti-based catalysts. On the other hand, ROMP of cycloalkenes occurs fairly readily with such systems, which are mostly based on $TiCl_4$ (or $TiBr_4$), activated either by $LiAlR_4$ ($R = C_7H_{15}$ or $C_{10}H_{21}$) or by R_3Al (R = Et, *i*-Bu), generally in two- or three-fold excess, and often used at low temperature (−20°C). Addition polymerization across the double bond can be minimized or avoided by inclusion of catalytic amounts of tertiary amines ($R_3N/Ti = 4$) (Tsujino 1964, 1965; Winstein 1977). The order of addition of components is crucial for the attainment of high yield. Thus, for dicyclopentadiene (M), 90% polymerization is obtained if the order of addition is M, Et_3Al, Et_3N, $TiCl_4$, but only 1% if the monomer is added last (Winstein 1977). Ring-opened polymers made with Ti-based catalysts usually contain more *trans* than *cis* double bonds, unless made at low temperature (−50°C) (Dall'Asta 1962).

Titanium compounds have also been used to enhance the activity of catalyst systems based on other transition metals, but the way in which they do this is not clear. Thus $TiCl_4$ in 150-fold excess acts as cocatalyst with $W[=C(OEt)R](CO)_5$ for the ROMP of cyclopentene (Chauvin 1976); the activity of this system is further increased by exposure to UV/vis radiation. Again, inclusion of TiO_2 in an MoO_3/Al_2O_3 catalyst increases its activity for the metathesis of propene and but-1-ene (Henrici-Olivé 1973), although not without a concurrent double-bond shift reaction. Dimethyltitanocene, Cp_2TiMe_2, used in conjunction with $WOCl_4$ or WCl_6 (Ti/W = 1.2), effects the metathesis of ethyl oleate in benzene at 70°C (Tsuji 1980). $TiCl_4$- and $Ti(O\text{-}i\text{-}Pr)_4$-treated highly dispersed lanthanide particles (Sm and Yb), prepared by metal vapour deposition, exhibit activity for propene metathesis at 100–200°C (Imamura 1989).

It is very significant that not only can bridged compounds of the type **17** be prepared from Cp_2TiCl_2 by reaction with 2 equivalents of Me_3Al, eqn. (8), but they

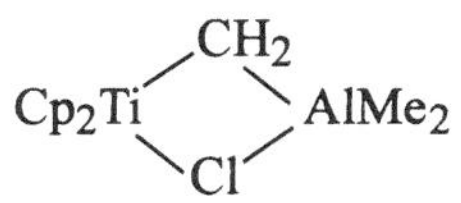

(17)

can also exchange methylene groups with isobutene, eqn. (9), and induce a degenerate-type exchange between isobutene and methylenecyclohexane (Tebbe

1978, 1979); eqn. (10). Compound **17** may therefore be regarded as a metal carbene, $Cp_2Ti{=}CH_2$, stabilized by coordination of Me_2AlCl but showing the

$$Cp_2TiCl_2 + 2\,Me_3Al \longrightarrow CH_4 + Cp_2TiCH_2(AlClMe_2) + Me_2AlCl \qquad (8)$$

$$Cp_2TiCH_2(AlClMe_2) + {}^{13}CH_2{=}CMe_2 \rightleftharpoons Cp_2Ti^{13}CH_2(AlClMe_2) + CH_2{=}CMe_2 \qquad (9)$$

(10)

typical reactions of a metal carbene. The exchange of methylene groups may then be written in the simplified form (11), according to the accepted metathesis

$$\begin{matrix} \overset{\delta+\ \ \delta-}{[Ti]{=}CH_2} \\ + \\ \overset{\delta-\ \ \delta+}{CH_2{=}CMe_2} \end{matrix} \longrightarrow \begin{matrix} [Ti] - CH_2 \\ | \quad\quad | \\ CH_2 - CMe_2 \end{matrix} \longrightarrow \begin{matrix} [Ti] \\ \| \\ CH_2 \end{matrix} + \begin{matrix} CH_2 \\ \| \\ CMe_2 \end{matrix} \qquad (11)$$

mechanism. The methylene ligand evidently behaves as a nucleophile in this system. Analogous reactions such as (12) occur with ketones, and it has been

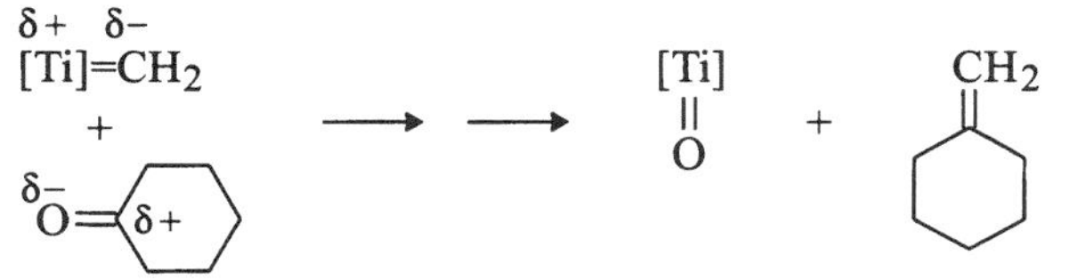

(12)

demonstrated by selective deuteration that the methylene entity is transferred intact. When **17** is treated in toluene at 25°C with 3,3-dimethylbut-1-ene in the presence of a tertiary amine to sequester Me_2AlCl, a stable titanacyclobutane (**18**) is formed, which has been well characterized by ^1H and ^{13}C NMR; reaction (13). On treatment

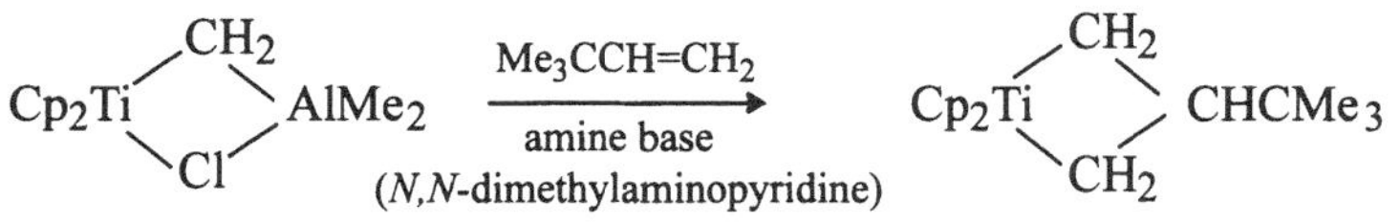

(**17**) (**18**) (13)

of the product with Me_2AlCl at −40°C, the reaction is reversed and the olefin regenerated quantitatively (Howard 1980a; Ott 1982). Reactions analogous to (13) occur with ethene, propene, styrene, cyclopentene, and norbornene. The product of reaction (13) can be equilibrated with *trans*-3-methylbut-1-ene-d_1 at 25°C without loss of stereoselectivity over 50 h; reaction (14).

No cyclopropanes are produced in any of the above reactions. However, the reaction with ethene in the presence of trimethylamine gives traces of

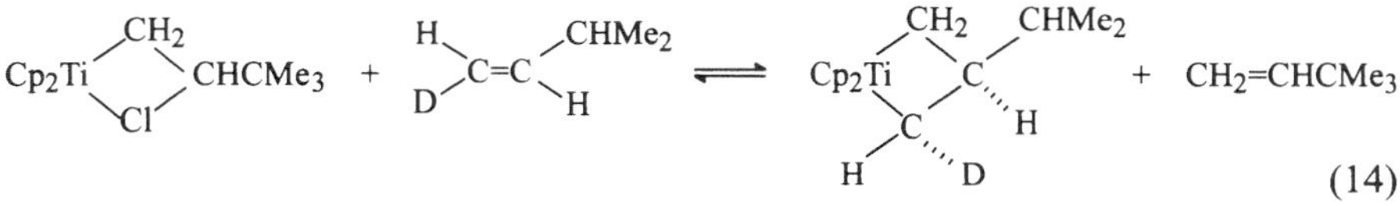

(14)

cyclopropane; in its absence the products are the homologues shown in reactions (15) and (16) (Tebbe 1978).

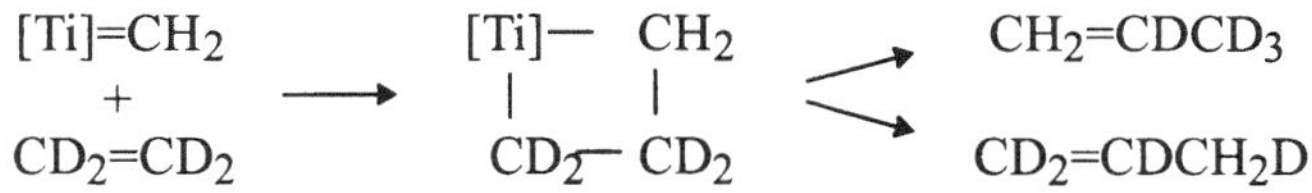

(15)

(16)

Titanacyclobutane complexes have been explored as catalysts for the metathesis of alk-1-enes (non-productive reaction) (Lee, J.B. 1982), alk-2-enes and alkadienes (poor yields) (Straus 1985). The titanacyclobutane **19**, formed by reaction of **17** with norbornene at 20°C, eqn. (17), initiates the ROMP of further norbornene when the temperature is raised to 65°C. The polymer formed has a narrow molecular

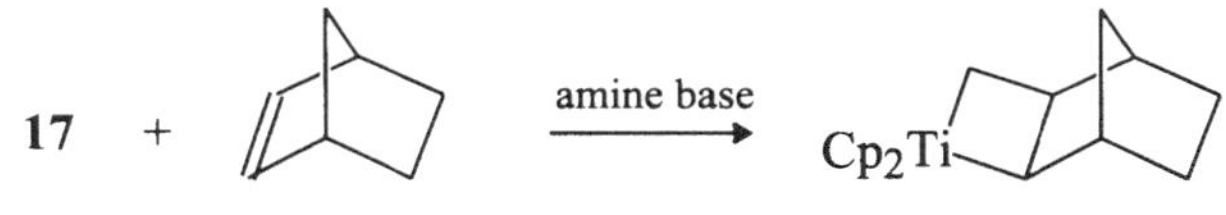

(17)

(19)

weight distribution (MWD); this reaction was the first well-documented example of 'living' ROMP (Gilliom 1986b).

Cp_2TiMe_2, $CpTiMe_3$, $CpTiMe_2Cl$, and $Cp_2Ti(CH_2SiMe_3)_2$ also catalyze the ROMP of norbornene. The rate-determining step is presumed to be the loss of CH_4 or $SiMe_4$ to form the initiating carbene complex (Petasis 1993).

2.2.2 Zirconium

$ZrCl_4/Et_3Al$ gives a low yield of mainly *trans*, ring-opened polymer of cyclopentene (Natta 1964a). An Al_2O_3-supported zirconium catalyst brings about the metathesis of pent-2-ene (Alkema 1970), as does $Zr(acac)_4/Me_3Al_2Cl_3$ (Kubicek 1972).

2.3 Group VA

2.3.1 Vanadium

Most V-based catalysts of the Ziegler–Natta type cause double-bond opening (addition polymerization) rather than ROMP of cycloalkenes. For example, $V(acac)_3/Et_2AlCl$ causes only Ziegler-Natta-type addition polymerization of cyclobutene (Natta 1963) and its copolymerization with ethene (Natta 1962).

Likewise VCl_4/Et_3Al causes addition polymerization of cyclobutene (Oshika 1968) and norbornene (Dall'Asta 1968a). ROMP is observed only when BuLi is used as cocatalyst (Dall'Asta 1968a). Thus VCl_4/BuLi (1/4) in toluene at −20°C causes ROMP of cyclobutene (45% yield) and of 3-methylcyclobutene (3% yield). $VOCl_3$/BuLi is also effective and gives a polymer containing comparable proportions of *cis* and *trans* double bonds.

The bis(neopentyl)vanadium(III) complex $V(CH_2CMe_3)_2(Cp)(PMe_3)$ catalyzes the ROMP of norbornene in benzene at 20°C. The initiating species is presumably a vanadium carbene complex formed by the decomposition of the catalyst. The vanadium(III) carbene complex $V(=CHCMe_3)(Cp)(dmpe)$, where dmpe = 1,2-bis(dimethylphosphino)ethane, is a poor initiator (Hessen 1993).

Supported vanadium oxide catalysts have some activity for the olefin metathesis reaction but are not very selective (Banks 1969). In combination with Re_2O_7 a very active catalyst system is obtained with alumina as the support (Nakamura, R. 1977b; Xu, X. 1985b). When V_2O_5/Al_2O_3 is preheated at about 575°C and treated with Me_4Sn it forms an effective catalyst system (V/Al = 3/97; Sn/V = 0.05–0.08) for propene metathesis at 25°C (Ahn, H-G. 1993).

2.3.2 Niobium and tantalum

$NbCl_5$ and $TaCl_5$, activated by Et_2AlCl or $EtAlCl_2$ plus an oxygen-containing compound, are effective for the ROMP of cyclopentene (Günther 1970; Dall'Asta 1971a) and norbornene derivatives (Kurosawa 1976b). The system $TaCl_5/Me_3SiCH_2Li$ is also active. The catalyst system itself leads to 35% Me_4Si, 12% Me_3SiCl, and 11% $(Me_3SiCH_2)_2$, based on Me_3SiCH_2Li (Ta/Li = 1/1) (Kershenbaum 1978). This presumably indicates the generation of metal carbenes by reactions such as those shown in Scheme 2.2.

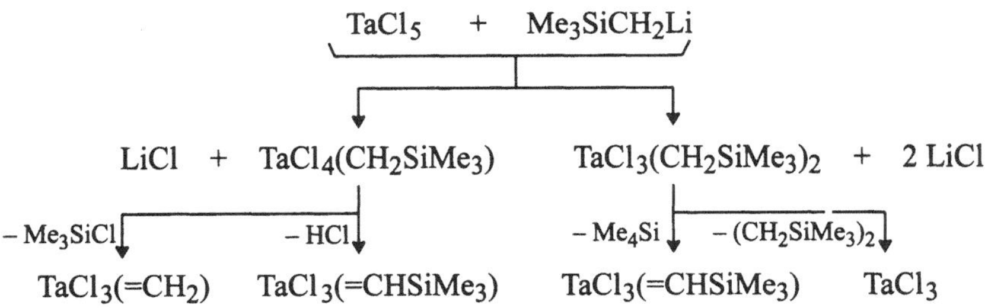

Scheme 2.2 Possible mechanism for generation of tantalum carbene complexes.

A number of stable $[Mt]=CHCMe_3$ complexes (Mt = Nb, Ta) have been prepared which react with terminal olefins in various ways depending on the nature of the ligands. Thus $Ta(=CHCMe_3)(Cl)_2(Cp)$ reacts with ethene to give only the homologation product, reaction (18), but $Ta(=CHCMe_3)(Cl)(OCMe_3)_2(PMe_3)$ reacts with but-1-ene to give stoichiometric metathesis products, Scheme 2.3.

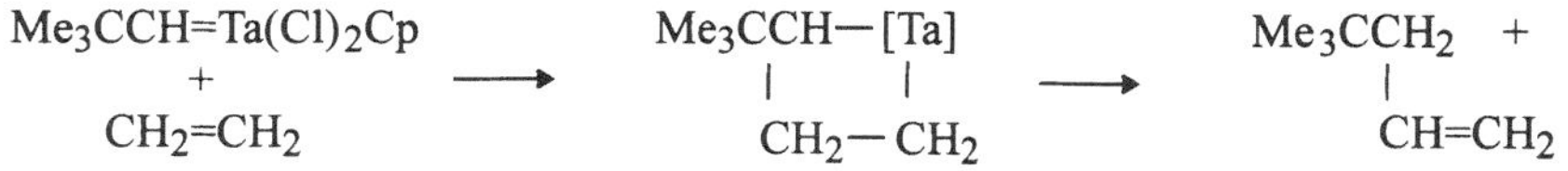

(18)

Note that there is a preferred direction of reaction with the original complex. The reaction is prevented from developing into a chain reaction by the propensity of the $[Ta]{=}CH_2$ complex to undergo the homologation reaction (McLain 1977; Schrock 1980; Rocklage 1981); see Abbenhuis (1994) for another example of a tantalum carbene complex which will undergo only stoichiometric metathesis. However, the complexes **2** and **3** (Table 2.2) are good initiators of ROMP as also is the tantalacyclobutane complex derived from **2** by the addition of norbornene. The complexes **1** and **2** are both effective for the metathesis of *cis*-pent-2-ene.

$Me_3CH{=}[Ta]$ + $CH_2{=}CHEt$ → $Me_3CCH{=}CHEt$ (18%) + $[Ta]{=}CH_2$
→ $Me_3CCH{=}CH_2$ (76%) + $[Ta]{=}CHEt$

$EtCH{=}[Ta]$ + $CH_2{=}CHEt$ → $EtCH{=}CHEt$ + $[Ta]{=}CH_2$ $\xrightarrow{CH_2=CHEt}$ $CH_2{=}CEtMe$ + [Ta]

Scheme 2.3 Reaction of $Ta({=}CHCMe_3)(Cl)(OCMe_3)_2(PMe_3)$ with but-1-ene (Rocklage (1981)

SiO_2-, Al_2O_3- and TiO_2-attached Nb catalysts, containing 2.0–2.3 wt% Nb, prepared by the reaction between $Nb(\pi\text{-allyl})_4$ and surface OH groups of the support followed by thermal treatment with H_2, are active for propene metathesis at 0–37°C. In particular the SiO_2-attached Nb catalyst at 25°C shows a high activity, nearly equal to that of a SiO_2-attached Mo-allyl catalyst, which is among the most active heterogeneous metathesis catalysts (see Section 2.4.2) (Nishimura 1986). A catalyst prepared by impregnation shows no activity under similar conditions. A further study of such catalysts provided one of the first pieces of spectroscopic evidence for the formation of metal carbenes on solid catalysts (Asakura 1989); see sequence (19).

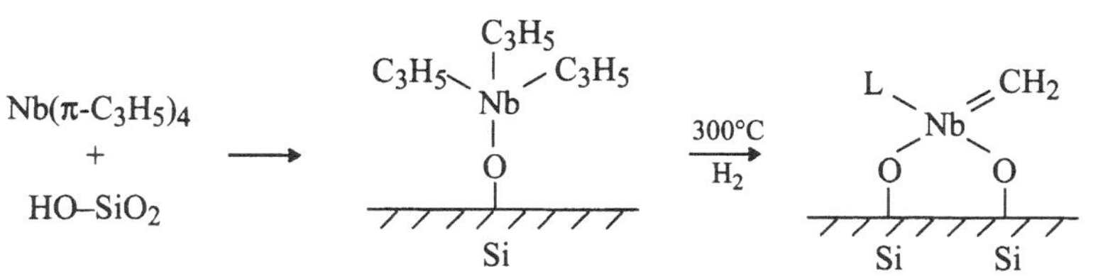

(19)

(L = C_1 or C_3 ligand)

Tantalum and niobium halides also initiate the polymerization of acetylenes; see Ch. 10.

2.4 Group VIA

2.4.1 Chromium

Although many stable chromium carbene complexes are known, attempts to initiate olefin metathesis reactions with Cr-based catalysts have generally failed. Rare successes are the ROMP of 2,3-dihydrofuran catalyzed by $Cr(=CPh_2)(CO)_5$ (Thu 1981b), the ROMP of 2,3,4,5-tetrahydrooxypin-2-yl acetate (Thu 1985) and the metathesis of vinyl ethers, see Ch. 7; also the reactions of pent-1-ene induced by $Bu_4N[CrCl(CO)_5]/MeAlCl_2$ (1/8) in chlorobenzene (Doyle 1973), and of *cis*-pent-2-ene initiated by $Cr(CO)_3$(mesitylene)/$EtAlCl_2/O_2$ (Leconte 1979a).

Chromium oxide catalysts, supported on Al_2O_3 and activated by $LiAlH_4$, have been claimed to cause ROMP of cycloalkenes but the yields are low (Eleuterio 1957). $Cr(CO)_3$(mesitylene) initiates the polymerization of alkynes, possibly via a metallacyclobutene intermediate (Farona 1974; Woon 1974); see Ch. 10.

2.4.2 Molybdenum

Apart from the molybdenum carbene complexes already listed in Tables 2.1 and 2.2 Mo-based catalysts are of three main types: (i) other Mo complexes, activated by a suitable cocatalyst; (ii) $MoCl_5$, also activated by a cocatalyst; and (iii) supported oxides, generated in various ways. For the metathesis of terminal olefins higher than propene, Mo-based catalysts are generally more effective than the corresponding W-based catalysts.

2.4.2.1 Catalyst systems based on $MoCl_5$ and Mo complexes

Many catalysts of these types are known; see also Ch. 6. $Mo(CO)_5(py)/EtAlCl_2/R_4NCl$ is unusual in that it is effective for terminal olefins but not for internal olefins (Farona 1976). $MoCl(CO)_2(PPh_3)_2(NO)/RAlCl_2$ is a very active long-lived catalyst which is readily prepared (Seyferth 1982a). The reaction of $MoCl_3(NO)(OPPh_3)_2$ and $MoCl_2(PPh_3)_2(NO)_2$ with $EtAlCl_2$ gives an [MoCl(NO)]-containing precursor to the active species. The system $MoCl(CO)_2(PPh_3)_2(NO)/EtAlCl_2$ is a highly active metathesis catalyst (Seyferth 1984b,d, 1985). The nitridomolybdenum(VI) complexes $MoNCl_3L_2$ ($L_2 = (OPPh_3)_2$ or Bu_4NCl) in combination with $EtAlCl_2$ have an initial metathesis activity comparable to that of dinitrosylmolybdenum complexes (Seyferth 1982b, 1985).

Many Mo-based catalysts cause the metathesis of *cis* olefins to proceed with a high degree of *cis* stereoselectivity, e.g. for pent-2-ene (Table 6.3). The ROMP of cyclopentene initiated by $MoCl_5/Et_3Al$ at $-30°C$ gives a polymer with 99% *cis* double bonds (Dall'Asta 1971a). This *cis* stereoselectivity also applies to certain degradation reactions; for example $(\pi\text{-}C_4H_7)_4Mo/EtAlCl_2$ with *cis*-1,4-polybutadiene gives short-chain polymer and cyclic oligomer, both having high *cis* content (Kropacheva 1972). Corresponding *trans* stereoselectivity is observed in the initial products of reaction of *trans*-pent-2-ene catalyzed by, for example, $MoCl_2(NO)_2(py)_2/EtAlCl_2$ (Hughes 1969). Unless the initial stereoselectivity is very high, it is usual for secondary metathesis reactions to cause a more or less rapid conversion to the equilibrium *cis/trans* ratio. The initial stereoselectivity can sometimes be opposite. Thus the ROMP of cycloheptene (which contains a *cis* double bond) initiated by $MoCl_5/Et_3Al$ gives a polymer with 93% *trans* content (Natta 1966a), while the ROMP of norbornene initiated by $MoCl_5/EtAlCl_2$ gives a polymer with 65% *trans* content (Ivin 1979c).

The organoisopolymolybdate $[(C_{13}H_{27})_3NH]_4[Mo_8O_{26}]$ with Et_2AlCl is a soluble catalyst combination developed to effect the smooth ROMP of dicyclopentadiene (Goodall 1993).

2.4.2.2 *Supported MoO_3 catalysts*

Supported molybdenum oxide catalysts have received much attention because they are widely used in industrial petrochemical processes, including metathesis (SHOP process, see Ch. 17). Their surface properties and catalytic activity are strongly influenced by the specific oxide support, surface Mo-oxide content, calcination temperature, etc. Such catalysts have generally been made in one of three ways: (i) by impregnation of the support (Al_2O_3, SiO_2, TiO_2) with molybdate solution; (ii) by treatment of the support with $Mo(CO)_6$; (iii) by treatment of the support with organomolybdenum compounds. In each case the supported Mo compound is subjected to an appropriate heat treatment coupled with oxidation (O_2) or reduction (CO, H_2, C_3H_6) to bring it into an active state. Examples of these catalysts are listed in Table 2.3.

The most well-defined supported catalysts are those in group (iii), e.g. supported $(\pi\text{-}C_3H_5)_4Mo$. The high activity of the catalysts in group (iii) is to be attributed to the fact that the Mo is present entirely as a surface compound in its optimum oxidation state. The molybdenum compound attaches itself to the support by reaction with two adjacent hydroxyl groups on the surface, evolving two molecules of propene, eqn. (20); (Candlin 1974; Iwasawa 1978). The resulting surface

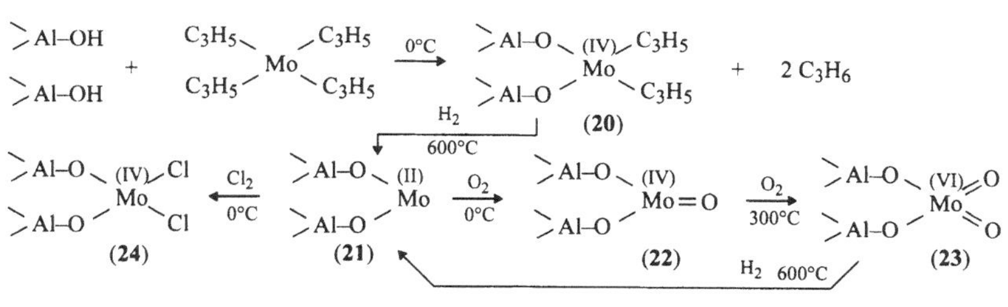

(20)

compound **20** can be converted into **21–24** by stoichiometric reactions with H_2, Cl_2, or O_2. Of these compounds, only those in oxidation state IV, namely **20**, **22**, and **24**, each characterized by X-ray photoelectron spectroscopy (XPS) binding energies of 235.3 and 232.5 eV, show appreciable metathesis activity for propene at 0°C.

Table 2.3 Examples of supported molybdenum catalysts for metathesis reactions[a]

Catalyst system	T/°C	Substrate	Reference
MoO_3/Al_2O_3	20–180	Pent-2-ene	Ismayel-Milanovic 1973
MoO_3/Al_2O_3	40	Propene	Engelhardt 1985a
$MoO_3/CoO/Al_2O_3$	150	Propene	Engelhardt 1982b
MoO_3/γ-$Al_2O_3/LiAlH_4$	20	Cycloalkenes	Eleuterio 1957
$MoO_3/Al_2O_3/Et_3Al$	20–100	Propene	Shmidt 1973
$MoO_3/Al_2O_3/Me_3Al$	60	Non-1-ene	Startseva 1987
$MoO_3/Al_2O_3/Bu_4Sn$		Hept-1-ene	Fridman 1977b
MoO_3/SiO_2	~400	Propene	Sodewasa 1979
MoO_3/β-TiO_2	20	Propene-d_6+d_0	Tanaka, K. 1979, 1981
MoO_3/β-TiO_2/Me_4Sn	25	Propene-d_6+d_0	Tanaka, K. 1984
$MoO_3/SnO_2/Me_4Sn$	25	Propene	Tanaka, K. 1988b
MoO_3/ZrO_2	25	Propene	Indovina 1993
$Mo(CO)_6/Al_2O_3$	69	Propene	Davie 1972b
$Mo(CO)_6/MgO$	25	Propene	Smith, J. 1974a
$Mo(CO)_6/SiO_2$	27	Propene	Howe 1974
$Mo(CO)_6/ZnO/O_2$	20	Propene	Tanaka, K. 1982b
$Mo(CO)_6$/HNa-Y zeolite	1	Propene	Yashima 1986
$Mo(CO)_6/TiO_2$	50	Propene	Masuyama 1988
$Mo(CO)_6$/carbon	0	Propene	Nakamura, T. 1992
$(\eta^6$-$C_6H_6)Mo(CO)_3/Al_2O_3$	50	Propene	Wagner 1987
$MoCl_5$/Na-mordenite	100	Propene	Johns 1988
$Mo(NMe_2)_4/SiO_2$	110	Ethene[c]	Zhuang 1990
$Mo_2(NMe_2)_6/SiO_2$	110	Ethene[c]	Zhuang 1990
$Mo_2(OAc)_4/Al_2O_3$[b]	20	Propene	Evans 1990
$Mo_2(OAc)_4/SiO_2$	25	Propene	Zhuang 1991; Ichikawa, M. 1993
$[Mo_2O_4(C_2O_4)_2(H_2O)_2]^{2-}/Al_2O_3$	150	Propene	Startsev 1990; Klimov 1990b
$[Mo_3O_4(C_2O_4)_3(H_2O)_3]^{2-}/Al_2O_3$	100	Propene	Klimov 1990a
$MoO_2(C_5H_7O_2)_2/\gamma$-$Al_2O_3$	100	Propene	Klimov 1991
$(\pi$-$C_3H_5)_4Mo/SiO_2$	90	Propene	Yermakov 1974, 1975/76; Kuznetsov 1975
$(\pi$-$C_3H_5)_4Mo/SiO_2$	20	Hex-1-ene	Candlin 1974
$(\pi$-$C_3H_5)_4Mo/SiO_2$-Al_2O_3	0	Propene	Iwasawa 1978, 1980, 1981
$(\pi$-$C_3H_5)_4Mo/Al_2O_3$	22	Propene, etc.	Ichinose 1978
$(\pi$-$C_3H_5)_4Mo/SiO_2$, etc.	350	Hex-2-yne, etc.	Mortreux 1980
$MoCl_5/SiO_2/R_4Sn$	20	Dec-1-ene	Bykov 1988
$MoOCl_4/SiO_2$-Al_2O_3/Et_3Al, Bu_4Sn	60	Pent-1-ene	Maksimowski 1991
$[Mo_2(MeCN)_8][BF_4]_4/SiO_2$	90	Norbornene	McCann 1994

[a] Catalysts suitably activated; see original papers for details. [b] Ac = CH_3CO. [c] $^{12}C/^{13}C$ exchange.

Neither **21** [Mo(II)] nor **23** [Mo(VI)] is active; nor do catalysts containing a few % Mo(V) show any correlation of activity with the electron spin resonance (ESR) signal due to Mo(V). The basic nature of the allyl ligands makes it understandable that reaction (20) is more facile when the support is more acidic. Thus, the following order is observed for the rate of propene metathesis: SiO_2-$Al_2O_3 > Al_2O_3 \gg SiO_2$, corresponding to the decreasing Brönsted acidity of these supports (Iwasawa 1978, 1981). It is a remarkable fact that the SiO_2 analogue of **23** is a highly active and stable catalyst for the metathesis of hex-2-yne at 350°C, with a selectivity of 90–100%, the activity being several times that of a catalyst prepared by the impregnation method (Mortreux 1980). In this case it is possible that the reactant itself effects a change of oxidation state before the reaction proper sets in. The increase of rate during the metathesis of propene on MoO_3/SiO_3 at 407°C (Thomas 1980) is likewise to be attributed to the formation of a more favourable, lower oxidation state ('break-in' phenomenon).

Surface compounds containing two adjacent Mo sites can be formed by reaction of $(\pi\text{-}C_3H_5)_4Mo_2$ with the surface hydroxyl groups of Al_2O_3 or SiO_2. Their activity is generally not much different from those of $(\pi\text{-}C_3H_5)_4Mo$, which suggests that the metathesis reaction proceeds on each of the Mo(IV) sites of the Mo_2 compound, although there may be some perturbation of the local structure of a molybdenum atom by an adjacent Mo structure (Iwasawa 1985). A SiO_2-grafted dinuclear molybdenum species can be derived from $Mo_2(OAc)_4$ by reaction with silanol groups on SiO_2 followed by thermal activation below 350°C. This Mo_2(V) species, which has been characterized by several spectroscopic techniques and which gives an ESR signal, is remarkably active for propene metathesis at 20°C (Zhuang 1991; Ichikawa, M. 1993). By thermal activation at >400°C, however, the Mo-Mo multiple bond is cleaved to give highly dispersed Mo(V) oxide which is not active in propene metathesis (Ichikawa, M. 1993).

For MoO_3/Al_2O_3 optimum selectivity is obtained if the MoO_3 is only slightly reduced during activation (Engelhardt 1982e). Others report that the metathesis activity is nearly independent of extent of reduction (Lombardo 1980), but using the same catalyst Indovina (1993) concludes that mononuclear Mo(V) species are the active site precursors.

A comparison has been made of the activity of Al_2O_3-supported Mo catalysts prepared in various ways. The order of activity for propene metathesis at 100°C is: $MoO_2(acac)_2 > [Mo_3O_4(C_2O_4)_3(H_2O)_3]^{2-} > [Mo_2O_4(C_2O_4)_2(H_2O)_2]^{2-} > Mo(C_3H_5)_4 > (NH_4)_6Mo_7O_{24}.4H_2O$, where the molybdenum compound is that used to prepare the catalyst, which is finally activated by reduction in CO or H_2 at 500°C (Startsev 1993); also see Klimov (1995a,b).

After reduction, $MoO_3/\gamma\text{-}Al_2O_3$ is active for both self-metathesis and polymerization of ethene. Nitric oxide is capable of poisoning the sites for both reactions (Olsthoorn 1980); that which is adsorbed on the active site for metathesis gives an IR absorption at 1796 cm^{-1}. The number of metathesis sites determined from poisoning experiments is only a small fraction of the molybdenum atoms present (Lombardo 1978; Hardee 1979, 1983). This may be one of the reasons that different values for the optimal oxidation state of supported Mo oxide catalysts have been reported; since only a small fraction of Mo atoms are active, the oxidation state of the active sites may be different from the average oxidation state. After activation of MoO_3/Al_2O_3 at the usual temperature (550°C) the activity for the metathesis of propene at 200°C shows an S-shaped dependence on Mo content, increasing sharply between 0.5 and 2 Mo atoms nm^{-2} (Thomas 1982). If activation is carried out at much higher temperature (870°C) under argon, the metathesis activity is greatly improved, especially at low Mo content; see Fig. 2.1 (Grünert 1992a). The low activity at low Mo content when the catalyst is activated at 547°C is attributed to the difficulty of reducing the extremely stable tetrahedrally coordinated Mo(VI), which is the dominant species at low Mo content (Grünert 1988, 1992a). At higher Mo content (> 4 wt% MoO_3) the Mo is mainly present in octahedrally coordinated form and is easier to reduce (Okamoto 1988). The fact that with high activation temperature the turnover frequency (TOF) actually decreases with increasing Mo content suggests that a second, more active site is produced from the tetrahedrally coordinated Mo(VI) atoms.

It is concluded that on a reduced MoO_3/Al_2O_3 surface, e.g. after activation in hydrogen, active sites are formed from both isolated Mo(IV) and Mo(VI) precursors, whereas after thermal treatment in inert gas the activity originates exclusively from active sites derived from Mo(VI) precursors. Mo-carbene formation might take place in a way similar to that for photocatalytic propene metathesis; see Section 2.7, eqn. (28).

The catalytic activity of MoO_3/SiO_2 also depends on the Mo content (Nakamura, R. 1972b; Vaghi 1976; Thomas 1980; Zhang, B. 1988): the TOF increases with Mo content up to about 1 Mo ion nm^{-2} corresponding to the formation of a monolayer

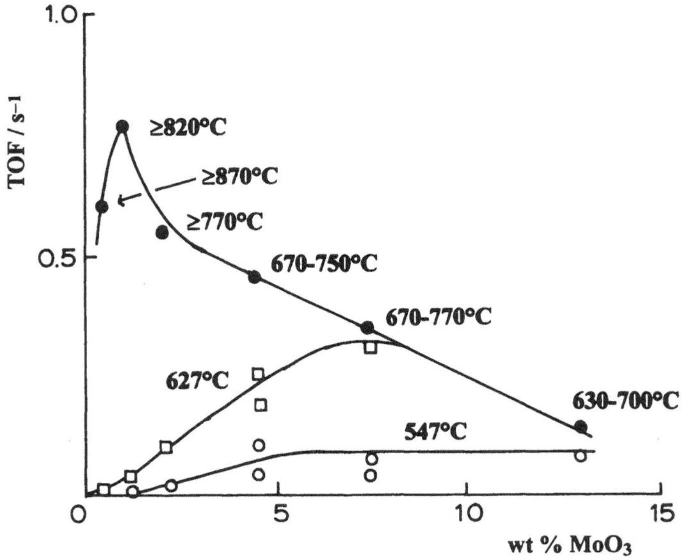

Fig. 2.1 Turnover frequencies (TOF) for the metathesis of propene (1 bar, 200°C) on MoO_3/Al_2O_3 as a function of MoO_3 content and catalyst activation temperature (shown on the figure) (Grünert 1992a).

of Mo species with maximum dispersion. At higher Mo concentrations, a decrease in activity is observed, owing to formation of inactive bulk MoO_3 (Mol 1990b). In a reduced MoO_3/SiO_2 impregnation catalyst, a complete parallelism is observed between the concentration of tetrahedrally coordinated Mo ions, as determined by a photoluminescence technique, and the rate of propene metathesis at 200°C (Ono 1986a,b). It is concluded that the tetrahedral Mo species (**25**) gives the active site precursor (**26**) after reduction. With increasing Mo content the fraction of inactive crystalline MoO_3 increases, thus decreasing the activity (Ono 1986b). In contrast, a catalyst prepared by vapour deposition of $MoCl_5$ on dehydrated SiO_2, containing only tetrahedral Mo ions, has a remarkably high activity, which is attributed to the isolated character of the Mo ions (Ono 1986a). The oxidation state of the active Mo site on silica is still a matter of debate. From an ESR study of a MoO_3/SiO_2 catalyst, a Mo(V) species with a distorted square pyramidal structure is suggested to be primarily responsible for metathesis (Zhang, B. 1988, 1991), although others have reported that mononuclear Mo(V) oxide/SiO_2 is not active for metathesis (Ichikawa, M. 1993).

Photoreduction in CO at room temperature of calcined MoO_3/SiO_2 catalysts, prepared by impregnation of silica with an ammonium paramolybdate solution, gives a more active metathesis catalyst than thermally reduced MoO_3/SiO_2 (Shelimov 1986). The photoreduction, using the full radiation of a high-pressure Hg lamp, is visualized to occur as shown in eqn. (21). It is reported that 50% (for catalysts with 1% Mo) to 70% (for catalysts with 0.1% Mo) of the molybdenum is transformed into coordinatively unsaturated surface Mo(IV) ions (characterized by UV/vis spectroscopy) (Pershin 1980). An even higher yield (70–80%) has been reported for catalysts with 1 wt% Mo (Kazansky 1993). Evidence for high homogeneity of Mo(IV) ions in photoreduced samples comes from microcalorimetric measurements (Elev 1988). These ions can react with olefins to generate the active molybdenum carbenes (Shelimov 1986).

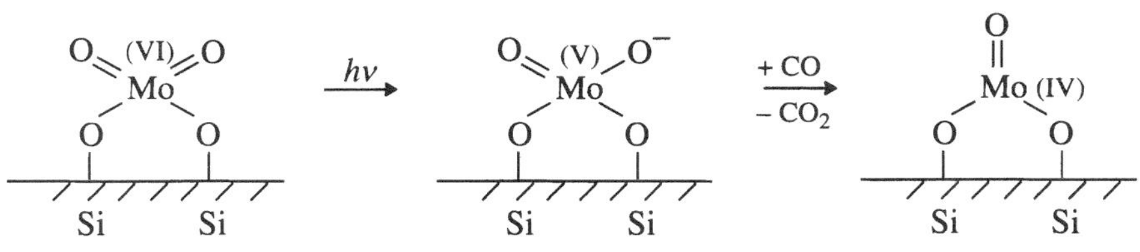

(**25**) (**26**)

(21)

If the photoreduced MoO_3/SiO_2 catalyst is treated with cyclopropane (Elev 1989a), methylcyclopropane (Vikulov 1991) or cycloheptatriene (Vikulov 1992b), followed by evacuation at 300–350°C, a much more active metathesis catalyst is obtained. Molybdenum carbenes are formed, as shown by IR and UV/vis spectroscopy (Vikulov 1989a,b), and are assumed to result from a sequence of reactions such as those shown for cyclopropane in Scheme 2.4 (Shelimov 1988;

Vikulov 1989b; Kazansky 1991a). A small proportion ($<5\%$) of the molybdacyclobutane complexes yields propene by reductive elimination at 350°C. Calorimetric measurements support this scheme (Vikulov 1992a). The reaction of

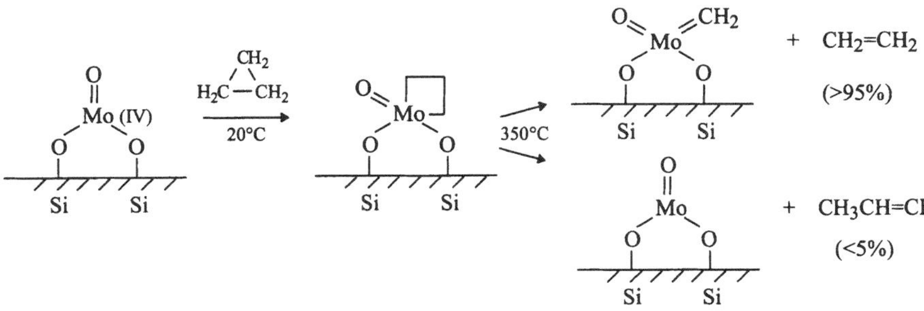

Scheme 2.4 Generation of molybdenum carbenes by the reaction of cyclopropane with Mo(IV)/SiO_2 (Shelimov 1988).

methylcyclopropane yields both [Mo]=CH_2 and [Mo]=CH−CH_3 (Vikulov 1991; Kazansky 1991b). The reaction of 1,3,5-cycloheptatriene (CHT) gives [Mo]=CH_2 via a rather complicated stepwise scheme (Vikulov 1992b; Kazansky 1993). The extent of Mo(IV) conversion into molybdenum carbene species is 90% for cyclopropane, 55% for methylcyclopropane and 10% for cycloheptatriene. The highest activity for propene metathesis is obtained with cyclopropane treatment (see Ch. 5). Photoreduction of MoO_3/SiO_2 in CO with a laser beam ($\lambda = 308$ nm) instead of a mercury lamp is at least as efficient. If the same laser-photoreduction procedure is applied to MoO_3/SiO_2-Al_2O_3 and followed by cyclopropane treatment, a catalyst of even higher activity for propene metathesis is obtained (Mol 1994a).

MoO_3/SiO_2 catalysts can also be activated by mechanical treatment at room temperature, first in hydrogen and then in helium. This results in the formation of Mo(IV) ions via reactions involving Si and H radicals. The activity is even better than that of photoreduced MoO_3/SiO_2. Subsequent cyclopropane treatment increases the activity by a factor of six (see Table 5.3) (Bobyshev 1992).

MoO_x/β-TiO_2 ($x = 2.3$–2.9) is remarkable in that the exchange between *trans*-but-2-ene-d_0 and -d_8 at 25°C is very stereoselective, giving 98% *trans*-CH_3CH=$CDCD_3$ in the initial products. The reaction of the *cis*-but-2-enes is rather less stereoselective ($>60\%$ *cis*-CH_3CH=$CDCD_3$) (Tanaka, K. 1980a). It appears that some oxygen must be coordinated to the Mo ion in order that the site shall be active (Tanaka, K. 1981). A catalyst for the metathesis of propene at 100°C can be obtained by reducing highly dispersed MoO_3/TiO_2 to MoO_x/TiO_2 ($x = 2$) in a gas flow of 5% H_2 in N_2. However, no metathesis activity is observed if $x > 2.5$ or $x < 1$ (Segawa 1988, 1991).

An ESR study of a reduced MoO_3/MgO catalyst (Mo/Mg = 0.5–0.7) indicates that Mo(V) ions with a coordination number less than six are the active sites for propene metathesis at 150°C with this catalyst (Hasegawa, S. 1992). For the catalyst MoO_3/ZrO_2, reduced by H_2, Mo(V) ions with special coordination features are thought to be the active sites for metathesis (Indovina 1993). For the metathesis of propene over a Mo/HNa-Y zeolite, it is concluded that the active sites are a slightly aggregated Mo(0) species (Komatsu, T. 1985, 1987).

Inactive fully oxidized MoO_3 samples supported on various kinds of oxides have been treated with Me_4Sn to convert them to active metathesis catalysts. For instance, MoO_3/TiO_2 (Tanaka, K. 1984, 1988c), MoO_3/SiO_2 (Tanaka, K. 1987a), MoO_3/Al_2O_3, MoO_3/ZrO_2 and MoO_3/SnO_2 (Tanaka, K. 1988b), which are inactive at room temperature, are successfully activated for propene metathesis by treatment with Me_4Sn, while the metathesis activity of partially reduced MoO_3, supported on TiO_2, SiO_2, Al_2O_3 or ZrO_2 is greatly enhanced by treatment with Me_4Sn (Tanaka, K. 1986a, 1990b); see also Table 5.1. CH_4 is formed in small amounts upon treatment of MoO_3/TiO_2 or MoO_x/TiO_2 with Me_4Sn and it is concluded that $[Mo]{=}CH_2$ species are formed on the catalyst surface.

An interesting observation has been made on the activity of the $Mo(CO)_6/SiO_2$ system. As the catalyst temperature is raised, CO is evolved and the activity for the metathesis of propene increases, reaching a peak at about 140°C. At temperatures exceeding 200°C most of the CO has been evolved and the metathesis activity falls to zero. Two observations indicate an increase in oxidation number of Mo by reaction with the support during the heating process. First, a certain amount of H_2, CO_2, and CH_4 is evolved during the heating period, and second, the evolution of CO is irreversible (Brenner 1979). Optimum conditions for metathesis are evidently reached before this process has gone too far.

Unlike SiO_2, which can hold an initial amount of only 0.1% $Mo(CO)_6$, Al_2O_3 can support up to 6%. The formation of active centres for propene metathesis is dependent on the density of surface hydroxyls on the alumina (Matyshak 1983). Activation of $Mo(CO)_6$ on partially dehydroxylated Al_2O_3 by heating in flowing helium or hydrogen leads to the loss of CO, beginning at approximately 25°C. By 100°C the catalyst can be represented as $Mo(CO)_3/Al_2O_3$ or $[Mo(CO)_3]_{ads}$. The formation of $[Mo(CO)_3]_{ads}$ at 100°C is indicated by diffuse reflectance spectroscopy (DRS) and thermal desorption (Kadushin 1982b). On prolonged heating in flowing helium or hydrogen, the species $[Mo(CO)_3]_{ads}$ appears to be very stable, in particular if the alumina is not extensively dehydroxylated. Heating in helium to higher temperatures leads to further liberation of CO starting at around 150°C; above 300°C complete and irreversible decarbonylation takes place, accompanied by a reaction of Mo with the hydroxyl groups of the support, causing oxidation of the molybdenum to Mo(II) (Burwell 1984). Further heating results in oxidation of Mo(II), and by 500°C the molybdenum is mostly present as Mo(VI). The highest initial metathesis activity is found when $[Mo(CO)_3]_{ads}$ is preheated in helium at 280°C (Brenner 1978). A mechanism involving reaction of the surface hydroxyl groups with the adsorbed $Mo(CO)_3$ to form surface $[Mo]{=}CH_2$ species has been proposed (Brown 1981). The metathesis activity of $[Mo(CO)_3]_{ads}$ is increased by a factor of 15 by exposure to oxygen at room temperature, $[Mo(CO)_2O_2]_{ads}$ supposedly being formed (Burwell 1975/76). $Mo(CO)_6$ adsorbed on Al_2O_3

pretreated at 250°C gives, after evacuation at 100°C, a catalyst that is active not only for metathesis of $C_2H_4 + C_2D_4$ but also for ethene polymerization and H/D exchange (Olsthoorn 1980).

Molybdenum oxide films prepared by sublimation *in vacuo* on to a glass wall are converted into an active catalyst for propene metathesis at 0.4 kPa and 20°C by treatment of the surface with ethene (or propene) and atomic hydrogen at −196°C (Kazuta 1987). Active sites can also be obtained by exposing the molybdenum oxide film to CH_2 radicals produced by the reaction of CH_2I_2 with Al or Mg metals deposited near the oxide film (Kazuta 1988, 1990).

At 600°C propene metathesis can also be catalyzed by metallic molybdenum in the form of an Mo(100) single crystal or a clean polycrystalline molybdenum foil, although the TOFs are considerably lower than for the supported oxide catalysts (Wang, L.P. 1990, 1993). Tests of various molybdenum oxides supported on an Mo foil, prepared both by oxidation of metallic molybdenum and reduction of MoO_3, suggest that the presence of an oxygen layer on the catalyst inhibits reaction and that the ranking of activity for propene metathesis at 600°C and 60 kPa is $MoO_2 > MoO_3 > Mo$, where MoO_2 is ~30 times more active than Mo metal (Bartlett 1993).

2.4.3 Tungsten

There are perhaps more known W-based catalysts for olefin metathesis than all others together. Many studies have been made on the catalyst systems themselves in an attempt to elucidate the nature of the active species and its mode of formation. They are especially effective for internal and cyclic olefins. Many examples will be found throughout this book. Here we will illustrate some of the main types.

W-based catalysts for the metathesis of terminal olefins are comparatively few in number. However, this is partly an illusion because systems such as $WCl_6/EtAlCl_2/EtOH$, although not effective in the sense of yielding ethene and an internal olefin, cause rapid non-productive metathesis in which the products can only be distinguished from the reactants by isotopic labelling; see Ch. 5.

Typical systems are listed in Tables 2.1, 2.2, and 2.4. The stereospecificity, especially for the ROMP of cycloalkenes, can vary widely according to the precise nature of the catalyst system. There is, however, a marked tendency towards retention of *cis* double bonds in certain cases: (i) if the catalyst is $W(=CPh_2)(CO)_5$ or $WF_6/RAlCl_2$; (ii) if the cocatalyst is a metal allyl compound; (iii) if certain additives are present, such as ethyl acrylate (Ivin 1979c), 2-*t*-butyl-*p*-cresol (Castner 1977b), or divinyldiphenylsilane (Goodyear 1977); and (iv) if the temperature is low. Such effects are associated with increased crowding at the site of reaction and hence greater steric control.

Electrochemical reduction of WCl_6, eqn. (22), or $MoCl_5$ in methylene chloride under controlled potential at a platinum cathode, with an aluminium anode, gives *in situ* formation of a species that catalyzes olefin metathesis. Good activity and high

$$WCl_6 + e^- \xrightarrow{CH_2Cl_2} Cl_nW{=}CH_2 \tag{22}$$

selectivity is maintained even after several charges of olefin, e.g. pent-2-ene, have undergone metathesis (Gilet 1983; Bages 1990a).

During the last decade different tungsten(VI) phenoxides of the type $WCl_{6-x}(OAr)_x$, $WOCl_{4-y}(OAr)_y$, or $W(=NAr)Cl_{4-y}(OAr)_y$ have been reported which in combination with a cocatalyst, e.g. R_xAlCl_{3-x} or an alkyltin hydride, form very active catalysts for the metathesis of linear as well as cyclic olefins (Bell, A. 1994b). They can be even more active for the metathesis of pent-2-ene when brought on to a support (Verpoort 1996a,b).

Tungsten(VI) carbyne complexes exhibit varying degrees of metathesis activity, depending on the ligands. For instance, $W(\equiv CCMe_3)(Cl)_3(dme)$ catalyzes the metathesis of linear and cyclic olefins at 20°C (Weiss 1986a, 1988c), whereas $W(\equiv CCMe_3)(OCMe_3)_3$ only becomes active for the metathesis of oct-1-ene in hexane at 69°C after reaction with SiO_2. The carbyne complex is presumed to react with the weakly Brönsted-acidic Si-OH protons, resulting in a Schrock-type tungsten carbene complex (Weiss 1989b,c); reaction (23). The surface W(VI) carbene complex is coordinatively unsaturated and the electrophilic character of the W centre is further increased by the coordination of siloxy groups. The heterogeneous tungsten(VI) systems $W(\equiv CCMe_3)Np_3/SiO_2$ (Np = neopentyl) and $W(\equiv CCMe_3)(Cl)_3(dme)/SiO_2$ also have high catalytic activities in the metathesis reactions of linear alkenes, α,ω-dienes and cycloalkenes (Weiss 1990b). For example, the first of these systems shows a TOF of 830 min^{-1} for oct-1-ene at 122°C (Weiss 1989c), 956 min^{-1} for dec-1-ene at 170°C (Weiss 1990b) and 200 min^{-1} for dec-4-ene at 122°C (Weiss 1989b).

$$W(\equiv CCMe_3)(OCMe_3)_3 + 2 \equiv Si{-}OH \longrightarrow (\equiv Si{-}O)_2W(=CHCMe_3)(OCMe_3)_2 + Me_3COH \tag{23}$$

Other supports with a stronger surface acidity can be used to generate very efficient metathesis catalysts. For example, in the metathesis of *cis*-pent-2-ene $W(\equiv CCMe_3)Np_3/Nb_2O_5$ (initial TOF = 370 min^{-1} at 20°C), and $W(\equiv CCMe_3)(Cl)_3(dme)/Nb_2O_5$ (TOF = 60 min^{-1}) are more active than $W(\equiv CCMe_3)Np_3/SiO_2$ (TOF = 15 min^{-1}) or $W(\equiv CCMe_3)(Cl)_3(dme)/SiO_2$ (TOF = 0). Al_2O_3 as the support also gives good activity. The highest conversions are observed with SiO_2-Al_2O_3 as the support, but the selectivities are poor as a result of double-bond shift reactions, followed by cross-metathesis (Buffon 1993, 1994).

WO_3/Al_2O_3 has a lower activity towards olefin metathesis than MoO_3/Al_2O_3 owing to the fact that WO_3/Al_2O_3 is more difficult to reduce (Thomas 1982). Likewise, after standard activation, the performance of WO_3/Al_2O_3 is much worse than that of WO_3/SiO_2 (Andreini 1985). The reduction behaviour of WO_3/Al_2O_3 catalysts has been studied by XPS and ESR (Grünert 1987, 1989a). When the WO_3 content approaches the monolayer capacity of the alumina, the reduction proceeds at temperatures of ⩽700°C; a W(IV) intermediate is detected, while very small amounts are converted to W(V), which is of no relevance to metathesis (Grünert 1989a). WO_3/Al_2O_3 catalysts with low metal oxide content (< 12 wt% WO_3) may

Table 2.4 Examples of W-based catalyst systems[a] for productive olefin metathesis (see also Tables 2.1 and 2.2)

Catalyst system[b]	*T*/°C	Olefin	Reference
Terminal olefins[c]			
WO_3/SiO_2	400	C_3	Heckelsberg 1968
$WO_3/SiO_2/MgO$	250–400	C_3	Banks 1985
WO_3/Al_2O_3	150	C_3	de Vries 1977
WO_3/TiO_2	240	C_3	Andreini 1985
WO_3/SiO_2-Al_2O_3	400	C_3	Andreini 1985
$W(CO)_6/SiO_2$	140	C_3	Brenner 1979
$W(CO)_3/Al_2O_3$	130	C_3	Okamoto 1988
$(\pi$-$C_3H_5)_4W/SiO_2$-Al_2O_3	20	1-C_6	Oreshkin 1971
$(\pi$-$C_4H_7)_4W/SiO_2$	250	C_3	Startsev 1975
(isoprene)$_3W/SiO_2$	100	C_3	Startsev 1986
Me_6W/Al_2O_3	25	C_3	Mowat 1974
$W(CO)_5[P(OPh)_3]/EtAlCl_2/O_2$	20	$PhCH_2CH{=}CH_2$	Bilhou 1977c
(PhMe)$W(CO)_3$	98	$PhCH{=}CH_2$	Lewandos 1971b
$WCp(CO)_3^-/Al_2O_3$	220	C_3	Otten 1992
$WCl_6/Et_3Al_2Cl_3/PhNH_2$	20	C_3	Menapace 1975
$WCl_6/Et_3Al_2Cl_3/X/PPh_3$[d]	20	1,4-C_5, etc.	Dall'Asta 1973
WCl_6/R_4Sn	50	1-C_6	Bespalova 1975
WCl_6/Ph_4Sn	22	1-C_7	Balcar 1983
$WCl_6/Bu_4Sn/MeCN$[e]	80	1-C_5	Ichikawa, K. 1976a
WCl_6/Et_3SiH	50	1-C_5	Nametkin 1973
$WCl_2(OAr)_4/EtAlCl_2$[f]	0	1-C_8, 2-C_5	Dodd 1982
$W(2\text{-}MeC_6H_4O)_4Cl_2/Et_3Al_2Cl_3$	20	1-C_8	Dodd 1988
$W(4\text{-}ClC_6H_4O)_4Cl_2/Bu_4Sn$	140	1-C_8	Dodd 1988
$WCl_6/Me_4Sn/EtOAc$	80	1-C_{13}	Gibson, T. 1981
WCl_6/Ph_4Pb	75	1-C_{10}	Startseva 1987
Internal acyclic olefins[c]			
$W(CO)_5(PPh_3)/EtAlCl_2/O_2$	20	*cis*-2-C_5	Bilhou 1977b
$[Et_4N][WCl(CO)_5]/ZrCl_4$		2-C_5	Szymańska-Buzar 1987
$W(CO)_4(MeCN)_2/AX_n$[g]	20	2-C_5	Szymańska-Buzar 1991
$WCl_2(CO)_3(PPh_3)_2/ZrCl_4$	20	2-C_5	Szymańska-Buzar 1991
$WCl_2(CO)_3(AsPh_3)_2$/norbornene/$AlCl_3$	80	2-C_5	Bencze 1985a
$R_nCl_{3-n}SnW(CO)_3Cp/i\text{-}BuAlCl_2/O_2$ [R = Me, Ph; *n* = 0–3]	20	2-C_5	Vanderyse 1988a,b
$[W(CO)_3Cp]_2/MeSnCl_3/i\text{-}BuAlCl_2/O_2$	20	2-C_5	van Ruyskensvelde 1995
$WOCl_4/EtAlCl_2$	20	2-C_5	Mocella 1976b
$WOCl_4/(t\text{-}BuCH_2)_2Mg$	25	2-C_5	Kress 1980
$WCl_6/EtAlCl_2/EtOH$	20	various	Calderon 1968
WCl_6/Al anode/e^-[h]	20	2-C_5	Gilet 1979

Table 2.4 Continued

Catalyst system[b]	T/°C	Olefin	Reference
WCl_6/BuLi (or PrLi)	25–130	2-C_5	Wang, J-L. 1973; van Dam 1974d
WCl_6/Ph_2SiH_2	20	2-C_6 + 2-C_4	Levisalles 1980a
WCl_6/Ph_4Sn	20	2-C_5	van Dam 1974a
WCl_6/Me_4Sn	20	2-C_5	Thorn-Csányi 1981
WCl_6/PhC≡CH	30	*cis*-2-C_5	Katz 1982
$WCl_{6-x}(OAr)_x/EtAlCl_2$[f,i,j]	25	*cis*-2-C_5	Quignard 1985a
$WCl_4(OC_6H_3\text{-}X_2\text{-}2,6)_2/R_4Mt$[k] [X = Me, Ph, Br, Cl, F]	85	*cis*-2-C_5	Quignard 1986
$WBr_5/EtAlCl_2$		*cis*-2-C_5	Vialle 1975
WO_3/SiO_2/*i*-$BuAlCl_2$	20	2-C_5	Verpoort 1995
$WCl_6/SiO_2/Me_4Sn$	22	2-C_5	van Roosmalen 1980b
$WCl_4(OC_6H_3\text{-}X_2\text{-}2,6)_2/SiO_2$/*i*-$BuAlCl_2$ [X = Me, *t*-Bu, Br, Cl]	20	2-C_5	Verpoort, 1996b
$WCl_4(OC_6H_3\text{-}X_2\text{-}2,6)_2/NbO_x/SiO_2$/*i*-$BuAlCl_2$ [X = *t*-Bu, Br]	20	2-C_5	Verpoort 1996a
$WCl(CO)_3Cp/SiO_2$/*i*-$BuAlCl_2$	20	2-C_5	Verpoort 1995
Cycloalkenes[l]			
$(mes)W(CO)_3/EtAlCl_2$/EPO[m]	20	NBE	Ivin 1981b
$(\pi\text{-}C_4H_7)_4W/AlBr_3$	30	COC	Kormer 1972
$WF_6/EtAlCl_2$	0	CPE	Ofstead 1980
$WCl_6/Et_3Al/(PhCOO)_2$[n]	−50	CPE	Minchak 1972
$WCl_6/Et_2O/Et_4Sn$	20	CPE	Lehnert 1974
$WOCl_4/(C_3H_5)_4Si$	−30	CPE	Pakuro 1982
WCl_6/R_4Sn	20	NBE	Ivin 1979c
WCl_6/Me_4Sn/siloxanes	45	COC	Guay 1995
WCl_6/*i*-Bu_3Al/phenols	20	CPE	Chen 1981
WCl_6/PhC≡CH	21	CPE	Katz 1982
$WOCl_4$	70	DCPD	Balcar 1992a
$WCl_5(OC_6H_4$-*t*-Bu-4)/Et_2AlCl, Bu_4Sn	20	DCPD	Balcar 1992a
$WCl_4[OCH(CH_2Cl)_2]_2/Et_2AlCl$	20	DCPD	Balcar 1992a
$WCl_{6-x}(OAr)_x/Bu_3SnH$ (x = 1,2)	90	DCPD	Sjardijn 1986, 1987
$WOCl_{4-y}(OAr)_y$[f,p]/R_3SnH[q]	32/80	DCPD	Bell, A. 1992
$W({=}NPh)Cl_4(OEt_2)/Et_2AlCl$	30/147	DCPD	Bell, A. 1993
$WCl_2(CO)_3(AsPh_3)_2$	80	NBE	Bencze 1985b
$[R_3NH]_x[XW_{12}O_{40}]$[r] (x = 3, X = P; x = 4, X = Si; x = 5, X = B)	60	DCPD	Goodall 1993

[a] Oxide-supported catalysts generally used with gaseous olefins; other catalysts used with olefins in solution (e.g. in hexane, benzene, chlorobenzene, toluene, or trichloroethene). [b] Oxide-supported catalysts require appropriate activation. [c] C_3 denotes propene, 1-C_4 but-1-ene, 1,4-C_5 penta-1,4-diene, 2-C_5 pent-2-ene, etc. [d] X is $MeCCl(OH)(CH_2Cl)$. [e] MeCN improves selectivity (94%). [f] OAr = various substituted phenoxides. [g] $AX_n = ZrCl_4$, $AlCl_3$, $TiCl_4$. [h] Electrolysis in a halogenated solvent; high selectivity (95–100%). [i] x = 2, 4. [j] The most active is $WCl_4(OC_6H_3\text{-}X_2\text{-}2,6)_2$ (X = Cl, Br). [k] Mt = Sn, Pb; R = Me, Bu; order of activity for a given cocatalyst: X = Me < Ph < F < Cl < Br; for a given precursor complex, the activity increases in the order $Me_4Sn < Bu_4Sn < Bu_4Pb$. [l] CPE = cyclopentene; COC = cyclooctene; NBE = norbornene; DCPD = dicyclopentadiene. [m] mes = mesitylene; EPO = *exo*-2,3-epoxybicyclo[2.2.1]hept-2-ene. [n] *Cis* content of polymer shows strong dependence on temperature: 100% and 10% at −50°C and 30°C, respectively. [p] y = 1–3. [q] R = alkyl or aryl. [r] R = C_{14-18} alkyl chains.

be activated at very high temperatures (up to 870°C for <4 wt% WO_3) in flowing argon and then exhibit specific activities for propene metathesis superior to those of catalysts with higher metal oxide content and comparable to WO_3/SiO_2, in particular at low W loadings (see Ch. 5). The inverse relationship between the specific activity and the reducibility, and the inability of hydrogen to improve the performance of a catalyst with ≤ 12 wt% WO_3, suggests that active sites can be formed not only from W(IV) but also from W(VI) precursors, and that W(VI) is the only active site precursor on WO_3/Al_2O_3 catalysts with WO_3 loadings far less than the monolayer capacity of alumina (Grünert 1988, 1989b).

SiO_2-supported WO_3 (and MoO_3) catalysts exhibit characteristic induction periods when they are brought into contact with the reactant olefin, e.g. propene; see Ch. 5.

The system $W(CO)_6/Al_2O_3$ is very similar in behaviour to $Mo(CO)_6/Al_2O_3$, both with respect to its surface chemistry during activation and its subsequent metathesis activity (Banks 1964; Smith, J. 1974b). Supported allyl and methallyl tungsten complexes will be discussed in Ch. 5.

2.5 Group VIIA

2.5.1 Technetium

The technetium complex $TcCl(CO)_3(PPh_3)_2$ in combination with $EtAlCl_2$ is moderately active for metathesis of pent-2-ene (Lorenz 1982).

2.5.2 Rhenium

Catalysts containing rhenium are generally based on Re_2O_7, $ReCl_5$, or a carbonyl or other complex compound. Some typical examples are listed in Table 2.5.

Supported rhenium oxide catalysts have attracted much attention because of practical applications (see Ch. 17). One of the potential advantages of rhenium catalysts is that they are more tolerant of functional groups than tungsten or molybdenum catalysts (see Ch. 7).

When the oxide is supported on Al_2O_3, the reaction may be carried out at 20–100°C, with the substrate olefin either in the gas phase or in solution, and takes place with high selectivity. This catalyst, when suitably activated, brings about the metathesis of all types of olefin, including unsaturated esters and allyl compounds; see Ch. 7. The oxide catalyst is typically prepared by impregnating γ-Al_2O_3 with ammonium perrhenate solution, drying at 110°C, and heating first in dry air at 550°C and then in nitrogen.

The surface structure of Re_2O_7/Al_2O_3 metathesis catalysts has been extensively studied by many techniques, including DRS (Edreva-Kardjieva 1976, 1985, 1992; Kadushin 1982a); ESR (Nakamura, R. 1982; Xu, X. 1985a; Xu, Y. 1986), extended X-ray absorption fine structure (EXAFS) (Ellison 1985, 1986, 1988), fast atom

Table 2.5 Examples of Re-based catalyst systems for olefin metathesis

Catalyst system	T/°C	Olefin	Reference
Re_2O_7/Al_2O_3[a]	20–100	Acyclic	Mol 1968, 1970
Re_2O_7/Al_2O_3	40–140	Cyclic	Sato 1977b[b]; Saito 1979
$Re_2O_7/Al_2O_3/Mt_xO_y$[c]	20–100	Acyclic	Nakamura, R. 1977b
$Re_2O_7/Al_2O_3/Bu_4Sn$	20	Hept-1-ene	Bashkirov 1974
$Re_2O_7/Al_2O_3/R_4Sn$	20	Propene	Fridman 1977b
$Re_2O_7/Al_2O_3/Et_4Pb$	35	Hex-1-ene	Finkel'shtein 1992a
$Re_2O_7/Al_2O_3/MoO_3/Bu_4Sn$	20	Hept-1-ene	Bogolepova 1979
Re_2O_7/SiO_2-Al_2O_3	25–50	Propene	Andreini 1986
Re_2O_7/SiO_2-Al_2O_3/R_4Sn	25	Propene	Andreini 1986
Re_2O_7/SiO_2-Al_2O_3/Et_3Al	20	Pent-2-ene	Bradshaw 1968
$Re_2O_7/TiO_2/Et_3Al/Me_4Sn$	40	Propene	Banks 1984b
Re_2O_3/unsupported and on SiO_2	200	Propene	Tsuda 1981
CH_3ReO_3/SiO_2-Al_2O_3	20	Pent-2-ene	Herrmann 1991
CH_3ReO_3/Nb_2O_5	20	Pent-2-ene	Buffon 1992a
CH_3ReO_3/R_nAlCl_{3-n}	20	Pent-2-ene	Herrmann 1991
$(ArO)_2Al(THF)[OAl(OAr)(THF)]$ $OReO_3$[d]	25	All types	Commereuc 1995
$ReCl_5$	20	Norbornene	Ivin 1977a
$ReCl_5/EtAlCl_2$	20	Norbornene	Ivin 1978a
$ReCl_5/Et_3Al/O_2$[e]	20	Acyclic	Uchida, Y. 1972
$ReCl_5/LiAlH_4$[f]	20	Pent-2-ene	Chatt 1972
$ReCl_5/Bu_4Sn$	20	Pent-2-ene	Moulijn 1971
$ReOCl_3(PPh_3)_2/EtAlCl_2$	20	Pent-2-ene	Rybak 1987
$ReOCl_3[P(OEt)_3]_2/EtAlCl_2$	20	Pent-2-ene	Rybak 1987
$Re(CO)_5X/EtAlCl_2$[g]	90	Acyclic	Farona 1975
$Na[(CO)_5Mt$-$Re(CO)_5]/$ $MeAlCl_2/R_4NCl$[h]	20	Pent-1-ene	Kroll 1972
$ReCl_4(PPh_3)/EtAlCl_2$	20	Pent-2-ene	Kittleman 1970
$ReBr_3(O)(PPh_3)$	20	Pent-2-ene	Kittleman 1970
Ph_2Ge-$Re(CO)_4(COMe)/$ i-$BuAlCl_2$	50–100	Hept-3-ene	Warwel 1983b

[a] Other supports tried include SiO_2 (Aldag 1977; Edreva-Kardjieva 1986), SnO (Bradshaw 1968) and MgO (Edreva-Kardjieva 1977). Catalytic activity on Al_2O_3 dramatically increased by O_2 (Nakamura, R. 1982). [b] The proportion of cyclic oligomers relative to high polymer in the product is increased by working at lower concentration. [c] Mt = W, V, etc. [d] Ar = C_6H_3-t-Bu_2-2,6. [e] No reaction in absence of O_2; with terminal olefins both double-bond migration and metathesis occur. [f] Best to premix the components for 2–3 h. [g] X = Cl, Me, Ph. [h] Mt = Mo, W.

bombardment mass spectrometry (FABMS) (Coverdale 1983; Ellison 1985), ^{1}H MAS NMR (Hietala 1994), IR (e.g. Olsthoorn 1976a; Nakamura, R. 1981; Wang, L. 1983; Sibeijn 1991b; Vuurman 1992), Raman spectroscopy (Kerkhof 1979; Wang, L. 1983; Hardcastle 1988; Williams 1990; Vuurman 1992), temperature-programmed reduction (TPR) (e.g. Arnoldy 1985; Xu, X. 1985a, 1986e; Xu, Y. 1986; Vuurman 1992), X-ray absorption near-edge spectroscopy (XANES) (Ellison 1986; Hardcastle 1988), X-ray diffraction (XRD) (Maksimov 1976; Kapteijn 1977; Nakamura, R. 1977b; Ellison 1985) and XPS (Xu, Y. 1986; Spronk 1993a); see also Moulijn 1988; Mol 1990b; Ellison 1990. Re_2O_7 forms a monolayer up to a surface coverage of approximately one Re atom per 0.35 nm^2 (~18 wt% Re_2O_7). Although not all the results are consistent, it is generally agreed that the rhenium oxide species which are catalyst site precursors have a monomeric tetrahedral structure, stabilized by the support via Al—O—Re bonds. It has been observed by Raman spectroscopy that, under dehydrated conditions at high loadings, two slightly different monomeric surface rhenium oxide species are present; both possess three terminal Re=O bonds and one bridging Re—O—Al moiety (Wang 1983; Vuurman 1992). According to certain authors some Re—O—Re bonds are also present at higher loadings. At present it is difficult to draw any definite conclusions with respect to the oxidation state of the metal under reaction conditions. However, partial reduction of Re(VII) to Re(VI) and Re(III) has been observed by XPS studies (Duquette 1984; Xu, Y. 1991).

For propene metathesis, it is best to pretreat the catalyst with H_2 or CO at 500°C followed by exposure to oxygen at room temperature. This treatment presumably brings the surface compound into the most favourable oxidation state for the formation of the initial metal carbene. The same result can be achieved by treatment first with propene at 450°C and then with oxygen at 80°C (Nakamura, R. 1982; Xu, Y. 1986). γ-Irradiation of Re_2O_7/Al_2O_3 in a reducing atmosphere gives an increase in reaction rate of at least a factor of 10 (Tarasov 1996).

Poisoning experiments with nitric oxide indicate that, with some catalysts, less than 0.3% of the rhenium sites are active when the support is covered with a complete layer of rhenium oxide (Olsthoorn 1976b). Kinetic studies (Aldag 1978; Kapteijn 1981) and 'chemical counting' of the initiating carbene species (Chauvin 1992, cf. Section 3.7) confirm that less than 2% of the Re ions form active sites.

The activity of Re_2O_7/Al_2O_3 in the region of lower loadings (0–10 wt % Re_2O_7) can be greatly enhanced by (i) the incorporation of a third oxide such as V_2O_5, MoO_3 or WO_3 (Nakamura, R. 1977b; Bogolepova 1979; Xu, X. 1985b; Spinicci 1989); (ii) treatment of the support with a mineral acid, e.g. HCl (Hietala 1994), or with phosphate (Sibeijn 1991b); (iii) the use of mixed support, e.g. SiO_2–Al_2O_3 (Xu, X. 1985c; Andreini 1986) or Al_2O_3–B_2O_3 (Xu, X. 1986b; Sheu 1992; Sibeijn 1994). Fourier-transform infrared (FTIR) spectra of pyridine adsorbed on various supported Re_2O_7 catalysts indicate a positive correlation between the catalyst activity and the Brönsted acidity of both the supports and their corresponding catalysts (Xu, X. 1986c,d). It is also possible that SiO_2–Al_2O_3 or Al_2O_3–B_2O_3 stabilize the Re species in the oxidation state which is most favourable for the formation of the initial metal carbene complex (Sheu 1992).

A spectacular improvement of the catalytic performance of Re_2O_7/Al_2O_3 is obtained by addition of tetraalkyltin or tetraalkyllead (Fridman 1977b, Andreini 1986, Moulijn 1988). Thus, Re_2O_7/Al_2O_3 is inactive for the metathesis of functionalized olefins, but becomes active when promoted with R_4Sn (R = Me, Et, Bu) (see Ch. 7). Again, the rate of metathesis of propene on Re_2O_7/Al_2O_3 is increased by 1–2 orders of magnitude by adding R_4Sn to the catalyst. When R = Me, some methane is formed, suggesting that new initiating metal carbene species are generated via double alkylation of the rhenium, followed by α-H elimination: $[Re](CH_3)_2 \rightarrow [Re]{=}CH_2 + CH_4$. The active sites generated by R_4Sn may be intrinsically different from those present on the unpromoted catalyst (Spronk 1991a; Moloy 1994).

If the support is changed from Al_2O_3 to SiO_2–Al_2O_3 the catalyst becomes 10–100 times more active for propene metathesis at 50°C, and if R_4Sn is now added the activity is increased by a further factor of 5–10 (Andreini 1986). For an NMR and Mössbauer study of the Sn species on the surface of Re_2O_7/SiO_2–Al_2O_3/Me_4Sn catalysts, see Buffon (1995).

$Re_2(CO)_{10}/Al_2O_3$ is active for propene metathesis after appropriate heat treatment (Danilyuk 1983, 1988). These treatments include: (i) O_2-adsorption at 120°C on partially decarbonylated $Re_2(CO)_{10}/Al_2O_3$; (ii) O_2-adsorption at 25°C on $Re_2(CO)_{10}/Al_2O_3$ which has been completely decarbonylated at 500°C; and (iii) partial reduction at 200–300°C of $Re_2(CO)_{10}/Al_2O_3$ preoxidized at 300°C. The processes of oxygen adsorption in the first two stages may be represented as in eqn. (24). The close similarity in the catalytic properties and in the conditions of the

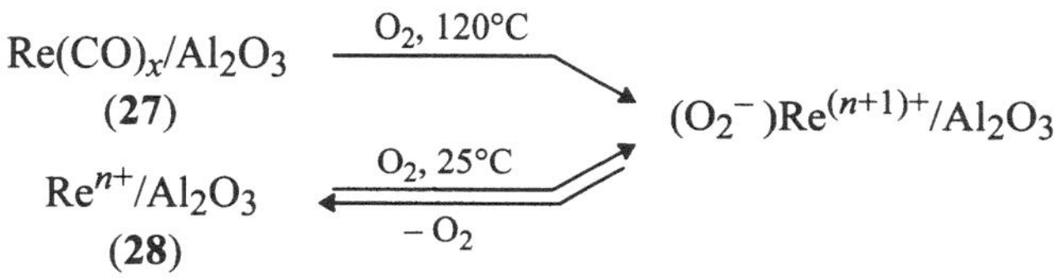

(24)

formation of metathesis-active sites suggests that both **27** and **28** contain the same type of active site. These active sites, or their precursors, are denoted by $(O_2^-)Re^{(n+1)+}/Al_2O_3$, where O_2^- is bonded to an Al^{3+} ion that is close to a reversibly oxidizing rhenium ion in a high oxidation state. In situation (iii), the oxidation results in a catalyst exhibiting low activity at room temperature, although treatment with hydrogen increases its activity. The carbene ligands in the active species in these catalysts are thought to be formed by reaction (25) (Danilyuk 1988).

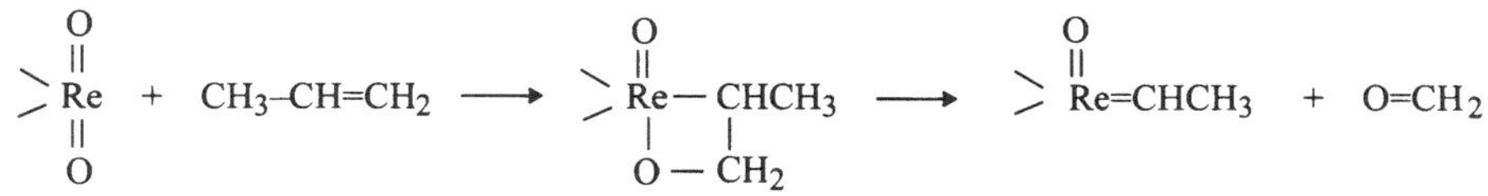

(25)

An active and stable supported rhenium catalyst for propene metathesis is formed by heating $[Re(CO)_3OH]_4/SiO_2$ under a propene–helium atmosphere at 150°C (Kirlin 1985a). $[Re(CO)_3OH]_4$ acts as a reservoir of rhenium from which catalytically active mononuclear rhenium complexes are formed (Kirlin 1985b). The reaction of $[ReBr(CO)_5]$ with γ-Al_2O_3 leads to the formation of a supported complex, which is the precursor of a catalyst with low activity for propene metathesis at 125°C (Nicolaides 1986).

Al_2O_3-supported (calcined) Re catalysts prepared from $Re(acac)_3$ have a higher activity for the metathesis of hex-1-ene than catalysts prepared from NH_4ReO_4, both alone and when promoted by various cocatalysts (Klimov 1993).

In solution, methyltrioxorhenium (CH_3ReO_3) is active for the metathesis of cyclic and acyclic alkenes, but only in the presence of a Lewis acid, viz. $AlCl_3$, and an alkylating agent such as Me_4Sn (Herrmann 1988), or in combination with R_nAlCl_{3-n} (R = Me, Et; $n = 1, 2$) (Herrmann 1991). When supported on calcined SiO_2–Al_2O_3 it becomes an efficient metathesis catalyst at room temperature without additives (Herrmann 1991). Another highly active system for the metathesis of acyclic olefins at room temperature is obtained when CH_3ReO_3 is chemisorbed on Nb_2O_5 (pretreated at 200–300°C). A correlation is observed between the catalytic activity and the Lewis acidity of the support, and between the catalytic activity and the presence of a methyl ligand (Buffon 1992a), although the initiating metal carbene is not formed from the methyl ligand (Buffon 1992b).

The most notable feature of $ReCl_5$ as a catalyst is its ability to stimulate the formation of all-*cis* polymers of norbornene and its derivatives; see Ch. 13. The system $ReCl(CO)_5/EtAlCl_2$ is remarkable in two respects: first, it can be kept for long periods at room temperature but only becomes active for metathesis of acyclic olefins when heated to 90°C (Farona 1975; Greenlee 1976a); second, in most cases it does not catalyze the ROMP of cycloalkenes but gives instead addition-type polymers (Tsonis 1979; Alonso 1988). However, it does effect the ROMP of norbornene at 100°C though not at 132°C, where only addition polymer is formed (Johnston 1992); see Ch. 4. A further point of interest is that stable rhenium carbene complexes such as $Re[{=}C(OH)Ph](Cl)(CO)_4$ can be isolated by hydrolysis of mixtures of $RePh(CO)_5/2EtAlCl_2$; although the original mixture is active for the metathesis of terminal olefins, the metal carbene derived from it is not, even when mixed with $EtAlCl_2$ (McKinney 1980).

The well-characterized four-coordinate Re(VII) alkylidene–alkylidyne complex, $Re({=}CHCMe_3)({\equiv}CCMe_3)[OCMe(CF_3)_2]_2$ brings about the metathesis of *cis*-pent-2-ene, but its activity is significantly lower than that of related Mo or W catalysts (Toreki 1990, 1993).

2.6 Group VIII

2.6.1 Cobalt

Goodall (1995) was the first to report the ROMP of norbornene initiated by a Co-based catalyst. Cobalt neodecanoate or acetylacetonate in combination with R_3Al (R = Et, *i*-Bu) in heptane at 20°C gives an 80% yield of an all-*cis* polymer in three days.

2.6.2 Ruthenium

The ROMP of cyclobutene and 3-methylcyclobutene (Natta 1964b, 1965a), bicyclo[4.2.0]oct-7-ene and norbornene (Michelotti 1965b; Lahouste 1976), but not cyclopentene, can all be initiated by $RuCl_3.3H_2O$, not only in alcoholic solution but also in aqueous emulsified systems at 20–100°C. In the case of norbornene, low-molecular-weight secondary alcohols may be detected and their formation interpreted in terms of side reactions of the intermediates, which would otherwise lead to the metal carbene that propagates polymerization (Laverty 1976a).

Ruthenium complexes, prepared by heating Ru–Cl_3 with cycloocta-1,5-diene in ethanol, also initiate the ROMP of norbornene, and the activity correlates with the strength of an IR band (1900–2100 cm^{-1}) attributed to Ru–H bonds (Laverty 1976b). Complexes of the type **29–32** are also effective for the polymerization of norbornene, but not of cyclopentene (Hiraki 1971; Porri 1974).

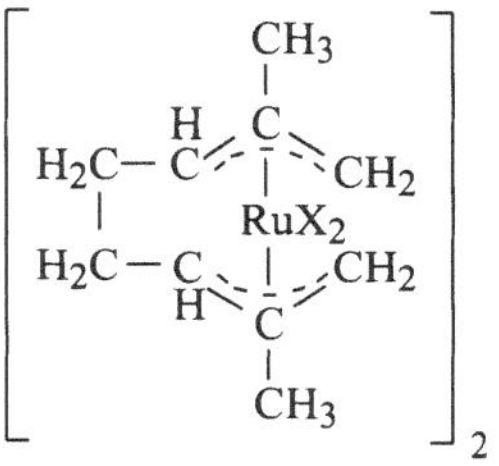

(**29**) X = Cl
(**30**) X = $OCOCF_3$

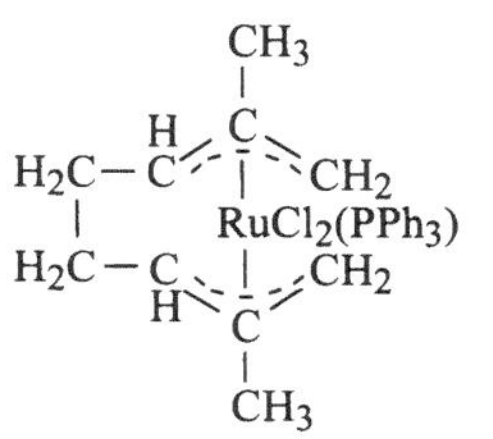

(**31**)

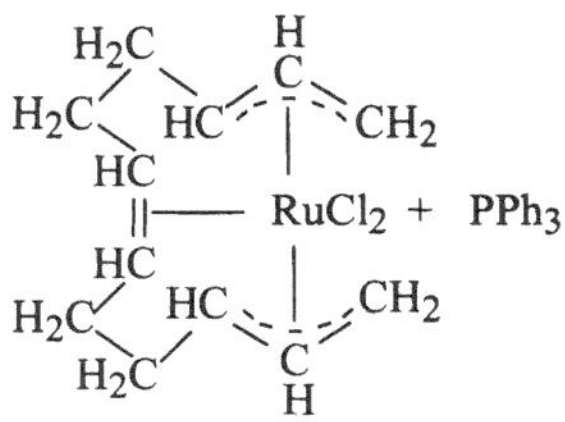

(**32**)

In order to obtain a catalyst capable of initiating ROMP of cyclopentene it is necessary to generate active [Ru]-H bonds. Passage of hydrogen through solutions of **29** or **30** is effective, the diene ligand being displaced in the process (Porri 1974). A number of other ruthenium complexes become active in the presence of excess trifluoroacetic acid (TFA), the most effective being $RuH_2(PPh_3)_4$/TFA (1/200) (Porri 1975). That the essential difficulty with cyclopentene lies not in propagating the polymerization but in initiation is confirmed by the fact that **30** readily initiates the copolymerization of norbornene with cyclopentene (Porri 1974). This shows that the active centres, after generation by the reaction of **30** with norbornene, can readily react with cyclopentene. Related complexes have been prepared by Wache (1995a,b); for example, the water-soluble complex $Ru(\eta^3:\eta^3\text{-}C_{10}H_{16})(OH_2)(O_3SCF_3)_2$ initiates the emulsion ROMP of norbornene to give a high-*cis* polymer.

The initiation of polymerization of norbornene by ruthenium complexes may well involve the participation of traces of oxygen. Thus, when oxygen is bubbled at room temperature through solutions of norbornene in chlorobenzene containing $RuCl_2(py)_2(PPh_3)_2$, the rate of ROMP is increased as much as 100-fold. Other ruthenium complexes such as $RuCl(PPh_3)_3$ and $RuHCl(PPh_3)_3$ behave in the same way. The reaction is accompanied by oxidation of norbornene to the corresponding epoxide, together with formation of some norbornanone. The epoxide can itself be used as a very effective cocatalyst with certain W-based catalysts. These observations have been interpreted in terms of formation of a metallaoxacyclobutane intermediate (Ivin 1981b); see Section 4.4.

Most of the polymers prepared using Ru-based catalysts have a high *trans* double bond content (Ivin 1979c; Porri 1974, 1975; Nguyen 1992), except when strongly chelating ligands are present; see Ch. 13.

The ROMP of functionalized monomers, such as 7-oxanorbornene and its derivatives can be performed successfully in organic solvents using $RuCl_3.3H_2O$ and $OsCl_3$ (Novak 1988a; Hamilton 1990b); reaction (26). Water acts as a

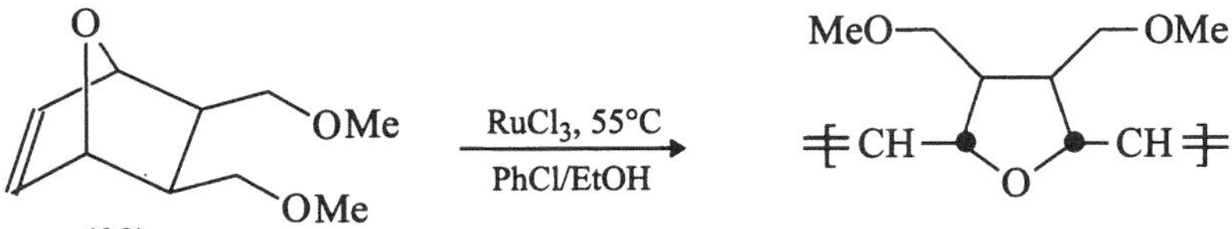

(33) (26)

cocatalyst and dramatically reduces the induction period which in the absence of water is typically 24 h for **33** (Novak 1988b). In water the polymerization of the 7-oxanorbornene monomers proceeds very rapidly under air in nearly quantitative yields with induction periods of 30–35 min. The aqueous ruthenium solution can be recycled at least 14 times without detectable loss of activity, and with progressively smaller induction period (Novak 1988b). When the complex $Ru(H_2O)_6(OTs)_2$ reacts with *n* equivalents of **33**, there is a much smaller induction period (55 s), and the product consists of 1 equivalent of mono-olefin adduct **34**, and polymer

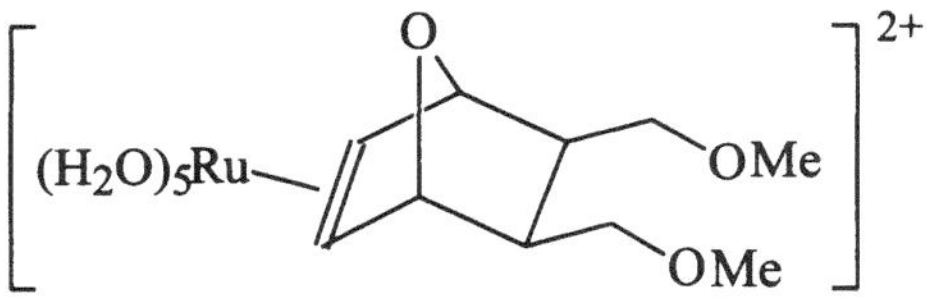

(34)

containing ($n - 1$) equivalents of monomeric units. The adduct **34** is highly active in subsequent polymerizations (Novak 1988b, 1992; McGrath 1991). The activity of $Ru(H_2O)_6(OTs)_2$ for ROMP can be extended to less reactive monomers, such as cyclopentene and cyclooctene, by the addition of ethyl diazoacetate (France 1993a); see also Section 13.5.1.

Reaction of 3,3-diphenylcyclopropene with either $RuCl_2(PPh_3)_3$ or $RuCl_2(PPh_3)_4$ produces the vinylcarbene complex $Ru(=CHCH=CPh_2)(Cl)_2(PPh_3)_2$ (**35**). This initiates the ROMP of norbornene in organic media both in the absence and presence of protic/aqueous solvents at room temperature (Nguyen 1992). The complexes **36** and **37** are more active than **35** as a result of increased σ-donation from the phosphine ligands. They catalyze the ROMP of both high- and low-

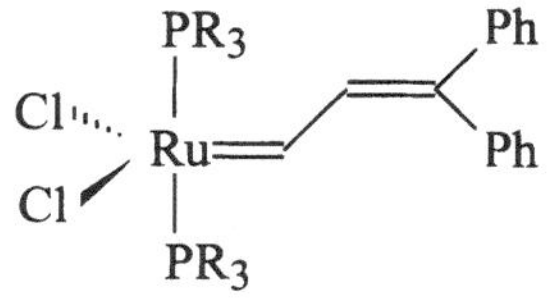

(**35**) R = Ph
(**36**) R = Cy
(**37**) R = *i*-Pr

strained cyclic olefins as well as the metathesis of acyclic olefins at room temperature. The propagating species are observed by ^{1}H NMR (Nguyen 1993). For some synthetic applications of these catalysts, see Ch. 8.

The use of diazoalkanes instead of 3,3-diphenylcyclopropene provides a series of air-stable alkylidene complexes, $Ru(=CHR')(Cl)_2(PPh_3)_2$ and $Ru(=CHR')(Cl)_2(PCy_3)_2$ (R′ = Me, Et, Ph, *p*-C_6H_4Cl), which are very efficient catalysts for the ROMP of norbornene and substituted cyclobutenes. Thus, for the $Ru(=CHPh)(Cl)_2(PPh_3)_2$-initiated ROMP of norbornene, $k_i/k_p = 9$ (see Table 11.4). Reaction of $Ru(=CHPh)(Cl)_2(PCy_3)_2$ with excess ethene gives $Ru(=CH_2)(Cl)_2(PCy_3)_2$, the first methylidene complex which has been isolated and shown to be an active metathesis catalyst (Schwab 1995, 1996).

$RuCl_2$(*p*-cymene)(PCy_3), when activated by a diazocompound, e.g. trimethylsilyldiazomethane, also gives a very active catalyst for promoting the ROMP of functionalized norbornenes and cyclooctenes (Stumpf 1995). Various Ru complexes, including $Ru_2(OAc)_4$ (Noels 1988), $Ru_2(O_2CCF_3)_4$ (Demonceau 1992), and various allylruthenium (IV) complexes (Herrmann 1996), can readily be activated towards ROMP by addition of a catalytic amount of a diazoester.

Ruthenium catalysts active for the metathesis of certain silicon-containing olefins are discussed in Ch. 7 and 9.

2.6.3 Osmium

Catalysis by osmium compounds is mainly confined to the ROMP of norbornene and its derivatives by $OsCl_3$, or related compounds. This was first observed for $OsCl_3.3H_2O$ in ethanol at 80°C (Michelotti 1965b). The reaction also occurs in benzene at 20°C to give a polymer containing 46% *cis* double bonds distributed at random in the chain (Ivin 1979c). The cyclooctadiene complex derived from $OsCl_3$ is also an active metathesis catalyst. Os–H bonds have been detected in this complex by both IR and ^{1}H NMR (δ −1.8) (Laverty 1976b). A solution of OsO_4 in chlorobenzene slowly polymerizes norbornene at 60°C. An oxametallacyclobutane is proposed as the intermediate leading to the propagating osmium carbene complex (Hamilton 1990a).

2.6.4 Iridium

Reaction between the cyclooctene complex $[(C_8H_{14})_2IrCl]_2$ and $\geqslant 4$ equivalent of silver fluoroacetate produces a highly active system for metathesis/isomerization of acyclic olefins, including methyl oleate; a lower molar ratio of silver fluoroacetate/$[(C_8H_{14})_2IrCl]_2$ results in extensive isomerization, but no metathesis (France 1994).

The ROMP of cyclopentene, cycloheptene, and cyclooctene at 20–60°C is brought about by $[(C_8H_{14})_2IrOCOCF_3]$ activated by TFA (Porri 1975). With $[(C_8H_{14})_2IrCl]_2$ or $[(C_8H_{14})_2IrOCOCF_3]$ as initiator it is possible to make copolymers with norbornene having up to 20–30% of units derived from cyclopentene or cyclooctene (Porri 1972, 1974). The ROMP of norbornene itself can be effected not only by $[(C_8H_{14})_2IrCl]_2$ and $[(C_8H_{14})_2IrOCOCF_3]$, but also by $IrCl_3.3H_2O$ (Michelotti 1965b), by $(NH_4)_2IrCl_6$/EtOH in aqueous emulsion (Rinehart 1965), by $IrCl(CO)(PPh_3)_2$ in emulsion, solution, or bulk (Rinehart 1969), and by cyclooctadiene complexes such as $[(C_8H_{12})IrCl]_2$ (Rinehart 1969; Ivin 1979c). Ir-H bonds have been detected in active catalysts derived from $[(C_8H_{12})IrCl]_2$, by both IR and ^{1}H NMR (Laverty 1976b).

Although acyclic olefins do not generally undergo metathesis with Ir-based catalysts, they can act as chain-transfer agents in the ROMP of norbornene (Rinehart 1965; Porri 1974; Ivin 1979c); also see Ch. 15. It appears therefore that the failure of most Ir-based catalysts to induce self-metathesis of acyclic olefins lies in their inability to generate an initial metal carbene from such olefins.

2.6.5 Rhodium

The rhodium complexes $Rh(PPh_3)_2LCl$ (L = CO or PPh_3) can effect the metathesis of electron-rich olefins such as **38** and **39**; reaction (27) (Cardin 1972b). Certain

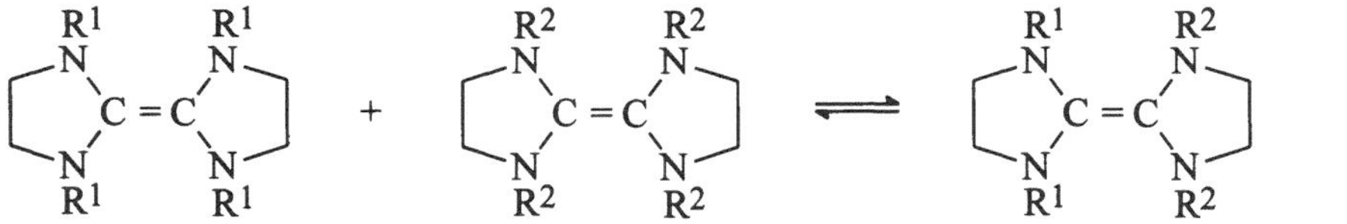

(38) **(39)**

(27)

other Rh-based catalyst systems are fairly active for metathesis but there is rapid concurrent isomerization (Hughes 1970b). Rhodium catalysts active for the metathesis of certain silicon-containing olefins are discussed in Ch. 7.

2.7 Photochemically activated catalysts

An indirect photochemical effect on a metathesis reaction was first noticed by Günther (1970) who found that the activity of $WCl_6/i\text{-}Bu_3Al$ as catalyst for the ROMP of cyclopentene could be enhanced if the WCl_6 was first irradiated in C_2Cl_4 solution at 100°C or in toluene solution at about 40°C. Since then it has been found that all types of olefin metathesis can be stimulated by photochemical means, normally using a mercury lamp as the source. Different results are sometimes obtained according to whether the olefin is present from the start (photoassisted reaction), or is added immediately after irradiation of the catalyst system (photocatalyzed reaction), or is added only after a period of incubation of the photochemical products (modified photocatalyzed reaction); this classification is due to Taniélian (1981). The effect of aging the irradiated solution before addition of substrate is particularly striking in the system $W(CO)_6/CCl_4/h\nu$ (Taniélian 1981). Some results using pent-2-ene as substrate are presented in Figs. 2.2–2.4. These show clearly that, for short irradiation periods (3 min), a very active species is produced with a half-life of only a few hours; with longer irradiation periods (60 min) a less active species with a much longer half-life is the outcome, one of these species being heterogeneous in nature. The initiating species in the $W(CO)_6/CCl_4/h\nu$ system appear to depend on the substrate. In the reaction with 1-ethoxynorbornene, the product contains the $=CCl_2$ entity, suggesting that $[W]=CCl_2$ is formed (Rudler 1980). However, in the reaction with oct-4-ene and tetradec-7-ene, no $RCH=CCl_2$ is detected during cross-metathesis (Karan 1989) and in the reaction with pent-2-ene there is again no evidence for metathesis initiation by $[W]=CCl_2$ (Schilder 1992). In the last case it appears that $[W]=CCl_2$ is destroyed before it can react with the substrate and the metathesis reaction is initiated from the surface of a precipitate; the nature of the intiating species is not clear.

Another case of a dramatic aging effect is in the photoassisted ROMP of cyclopentene by $WCl_6/h\nu$. Irradiation of a freshly made solution gives all-*cis*

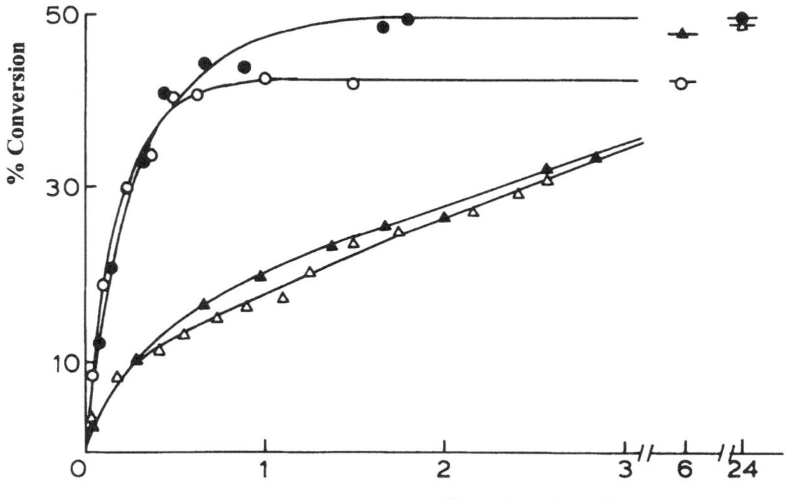

Fig. 2.2 Conversion of pent-2-ene (0.5 M) into metathesis products after irradiation in the presence of $W(CO)_6$ (10^{-2} M) in CCl_4 for 90 s (○), 3 min (●), 1 h (△), and 10 h (▲) (Taniélian 1981).

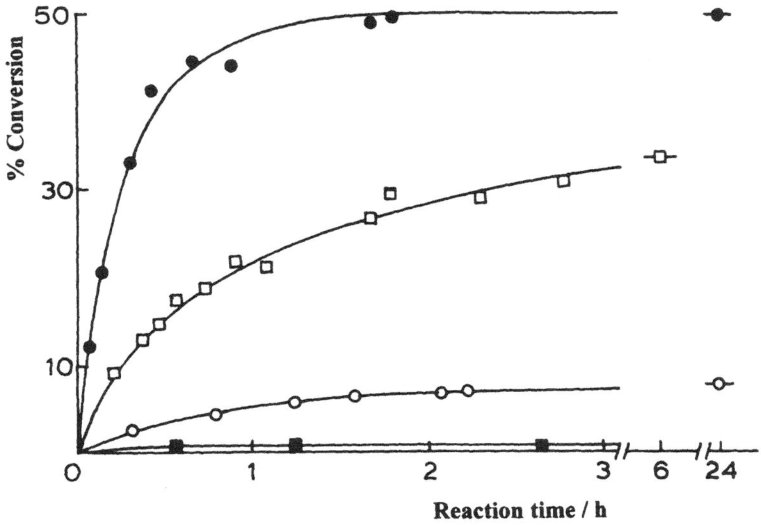

Fig. 2.3 Effect of incubation time on conversion into metathesis products. Three-minute irradiation of $W(CO)_6$ (10^{-2} M in CCl_4), followed by an incubation time of zero (●), 3 h (□), 8 h (○), and 10 h (■), before addition of pent-2-ene (0.5 M) (Taniélian 1981).

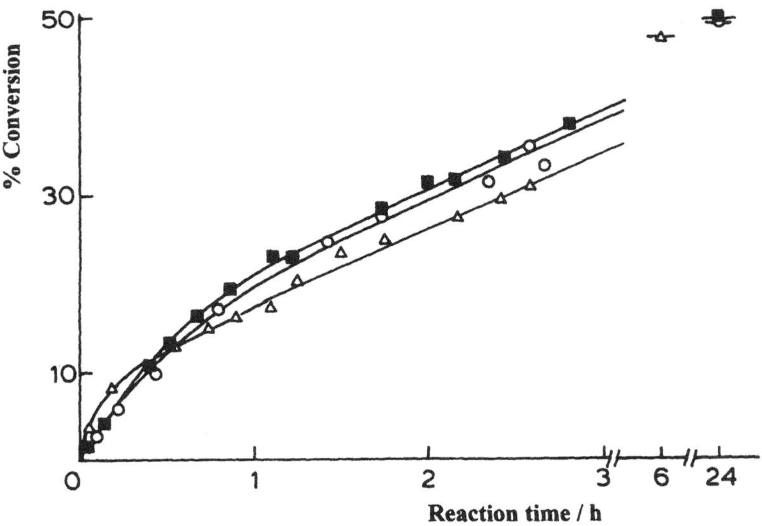

Fig. 2.4 As for Fig 2.3, but with 1 h irradiation followed by incubation times of zero (△), 3 h (■), and 10 h (○) before addition of pent-2-ene (Taniélian 1981).

polymer but, if the solution is kept for an hour and then irradiated, the *cis* content of the polymer is considerably reduced. Such effects have been attributed to progressive reduction from W(VI) to W(V) to W(IV) on standing, as indicated by the changes in absorption spectrum (Yakovleva 1977b). Irradiation of a solution in EtCl gives all-*trans* polymer (Yakovleva 1977a; Sheinker 1979). The ROMP of cyclopentene is also photocatalyzed by $W(=CRPh)(CO)_5/TiCl_4$ in benzene, where the $TiCl_4$ is present in 150-fold excess. Raising the temperature increases the *trans* content of the polymer from 50% at −25°C to 86% at 30°C (Chauvin 1976), an effect that is quite general for all manner of catalysts; see Ch. 12. The system WCl_6/2,6-di-*t*-butyl-*p*-cresol, when preirradiated for 30 min in cyclohexane, induces the ROMP of cycloocta-1,5-diene, giving all-*cis* polymer, this being attributed to the presence of bulky *t*-butyl groups close to the reaction site (Castner 1977a). Irradiated WCl_6 in benzene solution brings about the degradation of polybutadiene without affecting the total number of double bonds present (Zemtsov 1977).

The *cis/trans* equilibrium of but-2-ene is induced by $W(CO)_6/CCl_4/h\nu$ or by adding but-2-ene to preirradiated $W(CO)_6/CCl_4$; a comparison of the two types of experiment shows that the catalytic entity is continuously regenerated on irradiation (Krausz 1978). In the photoinduced metathesis of pent-2-ene, using the same system, very high stereospecificity is observed with respect to the *trans*-but-2-ene (> 95%) produced from *trans*-pent-2-ene, rather higher than that for *cis*-but-2-ene (78%) from *cis*-pent-2-ene; a binuclear tungsten intermediate has been proposed (Garnier 1979). A similar pattern is observed for the butenes formed from other olefins of the type $RCH{=}CHCH_3$ (R = Et, Pr, *i*-Pr, Bu, *i*-Bu). Other catalyst systems that have been shown to act as photosensitizers for metathesis of internal olefins are $W(CO)_6$/*i*-$BuAlCl_2$ (Warwel 1975), and *cis*-$WCl_2(CO)_4$ (Agapiou 1975; Borowczak 1982, 1984).

In the metathesis of pent-2-ene catalyzed by $W(CO)_6/TiCl_4/h\nu$ the initial rate shows a marked dependence on the pre-irradiation time (t_i) (Fig. 2.5). For longer t_i the conversion steadily increases reaching equilibrium conversion (50%) at $t_i = 60$ min. For $t_i > 60$ min the conversion–time curve remains the same as for $t_i = 60$ min (Vanwijnsberghe 1987). This behaviour should be contrasted with that for the $W(CO)_6/CCl_4/h\nu$ system (Fig. 2.2).

The role of the halide (AX_n) in the system $W(CO)_6/AX_n/h\nu$ has been extensively investigated; the most active systems are those containing strong Lewis acids, e.g. $ZrCl_4$ (Szymańska-Buzar 1987), $TiCl_4$ (Nagasawa 1978; Szymańska-Buzar 1987), BF_3 and $AlCl_3$ (Borowczak 1984).

Pent-1-ene metathesis is observed with $W(CO)_6/ZrCl_4/h\nu$ and $W(CO)_6/TiCl_4/h\nu$, but is accompanied by double-bond isomerization and cross-metathesis of pent-2-ene with pent-1-ene. In the series of $[WL(CO)_5]$-type complexes (L = Cl, CO, py, PPh_3), the most active for pent-2-ene metathesis is $(Et_4N)[WCl(CO)_5]$ (Szymańska-Buzar 1987).

A solution of $W[=C(OMe)Ph](CO)_5$ in CCl_4 becomes more active for the ROMP of cycloocta-1,5-diene when irradiated with UV (Kolesnikov 1981). There is evidence that the $W[=C(OMe)Ph](CO)_5/h\nu$-initiated ROMP of norbornene

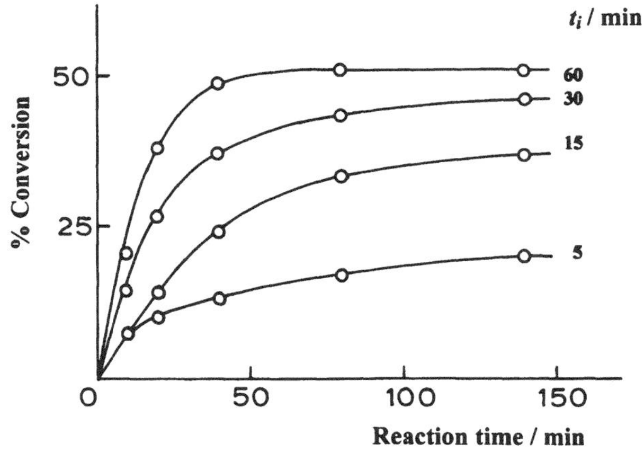

Fig. 2.5 Conversion of pent-2-ene (0.5 M) into metathesis products at 40°C for different pre-irradiation times (t_i) in the system $W(CO)_6$ (0.5×10^{-2} M)/$TiCl_4$ (0.5×10^{-2} M); solvent chlorobenzene (Vanwijnsberghe 1987).

proceeds not only via the removal of a CO ligand but also through the formation of a [W]-H species by transfer of hydrogen from the methoxy group (Bechera 1986). Another catalyst system that is sensitive to light is $(Me_3CCH_2)_3WBr(O)/AlBr_3$, the initiating metal carbene being generated by loss of Me_4C (Kress 1981).

$Mo(CO)_6$ and $W(CO)_6$ adsorbed on porous Vycor glass or NaY zeolite become active when irradiated with a mercury lamp (> 290 nm) at room temperature in the presence of propene (Wada 1988). They remain active after irradiation ceases, indicating that the catalytically active species is formed by photoreduction of the metal carbonyl on the surface. The activity of $Mo(CO)_6/Al_2O_3$ is enhanced by irradiation.

MoO_3, supported on porous Vycor glass and activated by heating in O_2 at 500°C followed by evacuation at 340°C and irradiation with UV (> 280 nm), catalyzes the metathesis of propene at 17°C (Anpo 1982a). The reaction ceases as soon as the radiation is cut off, and is retarded by O_2 and CO. Some ethanal is formed. The reaction is thought to be initiated as shown in eqn. (28). UV-irradiation of this catalyst at 17°C in the presence of ethene yields propene, but at 200°C small amounts of 2-C_4H_8, 1-C_4H_8 and HCHO are also formed, suggesting methylene formation and subsequent metathesis (Anpo 1982b).

$$Mo^{6+}{=}O^{2-} \xrightarrow{h\nu} Mo^{5+}{-}O^{-} \xrightarrow{CH_2=CHCH_3} \begin{array}{c} Mo — O \\ | \quad\quad | \\ CH_2—CHCH_3 \end{array} \longrightarrow \begin{array}{c} Mo \\ \| \\ CH_2 \end{array} + \begin{array}{c} O \\ \| \\ CHCH_3 \end{array} \tag{28}$$

The yield of photoinduced metathesis of propene over MoO_3/SiO_2 at 25°C is higher on catalysts obtained by reaction of gaseous $MoCl_5$ with surface silanol groups than on catalysts prepared by impregnation with ammonium paramolybdate (Mo content 0.01–0.37 wt%), owing to a higher molybdenum dispersion in the former case (Anpo 1987, 1988). For anchored catalysts, the TOF is highest at low loadings (e.g. 0.01 wt% Mo). At higher Mo content the formation of inactive octahedral Mo species reduces the rate. Similar results have been obtained for WO_3 and CrO_3 on porous Vycor glass, although the initial reaction rates are much lower (Anpo 1986). Photoreduction of supported MoO_3 in the presence of CO has already been discussed (see Section 2.4.2). A curiosity is that silica alone, activated by evacuation at high temperatures, is reported to catalyze metathesis of (deuterated) ethene and propene under photoirradiation (Yoshida, H. 1995).

Upon UV irradiation, complexes of the type $[(\eta^6\text{-arene}_1)Ru(\eta^6\text{-arene}_2)]^{2+}$ and $[Ru(NCR)_6]^{2+}$ are transformed into active catalysts for the ROMP of norbornene and 7-oxanorbornene derivatives (Karlen 1995).

Finally, one may note the curious behaviour of alkynes. If $Mo(CO)_6$/non-4-yne is irradiated at room temperature and an excess of 3-chlorophenol then added, there is rapid metathesis to give oct-4-yne and dec-5-yne with nearly 100% selectivity (Mortreux 1977). In contrast, $W(CO)_6/CCl_4/h\nu$ (350 nm) induces polymerization of hept-2-yne while causing metathesis of pent-2-ene present in the same reaction mixture (Stockel 1978). The difference in behaviour of the two systems presumably lies in the ability of $Mo(CO)_6$/non-4-yne/$h\nu$ to generate a metal carbyne on addition of the phenol, whereas $W(CO)_6/CCl_4/h\nu$ gives only a metal carbene. Photoassisted polymerization of terminal alkynes takes place with $WCl_6/h\nu$ in hydrocarbon solutions (Landon 1985); photocatalyzed polymerization of substituted alkynes is induced by $W(CO)_6/SnCl_4/h\nu$ (Tamura 1994). For related systems, see Ch. 10.

3

The Metal Carbene/Metallacyclobutane Mechanism

3.1 Introduction

The metal carbene/metallacyclobutane mechanism of olefin metathesis, as outlined in Section 1.3, was first proposed by Hérisson and Chauvin in 1971. By 1975 the evidence in its favour had become so compelling that the earlier 'pairwise' mechanism had been totally discarded. From 1980 onwards well-defined carbene complexes of Ta, Mo, W, Re, and Ru were discovered which would act as initiators without the need for activation by heat, light, or cocatalyst. This in turn led to the spectroscopic detection of the propagating metal–carbene complexes in many systems, to the detection of the intermediate metallacyclobutane complexes in a few cases, and in one case to the detection of the metal–carbene–olefin complex that precedes the formation of the metallacyclobutane complex. In no individual case have all three intermediates been detected: at most two have been observed, sometimes one, more often none. After 1980 metallacyclobutane complexes of Ti and Ta were found which would act as initiators at 60°C, but where the intermediate metal carbene complexes could not be detected.

For the many hundreds of 'first-generation' catalysts used before and since 1980, the precise nature of the metal carbene or metallacyclobutane intermediates remains a matter for speculation since their concentrations are too small to be detected except in a few cases. However, the structure of the initiating carbene can often be deduced from the initial products, or by using a carbene trap, or from the polymer end groups. The identity, number, and spatial arrangement of the permanent ligands in the propagating complexes operating in the first-generation systems and their effect on the stereochemistry of the metathesis reaction is generally not well understood. Closer investigation of systems in which well-defined metal carbene complexes act as initiators should allow a better understanding of these factors, but at present it is difficult to arrive at generalizations because quite small changes in the structure of the initiator or substrate can have a marked effect on the stereochemistry of the reaction.

In this chapter we present the historical development of the metal carbene/ metallacyclobutane mechanism of olefin metathesis.

3.2 Evidence from cross-metathesis reactions

The cross-metathesis of cyclopentene with unsymmetrical olefins is catalyzed by $WOCl_4/Bu_4Sn$ or $WOCl_4/Et_2AlCl$. Hérisson (1971) observed that the products of reaction with pent-2-ene consisted of three series of compounds: $Q^1M_nQ^1$, $Q^1M_nQ^2$, and $Q^2M_nQ^2$ ($n = 1$–4), where Q^1 = ethylidene, Q^2 = propylidene, and M_n represents n ring-opened units of cyclopentene M. These series were formed in the statistical ratio 1 : 2 : 1 even in the *initial* products. Similar results were obtained with cyclooctene, cycloocta-1,5-diene, and cyclododeca-1,5,9-triene in place of cyclopentene. It was these observations that led to the proposal of the metal carbene mechanism, since direct exchange between the double bonds of the reactant molecules would yield only the unsymmetrical series. The formation of the three series of compounds is accounted for in terms of reactions (1)–(6).

$$[Mt]{=}Q^1 + nM \longrightarrow [Mt]{=}M_nQ^1 \qquad (1)$$

$$[Mt]{=}M_nQ^1 + Q^1Q^2 \longrightarrow [Mt]{=}Q^2 + Q^1M_nQ^1 \qquad (2)$$

$$[Mt]{=}M_nQ^1 + Q^1Q^2 \longrightarrow [Mt]{=}Q^1 + Q^2M_nQ^1 \qquad (3)$$

$$[Mt]{=}Q^2 + nM \longrightarrow [Mt]{=}M_nQ^2 \qquad (4)$$

$$[Mt]{=}M_nQ^2 + Q^1Q^2 \longrightarrow [Mt]{=}Q^1 + Q^2M_nQ^2 \qquad (5)$$

$$[Mt]{=}M_nQ^2 + Q^1Q^2 \longrightarrow [Mt]{=}Q^2 + Q^1M_nQ^2 \qquad (6)$$

In a more detailed study of this type Katz (1977a) found that in the system cyclooctene/hex-2-ene, catalyzed by $MoCl_2(PPh_3)_2(NO)_2/Me_3Al_2Cl_3$, the initial ratio of the three products with $n = 1$, as determined by extrapolation back to zero time, was 1 : 3.2 : 1 (C_{12} : C_{14} : C_{16}). The C_{14}/C_{12} ratio drops from 3.2 to 2 as the reaction proceeds, indicating the occurrence of secondary metathesis reactions and emphasizing the need to use extrapolation procedures in order to obtain a true indication of the primary mechanism. The fact that the initial ratio is greater than two shows that hex-2-ene has a preferred mode of orientation at the reaction site. This effect is much more pronounced with terminal olefins, where $Q^1M_nQ^2$ is the dominant series of products. These reactions are considered in greater detail in Ch. 15.

A more stringent test is to react cyclooctene with a mixture of but-2-ene and oct-4-ene. If the metal carbene mechanism is correct, one may expect to find the C_{14} product of double cross-metathesis, i.e. $MeCH{=}CH(CH_2)_6CH{=}CHPr$, and an initial value of 4.0 for the product of the two ratios (C_{14}/C_{12}) and (C_{14}/C_{16}). The observed ratio is 4.05 ± 0.05 for *cis* reactants and 4.11 ± 0.09 for *trans* reactants (Katz 1977a).

Cross-metathesis reactions between undeuterated and deuterated dienes, such as octa-1,7-diene (Grubbs 1976) and 2,2′-divinylbiphenyl (Katz 1976b) have also been studied as a means of testing the metal carbene mechanism. The initial proportions of ethene-d_0, -d_2, and -d_4 formed in these reactions support this mechanism. Likewise for the cross-metathesis between *cis,cis*-deca-2,8-dienes containing either CD_3 or CH_3 end groups, the initial d_3/d_6 ratio of the but-2-enes formed is 2.4 ± 0.1, compared with the predicted value of 2.43 (Grubbs 1979).

3.3 Evidence from the stereochemistry of metathesis of internal olefins

If two *trans* olefin molecules were to exchange alkylidene moieties in pairwise fashion, either both products would be *cis* or both would be *trans*; similarly if the two reactant molecules were both *cis*, the products would either both be *cis* or both *trans*. Come what may, the *trans/cis* ratio in the two product molecules would be identical. Sometimes this is the case, or nearly so. For example, with the highly active catalyst **1**, *cis*-pent-2-ene gives almost pure *cis*-but-2-ene and *cis*-hex-3-ene at the beginning of reaction, while *trans*-pent-2-ene gives almost pure *trans* products (Couturier 1992); see Fig. 6.5.

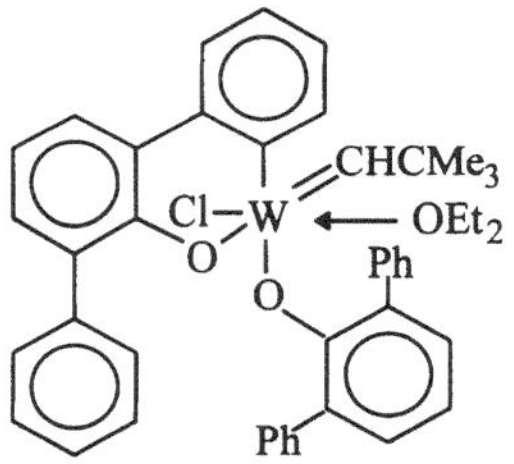

(1)

With $W(=CPh_2)(CO)_5$ as catalyst, *cis*-pent-2-ene gives but-2-ene that is initially $94.3 \pm 1.0\%$ *cis* and hex-3-ene that is initially $92.5 \pm 1.2\%$ *cis*, whereas *trans*-pent-2-ene gives but-2-ene that is $72.8 \pm 2.7\%$ *trans* and hex-3-ene that is $83.4 \pm 1.2\%$ *trans* (Katz 1977b). Again, the catalyst system $W(CO)_5(PPh_3)/EtAlCl_2/O_2$ with *cis*-pent-2-ene gives initially but-2-ene that is 58% *cis* and hex-3-ene that is 53% *cis* (Bilhou 1977a). For a series of catalyst systems, $WCl_2(OAr)_4/Et_3Al_2Cl_3$, the metathesis of *trans*-pent-2-ene always gives a higher proportion of *trans* isomer in the resulting hex-3-ene (79–53%) than in the accompanying but-2-ene (58–40%) (Dodd 1985).

The substrate pent-2-ene does not provide a very exacting test. Results with a more unsymmetrical substrate such as 4-methylpent-2-ene are much more revealing. In this case the initial products, using $WCl_6/Bu_4Sn/Et_2O$ as catalyst system, are as shown in eqns. (7) and (8) (Ofstead 1980). Although the but-2-ene tends to have the same configuration as the reactant, the 2,5-dimethylhex-3-ene is all-*trans* in both cases.

$$cis\text{-}Me_2CHCH{=}CHMe \longrightarrow \underset{67\%\ cis}{MeCH{=}CHMe} + \underset{100\%\ trans}{Me_2CHCH{=}CHCHMe_2} \qquad (7)$$

$$trans\text{-}Me_2CHCH{=}CHMe \longrightarrow \underset{90\%\ trans}{MeCH{=}CHMe} + \underset{100\%\ trans}{Me_2CHCH{=}CHCHMe_2} \qquad (8)$$

This remarkable result can be explained in terms of the metal carbene mechanism. The four possible propagation steps for the *cis* reactant are shown in Scheme 3.1, and for the *trans* reactant in Scheme 3.2. Steps 3 and 7 would lead to *cis*-2,5-dimethylhex-3-ene, and it is not unreasonable to suppose that these are strongly disfavoured by the 1,2-interactions of the isopropyl groups in the metallacycle which, if puckered, allows them at best to take up an axial/equatorial relationship. Steps 4 and 8, leading to *trans*-2,5-dimethylhex-3-ene, are therefore favoured since they can give a puckered ring with 1,2-diequatorial substituents. On the other hand, the relative probability of steps 1 and 2, and of 5 and 6, is more likely to be governed by the interactions of the 1,3-substituents, one of which is isopropyl and the other methyl. Step 6 can proceed through a puckered ring with all three substituents in equatorial positions, so it is strongly favoured over step 5; accordingly the but-2-ene has a high initial *trans* content when made from the *trans*

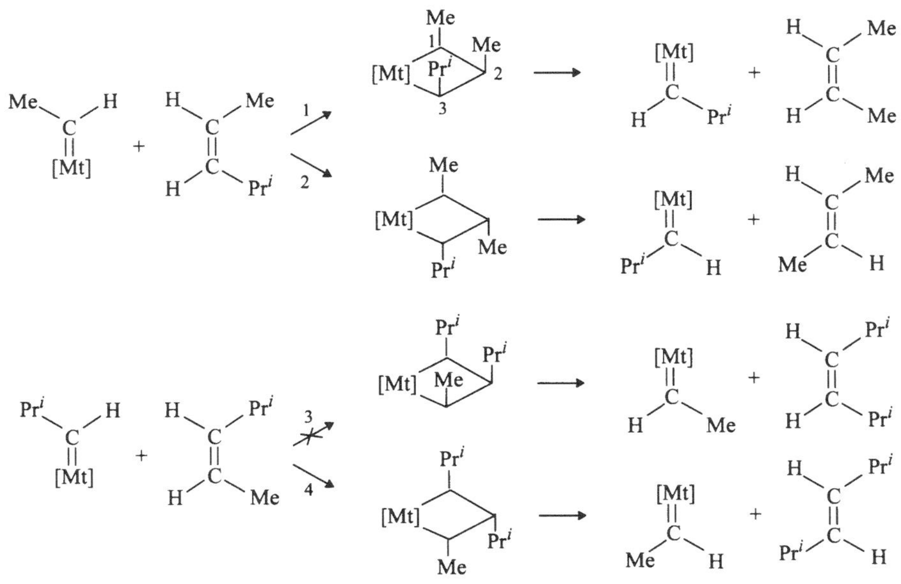

Scheme 3.1 Reactions of *cis*-4-methylpent-2-ene.

reactant. Step 1 can proceed through a puckered ring with the 1,3-substituents in equatorial positions and the 2-substituent in an axial position, whereas for step 2 the 1,3-substituents must necessarily have an axial/equatorial relationship. Thus one may expect step 1 to be favoured over step 2, leading to a preference for *cis*-but-2-ene in the initial products from the *cis* reactant, as observed. The postulate of a puckered ring in 1,3-substituted tungstametallacyclobutanes is supported by theoretical calculations (Tinland 1983) and by crystallographic data on certain stable complexes (Schaverien 1986; Feldman 1991).

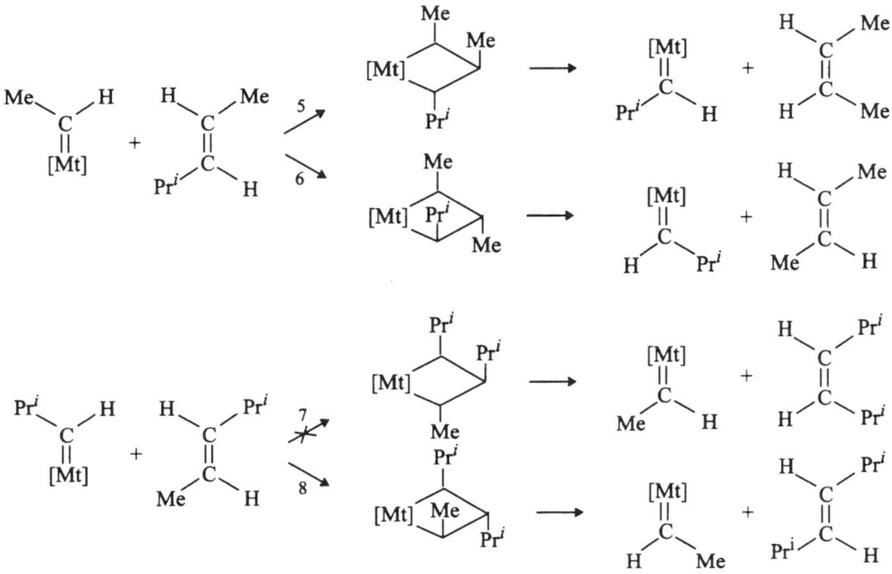

Scheme 3.2 Reactions of *trans*-4-methylpent-2-ene.

The catalyst system $W(CO)_6/CCl_4/h\nu$ is especially interesting in this connection, in that *trans* olefins of the type RCH=CHMe (R = Et, Pr, *i*-Pr , Bu, *i*-Bu) give but-2-ene having more than 95% *trans* content, while the corresponding *cis* olefins give but-2-ene having 75–79% *cis* content (Garnier 1979). Similar behaviour is observed in the exchange of *trans*-$CH_3CH{=}CHCH_3$ with *trans*-$CD_3CD{=}CDCD_3$ on $MoO_x/\beta\text{-}TiO_2$, giving $CD_3CD{=}CHCH_3$ with 98% *trans* content, while the corresponding *cis* olefins give $CD_3CD{=}CHCH_3$ with more than 60% initial *cis* content (Tanaka, K. 1980a). These stereoselectivities are much higher than those observed with other catalyst systems such as $WCl_6/EtAlCl_2/EtOH$ and $MoCl_2(NO)_2(py)_2/EtAlCl_2$, so that interactions of the alkyl substituents on the metallacycle with the permanent ligands on the metal must also be important in some cases.

3.4 Evidence from ring-opening metathesis polymerization (ROMP)

The ROMP of cycloalkenes shows the following features:

(1) The product generally consists of two distinct parts: a high-molecular-weight fraction, often having MW > 10^5, and a low-molecular-weight fraction consisting of a series of cyclic oligomers; for examples, see Table 11.3.

(2) Provided that the initial monomer concentration $[M]_o$ exceeds a certain value, high polymer is formed and an equilibrium between monomer, cyclic oligomers, and high polymer is eventually established. The same equilibrium composition can

be obtained starting from cyclic oligomers or from high polymer, or from a mixture of the two at the same overall concentration. The different routes to equilibrium can be represented on a triangular diagram as shown for the cycloocta-1,5-diene system in Fig. 3.1. In this case the equilibrium concentration of monomer is extremely small and the equilibrium mixture is represented by a point E very close to the polymer–oligomer line.

(3) Provided that $[M]_o$ exceeds the critical value required to produce high polymer, the concentration of oligomer at equilibrium is approximately constant. Hence, the greater the value of $[M]_o$, the greater is the proportion of high polymer at equilibrium. When $[M]_o$ is below the critical value, no high polymer is formed and the only products are cyclic oligomers. This is well understood in terms of the thermodynamics of addition polymerization, of which ROMP is a particular case (Jacobson 1950; Ivin 1974).

(4) Provided that $[M]_o$ exceeds the critical value, both cyclic oligomers and high polymer are formed *concurrently*, as indicated by path A in Fig. 3.1. This is not to say that they are formed initially at rates in proportion to their equilibrium concentrations. Sometimes there will be an initial bias towards cyclic oligomers, as in path A of Fig. 3.1, and towards particular *cis/trans* isomers of these oligomers especially for the smaller rings. In other cases there will be an initial bias towards

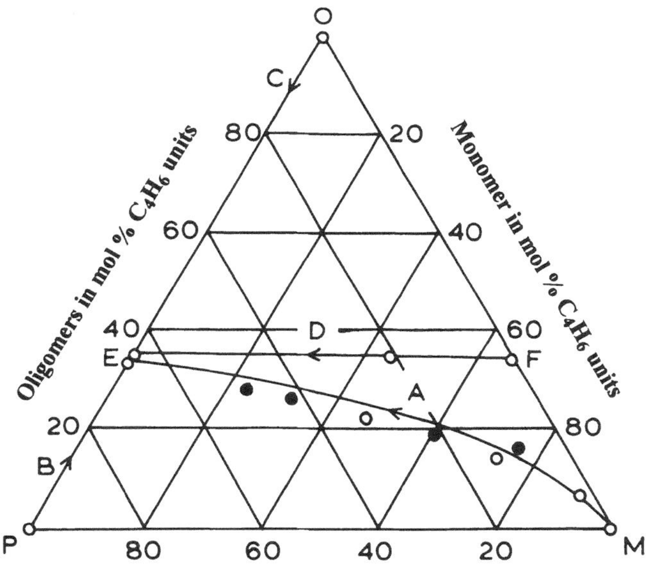

Fig. 3.1 Routes A, B, C, D to equilibrium E between cycloocta-1,5-diene, cyclic oligomers, and linear high polymer, starting from monomer M, polymer P, oligomers O, or a mixture F of monomer and oligomer, respectively. Starting concentration in all cases equivalent to 2.7 mol C_4H_6 units $litre^{-1}$. Catalyst: $W[=C(OMe)Ph](CO)_4(PPh_3)/TiCl_4$. Solvent: ○ = chlorobenzene; ● = benzene, 30°C (Chauvin 1977, 1978).

high polymer, and the viscosity of the system will pass through a maximum as the polymer/oligomer ratio moves towards equilibrium; this is observed for the ROMP of cyclopentene, cyclooctene, and cyclododecene, catalyzed by WCl_6/*i*-Bu_2AlCl (1/5) in benzene at 25°C; see Fig. 11.3 (Dolgoplosk 1977a).

(5) The relative proportions of cyclic oligomers, *c*-M_x, both initially and at equilibrium, form a continuously declining series for values of x larger than 3 or 4. In the case of cycloocta-1,5-diene, the series of cyclic oligomers have the formula $(C_4H_6)_x$ with no tendency for the even-numbered members of the series to predominate; see Fig. 3.2(a) (Kelly 1975; Kumobayashi 1976; Chauvin 1977, 1978).

(6) The catalyst/monomer ratio required to achieve equilibrium or high conversion can be extremely low, of the order of 10^{-5} or less. Even a 10^{-9} ratio is sufficient to cause significant polymerization of cyclopentene when the catalyst system is $W(\equiv CMe)Br(CO)_4/TiCl_4$ (Fischer, E.O. 1976). A chain mechanism is clearly indicated.

(7) When polymerizations are conducted in the presence of acyclic olefins, the MW is much reduced (Witte 1978) and, if sufficient alk-1-ene is used, the end groups derived therefrom (RCH= and =CH_2) may be detected in the ^{13}C NMR

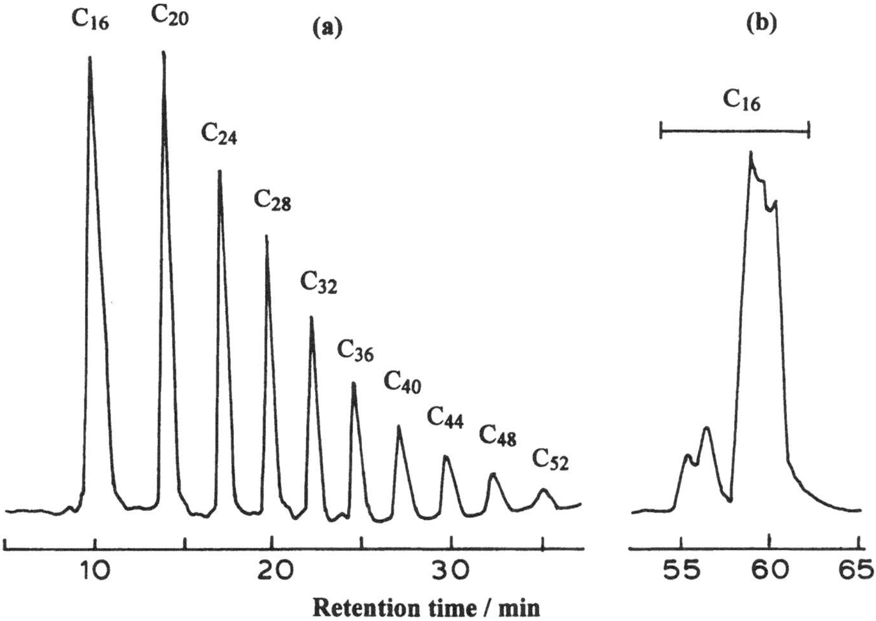

Fig. 3.2 (a) Gas chromatogram of cyclic oligomers $(C_4H_6)_x$ formed during the ROMP of cycloocta-1,5-diene catalyzed by Re_2O_7/Al_2O_3; oligomers extracted using hexane/propan-2-ol (vol. ratio 1/3). (The proportions of the different oligomers have not reached their equilibrium values.) (b) C_{16} fraction run on capillary column; peaks correspond to *cis/trans* isomers, of which there are six: *tttt*, *tttc*, *ttcc*, *tctc*, *tccc*, *cccc*, thought to be eluted in that order (Sato 1977b).

spectrum of the polymer; see Fig. 15.1 (Ivin 1978b, 1982a). In the limit the products are telomers of formula $Q^1M_nQ^1$, $Q^1M_nQ^2$, and $Q^2M_nQ^2$ as discussed in Section 3.2. The acyclic olefins clearly act as chain transfer agents and the polymer produced must be linear; see reactions (1)–(6).

(8) High polymer made from cyclopentene in the absence of deliberately added acyclic olefin gives a viscosity/MW relationship identical with that made in the presence of added acyclic olefin (Witte 1978). A cyclic polymer would be more compact in solution than a linear polymer of the same MW and would give a lower solution viscosity. There is thus no evidence from solution properties that the high polymer made by ROMP of cyclopentene is macrocyclic in nature.

Had the cyclopentene molecules reacted by a pairwise mechanism one would have observed: (i) a gradual increase in MW (as in condensation polymerization) via cyclic oligomers of increasing size; (ii) a marked change in the relative proportions of the various cyclic oligomers as the reaction progressed; and (iii) in the case of cycloocta-1,5-diene, a predominance of the even-numbered members of the series $(C_4H_6)_x$ in the initial stages of the reaction. None of these applies so that such a mechanism must be rejected.

The metal carbene mechanism provides a very satisfactory interpretation of all these observations. The simultaneous production of high polymer and cyclic oligomer is readily explained in terms of competition between the propagation reaction (9) and the intramolecular reaction (10). At low [M] the formation of cyclic oligomer $c\text{-}M_x$ is favoured, while at high [M] the formation of high polymer is preferred. The chance of a given double bond in the P_n chain coming close enough to [Mt] to undergo reaction (10) will clearly decrease with increasing x, so accounting both for the decreasing proportion of $c\text{-}M_x$ with increasing x and for the fact that there is little change in the proportion of the individual oligomers as the reaction proceeds.

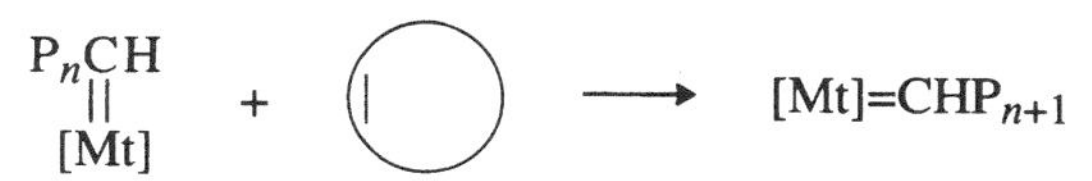

(9)

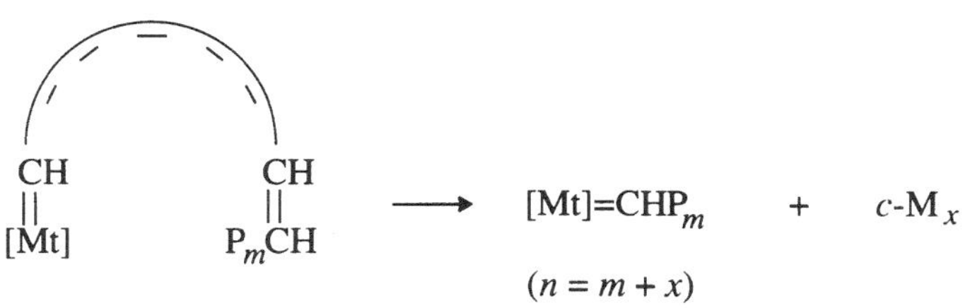

(10)

With cycloocta-1,5-diene, reaction (10) can take place at any double bond in the chain, so accounting for the continuous series of cyclic oligomers of formula $(C_4H_6)_x$. At the beginning of the reaction, alternate double bonds in the chain P_n will have *cis* configuration (being pre-formed in the monomer) but the other double bonds may be *cis* or *trans*. It appears that this is not a significant factor in determining the relative proportions of the larger cyclic oligomers. For any given value of x there will be a number of isomers containing different proportions of *cis* and *trans* double bonds, and in various possible sequences; see Fig. 3.2(b). The proportions of the different *cis/trans* isomers are kinetically controlled in the initial stages and may take some time to adjust to their equilibrium values, depending on the catalyst system. This is especially true of the *ttt* cyclic trimer $(C_4H_6)_3$, which is not present in the initial products, but which eventually appears as the dominant trimer (Thorn-Csányi 1995a).

3.5 Evidence from the reactions of well-defined metal carbene complexes

Stable metal carbene complexes, such as $W[{=}C(OMe)Me](CO)_5$, were first prepared by Fischer, E.O. (1964). These 18-electron complexes can be activated as catalysts for the metathesis of pent-1-ene or the ROMP of cycloalkenes by the use of a cocatalyst, or by heat or UV irradiation; see Table 2.1. For such complexes to become active as initiators of olefin metathesis it is necessary for a CO ligand to be displaced, allowing the substrate to enter the coordination shell and react with the metal carbene bond. For the ROMP of 1-methyl-*trans*-cyclooctene initiated by $W({=}CPh_2)(CO)_5$ at 50°C the $Ph_2C{=}$ end groups may be detected in the polymer by the UV absorption at 245 nm (Lee, S.J. 1976).

The most compelling evidence for the metal carbene mechanism comes from the discovery that metal carbene complexes having an electron count of less than 18, and with 4 or 5 rather than 6 ligands, can act directly as initiators of olefin metathesis. A vast number of such complexes have now been prepared; a small selection is given in Table 2.2. Early work concentrated on tungsten complexes. Preparative methods have steadily improved, and molybdenum and ruthenium complexes are now preferred. As one moves across the Periodic Table from Group V to Group VIII, or up a given Group, these complexes become progressively more stable towards polar groups (in turn towards esters, ketones, aldehydes, and alcohols) until with ruthenium-based initiators the reaction may be carried out in alcohol or in aqueous emulsion.

3.5.1 Detection of propagating metal carbene complexes

The reactions of metal carbene complexes with olefins can be readily followed by ^1H NMR spectroscopy. An early example was the reaction of $W({=}C\mathit{H}CMe_3)({=}O)(Cl)_2(PEt_3)_2/AlCl_3$ (δ 12.03) with but-1-ene to give $[W]{=}C\mathit{H}Et$ (δ 12.1) and $[W]{=}C\mathit{H}_2$ (δ 12.34 and 11.47) in the ratio 80 : 20 (Schrock 1980; Wengrovius 1980). $[W]{=}CH_2$ is thus more reactive than $[W]{=}CHEt$ so far as productive metathesis is concerned. The non-equivalence of the protons in $[W]{=}C\mathit{H}_2$ and the fact that there is a single chemical shift for $[W]{=}C\mathit{H}Et$ show that, in this case, there is a barrier to rotation but only one distinguishable orientation of the carbene ligand with respect to the other ligands. In other cases, as for compound **2**, there are distinct rotational isomers, **2a** and **2b**, eqn. (11), with different reactivity towards olefinic substrates, an important fact to which we will later return.

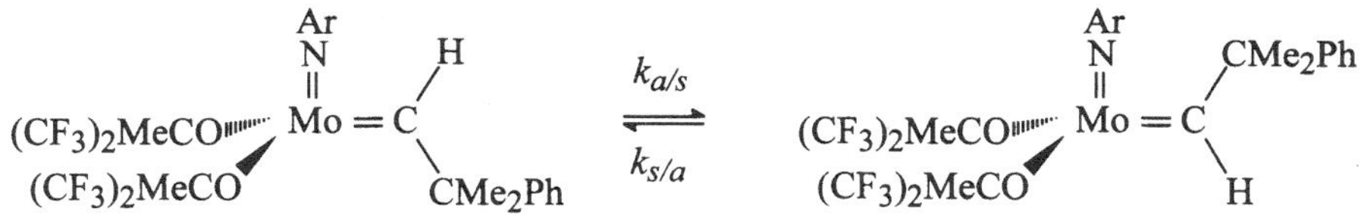

(**2a**) (*anti*) $K = k_{a/s} / k_{s/a}$ (**2b**) (*syn*) (11)

An early example of a metathesis equilibrium in which all the component metal carbene complexes may be seen simultaneously in the ^{1}H NMR spectrum is shown in Fig. 3.3. No reaction occurs in the absence of $AlBr_3$ but in its presence equilibrium is attained within 1 min. Complex formation with $AlBr_3$ has a deshielding effect on both the carbene carbon and hydrogen nuclei, but no significant effect on J_{CH}. The $AlBr_3$ thus interacts with a Br ligand and not directly with the carbene bond. Its activating effect is attributed to a reduction in electron density at the reaction site resulting in a stronger attraction for the π-electrons of the substrate olefin. In the initiation reaction, the products $Me_3CCH{=}CHMe$ and

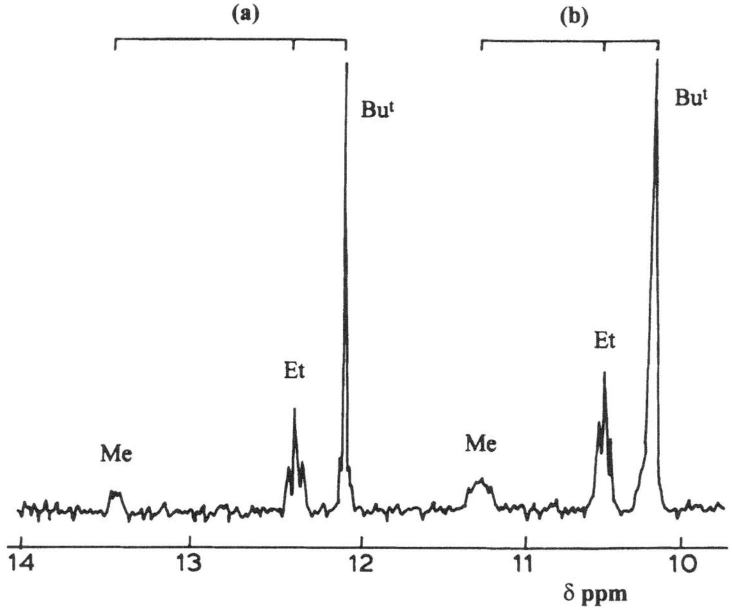

Fig. 3.3 ^{1}H NMR spectrum (carbene proton region) of a mixture of [W]=CHR (R = Me, Et, *t*-Bu) at −30°C, produced from a mixture of the species, (a) $W(=CHCMe_3)(Br)_2(ONp)_2/AlBr_3$ and (b) $W(=CHCMe_3)(Br)_3(ONp)/AlBr_3$, by reaction with *cis*-$MeCH{=}CHEt$ at − 10°C in C_6D_5Cl (Kress 1982).

$Me_3CCH{=}CHEt$ are formed in equal amount showing that there is no preferred orientation of the pent-2-ene in its reaction with initiator, but the propagating species [W]=CHEt and [W]=CHMe are present in about 4 : 1 ratio showing that the latter is somewhat more reactive in the propagation reactions. The broad signals for [W]=C*H*Me at −30°C (Fig. 3.3) suggest the occurrence of a dynamic exchange process (perhaps involving degenerate exchange with pent-2-ene). At −50°C this becomes sufficiently slow to show resolved quartets for these resonances.

Another clear example of an equilibrium metathesis reaction is the ROMP of cyclopentene initiated by the tungsten analogue of **2**. The reaction may be totally reversed by placing the system under vacuum thereby stripping off the monomer units from the 'living' metal carbene complexes (Schrock 1989).

When norbornene or its methyl derivatives are added to $W(=CHCMe_3)(Br)_2(OCH_2CMe_3)_2/GaBr_3$ the ^1H NMR spectra shown in Fig. 3.4 are obtained. All the monomers react completely within 1 min but not all the initiator reacts, indicating that propagation is faster than initiation. This is not surprising since the carbene ligand in the initiator has a tertiary alkyl substituent whereas the carbene ligands in the propagating species have mostly secondary substituents and will therefore offer less hindrance to the approach of olefin. For the case of 1-methylnorbornene, Fig. 3.4(d), both the head species **3** (singlet) and the tail species **4** (doublet) may be seen; it is the more hindered head species, with the tertiary alkyl substituent, that is present in higher concentration and therefore less reactive, as expected from the above argument.

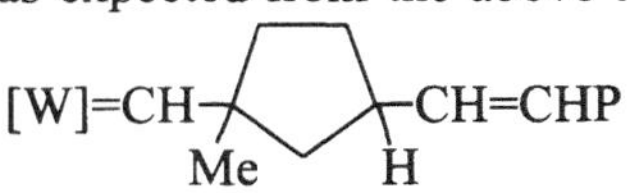

(3)

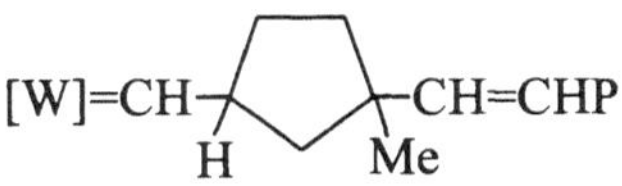

(4)

For norbornene, Fig. 3.4(a), only one carbene doublet is seen, the chemical shift being insensitive to the configuration of the nearest C=C bond or to other more remote structural variations. For the carbene adducts of monomers (b) and (c), however, two doublets are seen, which may be ascribed to head and tail structures. The weaker, having the chemical shift closest to that derived from norbornene, is assigned to the tail structure. Here too the head structure is the less reactive of the two, giving the higher final concentration. In the case of (e) the *anti* isomer reacts far more rapidly than the *syn* isomer, initially giving **5**, corresponding to the upfield doublet in Fig. 3.4(e). However, in time, the *syn* monomer adds to **5** to give **6**, which is responsible for the downfield doublet in Fig. 3.4(e).

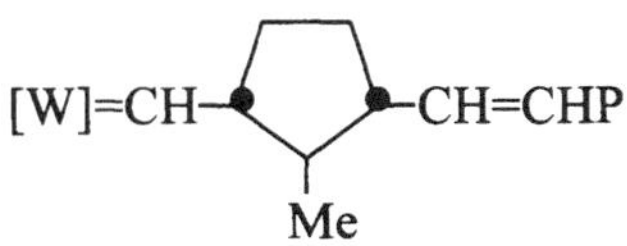

(5) (P = polymer chain)

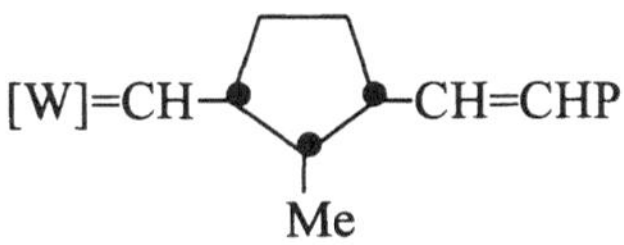

(6)

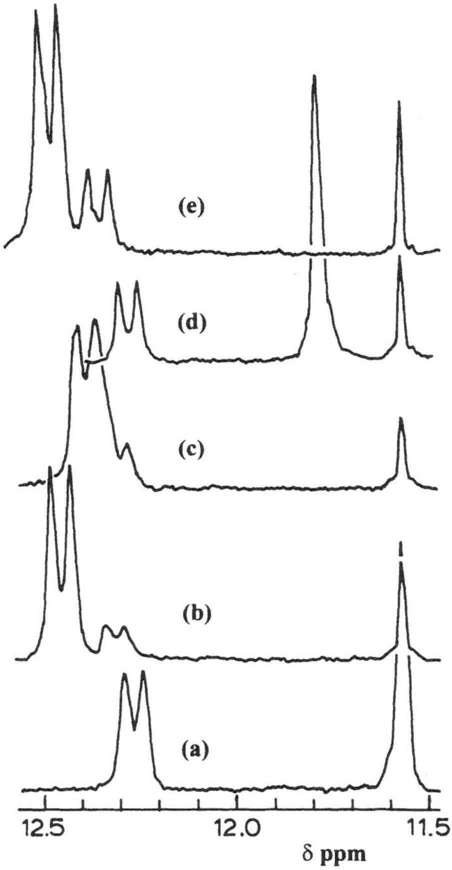

Fig. 3.4 ^{1}H NMR spectra (carbene proton region) after addition of 5 equivalents of (a) norbornene, (b) *endo*-5-methylnorbornene, (c) 5,5-dimethylnorbornene, (d) 1-methylnorbornene, (e) 1 : 1 mixture of *syn*- and *anti*-7-methylnorbornene to $W(=CHCMe_3)(Br)_2(ONp)_2/GaBr_3$ (1 : 0.5) at 20°C for 3 min and cooling to −33°C. All spectra are aligned on the singlet (δ 11.57) for the initiator in order to correct for small variations of δ with W : Ga ratio (Kress 1985).

Stable propagating metal carbene complexes may also be observed when certain initiators of the type listed in Table 2.2 are used *without* a Lewis acid cocatalyst. The propagating species are 'living' and addition of successive batches of different monomers can be used to make block copolymers; see Ch. 14. The conversion of the living polymer derived from the first monomer, into the propagating species of the second monomer can be readily followed by ^{1}H NMR.

When the reaction is sufficiently slow, separate signals for the first addition product and subsequent addition products are often seen, either for the carbene proton resonance (Kress 1987c; Feast 1994d) or elsewhere in the spectrum (Ivin 1988). The concentration of first-addition product $[P_1]$ passes through a maximum with increasing time. At this point the rates of formation and disappearance of P_1 are equal so that $k_i/k_p = [P_1]/[I]$, where [I] is the momentary concentration of initiator. An example is shown in Fig. 3.5. In this case initiation is nearly four times faster than propagation.

As mentioned earlier, initiator **2** has rotational isomers in which the alkylidene substituent is turned either towards the second multiple bond (*syn*) or away from it (*anti*); eqn. (11). In this case the equilibrium position is very much in favour of the *syn* rotamer ($K = 1450$ in toluene at 25°C), making it difficult to detect the *anti* rotamer in routine spectra. However, the equilibrium can be displaced by UV irradiation (366 nm) of the solution for several hours at -80°C to yield a mixture containing about 33% of the *anti* rotamer as determined from the H_α resonances: *syn*, δ 12.11, $J_{CH} = 120.3$ Hz; *anti*, δ 13.30, $J_{CH} = 153.3$ Hz (Oskam 1992, 1993a). On adding 0.33 equivalents of 2,3-bis(trifluoromethyl)norbornadiene to this solution and running the spectrum again at -80°C it is found that the *anti* rotamer has been completely consumed, giving the *syn* first-addition product, eqn. (12), while the *syn* rotamer has scarcely reacted at all. It is estimated that the

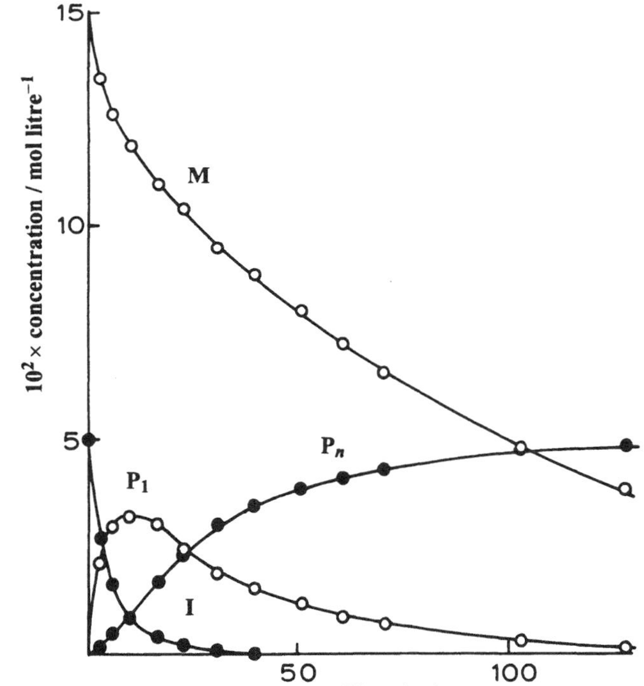

Fig. 3.5 Reaction of ~3.2 equivalents of *exo* adduct (M) of cyclopentadiene and maleic anhydride with 1 equivalent of $W[=C(CH_2)_3CH_2](Br)_2(OCH_2CMe_3)_2$ (I) to give the tungsten–carbene complex P_1 and then P_n ($n > 1$). Solvent CD_2Cl_2, 21°C. The values of $[P_1]$ and [I] at the maximum in $[P_1]$ give $k_i/k_p = 3.8$ (Ivin 1988).

anti rotamer is at least 100 times more reactive than the *syn* rotamer.

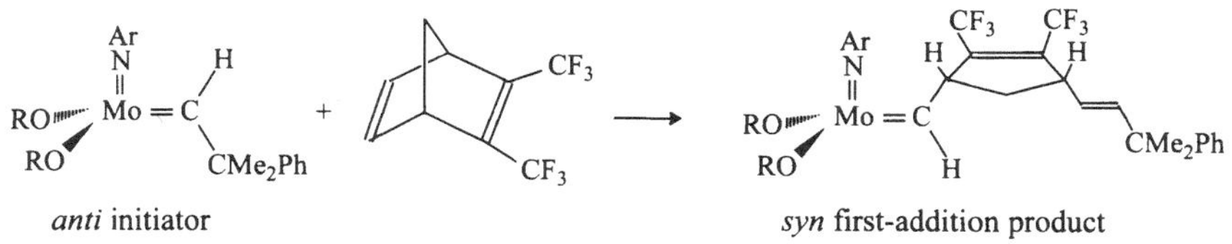

(12)

The C=C double bond formed in this reaction is *trans* on the basis of the value for J_{HH} (15.4 Hz). Further irradiation of the reaction mixture (containing 0.67 equivalents of *syn* initiator and 0.33 equivalents of *syn* first-addition product) results in the isomerization of some of each to the *anti* isomers. In THF the coordination of the solvent alters the position of equilibrium ($K = 23$ for **2** at 25°C) and reduces the rate of attainment of equilibrium but qualitatively the same effects are observed as in toluene; see Fig. 3.6.

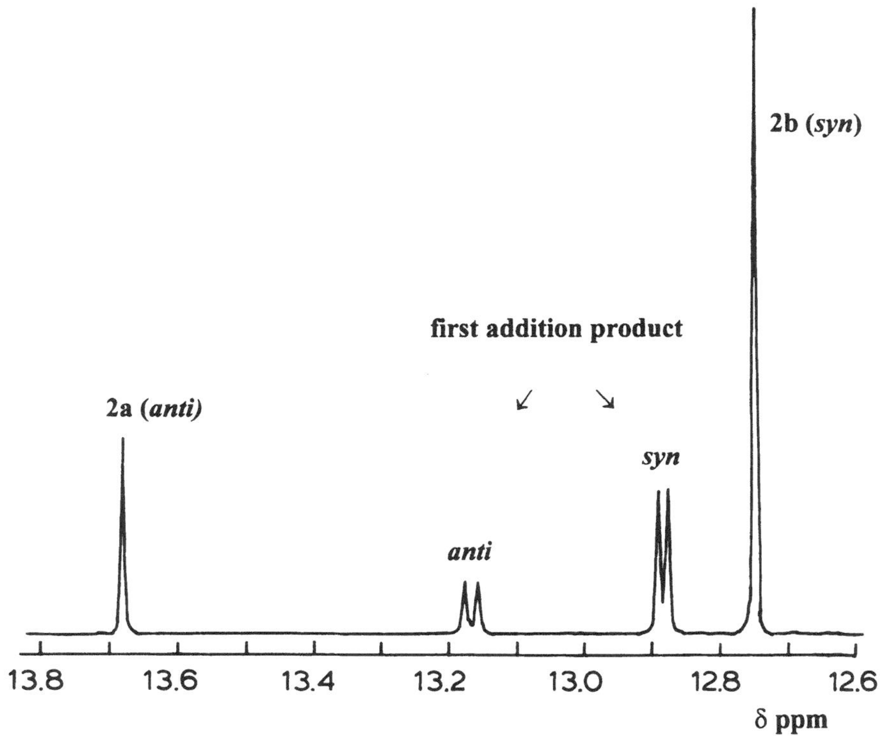

Fig. 3.6 500 MHz ^{1}H NMR spectrum at 35°C showing the alkylidene region after the addition of 0.33 equivalent 2,3-bis(trifluoromethyl)norbornadiene to irradiated **2** in THF-d_8, followed by further irradiation at low temperature (Oskam 1992).

Further work along these lines will clearly improve our understanding of the mechanism. For detailed studies on various Re carbene complexes, see Toreki (1993) .

The rate of reaction of $Mo(=CHCMe_3)(=NAr)[OCMe(CF_3)_2]_2$ (**2′**) with *anti*-7-methylnorbornene at $-13°C$ in CD_2Cl_2 is first-order with respect to **2′**, but zero-order with respect to monomer over a 4.5-fold change of concentration. The first addition of monomer gives a *trans* double bond, as judged by $J_{HH} = 15.6$ Hz. This is consistent with a mechanism in which the rate is controlled by the first-order conversion of the *syn* to the *anti* rotamer (3.5×10^{-3} s^{-1}) followed by rapid reaction of the *anti* rotamer with the monomer (Feast 1994c). Similarly, the rate of reaction of **2** with 1,7,7-trimethylnorbornene at 20°C in CD_2Cl_2 is first-order with respect to **2** but tends to zero-order in monomer at high concentration. The first-order rate constant, $k_{s/a} = 6.1 \times 10^{-5}$ s^{-1}, derived by extrapolation, is in good agreement with the value of 7×10^{-5} s^{-1} at 25°C in toluene determined from the rate of recovery of equilibrium (11) after a photochemical displacement (Oskam 1993a).

The rate of interconversion of the *syn* and *anti* rotamers is thus very dependent on the nature of the ligands, being much slower, for example, with $OCMe(CF_3)_2$ ligands than with $OCMe_3$ ligands (Oskam 1992, 1993a). Whether such factors come into play with any first-generation catalysts is not yet clear.

3.5.2 Detection of propagating metallacyclobutane complexes

In Section 3.6 we shall consider cases of initiation by metallacyclobutane complexes. In the present section we are concerned with the detection of metallacyclobutane complexes in systems initiated by metal carbene complexes.

We have seen from Fig. 3.4 that some initiator remains after all the monomer has reacted with $W(=CHCMe_3)(Br)_2(OCH_2CMe_3)_2/GaBr_3$. This is still the case when the initiator and monomer are mixed at $-78°C$ in CD_2Cl_2 confirming that this is not due to inefficient mixing. If the temperature is raised slowly, reaction begins at about $-53°C$ but the initial spectra are different from those observed at room temperature. The species formed at low temperature have been identified as intermediate *transoid* metallacyclobutane complexes **7**.

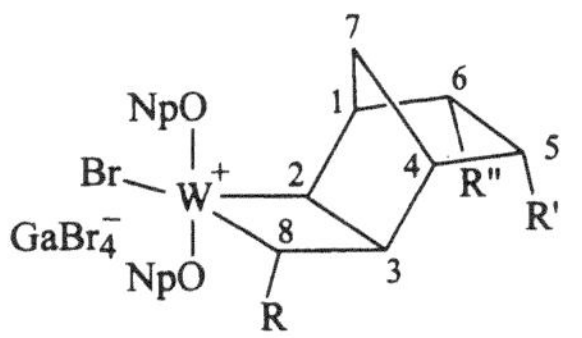

(**7a**): R = CMe_3, R' = R" = H
(**7b**): R = CMe_3, R' = R" = Me
(**7c**): R = CMe_3, R' = Me, R" = H
(**7d**): R = CMe_3, R' = H, R" = Me
(**7e**): R = $C_5H_8CH=[CHC_5H_8CH]_n=CHCMe_3$, ($n > 0$), R' = R" = H
(**7f**): R = CHMeEt, R' = R" = H
(**7g**): R = $C_5H_8CH=[CHC_5H_8CH]_n=CHCHMeEt$, ($n > 0$), R' = R" = H

(**7**) (Np = neopentyl, C_5H_8 = )

These complexes disappear when the temperature is raised to 20°C and are replaced by the metal carbene complexes seen in Fig. 3.4 (Kress 1987b,c, 1988, 1989, 1990). For **7a**, the initial adduct derived from norbornene, the ^{1}H NMR spectrum (-53°C, CD_2Cl_2) is characterized by doublets for H-2 (δ 6.77) and H-8 (4.91), each split by H-3 only, and a triplet for H-3 (0.53), split by H-2 and H-8 only; also by singlets for H-1 (3.33) and H-4 (3.51). The ^{13}C NMR shifts follow a similar pattern: C-2, δ 152.5; C-8, 141.6; C-3, 25.6; C-1, 46.6; and C-4, 47.5 (Kress 1987b). The shifts for the two α-protons H-2 and H-8 appear at much lower field than for those reported for the analogous bis(cyclopentadienyl)titanium compounds (Gilliom 1986b), no doubt because of the strong Lewis acidity of the tungsten centre. The relative positions of the chemical shifts for the nuclei in the α and β positions are as expected for trigonal–bipyramidal geometry (Feldman 1991). Differential nuclear Overhauser enhancement (NOE) experiments show that all the observed metallacyclobutane intermediates have a *transoid* C-8–C-3 bond. This becomes a *trans* double bond when the complex rearranges to a metal carbene complex. Since the polymers formed generally contain *cis* as well as *trans* double bonds it must be concluded that intermediate metallacyclobutane complexes with a *cisoid* C-8–C-3 bond are too unstable to be observed. When the $GaBr_3$ cocatalyst is omitted, no metallacyclobutane complexes of any sort are detected in these systems. The positive charge at the tungsten centre is thus a source of enhanced stability of the *transoid* species **7**.

The complex **7b**, produced from *endo,endo*-5,6-dimethylnorbornene, is an order of magnitude more stable than **7a**. It can be obtained in 70% yield from the initiator at -38°C which allows the kinetics of its first-order rearrangement to be studied at -28 to -18°C; Fig. 3.7. The Arrhenius parameters are $A = 7 \times 10^{11}$ s^{-1} and $E_a = 72$ kJ mol^{-1}. The extrapolated half-life at -38°C is 150 min compared with about 17 min for the rearrangement of **7a**.

From *endo*-5-methylnorbornene, two isomers, **7c** and **7d** are produced in equal proportion at -53°C, the *endo*-methyl group being located at C-5 and C-6, respectively. **7c** is the more stable, its rate of rearrangement at -23°C being similar to that of **7b**, while **7d** rearranges at -38°C at a rate that is close to that of **7a**. A significant degree of stability is thus conferred on the metallacycle by the presence of an *endo* group on C-5. This effect is specific and is not induced by methyl groups in any other position (*endo*-6, *exo*-5, *exo*-6, 1, or 7). It may be interpreted in terms of a repulsion between the *endo*-methyl group at C-5 and either the developing pπ orbital at C-3, or the C–H bond at C-3, which would raise the activation energy of the cleavage reaction.

Earlier we saw that the first-formed metal carbene complex P_1 could often be distinguished from the carbene complexes P_n ($n > 1$) formed by further addition of monomer. The same is true of the intermediate metallacyclobutane complexes X_1, such as **7a**, and X_n ($n > 1$), such as **7e**. The latter are formed when the reaction mixture is held at $-38°$ to -28°C, allowing X_1 to convert slowly to P_1 which then adds monomer to form X_2, and so on. The clearest indication of the formation of these higher homologues comes from the H-2 signal. Thus **7a** gives an H-2 doublet at δ 6.77 ($J = 9$ Hz) which is gradually replaced by two isomers of **7e** giving

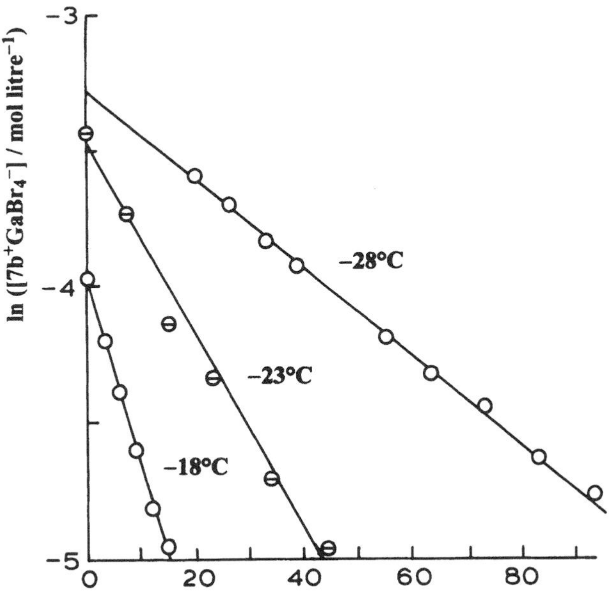

Fig. 3.7 First-order plots for the rearrangement of **7b** into the corresponding metal carbene complex in CD_2Cl_2 at three temperatures (Kress 1988).

distinct doublets at δ 6.92 (major component) and δ 6.87 (minor component) (Kress 1987b). If a racemic initiator is used, having a chiral substituent on the carbene ligand, then the *initially* formed metallacyclobutane complex X_1 is also present as two isomers; for example, **7f** gives H-2 doublets at δ 7.06 (42%) and 6.80 (58%). At −38°C, the isomers **7f** undergo further reaction to give **7g**, the spectrum of which shows exactly the same pair of H-2 doublets and with the same relative intensity as for **7e**. This is to be expected since the two structures differ only in the nature of the end group. The two isomers of **7f** must be diastereomers arising from the presence of chiral centres in R and at C-3. Differential NOE experiments on an analogue of **7e** and **7g** (differing only in the end group) have shown that the two isomers are both *transoid*; they too must therefore be diastereomers and are the precursors to the *m* and *r* dyads in the metal carbene complexes formed by cleavage of the W−C-8 and C-2−C-3 bonds (Kress 1989).

When the initiator has a doubly substituted carbene ligand the initially formed metallacyclobutane complex X_1 is not detected. This is because it has both *cisoid* and *transoid* substituents at C-8, and *cisoid* structures are never detected. In this case, for the reaction of, say, norbornene, the first product seen in the ^1H NMR spectrum is P_1, followed by the two *transoid* diastereomers of X_2. When the initiator has a monosubstituted carbene ligand, and the norbornene carries a *syn*-7-methyl substituent, again the X_1 species is not observed. One can expect from the structure of **7** that a *syn*-methyl group at C-7 will have the same sort of destabilizing effect as a *cisoid* substituent at C-8. This sort of interaction is not present in the case of an *anti*-7-methyl substituent and accordingly the X_1 species *is* observed in this case. When the monomer is unsymmetrical, as for 1-methylnorbornene, *endo*-5-methylnorbornene, 5,5-dimethylnorbornene, and *endo*-dicyclopentadiene, the X_1 species has head (H) and tail (T) forms while the X_2 species has the possibility of HH, HT, TH, and TT structures, each of which has two diastereomeric forms, i.e. eight distinct structures so far as the immediate neighbourhood of the metallacycle is concerned. In fact seven H-2 doublets are resolved for the adducts derived from *endo*-dicyclopentadiene, at δ 7.35, 7.23, 7.19, 7.09 (double intensity), 6.98, 6.95, and 6.81; and six for the adducts derived from *endo*-5-methylnorbornene (Kress 1989). The relief of ring strain is an important factor in the formation of these detectable metallacyclobutane intermediates; they are not observed for the metathesis reactions of acyclic olefins under similar conditions.

Intermediate molybdacyclobutane complexes have also been detected in the reactions of $Mo(=CHCMe_2R)(=NAr)(OCMe_3)_2$ ($Ar = C_6H_3$-*i*-Pr_2-2,6; R = Me, Ph) with **8–11** (Bazan 1991c).

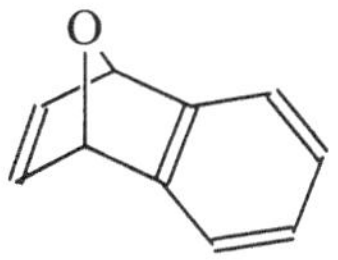

(**8**)

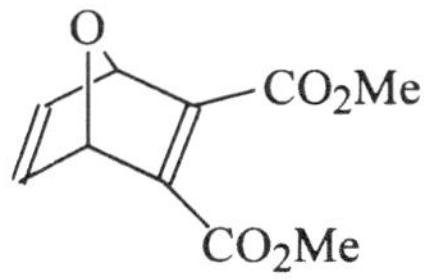

(**9**)

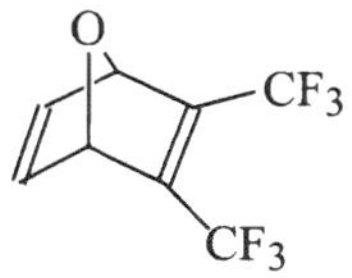

(**10**)

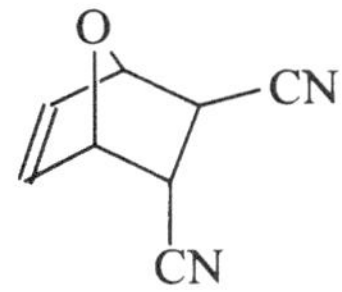

(**11**)

Only in the case of **8** is the ultimate product a long-chain polymer, but in all cases one may observe, at 0–60°C, a clean first-order rearrangement of the initial metallacyclobutane complex to the first metal carbene adduct, consisting of an equilibrium mixture of *syn* and *anti* rotamers in the ratio 9:1 (cf. Section 3.5.1). Except in the case of **8**, the metal carbene complexes do not survive for very long in the presence of excess monomer. For **8**, however, ROMP is propagated, and distinct ^{1}H NMR signals are seen for the longer-chain metal carbene complexes in both *syn* and *anti* forms.

In some cases the metallacyclobutane complexes can be isolated and their crystal structure determined.

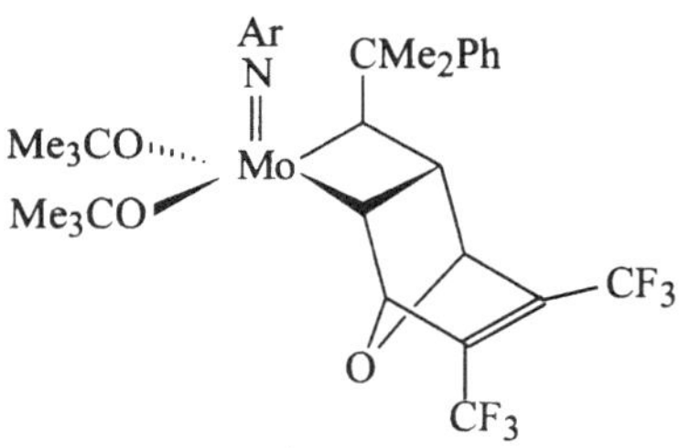

(**12**)

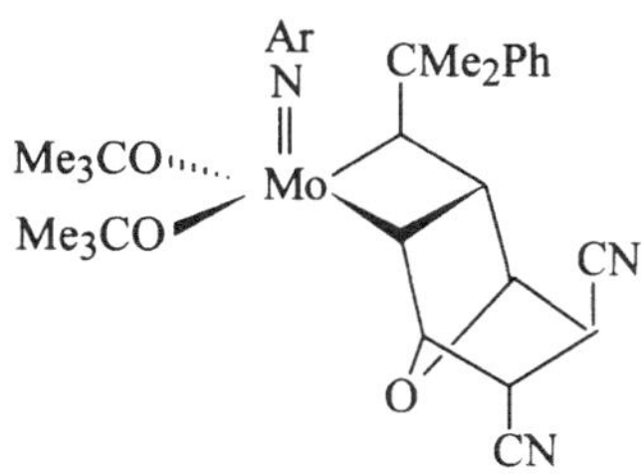

(**13**)

Thus, for **12** the geometry about the molybdenum atom is square-pyramidal (with NAr at the apex), the CMe_2Ph substituent is *trans* to the norbornene ring with the phenyl group directed towards this ring, and the MoC_3 ring is planar. The distance between Mo and O (3.32 Å) indicates that there is no significant bonding between them. In the reaction of **11** the ^1H NMR spectrum shows that two square-pyramidal *transoid* metallacycles are formed, in one of which the cyano group closest to CMe_2Ph is in the *endo* position, see **13** (65%), while in the other (not shown) it is in the *exo* position.

The half-life for the rearrangement of **12** to the metal carbene adduct in C_6D_6 at 35°C is 22 h, with an activation energy of 97 kJ mol^{-1}. When the substituent is CMe_3 in place of CMe_2Ph, the rate constant for the rearrangement increases five-fold. The rate goes up by another order of magnitude if the CF_3 groups are replaced by CO_2Me, and by yet another order of magnitude for the rearrangement of the metallacyclobutane derived from **8** (half-life ~1 h at 0°C). The stabilization brought about by the CF_3 groups is attributed to their inductive effect. Replacement of the Me_3CO ligands by Et_3CO ligands reduces the rate of rearrangement of the metallacyclobutane (R = Ph) derived from **9** by a factor of six (Bazan 1991c).

Square-pyramidal metallacycles are not observable at 25°C upon adding 7-oxa-norbornadiene derivatives to $Mo(=CHMe_2R)(=NAr)(OCMe_2CF_3)_2$ except as a transient red colour, rapidly changing to the characteristic orange colour of the living carbene complexes. However, the metallacycles can be observed at low temperature (Bazan 1991c).

It has sometimes been questioned whether metallacyclobutanes are true intermediates in olefin metathesis reactions (Wilker 1983; Hamilton 1984b). The above examples provide clear proof of their real existence.

3.5.3 Detection of propagating metal–carbene–olefin complexes

It has generally been assumed that in olefin metathesis reactions the olefin first coordinates to the metal carbene complex, *en route* to the formation of the intermediate metallacyclobutane complex, and that after cleavage of this intermediate the newly formed double bond is temporarily coordinated to the metal centre. A number of stable metal–carbene–olefin complexes are known; for references see McGeary (1985) and Toledano (1984a). They are mostly stabilized by chelation of the olefin and/or by heteroatom substituents on the carbene, although Schultz (1981) and Angermund (1984) have prepared some which enjoy neither of these modes of stabilization.

Two types of chelated complex have been detected as intermediates in the rearrangement of $W[=C(OCH_2CH_2CH=CH_2)(C_6H_4\text{-}Me\text{-}4)](CO)_5$ in benzene at 38°C, though the products, which are cyclopropane derivatives, are the result of reductive elimination from the metallacyclobutane and not of metathesis (Casey 1985). The major intermediate, containing four CO ligands, is characterized by ^1H NMR signals at δ 3.59 ($CH{=}CH_2$), 3.14 ($CH{=}CHH$), and 2.73 ($CH{=}CHH$) with $J_{trans} = 13.7$ Hz, $J_{cis} = 8.7$ Hz, and J_{gem} too small to observe. The upfield shift of

2.0–2.5 ppm of these vinyl resonances relative to their position in the uncomplexed alkene of the precursor carbene complex, as well as the absence of any geminal coupling, is consistent with π-coordination of the vinyl group to tungsten. The minor intermediate has a similar spectrum, with the terminal vinyl protons appearing at δ 2.86 and 2.78. There is good evidence to show that in the major intermediate the W=C and C=C bonds are parallel while in the minor intermediate they are perpendicular to one another. Surprisingly, they do not readily interconvert so that their relative ratio during their reactions is somewhat variable.

The complexes $W[=C(OCH_2CH_2CH=CHOMe)(C_6H_4\text{-}Me\text{-}4)](CO)_5$ (*cis* and *trans*), decompose in benzene at 22°C to give some *metathesis* product, 5-(4-methylphenyl)-2,3-dihydrofuran (44% yield from the *trans* isomer), together with certain amounts of cyclopropane derivatives formed by reductive elimination from the intermediate metallacyclobutane. If the original *trans* complex is first irradiated (350 nm) at −78°C in CD_2Cl_2, it is largely converted into the chelated metal–carbene–olefin complex containing four CO ligands. This has a half-life of 1.3 h at 5°C yielding metathesis and reductive elimination products in proportions comparable to those obtained in the decomposition of the original complex at 22°C. Similar results are obtained with the *cis* complex (Casey 1987). It is reasonable therefore to postulate that the metal–carbene–olefin complex is a precursor to the metallacyclobutane complex, the common intermediate for both metathesis and reductive elimination. The metathesis reaction is entirely suppressed if the reaction is carried out in a donor solvent such as MeCN or in the presence of a donor such as PPh_3. The relative ease of these two types of reaction is thus very sensitive to the coordination number of the metallacycle.

The only direct evidence for the presence of metal–carbene–olefin intermediates in catalytic metathesis systems comes from a study of the interaction of the tungsten cyclopentylidene complex **14** with cycloalkenes such as cycloheptene **15** in CD_2Cl_2. When these are mixed at −96°C and the temperature raised to between −53° and −28°C no polymerization occurs, but the ^{13}C NMR spectrum contains additional resonances which may be assigned to the metal–carbene–olefin complex **16**. The line intensities show that the equilibrium (13) moves to the right as the temperature is lowered (Kress 1992).

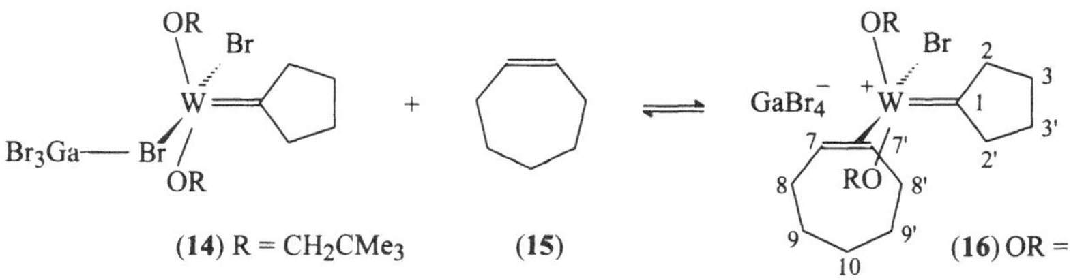

(**14**) R = CH_2CMe_3 (**15**) (**16**) OR = OCH_2CMe_3

(13)

The nature of **16** is established by the presence at −73°C of a signal at δ 355.3 ($^1J_{WC}$ = 142 Hz), which is assigned to the carbene carbon (C-1) of the cyclopentylidene unit, and of two signals at δ 124.4 and 107.8 ($^1J_{CH}$ = 165 Hz) corresponding to two non-equivalent olefinic carbons, upfield from the olefinic carbon signal of **15** at δ 132.4 ($^1J_{CH} = 154$ Hz) The assumption that $GaBr_4^-$ has left the coordination sphere of tungsten to yield a cationic complex is based on the large low-field shift of the alkylidene carbon and on the substantial decrease of $^1J_{WC}$, which suggest that the tungsten centre is significantly more electron-deficient in **16** than in **14** (δ 335.6 and $^1J_{WC} = 169$ Hz). The olefin occupies an apical site of the trigonal–bipyramidal geometry with its C=C axis aligned with the W=C axis and with its 'plane' parallel to that of the cyclopentylidene ligand. This particular conformation is the one that would most easily lead to a metallacyclobutane.

The ^{1}H NMR spectra are in keeping with this interpretation, showing that the two hydrogens of each β-methylene group of the cyclopentylidene ligand, as well as those of the neopentoxy methylene groups, are non-equivalent. However, the two signals for the olefinic hydrogens give only one multiplet at δ 5.64 which is little shifted from the corresponding signal (δ 5.76) in **15**. Such small changes, compared with those for non-alkylidene d^2 olefin complexes, tend to show that the complex **16** should be viewed as a d^0 metal complex in which the olefin is bound mainly through donation of its π electrons into an empty orbital of tungsten. A weak additional interaction of the π^* orbital with the π electrons of the W=C bond may, however, be at the origin of the parallel configuration and of the substantial barrier to rotation (see below).

The equilibrium constant for the formation of **16**, determined from the NMR spectrum at −38°C, is 4.5 M^{-1}; the temperature variation gives $\Delta H^\circ = -57$ kJ mol^{-1} and $\Delta S^\circ = -230$ J $K^{-1}mol^{-1}$. Raising the temperature to −33°C leads to the reversible coalescence of the olefinic carbon signals C-7/C-7′ (δ 124.4, 107.8). Likewise the pairs of peaks for C-3/C-3′ and C-5/C-5′ coalesce to singlets at −58° and −63°C respectively, while the two AB (2H) patterns due to the pairs of geminal protons attached to C-2 and C-2′ simplify to a single AB (4H) pattern above −48°C, and the same applies to the two AB patterns due to the two $-OCH_2-$ groups. Consideration of the splittings as a function of coalescence temperature show that they result from a single intramolecular dynamic process with an energy barrier of 44 kJ mol^{-1}. In this process, equilibration of the two neopentoxo ligands and of the two sides of both cycloheptene and cyclopentylidene ligands occurs, leaving their CH_2 protons non-equivalent. This can only be reconciled with the occurrence in **16** of olefin ligand rotation about the tungsten–olefin axis; eqn. (14).

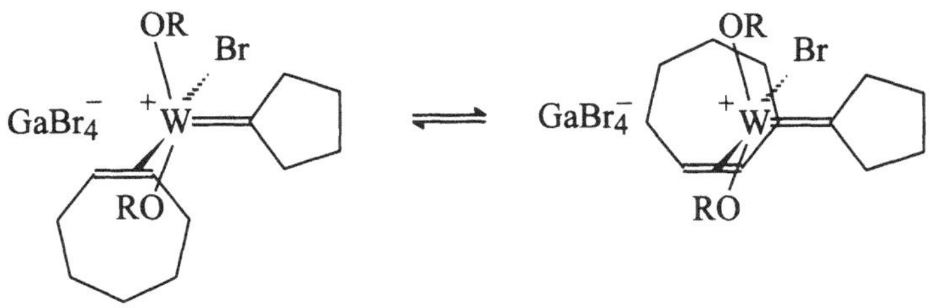

(14)

Such rotational barriers with one rotamer favoured at low temperature could clearly be of importance in determining the stereochemistry of propagation reactions in olefin metathesis.

When the temperature is raised above −18°C, the ROMP of cycloheptene begins to occur, indicating that **16** can be considered as a true intermediate in this system. No further intermediates are, however, detected and hardly any of the initiator is consumed, showing that propagation is very much faster than initiation.

It has further been found that replacing bromides by chlorides, cyclopentylidene by cyclohexylidene, or cycloheptene by cyclooctene leads to similar observations. However, when one bromo ligand is replaced by neopentoxo, or cyclopentylidene by neopentylidene, no intermediate can be detected, even though the ROMP of **15** occurs at −33°C. No interaction is found between **14** and cyclohexene, and no ROMP occurs, suggesting that ring-strain relief is involved in the formation of the cycloalkene adducts as well as in their actual ROMP. Replacing the cyclopentylidene ligand by *n*-pentylidene leads instead to a metallacyclobutane complex as the main observable intermediate and the ROMP of **15** starts at even lower temperature (−53°C), this behaviour being similar to that for norbornene and its derivatives (Section 3.5.2), and arising probably from further ring-strain relief on conversion of the metal–carbene–olefin complexes into the metallacyclobutanes (Kress 1992).

These results show that subtle changes in the nature of the metal carbene initiator or of the substrate can lead to important modifications in the relative energy levels of the three types of intermediates involved in catalytic olefin metathesis reactions.

3.6 Evidence from the reactions of metallacyclobutane complexes

Titanacyclobutane d^0 complexes with two η^5-cyclopentadienyl ligands (Cp), of the kind shown in eqn. (15), were first prepared by Howard (1980a).

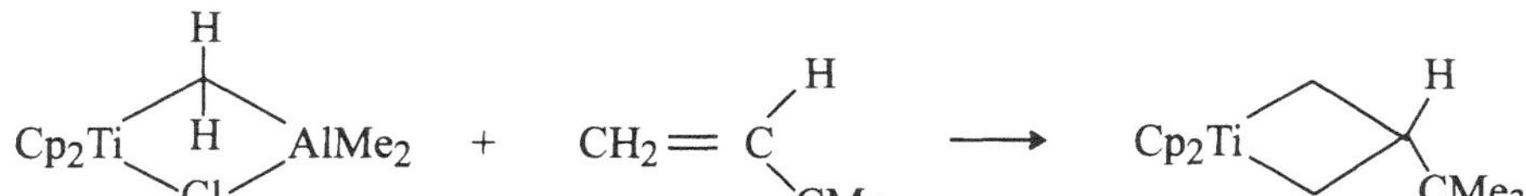

(17)

(15)

The complex **17** reacts with diphenylacetylene **18** in C_6D_6 at 40°C to give the titanacyclobutene complex **19** at a rate proportional to [**17**] but independent of [**18**], with an activation energy of 113 kJ mol^{-1} (Lee, J.B. 1982).

(17) (18) (19)

(16)

The reaction of **17** with Me_3CCHO or with Me_3SnCl proceeds at the same rate as reaction with **18**. It is therefore postulated that all these reactions proceed via slow dissociation of **17** to $Ti(=CH_2)Cp_2$, followed by rapid reaction with PhC≡CPh, Me_3CCHO or Me_3SnCl. These compounds thus behave as metal carbene traps.

The reaction of **17** with **18** is essentially irreversible, but the reaction of **17** with acyclic olefins is reversible and leads to the expected metathesis reactions, for example, the *cis/trans* isomerization of HDC=CMePh (Lee, J.B. 1982). Many isotopic labelling and kinetic experiments have been carried out in an attempt to discover whether a titanium–carbene–olefin complex plays a significant kinetic role in these reactions. The general conclusion is that this is unlikely and it is thought that complete dissociation to $Ti(=CH_2)Cp_2$ must occur before reaction takes place with an olefin or acetylene (Gilliom 1986a; Anslyn 1987; Hawkins 1988, see ref. 4 therein). If such a complex does have a finite existence, it probably corresponds only to a very shallow minimum in the energy profile for the reaction. Stereochemical evidence for this conclusion comes from a study of the isomerization reaction (17).

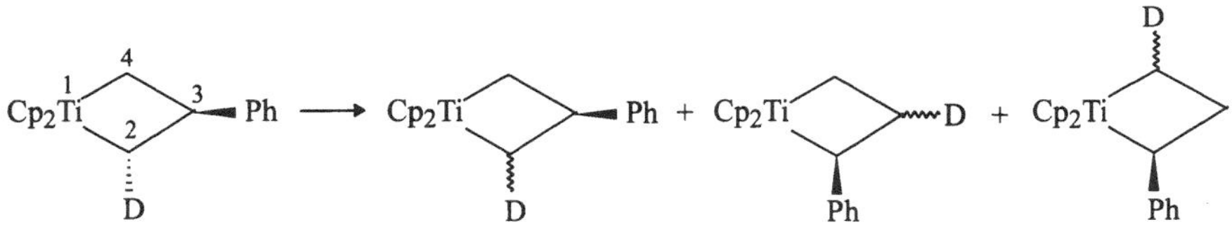

(**20**)(*trans*) (**20**)(*cis/trans*) (**21**)(*cis/trans*) (**22**)(*cis/trans*)

(17)

Cleavage of bonds 1−4 and 2−3, followed by rotation of the olefin in the titanium–carbene–olefin complex and re-formation of the titanacyclobutane complex, would have led only to *cis*-**22**-4-d_1, while cleavage of bonds 1−2 and 3−4, followed by rotation of the olefin, would have led only to *trans*-**21**-3-d_1. In fact both *cis* and *trans* isomers of **21** and **22** are formed, as well as the *cis* isomer of **20** (Ikariya 1985).

Another test is provided by reaction (18). Starting from a 4.7 : 1 mixture of **23a** : **23b** or from a 0.6 : 1 mixture results in the same 1.7 : 1 mixture of **24b** : **24a** at both early and late stages of reaction. Hence it is most likely to proceed by complete dissociation to $Ti(=C=CRR')Cp_2$ (Hawkins 1988).

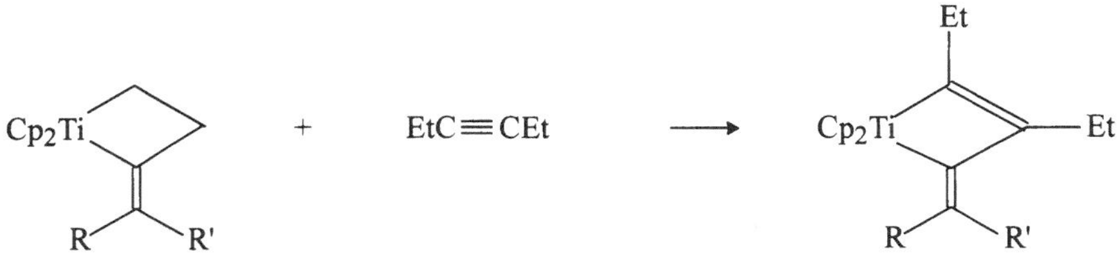

(23a) R = Me, R' = *i*-Pr
(23b) R = *i*-Pr, R' = Me

(24a) R = Me, R' = *i*-Pr
(24b) R = *i*-Pr, R' = Me

(18)

Titanacyclobutanes such as **25** and **26** can be made by adaptations of reaction (15), and then used to initiate the ROMP of norbornene at 60°C (Gilliom 1986b). The reaction is first-order in initiator and zero-order in substrate, again indicating that the conversion of the metallcyclobutane to the metal carbene complex is the rate-controlling step; the activation energy is 113 kJ mol^{-1}. Addition of the monomer to the metal carbene complex is so fast that the concentration of the latter does not reach detectable levels. Such titanium complexes have been used to initiate the ROMP of various other strained ring compounds (Swager 1987; Gilliom 1988).

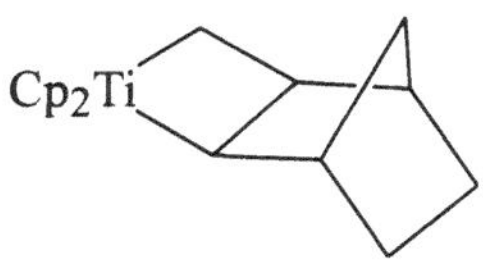

(25)

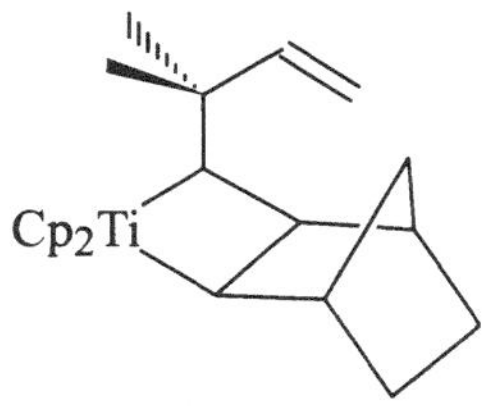

(26)

The tantalum carbene complex **27** reacts with norbornene at −30°C in ether, eqn. (19), to give the d^0 tantalacyclobutane complex **28**, which initiates the ROMP of norbornene at 60°C (Wallace 1986, 1987).

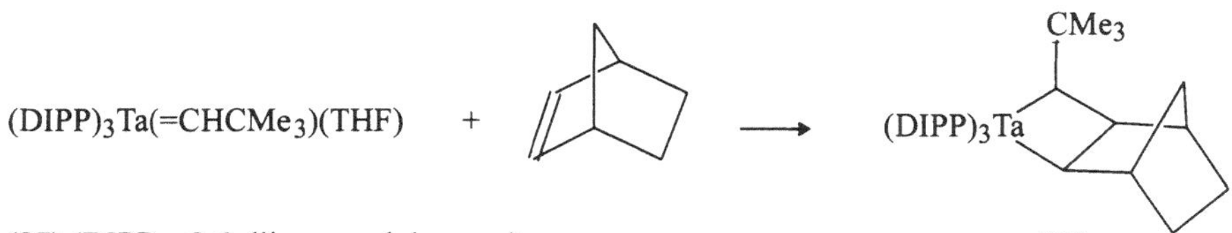

(27) (DIPP = 2,6-diisopropylphenoxy)

(28)

(19)

The kinetics of polymerization are remarkably similar to those for the reaction initiated by the analogous titanium complexes **25** and **26**: first-order in initiator, zero-order in monomer for the major part of the reaction, and activation energy of 105 kJ mol^{-1}. Propagation via tantalacyclobutane and tantalum carbene complexes is clearly implied.

Many stable tungstacyclobutane complexes are known, but few will initiate the metathesis of internal olefins or ROMP of cyclic olefins. Yet many will undergo exchange reactions with ethene or terminal olefins by a mechanism which must involve dissociation to a tungsten carbene complex. A great deal can therefore be learnt about the olefin metathesis mechanism from a study of such reactions. The following is a short summary.

Complexes **29** and **30** are decomposed photochemically to give the products expected as the latter stage of an olefin metathesis reaction, namely ethene from **29**, propene from **30**, and both ethene and propene from **31**, but their thermal decomposition at 80°C gives cyclopropane and propene from **29**, and butenes from **30** and **31** (Ephritikhine 1976, 1977; Adam 1980).

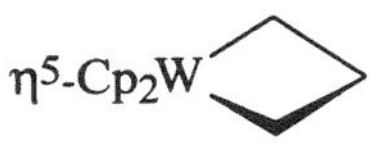

(29)

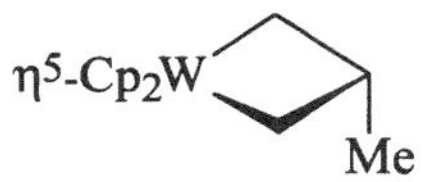

(30)

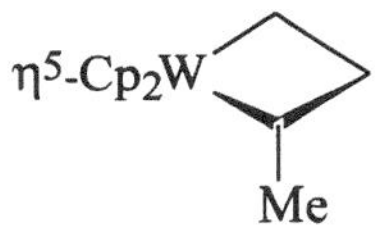

(31)

The difference in thermal and photochemical behaviour has been attributed to the fact that **29–31** are 18-electron compounds which cannot easily convert thermally to 20-electron tungsten–carbene–olefin species, such as $W(=CH_2)(C_2H_4)(\eta^5\text{-Cp})_2$, *en route* to loss of olefin. In the photochemical reaction, it is proposed that one η^5-Cp ligand is converted to an η^3-Cp ligand, so enabling the intermediate formation of an 18-electron species such as $W(=CH_2)(C_2H_4)(\eta^5\text{-Cp})(\eta^3\text{-Cp})$. These results suggest that for a metal carbene chain reaction to be sustained the ligands should be such that a 16-electron count is not exceeded in the metal carbene carrier, allowing the substrate olefin to be coordinated to give an 18-electron species which then rearranges to give a 16-electron metallacyclobutane complex.

Tungsten–carbene complexes of the type $W(=CHCMe_3)(=NAr)(OR)_2$ have a 12-electron count, and react reversibly with ethene and terminal olefins such as vinyltrimethylsilane to yield isolable tungstacyclobutane complexes in some cases. The position of equilibrium, the speed of reaction, and the geometry of the complex that is formed, are very dependent on the nature of the OR ligands and of the substituents in the ring (Feldman 1991). Thus, equilibrium (20) at 25°C is far to the left when $OR = OCMe(CF_3)_2$, but far to the right when $OR = OCMe_2CF_3$ (Schrock 1988b).

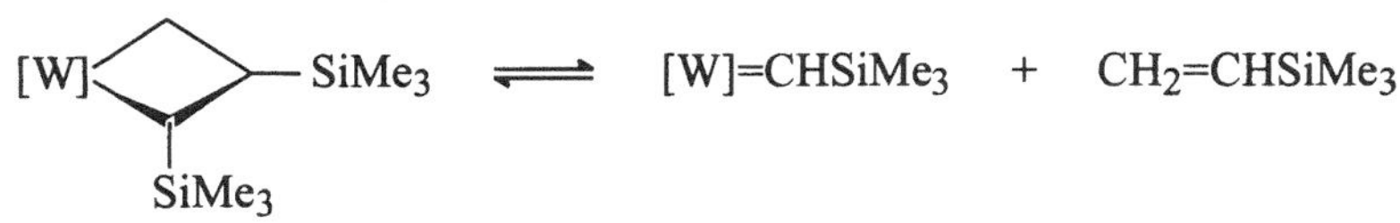

(20)

A contrast in speed is provided by reaction (21), which is very rapid at 25°C when OR = $OCMe_2CF_3$ but rather slow when OR = $OCMe_3$.

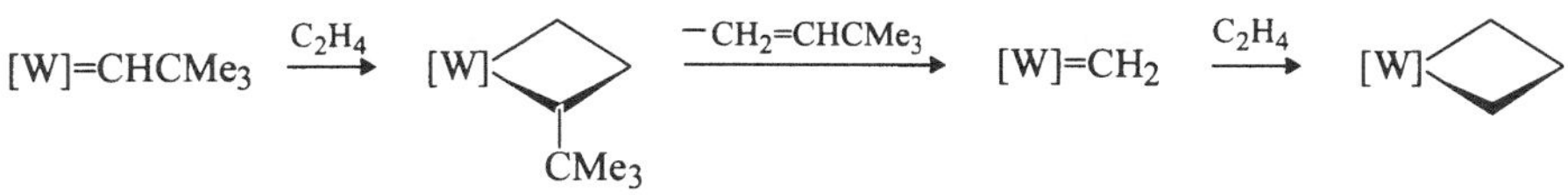

(21)

X-ray crystallographic studies show that when OR is strongly electron-withdrawing, for example, $OCMe(CF_3)_2$, the complexes are trigonal–bipyramidal, the metallacycle being planar when unsubstituted, **32**, and slightly puckered when substituted, **33** (Schrock 1988b). However, when OR is electron-donating, for example, $OCMe_3$, the complex assumes a square-pyramidal configuration with the imidoaryl ligand at the apex and the tungsten atom about 0.58 above the plane of the other four attached atoms, **34** (Feldman 1989a,b).

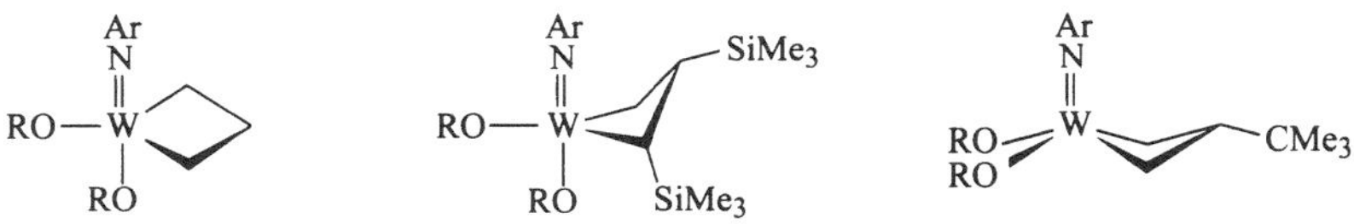

(**32**) OR = $OC(CF_3)_2(C_3F_7)$ (**33**) OR = $OCMe(CF_3)_2$ (**34**) OR = $OCMe_3$, $OCMe_2(CF_3)$, OAr

This geometry is thought to be favoured because the electron-donating OR ligand is then no longer competing directly with the NAr ligand, which is also a π-donor, for the same vacant tungsten d orbital. These two geometries can be conveniently distinguished by the chemical shifts of the nuclei in the ring. For the trigonal–bipyramidal structure the values of δ in ppm are 100–110 for C_α and −5.0–5.0 for C_β; 4.5–5.5 for H_α and about −1 for H_β; whereas for the square-pyramidal structure the values are 40–50 for C_α and 20–50 for C_β; 1–2 for H_α and 3–4 for H_β (Feldman 1991). When the OR ligands have intermediate electron-withdrawing power, for example, $OCMe_2CF_3$ and OAr, the two forms sometimes coexist in dynamic equilibrium.

The rate of reaction of **34** with ethene to give the unsubstituted tungstacyclobutane complex is independent of ethene concentration, indicating that here too the dissociation to $[W]{=}CH_2$ is the rate-controlling process; the activation energy is 82 $kJ\ mol^{-1}$.

The tungstacyclobutane complex formed by the addition of $W({=}CHCMe_3)({=}NAr)(OCMe_3)_2$ to 2,3-bis(trifluoromethyl)norbornadiene at 0°C has been characterized at low temperature and, as expected, has the *transoid* structure and square–pyramidal geometry. It rearranges to the tungsten carbene complex (*trans* double bond), but is unstable and does not polymerize the monomer smoothly (Bazan 1990).

3.7 Evidence of initiating species in systems with non-carbene catalysts

With every non-carbene catalyst system there is the question as to the nature of the initiating metal carbene complex and how it is formed from the catalyst/cocatalyst/(substrate olefin). When the cocatalyst contains an alkyl group, there is usually an obvious path whereby this becomes the source of the alkylidene group attached to the metal and there is much evidence from the initial products of reaction or from end groups in polymers formed by ROMP that can be brought to bear on this question.

As an example we may take the case of WCl_6/Me_4Sn. The ^1H NMR spectrum of the mixture in CD_2Cl_2 at 20°C develops a singlet at 10.52 ppm, reaching a maximum after 25 min, then slowly declining. When $(^{13}CH_3)_4Sn$ is used as cocatalyst this signal appears as a doublet ($J = 134$ Hz); a second, much weaker, doublet is seen at about 9 ppm. Corresponding peaks are seen in the ^{13}C NMR spectrum at 298.7 and 285.2 ppm, respectively. The formation of $(CH_3)_3SnCl$, $(CH_3)WCl_5$, $(CH_3)_2WCl_4$, and CH_4 is also observed. If 3 equivalents of norbornene are added to the mixture after 25 min the low-field signals disappear but are not replaced by any other low-field signals. The ^{13}C NMR spectrum of the recovered polymer shows peaks at 112.49 and 112.38 ppm, which may be assigned to methylene end groups near to *cis* and *trans* double bonds, respectively. Their intensity indicates that the polymer contains at least 150 monomer units from which the overall efficiency of initiation is estimated to be less than 0.7%. The low-field ^1H and ^{13}C signals are assigned to $[W]{=}CH_2$ by analogy with the shifts for known tungsten–carbene complexes. It seems likely that the main initiating species is $W({=}CH_2)Cl_4$ formed by elimination of methane from $(CH_3)_2WCl_4$ (Ivin 1987; also see Wei 1994; Isagulyants 1986). When this catalyst system is used with pent-2-ene as substrate, propene and but-1-ene can be detected initially in equal amounts. These are the expected products of reaction of the initiating species $[W]{=}CH_2$ with MeCH=CHEt, and are readily distinguished from the main products of olefin metathesis, namely but-2-ene and hex-3-ene (Soufflet 1973). Many similar examples have been listed for a variety of both homogeneous and heterogeneous catalysts (Ivin 1983, p. 62). The presence of two distinct initiating species in the WCl_6/Me_4Sn system, when used to catalyze the metathesis of pent-2-ene, is also implied by the way in which the yields of propene and but-2-ene change with time, but their precise nature has not been identified (Thorn-Csányi 1985).

When there is no cocatalyst the initiating metal carbene complex must result from a reaction between the catalyst and the substrate olefin. In the case of Re_2O_7/Al_2O_3 (activated at 550°C) its interaction with but-2-ene to form $[Re]{=}CHCH_3$ can be clearly demonstrated by its subsequent reaction with ethene (after removing the excess but-2-ene by evacuation) to form propene, the amount formed corresponding to about 1.8% of the total Re in the catalyst (Chauvin 1992). The nature of the initiating complex can also sometimes be deduced from the product of its reaction with a carbene trap. For example, the 7-coordinate, 18-electron complex $WCl_2(CO)_3(AsPh_3)_2$ catalyzes the ROMP of norbornene in benzene at 80°C, presumably through loss of a CO ligand followed by coordination of the norbornene, and rearrangement to a tungsten carbene complex which then propagates the ROMP. In the presence of benzaldehyde as carbene trap, polymerization is inhibited and the main product is 2-benzylidenenorbornane **37**. One may therefore conclude that its precursor **36**, produced by a 2,3-hydrogen shift in the tungsten–olefin complex **35**, is the initiating tungsten–carbene complex (Bencze 1985c).

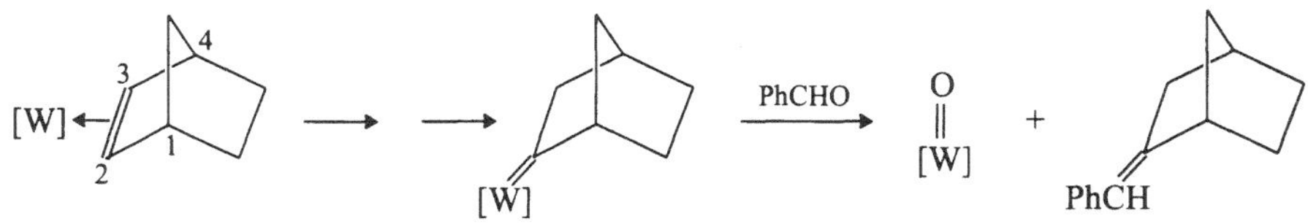

(35) **(36)** **(37)**

Similarly the carbene ligands generated in the systems $W(CO)_6/CCl_4/h\nu$ and $W(CO)_6/Ph_2CCl_2/h\nu$ have been identified as $=CCl_2$ and $=CPh_2$, respectively, from the products of reaction with 2-ethoxynorbornene, e.g. sequence (22) (Garnier 1980b).

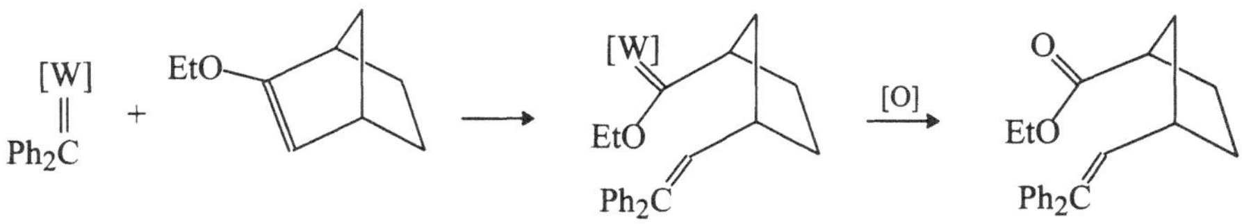

(22)

A novel approach to the determination of the nature of the active intermediate in the ROMP of cyclopentene initiated by $WOCl_4/EtAlCl_2$ was developed by Gassman (1985). This involves the isolation of the species on a porous polystyrene film and its characterization by XPS, which shows it to contain W : O : Al : Cl in the ratio 1 : 1 : 1 : 4–5. On treatment with Me_3P it is converted into a new tungsten complex (non-catalytic) with bonding energies for $W(4f_{7/2})$ and $W(4f_{5/2})$ very close to those for the known complex $W(=CHCMe_3)(=O)(Cl)_2(PEt_3)_2$.

The direct spectroscopic identification of the metal carbene species that propagate metathesis in heterogeneous systems is more difficult since NMR methods are not applicable. However, some success has been achieved using IR and UV/visible diffuse reflectance spectroscopy. Thus, when a MoO_3/SiO_2 catalyst has been photoreduced in CO at 20°C (Shelimov 1988), outgassed at 150°C, exposed to cyclopropane, and again outgassed at 20°C, four IR absorption bands are seen in the C−H stretching region, at 3080, 2985, 2945, and 2870 cm^{-1}. Outgassing the sample at 350°C causes the bands at 2985 and 2870 cm^{-1} to disappear, and the other two bands to decrease somewhat in intensity. On treating the sample with

C_2D_4 the bands at 3080 and 2945 cm^{-1} are completely replaced by bands at 2245 and 2160 cm^{-1}, characteristic of C–D stretching. Subsequent treatment of the sample with C_2H_4 restores the bands at 3080 and 2945 cm^{-1}, which are therefore attributed to [Mo]=CH_2. Concurrently one may observe a weak absorption band at 460–480 nm in the visible spectrum (Vikulov 1989a). Similar observations have been made when such a catalyst is exposed to cycloheptatriene (Vikulov 1992b). For other examples see Ch. 2. An attempt to detect [Mo]=CH_2 by inelastic electron-tunnelling spectroscopy, for a MoO_3/Al_2O_3 catalyst which had been exposed to ethene (0.27 kPa at 100°C), proved unsuccessful (Gajda 1987).

Isotopic labelling can be used to provide indirect evidence of the nature of the chain carrier and its mechanism of formation. For example, using a catalyst prepared from $Mo(\pi^3\text{-}C_3H_5)_4/Al_2O_3$ the initial rate of propene metathesis at 0°C is considerably slower for $CD_2{=}CHCH_3$ and $CD_2{=}CDCD_3$ than for $CH_2{=}CHCD_3$ and $CH_2{=}CDCH_3$. On this basis it is concluded that the initial carbene complex is formed by a 1,2-hydrogen shift mechanism, sequence (23), the isotope effect being associated with the rupture of a C–H bond in the methylene group (Iwasawa 1983).

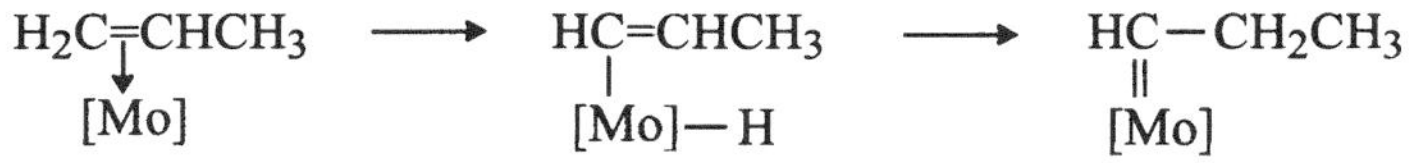

(23)

By studying the *cis/trans* ratios for the various products in the self-metathesis of *cis*-propene-1-d_1, and its cross-metathesis with propene-d_6, Tanaka, K. (1987b) has concluded that the main carrier for degenerate metathesis on a $MoO_3/TiO_2/Me_4Sn$ catalyst is [Mo]=$CHCH_3$ rather than [Mo]=CH_2; also see Tanaka, K. (1988a) and Section 5.3.2.

3.8 Theoretical treatments

Structural parameters and other data have been calculated by *ab initio* molecular orbital (MO) methods for models of the intermediates in the olefin metathesis reaction derived from the following species: for titanium, $Ti(=CH_2)(Cl)_2$ (Rappé 1982a), $Ti(=CH_2)(CH_3)(Cp)$ (Marynick 1985), $Ti(=CH_2)(Cp)_2$ (Upton 1985), $Ti(=CH_2)(L)_2$, where L = H, Cl, Cp (Gregory 1985); for niobium, $Nb(=CH_2)(H)_2(CH_3)$ (Ushio 1984), $Nb(=CH_2)(Cl)_2Cp$ (Taylor 1984); for chromium, $Cr(=CH_2)(Cl)_4$ (Rappé 1982a), $Cr(=CHOH)(CO)_5$ (Ushio 1984), $Cr[=C(OMe)Me](CO)_5$ (Marynick 1985); and for molybdenum and tungsten, $Mo(=CH_2)(CH_3)(Cl)(OAlH_3)$ (Nakamura, S. 1982), $Mo(=CH_2)(Cl)_4$ (Sodupe 1991; Rappé 1982a; Dedieu 1982), $Mt(=CH_2)(=O)(Cl)_2$, where Mt = Mo, W (Rappé 1982b; Cundari 1992), $Mo(=CHOH)(CO)_5$ (Taylor 1984), $Mt(=CHR)(=NH)(OH)_2$, where Mt = Mo, W (Cundari 1992), $Mo(=CH_2)(=X)(L)_2$, where X = O, NH, and L = Cl, OMe, OCF_3 (Folga 1993), $Mo(=CH_2)(=NH)(OR)_2$, where R = H, Me (Fox 1994c), and

$W(=CR^1R^2)Cl_4$ (Bencze 1995). Force field methods have been applied to the intermediates derived from various metal carbene initiators (Bencze 1992, 1994a,b,c).

The interconversion of transition-metal alkylidene-olefin complexes with corresponding metallacycles are formally $2_\pi + 2_\pi$ processes. It might have been expected, from the Woodward–Hoffmann and other rules (see Upton 1985 and Taylor 1984 for summaries), that these would have large activation barriers. Yet they generally proceed with remarkable facility and the first question to be answered is how these reactions manage to circumvent the rules that apply, say, to the formation of cyclobutane from two molecules of ethene. The answer, as determined from a detailed consideration of the reaction of $Ti(=CH_2)(Cl)_2$ with $CH_2=CH_2$, is that the participation of a 3d orbital allows the Pauli principle constraints to be satisfied in a unique way that avoids the unfavourable transition-state bonding interactions that are usually the source of the high barrier (Upton 1985; also see Cundari 1992).

On the question of the transitory existence of the metal–carbene–olefin intermediate, for which there is kinetic and spectroscopic evidence in certain systems (Anslyn 1987; Kress 1992), MO calculations do not reveal a potential energy well intermediate between the reactants $Ti(=CH_2)(Cl)_2 + CH_2=CH_2$ and the product metallacyclobutane, although the metal–carbene–olefin configuration does have an intermediate energy in the overall exothermic reaction (Rappé 1982a, 1984). Similar conclusions have been drawn for the reaction of $Mo(=CH_2)(Cl)_4$ with $CH_2=CH_2$ (Sodupe 1991).

The unsubstituted metallacyclobutane formed from $Ti(=CH_2)(Cl)_2 + CH_2=CH_2$ is calculated to have a planar but easily puckered ring. Even a substituent in the 2-position (opposite to Ti which is numbered 4) is known to cause very little puckering (Lee, J.B. 1981). However, in 1,3-disubstituted tungstacyclobutanes, extended Hückel calculations show that the ring has a puckered *ee* configuration, as required by the interpretation of the *cis/trans* stereoselectivity in the metathesis reactions of alk-2-enes (Tinland 1983); see Section 3.3.

An important conclusion from MO calculations on high oxidation state complexes such as $Mt(=CH_2)(Cl)_4$ is that the Mt−Cl bonds are quite ionic; the chlorines may therefore be pictured as chloride ions. In contrast, the Mt=C bond is quite covalent involving: (i) a dσ orbital on the metal paired with a hybridized carbon orbital (sp$^2\sigma$); and (ii) a dπ orbital on the metal paired with a pπ orbital on the carbon. Thus the bond is an ordinary double bond (analogous to that in an olefin) and the CH_2 part of the system is essentially identical with the triplet ground state of CH_2 (one electron in sp$^2\sigma$ and one in pπ). It is therefore better to regard the metal as in oxidation state (IV), although formally the metal in $Mt(=CH_2)(Cl)_4$ is in oxidation state (VI) (Rappé 1982a,b; also see Taylor 1984). Similar conclusions have been drawn from a consideration of the orbitals in $Mo(=CH_2)(=NH)(OH)_2$, where the highest occupied MO is the relatively non-polar Mo−C π bond, and the lowest unoccupied MO is the Mo−N π^* orbital in the alkylidene plane and largely Mo-centred on the COO face of the pseudo-tetrahedral complex. The reactivity of complexes such as $Mo(=CHR^1)(=NAr)(OR^2)_2$ is greatest when R^2 contains electron-withdrawing substituents. This may be attributed to a reduction in the extent of donation from the oxygen p orbitals into the Mo unfilled orbitals (Cundari 1992; Fox 1994c).

The bonding in the low-valent metal carbene complexes, such as $W(=CPh_2)(CO)_5$ is quite different from the high-valent cases discussed above. The CPh_2 ligand then has the character of a singlet carbene with an empty $p\pi$ orbital and a doubly occupied σ orbital ($sp^2\sigma$) pointing at the metal. Thus the bond here is basically a coordinate bond (not unlike that for CO) with possible $d\pi$ back-bonding (as also with CO) (Rappé 1982b). It is also important to note that when W is in a low oxidation state, e.g. with an additional two electrons in the d_{yz} orbital (d^4), the C=C bond of the olefin will orient perpendicular to W=C of the carbene during its initial approach, in order to maximize the back-bonding interaction with the metal and avoid competition with the carbene moiety for the filled d_{xy} orbital. Hence in this d^4 case metallacycle formation will require a relative rotation of the two double bonds about the line joining their centres, which may well involve a higher barrier than if they had been able to approach in a parallel configuration from the start (Youinou 1988).

The difference in bonding for the two types of complex just described means that for the high-valent complexes the metal–carbene bond is generally weakly polarized in the direction $Mt^{\delta+}-C^{\delta-}$ (more strongly for W–C than for Mo–C), while for the low-valent complexes the *net* polarization is in the opposite direction, $Mt^{\delta-}-C^{\delta+}$. The former are generally electron-deficient with respect to the metal (<18e) and very active, while the latter are initially 18-electron complexes and must lose a ligand in order to initiate olefin metathesis (Rappé 1982b; Marynick 1985; Cundari 1992).

Lewis acids often greatly enhance the activity of metal carbene catalysts by making the metal centre more electrophilic. An *ab initio* calculation on the model system $Mo(=CH_2)(=O)(Cl)(CH_3) + AlH_3$ predicts that the resulting complex should be pseudo-tetrahedral with the Lewis acid AlH_3 bound to the O ligand (Nakamura, S. 1982). Conversely Lewis bases such as pyridine usually slow down the reaction by attaching themselves directly to the metal centre, often in the position *trans* to the Mt=C bond, whence they may slowly migrate partially to a *cis* position. Metathesis is then likely to occur only after exchange of olefin with Lewis base in the *cis* position, or after exchange with Lewis base in the *trans* position followed by migration to the *cis* position (Youinou 1988).

The ease with which the geometry of the metal–carbene complexes can adjust to accommodate the incoming olefin may be an important factor in determining the rate and stereoselectivity in a given metathesis reaction (Lee, J.B. 1981). Extended Hückel MO calculations on $Ti(=CH_2)L_2$, where L = H, Cl, Cp, have shown that the completely planar molecule is easily distorted into a flattish pyramid with Ti at the apex, ready to receive the incoming donor olefin (Gregory 1985). Similarly, calculations employing the self-consistent-field-Xα-scattered-wave method on $Mo(=CH_2)(=NH)(OMe)_2$ show that the reaction with ethene at the COO face to form the metallacyclobutane is facilitated by twisting the $=CH_2$ ligand about the Mo=C bond (Fox 1994c).

Finally we may note two other features which emerge from MO calculations. First, in $Mo(=CH_2)(=NH)(OH)_2$ one may identify the agostic interaction in which the metal attracts electron density from the *anti* C−H bond (the one directed away from NH). Such interactions are manifest in real systems as a lengthening of the $=C-H_{anti}$ bond and a shortening of the Mt=C bond, for example, in tantalum carbene complexes (Schultz 1981). Second, oxo 'spectator' ligands, as in $Mo(=CH_2)(=O)(Cl)_2$, can assist the formation of the metallacyclobutane by being converted into triply bonded oxygen (as in C≡O) (Rappé 1982b).

The earlier MO treatments mostly involved model compounds somewhat removed from real life. Nevertheless they reveal factors and trends which are likely to be valid in real situations. The more recent force field (METMOD) treatments, e.g. that of Bencze, deal with actual initiators and give remarkably accurate predictions of geometry, rotational barriers, etc.

4

Related Reactions

4.1 Introduction

The chemistry of metal carbene and carbyne complexes has undergone very rapid development since the sixties and olefin metathesis reactions now form part of a much larger family of [2 + 2] metathesis reactions between multiply bonded compounds. In Section 4.2 we shall briefly consider this wider context. Some examples will be given of the use of stoichiometric metathesis reactions of such complexes in organic synthesis (Dötz 1984; Schubert 1989).

Olefin metathesis reactions are sometimes accompanied by side reactions and the question then arises as to whether the two types of reaction are related and proceed through the same intermediates, or are completely independent. An intriguing situation exists when metal alkyl intermediates, capable of propagating Ziegler–Natta type polymerization, are in equilibrium with metal alkylidene hydride intermediates capable of propagating ROMP. In principle this can lead to polymer chains containing units formed both by addition to the double bond and by cleavage of the double bond. This aspect will be discussed in Section 4.3. In living ROMP the metal carbene intermediate can be transformed by suitable reaction into another type of chain carrier, which can then propagate the polymerization of a different type of monomer, enabling the preparation of various kinds of block copolymer; some examples will be given in Section 14.5.2.

Finally we consider how three-membered ring compounds may be involved in both initiation and termination reactions in olefin metathesis; Section 4.4.

4.2 [2 + 2] reactions between compounds containing multiple bonds

Such reactions between compounds containing double bonds may be represented in shorthand form by eqn. (1).

$$\begin{array}{c} \mathrm{Mt{=}X} \\ + \\ \mathrm{Z{=}Y} \end{array} \longrightarrow \begin{array}{ccc} \mathrm{Mt} & - & \mathrm{X} \\ | & & | \\ \mathrm{Z} & - & \mathrm{Y} \end{array} \longrightarrow \begin{array}{ccc} \mathrm{Mt} & & \mathrm{X} \\ \| & + & \| \\ \mathrm{Z} & & \mathrm{Y} \end{array} \qquad (1)$$

The reactants may also contain triple bonds in place of one or both double bonds.

In some cases the reaction stops at the four-membered ring; in others the ring rearranges to give products different from those represented in eqn. (1). In Table 4.1 various types of observed reaction are summarized other than those involved in metathesis of olefins or acetylenes, or of metathesis polymerization.

Table 4.1 Some types of observed [2 + 2] reactions involving metal atoms attached to multiple bonds (other than those involved in olefin or acetylene metathesis); cf. eqn. (1)

Multiple bonds [a] Mt···X	Y···Z	Type of compound containing Y···Z	Notes	References
Ti=C	C=O	Ketone		Cannizzo 1987
Mo=C	C=O	Aldehyde,		Bazan 1990
		β- or γ-hydroxyketone	b	Fujimura 1995
W=C	C=O	Ketone, aldehyde		Kauffmann 1986
				Fu 1993a
				Fujimura 1995
W=C	C=N	Carbodiimide		Meisel 1986
		Imine		Weiss 1989a
W=P	P=P	Diphosphene		Dillon 1995
Os=C	S=O	Sulfur dioxide	c	Roper 1984
Ta=C	C≡O	Carbonyl complex	d	Proulx 1993
Zr=N	C=N	Imine		Meyer 1994
W=N	C=N	Carbodiimide		Meisel 1986
Zr=N	C≡C	Alkyne	e	Meyer 1994
W≡C	C=O	Ketone, aldehyde, formate, formamide		Freudenberger 1986
W≡C	C=N	Isocyanate, carbodiimide		Weiss 1986c, 1987
		isothiocyanate		Goller 1993
W≡C	C≡N	Nitrile		Freudenberger 1986
Mo≡C	P≡C	Phosphoalkyne		Hill 1991
W≡W	C≡C	Alkyne	f	Chisholm 1989
W≡W	C≡N	Nitrile		Schrock 1982

[a] The products correspond to the formation of an initial metallacycle $\overline{\text{MtXYZ}}$. [b] The hydroxyl group is essential for the reaction to succeed. [c] Reaction stops at the metallacycle. [d] Product formed by rearrangement rather than cleavage of the metallacycle. [e] Product is an unsaturated metallacycle. [f] See Section 10.2.3.

The carbonyl–olefination reaction, eqn. (2), is like a Wittig reaction in which the Mt=C bond takes the place of a P=C bond, and applies to the first three entries in Table 4.1. It is important both as a means of effecting clean termination of living ROMP reactions when initiated with Ti, Mo, or W complexes, and as a convenient means of converting C=O groups into $C{=}CH_2$ or C=CHR groups in organic synthesis.

$$[\text{Mt}]{=}\text{CHR} + \text{R}^1\text{R}^2\text{C}{=}\text{O} \longrightarrow [\text{Mt}]{=}\text{O} + \text{RCH}{=}\text{CR}^1\text{R}^2 \qquad (2)$$

While ketones can be used to terminate living Ti and W carbenes, they do not react so readily with Mo carbenes, except when carrying a β- or γ-hydroxy substituent, in which case the reaction becomes very stereoselective; e.g. *E/Z* $>99/1$ for the product of reaction (3) in benzene at 20°C. This effect is thought to be caused by coordination of the hydroxy substituent to the metal centre (Fujimura 1995).

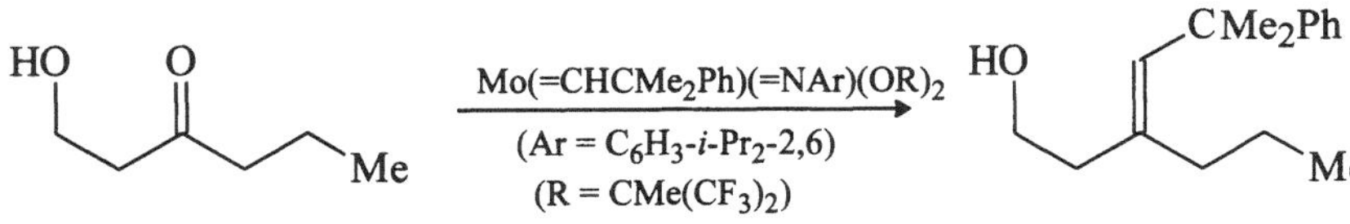

(3)

The analogous reaction using $W(=CHCMe_3)(=NAr)[OCMe(CF_3)_2]_2$ in toluene gives products with *E/Z* = 89/11 at 20°C and $>99/1$ at -78°C. The stereoselectivity is not so high when the HO group is replaced by $PhCH_2O$ or MeCOO, or if the HO group is moved to the α- or γ-positions. This suggests that the hydroxyl group is the only Lewis base functionality that is small enough or basic enough to coordinate strongly to the highly sterically hindered metal centre (Fujimura 1995). In passing it should be noted that living ruthenium carbene complexes are unreactive towards both aldehydes and ketones but can be terminated by stoichiometric metathesis with ethyl vinyl ether (Wu 1993) and other strained or functionalized olefins (Wu 1995b).

The complex $W(=CHCMe_3)(=NAr)[OCMe(CF_3)_2]_2$ can also be used in a stoichiometric metathesis sequence to effect the ring closure of unsaturated ketones so as to form 1-substituted cyclopentenes, cyclohexenes and cycloheptenes in good yield, e.g. eqn. (4).

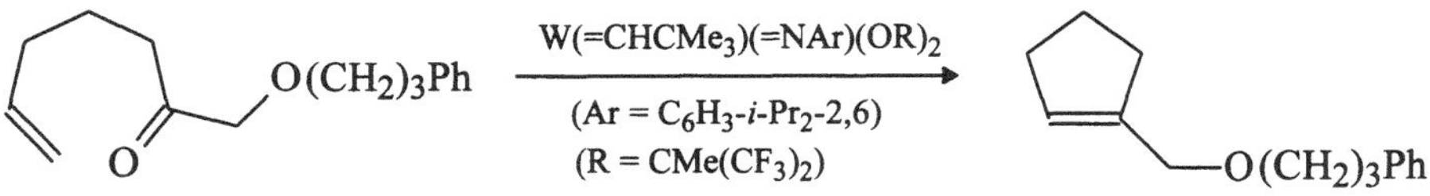

(4)

The C=C bond reacts first to give $[W]=CH(CH_2)_3CO(CH_2)O(CH_2)_3Ph$ which then undergoes an internal carbonyl-olefination reaction (Fu 1993a).

The same strategy has been used in the total synthesis of ($\pm$)-$\Delta^{9(12)}$-capnellene (**7**) from dimethyl-γ-butyrolactone (**1**), the key steps of which are shown in Scheme 4.1. The Tebbe reagent is used to cleave the C=C bond in **2** to give **4** via **3**, immediately followed by the carbonyl–olefination reaction to give **5**. This is then transformed to **6** by standard procedures and finally converted to **7**, again using the Tebbe reagent. This synthesis of **7** is the first to achieve the formation of all four asymmetric centres in a single step. The overall yield is 20% (Stille 1986, 1990). A similar strategy has been employed to synthesize polycyclic enol ethers (Nicolaou 1996).

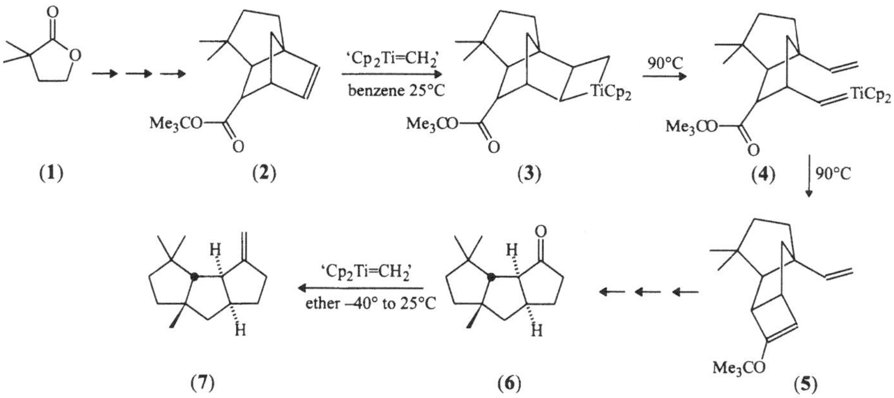

Scheme 4.1 Synthesis of (±)-$\Delta^{9(12)}$-capnellene involving metathesis reactions **2** → **4**, **4** → **5**, and **6** → **7**, facilitated by the Tebbe reagent '$Cp_2Ti=CH_2$'; see Section 2.2.1 (Stille 1986, 1990).

The WCl_6-assisted condensation polymerization of conjugated ketones such as benzylideneacetophenone (**8**) proceeds through the formation of oligomeric ketones PhCO(CH=CPh)$_n$CH=CHPh which can be detected ($n = 1$–6) in the early stages of the reaction (Schopov 1989; also see Schopov 1983, 1985; Jossifov 1991a,b, 1993a,b). By analogy with reaction (4) it is likely that the polymerization proceeds via alternate metathesis reactions of the C=C and C=O bonds with tungsten carbene species formed from the reactants; eqns. (5) and (6).

$$[W]{=}CHR + PhC(=O)CH{=}CHPh\ (\mathbf{8}) \longrightarrow [W]{=}CHC(=O)Ph + CHR{=}CHPh \qquad (5)$$

$$[W]{=}CHC(=O)Ph + PhC(=O)CH{=}CHPh\ (\mathbf{8}) \longrightarrow [W]{=}O + PhC(=O)CH{=}C(Ph)CH{=}CHPh \qquad (6)$$

Another example of a stoichiometric carbonyl–olefination reaction facilitated by a reagent which may be assumed to generate an intermediate tungsten carbene complex is shown in eqns. (7) and (8). The ketone can be added as soon as methane evolution has subsided (Kauffmann 1986).

$$WOCl_3(thf)_2 + 2\,MeLi \xrightarrow[-\,CH_4]{THF/45°C} W(=CH_2)(O)Cl \qquad (7)$$

$$W(=CH_2)(=O)Cl + Ph_2C=O \longrightarrow Ph_2C=CH_2 \qquad (8)$$

The tungsten alkyne complex $W(\equiv CCMe_3)(OC_6H_3\text{-}i\text{-}Pr_2\text{-}2,6)_3$, undergoes total metathesis with $MeC\equiv N$ to yield $MeC\equiv CCMe_3$, whereas with carbonyl compounds, such as acetone, benzaldehyde, ethyl formate and *N,N*-dimethylformamide, it reacts by double-bond metathesis to yield $W(=O)[C(CMe_3)(=CR^1R^2)](OC_6H_3\text{-}i\text{-}Pr_2\text{-}2,6)_3$ which may be readily hydrolyzed with 1 M KOH to give $Me_3CCH=CR^1R^2$ (Freudenberger 1986). Similarly $W(\equiv CCMe_3)Cl_3(dme)$ reacts with cyclohexyl isocyanate, CyN=CO, to give a product, the formation of which can be explained in terms of an initial double-bond metathesis reaction involving cleavage of the N=C bond to give an imido tungsten(VI) complex with a sigma-bonded ketenyl ligand. Into this bond a further molecule of CyNCO is then inserted, forming a chelating ligand (Weiss 1986c). Carbodiimides react in a similar way (Weiss 1987).

4.3 Relationship to Ziegler-Natta polymerization

A typical Ziegler–Natta polymerization (ZNP) is represented by eqn. (9).

$$CH_2{=}CH(CH_3) \xrightarrow{TiCl_4/Et_3Al} \left[CH_2{-}CH(CH_3) \right] \qquad (9)$$

The polymer may be isotactic, syndiotactic, or atactic according to the nature of the catalyst/cocatalyst system. The Cossee–Arlman mechanism for the ZNP of propene is depicted in Scheme 4.2.

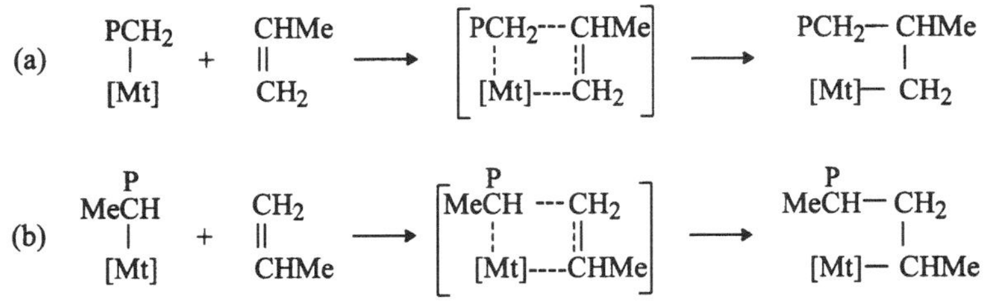

Scheme 4.2 Cossee-Arlman mechanism for (a) primary insertion and (b) secondary insertion in Ziegler-Natta polymerization of propene; P = polymer chain.

For isotactic polymer formation steric control is not lost even when some ethene units are introduced by copolymerization. This shows that the addition of the propene then occurs by 'primary insertion', with $[Mt]{-}CH_2P$ as chain carrier and the stereoregularity controlled by the chirality of the metal centre. On the other hand for syndiotactic polymer formation steric control is lost when ethene units are introduced by copolymerization. It may be concluded that the addition of the propene then occurs by 'secondary insertion', whereby the chain carrier is $[Mt]{-}CHMeP$ and the stereoregularity is controlled by the chiral centre at the carbon attached to the metal atom. The same two factors apply to the control of stereoregularity in the ROMP of chiral and prochiral monomers, as discussed in Section 11.6.2.2.

There is an analogy between the two modes of addition in Scheme 4.2 and the two possible mechanisms of degenerate olefin metathesis, Scheme 4.3, but, in the latter case [Mt]=CHMe is usually the predominant chain carrier; see Section 3.7.

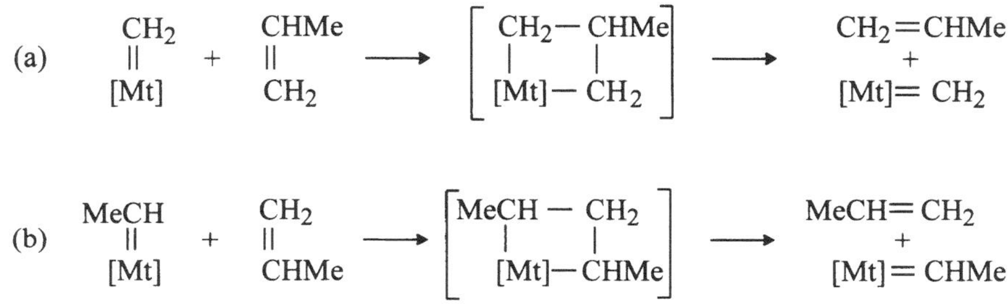

Scheme 4.3 The two possible mechanisms of degenerate olefin metathesis.

Ever since the discovery that Ziegler–Natta catalysts could sometimes bring about ROMP, the relationship between ZNP and ROMP has been a matter for considerable speculation (Dall'Asta 1968a; Olsthoorn 1976b; Ivin 1978c; Laverty 1983). In many cases the same catalyst system can induce ZNP of one olefin, and ROMP of another. For example, WCl_6/Bu_4Sn (1/10) in hydrocarbon solvents initiates ZNP of ethene (Gresham 1959) but ROMP of cyclopentene (Hein 1973). When the two olefins are reacted together in the presence of this catalyst, only the two homopolymers are formed, indicating the presence of two independent propagating species. Likewise $MoCl_5/(C_6H_{13})_3Al$ and $WCl_6/EtAlCl_2$ cause slow ZNP of ethene and rapid ROMP of cyclopentene, but from a mixture of the olefins only the ring-opened polymer of cyclopentene is formed, containing no trace of ethene units (Natta 1964a).

Firm evidence for an equilibrium between a tungsten(IV) methyl complex and a tungsten(VI) methylene hydride complex was first provided by Cooper (1974a,b, 1979).

$$[\mathrm{Mt}]-\mathrm{CH_2P} \rightleftharpoons \overset{\displaystyle \mathrm{H}\atop|}{[\mathrm{Mt}]}=\mathrm{CHP} \qquad (10)$$

If equilibrium (10) exists in a catalyst system for ZNP there is the possibility that ZNP actually proceeds via [2 + 2] addition of the olefin to the metal carbene complex by the mechanism shown in sequence (11).

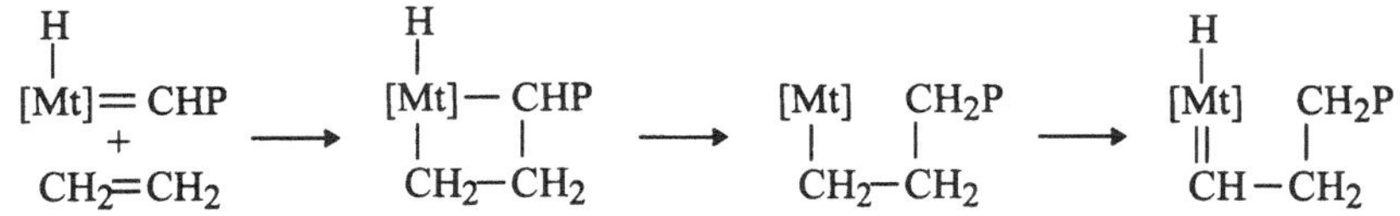

(11)

This would involve migration of hydrogen atoms to and from the α-carbons in the metallacycle and would be expected to exhibit a significant H/D isotope effect. No such effect has been detected in the copolymerization of C_2H_4 with C_2D_4 on $TiCl(C_2H_5)Cp_2/EtAlCl_2$ (Soto 1982), nor in reaction (12) which is a model for the propagation reaction (Clawson 1985).

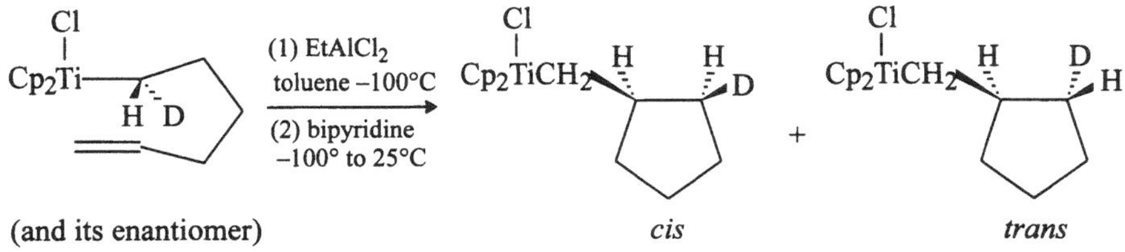

(12)

Racemic titanocene alkenyl chloride, of which one enantiomer is shown in eqn. (12), gives a 50 : 50 mixture of *cis* and *trans* isomers, as determined from the ^2H NMR spectrum of the *cis/trans* mixture of methylcyclopentanes obtained by hydrolysis of the product. However, the complex $Ta(=CHCMe_3)(H)(I)_2(PMe_3)_3$ adds ethene in the presence of PMe_3 to yield $Ta[=CH(CH_2)_nCMe_3](H)(I)_2(PMe_3)_3$. It is known that, for complexes of this type, α-hydrogen transfer is rapid compared with the rate of addition of ethene, so that this polymerization may well proceed by the mechanism shown in sequence (11) (Turner, H.W. 1983).

It must also be borne in mind that an alkyl ligand can be distorted significantly towards an alkylidene hydride system (agostic hydrogen), even when the metal has the d^0 configuration (Dawoodi 1982; Fellmann 1982). The implication is that an alkyl complex may sometimes behave as if it were an alkylidene hydride complex in the presence of certain reagents. If a cycloalkene can react with such a species in two different ways, or if it can react separately with the two distinct species portrayed in eqn. (10), the polymerization will result in a polymer containing units formed by both ZNP and ROMP. Consider, for example, the reaction of a metal alkyl complex $[Mt]-CH_2P$ with norbornene (NBE) by the following sequence of events: (i) insertion of one molecule of NBE into the Mt−C bond; (ii) migration of an α-hydrogen to the metal; (iii) addition of one molecule of NBE by metathesis with the Mt=C bond; (iv) migration of the hydrogen atom on the metal back to the α-carbon so as to reform the Mt−C bond; and (v) insertion of a further molecule of NBE into the Mt−C bond. The resulting structure **9** will contain two characteristic features at the junctions between the two types of unit, namely an olefinic carbon with no attached hydrogens (C-2), and a CH_2 group between the rings (C-6).

Two catalysts have been found which under limited conditions give polymers

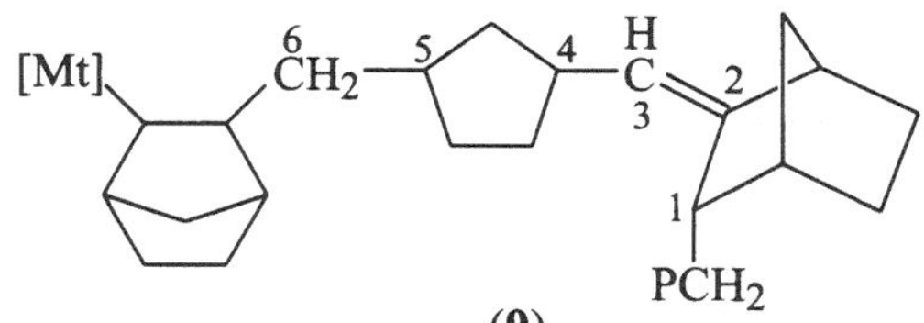

(9)

containing both these features. With $Mo(CO)_5(py)/EtAlCl_2/R_4NCl$ as catalyst the polymer made at 26°C has 100% ring-opened units, but when made at 110°C, only 31% such units; and with $ReCl(CO)_5/EtAlCl_2$ as catalyst the polymer made at 100°C contains 98.6% ring-opened units, falling to 5.1% when made at 110°C, and zero when made at 132°C (i.e. 100% ZNP). These figures are based on the olefinic to aliphatic proton intensity ratio in the ^1H NMR spectra, which is not in itself proof that the two types of unit are present in the same chain; but the fact that the gel permeation chromatogram (GPC) of the polymer exhibits a single, relatively narrow peak is a strong indication that this is so. The conclusive evidence is as follows (Johnston 1992). (i) The ^{13}C NMR spectra show four resonances in the olefinic region (131.8, 132.3, 132.5, and 132.8 ppm) which disappear in the DEPT spectrum showing that these carbons do not carry hydrogens and can therefore be assigned to C-2. The fine structure can be attributed to the four stereomers arising from the two possible configurations about C-3 (*E* and *Z*) and about C-1 (*endo* and *exo*). (ii) The product of ozonolysis of the polymer gives a resonance at 218.4 ppm consistent with the presence of an aliphatic ketone similar to that in norcamphor (216 ppm). (iii) When 2,3-dideuterionorbornene is used in place of NBE the hydrogens on C-6 in **9** are replaced by deuterium; C-6 can then be found in the $^{13}C\{^2H\}$ INEPT spectrum at 27 ppm.

Although most catalysts for the polymerization of NBE and other cycloalkenes give mainly ZNP or ROMP (Goodall 1995), quite a number are known where the IR spectrum or the olefin/aliphatic proton ratio indicates a mixed product (Ivin 1983, pp. 95–97). If subjected to examination along the above lines, some of these might be found to have both types of unit in the same chain. Similar questions arise in the reactions of acyclic olefins on heterogeneous supported oxide and other catalysts, where olefin metathesis is often in competition with polymerization or homologation reactions. An IR study of the polymerization of ethene on a sulfate-containing TiO_2 (anatase) sample shows evidence of on alkylidene surface end-group of the polymer, and for substantial perturbation of CH_2 groups of the polymer chain by interaction with the oxide surface (Al-Mashta 1983). However, for most Mo-based catalysts the evidence from isotopic labelling and selective poisoning experiments is that the different types of reaction proceed independently at distinct catalyst sites (Tanaka, K. 1986b; Suzuki, T. 1989a,d; Nakamura, T. 1992).

A method of converting metal carbene propagating species into metal alkyl propagating species for the purposes of making block copolymers is dealt with in Section 14.5.2.1.

4.4 Involvement of three-membered ring compounds in metathesis reactions

Cyclopropanes and oxiranes can play an important role in initiation of olefin metathesis reactions via the formation of four-membered metallacycles and their subsequent cleavage; sequences (13a, 13b). Conversely, the reverse of such reactions may sometimes be responsible for terminating a metathesis chain reaction.

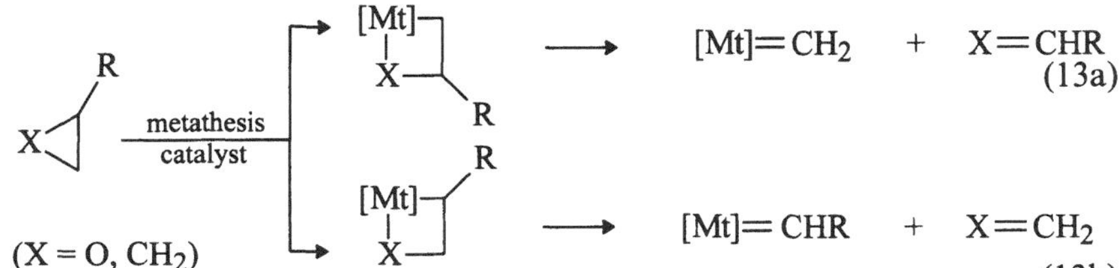

(13a)

(13b)

An example of this type of forward reaction has already been given in Scheme 2.4. Further examples are provided by: (i) the use of norbornene epoxide with W-based catalysts to enhance the ROMP of norbornene derivatives (Devine 1982); (ii) the use of oxygen with norbornene in Ru-based systems to generate such an epoxide (Ivin 1981b); (iii) the use of cyclopropane derivatives with Re_2O_7/Al_2O_3 (Anisimov 1991); see Ch. 2.

The reaction of cyclopropane with $PhWCl_3/AlCl_3$ at −78°C gives ethene but no propene. Likewise the alkylcyclopropanes (R = Et, *i*-Pr, Bu) are decomposed catalytically to give $RCH{=}CH_2$ and C_2H_4 with an overall conversion of 80% in 2.5 h at 20°C and no isomerization products. It seems likely that the reaction proceeds mainly as in sequence (13a) (Gassman 1976a). The metal carbenes formed in these reactions can be scavenged using suitable unsaturated compounds and identified from the nature of the products. In the reaction of ethylcyclopropane with $PhWCl_3/AlCl_3$ or $PhWCl_3/EtAlCl_2$, in the presence of a scavenger such as ethyl crotonate, Gassman (1976b) observed 7–14% yields of the methylidene

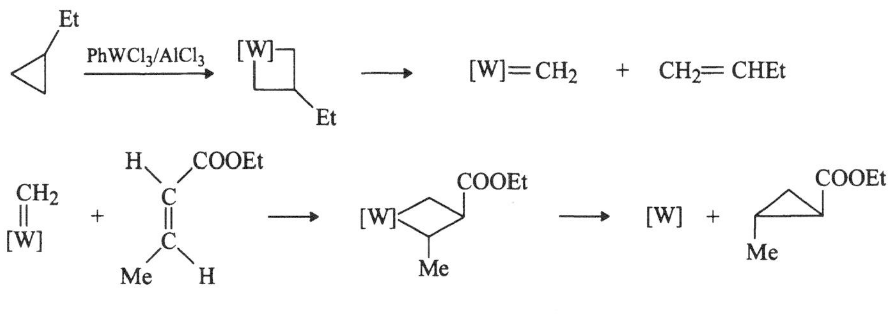

Scheme 4.4 Cross-reaction between ethylcyclopropane and ethyl crotonate.

adducts, but none of the propylidene adducts. The stereochemistry is retained in the product. Scheme 4.4 accounts for these observations.

This result shows that [W]=CH_2 is an intermediate; sequence (13a). For the same reaction at −20°C, using RCH=CH_2 (R = Et, Pr, Bu) as scavenger, Gassman (1977) observed the formation of EtCH=CHR, showing that [W]=CHEt is also an intermediate; sequence (13b). Although scavengers such as Michael acceptors thus quench the metathesis reactions of acyclic olefins and react with Fischer-type metal carbene complexes to give cyclopropane derivatives (Dötz 1972; Casey 1974, 1977) they do not quench the ROMP of norbornene. However, they may slow it down and enhance the *cis* double-bond content of the polymer (Ivin 1977b, 1979c). The variable effect of Michael acceptors on the olefin metathesis reaction is probably governed by the relative complexing ability of the olefin and Michael acceptor with the active site.

When the ROMP of norbornene is conducted in the presence of styrene and initiated by $Ru_2(OAc)_4/N_2CHCO_2Et$, the products contain both norbornene telomers and cyclopropane derivatives. This may be explained in terms of Scheme 4.5 in which the initially formed metallacyclobutane complex may rearrange either by metathesis-type cleavage to give the initiating species [Ru]=CHPh, or by reductive elimination to form a cyclopropane derivative (Noels 1988). The reaction of ethylene oxide on the palladium-(111) face yields ethene and acetaldehyde which can again be accounted for terms of competing reactions of an intermediate oxymetallacycle (Lambert 1994).

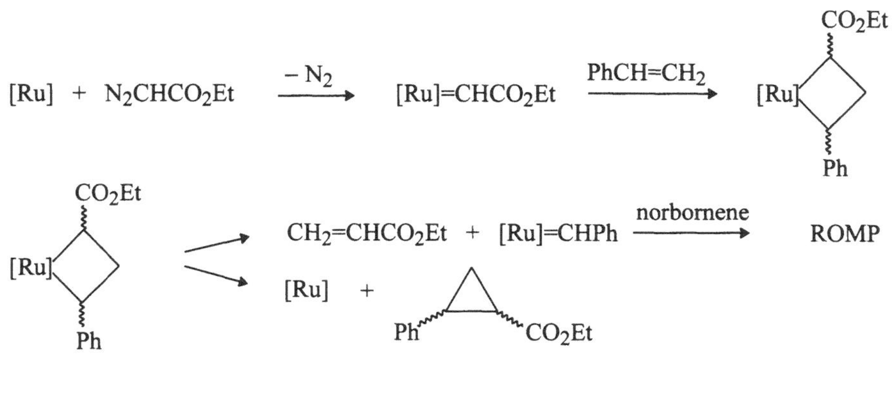

Scheme 4.5 Competition between metathesis and reductive elmination.

Metal carbene complexes bearing an alkenyl substituent can also undergo an internal cyclopropanation reaction to yield a bicyclo[*n*.1.0] compound (Parlier 1987).

5

Ethene and Terminal Alkenes

5.1 Introduction

The simplest of all olefin metathesis reactions is that of isotopically labelled ethene. With higher terminal olefins, productive metathesis can be observed in the presence of suitable catalysts but is generally accompanied by a much faster degenerate metathesis which can be detected by isotopic labelling (see Sections 1.1 and 1.3). With some catalysts productive metathesis can scarcely be detected, but with others, particularly $Mo(=CHCMe_2Ph)(=NAr)[OCMe(CF_3)_2]_2$, high conversions can be achieved (Fox 1994a). Indeed with dienes, this reaction is so efficient that its repeated occurrence leads to the formation of high polymers; see Ch. 8. Some of the more acidic catalyst systems can give rise to isomerization reactions in competition with metathesis; in such cases the products can become very complex.

In this chapter we take the terminal olefins in order to chain length, dwelling in particular on the important industrial case of propene.

5.2 Ethene

The forward reaction (1) has been investigated: (i) using Re_2O_7/Al_2O_3 as catalyst at 20°C after activation by exposure to higher olefins (Olsthoorn 1976b); (ii) using

$$CH_2{=}CH_2 \quad + \quad CD_2{=}CD_2 \quad \rightleftharpoons \quad 2\,CH_2{=}CD_2 \qquad (1)$$

$Mo(CO)_6/Al_2O_3$ at 20°C after activation in various ways (Olsthoorn 1980); (iii) using photoreduced MoO_3/SiO_2 (Elev 1987); and (iv) using MoO_x/TiO_2 ($+Me_4Sn$) (Tanaka, K. 1988c). The back reaction has been studied: (i) using MoO_3/Al_2O_3 at 160°C (Lewandos 1971a); (ii) Re_2O_7/Al_2O_3 at 75°C (Lin, C.J. 1976; Aldag 1978); and (iii) WCl_6/RLi in aromatic solvents (Grubbs 1975). In all cases the methylene fragment retains its integrity during exchange: very little $CHD{=}CHD$ is found. However, on MoS_2 some scrambling occurs (Okuhara 1976).

Good catalytic activity of Re_2O_7/Al_2O_3 for this exchange is obtained when it has been previously exposed to a higher olefin, e.g. propene (Olsthoorn 1976b) or *cis*-pent-2-ene (McCoy 1991), but no activity is generated by ethene itself. This might have been thought to indicate that the activation of the catalyst by propene occurs by the π-allyl mechanism (see Scheme 2.1). However, in the metathesis of $^{14}CH_3CH{=}CH_2$ on this catalyst the methyl fragment preserves its identity (Mol 1968). Hence if any activation does occur by the π-allyl mechanism it must involve the consumption of an extremely small fraction of the propene. If the catalyst is activated by contact with but-2-ene-d_8 a surface layer containing CD_2 and CD_3 groups may be detected by IR, but these groups are not removed during the subsequent metathesis of propene. Poisoning experiments using NO show that less than 0.3% of the Re atoms act as sites for olefin metathesis (Olsthoorn 1976b).

The metathesis rate of a mixture of $^{12}C_2H_4$ and $^{13}C_2H_4$ over a reduced MoO_3/SiO_2 catalyst at 25°C is enhanced in the presence of CO by a factor of 2.4 (Suzuki, T. 1989b).

The reactions of ethene with titanium and tantalum carbene complexes are described in Sections 2.2.1 and 2.3.2, respectively.

5.3 Propene

5.3.1 General remarks

The importance of being able to convert surplus propene into ethene has led to a great deal of industrially oriented research on the metathesis of propene, reaction (2). The forward reaction is slightly endothermic ($\Delta H = 1.7\ \mathrm{kJ\,mol^{-1}}$ at 25°C) and the equilibrium proportion of propene is 65.9%, 62.6%, 59.4%, and 56.6% at 25, 125, 225, and 325°C, respectively (Kapteijn 1983b).

$$2\,MeCH{=}CH_2 \rightleftharpoons CH_2{=}CH_2 + MeCH{=}CHMe \qquad (2)$$

The reaction takes place at a convenient speed in a flow system at 400–600°C on SiO_2-supported catalysts, or at 25–200°C on Al_2O_3-supported catalysts, and is measurable even at 0°C. For example, with $Re_2O_7/Al_2O_3/Et_4Sn$ at 20°C/1.5 bar, equilibrium conversion can be achieved with 100% selectivity at a throughput of $25\ \mathrm{dm^3\,h^{-1}}$ (g catalyst)$^{-1}$ (Spronk 1991a).

The metathesis of propene has frequently been used as a probe of catalyst activity; many references to this reaction will be found in Ch. 2.

5.3.2 Reactions of isotopically labelled propenes

Studies with $CH_3CD{=}CH_2$ (Mol 1970), $CH_3{}^{14}CH{=}CH_2$, $CH_3CH{=}{}^{14}CH_2$, and $^{14}CH_3CH{=}CH_2$ (Mol 1968) on Re_2O_7/Al_2O_3 (activated in dry air at 580°C) show that the initial metathesis reaction occurs cleanly at 85°C, but the product but-2-ene undergoes isomerization to but-1-ene, which then undergoes further metathesis reactions (Olsthoorn 1976b). On $MoO_3/CoO/Al_2O_3$, experiments with $CH_3CH{=}^{14}CH_2$ show that the metathesis reaction again occurs cleanly at 60°C (Clark 1969), but at 160°C the product but-2-ene is radioactive to the extent that nearly half the propene must have isomerized to $CH_2{=}CH^{14}CH_3$ before undergoing metathesis (Clark 1969; Woody 1969).

Some results on the reaction of a 1 : 1 mixture of $CH_3CH{=}CH_2$ and $CD_3CD{=}CD_2$, over MoO_x/β-TiO_2, with $2.3 < x < 2.9$ (with or without Me_4Sn treatment) at room temperature are shown in Table 5.1. The catalyst was prepared by impregnating β-TiO_2 with ammonium paramolybdate, then dried at 120°C and oxidized with O_2 at 500°C, reduced with H_2 or CO at 500°C for 1 h, treated with an equimolar mixture of N_2O and H_2 at 200°C for 1 h, before finally evacuating at 500°C. N_2O and H_2 react on the surface, causing partial reoxidation. These results show that the degenerate metathesis, yielding $CH_3CH{=}CD_2$ and $CD_3CD{=}CH_2$, is about one order of magnitude faster than productive metathesis, yielding $CH_2{=}CH_2$, $CH_2{=}CD_2$, and $CD_2{=}CD_2$. For the reverse reaction between $CD_2{=}CD_2$ and $CH_3CH{=}CHCH_3$, the product propene is entirely $CH_3CH{=}CD_2$ with no accompanying isomerization or hydrogen scrambling (Tanaka, K. 1979). In experiments using mixtures of $CD_3CD{=}CD_2$, $CH_3CH{=}CH_2$, $CH_2{=}CH_2$, and $^{13}CH_2{=}^{13}CH_2$ the predominant chain carrier for the degenerate metathesis of propene on these catalysts has been shown to be $[Mt]{=}CHCH_3$ rather than $[Mt]{=}CH_2$ (Sasaki, M. 1986). The same conclusion was reached on the basis of an analysis of the reaction products in the cross-metathesis between *cis*-$CH_3CH{=}CHCH_3$ and $CD_3CD{=}CD_2$ (Tanaka, K. 1987b).

On Re_2O_7/Al_2O_3 the ratio of degenerate/productive metathesis increases as the temperature decreases (~1000 at −40°C) (Gudkov 1981). On photoreduced MoO_3/SiO_2 the rate of productive metathesis of propene is 4–5 times faster than the rate of cross-metathesis between C_2H_4 and C_2D_4, while the rate of degenerate metathesis of propene ($C_3H_6 + C_3D_6$) is (5–10) × 10^3 times higher still. Here too it is concluded that degenerate propene metathesis occurs via $[Mo]{=}CHCH_3$ (Elev 1987). The presence of CO enhances the rate of degenerate metathesis at 25°C

Table 5.1 Turn-over frequencies (TOF) and V_D/V_P[a] ratio in the metathesis of propene at room temperature (Tanaka, K. 1984)

Catalyst[b]	TOF[c]	V_D/V_P
MoO_3/β-TiO_2	0	–
MoO_3/β-$TiO_2 + Me_4Sn$	9.3×10^{-5}	10.0
MoO_x/β-TiO_2	2.0×10^{-7}	6.7
MoO_x/β-$TiO_2 + Me_4Sn$	5.4×10^{-4}	27.0

[a] V_D initial rate for degenerate metathesis; V_P initial rate for productive metathesis. [b] 0.5 g catalyst, 6.7 wt% MoO_3, pressure of propene 3.3 kPa. [c] The rate of formation of ethene (mol C_2H_4/mol Mo/s) assuming that all Mo atoms are active sites.

($^{13}C_3H_6 + {}^{12}C_3H_6$) over a $MoO_{2.2}/SiO_2$ catalyst, whereas it has no effect on the rate of productive metathesis (Suzuki, T. 1989c).

5.3.3 Kinetics of reaction

Most of the kinetic studies are concerned with the metathesis of propene over solid catalysts such as $MoO_3/CoO/Al_2O_3$, $Mo(CO)_6/Al_2O_3$, WO_3/SiO_2, and Re_2O_7/Al_2O_3. Before the advent of the metal carbene mechanism a Langmuir–Hinshelwood model appeared to be the most appropriate, viz. a dual-site mechanism, with the surface reaction between two adjacent chemisorbed propene molecules as the rate-controlling process (Lewis 1969, 1971; Moffat 1970a; Davie 1972b; Luckner 1973a; Hattikudur 1974; Lin, C.J. 1976). This leads to eqn. (3) for

$$r_0 = kK^2p^2/(1 + Kp)^2 \tag{3}$$

the initial reaction rate r_0, where k is the rate constant, K is the adsorption equilibrium constant for propene, and p is the propene partial pressure. Later, a model was proposed in which it was assumed that only one active site is involved, at which two propene molecules arrive in turn and then react (Mol 1975; see also El-Sawi 1981, 1982). Eqn. (4) can then be derived for the case where the surface

$$r_0 = kK_1K_2p^2/(1 + K_1p + K_1K_2p^2) \tag{4}$$

reaction is rate-controlling, K_1 and K_2 being the adsorption equilibrium constants for the first and second propene molecules, respectively. In all of these early models the assumption was made that reaction occurred between two alkene molecules bound to the active centre of the catalyst.

When it became clear that carbene complexes might well be involved, van Rijn (1977) presented other kinetic models, based on alkene and alkylidene surface complexes as the reaction intermediates. The derived rate expressions were tested with data from the literature and it was shown that these models described the data as well as if not better than the Langmuir–Hinshelwood model.

An extensive kinetic study of the metathesis of propene over Re_2O_7/Al_2O_3 provides a closer analysis of further models as well as the ones mentioned above. Only eqn. (5) fits the experimental data adequately (Fig. 5.1). On this basis the

$$r_0 = kKp/(1 + Kp) \tag{5}$$

models giving eqns. (3) and (4) can be eliminated. Further results using a mixed feed (propene/ethene or propene/but-2-ene) fit best an expression derived from a metal carbene mechanism. It is concluded that product-desorption processes are the rate-determining steps. Less than 1% of the surface Re sites are involved in the metathesis reaction (Kapteijn 1981). Other authors (Nekrasov 1982; Spinicci 1985a) have confirmed that the metal carbene mechanism gives the best kinetic description for the metathesis of propene over a Re_2O_7/Al_2O_3 catalyst. These conclusions are supported by the results of temperature-programmed desorption measurements (Matulewicz 1983; Spinicci 1983, 1985b).

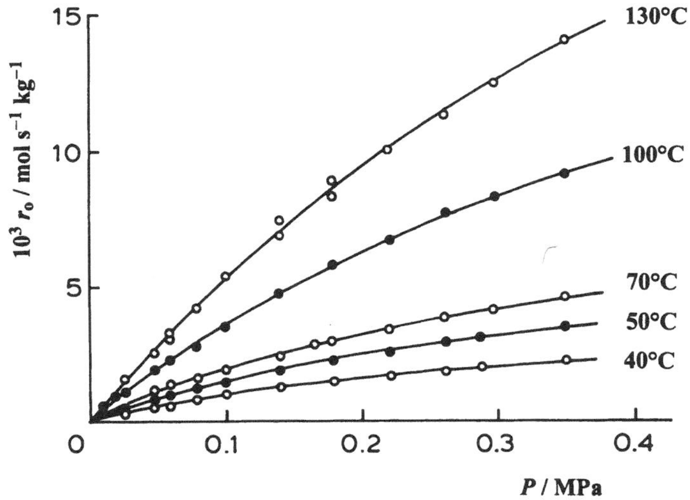

Fig. 5.1 Initial rate of propene metathesis, r_0, on a 5.8 wt% Re_2O_7/Al_2O_3 catalyst in a flow system, as a function of temperature and propene pressure *P*. The lines represent the rates calculated from eqn. (5) (Kapteijn 1981).

Other kinetic aspects are dealt with under specific catalyst systems either here or in Ch. 2.

5.3.4 Molybdenum-based catalysts

$Mo(CO)_6$, $(NH_4)_6Mo_7O_{24}.4H_2O$, or $(\pi\text{-}C_3H_5)_4Mo$ are among the normal starting compounds for the preparation of supported molybdenum catalysts, although a variety of others have been found effective (Smith, J. 1974b; see also Section 2.4.2). After normal activation the TOF for Al_2O_3-supported Mo-based catalysts is very small for low Mo loading, increases with increasing loading, and then passes through a maximum (Thomas 1982; Klimov 1990b; Startsev 1993); the optimum Mo/Al ratio is generally about 1/25 (Fig. 5.2). In the case of supported catalysts prepared from molybdenum oxalate or acetylacetonate, this has been explained as follows. At low Mo content the activity is low because the Mo precursor reacts first with the most basic surface OH groups on γ-Al_2O_3 giving inactive surface species. At higher Mo content the neutral surface OH groups are also consumed, resulting in highly active sites after reduction in CO or H_2. At still higher Mo loadings the more acidic OH groups react with the Mo precursor complex, which results in poorly active sites (Startsev 1993). Not only the Mo precursor complex, but also the nature of the support and the pretreatment conditions have an important effect on the ultimate metathesis activity as indicated in Table 5.2 (see also Mol 1990b). Catalysts made from $(\pi\text{-}C_3H_5)_4Mo/Al_2O_3$ are especially active when the molybdenum ions are brought into oxidation state IV by appropriate treatment (see Section 2.4.2).

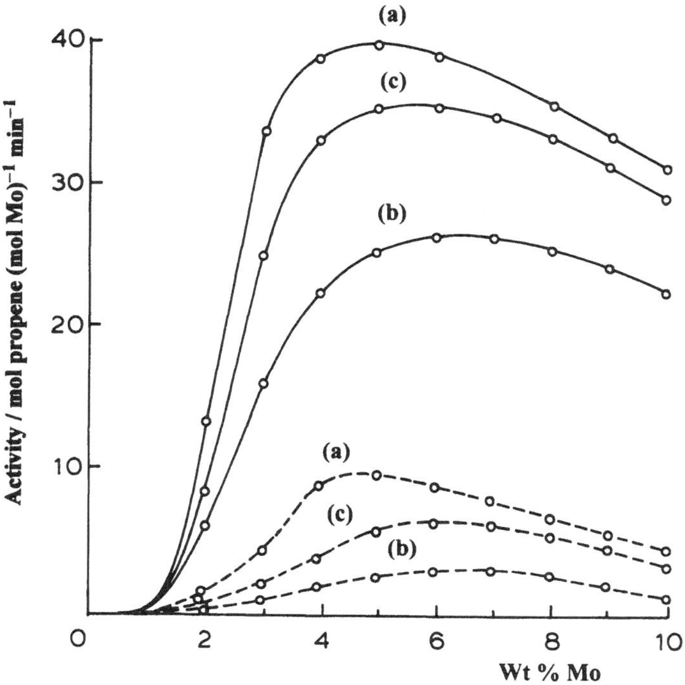

Fig. 5.2 Activity in the metathesis of propene (16 kPa, 100°C) over Al_2O_3-supported catalysts as a function of the Mo content, after pretreatment with CO (—) or H_2 (- - -) at 500°C. (a) $MoO_2(C_5H_7O_2)_2/Al_2O_3$, (b) $[Mo_2O_4(C_2O_4)_2(H_2O)_2]^{2-}/Al_2O_3$, (c) $[Mo_3O_4(C_2O_4)_3(H_2O)_3]^{2-}/Al_2O_3$ (Startsev 1993).

Al_2O_3-supported MoO_3 catalysts are rather prone to cause side reactions above 120°C, e.g. hydrogenation, isomerization (Ichinose 1978), polymerization, and deposition of carbon (Rodríguez-Ramos 1995), but these can be suppressed by the inclusion of small amounts of KOH or NaOH in the catalyst (Bradshaw 1967; Alkema 1968). As the temperature is raised, the metathesis rate passes through a

Table 5.2 Optimum conditions for preparation of catalysts based on $Mo(CO)_6$ (Smith, J. 1974b)

Support	Pretreatment temperature of support/°C	Catalyst activation temperature/°C	Initial rate of propene metathesis[a]/ % min^{-1} m^{-2}
γ-Al_2O_3	300	200	0.06
SiO_2	25	200	0.12
SiO_2-Al_2O_3	650	550	0.14
MgO	750	300	0.29

[a] 25°C/3.33 kPa.

maximum at 150°C for Al_2O_3-supported catalysts prepared from molybdenum oxalate or acetylacetonate (Startsev 1993), at 160–170°C for MoO_3/Al_2O_3 prepared via impregnation of Al_2O_3 with ammonium molybdate (Aliev 1977), or at 210°C for $MoO_3/CoO/Al_2O_3$ (Moffat 1970a), an effect that has been attributed to reversible deactivation of catalytic sites. Another feature of the MoO_3/Al_2O_3 catalyst is that its activity for the metathesis of propene at 80°C correlates with its activity for the decomposition of diazomethane in ether at 20°C (O'Neill 1972). It is generally found that the metathesis activity of Al_2O_3-supported catalysts increases up to a certain point with successive exposures to propene (Davie 1972b, Ogata 1974), and the catalyst life and product yield are both improved if hydrogen is present during metathesis (Alkema 1968). The rate of deactivation of MoO_3/Al_2O_3 using a liquid propene stream at low temperature (0°C) is lower than for a gas-phase stream. In the former case soluble oligomers and polymers formed as by-products from propene and especially ethene are more readily carried away in solution from the catalyst surface (Ogata 1973). A solvent such as heptane is even better for this purpose (Ogata 1976). However, the catalyst tends to deactivate more rapidly with increasing propene pressure (Sodewasa 1977). The reaction on a catalyst made from $Mo(CO)_6/Al_2O_3$ is somewhat stereoselective at 25°C. In the early stages of reaction the product but-2-ene has a *cis/trans* ratio of 67/33, but this very quickly changes to 27/73 as a result of secondary metathesis, coming close to the equilibrium ratio of 24/76 (Davie 1972a). The activity is greatly enhanced by pretreatment of the catalyst with trichloroethene (Davie 1971).

SiO_2-supported Mo catalysts show somewhat similar behaviour to those supported on Al_2O_3. Table 5.3 indicates relative activities of some catalysts

Table 5.3 Activities for propene metathesis of catalysts containing surface molybdenum complexes in different oxidation states (Yermakov 1975/76)

Surface complex	Oxidation state of Mo	TOF/s^{-1} [a]
$(\equiv SiO)_2Mo(C_3H_5)_2$ [b]	IV	0.167
$(\equiv SiO)_2Mo;$ [c]	II	0.007
$(\equiv SiO)_2MoO_2$ [d]	VI	0
$(\equiv SiO)_2MoO$ [e]	IV	0.127

[a] At 90°C/16 kPa. [b] SiO_2 dehydrated at 400°C, treated with $(\pi\text{-}C_3H_5)_4Mo$ in pentane at 20°C, and solvent removed at 20°C. [c] $(\equiv SiO)_2Mo(C_3H_5)_2$ reduced with H_2 at 600°C. [d] $(\equiv SiO)_2Mo$ or $(\equiv SiO)_2MoO$ oxidized with O_2 at 400°C. [e] $(\equiv SiO)_2Mo$ oxidized with O_2 at 25°C.

prepared from $(\pi\text{-}C_3H_5)_4Mo/SiO_2$ in various ways; those in state Mo(IV) are much the most active; Mo(VI) is inactive; Mo(II) is virtually inactive – its activity is probably due to the presence of some Mo(IV) ions. Surface Mo(IV) ions in an oxygen environment and with coordination number 6, prepared via $[Mo(OEt)_5]_2$, are not active in propene metathesis. It is clearly necessary for the surface Mo(IV) species to have a vacant position for coordination of olefin (Kuznetsov 1980). This is also indicated by the fact that oxygen is a poison for the catalyst $(\equiv SiO)_2Mo(C_3H_5)_2$, reducing the rate to zero when the ratio of adsorbed oxygen atoms to molybdenum atoms reaches 1 (Fig. 5.3).

Photoreduced SiO_2-supported MoO_3 is a very active catalyst, particularly when pretreated with an activator such as cyclopropane (see Section 2.4.2). After propene has been flowing over the catalyst for several hours it loses its original high activity. The deactivation is due to the isomerization of the Mo-cyclobutane intermediate to form an inactive π-complex of the corresponding olefin with an Mo(IV) ion (Vikulov 1994). A comparison of different SiO_2-supported molybdenum catalysts is given in Table 5.4. Cyclopropane-treated, photoreduced MoO_3 supported on $SiO_2\text{-}Al_2O_3$ is even more active in propene metathesis than when supported on SiO_2 (Mol 1994a).

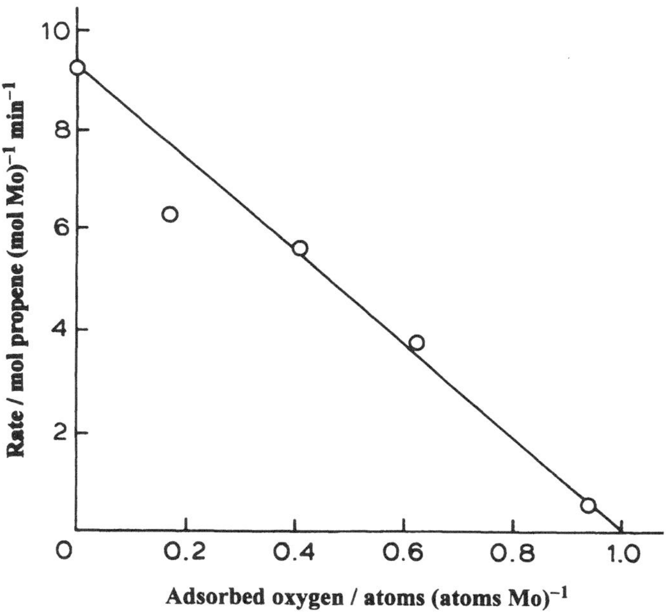

Fig. 5.3 Decrease in the activity of $(\equiv SiO)_2Mo(\pi\text{-}C_3H_5)_2$ for the metathesis of propene (16 kPa, 90°C) when poisoned by O_2; 3.8 wt% Mo (Yermakov 1975/76).

Table 5.4 Comparison of turnover frequencies (TOF) for propene metathesis on different MoO_3/SiO_2 catalysts

Catalyst precursor	Pretreatment[a]	T/°C	P/kPa	TOF/s^{-1}	Ref.[b]
MoO_3/SiO_2					
1 wt% Mo	Thermal reduction in CO at 500°C	20	8.5	0.07	(1)
	Idem + CP treatment	20	10.4	1.25	(2)
	Thermal reduction in H_2 at 500°C	20	6.4	0.003	(1)
	Idem + CP treatment	20	9.3	0.30	(2)
	Thermal reduction in H_2 at 800°C	20	9.7	0.03	(2)
	Idem + CP treatment	20	14.7	1.9	(2)
	Photoreduction in CO at 20°C	20	11.3	0.42	(1)
	Idem + CP treatment	20	20.9	5.05	(3)
	As above, 0.1 wt% Mo	20	20.0	12.8	(3)
	Photoreduction in CO + MCP treatment	20	22.7	2.14	(3)
	Photoreduction in CO + CHT treatment	20	11.1	1.0	(4)
	Conventional impregnation and calcination	400	100.0	0.004	(5)
2.5 wt% Mo	Mechanical activation in H_2 + He (20°C)	20	6.7	0.56	(6)
	Idem + CP treatment	20	26.7	3.6	(6)
$Mo(\pi\text{-}C_3H_5)_4/SiO_2$	No pretreatment	90	16.0	0.17	(7)
$Mo_2(OAc)_4/SiO_2$	Evacuation at 350°C	20	6.7	0.15	(8)
$Mo(isoprene)_3/SiO_2$ (2.66 wt% Mo)	Treated in O_2 (400°C) and H_2 (500°C)	100	16.0	0.073	(9)

[a] CP = cyclopropane; MCP = methylcyclopropane; CHT = 1,3,5-cycloheptatriene. [b] (1) Shelimov 1986; (2) Elev 1989a; (3) Vikulov 1991; (4) Vikulov 1992b; (5) Thomas 1980; (6) Bobyshev 1992; (7) Yermakov 1975/76; (8) Ichikawa, M. 1993; (9) Startsev 1986.

5.3.5 Tungsten-based catalysts

WO_3/SiO_2 is the most commonly used amongst many tungsten-based catalysts for propene metathesis (Smith, J. 1974b). Although a high temperature is needed the reaction proceeds with high selectivity and is not so vulnerable to common poisons or to coke formation as with Al_2O_3 supports (Wills 1977). The catalyst is generally activated in dry air at 550°C and the XRD pattern of freshly activated catalyst shows the presence of crystalline WO_3 only if the catalyst contains more than 10% WO_3 (Fig. 5.4). Below this level the tungsten appears to be present largely as an amorphous surface compound which is difficult to reduce. Maximum activity is obtained with a composition of about 20% WO_3 and an overall extent of reduction (by H_2, CO, C_2H_4, or $CH_3CH{=}CH_2$), which lies at $WO_{2.95}$ (Kerkhof 1977; Westhoff 1977).

The catalyst WO_3/SiO_2 has two distinctive features. First, it is unusually prone to restriction of the rate by diffusional effects (interphase mass transfer), in spite of calculations which predict that no limitation should occur (Moffat 1970b). These effects may be seen as a dependence of the rate on mass of catalyst at a given ratio of catalyst mass to flow rate (Fig. 5.5a). This does not occur with $MoO_3/CoO/Al_2O_3$ (Fig. 5.5b) or with Re_2O_7/Al_2O_3. Such effects have been accounted for by assuming that reaction occurs on a small number of very active sites that are widely separated

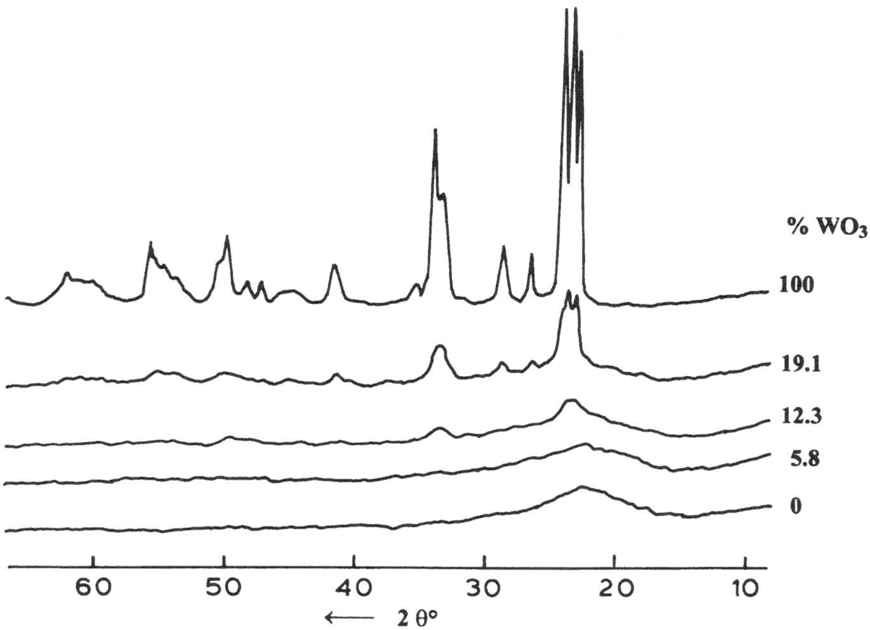

Fig. 5.4 X-ray diffraction pattern of fresh WO_3/SiO_2 catalysts (Kerkhof 1977).

on the catalyst surface, the activity being limited by localized interphase diffusional effects (Moffat 1971). However, this effect might be caused by fluctuations (velocity changes) in the feed purification unit instead of changes in the reactor (Banks 1979). This interphase mass-transfer limitation is negligible when the catalyst is activated at or above 600°C for more than 2 h (Basrur 1991). Secondly, if WO_3/SiO_2 is activated in air at 550°C and then used without further treatment, the rate of metathesis builds up very considerably over a period of minutes or hours as propene is passed through the catalyst (Luckner 1973b). This phenomenon is known as 'break-in' (Gangwal 1977; Wills 1977). The break-in time is reduced, but not eliminated, if the catalyst is pretreated with H_2 or CO (Luckner 1973b). At the same time there is a colour change from yellow to blue indicating reduction of WO_3. An injection of ammonia or certain amines into the gas stream produces a temporary burst of enhanced activity (Wills 1977; Gangwal 1978; Fathikalajahi 1980). High-temperature activation under inert gas greatly increases both the break-in rate and the activity (Andreini 1981). The break-in process has been attributed to the formation of a steady-state population of active sites, essentially exposed W ions in an appropriate oxidation state. The necessary slight loss of oxygen (see also Choung 1983) is accompanied by the production of trace quantities of acetone and acetaldehyde, formed by reaction of lattice oxygen from the WO_3 with propene, probably in two steps: (i) reduction at the catalyst yielding acetone and reduced metal oxide; and (ii) formation of a tungsten–carbene and acetaldehyde (Basrur

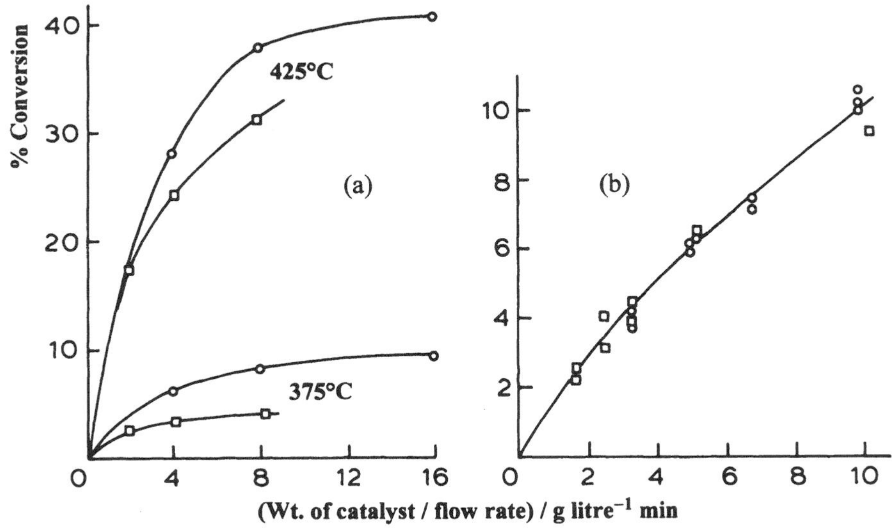

Fig. 5.5 Rate of propene metathesis on (a) WO_3/SiO_2 at 375°C and 425°C: squares 0.4 g, circles 0.8 g catalyst. (b) $MoO_3/CoO/Al_2O_3$ at 150°C: squares 0.3 g, circles 0.6 g catalyst. Note that (a) shows diffusional effects at low flow rates whereas (b) does not (Moffat 1971).

1991). Such complications make it difficult to interpret the kinetics of reaction on this catalyst (Maatman 1979).

The metathesis of propene on WO_3/SiO_2 is speeded up by pretreatment of the catalyst with HCl (Pennella 1974; Aliev 1977) but the product but-2-ene undergoes considerable isomerization to but-1-ene (Aliev 1978). The inclusion of 1% cycloocta-1,5-diene (COD) in the propene stream also increases the rate of metathesis and reduces the break-in time from 20 min to less than 5 min at 500°C. The latter effect disappears when the additive is removed, so it is not due to reduction of the catalyst. The effects of both HCl and COD have been attributed to a favourable modification in the metal d-orbital levels as the result of the presence of new ligands (Pennella 1973, 1974). Pretreatment with hexamethyldisilazane (HMDS) at 250°C also has a remarkable effect on the activity, increasing it as much as 140 times for the metathesis of propene at 427°C. The same treatment of silica alone completely eliminates its capacity to isomerize *trans*-but-2-ene at 427°C, so it is concluded that the Brönsted acidic hydroxyl groups, poisoned by HMDS, are not likely to be the precursors for the active sites in propene metathesis over WO_3/SiO_2 (van Roosmalen 1980a, 1982).

A WO_3/SiO_2 catalyst can be promoted by adding minor amounts of elemental Na, S, Si, Mg, Ba, Zn, Sb, or W to the catalyst, and treating the admixture at elevated temperatures under inert atmosphere (Banks 1985). The enhanced effect is attributed to a partial reduction of the supported tungsten catalyst by the additive. Combination of WO_3/SiO_2 with MgO produces a dramatic increase in its metathesis activity, possibly caused by allyl radicals generated on the surface of

MgO (Banks 1985). Addition of C_2H_2 to a WO_3/SiO_2 catalyst increases the conversion of propene at 400°C five-fold (Banks 1986b).

SiO_2-supported catalysts have also been made from $W(CO)_6$ (Brenner 1979), $(Me_3CCH_2)_6W_2$, Me_6W (Smith, J. 1974b), and from $(\pi\text{-}C_3H_5)_4W$ and $(\pi\text{-}C_4H_7)_4W$ (Startsev 1975; Kuznetsov 1980). The latter pair are active at room temperature and are much more active than the catalyst prepared by impregnation of SiO_2 with $(NH_4)_2WO_4$. The precursor with the highest activity is $(\equiv SiO)_2W(\pi\text{-}C_4H_7)_2$. These tungsten catalysts behave analogously to the molybdenum complexes (Table 5.3): an oxidation state of IV leads to the highest activity and coordinative unsaturation is a prerequisite for catalytic activity. The activity decreases linearly with adsorbed oxygen, falling to zero at an O/W ratio of 2; this observation suggests that the metathesis centres are uniformly active.

For WO_3/Al_2O_3, activated at 550°C in air, the TOF at 400°C shows a considerable increase with the tungsten content, passing through a maximum at about 2 mol W nm^{-2} (Thomas 1982). The activity of the catalyst is increased if the activation is conducted at 550°C in argon rather than air. The activity is also higher when the catalyst is activated at higher temperatures (650°C–870°C), but depends on the WO_3 loading (Grünert 1989b) (Fig. 5.6). A break-in is not observed for WO_3/Al_2O_3 after activation under standard conditions (550°C) in helium. A catalyst with 6 wt% WO_3 shows maximum activity at a reaction temperature of about 360°C; with 10.6 wt% WO_3 the maximum occurs at 425°C (Shmidt 1979; Andreini 1981).

Al_2O_3-supported catalysts made from $W(CO)_6$ have only one thousandth of the activity of those made from $Mo(CO)_6$, so they have not attracted much attention (Davie 1972b), but those made from Me_6W/Al_2O_3 and Me_6W/SiO_2, and activated at 200–300°C, have comparable activity at 20°C (Mowat 1974).

A 6 wt% WO_3/TiO_2 catalyst shows maximum activity at 217–247°C and can be more active than a WO_3/Al_2O_3 catalyst (Andreini 1985).

Unsupported WO_3 is an insulator and does not catalyze the metathesis of propene at 200°C; instead it catalyzes the dimerization (Tsuda 1985). Unsupported $Na_{0.25}WO_3$ is a conductor and does catalyze the metathesis reaction (Kosaka 1986; Mori, T. 1986). On a series of unsupported non-stoichiometric samples of Na_xWO_3 and K_xWO_3 $(0 < x < 0.8)$ the maximum activity and selectivity for propene metathesis at 400°C is found at $x \approx 0.2$, corresponding to the conductor/insulator transition (Stevenson 1991). However, both metathesis and dimerization occur over conducting $K_{0.3}MoO_3$ and insulating $K_{0.33}MoO_3$. Below 127°C metathesis is dominant over both oxides. In contrast to Na_xWO_3 there seems to be no correlation between the selectivity (metathesis *vs* dimerization) of the reaction and the conductivity of the catalyst (Suzuki, M. 1987).

$WBr_5/SiO_2/Et_3Al$ is active at 80°C, the catalytic activity increasing with the Al/W ratio up to a value of nine (Maksimowski 1991). In chlorobenzene solution at 25°C, propene metathesis can be brought about using $W(CO)_5(PPh_3)/EtAlCl_2/O_2$ as catalyst; the *cis*/*trans* ratio in the product but-2-ene extrapolated to zero % conversion, is 1.37 ± 0.05 (Basset 1979).

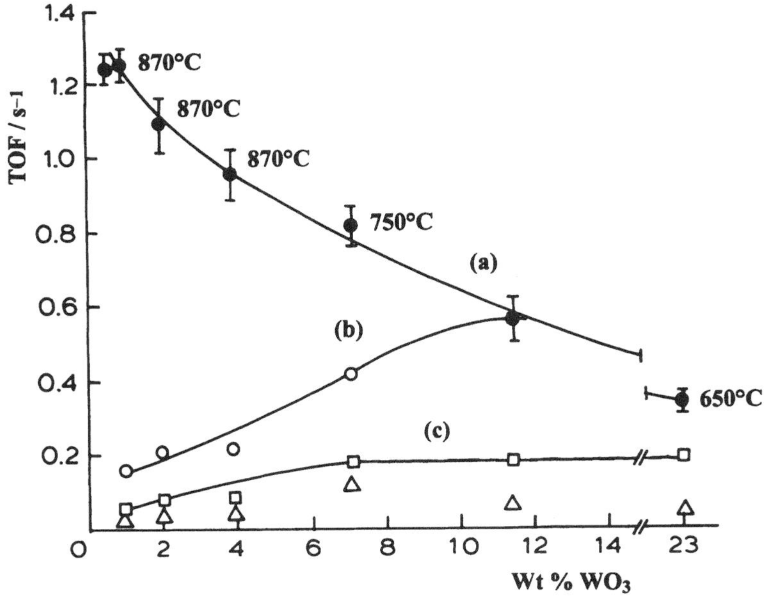

Fig. 5.6 Dependence of the propene metathesis activity at 300°C on the WO_3 content of WO_3/Al_2O_3 catalysts activated in argon. (a) Activation temperatures as shown on the figure; vertical bars indicate standard deviations. (b) Activation at 700°C. (c) Activation at 550°C. Δ = activation in air at 550°C (Grünert 1989b).

5.3.6 Rhenium-based catalysts

Re_2O_7/Al_2O_3, activated at 500°C in dry air, is an effective catalyst for propene metathesis at room temperature; see Section 2.5.2. A remarkable dependence of rate of metathesis on the percentage of Re_2O_7 is observed (Fig. 5.7). At low Re content, the catalytic activity is extremely low. The kinetics fit eqn. (5) but both k and K are markedly dependent on the percentage of Re_2O_7. An explanation for this behaviour is based on an examination of the surface OH groups by FTIR. At low rhenium loadings ReO_4^- ions have reacted mainly with the basic surface alumina OH groups during deposition of the rhenium compound, while at higher loadings the neutral and more acidic OH groups have also reacted, the latter resulting in the most active sites (Sibeijn 1991b).

The activity of Re_2O_7/Al_2O_3 catalysts can be increased by using a calcination temperature of 830–930°C, which is much higher than that normally employed (500–550°C). The optimum calcination temperature depends upon the wt% Re_2O_7 in the catalyst. The improvement is explained in terms of redistribution of ReO_4 groups over the alumina surface (Spronk 1993b).

Besides good activity and high selectivity, the lifetime, or stability, of a catalyst

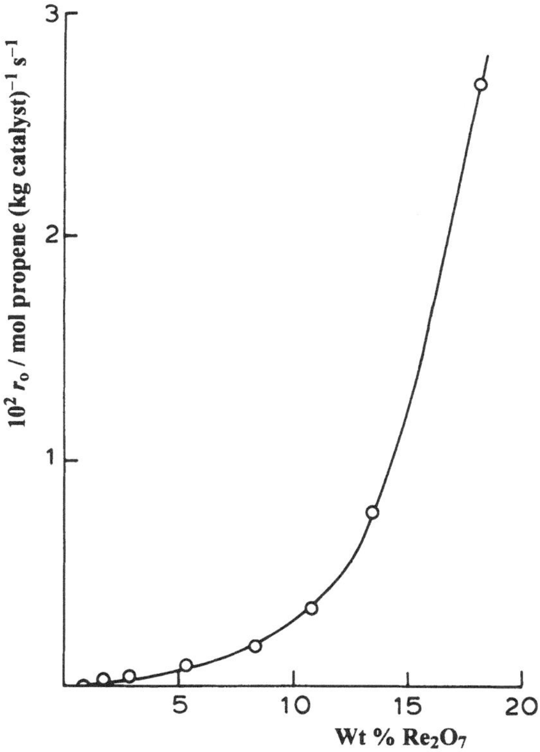

Fig. 5.7 Initial rate of propene metathesis, r_0, on Re_2O_7/γ-Al_2O_3 at 50°C as a function of Re_2O_7 content (Kapteijn 1977).

under working conditions is often crucial in industrial applications. Polar compounds, such a H_2O, which might be present as an impurity in the reactant(s), are catalyst poisons (Moulijn 1988; Amigues 1991). Other possible routes for the deactivation of rhenium-based catalysts are: (i) reduction of the rhenium below its optimum oxidation state; (ii) adsorption of polymeric by-products on the surface of the catalyst, blocking the active sites or even the catalyst pores; (iii) termination caused by the reverse of the initiation steps; and (iv) reductive elimination of the metallacyclobutane intermediate. Although the activity of Re_2O_7/Al_2O_3 is greatly promoted by treatment with R_4Sn, it falls off more quickly than the activity of an unpromoted catalyst; deactivation also occurs more readily at 80°C than at − 10°C (Spronk 1991a). Deactivation of the promoted catalyst is probably caused mainly by reductive elimination, whereas deactivation of the unpromoted catalyst is attributed to a decrease of the mean intrinsic activity of the active sites rather than a decrease in their number (Amigues 1991; Chauvin 1992).

Unpromoted supported rhenium oxide catalysts can be regenerated many times without any loss of activity. However, the activity of a deactivated tetraalkyltin-promoted Re_2O_7/Al_2O_3 catalyst can only be partly restored by heating in a stream of oxygen, followed by further addition of promoter. The partial loss of activity is attributed to accumulation of tin on the catalyst surface (Spronk 1991c).

Re_2O_7/SiO_2 is not active below 120°C (Aldag 1977). This is perhaps because SiO_2 does not stabilize the necessary intermediate oxidation state of rhenium. ReO_3 is catalytically active, either unsupported or on SiO_2, at 200°C. In contrast to WO_3, ReO_3 is a conductor, which favours metathesis relative to dimerization (Tsuda 1981, 1985).

A remarkable observation is that Me_4Sn deposited on dehydroxylated alumina brings about propene metathesis at 25°C in a pulse reactor with an activity comparable to that of Re_2O_7/Al_2O_3 (Ahn, H-G. 1992). This is one of the rare cases where metathesis appears to proceed without the mediation of a transition metal complex.

5.4 But-1-ene and its derivatives

5.4.1 But-1-ene

Brief reference has been made to metathesis of but-1-ene and reactions of but-1-ene with tantalum carbene complexes in Ch. 2. On catalysts such as MoO_3/Al_2O_3 (Engelhardt 1981, 1982a; Pantoja 1993), $MoO_3/CoO/Al_2O_3$ (Banks 1964), $MoO_3/TiO_2/Al_2O_3$ (Henrici-Olivé 1973), $Mo(CO)_6/Al_2O_3$ (Davie 1972a), $(\pi\text{-}C_3H_5)_4Mo/Al_2O_3$ (Ichinose 1978), Re_2O_7/Al_2O_3 (Edreva-Kardjieva 1976), Re_2O_7/MgO (Edreva-Kardjieva 1977), Re_2O_7/SiO_2 (Edreva-Kardjieva (1986), and $Re_2(CO)_{10}/Et_3Al_2Cl_3$ (Yamaguchi 1974), but-1-ene isomerizes to some extent to but-2-ene, which then undergoes cross-metathesis with but-1-ene, followed by further isomerization reactions to give a great variety of products. At 0–140°C the activity and selectivity of Re_2O_7/Al_2O_3 for but-1-ene metathesis is higher than that of MoO_3/Al_2O_3 (Sodewasa 1977). Competing isomerization reactions tend to increase in the following order of catalysts: $Re_2O_7/Al_2O_3 < WO_3/Al_2O_3 < MoO_3/Al_2O_3$ (Engelhardt 1982a,c). Deactivation of MoO_3/Al_2O_3 towards metathesis is accompanied by coke deposition on the catalyst surface (Grupp 1991). Both the rate of isomerization and the Brönsted acidity of the catalyst increase with increasing Mo content. A π-allyl mechanism is proposed for the isomerization over a 1 wt% Mo catalyst, while a Brönsted-acid mechanism is suggested for the isomerization over catalysts with a Mo loading of $\geqslant$3.9 wt% (Pantoja 1993).

One of the few catalysts to give reasonably selective metathesis to form ethene and hex-3-ene, reaction (6), is MoO_3/Al_2O_3 doped with 2% alkali metal ions (Bradshaw 1967; Alkema 1968) or thallium ions (Kobylinski 1972). The suppression of the isomerization is more effective the larger the cation: $Cs^+ > Tl^+ > Rb^+ > K^+ > Na^+ > Li^+$; it also correlates with a diminished ability to adsorb ammonia. The polarizability of the cation thus appears to be an important factor in reducing the surface acidity which is the cause of isomerization.

$$2\ EtCH{=}CH_2 \longrightarrow EtCH{=}CHEt + CH_2{=}CH_2 \qquad (6)$$

The reaction of $CH_2{=}CHC_2H_5$ with $CD_2{=}CDC_2D_5$ on MoO_x/TiO_2 at room temperature shows that degenerate metathesis is about 14 times faster than productive metathesis (Tanaka, K. 1979). When pretreated with Me_4Sn the TOF for metathesis of but-1-ene is enhanced by a factor of three (Tanaka, K. 1988c), but degenerate metathesis is much faster. UV irradiation of MoO_3 on porous Vycor glass in the presence of but-1-ene results in the formation of but-2-ene, followed by cross-metathesis to yield propene and pentene (Anpo 1984).

Catalyst systems suitable for metathesis of but-1-ene in solution are as listed for pent-1-ene in Table 5.5.

5.4.2 3-Methylbut-1-ene

Conversion of 3-methylbut-1-ene into 2,5-dimethylhex-3-ene, reaction (7), has been achieved using a Re_2O_7/Al_2O_3 catalyst (2 h at 20°C) with good yield (58%)

$$2\ Me_2CHCH{=}CH_2 \longrightarrow Me_2CHCH{=}CHCHMe_2 + CH_2{=}CH_2 \qquad (7)$$

and selectivity (Warwel 1983a). Catalysts of the type $Bu_4N[MoCl(CO)_5]/MeAlCl_2$ in chlorobenzene are also active at 20°C (Ruhle 1971; Doyle 1973).

5.4.3 3,3-Dimethylbut-1-ene

Metathesis of 3,3-dimethylbut-1-ene (neohexene) has not been reported (see Warwel 1983a), although degenerate cross-metathesis with 3-methylbut-1-ene has been observed (Howard 1980a).

5.4.4 Vinylcyclopropane

Metathesis of vinylcyclopropane over $Re_2O_7/Al_2O_3/Bu_4Sn$ proceeds with 2–3% conversion in the liquid phase at 35°C (Finkel'shtein 1992a); reaction (8). By continuous removal of ethene the yield can be increased to 60–65% after 3–4 h with a selectivity of $\geqslant$95%.

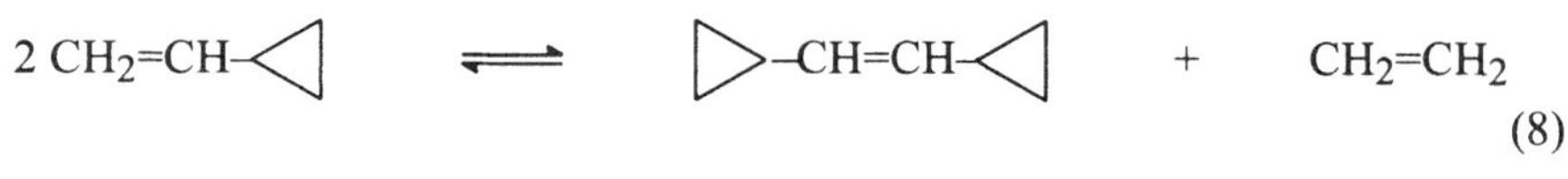

(8)

5.5 Pent-1-ene and its derivatives

5.5.1 Pent-1-ene

The behaviour of pent-1-ene is typical of that of terminal olefins. Its metathesis to form oct-4-ene and ethene, reaction (9), is most readily achieved with high

$$2\ PrCH{=}CH_2 \longrightarrow PrCH{=}CHPr + CH_2{=}CH_2 \qquad (9)$$

selectivity using one of the molybdenum nitrosyl or carbonyl complexes shown in Table 5.5. The corresponding chromium or tungsten complexes are generally less effective and may cause isomerization (Heenop 1978). Suitable catalysts can readily be generated and used *in situ* by contacting a variety of molybdenum compounds such as $MoCl_5$ in solution with NO (Zuech 1970). The formation of the initial metal carbene in these catalysts systems requires the displacement or reaction of one of the carbonyl (Motz 1977) or nitrosyl ligands (Hughes 1974), or the arene ligand in the system $W(CO)_3(arene)/EtAlCl_2/O_2$ (Korda 1982). The presence of the Bu_4N^+ ion is vital to the success of the molybdenum carbonyl catalysts but its role is unknown. $W[{=}C(C_6H_4Me)_2](CO)_5$ reacts with pent-1-ene to produce 36% $(MeC_6H_4)_2C{=}CH_2$ (and 27% cyclopropane derivative) but the metal carbene $W({=}CHPr)(CO)_4$, which may be presumed to be the other product, is apparently unable to propagate the metathesis reaction (Casey 1976).

Catalysts based on $MoCl_5$, WCl_6, or $ReCl_5$ generally lead to extensive alkylation (Graham 1971; Uchida, A. 1971) or isomerization (Uchida, Y. 1972; Uchida, A. 1977), with little or no apparent metathesis. Some exceptions to this rule are the

Table 5.5 Carbonyl and nitrosyl catalyst systems suitable for the metathesis of pent-1-ene (and terminal olefins in general)

Catalyst system	Solvent	Typical yield[a]	Reference
$MoCl_2(PPh_3)_2(NO)_2/Me_3Al_2Cl_3$	PhCl/5°C	48%/50 min	Zuech 1968, 1970; Hughes 1974
$MoCl_2(PPh_3)_2(NO)_2/EtAlCl_2$[b]	PhCl/25°C		Leconte 1979b
$Bu_4N[Mo(COPh)(CO)_5]/MeAlCl_2$[c]	PhCl/20°C	62%/4 h	Kroll 1971
$W(CO)_3(arene)/EtAlCl_2/O_2$[d]	PhCl/20°C	49%/15 min	Korda 1980, 1982
$W[{=}C(OMe)Et](CO)_5/Bu_4NCl$	PhCl/20°C	98%/4 h	Kroll 1971
$W[{=}C(OMe)Ph](CO)_5$/halides[e]	C_6H_{14}/55°C	40%/20 h	Banks 1982
$Mo(CO)_5(py)/Me_3Al_2Cl_3/Bu_4NCl$	PhCl/70°C	50%/30 min	Ruhle 1971 Motz 1977
$Mo(O_2CMe)_2(NO)_2/EtAlCl_2$	PhCl/20°C	Equilib./1 min	Keller 1989a
$[Mo(HSO_4)_2(NO)_4](HSO_4)_2/EtAlCl_2$	PhCl/20°C	40%/3 h	Keller 1990a
$Bu_4N[Mo_2(CO)_{10}]/MeAlCl_2$[f]	PhCl/20°C	60%/2 h	Kroll 1972
$Bu_4N[MoCl(CO)_5]/MeAlCl_2$	PhCl/20°C	90%/2 h	Doyle 1973

[a] Oct-4-ene; the forward reaction is favoured by allowing the ethene to escape. [b] The oct-4-ene has an initial *trans*/*cis* ratio of 2.0 ± 0.2; see Section 6.7. [c] The yields are less when Mo is replaced by W, or Ph by Me, or Bu_4N by Me_4N or Li. [d] The selectivity improves at lower temperatures. [e] Such as $TiCl_4$, $SnCl_4$, CCl_4, $MoOCl_4$. [f] The MoRe or MoMn complexes are not as effective as the Mo_2 complex.

following. $WCl_6/Bu_4Sn/X$, where X is a catalytic amount of propyl acetate or acetonitrile, in trichloroethene at 80°C gives a good yield of metathesis products (Ichikawa, K. 1976a). With diethyl ether as third component ($W/Sn/Et_2O/$alkene = 1/2/1/1000) a conversion of 84% is achieved in 2 h with 99% selectivity in $CHCl_3$ at 50°C (Warwel 1983a). With this system the thermodynamic *cis/trans* product ratio of the internal alkene is obtained. WCl_6 is also effective when activated with Et_3SiH or Ph_2SiH_2 at 50°C (Nametkin 1973). Moderate yields are reported using Ph_4Sn as cocatalyst (Stockel 1978). WCl_6/SiO_2-Al_2O_3 and $WOCl_4/SiO_2$-Al_2O_3, activated with Et_3Al, Me_3Al, or Bu_4Sn (cocat/W = 3/1) are effective, while $MoOCl_4$ supported on SiO_2-Al_2O_3 is active after addition of Bu_4Sn or Et_3Al (cocatalyst/Mo = 6/1) (Maksimowski 1991).

Photochemical systems have been discussed in Section 2.7. Re_2O_7/Al_2O_3 catalysts give high conversions with high selectivity, especially when treated with Cs (Warwel 1983a; Kawai 1988). Thus, at 28°C liquid pent-1-ene is converted into oct-4-ene and ethene with 99% selectivity.

5.5.2 Methylpent-1-enes

The metathesis of 4-methylpent-1-ene, reaction (10), can be effected by the

$$2\,Me_2CHCH_2CH{=}CH_2 \longrightarrow Me_2CHCH_2CH{=}CHCH_2CHMe_2 + CH_2{=}CH_2 \qquad (10)$$

molybdenum carbonyl complexes in Table 5.5 (Kroll 1971) or by the system $WCl_6/Bu_4Sn/Et_2O$. The latter system is also active for the metathesis of 3-methylpent-1-ene (Warwel 1983a). Re_2O_7/Al_2O_3 and $Re_2O_7/CsNO_3/Al_2O_3$ are active for these substrates, as well as for 4,4-dimethylpent-1-ene (Warwel 1983a; Kawai 1989). 4,4-Dimethylpent-1-ene also undergoes metathesis over $Mo(CO)_6/\gamma$-Al_2O_3 (Farona 1980).

5.6 Hex-1-ene and its derivatives

5.6.1 Hex-1-ene

Hex-1-ene undergoes metathesis under conditions essentially the same as those for pent-1-ene. Mo carbene complexes are quite effective at room temperature (Fox 1994a). The activity of Re_2O_7/Al_2O_3 at 20°C can be enhanced by a factor of more than 100 by the incorporation of other oxides such as V_2O_5 (Nakamura, R. 1977b). Incorporation of 3 wt% Re_2O_7 into a MoO_3/Al_2O_3 catalyst can increase the activity for hex-1-ene metathesis (at 200°C), but a large number of products is formed (C_2–C_{10}) as a result of isomerization (mainly on Mo) and subsequent metathesis (mainly on Re) (Guo 1988). The selectivity of Re_2O_7/Al_2O_3 can again be improved by incorporating a small amount of $CsNO_3$ (Kawai 1983).

For the vapour-phase metathesis of hex-1-ene at 150°C on $MoO_3/CoO/Al_2O_3$, a catalyst containing 4.5% CoO and 15.7% MoO_3 gives maximum conversion, but a complex mixture of products is formed (C_2–C_{12}). CoO slightly increases the selectivity for metathesis over MoO_3/Al_2O_3 (Ignatov 1986). MoO_3/SiO_2, photo-reduced in CO and treated with cyclopropane exhibits high initial catalytic activity (1 wt% Mo, TOF = ~20 mol hex-1-ene/mol Mo/s) and selectivity (~95%) at 20°C (Elev 1989b).

WCl_6 is effective in conjunction with Et_4Sn and $(Me_3SiCH_2)_4Sn$, as judged by evolution of ethene (Bespalova 1975); also using **1**, **2**, or **3** as cocatalysts

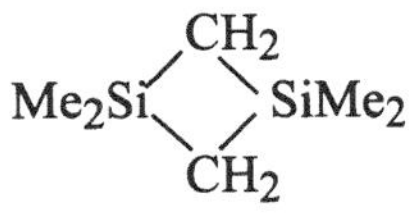

(1)

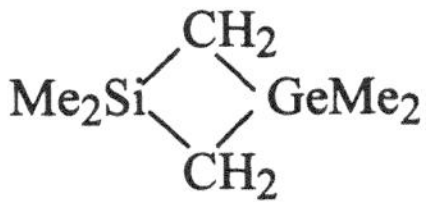

(2)

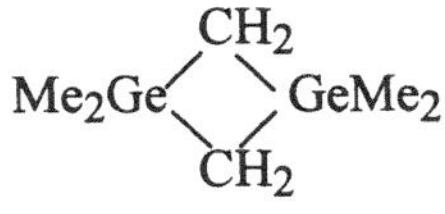

(3)

(Bespalova 1975; Nametkin 1975; Babich 1977). With **1** the evolution of ethene is preceded by an induction period, but with **2** and **3** reaction begins immediately. Analogous open-chain germanium compounds, such as Et_4Ge, exhibit no cocatalytic ability. The system WCl_6/1,1-dialkyl-3,3,5,5-tetramethyl-1-stanna-3,5-disila-4-oxacyclohexane is more stable than the corresponding WCl_6/R_4Sn systems; replacing WCl_6 by $WOCl_4$ gives a substantial increase in activity (Babich 1976). For $WCl_6/Bu_4Sn/Et_2O$ in $CHCl_3$ 79% conversion to decene and ethene is obtained after 20 min at 60°C with 98% selectivity (Warwel 1983a). For WCl_6/Bu_4Sn the conversion of hex-1-ene and hept-1-ene to metathesis products is increased four-fold if WCl_6 is immobilized on polystyrene (Liao 1989; Yao 1994).

5.6.2 Methylhex-1-enes

For a series of methylhex-1-enes, the reactivity decreases as the substituent(s) are placed closer to the double bond, no matter whether the catalyst is $Re_2O_7/CsNO_3/Al_2O_3$ (Kawai 1989), WCl_6/Me_4Sn (Kawai 1994), or $W(\equiv CCMe_3)(Cl)_3(dme)$ (Weiss 1988c). 4,4-Dimethylhex-1-ene has a reactivity intermediate between that of 4-methyl- and 3-methylhex-1-ene (Kawai 1989). Double-bond shift reactions and dimerization tend to predominate when the substituent is close to the double bond (Kawai 1992, 1994).

5.7 Higher acyclic terminal alkenes

5.7.1 Linear alkenes

The catalyst systems already described have been used to prepare reasonable to good yields of dodec-6-ene from hept-1-ene (Ichikawa, K. 1976a; Maruyama 1980; Warwel 1983a; Kawai 1988, 1992, 1994), tetradec-7-ene from oct-1-ene (Zuech 1968, 1970; Crain 1969; Ichikawa, K. 1976a; Warwel 1983a; Kawai 1988; Mol 1988;

Spronk 1991b), hexadec-8-ene from non-1-ene (Kawai 1988), octadec-9-ene from dec-1-ene (Ichikawa, K. 1976a; Warwel 1983a; Spronk 1991b; Fox 1994a), docos-11-ene from dodec-1-ene (Hughes 1977), triacont-15-ene from hexadec-1-ene (Maruyama 1980), symmetrical internal C_{24}–C_{42} alkenes from C_{13}–C_{22} alk-1-enes (Gibson, T. 1981), and octatriacont-19-ene from eicos-1-ene (Doyle 1973). For example, by refluxing oct-1-ene through a column of Re_2O_7/Al_2O_3 one may obtain a conversion of 86% (selectivity 92%) into tetradec-7-ene (Warwel 1983a).

In the reaction of oct-1-ene, catalyzed by $WCl_6/EtAlCl_2$ in chlorobenzene, the addition of $Ni[P(OPh)_3]_4$ suppresses polymerization, and allows metathesis accompanied by isomerization to yield a mixture of C_{10}, C_{11}, C_{12}, and C_{14} olefins (du Plessis 1979). $WOCl_4/EtAlCl_2$ also brings about a combined isomerization and metathesis reaction. Addition of the isomerization catalyst $Ni[P(OPh)_3]_4$ increases the formation of oct-2-ene, but does not change the distribution of the metathesis products (du Plessis 1980). *trans*-$WCl_2(OPh)_4/EtAlCl_2$ (1/6) is effective at 0°C, but causes some isomerization (Dodd 1982). Tungsten phenoxo compounds such as $WCl_2(OC_6H_4Me\text{-}2)_4$ with $Et_3Al_2Cl_3$ as cocatalyst (at room temperature), or $WCl_2(OC_6H_4Cl\text{-}4)_4/Bu_4Sn$ (at 140°C) are active for oct-1-ene metathesis in chlorobenzene, but some isomerization also takes place (Dodd 1988). During the interaction of oct-1-ene or oct-2-ene with $[Mo(CO)_3Cp]_2$ or $[Mo(CO)_3Cp]_2/NO$ and $EtAlCl_2$ in chlorobenzene at temperatures above 120°C, metathesis, isomerization, and oligomerization of olefins as well as alkylation of the solvent by short-chain olefins occur (du Plessis 1994).

A reduced Phillips polymerization catalyst, viz. Cr(II) surface-bound to SiO_2, reacts with the Fischer-type carbene complexes $Mt[{=}C(OMe)Ph](CO)_5$ (Mt = W or Mo) to yield active catalysts for the metathesis of oct-1-ene in hexane at 69°C (Weiss 1988a). Other substituted Fischer-type tungsten carbene complexes of the general formula $W({=}CR^1R^2)(CO)_5$ (R^1 = OMe, Ph; R^2 = Ph, C_6H_4Me), which are inactive in homogeneous solution, also become active for metathesis and double-bond shift reactions of oct-1-ene at 69°C by reaction with $Cr(II)/SiO_2$ (Weiss 1988d). Reactions of tungsten carbyne complexes $W({\equiv}CPh)(X)(CO)_4$ (X = Cl, Br, I) with the above-mentioned reduced Phillips catalyst yield very active heterogeneous bimetallic metathesis catalysts, as demonstrated for the metathesis of oct-1-ene. The original polymerization activity toward alk-1-enes of the surface-bound Cr(II) atoms disappears after reaction with carbyne complexes (Weiss 1988b). The catalyst $W({\equiv}CPh)(Cl)(CO)_4/Cr(II)/SiO_2$ is more active for oct-1-ene metathesis (at 122°C) than the carbene complex $W[{=}C(OMe)Ph](CO)_5/Cr(II)/SiO_2$ (Weiss 1990a).

The W-W single bond complex tetrachlorobis-(μ-ethoxy)tetraethoxyditungsten and, to some extent, the double-bonded tungsten–tungsten alkoxide dimer tetrachloro(μ-methoxy)bis(methanolo)dimethoxyditungsten are catalysts for the metathesis of oct-1-ene when Et_2AlCl is used as a cocatalyst. The reaction occurs under mild conditions, although the conversion to tetradecene is not high (Savard 1988). The photoassisted metathesis of oct-1-ene by $W(CO)_6/CCl_4/h\nu$ can be performed at 25°C. The reaction rate is very low compared with other catalyst systems; besides chlorinated hydrocarbons, small amounts of secondary metathesis products are also formed (Imamoğlu 1986).

A catalyst prepared by attaching $W(CO)_6$ to a strong anion-exchange resin (Amberlyst A-27) to form [resin]-$Me_3N[WCl(CO)_5]$ polymerizes oct-1-ene after addition of $EtAlCl_2$, but isomerization (to oct-2-ene) and cross-metathesis are also observed (du Plessis 1982). The combined metathesis and isomerization of oct-1-ene in the pesence of $RMe_3N[WCl(CO)_5]/EtAlCl_2$ (R = Me or Amberlyst) takes place at a molar ratio of Al/W > ~2 for R = Me, and Al/W > ~7 for R = resin, while the presence of small amounts of HCl or H_2O is necessary to activate the system (Pienaar 1985; du Plessis 1988).

A kinetic study has been made of the liquid-phase metathesis of oct-1-ene over Re_2O_7/Al_2O_3 in a flow reactor; Fig. 5.8 shows the fractional conversion X at four temperatures as a function of the contact time expressed in terms of W/F as defined in the caption. The data are best interpreted in terms of a model in which either product desorption or interconversion of the alkene/carbene complex is rate-determining. This model leads to eqn. (11), where r is the reaction rate per unit weight of catalyst, and from which the curves in Fig. 5.8 have been calculated to give the best fit to the data (Spronk 1992).

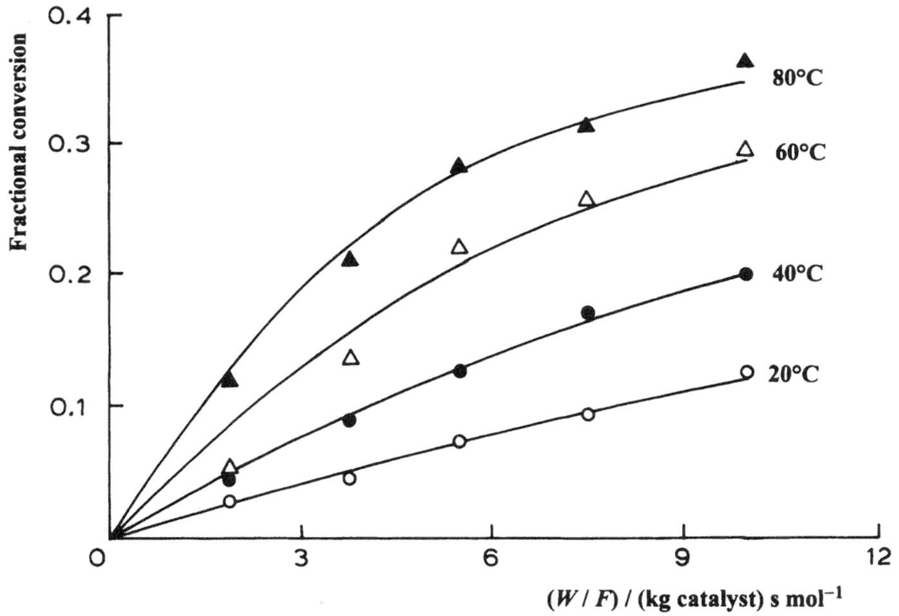

Fig. 5.8 Fractional conversion of oct-1-ene (6.4 M) as a function of contact time at four temperatures. Markers indicate the measured conversion and the lines represent the fits using eqn. (11). W = weight of catalyst (kg), F = reactant flow rate ($mol\ s^{-1}$); pressure held at 9 bar to keep the product ethene in solution (Spronk 1992).

$$r = k[1 - 2X + (1 - 1/4K)X^2] \tag{11}$$

For $Re_2O_7/CsNO_3/Al_2O_3$ the relative reactivities for productive metathesis of 1-C_6, 1-C_7, 1-C_8, and 1-C_9 are 1.0, 0.90, 0.66, and 0.44, respectively, the decrease being attributed to an increasing ratio of the rate of non-productive to productive metathesis with increasing chain length (Kawai 1988). In another piece of work using Re_2O_7/Al_2O_3, the relative reactivities for 1-C_6, 1-C_8, and 1-C_{10}, calculated from the initial rates, are 1.0, 0.5, and 0.1, respectively (see Fig. 5.9). In this case a decreasing rate of product desorption with increasing chain length has been offered as the explanation for the variation in rate of reaction (Spronk 1991b).

A method for the prediction of the equilibrium position for the liquid-phase metathesis of alk-1-enes has been developed and tested by comparison with experimental data for the metathesis of oct-1-ene at a pressure of 9 bar in the temperature range of 22–80°C (Dekker 1993).

Re_2O_7/SiO_2 shows no activity for oct-1-ene metathesis; only isomerization and dimerization occur (Duquette 1984; Mol 1988). Inorganic chlorocomplexes, such as $K_2[ReCl_6]$ and Re_2Cl_{10} on a SiO_2-Al_2O_3 support catalyze both double-bond shift and metathesis of oct-1-ene at room temperature (Brégeault 1988). In the presence of $MoCl_5/SiO_2/Me_4Sn$ the stereoselectivity of the metathesis is almost independent of the chain length of the alk-1-ene, and metathesis of 1-C_6, 1-C_8, 1-C_{10}, and 1-C_{15} results in the corresponding symmetrical internal alkene (*cis*/*trans* = 16/84) (Bykov 1990).

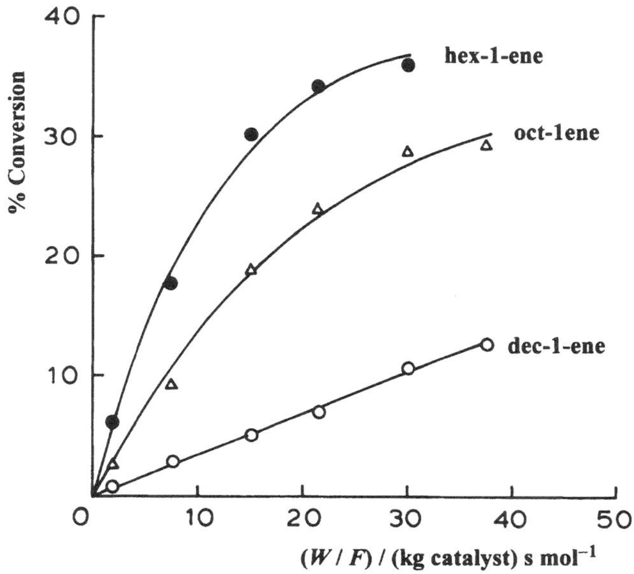

Fig. 5.9 Alk-1-ene conversions as a function of contact time over an Re_2O_7/Al_2O_3 catalyst at 20°C. *W* and *F* as in Fig. 5.8 (Spronk 1991b).

In the presence of $WOCl_4$/MeLi (1 : 2) in THF, believed to produce $W[=CH_2(O)Cl_2]$, a ten-fold excess of hexadec-1-ene is converted into triacont-15-ene and ethene in cyclohexane/THF (9/1) at 80°C (yield 45% after 4 h); the metathesis of hexadec-1-ene is almost completely suppressed when the solvent is THF (Kauffmann 1986).

5.7.2 Substituted alkenes

Vinylcyclohexane is metathesized by Re_2O_7/Al_2O_3 (Finkel'shtein 1992a) as also is allylcyclohexane (Warwel 1983a). 4-Vinylcyclohexene undergoes metathesis over MoO_3/Al_2O_3 (Crain 1972); the double-bond in the ring is not affected. With Re_2O_7/Al_2O_3 a conversion of 74% after 2 h at 50°C is reported (Warwel 1983a); reaction (12).

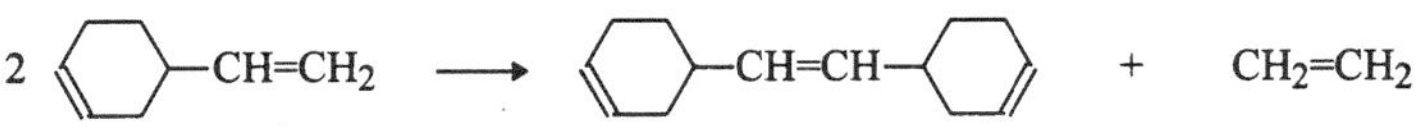

(12)

Styrene can be converted into stilbene, reaction (13), using $W(CO)_3(PhMe)$ in

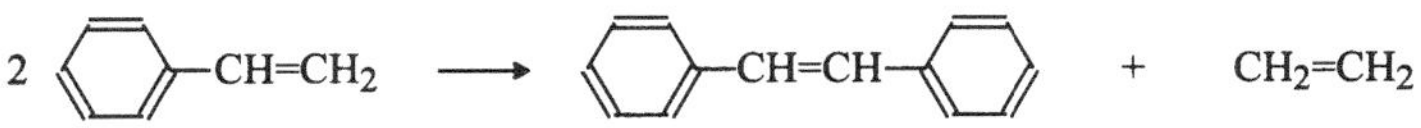

(13)

heptane at 98°C, provided that the styrene concentration is not too high and the displaced carbon monoxide is allowed to escape (Lewandos 1971b). Metathesis of styrene over Re_2O_7/Al_2O_3 results in 42% converion after 2 h at 50°C (Warwel 1983a). Styrene is efficiently transformed to *trans*-stilbene when using $Mo(=CHCMe_2Ph)(=NC_6H_3\text{-}Me_2\text{-}2,6)(OCMe_2CF_3)_2$ in the presence of dimethoxyethane (Fox 1994a). $MoCl_2(PPh_3)_2(NO)_2/EtAlCl_2$ in chlorobenzene causes rapid polymerization of styrene at 25°C but brings about the metathesis of the related olefins (**4**); reaction (14). The reaction may be driven to high conversion by continuous

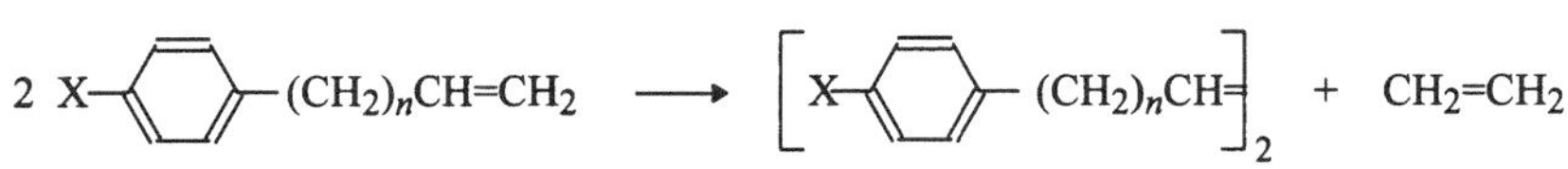

(14)

(**4**) (X = H, Me, Me_3C, F, Cl, Br, not MeO; $n \geq 1$)

removal of ethene under vacuum (Chevalier 1975). The rate is highest when $n = 2$. This has been attributed to favourable geometry for coordination of the benzene ring to the Mo site thereby allowing easier access for the incoming olefin. The metathesis of allylbenzene (**4**, X = H, $n = 1$) on Re_2O_7/Al_2O_3 gives 77% conversion after 2 h at 80°C (Warwel 1983a). The metathesis of allylbenzene is also catalyzed by $W(CO)_5L/EtAlCl_2/O_2$ in chlorobenzene at 20°C, the rate decreasing in the order $L = P(OPh)_3 > CO > PPh_3 > PBu_3$, i.e. with increasing basicity of L. However, the catalysts with the ligands of higher basicity are longer-lived and give higher ultimate yields (Bilhou 1977c).

$Bu_4N[MoCl(CO)_5]/MeAlCl_2$ does not metathesize allylbenzene but does cause metathesis of 4-phenylbut-1-ene (**4**, X = H, $n = 2$) to give good yields of 1,6-diphenylhex-3-ene (Doyle 1973).

Degenerate cross-metathesis between $C_6H_{13}CH{=}CD_2$ and *cis*-$C_8H_{17}CH{=}CHD$ shows that the *cis* configuration is partially maintained in the product $C_6H_{13}CH{=}CHD$, no matter what catalyst is used (Casey 1978); see Ch. 9.

6

Acyclic Disubstituted and Trisubstituted Ethenes

6.1 Introduction

Only a few 1,1-disubstituted ethenes have been reported to undergo metathesis; such reactions are not favoured thermodynamically but can be encouraged by removal of the ethene. Most of this chapter is therefore concerned with 1,2-disubstituted ethenes of which but-2-ene is the simplest example.

Such olefins exist in *cis* and *trans* forms, and can undergo *cis/trans* isomerization by the metathesis reaction. Indeed when the olefin is symmetrical this is the only observable change, provided that there is no concomitant double-bond shift reaction. When the olefin is unsymmetrical there is an added point of interest, namely the extent to which a *cis* reactant gives rise to *cis* products, and a *trans* reactant to *trans* products. We have already touched on this important question of stereoselectivity in Section 3.3, and here we shall elaborate further in Section 6.7, after summarizing the *cis/trans* ratio observed for the products of metathesis of both alk-1-enes and alk-2-enes. The selectivities in metathesis of internal olefins are usually very high.

Examples of suitable catalyst systems for the metathesis of internal olefins have been given in Ch. 2.

6.2 *Cis/trans* isomerization

Given time, both reactants and products of the metathesis of an internal olefin will reach *cis/trans* equilibrium. WCl_6/$EtAlCl_2$/EtOH (1/4/1) is particularly effective in bringing about rapid equilibrium, and some experimental *cis/trans* equilibrium data are shown in Table 6.1. *Cis/trans* equilibrium constants have been determined for both but-2-ene and pent-2-ene from −20°C to 550°C (using Re_2O_7/Al_2O_3 and WO_3/SiO_2 catalysts), thereby allowing improved values to be obtained for the relative thermodynamic functions of the *cis/trans* isomers (Kapteijn 1983a).

The exchange of alkylidene moieties in but-2-ene was first demonstrated using

Table 6.1 Equilibrium *cis* isomer content for internal olefins involved in the metathesis reaction[a] (Calderon 1968; Zowade 1973)

	Cis content of olefin in equilibrated mixture (%)							
Reactant[b]	2-C_4	2-C_5	2-C_6	3-C_6	2-C_7	3-C_7	4-C_8	5-C_{10}
cis-2-C_4	28							
cis-2-C_5[c]	29	18		13				
trans-2-C_5[c]	28	20		15				
trans-2-C_6	25		19				16	
2-C_7[d,e]	27				21			18
3-C_7[e]				14		15	17	
trans-5-C_{10}								17
Averages	27	19	19	14	21	15	17	18

[a] Catalyst: $WCl_6/EtAlCl_2/EtOH$ (1/4/1) in benzene at about 25°C. [b] *cis*-2-C_4 denotes *cis*-but-2-ene, and so on. Concentration 1 M. [c] Similar values observed in chlorobenzene at 0°C (Hughes 1969). [d] Similar values observed in 1,2,4-trichlorobenzene at 20°C (Ichikawa, K. 1976f). [e] Mixture of *cis* and *trans* isomers.

the catalyst system $WCl_6/EtAlCl_2/EtOH$ (1/4/1) (Calderon 1968); reaction (1). This

$$\begin{array}{c} CH_3CH{=}CHCH_3 \\ + \\ CD_3CD{=}CDCD_3 \end{array} \rightleftharpoons \begin{array}{c} CH_3CH \\ \| \\ CD_3CD \end{array} + \begin{array}{c} CHCH_3 \\ \| \\ CDCD_3 \end{array} \quad (1)$$

exchange can be probed more deeply by starting from a mixture of *cis*-but-2-ene and *cis*-but-2-ene-d_8, and examining the labelling pattern of both the *trans*-but-2-ene that is formed, and the residual *cis*-but-2-ene. On MoO_x/β-TiO_2 at room temperature degenerate metathesis, reaction (2a), is rather more probable than *cis/trans* isomerization, reaction (2b). For *trans* reagents the degenerate metathesis is dominant (Tanaka, K. 1980b, 1982a). Activation of MoO_x/β-TiO_2 with Me_4Sn increases both the activity and the preference for reaction (2a) (Tanaka, K. 1984). The reactions of the *cis* olefins proceed through the metallacyclobutanes **1** and **2**, respectively, and, at first sight, it might have been expected that **2**, and therefore reaction (2b), would be favoured because of the interactions between the substituents in the 1,3 and 1,2 positions. However, this takes no account of the fact that the ring is likely to pucker so as to relieve these interactions, giving **1a** and

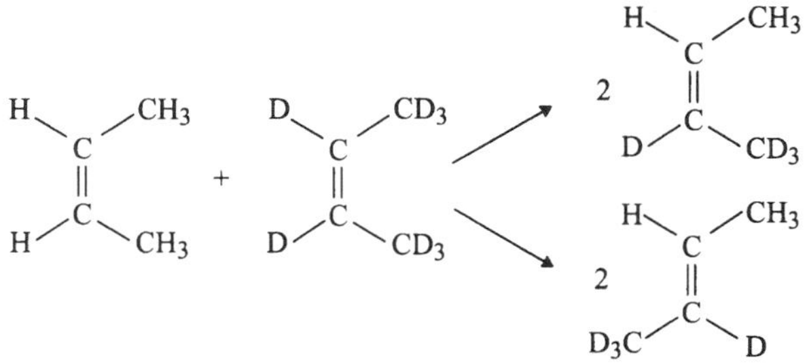

(2a)

(2b)

2a. There is then less difference between the likelihood of formation of **1a** and **2a**; indeed **1a** is somewhat favoured. In this case the important interaction seems to be that between the 1,3-substituents, *ee* being preferred to *ea*. This theme will be elaborated in Section 6.7.

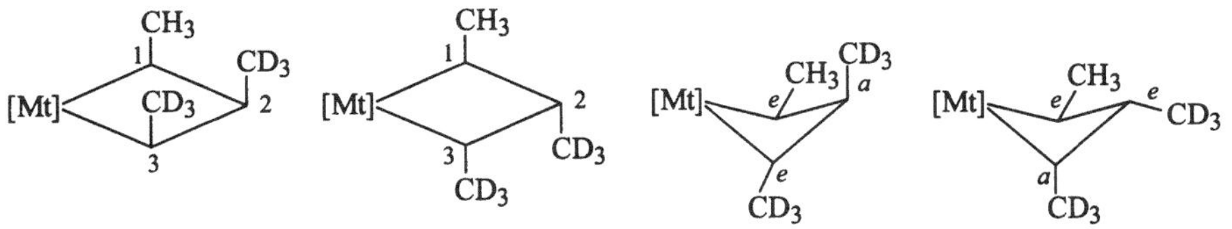

(**1**) (**2**) (**1a**) (**2a**)

On MoO_3/γ-Al_2O_3 the *cis/trans* isomeratization of but-2-ene follows second-order kinetics, but is accompanied by isomerization to but-1-ene and secondary cross-metathesis reactions (Davie 1972a; Engelhardt 1982e).

6.3 Pent-2-ene and 4-substituted derivatives

Pent-2-ene is the simplest internal olefin to undergo productive metathesis, reaction (3), and it has therefore been a favourite proving ground for a wide range of

$$2\,MeCH{=}CHEt \rightleftharpoons MeCH{=}CHMe + EtCH{=}CHEt \qquad (3)$$

catalytic systems. The reaction is almost athermal ($\Delta H = -0.58\,kJ\,mol^{-1}$) and the equilibrium mixture consists of but-2-ene, pent-2-ene, and hex-3-ene in the approximate ratio 1 : 2 : 1, pent-2-ene being favoured by the statistical factor of 2. The ultimate equilibrium *cis/trans* ratios are indicated in Table 6.1. Reaction (3) is catalyzed by metal carbene complexes and many other catalyst systems as described in Ch. 2. Further catalyst systems are listed by Ivin (1983). More recently an extensive series of molybdenum nitrosyl complexes have been prepared which are very active in the presence of a cocatalyst, e.g. $[Mo(NO)_2(OEt)_2]_n/EtAlCl_2$ (Keller 1989b); see also Taube (1977), Seyferth (1982a,b, 1984c), Keller (1987a,b, 1989a, 1990a,b,c, 1991a,c, 1992a,b). Various tungsten complexes have also been studied (Seyferth 1984a; Davidson, J.L. 1989).

The rate of metathesis is dependent on both the inherent activity (rate constant) of the chain carriers and their concentration. It is not generally possible to disentangle these two factors, although it is likely that the presence of more than one vacant site at the metal centre favours high activity but reduced stereoselectivity. Many of the catalyst systems depend for their activity on the prior dissociation or abstraction of one or more of the neutral ligands so as to leave vacancies for the coordination of substrate and the formation of the carbene ligand. It is thought that the role of oxygen in catalysts such as $W(CO)_5(PPh_3)/EtAlCl_2/O_2$

is to bond indirectly to the metal site through the aluminium cocatalyst so as to weaken the bond to the *trans*; ligand thereby permitting it to be readily removed. In some cases the reaction of pent-2-ene is markedly retarded by the presence of terminal alkenes, an effect that is attributed to their stronger coordination to the active site.

An unusual feature of the catalyst systems $W(CO)_5(PPh_3)/EtAlCl_2/O_2$ and $W(CO)_3(mes)/EtAlCl_2/O_2$ is that they initiate metathesis of *trans*-pent-2-ene only with difficulty whereas that of *cis*-pent-2-ene proceeds very readily (Bilhou 1977b). However, a trace (0.2%) of *cis* or terminal olefin is sufficient to set the reaction going after an induction period. This observation seems to indicate that the olefin is directly involved in the formation of the initial metal carbene from these catalyst systems and that for steric reasons *trans*-pent-2-ene finds it difficult to fulfil this role. The catalyst $WCl_6/Et_3Al_2Cl_3/(CH_2Cl)_2CHOH/PPh_3$, acting on a 27.5% *cis*/72.5% *trans* mixture of pent-2-ene, appears to convert 80% of the *cis* isomer but only 4% of the *trans* isomer in a given time; however, this is probably indicative of concurrent *cis/trans* isomerization of the reactant rather than a failure of the *trans* isomer to react (Dall'Asta 1973/74).

$W(SnCl_{3-n}R_n)(CO)_3Cp$ (R = Me, Ph; $n = 0$–3), activated with *i*-$BuAlCl_2$ in the presence of O_2 in an optimum $W/Al/O_2$ ratio of 1/10/5, gives high conversions and > 90% selectivity in pent-2-ene metathesis. With different tin ligands the order of activity is $SnCl_3 > Me_2SnCl > Me_3Sn \gg Ph_3Sn$ (Vanderyse 1988a). $WX(CO)_3Cp$, (X = Cl, I), in combination with R_nSnCl_{4-n} (R = Me, Ph; $n = 0$–4) and *i*-$BuAlCl_2/O_2$, also catalyze the rapid metathesis of pent-2-ene with good selectivity. For X = Cl the trend in reactivity is $Me_3SnCl > Me_2SnCl_2 \approx Me_4Sn > MeSnCl_3 > SnCl_4$. For X = I the initial reaction rate and conversion decrease in the order $MeSnCl_3 > Me_2SnCl_2 > Me_3SnCl > Me_4Sn$; when R = Ph, the initial rates obtained for $n = 1$ and $n = 2$ are about the same as for the methyltin chlorides (Verdonck 1992).

The activity of Mo complexes of the type $Mo(=CHR')(=NAr)(OR)_2$ ($OR = OCMe(CF_3)_2$, $OCMe_2CF_3$, $OCMe_3$; $Ar = C_6H_3$-*i*-Pr_2-2,6) roughly correlates with the electron-withdrawing ability of OR, being high when $OR = OCMe(CF_3)_2$ and low to negligible when $OR = OCMe_3$. $Mo(=CHCMe_3)(=NAr)[OCMe(CF_3)_2]_2$ metathesizes *cis*-pent-2-ene in toluene to an equilibrium mixture of *cis*- and *trans*-but-2-enes, pent-2-enes and hex-3-enes at 25°C with a turnover frequency (TOF) of about 250 min^{-1}, while for $OR = OCMe_2CF_3$ and $OCMe_3$ the TOFs are 5 h^{-1} and 1–2 day^{-1}, respectively. The analogous W complex gives a TOF of about 1000 min^{-1}; to a first approximation the W catalyst is an order of magnitude more active than the analogous Mo catalyst for the metathesis of *cis*-pent-2-ene (Schrock 1990b).

The metathesis of 4-methylpent-2-ene, reaction (4), occurs with $Bu_4N[MoCl(CO)_5]/MeAlCl_2$ (Doyle 1973), $W(CO)_5(PPh_3)/EtAlCl_2/O_2$ (Leconte 1978, 1979b)

$$2\,MeCH{=}CHCHMe_2 \rightleftharpoons MeCH{=}CHMe + Me_2CHCH{=}CHCHMe_2 \tag{4}$$

and $WCl_6/Me_4Sn/Et_2O$ (Ofstead 1980). With the last catalyst the *cis* reactant gives *i*-PrCH=CH-*i*-Pr, which is practically 100% *trans*; the remarkable stereoselectivity of this reaction has already been discussed in Section 3.3. The former catalyst is also effective with 4,4-dimethylpent-2-ene; reaction (5).

$$2\,\mathrm{MeCH{=}CHCMe_3} \rightleftharpoons \mathrm{MeCH{=}CHMe} + \mathrm{Me_3CCH{=}CHCMe_3} \qquad (5)$$

6.4 Hex-2-ene and 4-methylhex-2-ene

The conversion of hex-2-ene into but-2-ene and oct-4-ene, reaction (6), is readily

$$2\,\mathrm{MeCH{=}CHPr} \rightleftharpoons \mathrm{MeCH{=}CHMe} + \mathrm{PrCH{=}CHPr} \qquad (6)$$

achieved. Two catalyst systems deserve mention. First, by irradiating a solution of $W(CO)_6$/*i*-$BuAlCl_2$/*trans*-hex-2-ene (1/5/850) with UV radiation for 2 h, 50% of the hex-2-ene may be converted into metathesis products, while the remainder is partially isomerized to the *cis* isomer; the same occurs starting from *cis*-hex-2-ene (Warwel 1975). Secondly, the combination $W(MtR_3)(CO)_3Cp$/*i*-$BuAlCl_2/O_2$ (Mt = Ge, Sn; R = Me, Ph) in chlorobenzene is slow-acting but selective if O_2 is added after the olefin, but fast-acting and accompanied by double-bond shifts if the olefin is added last (Warwel 1977b). Metathesis of *cis*-hex-2-ene proceeds easily with 96% selectivity over $Re_2O_7/CsNO_3/Al_2O_3$ at 31°C; only small quantities of products of cross-metathesis between hex-2-ene and hex-3-ene are produced. Similar observations have been made with other alk-2-enes (C_5–C_9) (Kawai 1988). Further data are given in Section 6.7.

The metathesis of (*S*)-4-methylhex-2-ene catalyzed by WCl_6/Bu_4Sn at 20°C, reaction (7), proceeds with low yield (3%) and some double-bond shift in the

$$2\,\mathrm{Et}\underset{\substack{|\\ \mathrm{Me}}}{\overset{*}{\mathrm{C}}}\mathrm{HCH{=}CHMe} \rightleftharpoons \mathrm{MeCH{=}CHMe} + \mathrm{Et}\underset{\substack{|\\ \mathrm{Me}}}{\overset{*}{\mathrm{C}}}\mathrm{HCH{=}CH}\underset{\substack{|\\ \mathrm{Me}}}{\overset{*}{\mathrm{C}}}\mathrm{HEt} \qquad (7)$$

substrate, but with preservation of chirality, as judged by the optical purity of the recovered substrate, as well as by a study of the product in the related metathesis of (*S*)-6-methyloct-2-ene (Rossi 1976d). Retention of chirality is, of course, to be expected for the metal carbene mechanism.

6.5 Hept-2-ene and hept-3-ene

Hept-2-ene has been converted into an equilibrium mixture with but-2-ene and dec-5-ene, reaction (8), using various WCl_6-based catalyst systems (Takagi 1972, 1975; Ichikawa, K. 1976c,d,e). Polar additives inhibit alkylation of the solvent

$$2\,MeCH{=}CHBu \rightleftharpoons MeCH{=}CHMe + BuCH{=}CHBu \quad (8)$$

(Nishiguchi 1981). Metathesis of *cis*-hept-2-ene on $Re_2O_7/CsNO_3/Al_2O_3$ proceeds with 100% selectivity at 31°C (Kawai 1988).

Hept-3-ene can be converted into hex-3-ene and oct-4-ene with the usual catalysts; reaction (9). With $Bu_4N[MoCl(CO)_5]/MeAlCl_2$ the *trans* isomer is

$$2\,EtCH{=}CHPr \rightleftharpoons EtCH{=}CHEt + PrCH{=}CHPr \quad (9)$$

rapidly converted into an equilibrium mixture with the *cis* isomer, but the initially formed metathesis product oct-4-ene is 95% *trans* (Doyle 1973). Metathesis also occurs readily on the polystyrene-bound complex (pol)-$CH_2(\eta^5\text{-}C_5H_4)W(CO)_3R$ (R = Cl, H, Me) in conjunction with *i*-$BuAlCl_2$ (Heveling 1990).

6.6 Higher acyclic internal olefins

The catalysts already described for the metathesis of the lower internal olefins work equally well for oct-2-ene (Crain 1969; Uchida, A. 1985; Kawai 1988; Zümreoğlu-Karan 1991), 6-methyloct-2-ene (Rossi 1976d), *cis*-oct-3-ene, *cis*-non-3-ene (Kawai 1988), non-4-ene (Hughes 1970a; Heckelsberg 1983; Kawai 1988), and dec-2-ene (Grubbs 1979). The photoassisted and photocatalyzed metathesis of oct-2-ene in the presence of the system $W(CO)_6/CCl_4/h\nu$ proceeds with low activity at room temperature; secondary metathesis products and chlorinated hydrocarbons are also formed, although to a much lesser degree than with oct-1-ene as the substrate (Zümreoğlu-Karan 1991). The clean metathesis of non-4-ene on Re_2O_7/Al_2O_3 is an example of the use of olefin metathesis as a method for the rapid and accurate determination of double-bond locations and isomer purity in olefins (Heckelsberg 1983).

In the electrochemical system $WCl_6/e^-/Al/CH_2Cl_2$ the trend in metathesis reactivity is pent-2-ene > hex-2-ene > non-4-ene (Gilet 1983). With $Re_2O_7/CsNO_3/Al_2O_3$ as catalyst the general order of reactivity of linear olefins (C_5–C_9) is: alk-2-ene > alk-3-ene > alk-4-ene > alk-1-ene (Kawai 1988).

trans-1-Phenylbut-2-ene, $PhCH_2CH{=}CHMe$, is interesting in that with $MoCl_2(PPh_3)_2(NO)_2/EtAlCl_2$ the initial but-2-ene has a very high *trans* content (97%); this is similar to the behaviour of *trans*-pent-2-ene with the same catalyst (Chevalier 1976). It also gives a better yield of $PhCH_2CH{=}CHCH_2Ph$ than is obtainable by the metathesis of allylbenzene.

6.7 Stereoselectivity in the metathesis of acyclic olefins

Most of the available data on stereoselectivity of metathesis of acyclic olefins relate to *cis*- and *trans*-2-olefins, but it is convenient here also to consider some results for propene, but-1-ene, and pent-1-ene, also *cis,cis*- and *trans,trans*-deca-2,8-diene.

6.7.1 Necessity for use of extrapolation procedures

The occurrence of rapid secondary metathesis reactions makes it absolutely essential to extrapolate results back to zero time if they are to be properly interpreted. The changes in *trans/cis* ratio of reactants and products with time also carry useful kinetic information, although the mechanism then becomes so complex that any attempt at interpretation requires a number of assumptions.

Some examples of extrapolation procedures are shown in Figs. 6.1–6.4. As the abscissa, one may use time, percentage conversion of reactant, or *trans/cis* ratio for a *cis* reactant (the inverse for a *trans* reactant). The third method, illustrated in Fig. 6.2, is expected, on the basis of a simplified theory, to give a linear extrapolation, as found experimentally. This is the best method, provided that the rate of isomerization of the reactant is not much faster than the rate of productive metathesis, which, however, is the case with branched olefins (Fig. 6.3).

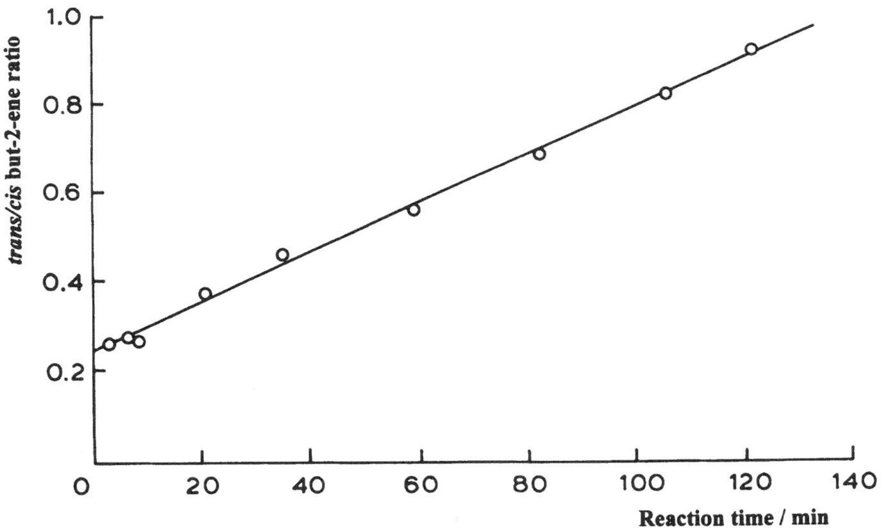

Fig. 6.1 Change in the *trans/cis* but-2-ene ratio with time during the metathesis of *cis,cis*-deca-2,8-diene. Catalyst: $MoCl_2(PPh_3)_2(NO)_2/Me_3Al_2Cl_3$ (1/3.5) in chlorobenzene (Grubbs 1979).

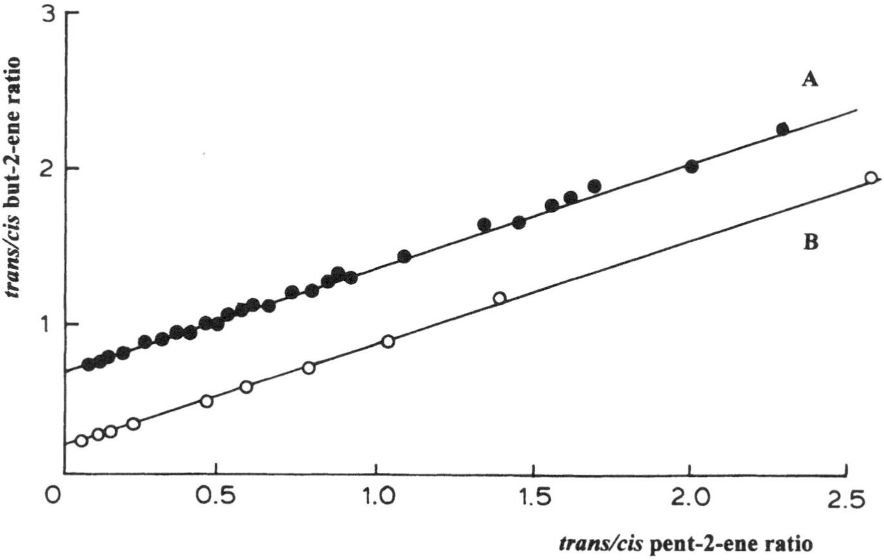

Fig. 6.2 Change in the *trans/cis* but-2-ene ratio with *trans/cis* ratio of pent-2-ene during the metathesis of *cis*-pent-2-ene. Catalyst A: $W(CO)_5(PPh_3)/EtAlCl_2/O_2$ (1/4/0.5) in chlorobenzene; catalyst B: $MoBr_2(PPh_3)_2(NO)_2/EtAlCl_2$ (1/6) in chlorobenzene (Leconte 1979b).

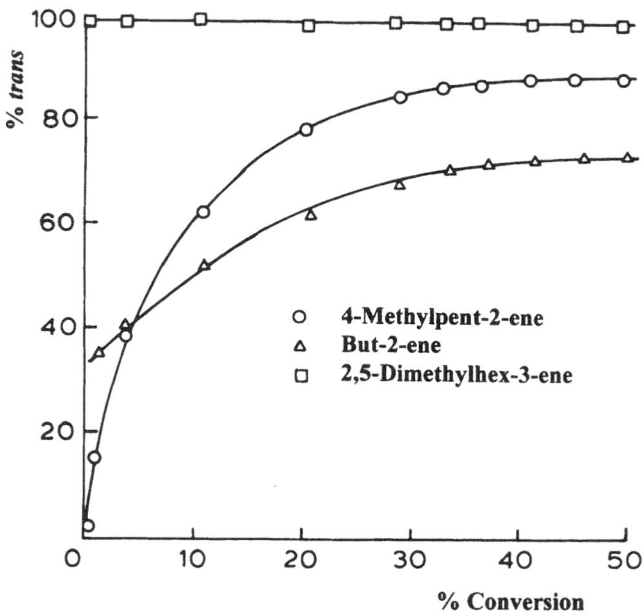

Fig. 6.3 Change in the *trans* content of reactant and products with % conversion during the metathesis of 4-methyl-*cis*-pent-2-ene. Catalyst: $WCl_6/Bu_4Sn/Et_2O$ (1/2/2) in toluene (Ofstead 1980).

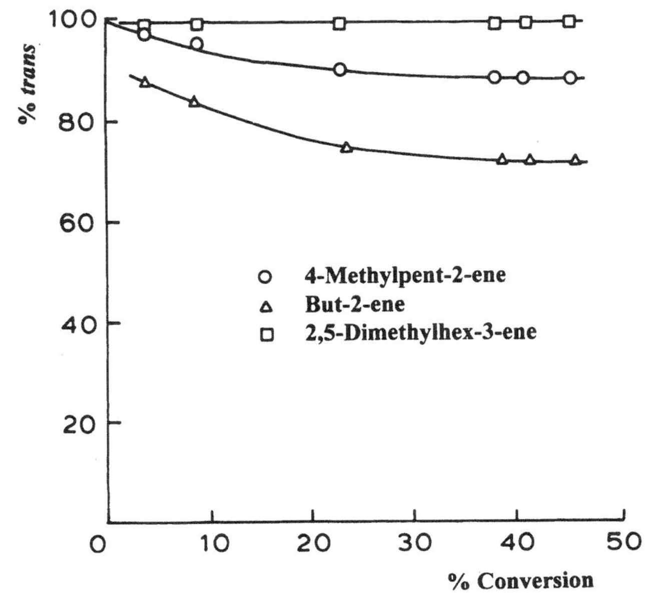

Fig. 6.4 As for Fig. 6.3 but starting from 4-methyl-*trans*-pent-2-ene (Ofstead 1980).

6.7.2 Experimental data

A selection of the available data is given in Tables 6.2–6.4. With the exception of 3-methylhex-1-ene, the alk-1-enes higher than but-1-ene give $75 \pm 8\%$ *trans* products (Table 6.2). The lower percentage *trans* in the product from 3-methylhex-1-ene is attributed to repulsive interactions involving the α-methyl group in the intermediate *trans*-metallacyclobutane (Kawai 1989).

For the metathesis of alk-2-enes, RCH=CHMe (Table 6.4), the product but-2-ene generally shows a *cis* preference when the reactant is *cis*, and *trans* when the reactant is *trans*. However, for the reaction of $Me_2CHCH{=}CHMe$, the product $Me_2CHCH{=}CHCHMe_2$ is fully *trans* no matter whether the reactant is *cis* or *trans* (see Section 3.3). For the metathesis of *cis*-pent-2-ene catalyzed by $WCl_2(OAr)_4$/$Et_3Al_2Cl_3$ the *cis/trans* ratios in the products are similar for catalysts with various substituted phenoxide ligands, but for the metathesis of *trans*-pent-2-ene, catalysts bearing *ortho* substituents on the phenoxide ligand tend to give products with a higher *cis* content, suggesting a specific ligand interaction (Dodd 1985).

In going from Cr to Mo to W in a given type of catalyst system there is some decrease in stereoselectivity with respect to the formation of *cis*-MeCH=CHMe from *cis*-EtCH=CHMe. Thus the *cis* content of the but-2-ene decreases as follows (Leconte 1979b, 1980):

(i) $Mt(CO)_3(mes)/EtAlCl_2/O_2$: Cr 69%, Mo 63%, W 56%;
(ii) $Mt(CO)_5(PPh_3)/EtAlCl_2/O_2$: Mo 63%, W 58%;
(iii) $MtCl_2(PPh_3)_2(NO)_2/EtAlCl_2$: Mo 83%, W 69%.

Although, on the whole, Mo-based catalysts tend to be more stereoselective than W-based catalysts, it should nevertheless be noted that some of the highest stereoselectivities are observed with W-based catalysts.

In the liquid-phase metathesis of linear internal olefins RCH=CHR′ on $Re_2O_7/CsNO_3/Al_2O_3$, the *trans* content of the products increases with the size of R or R′. The *trans* content in the dec-5-ene produced in the metathesis of BuCH=CHR decreases from 71% to 67% to 64% as R is changed from H to Me to Et (Kawai 1988).

6.7.3 General consideration of the mechanism

In the metathesis of an unsymmetrical olefin there are two propagating species, which for pent-2-ene may be represented as [Mt]=CHMe and [Mt]=CHEt. The reactant may present itself to the metal carbene in one of four ways, making eight propagation steps in all. These are depicted below for *cis*-pent-2-ene. It should be noted that the two processes, (12) and (16), leading to *cis* → *trans* 'isomerization' involve an exchange of alkylidene fragments; they are not isomerizations in the sense that there is a rearrangement of atoms within a single molecule.

trans-Pent-2-ene can be formed in both reactions (12) and (16), whereas *trans*-

Table 6.2 Initial percentage *trans* content of the products RCH=CHR of the metathesis of terminal olefins $RCH=CH_2$ at 25°C

R	Catalyst system[a]	% *trans*	Ref.[b]	R	Catalyst system[a]	% *trans*	Ref.[b]
Me	A	52 ± 2^c	(1,2)	$Me(CH_2)_3$	I	71	(7)
	B	42 ± 1	(3)		J	73	(8)
	C	33	(4)	$Me(CH_2)_4$	I	71	(7)
	D	44	(5)	$Me(CH_2)_5$	I	72	(9)
	E	49	(5)	$Me(CH_2)_7$	K	83	(8)
	F	53	(5)	Me_2CHCH_2	I	71	(9)
	G	69^d	(6)	$Me_2CH(CH_2)_2$	I	72	(9)
	H	83^d	(6)	$Et(Me)CHCH_2$	I	71	(9)
Et	A	59 ± 1^e	(1)	Pr(Me)CH	I	46	(9)
$Me(CH_2)_2$	A	67 ± 2^f	(1)	Me_3CCH_2	I	71	(9)
	I	67	(7)	$EtMe_2CCH_2$	I	71	(9)

[a] A: $MoCl_2(PPh_3)_2(NO)_2/EtAlCl_2$. B: $W(CO)_5(PPh_3)/EtAlCl_2/O_2$. C: $W(CO)_6/CCl_4/h\nu$ (pre-irradiated). D: $Re_2O_7/\gamma\text{-}Al_2O_3$ (independent of wt fraction of active phase). E: WO_3/SiO_2. F: MoO_3/SiO_2. G: $MoO_3/TiO_2/Me_4Sn$. H: $MoO_{3-x}/TiO_2/Me_4Sn$. I: $Re_2O_7/CsNO_3/Al_2O_3$. J: $Mo(=CHCMe_2Ph)(=NC_6H_3\text{-}i\text{-}Pr_2\text{-}2,6)(OCMe_3)_2/DME$. K: $Mo(=CHCMe_2Ph)(=NC_6H_3\text{-}i\text{-}Pr_2\text{-}2,6)[OCMe(CF_3)_2]_2/DME$. [b] (1) Leconte (1979b); (2) Basset (1979); (3) Leconte (1978); (4) Garnier (1980a); (5) Kapteijn (1982); (6) Tanaka, K. (1987b); (7) Kawai (1988); (8) Fox (1994a); (9) Kawai (1989). [c] Equilibrium 73%. [d] Reactant: *cis*-CDH=CHMe. [e] Equilibrium 86%. [f] Equilibrium 83%.

Table 6.3 Examples of stereoselectivities in the metathesis of *cis*- and *trans*-pent-2-ene with Cr-, Mo-, W-, and Re-based catalysts

Catalyst system[a]	*cis*-pent-2-ene: % *cis* in product[b] 2-C_4	*cis*-pent-2-ene: % *cis* in product[b] 3-C_6	*trans*-pent-2-ene: % *trans* in product[b] 2-C_4	*trans*-pent-2-ene: % *trans* in product[b] 3-C_6	Reference
$Cr(CO)_3(mes)/EtAlCl_2/O_2$	69				Leconte 1979b, 1980
$MoCl_2(NO)_2(py)/EtAlCl_2$[c]	~89	~73	~98	~100	Hughes 1969
$MoCl_2(PPh_3)_2(NO)_2/EtAlCl_2$	83	69	92.5	96	Leconte 1979b, 1980
$MoBr_2(PPh_3)_2(NO)_2/EtAlCl_2$	83	70			Leconte 1979b, 1980
$MoCl_2(NO)_2(py)/Me_3Al_2Cl_3$	81	71			Leconte 1979b, 1980
$Mo(CO)_5(PPh_3)/EtAlCl_2/O_2$	63				Leconte 1979b, 1980
$MoCl_5/Me_3Al_2Cl_3$	82	73			Taghizadeh 1982
$W(=CPh_2)(CO)_5$[d]	98[e]	96[e]	73	83	Katz 1977b
$W(CO)_6/CCl_4/h\nu$[f]	79		>95		Garnier 1980a
$WCl_2(PPh_3)_2(NO)_2/EtAlCl_2$	69	57	86	92	Leconte 1979b
WCl_6/R_4Sn (R = Me, Bu, etc.)	64	50	81	82	Bosma 1982
$WCl_6/EtAlCl_2/EtOH$[g]	62	50	54	61	Calderon 1968
$WCl_2(OC_6H_{5-n}X_n)_4/Et_3Al_2Cl_3$					
(X = H, 4-Cl, 4-Me, 4-MeO)	58	45	58	79	Dodd 1985[h]
(X = 2-Me, 2,4-Me_2, 2,4,6-Me_3)	58	45	39	60	Dodd 1985
[W]=$CHCMe_3$ (**1** in Ch. 3)	99	100	99.6	99.6	Couturier 1992
Re_2O_7/γ-Al_2O_3	72	50	77	89	Kapteijn 1982
Equilibrium values	27	14	73	86	See Table 6.1

[a] Solvent: chlorobenzene at 25°C except where indicated. [b] Initial values. [c] 0°C. [d] In octane. [e] Corrected for 4% *trans* isomer in the initial pent-2-ene. [f] Pre-irradiated. [g] In benzene. [h] Also see Quignard 1985a, 1986.

but-2-ene is formed only by reaction (11). Other things being equal, one might therefore expect the rate of formation of *trans*-pent-2-ene to be double that of formation of *trans*-but-2-ene. In fact the ratio is 2.6 ± 0.4 for a number of catalyst systems (Bilhou 1977b). For the metathesis of 4-methyl-*cis*-pent-2-ene the ratio is very much larger (Fig. 6.3).

As the metathesis reaction proceeds it is necessary to take account of the secondary reactions of the products in order to understand the evolution of the composition of the reaction mixture with time. There will then be another eight reactions involving *trans*-pent-2-ene, and four each for *cis*- and *trans*-but-2-ene and *cis*- and *trans*-hex-3-ene, making 32 propagation reactions in all. By assuming that the concentrations of [Mt]=CHMe and [Mt]=CHEt are the same, and by equating various ratios of rate constants, it is possible to show that a linear relationship of the type shown in Fig. 6.2 is to be expected (Leconte 1979b).

6.7.4 Interpretation of the stereoselectivities

Most of the observations on stereoselectivities in these reactions can be rationalized by assuming that the metallacycle transition state is slightly puckered so as to minimize its energy (Leconte 1979b; Tinland 1983). Puckering may occur in two ways and may place the substituents in either axial or equatorial positions. The four

Table 6.4 Stereoselectivities in the metathesis of some alk-2-enes RCH=CHMe other than pent-2-ene

		cis-RCH=CHMe		*trans*-RCH=CHMe		
		% *cis* in product[b]		% *trans* in product[b]		
R	Catalyst[a]	2-C_4	RCH=CHR	2-C_4	RCH=CHR	Reference
Propyl	A	82		92		Leconte 1980
	B	56		56		Leconte 1980
	D	77		>95		Garnier 1980a
	F		38			Kawai 1988
i-Propyl	A	70				Leconte 1980
	B	50		54		Leconte 1980
	C′	68	<1	>92	>99	Ofstead 1980
	D	77		>95		Garnier 1980a
Butyl	A			93		Leconte 1980
	B			56		Leconte 1980
	D	75		>95		Garnier 1980a
	F		33			Kawai 1988
i-Butyl	D	79		>95		Garnier 1980a
t-Butyl	A	69		89		Leconte 1980
	B	50		52		Leconte 1980
Pentyl	A			92		Leconte 1980
	B			56		Leconte 1980
	C″		44[c]		59[c]	Uchida, A. 1981
	E		0[c]		100[c]	Uchida, A. 1981
	F		31			Kawai 1988
Hexyl	F		30			Kawai 1988
Heptyl	A′	83		68		Grubbs 1979
	C	62				Grubbs 1979
Phenyl	A			89		Leconte 1980
Benzyl	A			89		Leconte 1980
	B			55		Leconte 1980
cis-Hept-5-enyl	A′	80				Grubbs 1979
	C	62				Grubbs 1979
trans-Hept-5-enyl	A′			67		Grubbs 1979
	C			67		Grubbs 1979

[a] A: $MoCl_2(PPh_3)_2(NO)_2/EtAlCl_2/O_2$. A′: $MoCl_2(PPh_3)_2(NO)_2/Me_3Al_2Cl_3/O_2$. B: $W(CO)_5(PPh_3)/EtAlCl_2/O_2$. C: WCl_6/Me_4Sn. C′: $WCl_6/Bu_4Sn/Et_2O$. C″: WCl_6/Ph_4Sn. D: $W(CO)_6/CCl_4/h\nu$ (pre-irradiated). E: $W(OPh)_6/EtAlCl_2$. F: $Re_2O_7/CsNO_3/Al_2O_3$. [b] Initial values. [c] As observed during cross-metathesis with oct-1-ene.

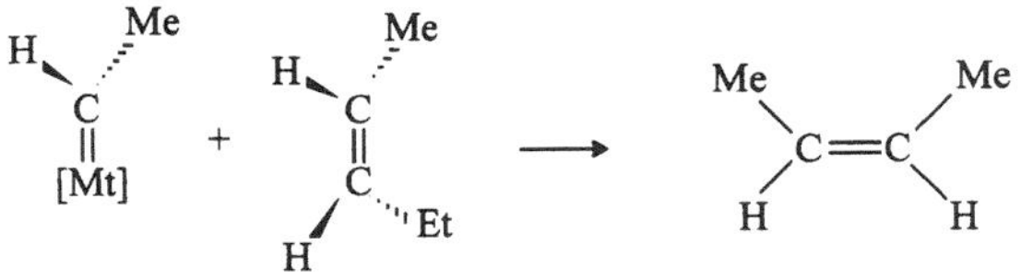

productive metathesis, *cis* product (10)

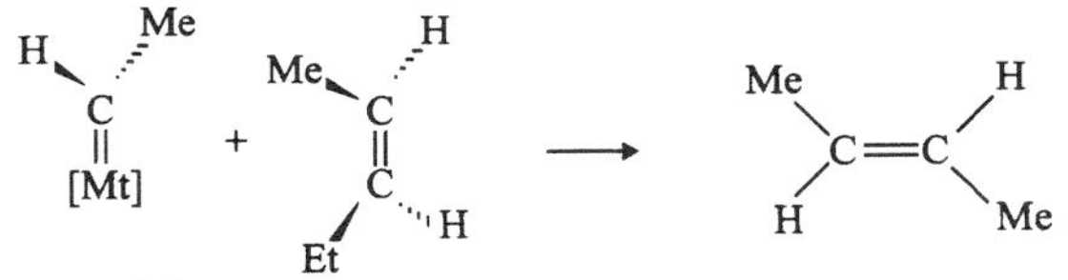

productive metathesis, *trans* product (11)

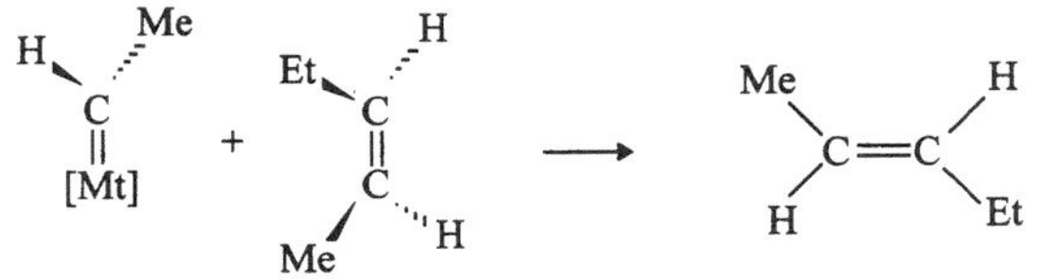

isomerization metathesis, *cis* → *trans* (12)

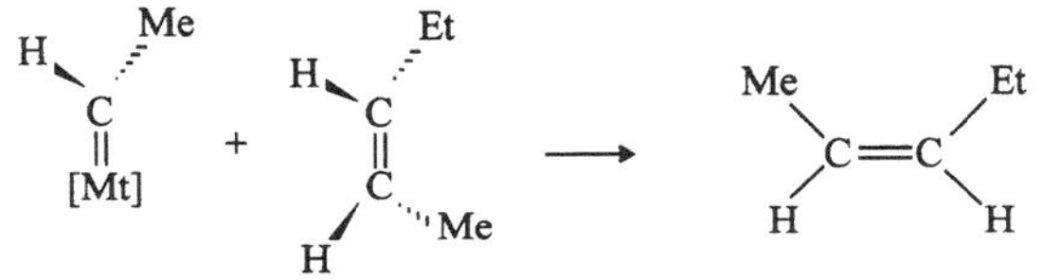

degenerate metathesis, *cis* → *cis* (13)

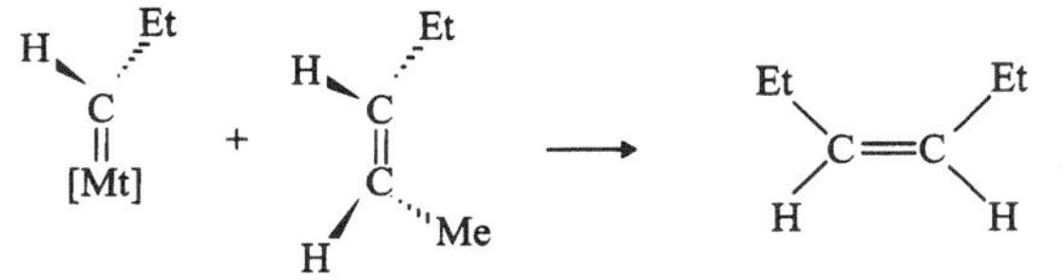

productive metathesis, *cis* product (14)

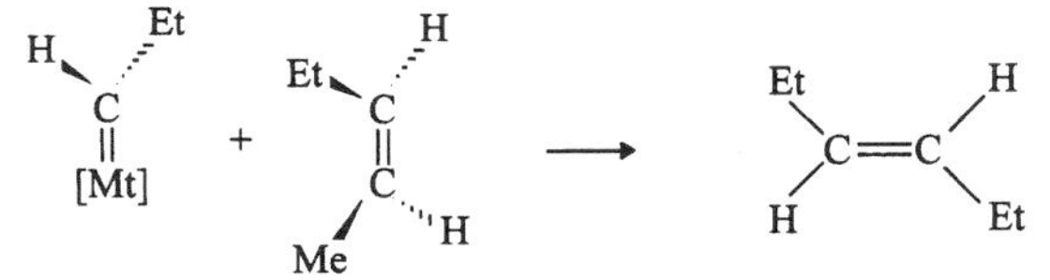

productive metathesis, *trans* product (15)

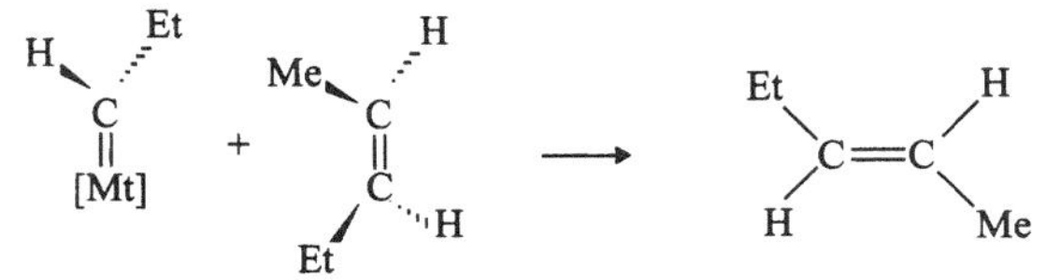

isomerization metathesis, *cis* → *trans* (16)

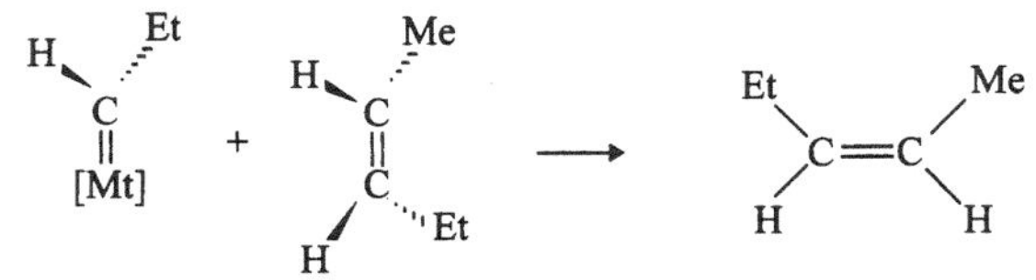

degenerate metathesis, *cis* → *cis* (17)

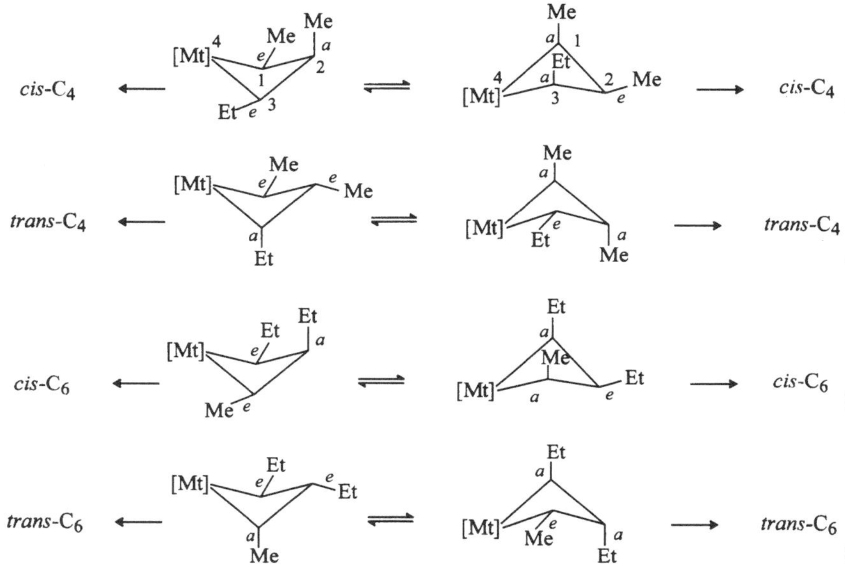

(10′)

(11′)

(14′)

(15′)

Scheme 6.1 Possible structures of puckered metallacycle intermediates in the productive metathesis reactions of *cis*-pent-02-ene.

possible pairs of structures corresponding to reactions of *cis*-pent-2-ene, (10′), (11′), (14′), (15′), and leading to *cis* or *trans* products, are shown in Scheme 6.1. In each case C^1 derives from the carbene carbon of the reacting metal carbene, while C^2 and C^3 are derived from the olefinic carbons of the reactant; the transition metal is numbered 4 in the cycle. The product olefinic carbons are C^1 and C^2. It should be noted that *ea* or *ae* substituents in the C^1,C^2 positions lead to *cis* products, while *ee* or *aa* lead to *trans* products. The corresponding pairs of structures for the reactions of *trans*-pent-2-ene leading to productive metathesis are shown in Scheme 6.2.

We may now proceed to the interpretation of the various *t/c* ratios for the products of these and related metathesis reactions. In doing so we have to consider the likely relative magnitudes of the various possible interactions, in particular those between the 1,2 substituents and those between the 1,3 substituents, and also the 2,4 and 3,4 interactions in so far as the ligands on Mt may exert an influence, as witnessed by the variation of *t/c* ratios with catalyst system. For 1,2 interactions, *ee* and *aa* will be of lower energy than *ea* or *ae*, but for 1,3 interaction the energies will decrease in the order *aa* > *ae* > *ee* for a given pair of substituents.

First consider the reactions of terminal olefins, with H instead of Et in Scheme 6.1, assuming negligible 1,3 interactions. The progressive increase in *t/c* ratio as the substituents in the 1,2 positions are changed from Me to Et to Pr (Table 6.2, catalyst system **A**) shows that 1,2-*ee* or -*aa* structures are increasingly favoured relative to *ae* or *ea*.

Now consider the case of RCH=CHMe, assuming that the 1,3 interaction is

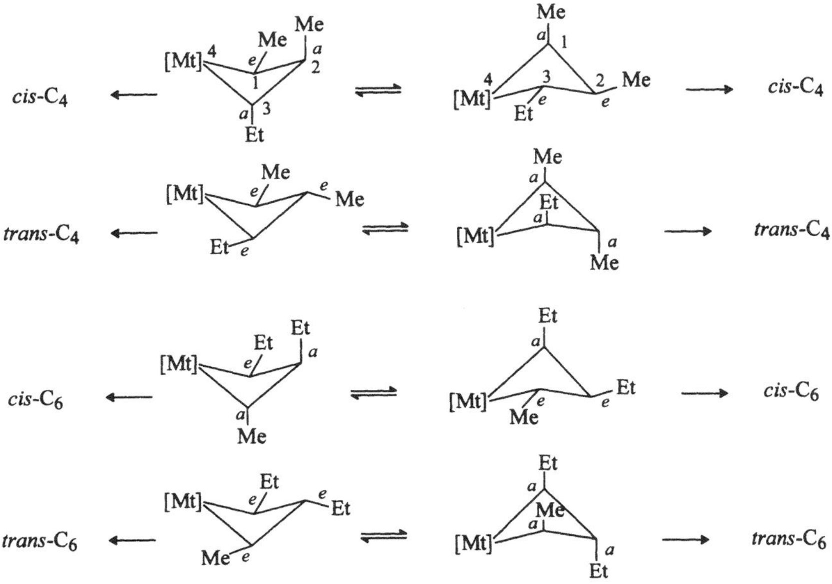

(18′)
(19′)
(20′)
(21′)

Scheme 6.2 Possible structures of puckered metallacycle intermediates in the productive metathesis reactions of *trans*-pent-2-ene.

dominant. In this event the *cis* olefin can react via the 1,3-*ee* structure to give *cis*-but-2-ene, whereas to form *trans*-but-2-ene it must react via the higher energy *ea* or *ae* structure. On the other hand the *trans* olefin can give *trans*-but-2-ene via the 1,3-*ee* structure whereas to form *cis*-but-2-ene it must react via the higher energy *ea* or *ae* structure. The general tendency for the configuration of the reactant to persist in the product but-2-ene can thus be explained. However, with some catalysts and substrates the *t*/*c* ratio is close to unity and this effect can evidently be nullified by other more dominant effects such as interaction of the bulky R group with the ligands.

Extremely high stereoselectivity in the metathesis of *cis*-pent-2-ene catalyzed by the [W]=$CHCMe_3$ complex (see Table 6.3) is maintained up to near-equilibrium conversion (Fig. 6.5). This can be explained either in terms of additional stabilization of the 1,3-*ee* structure of the tungstacyclobutane intermediate by the bulky ligands on tungsten, or by the rigidity generated by the cyclometallated ligand, or both (Couturier 1992).

We have already seen that 1,2 interactions are significant in the metathesis of the higher terminal alkenes. The fact that *t*/*c* ratios are always higher for hex-3-ene than for but-2-ene in the products of metathesis of pent-2-enes (Table 6.3) is also evidence for 1,2 interactions. Thus by examining Scheme 6.1 it will be seen that for the transition states leading to the formation of *cis*-C_4 and *cis*-C_6, the 1,3 interactions are the same (methyl–ethyl, *ee* or *aa*); this is also the case for the transition states leading to the formation of *trans*-C_4 and *trans*-C_6 (methyl–ethyl, *ea*

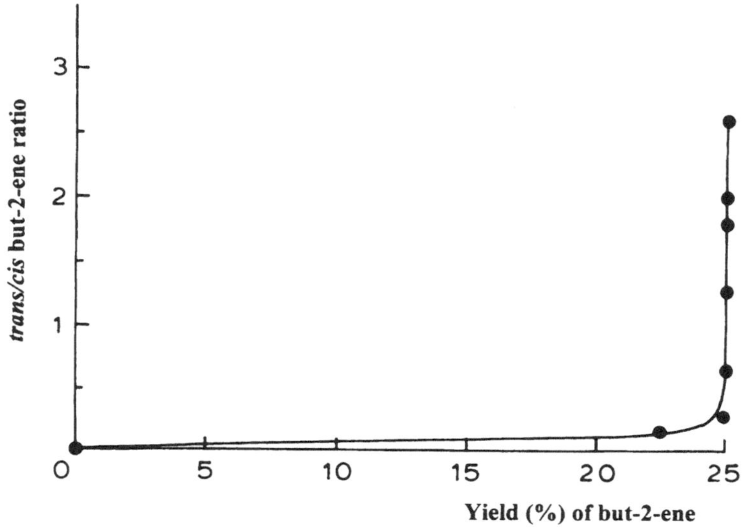

Fig. 6.5 Metathesis of *cis*-pent-2-ene catalyzed by [W]=$CHCMe_3$ (**1** in Ch. 3). Plot of the *trans/cis* ratio of the product but-2-ene versus the yield of but-2-ene. *cis*-Pent-2-ene/W molar ratio = 500/1; in chlorobenzene at 25°C (Couturier 1992).

or *ae*). The larger *t/c* ratios for the product hex-3-ene must therefore be attributed to the influence of the substituent at position 2, with the ethyl group exercising a stronger influence than the methyl group. Where this influence exists it must favour the 1,2-*aa* and -*ee* structures relative to *ae* or *ea*, so the *trans* bias in the hex-3-ene, relative to but-2-ene, can be explained. This effect becomes completely dominant in the metathesis of 4-methylpent-2-ene [*i*-Pr in place of Et in (14′), (15′), (20′), (21′)], the product 2,5-dimethylhex-3-ene being entirely *trans* (Ofstead 1980).

The generally greater stereoselectivity observed with molybdenum complexes compared with corresponding tungsten complexes may be due to a shorter Mt–C distance in the metallacycle, thereby increasing the various possible interactions and so making for greater discrimination between the competing propagation reactions.

6.8 1,1-Disubstituted olefins

It may be calculated from thermodynamic data that the metathesis of isobutene in a closed system at 25°C can yield only 1.4% each of ethene and 2,3-dimethylbut-2-ene; reaction (22) (Mol 1975). At 200°C the estimated equilibrium conversion is still less than 6% of each product. It is clear that to drive reaction (22) from left to right it is necessary to remove the ethene as fast as it is formed. The reaction is catalyzed by Me_4Sn-treated MoO_x/TiO_2 (x = 2.3–2.9) at 20°C (TOF =

$$CH_2{=}CMe_2 \rightleftharpoons CH_2{=}CH_2 + Me_2C{=}CMe_2 \qquad (22)$$

$3.3 \times 10^{-5}\,s^{-1}$); likewise the metathesis of 2-methylbut-1-ene (TOF = $7.2 \times 10^{-6}\,s^{-1}$) (Tanaka, K. 1990a). 2-Methylbut-1-ene also reacts in the presence of $WO(OPh)_4/AlCl_3/Et_3Al$ (or $EtAlCl_2$) evolving ethene with the formation of a *cis/trans* mixture of 3,4-dimethylhex-3-ene in 95% yield; reaction (23) (Knoche 1971). Likewise the metathesis of 2-methylpent-1-ene and 2-methylhept-1-ene can be induced by $W({=}CPh_2)(CO)_5$ (McGinnis 1976).

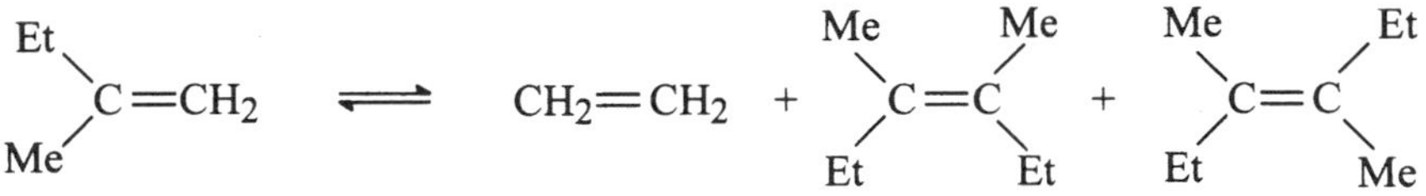

(23)

Re_2O_7/Al_2O_3 is very effective for the metathesis of methylenecyclobutane, particularly if the reaction is run at 35°C in the liquid phase under argon with continuous removal of ethene. A 70% yield of dicyclobutylidene can be obtained in 4 h; reaction (24) (Popov 1973; Strel'chik 1976). At 80°C the initial rate of

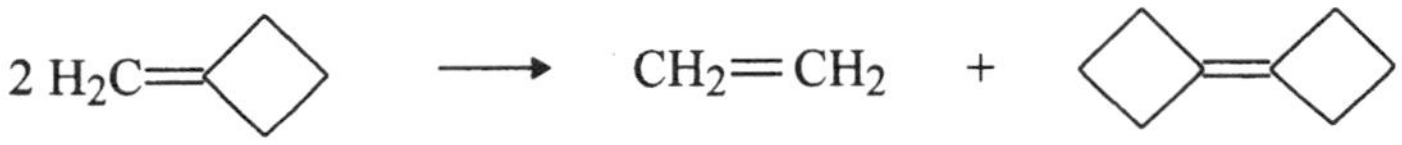

(24)

metathesis is 23 times lower than that of pent-1-ene. Under similar conditions methylenecyclopentane and methylenecyclohexane do not undergo significant metathesis, but instead isomerize to the corresponding 1-methylcycloalkene. Methylenecyclopropane also fails to undergo metathesis in the presence of various catalysts (Finkel'shtein 1992a). Degenerate exchange of the methylene group of methylenecyclohexane and isobutene is, however, catalyzed by a titanium-based catalyst (Tebbe 1978, 1979); see Section 2.2.1. Methylenecyclohexane will exchange with $W({=}CHCMe_2Ph)({=}NAr)[OCMe(CF_3)_2]_2$ to give $CH_2{=}$ $CHCMe_2Ph$ but the reaction does not proceed further (Konzelman 1995). 1,3-Dimethylenecyclobutane can be converted into a polycyclic hydrocarbon, reaction (25), in the presence of Re_2O_7/Al_2O_3 promoted by Bu_4Sn (Finkel'shtein 1992a).

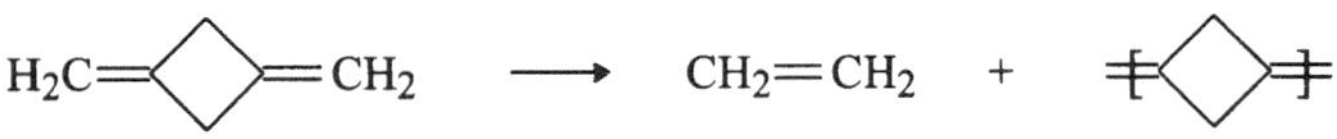

(25)

For a number of 1,1-disubstituted alkenes metathesis catalysts first bring about conversion to isomers which then undergo cross-metathesis with the remaining substrate (Strel'chik 1976; Usov 1983; Kawai 1990).

6.9 Trisubstituted ethenes

Metathesis of ethylidenecyclobutane can be performed over Re_2O_7/Al_2O_3 at 35°C giving 30% dicyclobutylidene in 4 h (Finkel'shtein 1977), and if promoted with Bu_4Sn the yield is close to 75% (Finkel'shtein 1992a). 2-Methylbut-2-ene undergoes metathesis in the presence of Mt(=$CHCMe_2Ph$)(=NC_6H_3-*i*-Pr_2-2,6)[$OCMe(CF_3)_2$]$_2$ (Mt = Mo or W), eqn. (26), reaching equilibrium conversion

$$MeCH{=}CMe_2 \rightleftharpoons MeCH{=}CHMe + Me_2C{=}CMe_2 \qquad (26)$$

(16 ± 1%) in 1 week. All attempts at metathesizing trisubstituted olefins containing an alkyl group larger than methyl have failed with these catalysts (Konzelman 1995).

With the catalyst $Re_2O_7/CsNO_3/Al_2O_3$ at 20–70°C, 2-methylpent-2-ene and 2-methylhex-2-ene undergo slow isomerization to 2-methylpent-1-ene and 2-methylhex-1-ene, respectively, which then rapidly cross-metathesize with the starting olefin. The major reaction products 4-methylhept-3-ene and 5-methylnon-4-ene respectively, stem from the cross-metathesis reactions, but self-metathesis of the starting olefins also occurs. With 2-methylpent-2-ene as substrate, the selectivity for the cross-metathesis product 4-methylhept-3-ene is 75%, while that for the self-metathesis products 2,3-dimethylbut-2-ene and hex-3-ene is 9%. The conversion of 3-methylpent-2-ene is lower than that of 2-methylpent-2-ene. It isomerizes slowly to 2-ethylbut-1-ene, which then cross-metathesizes with the starting olefin to give principally 3-ethylpent-2-ene and trace amounts of 3-methyl-4-ethylhex-3-ene, see Scheme 6.3. The selectivity for cross-metathesis is 80% and for self-metathesis of 3-methylpent-2-ene 21% at 40°C (Kawai 1990).

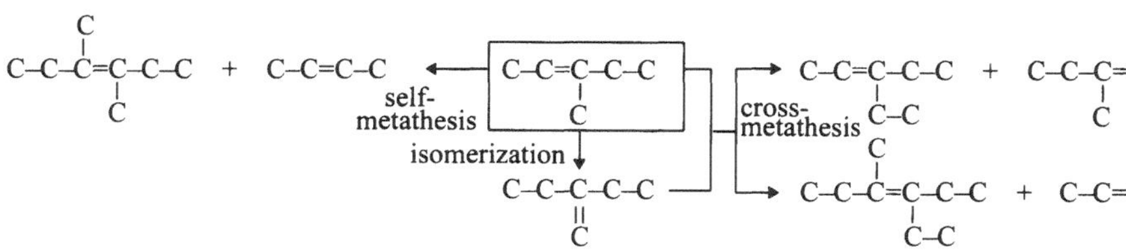

Scheme 6.3 Routes to the products of metathesis and cross-metathesis of 3-methylpent-2-ene.

7

Acyclic Functionalized Alkenes

7.1 Introduction

At one time it was thought it would be difficult to extend the olefin metathesis reaction to compounds containing functional groups. The classical catalyst system $WCl_6/EtAlCl_2/EtOH$ rapidly loses its activity in the presence of an excess of such compounds. A major step forward was the discovery that the metathesis of unsaturated esters could be brought about by WCl_6/Me_4Sn at 70°C (van Dam 1972). Later it was found possible to induce metathesis of unsaturated ethers, amines, nitriles, chlorides, etc. The occurrence of reactions, such as (1) (Cardin 1972a) and (2) (Fischer, E.O. 1972), shows that there is no inherent conflict

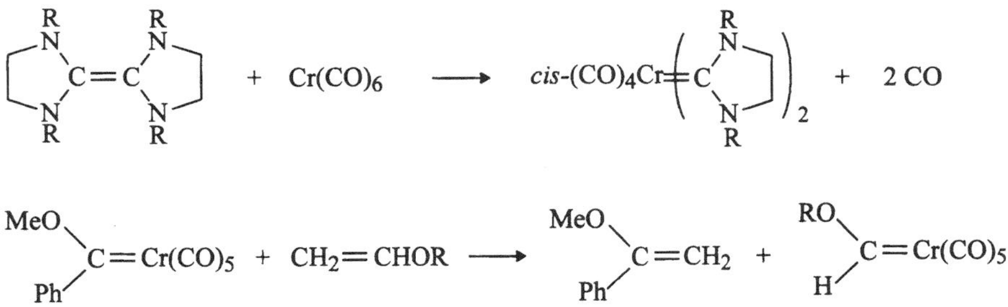

(1), (2)

between functional groups and the metal carbene bond. The reaction works best if the functional group is not too close to the double bond. There are indications that when the functional group is directly attached to the double bond, failure to undergo self-metathesis may be due to a strong propensity for degenerate metathesis, no doubt enhanced by the greater polarity of the double bonds. This area has been reviewed by Mol (1990a, 1991).

In the metathesis of acyclic functionalized olefins, catalysts are not nearly as active as they are for normal olefins. Much higher catalyst/substrate ratios must therefore be used than are required for normal olefins. The functional group may interfere: (i) by destruction of the catalyst; (ii) by interaction with the cocatalyst, if present; and (iii) by competition with the C=C bond of the olefin for complexation with the metal centre. Impurities in the substrate(s) may also be a reason for catalyst deactivation. All compounds containing those polar groups which totally deactivate the catalyst (e.g. alcohols, free acids, peroxides) must therefore be removed. When using heterogeneous catalysts adsorbed reaction products can block the active sites or even the pores of the catalyst; see also Mol (1994b).

7.2 Esters

Since the original discovery of the metathesis of methyl oleate in 1972, eqn. (3),

$$2\,Me(CH_2)_7CH{=}CH(CH_2)_7COOMe \rightleftharpoons \left[Me(CH_2)_7CH{=}\right]_2 + \left[MeOOC(CH_2)_7CH{=}\right]_2 \qquad (3)$$

unsaturated esters have been extensively tested for ability to undergo the reaction, as indicated in Table 7.1. Effective catalyst systems for such reactions are listed in Table 7.2. Interest in unsaturated esters was stimulated by the availability of natural unsaturated esters, e.g. in sunflower oil, olive oil, linseed oil, soybean oil, and tallow, which, through the metathesis reaction, may be converted into intermediates for the production of useful perfumes, polymers, etc. For example, the unsaturated diester from the reaction of ethyl oleate can be converted into civetone (**22**), an important constituent of many perfumes (Verkuijlen 1976; Tsuji 1980, 1981).

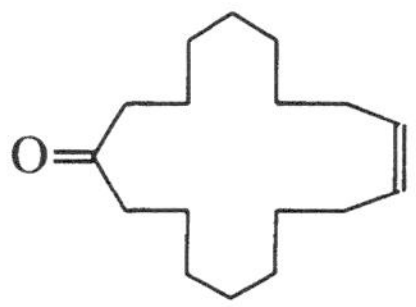

(**22**) mixture of *cis* and *trans*

Many of the metathesis reactions listed in Table 7.1 proceed best without the use of a solvent, and in some cases the ester group actually appears to have a beneficial effect on selectivity (Verkuijlen 1977a). A related observation is that saturated esters enhance the selectivity in the metathesis of alk-1-enes catalyzed by WCl_6/R_4Sn (Ichikawa, K. 1976a). Unreactive unsaturated esters such as **6** do not interfere with the reaction of other compounds such as **7**. The non-reactivity of methyl methacrylate (**4**) has been ascribed to degenerate metathesis, rather than poisoning of the catalyst, although this has not been checked by isotopic labelling. Its ready cross-metathesis with hex-3-ene, reaction (4), is consistent with such an

$$EtCH{=}CHEt + CH_2{=}C(Me)COOMe \xrightarrow{WCl_6/Me_4Sn} EtCH{=}CH_2 + EtCH{=}C(Me)COOMe \qquad (4)$$

Table 7.1 Unsaturated esters that have been tested for the metathesis reaction[a,b]

	Ester[c]	Notes	Reference[d]
1	$C{=}CCOOMe$	No reaction	(1–3)
2	$C_2{=}CCOOMe$	No reaction	(1–3)
3	$C{=}C_2COOR$	Good yield; high selectivity	(1–4)
4	$C{=}C(C)COOMe$	No self-metathesis, but cross-metathesizes with hex-3-ene	(2,3)
5	$C_2{=}C(C)COOMe$	As for **4**	(3)
6	$Me_2C{=}CCOOMe$	No reaction; does not cross-metathesize with **7**	(2,3)
7	$C_2{=}C_2COOMe$	Good yield of mainly *trans* diester; reaction not retarded by **6**	(2)
8	$C{=}C_3COOR$	Good yield; selectivity > 99% using $Re_2O_7/Al_2O_3/Me_4Sn$ (ref. 5); mainly *trans* diester formed	(2–9)
9	$C{=}C_7COOMe$	Good yield; 93% selectivity	(7)
10	$C{=}C_9COOMe$	High yield with 100% selectivity (ref. 10)	(4,6–11)
11	*cis*-$C_9{=}C_8COOR$	Good yield; high selectivity; equilibrium readily established; mainly *trans* isomer in both products	(2–4,6,10–13,18,19)
12	*trans*-$C_9{=}C_8COOMe$	As for **11**	(10,12)
13	*cis,cis*-$C_6{=}C_3{=}C_8COOMe$	Products contain $C_6{+}C_3{+}_nC_6$, $C_6{+}C_3{+}_nC_8COOMe$, and $MeOCOC_8{+}C_3{+}_nC_8COOMe$, with $n = 0$–4 in each case; also cyclohexadiene and higher cyclic polymers (see Table 7.3)	(1,10,11,14,15,20)
14	*cis,cis,cis*-$C_3{=}C_3{=}C_3{=}C_8COOMe$	Similar behaviour to **13**	(10,11,14)
15	*cis*-$C_9{=}C_{12}COOMe$	Good conversion; 100% selectivity	(8–11)
16	$C{=}COCOMe$	No self-metathesis, but cross-metathesizes with hex-3-ene	(3)
17	$C{=}C_2OCOMe$	17% conversion; 96% selectivity	(17)
18	$C{=}C_3OCOMe$	44–60% conversion; 95% selectivity	(3,21)
19	$C{=}C_9OCOMe$	41% conversion; 88% selectivity	(21)
20	*cis*-$C_9{=}C_9OCOMe$	Good conversion	(4,6)
21	$C_9{=}C_6$ | | O–CO	Polymerizes to unsaturated polyester, MW 95 000	(16)

[a] Olive oil (75% glyceryl trioleate), linseed oil, and soybean oil (glyceryl esters of oleic, linoleic, and linolenic acids) also undergo the reaction to yield substantial quantities of octadec-9-ene and cyclohexa-1,4-diene as well as higher polyesters. [b] For typical catalysts see Table 7.2. [c] $C{=}CCOOMe$ denotes $CH_2{=}CHCOOMe$, i.e. methyl acrylate; *cis*-$C_9{=}C_8COOMe$ denotes *cis*-$CH_3(CH_2)_7CH{=}CH(CH_2)_7COOMe$, i.e. methyl oleate; **21** is ambrettolide. [d] (1) Verkuijlen (1976); (2) Verkuijlen (1977a); (3) Bosma (1983); (4) Nakamura, R. (1976a); (5) Verkuijlen (1977b); (6) Nakamura, R. (1976b); (7) Baker (1977); (8) Ellison (1983); (9) Ellison (1985); (10) van Dam (1974b); (11) van Dam (1974c); (12) van Dam (1972); (13) Ichikawa, K. (1976b); (14) Verkuijlen (1974); (15) Ast (1976a); (16) Ast (1976b); (17) Mol (1979); (18) Tsuji (1980, 1981); (19) van Roosmalen (1980b); (20) Nishiguchi (1980); (21) Levisalles (1980b).

explanation (Verkuijlen 1977a). The failure of **1**, **2**, and **6** to undergo self-metathesis is probably likewise due to the relatively high polarity of the double bond, favouring the degenerate reaction. However, attempted cross-metathesis between **4** and methyl crotonate (**2**), eqn. (5), was unsuccessful. Therefore, the non-

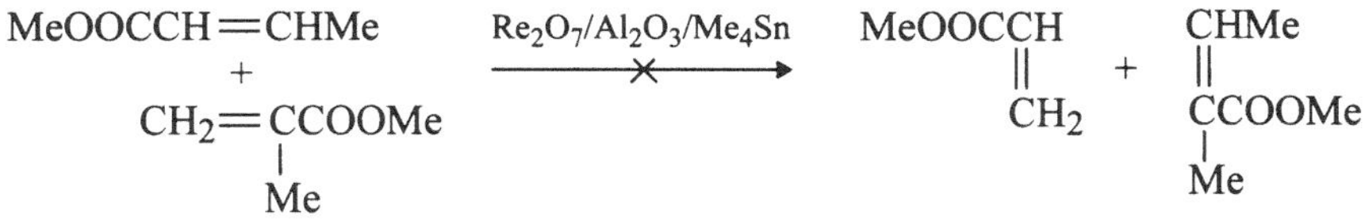

(5)

reactivity of α,β-unsaturated esters might instead be due to the difficulty of forming the propagating metal carbene species, or to the deactivating effect of an ester group as a result of its close proximity to the metal (Bosma 1983). The presence of one or more CH_2 groups between the double bond and the functional group is sufficient to ensure that productive metathesis takes place. No evidence for stereoselectivity has been found in these reactions, probably because the conditions have generally been such as to favour the establishment of *trans/cis* equilibrium, in which *trans* is favoured (Brégeault 1988).

Two types of side reaction have been noted. The catalyst system WCl_6/Me_4Sn generates a certain amount of HCl, which adds to terminal double bonds to the extent of about 10% of the metathesis reaction (Baker 1977). The catalyst system WCl_6/Et_3B causes some double-bond shift in the substrate prior to metathesis; with methyl undec-10-enoate (**10**) a series of products of the type MeOCO $(CH_2)_8CH{=}CH(CH_2)_nCOOMe$ are formed, the yield decreasing as n decreases from 8 (the direct metathesis product) to 7, 6, and 5 (Nakamura, R. 1976a).

When WCl_6/Me_4Sn is used as catalyst system for the metathesis of unsaturated esters some enhancement of reactivity is observed if WCl_6 is placed in contact with the ester before the addition of Me_4Sn (Balcar 1984, 1985). Spectroscopic evidence points to the formation of a $WOCl_4$–ester complex via a WCl_6–ester complex, accompanied by a partial reduction of WCl_6 to WCl_5 (Balcar 1986a,b), as previously suggested by van Roosmalen (1980b). Commercial unsaturated esters often contain free carboxylic acid. The effect of such acids on the metathesis of unsaturated esters by WCl_6/Me_4Sn has been examined by Balcar (1987).

When $WOCl_4/Me_4Sn$ is used as a catalyst system and the ester is added before Me_4Sn, an induction period is observed. However, if $WOCl_4$ is allowed to react with Me_4Sn before the ester addition there is no induction period (Balcar 1989).

Heterogeneous catalysts for the metathesis of acyclic unsaturated esters are mainly based on rhenium. The rhenium oxide catalysts (Table 7.2) not only have the advantage that they are active at room temperature, but they are also highly selective, and can be easily recovered (Mol 1991). MoO_3/SiO_2, that has been photoreduced in a CO atmosphere and subsequently treated with cyclopropane, has an even higher activity (Berezin 1991, Mol 1994b). Several well-defined metal

Table 7.2 Catalyst systems suitable for inducing the metathesis of esters of acyclic unsaturated acids (M)[a]

Catalyst system	Solvent[b] temp. (°C)	Olefin[c]	Notes	Reference
Homogeneous				
WCl_6/Me_4Sn	C/110	MeOl	W/Sn/M = 1/1/75	van Dam 1972, 1974b
WCl_6/Et_3B	C/32	MeOl	W/B/M = 1/3/20	Nakamura, R. 1976a
$WCl_6/Me_3Al_2Cl_3$	C/28	MeOl	W/Al/M = 1/2/30	Nakamura, R. 1976b
WCl_6/Ph_2SiH_2, $WOCl_4/Ph_2SiH_2$	–/75	MeUn	W/Si/M = 1/2/50	Levisalles 1984
WCl_6/1,1,3,3-tetramethyl-1,3-disilacyclobutane (T)	B/80	EtPe	W/T/M = 1/2/50	Vdovin 1986
	B/70	EtPe	W/T/M = 1/2/150	Bespalova 1990
$WOCl_4/Me_4Sn$[d]	–/70	MeOl	W/Sn/M = 1/1/50	Verkuijlen 1980
$WOCl_4/Cp_2TiMe_2$	B/70	EtOl	W/Ti/M = 1/1.2/10	Tsuji 1981
$WOCl_4$/1,1,3,3-tetramethyl-1,3-disilacyclobutane (T)	B/70	EtPe	W/T/M = 1/2/150	Bespalova 1992
$WCl_4(OC_6H_3\text{-}X_2\text{-}2,6)_2/R_4Mt$ (X = Cl,Br; Mt = Sn,Pb; R = Me,Bu)	C/85	EtOl	W/Mt/M = 1/2/50	Quignard 1986
trans-$WOCl_2(OC_6H_3Br_2\text{-}2,6)/Et_4Pb$	–/130	MePe-3	W/M = 1/50	Nugent 1995
$W[{=}C(OMe)Ph](CO)_5/SnCl_4$	H/55	MeUn	W/Sn = 1/9	Banks 1982
$W({=}CHCMe_3)(CH_2CMe_3)(Cl)(OC_6H_3Ph_2\text{-}2,6)_2(OEt_2)$	C/85	EtOl	W/M = 1/100	Quignard 1985b
$W({=}CHCMe_3)({=}NC_6H_3\text{-}i\text{-}Pr_2\text{-}2,6)[OCMe(CF_3)_2]_2$	P/25	MeOl	W/M = 1/300	Schaverien 1986; Schrock 1987a
$[W]{=}CHCMe_3$ (see **1** in Ch. 3)	C/25	EtOl	W/M = 1/500	Couturier 1992
$MoCl_3(OEt_2)/Et_3B$	C/32	MeOl	Mo/B/M = 1/1.3/10	Nakamura, R. 1976a
$Mo(OR_2)(NO)_2/EtAlCl_2$ (R = Et,*i*-Pr)	C/20	EtBu	Mo/Al/M = 1/6/100	Keller 1991b
$Mo({=}CHCMe_2Ph)({=}NC_6H_3\text{-}i\text{-}Pr_2\text{-}2,6)[OCMe(CF_3)_2]_2$	DME/20	MePe	Mo/M = 1/1000	Fox 1994a
$[Mo]{=}CHCMe_2Ph$[e]	B/25	MeOl	Mo/M = 1/400	Feher 1994
$Re({\equiv}CCMe_3)({=}CHCMe_3)[OCMe(CF_3)_2]_2$	D/20	MeOl	Re/M = 1/100	Toreki 1993
$Ru({=}CHCH{=}CPh_2)(Cl)_2(PCy_3)_2$	DB/20	MeOl	Ru/M = 1/550	Nguyen 1996
$[(C_8H_{14})_2IrCl]_2/AgO_2CCF_3$	T/25	MeOl	Ir/Ag/M = 1/6/10	France 1994

Heterogeneous				
$WCl_6/SiO_2/Me_4Sn$[f]	B/72	MeOl	W/Sn/M = 1/1/25	van Roosmalen 1980b
$WCl_6/SiO_2/Me_4Sn$[g]	DE/80	MePe	W/Sn/M = 1/3/25	Cameron 1983
$MoCl_5/SiO_2/Me_4Sn$	–/90	EtOl + dec-5-ene	Mo/Sn/M(total) = 1/4/110	Bykov 1988
$MoCl_5/SiO_2/Me_4Sn/Si(or\ Ge)Cl_4$	–/80	Fatty oils + alkene	Mo/Sn/Si/oil/alkene = 1/40/40/5/200	Bykov 1991
$MoO_3/SiO_2/CO(h\nu)$/cyclopropane	Hp/50	EtOl	Mo/M = 1/250	Berezin 1991
$Re_2O_7/Al_2O_3/Me_4Sn$	CT/50	MePe	Re/Sn/M = 1/0.04/9	Verkuijlen 1977b
$Re_2O_7/Al_2O_3/R_4Sn$ (R = Me, Et, Bu, Oct, etc.)	H/20	MeOl	Re/Sn/M = 1/0.1/8	Bosma 1983
$Re_2O_7/Al_2O_3/Me_4Pb$	Hp/20	MeUn	Re/Pb/M = 1/0.05/46	Chevalier-Seite 1983
$Re_2O_7/Al_2O_3/Me_3Al_2Cl_3$	H/20	MeOl	Re/Al/M = 1/0.2/5	Bosma 1983
$Re_2O_7/V_2O_5/Al_2O_3/R_4Mt$ (R = Et, Bu; Mt = Sn, Pb)	H/20	MeOl	V/Re/Sn/M = 10/1/0.6/60	Xu, X. 1985b; Mol 1986
$Re_2O_7/MoO_3/Al_2O_3/R_4Mt$ (R = Et, Bu; Mt = Sn,Pb)	H/20	MeOl	Mo/Re/Sn/M = 6/1/0.6/60	Xu, X. 1985b; Mol 1986
$Re_2O_7/WO_3/Al_2O_3/R_4Mt$ (R = Et, Bu; Mt = Sn,Pb)	H/20	MeOl	W/Re/Sn/M = 8/1/0.6/60	Xu, X. 1985b; Mol 1986
$Re_2O_7/AlPO_4/Al_2O_3/Me_4Sn$	Hp	MePe	Flow system	Lawrenson 1984
$Re_2O_7/AlPO_4/Al_2O_3/Et_4Sn$	H/20	MeOl	Re/M = 1/120	Mol 1991
Re_2O_7/Al_2O_3-B_2O_3/Et_4Sn	H/20	MeOl	Re/Sn/M = 1/0.6/60	Xu, X. 1986b
Re_2O_7/SiO_2-Al_2O_3/R_4Mt (R = Me, Et, Bu; Mt = Sn,Pb)	H/20	MeOl	Re/Sn/M = 1/0.6/120	Xu, X. 1985c, 1986a; Sibeijn 1991a
$Re_2O_7/B_2O_3/SiO_2$-Al_2O_3/Bu_4Sn	P/50	MeUn	Re/Sn/M = 1/0.8/350	Warwel 1992a
$Re_2O_7/B_2O_3/SiO_2$-$Al_2O_3/(AlMeO)_n$ ($n \approx 15$)		MeOl + oct-4-ene	Re/Al/M/O = 1/2/125/250	Warwel 1991
CH_3ReO_3/SiO_2-Al_2O_3	D/25	MeOl	Re/M = 1/50	Herrmann 1991
$CH_3ReO_3/B_2O_3/SiO_2$-Al_2O_3	Do/50	MeUn	Re/M = 1/500	Warwel 1992b

[a] For the reactions of cyclic monomers containing ester groups, see Ch. 13. [b] B = benzene; C = chlorobenzene; CT = carbon tetrachloride; D = dichloromethane; DB = dichlorobenzene; DE = dichloroethane; DME = 1,2-dimethoxyethane; Do = dodecane; H = hexane; Hp = Heptane; P = pentane; T = toluene. [c] MeOl = methyl oleate; EtOl = ethyl oleate; MeUn = methyl undec-10-enoate; EtPe = ethyl pent-4-enoate; MePe = methyl pent-4-enoate; MePe-3 = methyl pent-3-enoate; EtBu = ethyl but-3-enoate. [d] The presence of Me_3SnCl, formed from $WOCl_4/Me_4Sn$, is essential to catalytic activity. [e] A silesquioxane analogue of the catalyst immediately above. [f] Low activity. [g] Only active for ω-unsaturated esters.

Table 7.3 Composition of metathesis products of unsaturated C_{18} esters at equilibrium (catalyst system WCl_6/Me_4Sn) (Verkuijlen (1974, 1976)

Starting ester	Conversion (%)	Polyenes	Monoesters	Diesters	Cyclo-polyene
Methyl oleate	50	25	50	25	–
Methyl linoleate	84	23	46	23	8 (6)[a]
Methyl linolenate	95	15	30	15	40 (30)[a]

[a] Cyclohexa-1,4-diene.

carbene complexes are also effective and do not require a cocatalyst; see Table 7.2.

Polyunsaturated compounds, such as methyl linoleate (**13**) and methyl linolenate (**14**) offer some interesting possibilities. There are three main sets of products (hydrocarbons, monoesters and diesters), identified by mass spectroscopy (MS) and listed in Table 7.3 (Ast 1976a; Verkuijlen 1974, 1976). Cyclohexa-1,4-diene and traces of higher cyclopolyenes are formed. These must clearly result from secondary intramolecular metathesis reactions such as (6).

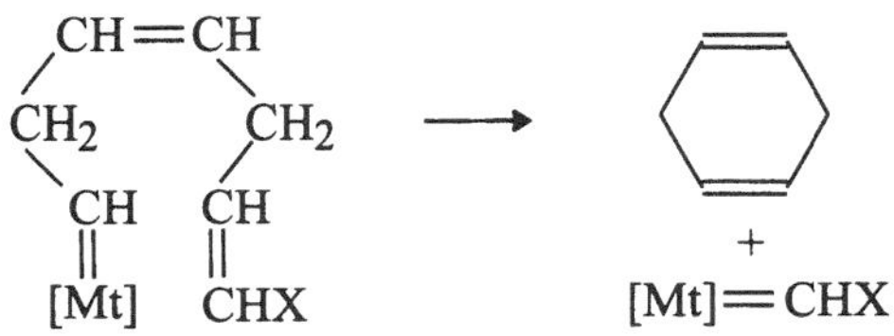

(6)

$[X = (CH_2)_4CH_3, (CH_2)_7COOMe, \text{etc.}]$

Methyl linolenate (**14**) has three C=C bonds and should be capable of forming cyclohexa-1,4-diene directly after its reaction with the propagating metal carbene at either of the outer double bonds. It is not surprising therefore that a much higher proportion (30%) of cyclohexa-1,4-diene is found in the products, together with a substantial proportion (10%) of higher cyclic polyenes (Verkuijlen 1974). Starting from linseed oil, consisting of 60% linolenate, 16% linoleate, 16% oleate, and 8% saturated esters, the yield of cyclohexa-1,4-diene can be as high as 56% when the reaction is catalyzed by WCl_6/Bu_4Sn in chlorobenzene at 80°C under optimum conditions (W/Sn/olefin = 1/4/30) (Nishiguchi 1980).

Glyceryl trioleate gives 15% octadec-9-ene when treated with WCl_6/Me_4Sn at 70°C (van Dam 1974c; Verkuijlen 1976). This may be formed by both intermolecular and intramolecular metathesis, as depicted in (7). The residue is of substantially higher viscosity and MW than the starting material, showing that most of the reaction proceeds by the intermolecular route. This reaction offers the possibility of converting drying and semi-drying oils, such as linseed oil and soy-bean oil, into viscous oils, so-called stand oils, with drying properties that are more pronounced than those of thermally polymerized oils where the polymerization

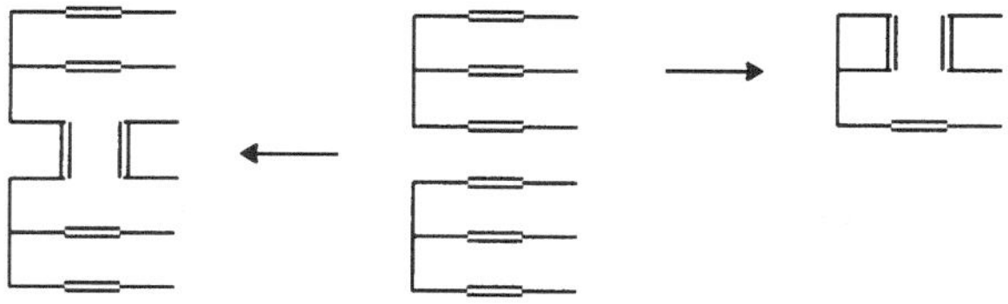

(7)

process has already reduced the number of double bonds available for cross-linking during the drying process.

The metathesis of fatty-acid esters and fatty oils has been reviewed by Boelhouwer (1985). More recent results in this field have been summarized by Mol (1994b).

7.3 Other carbonyl-containing compounds

With $WCl_6/Me_3Al_2Cl_3$ as catalyst allyl acetone, $CH_2{=}CH(CH_2)_2COMe$, gives a 10% yield of metathesis products at 60°C (Nakamura, R. 1976b). On $Re_2O_7/Al_2O_3/Me_4Sn$ the conversion of allyl acetone is 35% at room temperature with 97% selectivity (Mol 1979). Modification of a $Re_2O_7/Al_2O_3/Bu_4Sn$ catalyst by treatment of the Al_2O_3 support with trialkylborate increases its activity for the metathesis of allyl acetone 3–15-fold (Bogolepova 1992). If the carbonyl group is protected by silylation to give its silylenol ether, or by acetalization to the 1,3-dioxolane derivative, the reactivity is increased (Warwel 1989c).

The functional groups COOH, CHO, $CONH_2$, or OH deactivate most catalysts. Compounds containing OH groups can first be protected by esterification or silylation of the OH group. The protected compound will then undergo metathesis (Warwel 1987a). The first catalyst reported to bring about the self-metathesis of a COOH-containing olefin is $Ru({=}CHCH{=}CPh_2)(Cl)_2(PCy_3)_2$: 200 equivalents of oleic acid are metathesized by this catalyst at 20°C in 20 h (Nguyen 1996).

7.4 Ethers

Vinyl methyl ether and allyl methyl ether do not undergo self-metathesis in the presence of WCl_6/Me_4Sn (Ast 1977a; Levisalles 1979). However, on $Re_2O_7/Al_2O_3/Me_4Sn$ at 20°C, allyl ethyl ether in carbon tetrachloride gives an immediate evolution of ethene and, after 6 h, a 40% yield of the unsaturated diether (*trans*/*cis* = 2.6) with 98% selectivity; reaction (8) (Mol 1979). With WCl_6/Me_4Sn as catalyst but-3-enyl butyl ether and hex-3-enyl butyl ether give moderate yields of metathesis products, while pent-4-enyl butyl ether and undec-10-enyl butyl ether give rather better yields (Ast 1977a). Thus with the homogeneous tungsten catalyst, for metathesis to occur it is essential to have more than one methylene unit between the double bond and the ether function. For example, *cis*-$RO(CH_2)_nCH{=}CH$-

$$\begin{array}{c} CH_2{=}CHCH_2OEt \\ + \\ CH_2{=}CHCH_2OEt \end{array} \xrightarrow{Re_2O_7/Al_2O_3/Me_4Sn} \begin{array}{c} CH_2 \\ \| \\ CH_2 \end{array} + \begin{array}{c} CHCH_2OEt \\ \| \\ CHCH_2OEt \end{array} \quad (8)$$

$(CH_2)_nOR$ ($R = t$-$BuMe_2Si$) does not undergo metathetical *cis–trans* isomerization in the presence of a well-defined W carbene catalyst if $n = 1$, whereas for $n = 2$ isomerization from >95% *cis* to 75% *trans* is observed (Hillmyer 1995c).

o-Allylphenyl propyl ether is converted into ethene and 1,4-bis-(*o*-propoxyphenyl)but-2-ene with high yield and ~95% selectivity on $Re_2O_7/Al_2O_3/Bu_4Sn$ at 22°C (Bogolepova 1991). Oleyl ethyl ether undergoes metathesis in the presence of $W[{=}C(OMe)Ph](CO)_5/SnCl_4$ (W/Sn = 1/18) forming 1,18-diethoxyoctadec-9-ene with a yield of 11% after 20 h (Banks 1982).

cis-Propenyl ethyl ether and vinyl butyl ether (molar ratio 1/1), undergo cross-metathesis at 40°C in the presence of $Cr({=}CPh_2)(CO)_5$ (molar ratio ether/carbene = 50/1), but only propenyl butyl ether and vinyl ethyl ether are obtained, and no diethers or alkenes, i.e. only a head-to-tail reaction takes place; reaction (9) (Thu 1985).

$$\begin{array}{c} MeCH{=}CHOEt \\ + \\ BuOCH{=}CH_2 \end{array} \xrightarrow{Cr({=}CPh_2)(CO)_5} \begin{array}{c} MeCH \\ \| \\ BuOCH \end{array} + \begin{array}{c} CHOEt \\ \| \\ CH_2 \end{array} \quad (9)$$

ω-Unsaturated glucosides bearing protecting groups at the glucose unit, viz. acetyl esters or silyl ethers, undergo metathesis in the presence of certain homogeneous aryloxotungsten catalysts, affording bolaamphiphiles in good yields; reaction (10). Thus, in the presence of $[W]{=}CHCMe_3$ (**1** in Ch. 3) and for R = acetyl, a yield of 64% is obtained at 80°C in chlorobenzene, while a yield of 92% is obtained for $R = t$-$BuMe_2Si$ at 65°C (Descotes 1994; Ramza 1996).

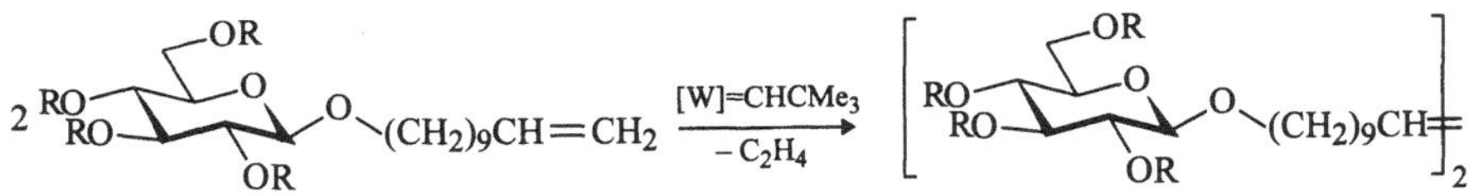

(10)

7.5 Amines

Catalysts such as $W(CO)_6/CCl_4/h\nu$, WCl_6/Me_4Sn, and $W({=}CPh_2)(CO)_5$ are without effect on unsaturated amines. Catalysts such as $W(CO)_5L$ ($L = PPh_3$, $={C(OMe)Me}$, CO) or $W(CO)_3$(arene) (arene = benzene, toluene, mesitylene) are effective at 25°C but only in the presence of a large excess of $EtAlCl_2$ and O_2 (Laval 1977; Nouguier 1977; Edwige 1980). Although O_2 is essential, the lifetime of the catalyst is inversely related to the amount of O_2 in the system, suggesting that it is also involved in termination of the chains. The combination $MtX_2(NO)_2L_2/EtAlCl_2$ (Mt = Mo, W; X = Cl, Br) is moderately active at 25°C provided that an excess of $EtAlCl_2$ is present. Such catalysts are less active but of longer life than the carbonyl catalysts (Edwige 1980).

Only unsaturated secondary and tertiary amines, and quaternary amine salts undergo metathesis. The yields of ω-unsaturated amines and amine salts are generally low, as are the turnover numbers (TONs) ($\leqslant 5$). Adverse factors are a low basicity of the amine, as in primary amines (inactive), bulky substituents on the nitrogen, as in $CH_2{=}CH(CH_2)_3N\text{-}i\text{-}Pr_2$ (inactive), and a small number of CH_2 groups between the double bond and the nitrogen. Best yields are achieved with three intervening carbons and a single alkyl substituent on the nitrogen; for example, $CH_2{=}CH(CH_2)_2CHMeNH\text{-}i\text{-}Pr$ gives 52% yield. The effects of basicity and substituents at the nitrogen, together with the necessity for an excess of $EtAlCl_2$, can be understood in terms of the position of equilibrium (11). This

$$CH_2{=}CHCH_2NRR' + \tfrac{1}{2}(EtAlCl_2)_2 \rightleftharpoons \underset{\displaystyle EtAlCl_2}{\underset{\downarrow}{CH_2{=}CHCH_2NRR'}} \qquad (11)$$

equilibrium will be driven to the right by high basicity of the amine and low steric hindrance, so it is reasonable to conclude that it is the complexed amine on the right-hand side of eqn. (11) that is the effective species undergoing metathesis. The optimum spacing of the double bond from the nitrogen suggests that intramolecular coordination of the nitrogen to the metal site may be an important factor in stabilizing the metal carbene intermediate, preventing its chain termination by O_2 or by oxygen-containing species or by the free amine itself; eqn. (12). However, the

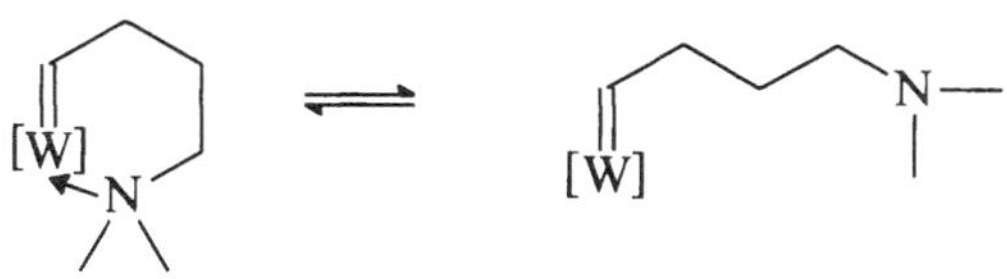

(12)

stereoselectivity for reaction of *trans*-$CH_3CH{=}CH(CH_2)_3NHMe$ is no different from that for reaction of *trans*-$CH_3CH{=}CH(CH_2)_4Me$ with similar catalysts, suggesting that at the moment of reaction the metal carbene is in its decoordinated state as on the right-hand side of eqn. (12).

We have seen that coordination of the amine with a Lewis acid ($EtAlCl_2$) increases the metathesis activity. Likewise coordination with a protonic acid or with an alkyl halide to give an amine salt also enhances the ability to undergo the metathesis reaction. Compare, for example, $CH_2{=}CHCH_2NH_2$ and $CH_2{=}CHCH_2NMe_2$ (0% yield) with $CH_2{=}CHCH_2NH_2{,}HCl$ (12% yield), $CH_2{=}CHCH_2NMe_2{,}HCl$ (22% yield), $CH_2{=}CHCH_2NMe_3Br$ (11% yield), and $CH_2{=}CHCH_2NMeEt(C_6H_{11})I$ (69% yield); catalyst system: $W(CO)_3$(mes)/

$EtAlCl_2/O_2$ in chlorobenzene, $W/Al/O_2$/amine = 1/40/30/10 (Nouguier 1977; Edwige 1980).

In the pesence of $Mo(=CHCMe_2Ph)(=NAr)[OCMe(CF_3)_2]_2$ the *t*-butoxycarbonyl (Boc)-protected amine, $CH_2=CHCH_2NHC(O)OCMe_3$ (2 equiv in toluene), gives a 60% yield of the bis(Boc) amine after 8 h (Marmo 1994). Attempts to metathesize olefinic amides with WCl_6-based catalysts were unsuccessful (Levisalles 1984).

7.6 Nitriles

The metathesis of ω-unsaturated nitriles is represented by eqn. (13). Yields

$$2\,CH_2{=}CH(CH_2)_nCN \longrightarrow NC(CH_2)_nCH{=}CH(CH_2)_nCN + CH_2{=}CH_2 \qquad (13)$$

obtained with three catalyst systems are shown in Table 7.4. No reaction is observed for acrylonitrile ($n = 0$) with any of these catalysts, and only very low yields for allyl cyanide ($n = 1$) (van den Aardweg 1983; Bosma 1985). However, metathesis of the latter can be accomplished with the homogeneous catalyst WCl_6/1,1,3,3-tetramethyl-1,3-disilacyclobutane. Thus with a substrate/WCl_6/cocatalyst ratio of 50/1/2 the conversion after 10 h at 60°C is 53%, with a selectivity of 82% (Bespalova 1988). For the longer-chain compounds ($n \geqslant 2$) good yields are obtained with the heterogeneous Re_2O_7-based catalysts (selectivity > 98%); and for $n \geqslant 5$ also with the homogeneous WCl_6/Me_4Sn system (Table 7.4).

Table 7.4 Self-metathesis of unsaturated nitriles of the general formula $CH_2=CH(CH_2)_nCN$ (van den Aardweg 1983; Bosma 1985)

	Conversion (mol%)		
	WCl_6/Me_4Sn[a]	$Re_2O_7/Al_2O_3/Me_4Sn$[b]	$Re_2O_7/Al_2O_3/Et_4Sn$[b]
0[c]	0	0	0
1	0	< 1	< 1
2	0	83	91
3	0	91	98
4	0	91	98
5	47	80	86
8	48	70	
9	50	70	

[a] Molar ratio nitrile/WCl_6/Me_4Sn = 10/1/2, 100°C. [b] Molar ratio nitrile/Re_2O_7/R_4Sn = 10/1/0.2, 20°C.
[c] For cross-metathesis reactions of acrylonitrile, see Ch. 9.

7.7 Chlorides and bromides

The metathesis of allyl bromide, reaction (14), is reported to proceed on $Re_2O_7/Al_2O_3/Me_4Sn$ with a selectivity of 95%, reaching 50% conversion at 20°C in

$$2\,CH_2{=}CHCH_2Br \xrightarrow{Re_2O_7/Al_2O_3/Me_4Sn} BrCH_2CH{=}CHCH_2Br \quad + \quad CH_2{=}CH_2 \qquad (14)$$

30 min (Mol 1979). Similar results are obtained with other low-molecular-weight halogen-substituted alkenes (Fridman 1977a) A $Re_2O_7/Al_2O_3/Bu_4Sn$ catalyst modified with triethylborate converts allyl chloride into 1,4-dichlorobut-2-ene and ethene within 10 min at 20°C (75% conversion, >95% selectivity) (Bogolepova 1992). Olefins containing a halogen at the double-bond display the lowest reactivity, while metathesis of other halogenated olefins proceeds at rates close to the rates of simple olefins. The slow rate for metathesis of vinyl chloride can be attributed to the difficulty of formation of an intermediate metallacyclobutane complex in which the two halogen atoms occupy adjacent positions; accordingly cross-metathesis with simple olefins proceeds rapidly (Fridman 1977a, 1979). The metathesis of oleyl chloride and its cross-metathesis with simple olefins has also been described (Nakamura, R. 1977a).

7.8 Sulfides and sulfonates

Metathesis of unsaturated sulfides proceeds to a substantial extent with the tungsten complex $[W]{=}CHCMe_3$ (**1** in Ch. 3). In its presence $CH_2{=}CHCH_2SMe$ ([olefin]/[catalyst] = 25/1) is converted to ethene and a *trans/cis* (9/1) mixture of $MeSCH_2CH{=}CHCH_2SMe$ at 80°C in chlorobenzene (40% conversion in 15 h) (Couturier 1993b).

Alkenyl tosylates of the type $RHC{=}CH(CH_2)_nOTs$ ($n = 7$–9) undergo metathesis when catalyzed by WCl_6/Me_3SnCl (Daly 1982).

7.9 Silanes and germanes

The metathesis of alkenylsilanes is represented by eqn. (15). Conversions of 4–15% have been reported for the metathesis of vinyltrimethylsilane ($n = 0$, R = Me) on

$$2\,CH_2{=}\,CH(CH_2)_nSiMe_2R \longrightarrow RMe_2Si(CH_2)_nCH{=}\,CH(CH_2)_nSiMe_2R \; + \; CH_2{=}CH_2 \qquad (15)$$

MoO_3/Al_2O_3 or on $Re_2O_7/Al_2O_3/Bu_4Sn$ (Fridman 1971, 1977b; Finkel'shtein (1992b). However, good yields of $Me_3SiCH{=}CHSiMe_3$ have been obtained with a number of ruthenium catalysts: $Ru_3(CO)_{12}/HSiPh_3$ (1/10) in benzene at 80°C, 75%

yield of *trans* product (Seki, Y. 1989); $[RuCl_2(CO)_3]_2$ in benzene at 130°C, 48% yield of *trans* product (Marciniec 1994c); $Ru(H)(Cl)(CO)(PPh_3)_3$ at 140°C, 38% yield of *trans/cis* (44/56) product (Wakatsuki 1991). With oxygenated benzene solutions of $RuCl_2(PPh_3)_3$ or $RuH_2(PPh_3)_4$ the yield is 58% but the product contains about 15% of the isomer $(Me_3Si)_2C{=}CH_2$ and other by-products (Marciniec 1994c). With $Ru(SiMe_3)(Cl)(CO)(PPh_3)_2$ in benzene at 130°C a yield of 50% *trans*-$Me_3SiCH{=}CHSiMe_3$ and 30% $(Me_3Si)_2C{=}CH_2$ has been reported (Marciniec 1995b). The formation of the two isomers has been interpreted in terms of a non-(metal-carbene) mechanism in which the C=C bond of the substrate can insert into the Ru-Si bond in both directions (Marciniec 1995b).

$RuCl_2(PPh_3)_3$ is also an efficient catalyst for the metathesis of vinyltri(phenyl, methyl)silanes; reaction (16). A yield of >75% is obtained after 48 h (for $n = 1$) or 96 h ($n = 2$) at 120°C in air. $RhCl(PPh_3)_3$ is less active (Marciniec 1991).

$$2\,CH_2{=}CHSiMe_{3-n}Ph_n \longrightarrow Ph_nMe_{3-n}SiCH{=}CHSiMe_{3-n}Ph_n + CH_2{=}CH_2 \quad (n = 1\text{–}3) \qquad (16)$$

The metathesis of vinyltri(alkoxy,methyl)silanes, reaction (17), proceeds smoothly at 60–130°C in the presence of ruthenium chloride complexes such as $RuCl_2(PPh_3)_3$, or in some cases rhodium chloride complexes such as $RhCl(PPh_3)_3$,

$$2\,CH_2{=}CHSiMe_{3-n}(OR)_n \longrightarrow (RO)_nMe_{3-n}SiCH{=}CHSiMe_{3-n}(OR)_n + CH_2{=}CH_2 \quad (n = 1\text{–}3;\ R = Me, Et, Pr, i\text{-}Pr) \qquad (17)$$

as well as in the presence of catalyst systems comprising ruthenium and rhodium chlorides with a Lewis acid and a hydride donor as cocatalyst. The reaction is accelerated by traces of dioxygen (Marciniec 1984, 1988, 1989). $Ru_3(CO)_{12}$ catalyzes the metathesis of vinyldimethylethoxysilane (eqn. (17), $n = 1$, R = Et) and vinyltriethoxysilane ($n = 3$, R = Et) in the presence of $HSiPh_3$, giving yields of 91% and 25%, respectively, after 4 h at 80°C (Seki, Y. 1989). $Ru(H)(Cl)(CO)(PPh_3)_3$ catalyzes the metathesis of vinyldimethylethoxysilane with a yield of 49% after 14 h at 140°C (Wakatsuki 1991).

The complex $Ru(SiMe_3)(Cl)(CO)(PPh_3)_2$ brings about the conversion of the vinyltrisubstituted silanes $CH_2{=}CHSiMe_2Ph$ and $CH_2{=}CHSi(OEt)_3$ to a mixture of two isomers, viz. *trans*-1,2-bis(silyl)ethene and 1,1-bis(silyl)ethene, similar to the reaction of vinyltrimethylsilane with this catalyst (Marciniec 1995b).

Allyltrimethylsilane, eqn. (15), $n = 1$, R = Me, undergoes metathesis in the presence of $WOCl_4$ without a cocatalyst (yield 10–12% at a substrate/W molar ratio of 100/1). The role of the cocatalyst is taken over by the substrate, which alkylates the tungsten compound (Bespalova 1990). The metathesis of allylsilanes containing different substituents at the Si atom (e.g. R = Ph, *p*-tol, $PhCH_2$, ferrocenyl) proceeds quite effectively in the presence of $Re_2O_7/Al_2O_3/R_4Mt$ (Mt = Sn, Pb; R = Et, Bu) at 35–75°C (Vdovin 1984c; Finkel'shtein 1989; Ushakov 1981, 1991).

With certain ruthenium complexes such a $[RuCl_2(p\text{-cymene})]_2$, allyltrisubstituted silanes (allyltrimethylsilane, allyltriethoxysilane) undergo isomerization to the corresponding 1-propenyltrisubstituted silane followed by cross-metathesis between the two isomers (Marciniec 1994a).

1-Hexenylmethylbis(trimethylsiloxy)silane undergoes metathesis with $>50\%$ yield in the presence of WCl_6/Me_4Sn (molar ratio = 1/4). A molar ratio of 1/2 does not, however, catalyze this reaction, in contrast to the metathesis of hex-1-ene, where the opposite is true (Stein 1994).

The metathesis of allyltrimethylgermane catalyzed by $Re_2O_7/Al_2O_3/Bu_4Sn$ proceeds as for its silicon analogue (Vdovin 1984b).

7.10 Phosphanes

The carbene complex $[W]{=}CHCMe_3$ (**1** in Ch. 3) brings about the metathesis of allyldiphenyl phosphane, eqn. (18), giving a conversion of 50% after 25 h at 80°C in benzene (Leconte 1995a).

$$2\,CH_2{=}CHCH_2PPh_2 \longrightarrow Ph_2PCH_2CH{=}CHCH_2PPh_2 + CH_2{=}CH_2 \qquad (18)$$

8

Acyclic Dienes

8.1 Introduction

Dienes can undergo olefin metathesis reactions of two types: (i) intermolecular, and (ii) intramolecular, as illustrated by the reactions of hexa-1,5-diene and octa-1,7-diene, respectively, eqns. (1) and (2). If necessary, reaction can be driven from left to right by the removal of ethene.

$$2\,CH_2{=}CH(CH_2)_2CH{=}CH_2 \longrightarrow [CH_2{=}CH(CH_2)_2CH{=}]_2 + C_2H_4 \qquad (1)$$

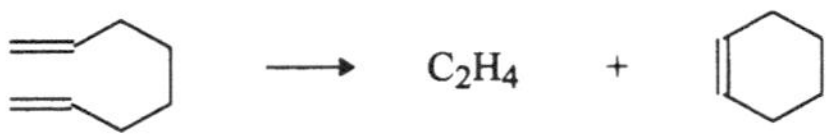

(2)

The linear product of reaction (1) can undergo further metathesis condensation reactions (Fig. 8.1), eventually leading to high polymer. This is known as ADMET (*a*cyclic *d*iene *met*athesis) polymerization. The evolution of the MWD in such reactions has been calculated by computer simulation (Martl 1991). Catalysts based on WCl_6 or Re_2O_7 usually give only low-molecular-weight products but metal carbene complexes are capable of yielding high polymers from monomer in bulk or in solution. Details are given in Tables 8.1–8.8. All the catalyst systems referred to in these tables are collected for convenience in the footnotes to Table 8.1.

Ring-closing metathesis (RCM) reactions, as in eqn. (2), occur with great readiness whenever the product is a 6-membered ring. They are also often favoured for the production of 5-, 7-, and 8-membered rings, depending on the nature, number and location of any substituents. The formation of much larger rings can also be achieved sometimes by imposing conformational restraints on the bonds lying between the two C=C bonds; examples are given later. Whether polymer or cyclic compound is formed from any given diene is usually determined by thermodynamic rather than kinetic factors. When the free energy of polymerization of the cyclic compound is close to zero, high concentration of substrate will favour polymer formation, while low concentration will favour the formation of the cyclic compound. For a discussion of the thermodynamics of polymerization of cyclic compounds, see Ivin (1991).

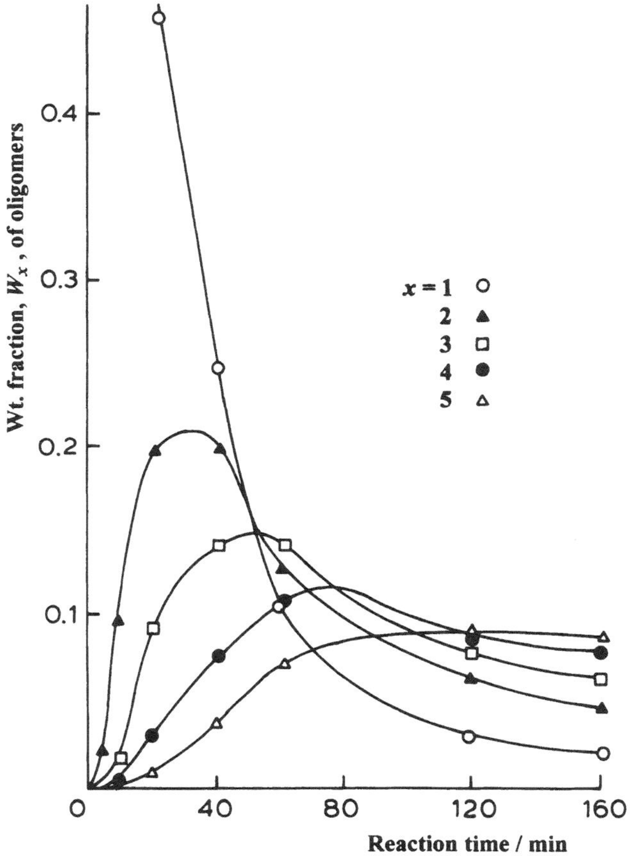

Fig. 8.1 Weight fraction of the oligomers $CH_2[=CHCH_2CH_2CH=]_xCH_2$ produced by the metathesis of hexa-1,5-diene (M), as a function of time. Catalyst: $WCl_6/Et_3Al_2Cl_3/MeCCl(OH)CH_2Cl/PPh_3$ (1/2/1/0.5); W/M = 1/375. Solvent: toluene, 25°C (Dall'Asta 1973).

8.2 Double bonds linked only by C atoms

8.2.1 Hydrocarbons

Examples of clean metathesis reactions for dienes in which the reactive double bonds are separated by x bonds (x = 1–11) are listed in Table 8.1. In favourable cases, for example with the carbene initiators **8** and **17**, high polymer is eventually formed. This may be accompanied by the formation of a range of cyclic species as a result of backbiting reactions, eqn. (3). In some cases a monomeric species may be

Table 8.1 Some metathesis reactions of diene hydrocarbons[a]

x^b	Substrate[c]	Products[d]	Catalyst[e]	Notes[f]	Reference
1	2,4-C_6	2,4,6-C_8 + 2,4,6,8-C_{10}	**4**	$C_6:C_8:C_{10} = 26:8:1$	Woerlee 1984
		Linear oligomer	**8**	$n = 10–20$	Tao 1994
2	1,4-C_5	1,4,7-C_8	**6**	70% yield	Doyle 1973
2	1,3-Dimethylenecyclobutane	Polymer	**5**		Finkel'shtein 1992a
2	1,4-C_6	1,4-CHX	**2**	50% yield per pass	Kawai 1986
2	1,4-C_8	1,4-CHX	**2**	Similar to 1,4-C_6	Kawai 1987
3	1,5-C_6	Linear oligomer	**2**	n up to 5; cf. Fig. 8.1	Kawai 1984a
3	1,5-C_6	Polymer	**17**	$n > 500$	Wagener 1991c
		Polymer	**10**	$n = 55^g$	Nubel 1994
3	2-Me-1,5-C_6	2,9-Me_2-1,5,9-C_{10}	**17**	$n = 2$	Konzelman 1995
		1,4-Polyisoprene[h]	**8**	$Z/E = 35/65$	Konzelman 1995
4	1,6-C_7	CPE(85%)[i] + linear oligomer	**2**	$n = 2$ (7%); $n = 3$ (1%)	Kawai 1984a
4	2,6-Me_2-2,7-C_8	(*S*)-3-Me-CPE	**8**	(+)-Substrate	Sita 1995
			15		Nugent 1995
5	4,8-Me_2-1,7-C_9	(*S*)-4-Me-CHX	**15**	(−)-Substrate	Nugent 1995
5	1,7-C_8	CHX	**2**	99% yield	Kawai 1984a
5	*cc*-2,8-C_{10}	CHX	**7,7a**		Grubbs 1979,1980
5	1,4-Divinylbenzene[j]	Linear oligomer	**17**	$n = 6$	Kumar, A. 1992
			8	$n = 2–4$	Fox 1994a
			16	$n = 2–720$	Thorn-Csányi 1994b
6	1,8-C_9	CHP (30%)[i] + linear oligomer	**2**	$n = 2$ (48%); $n = 3$ (15%)	Kawai 1984a
		1,8,15-C_{16}	**1**	$n = 2$ (80% yield)	Warwel 1989a
7	1,9-C_{10}	Linear oligomer[k]	**2**	$n = 2$ (75%)	Kawai 1984a
		Polymer	**17**	$M_w > 108\,000$	Wagener 1991c
		Polymer	**8**	$n = 130$	O'Gara 1993b
7	6-Me-1,5,9-C_{10}	Polymer	**8**	$M_w = 14\,000$	Wagener 1991d
9	1,4-Bis(homoallyl)cubane[l]	Linear oligomer[m]	**8**	$M_n \sim 1150$	Chauvin 1996
11	1,13-C_{14}	1,13,25-C_{26}	**1**	$n = 2$	Heckelsberg 1983

[a] For earlier references see Ivin (1983); also see Kawai (1987) for $x = 2$; Nel (1989) for $x = 3$; Bogolepova (1982) for $x = 3, 5$; Crease (1982), Davidson (1989) for $x = 5$; Lindmark-Hamberg (1987) for $x = 3, 7$; Blosch (1992b), Warwel (1989a) for $x = 7$.

[b] x = number of bonds separating the reacting double bonds.

[c] 2,4-C_6 denotes hexa-2,4-diene; similarly for the other abbreviations.

[d] CPE = cyclopentene; CHX = cyclohexene; 1,4-CHX = cyclohexa-1,4-diene; 2,4,6-C_8 = octa-2,4,6-triene. The small molecules formed, such as ethene or propene, are not listed.

[e] Catalyst systems for Tables 8.1–8.8: **1** Re_2O_7/Al_2O_3; **2** $Re_2O_7/Al_2O_3/CsNO_3$; **3** Re_2O_7/SiO_2-Al_2O_3/Bu_4Sn; **4** $Re_2O_7/Al_2O_3/Me_4Sn$; **5** $Re_2O_7/Al_2O_3/Bu_4Sn$; **6** $Bu_4N[MoCl(CO)_5]/MeAlCl_2$; **7** $MoCl_2(PPh_3)_2(NO)_2/Me_3Al_2Cl_3$; **7a** $MoO_3/CoO/Al_2O_3$; **8** Mo(=CHCMe$_2$Ph)(=NC$_6$H$_3$-*i*-Pr$_2$-2,6)[OCMe(CF$_3$)$_2$]$_2$; **9** WCl_6/Me_4Sn; **10** WCl_6/Me_4Sn/PrOAc; **11** WCl_6/Cp_2TiMe_2; **12** WCl_6/1,1,3,3-Me$_4$-1,3-disilacyclobutane; **13** $WOCl_4/Me_4Sn$; **14** $WOCl_4/Ph_2SiH_2$; **14′** $WOCl_4$; **15** *trans*-WOCl$_2$(OC$_6$H$_3$-Br$_2$-2,6)$_2$/Et$_4$Pb; **16** W(=CHC$_6$H$_4$-OMe-2)(=NC$_6$H$_3$-Me$_2$-2,6)[OCMe(CF$_3$)$_2$]$_2$(thf); **17** W(=CHCMe$_2$Ph)(=NC$_6$H$_3$-*i*-Pr$_2$-2,6)[OCMe(CF$_3$)$_2$]$_2$; **18** Ru(=CHCH=CPh$_2$)(Cl)$_2$(PCy$_3$)$_2$. See also Brzezinska (1996) for ADMET reactions catalyzed by Ru(=CHPh)(Cl)$_2$(PCy$_3$)$_2$.

[f] n = Number of units (average for large n).

[g] In the presence of 5-acetoxypent-1-ene as chain-transfer agent.

[h] Via initial formation of 2,9-Me$_2$-1,5,9-C$_{10}$.

[i] Plus trace of cyclic dimer.

[j] Also the 2,5-diheptyl derivative with **8** as catalyst giving product with $n = 10$ (Thorn-Csányi 1995b)

[k] Plus trace of cyclooctene.

[l] Also forms a copolymer with hexa-1,5-diene.

[m] All-*trans* double bonds.

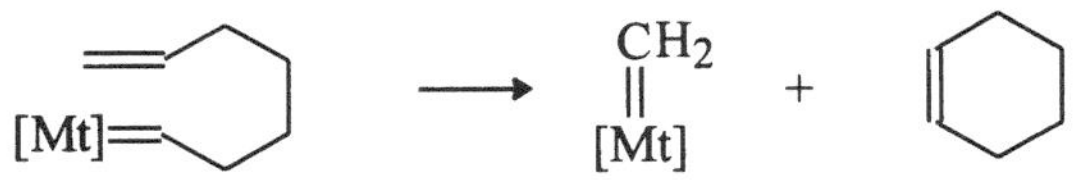

(3)

the dominant product, for example, the production of cyclohexene from octa-1,7-diene.

Isopropenyl groups are relatively unreactive towards metathesis. 2,5-Dimethylhexa-1,5-diene, with an isopropenyl group at both ends, does not react at all in the presence of either **8** or **17**. However, while 2-methylhexa-1,5-diene reacts only at the vinyl end in the presence of the tungsten complex **17**, to give the 'dimer' ($n=2$ in Table 8.1), the corresponding molybdenum complex **8** is much more active and the end product is perfectly head–tail 1,4-polyisoprene, formed by alternate reactions of an isopropenyl groups with $[Mo]{=}CH_2$ and of a vinyl group with [Mo]=CMeP where P is the polymer chain (Konzelman 1995).

The RCM of (+)-β-citronellene (2,6-dimethylocta-2,7-diene) in toluene (0.75 M), induced by **8**, proceeds quantitatively at 20°C to yield isomerically pure (*S*)-3-methylcyclopentene (Sita 1995). The same result can be achieved with *trans*-$WOCl_2(OC_6H_3Br_2\text{-}2,6)_2/Et_4Pb$ as catalyst (Nugent 1995). This is a remarkable improvement on the previous very difficult synthesis of this compound. In concentrated solution in toluene (5 M) it undergoes ROMP when initiated by **8** at − 30°C (Sita 1995). (*S*)-4-methylcyclohexene can be made in a similar way, as can 3,3-dimethylcyclohexene and 2-methylcyclohexene (Nugent 1995).

Octa-1,7-diene has been a favourite molecule for testing catalytic activity and for carrying out labelling experiments, because reaction (2) is generally highly selective. The reaction of 1 : 1 mixtures of normal and 1,1,8,8-d_4-octa-1,7-diene gives ethene, the isotopic composition of which ($d_0 : d_2 : d_4$) corresponds closely to that expected from the metal carbene mechanism (1 : 2 : 1), for both $MoCl_2(PPh_3)_2(NO)_2/Me_3Al_2Cl_3$ and $PhWCl_3/AlCl_3$ as catalyst systems. With the former catalyst, some $CH_2{=}CH(CH_2)_4CH{=}CD_2$ is found in the initial products, showing the occurrence of the degenerate exchange of the methylene groups, but with the latter this is relatively slight. Isotopic labelling experiments with 2,2′-divinylbiphenyl and its end-labelled d_4 analogue give very similar results so far as $d_0 : d_2 : d_4$ composition is concerned, with very little degenerate exchange of methylene groups (Katz 1976b).

Degenerate metathesis of alk-1-enes could, in principle, be propagated either by $[Mt]{=}CH_2$ or [Mt]=CHR. In the reactions of octa-1,7-diene it is clear that $[Mt]{=}CH_2$ must be formed when cyclohexene is eliminated by reaction (3), so providing an opportunity to study its subsequent reactions in the presence of other olefins such as cyclopentene. Reaction (3) is not quenched even in the presence of a 20-fold excess of cyclopentene, using $WCl_2(CO)_3(AsPh_3)_2/EtAlCl_2$ or WCl_6/Me_4Sn as catalyst system (Bencze 1980). Hence most of the products containing cyclopentene units must result from the initial addition of $[Mt]{=}CH_2$ to cyclopentene. The relative proportions of products, $C_{13}/C_7 = 8.5$, $C_{18}/C_{12} = 7$, then give a good indication of the relative rates of reactions (4) and (5), for $n = 1$ and 2, respectively.

$$[Mt]{+}CH(CH_2)_3CH{+}_nCH_2 \quad + \quad CH_2{=}CH(CH_2)_4CH{=}CH_2$$

$$\longrightarrow [Mt]{=}CH_2 \quad + \quad CH_2{=}CH(CH_2)_4CH{+}CH(CH_2)_3CH{+}_nCH_2 \qquad (4)$$

$$\longrightarrow [Mt]{=}CH(CH_2)_4CH{=}CH_2 \quad + \quad CH_2{+}CH(CH_2)_3CH{+}_nCH_2 \qquad (5)$$

However, the yields of C_7 and C_{12} are still only 20% and 12%, respectively, relative to cyclohexene, so that most of the $[Mt]{=}CH_2$ species must react preferentially with octa-1,7-diene to give ethene by reaction (6). One may conclude that there is no impediment to reaction (6) and that therefore in the metathesis of alk-1-enes the carrier is mainly $[Mt]{=}CHR$; cf. Section 5.3.2.

$$\begin{array}{c} [Mt]{=}CH_2 \\ + \\ CH_2{=}CH(CH_2)_4CH{=}CH_2 \end{array} \longrightarrow \begin{array}{c} [Mt] \\ \| \\ CH_2{=}CH(CH_2)_4CH \end{array} + \begin{array}{c} CH_2 \\ \| \\ CH_2 \end{array} \qquad (6)$$

Asymmetric RCM of substituted 1,6- and 1,7-dienes, taken to partial conversion using a chiral molybdenum carbene complex as catalyst, results in residual reactant having 19–84% excess of one enantiomer (Fujimura 1996a,b).

8.2.2 Ketones, side-chain esters, side-chain ethers, etc.

Examples of such reactions are shown in Table 8.2. The main thing to notice is how, for a given value of x, the introduction of substituents, or a change in the nature of the substituent, e.g. from $COCMe_3$ to COPh, can swing the reaction from being mainly intermolecular giving oligomer or polymer, to intramolecular giving cycloalkene as the main product. The catalyst also sometimes has a marked influence on the direction of the reaction, which must therefore be kinetically controlled (Maier 1995).

An efficient method for the synthesis of 5- and 6-membered cyclic enol ethers from unsaturated esters is first to convert the esters into acyclic olefinic enol ethers, which are then transformed to the desired products by RCM, catalyzed by **8** (Fujimura 1994). Further examples will be found in Table 8.5.

8.3 Double bonds linked by C and Si, Ge or Sn atoms

Examples of such reactions are shown in Table 8.3. The fact that dimethyldivinylsilane will not undergo self-metathesis, while it can be cross-metathesized with deca-1,9-diene, indicates that the propagation reaction (7) is prohibited, but that

Table 8.2 Some metathesis reactions of diene ketones, side-chain esters, side-chain ethers, etc.

x [a]	Substrate [b]	Product [c]	Catalyst [d]	Notes	Reference
	Ketones				
4		Oligomers	**8**		Forbes 1992
4	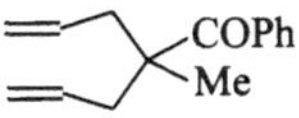		**8**	55% yield	Forbes 1992
6	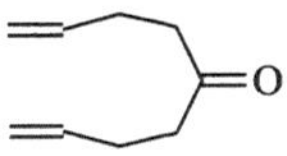	Oligomers	**8**		Forbes 1992
6	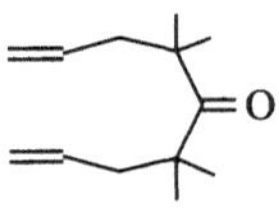	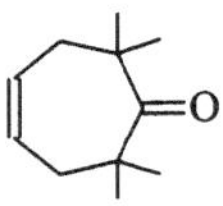	**8**	95% yield	Forbes 1992
6	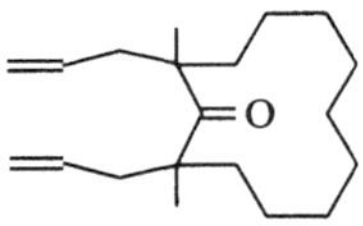	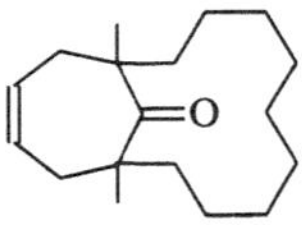	**8**	95% yield	Forbes 1992
10	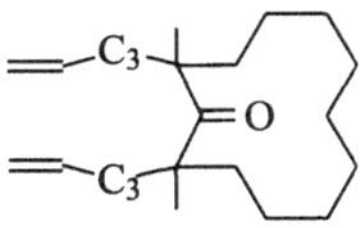	Polymer	**8**		Forbes 1992
10	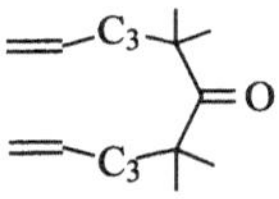	Polymer	**8**		Forbes 1992
16	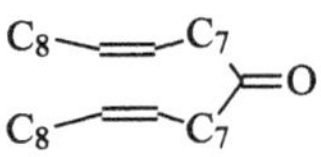	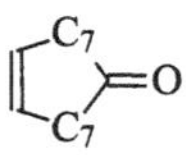	**3**	Reaction at 10^{-3} M [e]	Plugge 1991
	Esters				
4	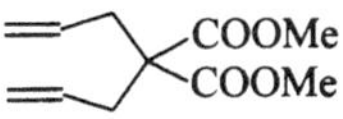	Oligomers	**8**		Forbes 1992
4		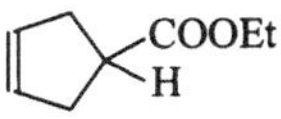	**12** **15**	yield >80% yield 74%	Bespalova 1994b Nugent 1995
4	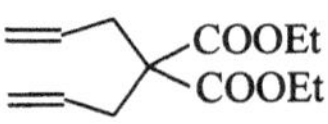	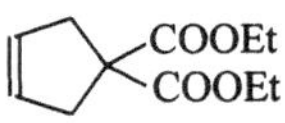	**15**	yield 86%	Nugent 1995

Table 8.2 (*continued*)

x [a]	Substrate [b]	Product [c]	Catalyst [d]	Notes	Reference
5	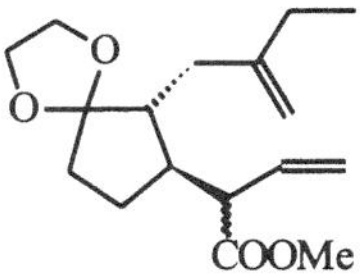	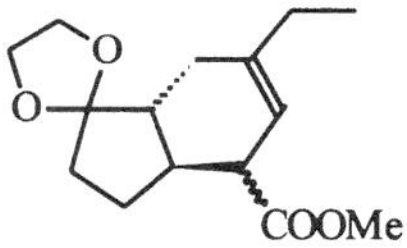	**8**	91% yield[g]	Holder 1996
	Ethers				
4	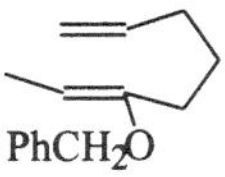	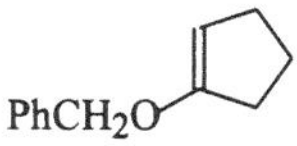	**8**	88% yield	Fujimura 1994
4	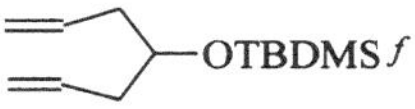	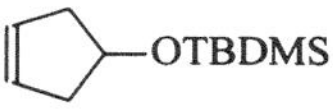	**8** **18** **15**	91% yield 85% yield 81% yield	Fu 1993a Fu 1993b Nugent 1995
4	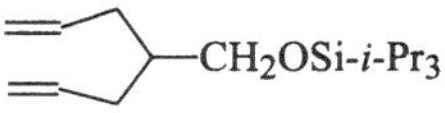	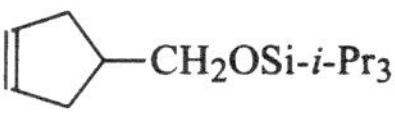	**15**	91% yield	Nugent 1995
4	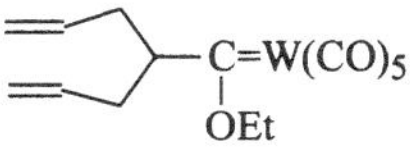	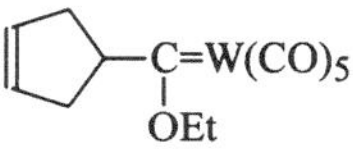	**14**		Toledano 1984b
5	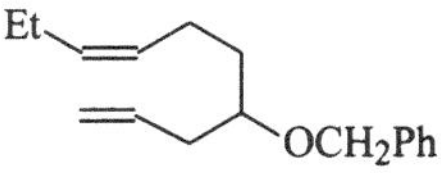	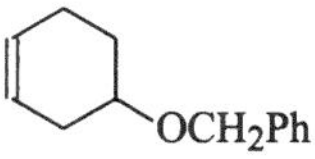	**8**	89% yield	Fu 1993a
5	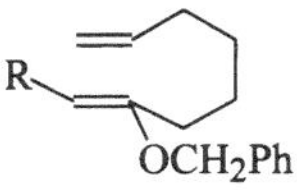	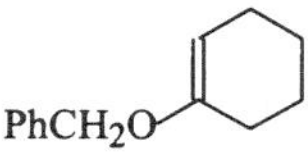	**8** **8**	R = H, 80% yield R = Me, 84% yield	Fujimura 1994
6	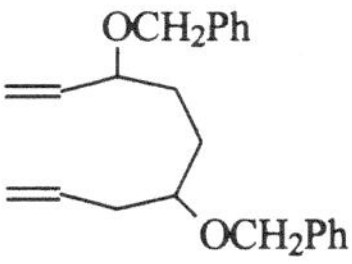	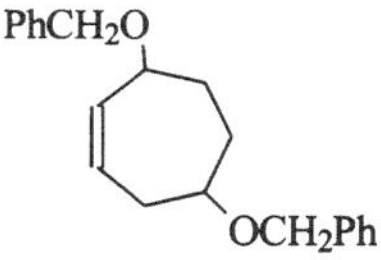	**18** **8**	40% yield nil (linear dimer only)	Maier 1995
	Miscellaneous				
5	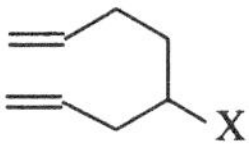	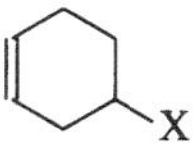	**18**	X = CH_2OH 88% yield X = CHO 82% yield X = COOH 87% yield	Fu 1993b
5	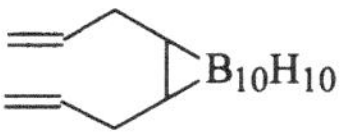	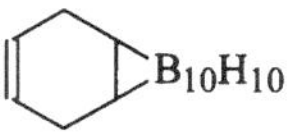	**13**	30% yield	Vdovin 1984a

[a] x = number of bonds separating the reacting double bonds. [b] C_y represents a chain of y carbon atoms. [c] The eliminated olefin, such as ethene, is not listed. [d] See Table 8.1. [e] Product is civetone, *cis/trans* = 50/50. [f] OTBDMS = *tert*-butyldimethylsilyloxy; see Junga (1993) for RCM of analogous compounds with x = 5,6. [g] Readily converted to coronafacic acid.

Table 8.3 Some metathesis reactions of dienes linked through Si, Ge or Sn atoms

x [a]	Substrate	Product [b]	Catalyst [c]	Notes [d]	Reference
2		Nil	**17**	Copolymer [e]	Wagener 1991f
4			**5**	Mainly 'dimer'	Finkel'shtein 1981
		+ linear oligomer	**14'**	28% cyclic	Bespalova 1990
	(Neat)	Polymer, 19% *cis*	**17**	$n = 100$	Anhaus 1993
	(Neat)	Polymer, 25% *cis*	**8**		Anhaus 1993
	(Soln.) [f]	Cyclic 'monomer' and 'dimers'	**8**	73% *cc* 'dimer'	Anhaus 1993
4	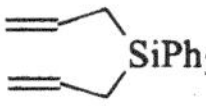 (neat)	Polymer, 10% *cis*	**8**	$n = 20$	Anhaus 1993
4		Linear 'dimer' (55%) + cyclic 'monomer' (12%)	**5**		Vdovin 1984c
4	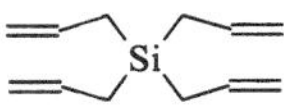	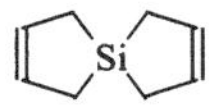 + oligomer	**5**		Ushakov 1989
4	 (neat)	Linear oligomer (57%) + cyclic 'monomer' (38%)	**8**	84% *trans*	Cummings 1995
4 5 6	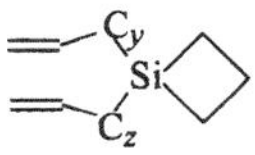	Linear oligomer + cyclic 'monomer' [g]	**5**	(y, z = 1 or 2)	Ushakov 1989
4 5 6	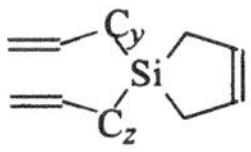	Linear oligomer + cyclic 'monomer' [h]	**5**	(y, z = 1 or 2)	Ushakov 1989
5	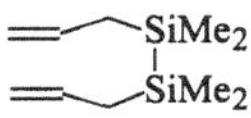		**14**	60% yield	Ushakov 1981
7	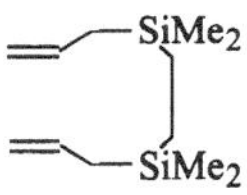	Polymer, 27% *cis*	**17**	$n = 240$ [i]	Wagener 1991f
9	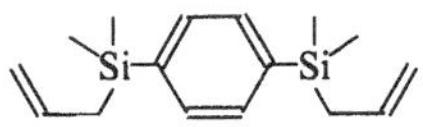	Polymer, 53% *cis*	**17**	$n = 6$	Wagener 1991f
4		 + linear 'dimer'	**5**		Vdovin 1984b
4	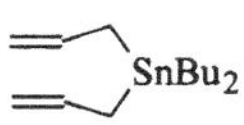		*j*	Side reactions	Lefebvre 1995

[a] x = number of bonds separating the reactive double bonds. [b] Not including the eliminated olefin such as ethene. [c] See Table 8.1. [d] n = number of units in polymer (average for large n). [e] Copolymer, formed by reaction with deca-1,9-diene, contains isolated ╪CHSiMe$_2$CH╪ units. [f] Also see Leconte (1995b): 90% yield of cyclic 'monomer' using complex **1** of Ch. 3. [g] The highest proportion of intramolecular metathesis (60%) is observed for the formation of 6-membered rings ($y = 1, z = 2$). [h] As for footnote *g* (34%). [i] See Fig. 8.2. [j] [W]=CHCMe$_3$ complex **1** of Ch. 3.

$$\begin{array}{l} [W]{=}CHSiMe_2CH{=}CH_2 \\ \quad + \\ CH_2{=}CHSiMe_2CH{=}CH_2 \end{array} \not\longrightarrow \begin{array}{c} [W] \\ \| \\ CH_2 \end{array} + \begin{array}{l} CHSiMe_2CH{=}CH_2 \\ \| \\ CHSiMe_2CH{=}CH_2 \end{array} \qquad (7)$$

$$\begin{array}{l} [W]{=}CHSiMe_2CH{=}CH_2 \\ \quad + \\ CH_2{=}CH(CH_2)_6CH{=}CH_2 \end{array} \longrightarrow \begin{array}{c} [W] \\ \| \\ CH_2 \end{array} + \begin{array}{l} CHSiMe_2CH{=}CH_2 \\ \| \\ CH(CH_2)_6CH{=}CH_2 \end{array} \qquad (8)$$

reaction (8) will occur; also that $[W]{=}CH_2$ can react with the vinyl group in both molecules so as to eliminate ethene.

Dimethyldiallylsilane is a good example of a reaction in which there is a very delicate balance between the intramolecular and intermolecular processes. The balance may be swung either way by adjusting the substrate concentration. Even more remarkable is the formation of all three cyclic 'dimers' (*cc, ct,* and *tt*) at an overall concentration which exceeds that of the residual cyclic monomeric species.

The ^{13}C NMR spectrum of one of these polymers, shown in Fig. 8.2, illustrates the very clean nature of the ADMET reaction.

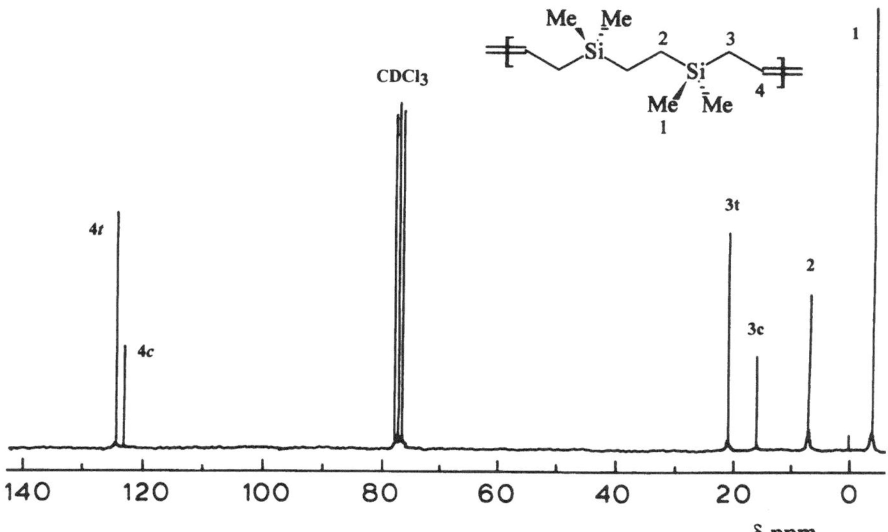

Fig. 8.2 50 MHz ^{13}C NMR spectrum of the product of metathesis condensation polymerization of 4,4,7,7-tetramethyl-4,7-disiladeca-1,9-diene (Wagener 1991f).

8.4 Double bonds linked by C and N atoms

Dienes in which the double bonds are separated by 4, 5, or 6 bonds, and linked through amine or amide functions, readily undergo RCM in the presence of various initiators to yield substituted unsaturated cyclic amines or amides (pyrrolidinones, piperidinones, piperidines); see Table 8.4.

Table 8.4 Some intramolecular metathesis reactions of dienes linked through N atoms[a]

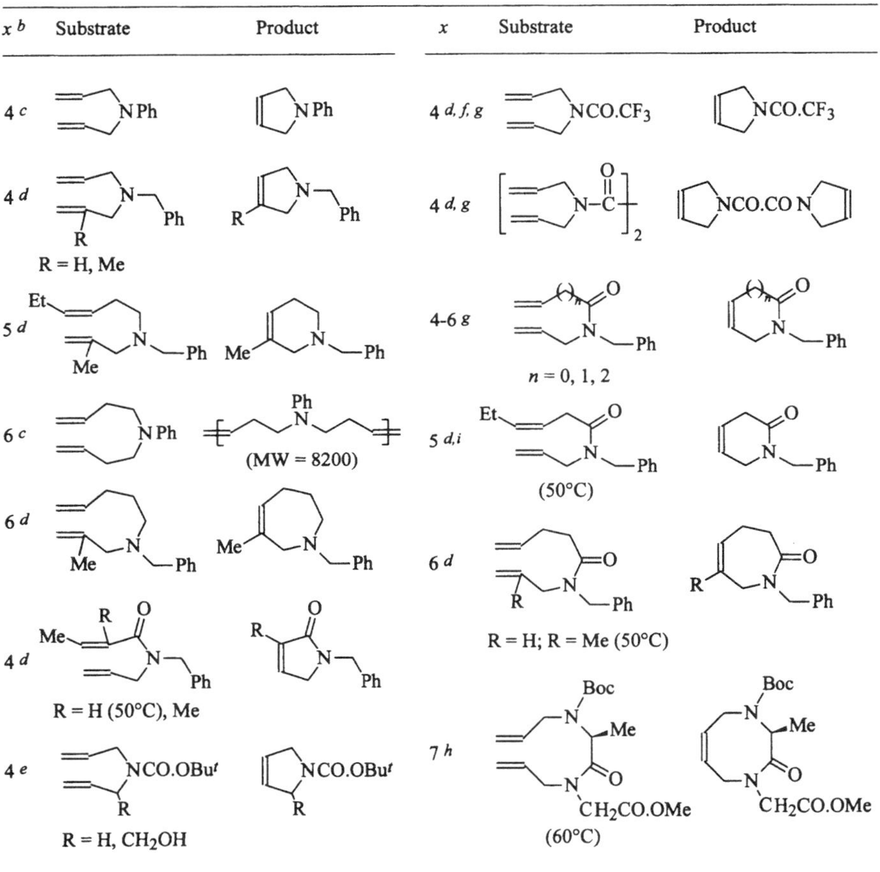

[a]Catalysts as specified in Table 8.1. Temperature: 20°C except where indicated. Yields generally 70–90% in less than 3 h.
[b]x = number of bonds separating the reactive double bonds.
[c]Catalyst **8** (Portmess 1995, 1996).
[d]Catalyst **8** (Fu 1992b).
[e]Catalyst **18** (Fu 1993b; Huwe 1995, 1996).
[f]Catalyst **15** (Nugent 1995).
[g]Catalyst **18** (Fu 1993b).
[h]Catalyst **18** (Miller 1995a).
[i]See Garro-Hélion (1996) for a related example using catalyst **18**.

The bis(*N*-allyl) dipeptide ($x = 7$) shown in Table 8.4 gives a 51% yield of the cyclic dipeptide whereas an analogous *N,O*-bis(allyl) ester ($x = 7$) fails to undergo RCM. The success or failure of RCM of such acyclic dienes thus depends critically upon the preferred conformations about the seven intervening bonds and upon the potential strain energy in the 8-membered ring to be formed. When one of these bonds is constrained by being part of a ring, as in **19**, the chance of effecting RCM is enhanced, eqn. (9). The product **20** can be converted to **21**, which is the core fragment of the anti-cancer agent FR-900482 (Miller 1995a); also see Martin (1995). The same strategy can be used for the preparation of fused nitrogen heterocycles such as pyrrolizidines, indolizidines, quinolizidines, pyrrolidinoazocines, and piperidinoazocines (Martin 1994b, 1996). A synthesis of castanospermine, in which the key step is the RCM of a diene to yield a bicyclic lactam, has been reported (Overkleeft 1996).

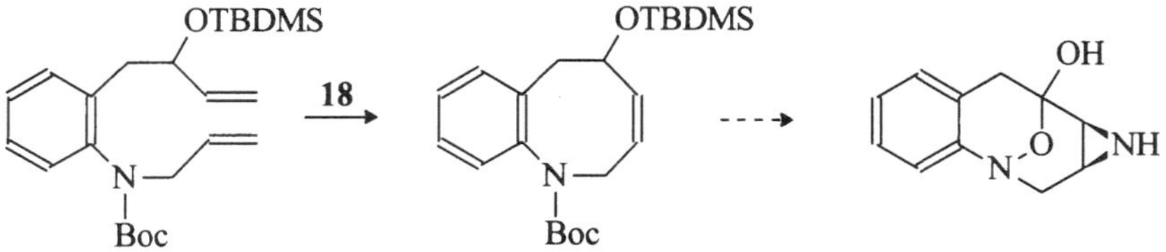

(19) **(20)** **(21)** (9)

An even more remarkable case of RCM assisted by conformational restraint is that shown in eqn. (10). This involves a substrate in which the two double bonds are separated by 13 single bonds only one of which forms part of a ring, but in which an additional constraint is imposed by the hydrogen bond. When a mixture of the four stereoisomers denoted by **22** is treated with catalyst **18** (20 mol %, 0.002 M in CH_2Cl_2, 40°C) only one of the isomers undergoes RCM to give (*S,S,S*)-**23**, showing that in this case the configuration with respect to the chiral centres *a* and *b* is also crucial (Miller 1995b).

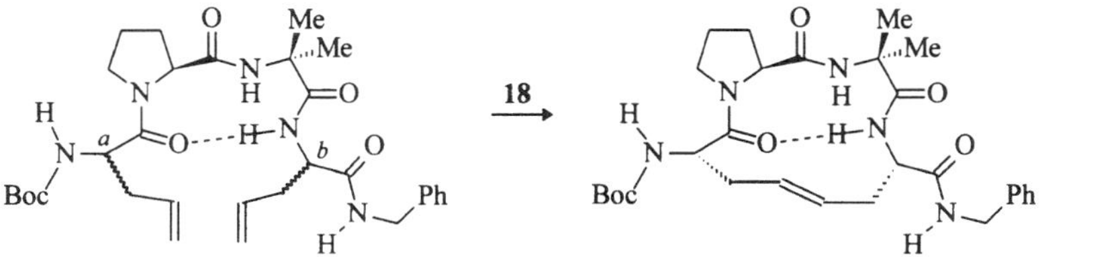

(10)

(**22**) (mixture of 4 diastereomers) (*S,S,S*)-(**23**) formed from one diastereomer

With catalyst **18** one can also carry out RCM in compounds containing unprotected peptidic structures, as illustrated by eqn. (11) (Miller 1995b).

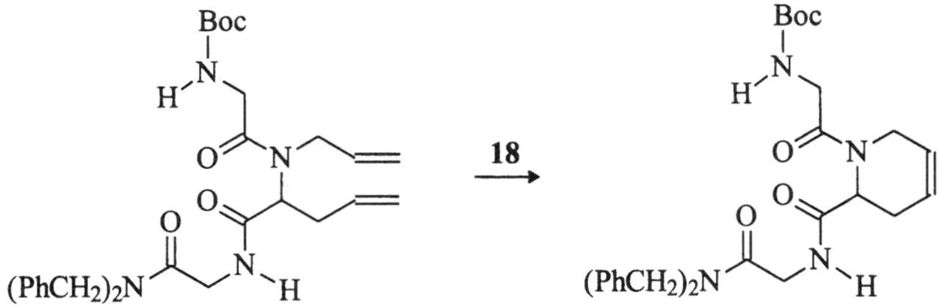

(11)

8.5 Double bonds linked by C and O atoms

8.5.1 Ethers and formals

Examples of such reactions are shown in Table 8.5. Divinyl ether is unreactive but diallyl ether, when treated in bulk with **8**, gives an equilibrium mixture of 63% 2,5-dihydrofuran and 37% of its all-*trans*, ring-opened polymer, albeit of low MW (Fig. 8.3). The analogous tungsten complex is not a good initiator for the ADMET polymerization of diallyl ether.

For any substrate with $x = 4$, 5, or 6, the presence of substituents tips the balance totally in favour of the cyclic product; for larger values of x the situation is reversed

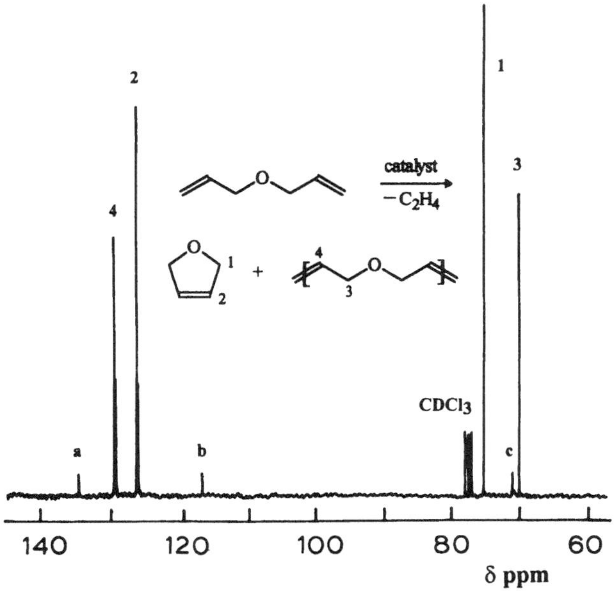

Fig. 8.3 50 MHz ^{13}C NMR spectrum of the equilibrium products of metathesis reactions of diallyl ether catalyzed by $Mo(=CHCMe_3)(=NC_6H_3\text{-}i\text{-}Pr_2\text{-}2,6)[OCMe(CF_3)_2]_2$. Peaks 1 and 2: 2,5 dihydrofuran. Peaks 3 and 4: linear polymer. Peaks a, b and c: end-group carbons, a $-CH_2CH{=}CH_2$, b $-CH_2CH{=}CH_2$, and c $-CH_2CH{=}CH_2$. Solvent $CDCl_3$ (Wagener 1992).

Table 8.5 Metathesis reactions of diene ethers and formals

x [a]	Substrate	Product [b]	Catalyst [c]	Notes [d]	Reference
2	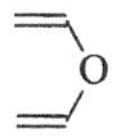	nil	**17**, **8**		Wagener 1991e Wagener 1992
4	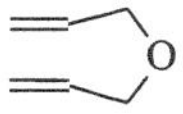	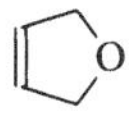	**1**, **5** [e,f]	Inhibited by product	Bogolepova 1978, 1984, 1986, 1987, 1989, 1990
	(Neat)	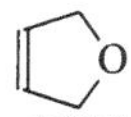 + polymer (63%) (37%)	**8**	Polymer, $n = 10$, 100% *trans* (Fig.8.3)	Wagener 1992
4	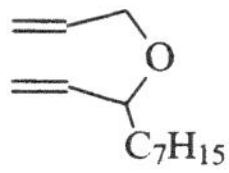	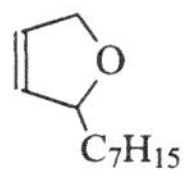	**15**	(*R*)-Substrate and product	Nugent 1995
4	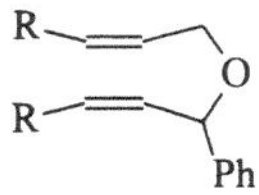	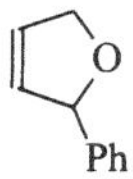	**8** **18**	R = H, 84% yield R = Me, 92% yield	Fu 1993b Fu 1992a
4	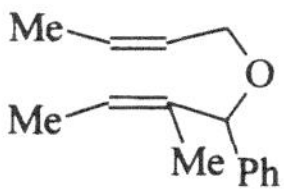	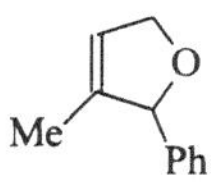	**8**	89% yield	Fu 1992a
4	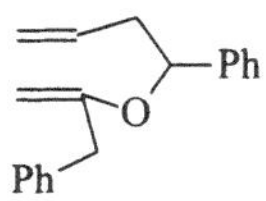	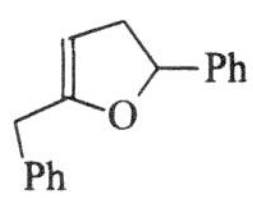	**8**	81% yield	Fujimura 1994
4	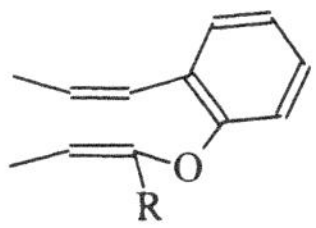	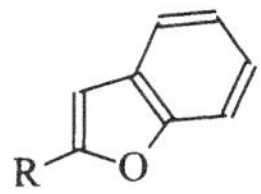	**8**	R = Pr, 87% yield R = $PhCH_2$, 87% yield	Fujimura 1994
5	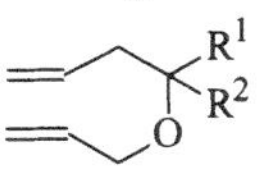	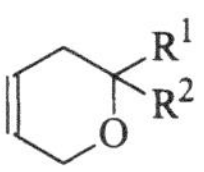	**5**	$R^1 = R^2$= H, Me R^1 = H; R^2 = Me, Ph	Bogolepova 1987
5	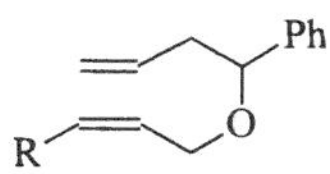	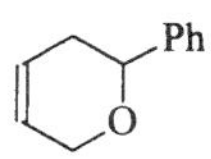	**8**	R = Me, 92% yield R = H, 86% yield	Fu 1992a Fu 1993b
5	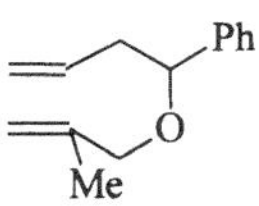	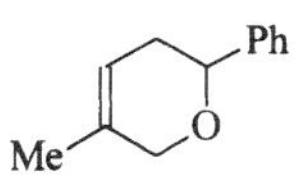	**8**	92% yield	Fu 1992a
5 [g]	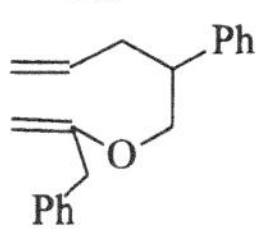	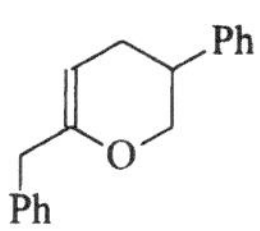	**8**	84% yield	Fujimura 1994
6	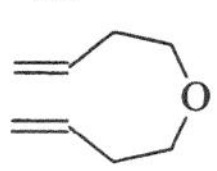	Low MW	**17**		Wagener 1991e

Continued overleaf

Table 8.5 *(continued)*

6	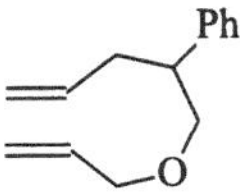	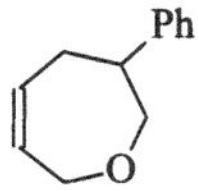	**18**	72% yield	Fu 1992b
6	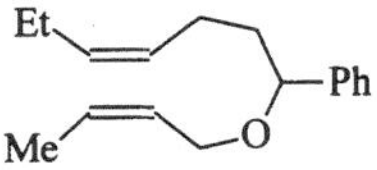	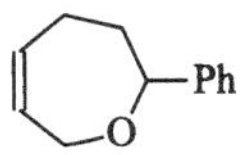	**8**	92% yield	Fu 1992a
6	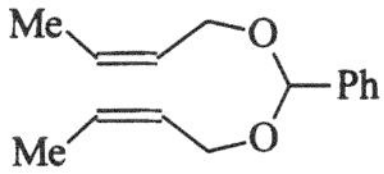	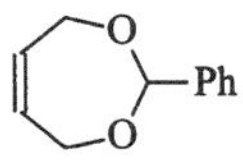	**8**	89% yield	Fu 1992a
6	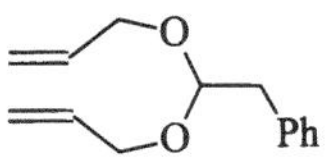	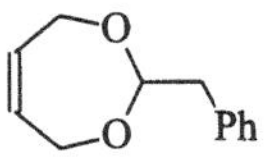	**18**	87% yield	Fu 1993b
7	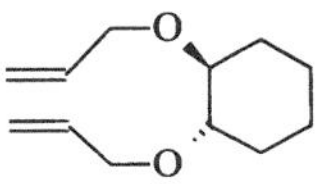	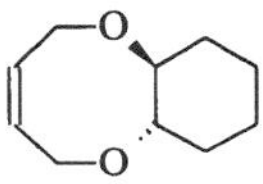	**18**	60% yield/2 h/55°C	Miller 1995a
7	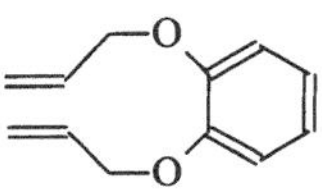	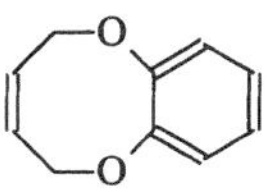	**18**	75% yield/3 h/55°C	Miller 1995a
8	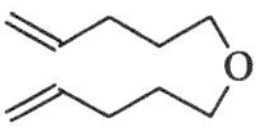	Polymer, 20% *cis*	**8**	96% yield, $n = 140$	Wagener 1991e
10	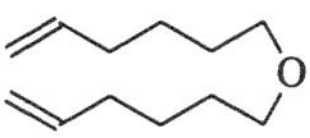	Polymer	**8**	99% yield, $n = 100$	Wagener 1991e

[a] x = number of bonds separating the reacting double bonds.
[b] Not including the eliminated olefin.
[c] See Table 8.1.
[d] n = average number of repeat units in polymer.
[e] Pretreatment of the Al_2O_3 support with $(RO)_3B$ increases the reactivity and allows the Re_2O_7 content to be decreased by a factor of 10.
[f] See also Leconte (1995b).
[g] See Nicolaou (1996) for related examples.

and long-chain polymer is formed except when there is a conformational restraint, for example, when one of the bonds between the two reacting double bonds forms part of a ring system. The probability of RCM is then much better for compounds with $x = 7$ or more. Thus the catechol derivative shown in Table 8.5 gives a good yield of bicyclic product, as also does the corresponding *trans*-1,2-cyclohexane derivative; but the situation is less favourable for the *cis*-1,2-cyclohexane derivative which gives only 20% yield of bicyclic product (Miller 1995a).

Substituted unsaturated pyrans prepared by RCM using **18** as catalyst can be immediately submitted to zirconium-catalyzed kinetic resolution of the racemic product at 70°C. This provides a new route to medicinally important agents containing 6-membered cyclic ethers. A one-pot synthesis can give 63% conversion with >99% enantiomeric purity (Morken 1994). Dienes of the type $CH_2{=}CH(CH_2)_3O(CH_2(H_2O)_nCH_2)_3CH{=}CH_2$ ($n = 2$–4) readily undergo ADMET polymerization in the presence of catalyst **8** (Qiao 1995).

8.5.2 Esters and carbonates

Villemin (1980) first demonstrated the occurrence of the RCM reaction (12) to form a 16-membered unsaturated lactone.

$$\begin{array}{l} CH_2{=}CH(CH_2)_8C{=}O \\ \qquad\qquad\qquad\quad | \\ CH_2{=}CH(CH_2)_4O \end{array} \xrightarrow[80^\circ C]{WCl_6/Me_4Sn} \begin{array}{l} CH_2 \\ \| \\ CH_2 \end{array} + \begin{array}{l} CH(CH_2)_8C{=}O \\ \| \qquad\qquad\quad | \\ CH(CH_2)_4O \end{array} \qquad (12)$$

ADMET polymerization reactions of this and other diene esters, also carbonates, are listed in Table 8.6. As with monoene esters, for reaction to occur there must be more than one methylene group between the double bond and the ester (or carbonate) group. The inhibition by a proximate ester group may be a polarization effect hindering the formation and/or the rearrangement of the intermediate metallacycle, or it may result from a coordination of the carbonyl oxygen to the metal centre (Patton, J.T. 1992).

Main-chain ferroelectric liquid crystal oligomers (MW ~ 6000) have been made by the ADMET polymerization of $H_2C{=}CH(CH_2)_7OC_6H_4C_6H_4COOC_6H_3(NO_2)OCHMe(CH_2)_nCH{=}CH_2$ ($n = 2$–4) using **18** at 80°C as a catalyst (Walba 1996).

8.6 Double bonds linked by C and P atoms

The RCM of diallylphenyl phosphane, eqn. (13), proceeds smoothly at 80°C in chlorobenzene (95% conversion in 5 h) when catalyzed by complex **1** of Ch. 3. The reaction occurs more readily than the self-metathesis of allyldiphenyl phosphane (10% conversion in 5 h). Steric hindrance evidently prevents the phosphorus atom from competing seriously with the olefinic bond for the vacant coordination position at the metal site (Leconte 1995a,b).

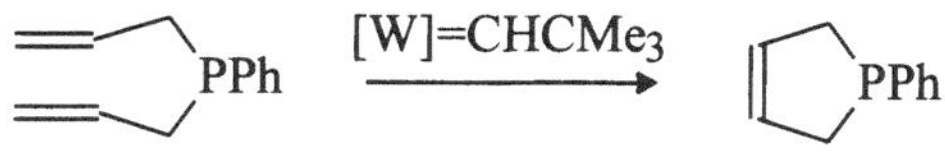

(13)

Table 8.6 Metathesis reactions of diene esters and carbonates

x	Substrate skeleton	Product	Catalyst [a]	Notes	Reference
	Esters				
8	C=CC$_4$O | C=CCC=O	Nil	**8**		Patton, J.T. 1992
9	C=CC$_4$O | C=CC$_2$C=O	Polymer, 10–20% *cis*	**8**	$n = 101$	Patton, J.T. 1992
	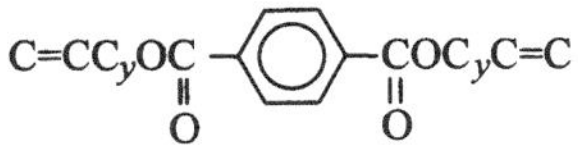				
11	$y = 1$	Nil	**8**		Patton, J.T. 1992
13	$y = 2$	Polymer, 10–20% *cis*	**8**	$n = 22$	Patton, J.T. 1992
15	$y = 3$	Polymer, 10–20% *cis*	**8**	$n = 45$	Patton, J.T. 1992
17	$y = 4$	Polymer, 10–20% *cis*	**8**	$n = 39$	Patton, J.T. 1992
13	C C=CC$_7$O | C=CC$_3$C=O	C CC$_7$O || | CC$_3$C=O	**18**	75% yield [b,c]	Fürstner 1996
15	C=CC$_4$O | C=CC$_8$C=O	CC$_4$O || | CC$_8$C=O	**9** **18**	65% yield [c] 79% yield [c]	Villemin 1980 Fürstner 1996
15	C$_8$C=CC$_8$O | C$_8$C=CC$_7$C=O	CC$_8$O || | CC$_7$C=O	**11**	18% yield	Tsuji 1980
20	C=CC$_9$O | C=CC$_8$C=O	CC$_9$O || | CC$_8$C=O	**18**	71% yield [c]	Fürstner 1996
23	(C=CC$_8$CO.O.C)$_2$	Polymer, 20% *cis*	**17**	$n = 15$	Bauch 1991
	Carbonates				
	(C=CC$_y$O)$_2$CO				
6	$y = 1$	Nil	**8**		Wagener 1993a
8	$y = 2$	Polymer	**8**	$n = 58$	Wagener 1993a
10	$y = 3$	Polymer	**8**	$n = 51$	Wagener 1993a
12	$y = 4$	Polymer	**8**	$n = 52$	Wagener 1993a
20	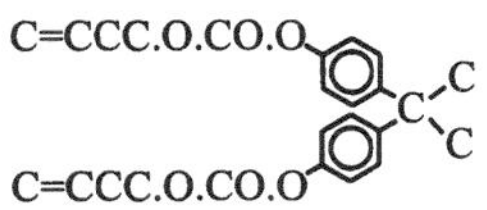	Polymer	**8**	$n = 40$	Wagener 1993a

[a] See Table 8.1. [b] Can be hydrogenated to exaltolide, a perfume ingredient. [c] *cis/trans* mixture.

8.7 Double bonds linked by C and S atoms

The sulfides $[CH_2{=}CH(CH_2)_y]_2S$ ($y = 1$–4) have been studied using **8** as initiator; also **1** of Ch. 3 (Leconte 1995b). Diallyl sulfide reacts best in benzene solution (0.36 M) at 20°C to give a quantitative yield of 2,5-dihydrothiophene (Couturier 1993b). Surprisingly, the yield is much lower when carried out in bulk, perhaps because the initiator is then more easily poisoned by the product. Allyl isopropenyl sulfide also undergoes RCM but not diisopropenyl sulfide (Leconte 1995b; Lefebvre 1995). The higher homologues of diallyl sulfide ($y = 2$, 3, 4) give polythioethers (MW ~ 3000-30 000) (O'Gara 1993a).

8.8 Double bonds linked by C, Si, and O, or C, B and O atoms

Examples are listed in Table 8.7 for various numbers of bonds (x) between the double bonds. For the compounds with $x = 6$, the formation of the 7-membered ring is the preferred reaction. For $x > 6$, the polymer is the favoured product. For $x = 4$ there is a remarkable variation in behaviour with the catalyst: no reaction is observed with the molybdenum carbene catalyst, but with the rhodium complex there is 86% conversion of substrate in 72 h to products consisting of about 5% of cyclic 'dimer', 4% of cyclic 'trimer' and 91% of linear oligomers ($M_n = 1815$). In the early stages of reaction the products are mainly the cyclic species but these undergo ROMP once their equilibrium concentration has been exceeded. With the ruthenium complex as initiator the kinetics of ROMP are less favourable and the products after 72 h consist of 25% cyclic dimer, 17% cyclic trimer and 58 % of linear oligomers (Marciniec 1995a).

'Dienes containing the boronate functionality, with $x = 9$ or 11, undergo ADMET polymerization in the presence of $Ru(=CHPh)(Cl)_2(PCy_3)_2$ (Wolfe 1996).'

8.9 Divinylferrocene

1,1′-Divinylferrocene in toluene solution undergoes metathetical condensation in the presence of **17** to give oligomers averaging 4 units, and 30% *cis* content. The low average degree of polymerization (DP) is attributed to the poor solubility of the oligomers (Gamble 1992).

8.10 Some further applications in organic synthesis

A number of examples have already been given of the application of the RCM of α,ω-dienes to the synthesis of complex molecules. Some further examples are the following.

Table 8.7 Metathesis reactions of carbodisiloxadienes

x [a]	Substrate	Catalyst [b]	Product	Notes	Reference
	$[CH_2{=}CH(CH_2)_ySiMe_2]_2O$				
4	$y = 0$	Mo-1	Nil		Smith, D.W. 1993a
		Rh-1	Linear oligomer (91%) + cyclic species (9%)	86% conversion in 72 h/130°C	Marciniec 1995a
		Ru-1	Linear oligomer (58%) + cyclic species (42%)	78% conversion in 72 h/130°C	Marciniec 1995a
6	$y = 1$	Mo-1	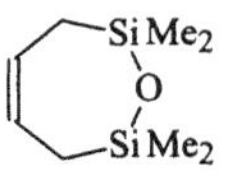	100% yield	Forbes 1992 Smith, D.W. 1993a
10	$y = 3$	Mo-1	Polymer	$n = 128$	Smith, D.W. 1993a
8	$(CH_2{=}CHCH_2SiMe_2O)_2SiMe_2$	Mo-1	Polymer	$n = 55$	Smith, D.W. 1993a
13	$CH_2{=}CHCH_2SiMe_2OSiMe_2$  $CH_2{=}CHCH_2SiMe_2OSiMe_2$	Mo-1	Low polymer	$n = 3$	Smith, D.W. 1993a
6	$(MeCH{=}CHCH_2O)_2SiMePh$	Mo-1	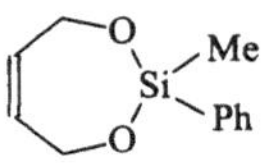	71% yield	Fu 1992a

[a] x = number of bonds separating the reacting double bonds. [b] Catalysts: Mo-1, $Mo({=}CHCMe_2Ph)({=}NC_6H_3\text{-}i\text{-}Pr_2\text{-}2,6)[OCMe(CF_3)_2]_2$/50°C; Rh-1, $[RhCl(COD)]_2$ where COD = cycloocta-1,5-diene; Ru-1, $RuCl_2(PPh_3)_3$.

(i) The synthesis of compound **24**, an intermediate in the synthesis of manzamine A (a novel anti-tumour alkaloid) by reaction (14), catalyzed by **18** (Table 8.1) (Borer 1994; also see Martin 1994a). The total synthesis of the ring system of manzamine A, with correct absolute stereochemistry at the chiral

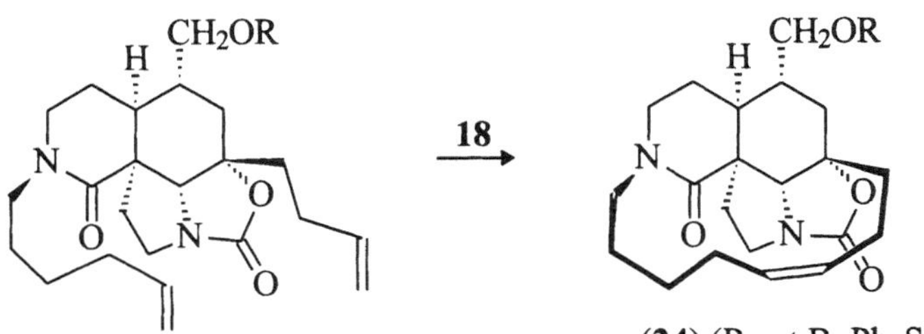

(**24**) (R = *t*-BuPh$_2$Si) (14)

centres, has been achieved through an extension of this work, using a second RCM reaction to form the azocine ring (not present in **24**); (Pandit 1996).

(ii) The synthesis of some hydroazulenes, such as **25** by reaction (15) catalyzed by $CH_3ReO_3/SiO_2\text{-}Al_2O_3$ (Junga 1993), or better still by **18** (Schneider, M.F. 1995). Such ring systems occur in many natural products of pharmacological interest.

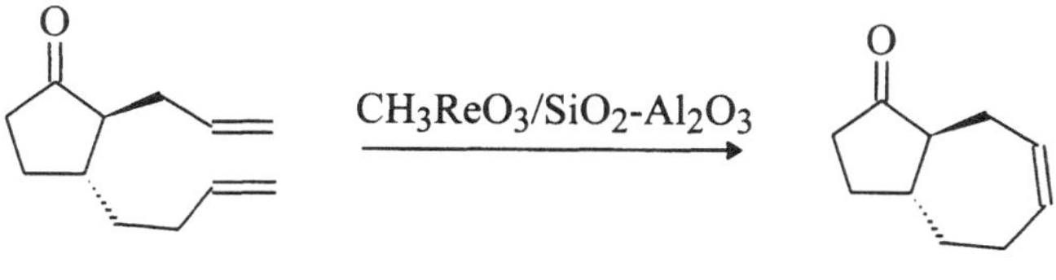

(15)

(**25**)

(iii) The synthesis of compound **26** by reaction (16) is remarkable in that there are 13 single bonds between the reacting double bonds, with only a single peptide link to provide a measure of conformational restraint. A 60% yield of a single stereoisomer (> 98% *Z*) is achieved when carried out in 0.01 M solution in THF, using 25 mol% **8** as catalyst. After hydrogenation and removal of the silyl protecting group one obtains a direct relative of the anti-fungal agent Sch 38516 (fluvirucin B_1) (Houri 1995).

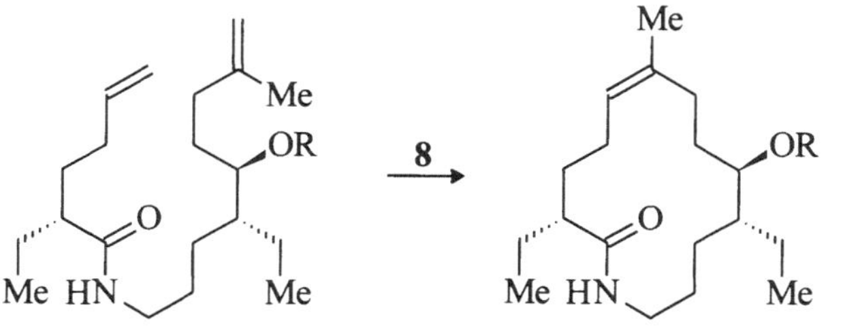

(16)

(**26**) (R = *t*-$BuMe_2Si$)

(iv) The synthesis of compound **27** by reaction (17) is catalyzed by **8**. After removal of the benzyl protecting groups one obtains the anti-fungal phytoalexine, *Sophora* compound I (Fujimura 1994).

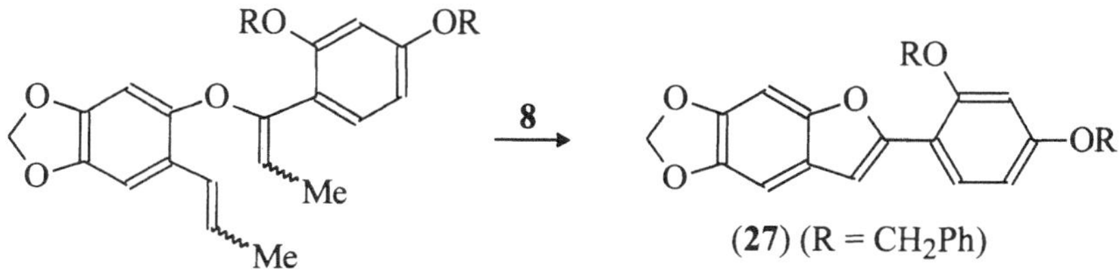

(17)

(**27**) (R = CH_2Ph)

(v) Some diene derivatives of glucosides undergo RCM under the influence of $WCl_4(OC_6H_3\text{-}Ph_2\text{-}2,6)_2/Bu_4Pb$ at 80°C to give bicyclic products containing 12- and 14-membered rings (Descotes 1996).

(vi) The eight-residue cyclic polypeptide *cyclo*[-(L-Phe-D-^{Me}NAla-L-Hag-D-^{Me}N-Ala)$_2$-], containing two L-homoallylglycine (Hag) residues, self-assembles to form two interconverting hydrogen-bonded dimers, one of which is represented by **28** and contains two pairs of double bonds in sufficiently close proximity that each pair undergoes RCM in the presence of **18** to give **29** with a conversion of 65% (mixture of *cc*, *ct*, and *tt* isomers); eqn. (18). Such a strategy may be useful in stabilizing kinetically labile α-helical and β-sheet peptide secondary structures (Clark, T.D. 1995).

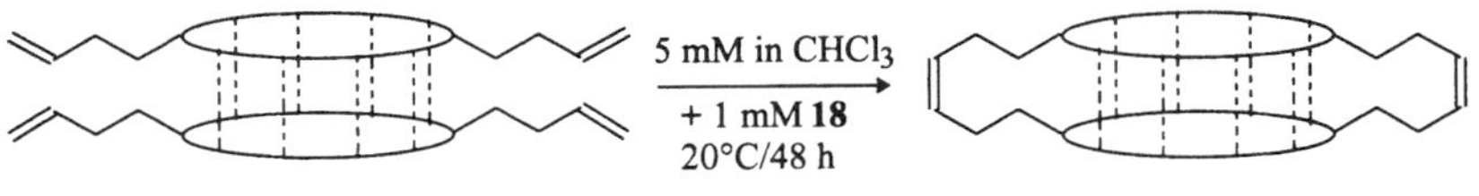

(28) **(29)** (18)

(vii) RCM is also possible between pairs of double bonds contained in the side chains of polymers, for example, atactic 1,2-polybutadiene (**30**); eqn (19).

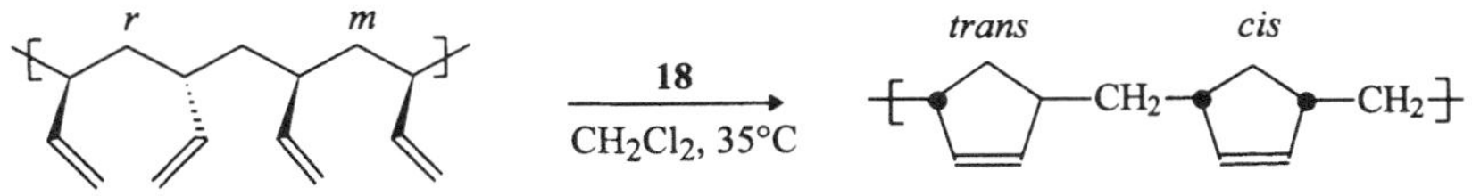

(30) **(31)** (19)

The reaction proceeds to 90% conversion in 30 min and then much more slowly to 97% conversion in 200 min. The first stage corresponds to the random reaction of adjacent double bonds in the chain, leaving 13.5% of isolated vinyl groups. These react more slowly by secondary metathesis with the double bonds in the neighbouring cyclopentene rings, thereby causing the vinyl group in effect to move along the chain until it meets another isolated vinyl group with which it can undergo RCM. The product (**31**) contains two types of repeat unit, *trans* and *cis*, according to whether the reacting dyad was *r* or *m*, giving rise to distinct olefinic ^{1}H NMR signals (Coates 1996).

For a review of this important, rapidly developing area, see Schmalz (1995) and Grubbs (1995); see also McKervey (1996) for an application to the synthesis of bridged- and oligo-calix[4]arenes.

8.11 Copolymers by metathesis condensation

The results of attempted copolymerizations, mostly successful, of deca-1,9-diene (M_1) with other dienes (M_2) are summarized in Table 8.8.

In system 1 the product has a statistical distribution of [$=CH(CH_2)_2CH=$] and [$=CH(CH_2)_6CH=$] units as shown by the olefinic region of the ^{13}C NMR spectrum, where the carbons in the M_1M_2 junctions are clearly resolved from those in the M_1M_1 and M_2M_2 junctions (see Fig. 8.4). The double bonds in the polymer chain

Table 8.8 Acyclic diene metathesis copolymerizations between deca-1,9-diene (M_1) and other dienes (M_2)

System	M_2	Catalyst	Notes	Reference
1	CH_2=CHCH$_2$ | CH_2=CHCH$_2$	**17**	Statistical copolymer, same composition as feed [a]	Wagener 1990
2	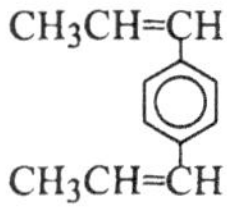	**17**	M_1 slightly less reactive than M_2 at high $[M_1]/[M_2]$; MW up to 3500	Wolf 1991
3	CH_2=CH SiMe$_2$ CH_2=CH	**17**	Feed of $[M_1]/[M_2]$ gives copolymer with 6% isolated units derived from M_2	Wagener 1991f
4	CH_2=CHCH$_2$CMe$_2$ CO CH_2=CHCH$_2$CMe$_2$	**8**	Reaction gives only homopolymer from M_1 and cyclohepten-5-one derivative from M_2	Forbes 1992
5	CH_2=CH(CH$_2$)$_3$CMe$_2$ CO CH_2=CH(CH$_2$)$_3$CMe$_2$	**8**	Copolymer, same composition as feed (1:1); MW = 30 600 [b]	Forbes 1992
6	[CH_2=CH(CH$_2$)$_4$]$_2$O	**17**	Statistical copolymer (M_1/M_2 = 1.95), 94% yield; MW = 6000	Wagener 1991e
7	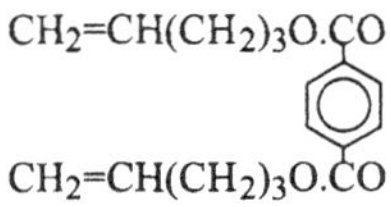	**8**	Statistical copolymer; MW = 4100	Patton, J.T. 1992
8	[CH_2=CH(CH$_2$)$_4$]$_2$S	**8**	Statistical copolymer, same composition as feed (1 : 1); MW = 35 000	O'Gara 1993a
9	1,1'-Divinylferrocene	**17**	Copolymer containing one or two units derived from M_2 per chain; MW = 3000	Gamble 1992
10	CH_2=CH(CH$_2$)$_x$CR CO (CH$_2$)$_9$ CH_2=CH(CH$_2$)$_x$CR	**8**	$x = 3$, R = Me: $M_n = 5000$ $x = 1$, R = H: $M_n = 640$ $x = 1$, R = Me: $M_n = 4900$	Wagener 1993b

[a] The same result can be achieved by exposing a mixture of homopolymer and the other monomer to the metathesis catalyst. [b] Low–molecular-weight polymers can also be obtained from diene derivatives of cyclododecanone.

readily undergo secondary metathesis reactions as indicated by the formation of copolymers by reaction of a homopolymer with another monomer in the presence of a tungsten carbene catalyst.

For system 2, 1,4-dipropenylbenzene (M_2) by iself gives only 'dimer' and a trace of 'trimer', but in the presence of M_1 it gives a copolymer having a composition not

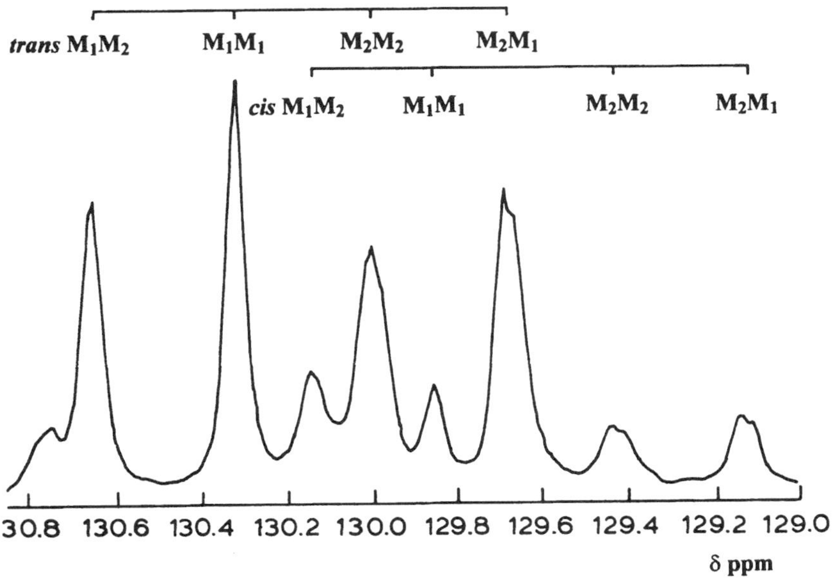

Fig. 8.4 ^{13}C NMR spectrum (olefinic region) for copolymer obtained by the metathesis condensation of deca-1,9-diene (M_1) with hexa-1,5-diene (M_2) in the presence of $W(=CHCMe_3)(=NC_6H_3\text{-}i\text{-}Pr_2\text{-}2,6)[OCMe(CF_3)_2]_2$. M_1M_2 denotes the italicized carbon in the dyad $=CH(CH_2)_6$*CH*$=CH(CH_2)_2CH=$; the adjacent olefinic carbon is denoted by M_2M_1 (Wagener 1990).

very different from that of the feed. For system 3, divinyldimethylsilane (M_2) is unreactive by itself, but with M_1 gives a copolymer, $M_n = 6400$, containing about three isolated units of M_2 per chain.

Systems 4 and 5 provide an interesting contrast. The thermodynamic stability of the cyclohepten-5-one derivative produced in system 4 effectively prevents any units of M_2 from being incorporated in the polymer of M_1. The carbonyl group is evidently sufficiently protected to prevent it from undergoing a Wittig-like reaction with the metal carbene complex.

In all cases where copolymer is formed, the product gives a single GPC peak, showing that it is not admixed with an independently formed homopolymer. The ability to form copolymers by ADMET reactions of unsaturated homopolymers with a diene monomer is not confined to system 1 and is probably quite general. For an example of insertion of hydrocarbon units into an unsaturated polysiloxane, see Smith, D.W. (1993a).

In summary, three types of behaviour have been observed: (i) both M_1 and M_2 undergo ADMET polymerization separately and form statistical copolymers when reacted together (systems 1, 5–9 in Table 8.8); (ii) M_2 by itself does not give long-chain polymer under ADMET conditions, but forms a copolymer with M_1 (systems 2 and 3); (iii) M_2 by itself gives a cyclic species under ADMET conditions, and still does so in the presence of M_1 (system 4).

9

Cross-metathesis between Acyclic Compounds

9.1 Introduction

Some examples of cross-metathesis between isotopically labelled alkenes have been given in Chs. 3 and 5. They demonstrate not only the fission of the double bond but also the occurrence of non-productive metathesis. Here we summarize the information on all types of cross-metathesis reaction involving only acyclic compounds. Cross-metathesis between cyclic and acyclic compounds is discussed in Ch. 15.

The simplest example of a productive cross-metathesis reaction between acyclic olefins is that between ethene and but-2-ene; reaction (1). In this case only one product is possible, apart from *cis/trans* isomerization of the but-2-ene; the equilibrium mixture thus consists of four compounds. At the other extreme, the reaction of two unsymmetrical olefins, $R^1CH{=}CHR^2$ and $R^3CH{=}CHR^4$, with R^1, R^2, R^3, R^4 all different, can produce *cis/trans* isomers of four different unsymmetrical olefins by cross-metathesis as well as four symmetrical olefins by self-metathesis. Counting the *cis/trans* isomers of the reactants as well, this means that the equilibrium mixture will contain 20 different compounds. Side reactions, such as double-bond shift reactions, will complicate the situation still further. The main value of cross-metathesis reactions, apart from their use in the proof of mechanism, lies in their application to the synthesis of olefins that are otherwise expensive or difficult to prepare. A number of higher olefins, useful as insect sex attractants, have been made in this way.

$$\begin{array}{c} CH_2{=}CH_2 \\ + \\ MeCH{=}CHMe \end{array} \rightleftharpoons \begin{array}{c} CH_2 \\ \| \\ MeCH \end{array} + \begin{array}{c} CH_2 \\ \| \\ CHMe \end{array} \qquad (1)$$

An indication of equilibrium positions for various types of cross-metathesis reaction is shown in Fig. 9.1. It will be seen that, in general, the equilibrium favours the less substituted olefins. However, it is sometimes possible to drive the equilibrium in the opposite direction by continuous removal of the more volatile components of the equilibrium mixture (see Ch. 5). In the rest of this chapter we

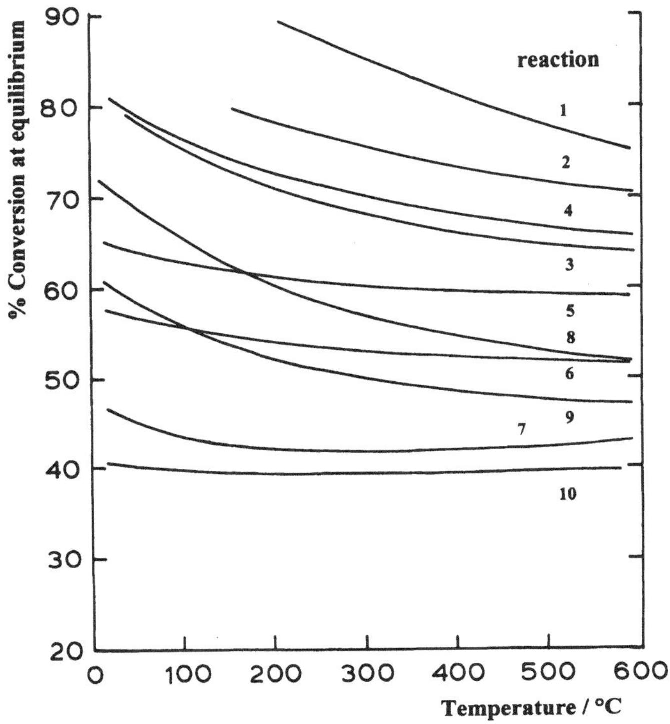

Fig. 9.1 Predicted equilibrium conversions for some cross-metathesis reactions, starting from equal proportions of reactants (adapted from Banks 1971).

1. $C_2H_4 + Me_2C{=}CMe_2 \rightleftharpoons 2Me_2C{=}CH_2$
2. $C_2H_4 + EtMeC{=}CHMe \rightleftharpoons EtMeC{=}CH_2 + MeCH{=}CH_2$
3. $C_2H_4 + Me_2C{=}CHEt \rightleftharpoons Me_2C{=}CH_2 + EtCH{=}CH_2$
4. $C_2H_4 + Me_2C{=}CHMe \rightleftharpoons Me_2C{=}CH_2 + MeCH{=}CH_2$
5. $C_2H_4 + EtCH{=}CHMe \rightleftharpoons EtCH{=}CH_2 + MeCH{=}CH_2$
6. $C_2H_4 + MeCH{=}CHMe \rightleftharpoons 2MeCH{=}CH_2$
7. $C_2H_4 + EtCH{=}CHEt \rightleftharpoons 2EtCH{=}CH_2$
8. $MeCH{=}CH_2 + EtMeC{=}CHMe \rightleftharpoons EtMeC{=}CH_2 + MeCH{=}CHMe$
9. $MeCH{=}CH_2 + Me_2C{=}CHMe \rightleftharpoons Me_2C{=}CH_2 + MeCH{=}CHMe$
10. $MeCH{=}CH_2 + EtCH{=}CHMe \rightleftharpoons EtCH{=}CH_2 + MeCH{=}CHMe$

shall consider some specific examples, taking the olefins in order of increasing carbon number of the lower-molecular-weight reactant, except for functionalized olefins, which are discussed separately at the end.

9.2 Ethene

The cross-metathesis of ethene with higher olefins has been termed 'ethenolysis' (Bradshaw 1967), and provides a useful means of reducing the extent of substitution of an olefin feedstock and of converting cyclic olefins into linear dienes, trienes, etc. Some examples for acyclic olefins are shown in Table 9.1. The reactions are best done at 50 bar in order to drive the reaction to the right and to minimize self-metathesis of the substrate.

Ethenolysis can also be applied to difficult *analytical* problems: it provides a simple and accurate method for locating double bonds and determining isomer purity of internal (poly)olefins with or without functional groups, via analysis of the terminal alkenes produced. Direct chromatographic separation of positional isomers is often impossible, while other methods for determining the double-bond position, such as ozonization followed by subsequent reduction, are far more complicated.

9.3 Propene

The reaction with isobutene (Fig. 9.1, reverse of reaction 4) proceeds readily on WO_3/SiO_2 at 450°C. Equilibrium can be achieved in a flow system at 20 bar. The reaction provides a useful route to 2-methylbut-2-ene, and thence to isoprene and polyisoprene (Banks 1971, 1984a). Reaction with butadiene likewise gives penta-1,3-diene (Heckelsberg 1969b).

Cross-metathesis of propene with pent-2-ene (Fig. 9.1, reaction 10), when catalyzed by $WCl_6/Et_3Al_2Cl_3/EtOH$, is accompanied by loss of about two-thirds of the propene, presumably by addition polymerization; only traces of ethene are detected (Zowade 1973). Metathesis of a 1 : 2 mixture of propene and methylenecyclobutane over Re_2O_7/Al_2O_3 at 20°C gives a 20% yield of ethylidenecyclobutane, eqn. (2); but-2-ene and dicyclobutylidene are formed as by-products (Finkel'shtein 1977). When the Re_2O_7/Al_2O_3 catalyst is promoted with Bu_4Sn, the reaction of methylenecyclobutane with hex-1-ene and hept-1-ene gives 60–65% of the cross-metathesis product (Finkel'shtein 1992a).

(2)

9.4 Butenes

Reactions with the lower olefins are described in the preceding sections. The cross-metathesis of but-1-ene with *cis*- or *trans*-but-2-ene on Re_2O_7/Al_2O_3 at 30–40°C is accompanied by *cis/trans* isomerization of the but-2-ene, but not by self-metathesis

Table 9.1 Ethenolysis of acyclic olefins

Reaction[a]	Catalyst[b]	Temp./°C	Reference
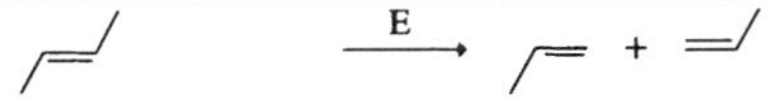	$MoO_3/CoO/Al_2O_3/NaOH$ Re_2O_7/Al_2O_3	120–240 35	Bradshaw 1967 Amigues 1991
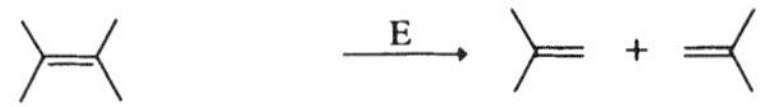	$MoO_3/Al_2O_3/KOH$	165	Crain 1969
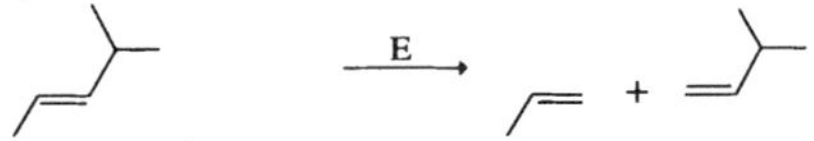	$MoO_3/CoO/Al_2O_3/NaOH$	120–240	Bradshaw 1967
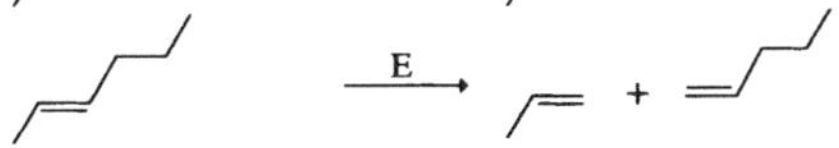	WO_3/SiO_2	450	Banks 1971
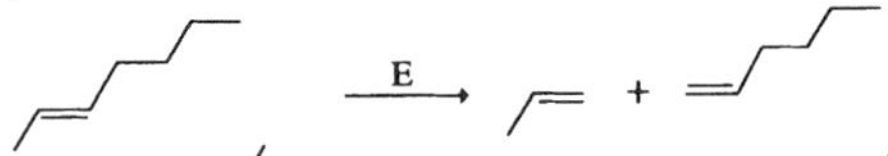	$MoCl_2(NO)_2(PPh_3)_2$[c]$/Me_3Al_2Cl_3$	0–20	Zuech 1970
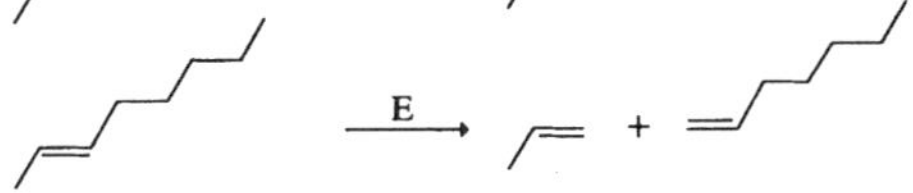	$MoO_3/Al_2O_3/KOH$	165	Crain 1969
$C_6H_{13}CH{=}CHC_6H_{13} \xrightarrow{E} 2\ CH_2{=}CHC_6H_{13}$	Re_2O_7/Al_2O_3	80	Warwel 1982a
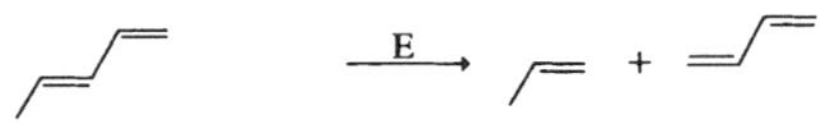	$Re_2O_7/Al_2O_3/R_4Sn$	20	Fridman 1978
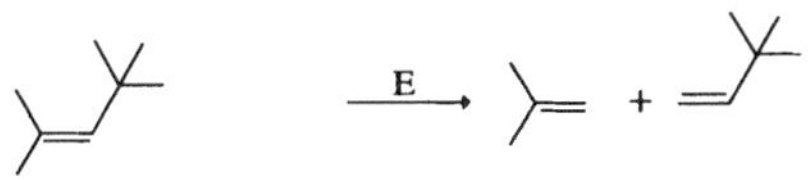	WO_3/SiO_2	370	Hughes 1977 Banks 1982
$PhCH = CHPh \xrightarrow{E} CH_2{=}CHPh$	$WO_3/SiO_2 + K_2O$[d]	480	Montgomery 1976

[a] E = ethene. [b] Double-bond shift reactions are sometimes a complication with MoO_3- and WO_3-based catalysts. [c] Many Mo and W compounds can be activated by exposure to NO for 30 min (suspended in chlorobenzene) followed by removal of excess NO by evacuation and then addition of dry pyridine. [d] See Ch. 17.

of but-1-ene, which is evidently inhibited by the presence of but-2-ene. Reaction with the *trans* isomer proceeds without an induction period and more rapidly than reaction with the *cis* isomer, which is preceded by an induction period. The rate of the cross-metathesis reaction relative to *cis/trans* isomerization is higher on Re_2O_7/Al_2O_3 than on MoO_3/Al_2O_3; also the proportion of *trans*-pent-2-ene in the products is higher for reaction of *trans*-but-2-ene than for reaction of *cis*-but-2-ene. In all cases more propene is formed than pent-2-ene (Engelhardt 1982d, 1985b). A mixed catalyst based on $Ni(acac)_2$ and WCl_6 has been used to effect ethene dimerization coupled with cross-metathesis of the resulting but-1-ene with but-2-ene (Muthukumaru Pillai 1992).

$$\begin{array}{c} CH_2{=}CH{-}CMe{=}CH_2 \\ + \\ MeCH{=}CHMe \end{array} \longrightarrow \begin{array}{c} CH_2 \\ \| \\ MeCH \end{array} + \begin{array}{l} CH{-}CMe{=}CH_2 \\ \| \\ CHMe \end{array} + \begin{array}{r} CH_2{=}CH{-}CMe \\ \| \\ MeCH \end{array} \qquad (3)$$

Reaction of but-2-ene with isobutene (Fig. 9.1, reverse of reaction 9) on WO_3/SiO_2 at 450°C (Banks 1971) or on MoO_3/Al_2O_3 (Nakamura, R. 1972a), or Re_2O_7/Al_2O_3 (Nakamura, R. 1972a, 1976c) at 150°C provides another route to 2-methylbut-2-ene. Incorporation of KOH into the MoO_3/Al_2O_3 (K/Mo = 1/7.7) prevents dimerization of the isobutene at the acidic sites. $WCl_4[OCH(CH_2Cl)_2]_2$/*i*-Bu_2AlCl catalyzes the cross-metathesis of but-2-ene with pent-1-ene without inducing any self-metathesis of pent-1-ene (Dolgoplosk 1977b). But-2-ene reacts with isoprene on WO_3/SiO_2 at 538°C to give a low yield (2%) of the various possible cross-metathesis products; reaction (3). Likewise isobutene and butadiene give 4-methylpenta-1,3-diene (Heckelsberg 1969b). Reaction between but-2-ene-d_8 and hex-3-ene on $WCl_6/EtAlCl_2/EtOH$ gives a *cis/trans* mixture of pent-2-ene having mass 74 as sole products, indicating preservation of the alkylidene moieties; reaction (4) (Calderon 1968). Reaction of but-2-ene with oct-4-ene to give an equilibrium mixture with hex-2-ene also occurs very readily (Kroll 1972; Doyle 1973). Cross-metathesis of a two-fold excess of a symmetrical internal alkenes (but-2-ene, dec-5-ene, etc.) with 4-vinylcyclohexene, reaction (5), in the presence of the catalyst $WCl_6/Bu_4Sn/Et_2O$ gives a conversion of > 80% at 50°C with a high selectivity (84–92%) for the formation of the 4-alkenylcyclohexenes (Warwel 1982b). The double bond in the ring is not involved; its participation in the metathesis reaction is thermodynamically unfavourable.

$$\begin{array}{c} CD_3CD{=}CDCD_3 \\ + \\ C_2H_5CH{=}CHC_2H_5 \end{array} \rightleftharpoons \begin{array}{c} CD_3CD \\ \| \\ C_2H_5CH \end{array} + \begin{array}{c} CDCD_3 \\ \| \\ CHC_2H_5 \end{array} \qquad (4)$$

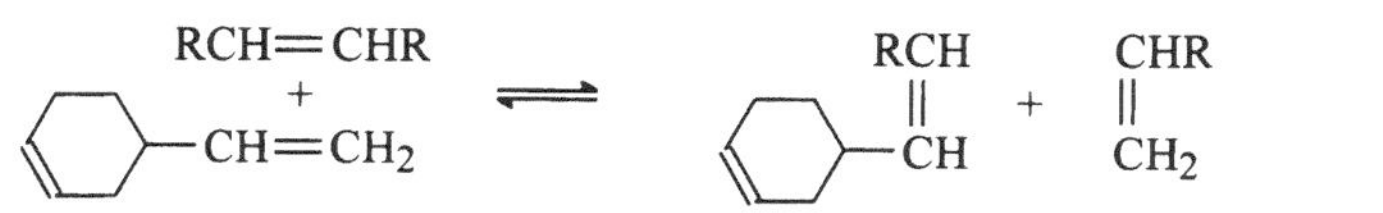

(5)

Cross-metathesis of isobutene (isobutenolysis) with higher terminal alkenes or symmetrical internal alkenes on $Re_2O_7/Al_2O_3/Me_4Sn$ leads to 2-methylalk-2-enes with conversions of 70–80%. However, in both types of reaction, products of self-metathesis of the substrate alk-1-ene or of the alk-1-ene formed in the isobutenolysis of the symmetrical alkene, are also formed (Warwel 1988b). Isobutenolysis of hexa-1,5-diene gives 6-methylhepta-1,5-diene (an intermediate in the synthesis of vitamins and carotenoids); reaction (6). A yield of $>20\%$ at 40°C is obtained using the catalyst $Re_2O_7/Al_2O_3/Bu_4Sn$ (Fitton 1974).

$$\begin{array}{c} CH_2{=}CH(CH_2)_2CH{=}CH_2 \\ + \\ Me_2C{=}CH_2 \end{array} \longrightarrow \begin{array}{c} CH_2{=}CH(CH_2)_2CH \\ \| \\ Me_2C \end{array} + \begin{array}{c} CH_2 \\ \| \\ CH_2 \end{array} \qquad (6)$$

9.5 Pentenes

The reactions with lower olefins are described above. In cross-metathesis of pent-1-ene with pent-2-ene, catalyzed by $MoO_3/CoO/Al_2O_3$, $WCl_6/EtAlCl_2/EtOH$, or $Bu_4N[MoCl(CO)_5]/MeAlCl_2$, all possible products are observed, including ethene and oct-4-ene. However, the non-productive metathesis of pent-1-ene occurs 10–1000 times faster than productive metathesis, depending on the initiator (Kelly 1975; Mocella 1976a). The rate of self-metathesis of pent-2-ene is sharply reduced by small amounts of terminal olefins such a pent-1-ene (Calderon 1976) or non-1-ene (Mocella 1976a), suggesting that the terminal olefins can coordinate more strongly than internal olefins at the active site. However, the presence of traces of terminal olefins is beneficial for the initiation of reactions of *trans* alk-2-enes on some catalysts (Bilhou 1977b). In the presence of $Mo(CO)_6/ZrCl_4/h\nu$ pent-1-ene undergoes substantial isomerization to pent-2-ene, which then cross-metathesizes with pent-1-ene without itself undergoing appreciable self-metathesis (Szymańska-Buzar 1987). In the reaction of pent-1-ene with 4-methylpent-2-ene, the cross-products, hex-2-ene and 2-methylhept-3-ene, predominate over the products of self-metathesis. A curious feature of the catalyst $Mo(CO)_5(py)/EtAlCl_2/Bu_4NCl$ is that, although it will metathesize terminal olefins, it will not induce their cross-metathesis with alk-2-enes (Motz 1977).

The reaction of *cis*-pent-2-ene with 4-methylpent-1-ene has been investigated in some detail under conditions that allowed most of the ethene and propene to escape; reaction (7). As may be seen from Fig. 9.2, the reaction is very rapid when catalyzed by $Bu_4N[MoCl(CO)_5]/MeAlCl_2$ and reaches a steady state after 8 min. The cross-metathesis products, 6-methylhept-3-ene and 5-methylhex-2-ene, are formed in somewhat larger amount than the self-metathesis products, hex-3-ene and 2,7-dimethyloct-4-ene. The initial and final *trans/cis* ratios, obtained by plotting Doyle's data according to the method of Fig. 6.2, are (final values shown in brackets): but-2-ene 0.9 (2.5), 5-methylhex-2-ene 1.6 (3.3), 6-methylhept-3-ene 2.5

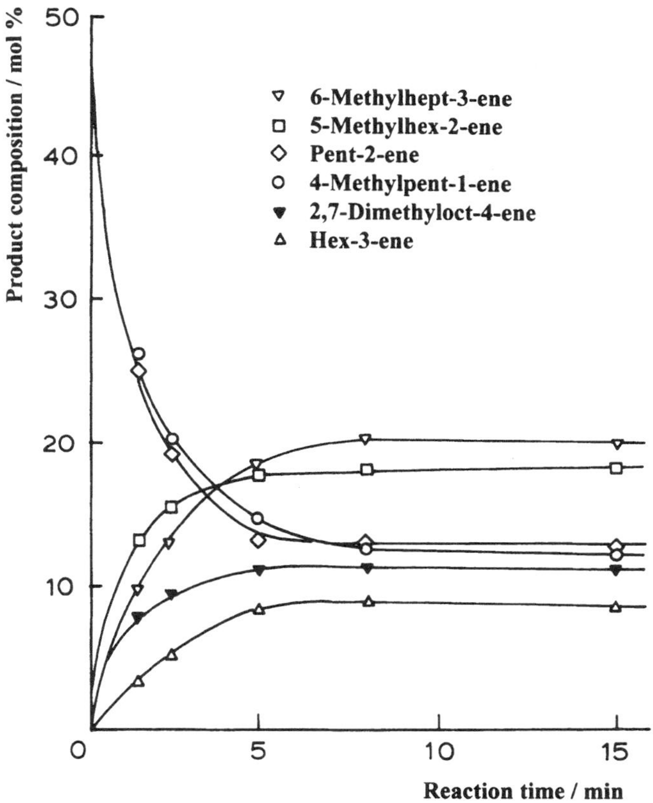

Fig. 9.2 Composition of reaction mixture as a function of time in the cross-metathesis of a 1 : 1 mixture of *cis*-pent-2-ene with 4-methylpent-1-ene. Catalyst: $Bu_4N[MoCl(CO)_5]$/ $MeAlCl_2$ (Doyle 1973).

(5.0), 2,7-dimethyloct-4-ene 2.1 (5.0). For the olefins with the bulkier substituents, there is thus a bias towards the *trans* isomer from the start of reaction, which may be attributed to the dominance of the 1,2 interactions in the metallacycle transition state (see Section 6.7).

EtCH=CHMe + *i*-BuCH=CH_2 →

→ EtCH=CHEt + MeCH=CHMe

→ *i*-BuCH=CH-*i*-Bu + CH_2=CH_2

→ *i*-BuCH=CHEt + MeCH=CH_2 (7)

→ *i*-BuCH=CHMe + EtCH=CH_2

The reaction of pent-2-ene with dodec-6-ene leads to an equilibrium mixture of *cis/trans* isomers of six different internal olefins which can be readily analyzed by GC. The equilibrium proportions of the olefins are very close to those expected for a random exchange of alkylidene moieties (Fig. 9.3).

9.6 Hexenes

Cross-metathesis reactions with lower olefins are discussed in the earlier sections. The reaction of hex-1-ene with hept-1-ene-1,1-d_2 (Mocella 1976a) and with oct-1-ene-1,1-d_2 (McGinnis 1976) results both in exchange of methylene groups and in productive metathesis; reaction (8). The non-productive reaction always far outstrips the productive reaction ($k/k' \gg 1$), to an extent that depends on the catalyst (see Table 9.2).

The reaction of hex-1-ene with higher alk-1-enes, catalyzed by Re_2O_7/Al_2O_3 (Spronk 1991b) or $Re_2O_7/CsNO_3/Al_2O_3$ (Kawai 1988), results in the products

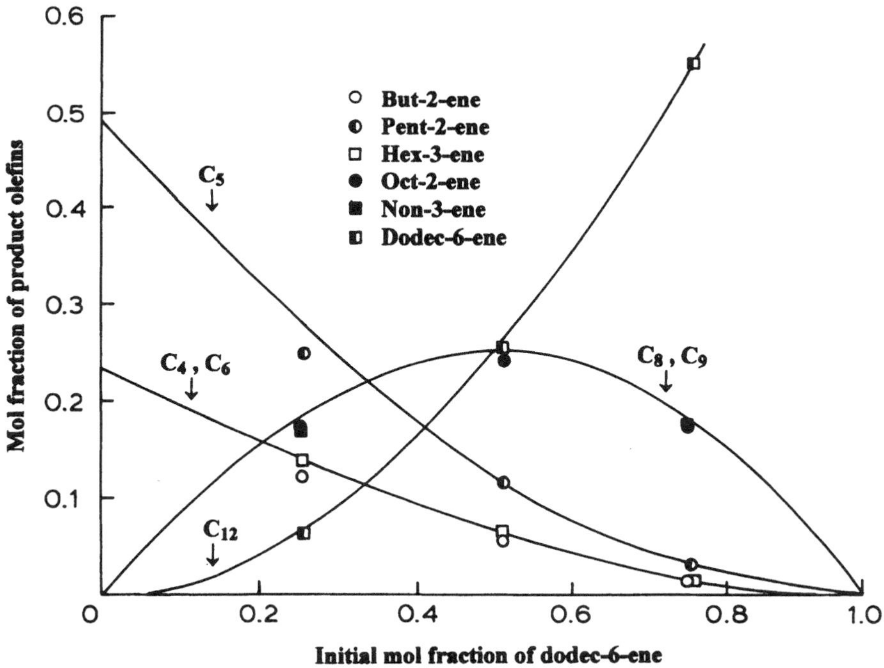

Fig. 9.3 Metathesis of pent-2-ene with dodec-6-ene. Equilibrium proportions of reactants and products at different starting ratios of the two reactants. Solid lines correspond to random exchange of alkylidene moieties (thermoneutral reaction) (Calderon 1968).

Table 9.2 Ratio of non-productive to productive metathesis in the reaction of hex-1-ene with oct-1-ene-1,1-d_2[a] (McGinnis 1976)

Catalyst system	Temp./°C	k/k'
$MoCl_2(PPh_3)_2(NO)_2/Me_3Al_2Cl_3$	0	26
$W(=CPh_2)(CO)_5$	50	75
$WCl_6/BuLi$	20	80
WCl_6/Ph_3EtSn	20	103
$WCl_6/EtAlCl_2/EtOH$	20	155

[a] Derived from the products measured as a function of time.

$$R^1CH{=}CD_2 + R^2CH{=}CH_2 \xrightarrow{k} R^1CH{=}CH_2 + R^2CH{=}CD_2$$

$$R^1CH{=}CD_2 + R^2CH{=}CH_2 \xrightarrow{k'} \begin{cases} R^1CH{=}CHR^1 + CD_2{=}CD_2 \\ R^1CH{=}CHR^2 + CH_2{=}CD_2 \\ R^2CH{=}CHR^2 + CH_2{=}CH_2 \end{cases} \qquad (8)$$

of both self-metathesis and cross-metathesis, reaction (9). For a 1 : 1 mixture of hex-1-ene and oct-1-ene ($n=5$), the ratio of products, $C_{10}:C_{12}:C_{14}$, is about 1:2:1, with C_{10} slightly favoured over C_{14} (see Fig. 9.4). This near-random exchange of alkylidene groups is in contrast to the large decrease in reactivity with increasing chain length observed for the separate self-metathesis of the alk-1-enes. This observation has its parallel in copolymerization, where a monomer that is relatively unreactive towards homopolymerization can readily enter into copolymerization with a more reactive monomer as a result of favourable cross-propagation reactions. Product desorption effects may also be a factor in the cross-metathesis of acyclic olefins on solid catalysts.

$$\begin{matrix} CH_2{=}CH(CH_2)_3CH_3 \\ + \\ CH_2{=}CH(CH_2)_nCH_3 \end{matrix} \rightleftharpoons \begin{matrix} CH_2 \\ \| \\ CH_2 \end{matrix} + \begin{matrix} CH(CH_2)_3CH_3 \\ \| \\ CH(CH_2)_nCH_3 \end{matrix} \qquad (9)$$

The reaction of hex-1-ene with *cis*-alk-2-enes, catalyzed by $Re_2O_7/CsNO_3/Al_2O_3$ at 31°C, results in the products of both self-metathesis and cross-metathesis, the latter being represented by eqns. (10) and (11). For reaction with *cis*-hex-2-ene ($n=4$), the initial ratio of non-4-ene to hept-2-ene produced in reactions (10) and (11), repectively, is 0.62, but the self-metathesis of *cis*-hex-2-ene to give oct-4-ene is six times faster than that of hex-1-ene to give dec-5-ene (Kawai 1988). However, like pent-2-ene, hex-2-ene fails to cross-metathesize with terminal olefins on $Mo(CO)_5(py)/EtAlCl_2/Bu_4NCl$ (Motz 1977).

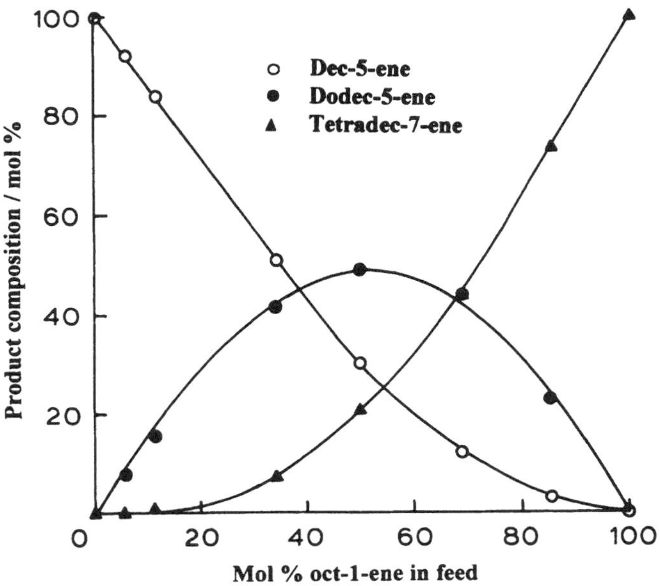

Fig. 9.4 Product distribution for the cross-metathesis of hex-1-ene and oct-1-ene on Re_2O_7/Al_2O_3 at 20°C and 9 bar, as a function of oct-1-ene content in the feed (Spronk 1991b).

$$C_1{=}C_5 + C_2{=}C_n \rightleftharpoons C_1{=}C_2 + C_5{=}C_n \tag{10}$$

$$C_1{=}C_5 + C_n{=}C_2 \rightleftharpoons C_1{=}C_n + C_5{=}C_2 \tag{11}$$

Cross-metathesis of hex-1-ene with an equimolar amount of tetradec-7-ene in the presence of $WCl_6/Bu_4Sn/Et_2O$ at 50°C results in a 76% conversion of hex-1-ene and a selectivity of 68% for the cross-metathesis product dodec-5-ene (Warwel 1983a). Using a four-fold excess of the symmetrical alkene, the conversion of the alk-1-ene is 90% with a selectivity of 90% for the cross-metathesis product.

4-Methylpent-1-ene reacts with dodec-1-ene (Küpper 1976) and tetradec-1-ene (Rossi 1975) on Mo-based catalysts. The latter, reaction (12), provides a route to 2-methylheptadecane, a sex pheromone for at least nine species in the family of *Arctiidae*. 4-Methylpent-1-ene undergoes cross-metathesis with a symmetrical alkene in the same way as linear alk-1-enes such as hex-1-ene. 3-Methylpent-1-ene, which is less reactive in self-metathesis, reacts readily via cross-metathesis with an equimolar amount of a symmetrical alkene (dec-5-ene, tetradec-7-ene, octadec-9-ene) (Warwel 1983a). 3,3-Dimethylbut-1-ene (neohexene), which is inactive in self-metathesis, undergoes cross-metathesis with internal olefins to high conversion in the presence of $WCl_6/Bu_4Sn/Et_2O$ or Re_2O_7/Al_2O_3 (Warwel 1983a).

$$\begin{array}{c} CH_2{=}CHCH_2CH(CH_3)_2 \\ + \\ CH_2{=}CHC_{12}H_{25} \end{array} \rightleftharpoons \begin{array}{c} CH_2 \\ \| \\ CH_2 \end{array} + \begin{array}{c} CHCH_2CH(CH_3)_2 \\ \| \\ CHC_{12}H_{25} \end{array} \quad (12)$$

A selective synthesis of 2-methylalk-2-enes succeeds via cross-metathesis of 2,3-dimethylbut-2-ene with symmetrical internal alkenes (oct-4-ene, dec-5-ene, etc.) on $Re_2O_7/Al_2O_3/Me_4Sn$; eqn. (13) (Warwel 1988b). Cross-metathesis of 2,3-dimethylbut-2-ene with hexa-1,5-diene in the presence of $Re_2O_7/CsNO_3/Al_2O_3$ at 25°C provides another route to 6-methylhepta-1,5-diene. The conversions of 2,3-dimethylbut-2-ene and hexa-1,5-diene are 37% and 60% respectively, with 32% selectivity for the product (Kawai 1984b).

$$RCH{=}CHR \quad + \quad Me_2C{=}CMe_2 \longrightarrow 2\,RCH{=}CMe_2 \quad (13)$$

9.7 Higher olefins

Some cross-metathesis reactions of higher olefins are collected in Table 9.3; also see Kawai (1989).

C_{10}–C_{18} olefins (feedstock for the manufacture of surfactants) can be obtained from lower-molecular-weight alkenes, such as octenes by combining self-metathesis, cross-metathesis and isomerization reactions using the homogeneous bifunctional catalyst system $Ph_3Ge\text{-}W(CO)_3(Cp)/i\text{-}BuAlCl_2/O_2$ or $W(H)(CO)_3(Cp)/i\text{-}BuAlCl_2/O_2$ (at 20°C) or the heterogeneous bifunctional catalyst $Re_2O_7/Pt/Al_2O_3$ (at 120–160°C) (Warwel 1982a).

1-Phenyl-3-methylbut-2-ene does not cross-metathesize with 1-phenylbut-2-ene but reduces the rate of self-metathesis of the latter (Chevalier 1976). In the cross-meta-thesis of 1-phenylbut-2-ene with allylbenzene, the light olefins initially produced contain only ethene and propene, and no but-2-ene. Reaction (14) is evidently much preferred to reaction (15). But-2-ene forms as the result of a secondary metathesis reaction of propene and eventually reaches its equilibrium concentration.

$$\begin{array}{c} [Mt]{=}CHMe \\ + \\ PhCH_2CH{=}CH_2 \end{array} \longrightarrow \begin{array}{c} [Mt] \\ \| \\ PhCH_2CH \end{array} + \begin{array}{c} CHMe \\ \| \\ CH_2 \end{array} \quad (14)$$

$$\begin{array}{c} [Mt]{=}CHMe \\ + \\ PhCH_2CH{=}CHMe \end{array} \longrightarrow \begin{array}{c} [Mt] \\ \| \\ PhCH_2CH \end{array} + \begin{array}{c} CHMe \\ \| \\ CHMe \end{array} \quad (15)$$

The reaction of 2,4,4-trimethylpent-2-ene with 4-vinylcyclohexene, eqn. (16), proceeds with a selectivity of $> 75\%$ in the presence of Re_2O_7/Al_2O_3 if the catalyst is pretreated with $Cu(NO_3)_2$ to suppress side reactions (Lyons 1988).

Table 9.3 Examples of cross-metathesis of higher olefins with each other

Reactants[a]		Catalyst[b]	Notes	Reference
2-C_7	2-C_8	Re-based	No side reactions	Kawai 1988
3-C_7	2-C_8	Re-based	No side reactions	Kawai 1988
1-C_8	*cis*-2-C_8 *trans*-2-C_8	WCl_6/Ph_4Sn	Initial rates of formation: 2-C_9 > 6-C_{13} > 7-C_{14}	Uchida, A. 1981
1-C_8	*trans*-2-C_8	$W(OPh)_6/EtAlCl_2$	Initial products all *trans* for both *cis* and *trans* reactants	Uchida, A. 1981
1-C_8	4-C_9	Re-based	No side reactions	Kawai 1988
1-C_8-1-d_2	*cis*-1-C_{10}-1-d_1	Mo-, W-, or Re-based	$c/t > 1$ for the product $C_6H_{13}CH{=}CHD$; see Ch. 6	Casey 1978
1-C_8	1-C_{10}	Re-based	The mixture of C_{14}–C_{18} alkenes is dimerized to lube oil range hydrocarbons	Nelson 1983
1-C_8	1-C_{10}	Re-based	Statistical product distribution	Spronk 1991b
1-C_8	5-C_{10}	Re-based	Example of check of isomer purity of a symmetrical alkene	Heckelsberg 1983
1-C_8	5-C_{10}	$WCl_6/Bu_4Sn/Et_2O$	Results comparable to those of cross-metathesis of 1-C_6 with 7-C_{14} (Section 9.6)	Warwel 1983a
2-C_8	3-C_8	Re-based	No side reactions	Kawai 1988

$4\text{-}C_8$	4-Vinylcyclohexene	$WCl_6/EtAlCl_2$	Product is 4-pentenylcyclohexene	Pinazzi 1977a
$4\text{-}C_8$	$7\text{-}C_{14}$	$W(CO)_6/CCl_4/h\nu$	56% yield of C_{11} at 25°C; selectivity 90%	Karan 1989
$7\text{-Me-}1\text{-}C_8$	$1\text{-}C_{12}$	Mo-based	22% yield of $2\text{-Me-}7\text{-}C_{18}$[c]	Küpper 1975, 1976
$1\text{-}C_9$	$2,5\text{-}Me_2\text{-}2,4\text{-}C_6$	Re-based	20% yield of $2\text{-Me-}2,4\text{-}C_{12}$	Woerlee 1984
$1\text{-}C_{10}$	4-Vinylcyclohexene (4-VCH)	Re_2O_7/Al_2O_3 50°C	82% conversion of 4-VCH, selectivity for 4-decenylcyclohexene 78% ($1\text{-}C_{10}$/4-VCH ratio = 4/1)	Warwel 1982a
$1\text{-}C_{10}$	$1\text{-}C_{15}$	Mo-based	26% yield of $9\text{-}C_{23}$[d]	Rossi 1975
$1\text{-}C_{10}$	$1\text{-}C_{15}$	$WCl_4(OC_6H_3\text{-}Cl_2\text{-}2,6)_2/Me_4Sn$	25% yield of $9\text{-}C_{23}$[d]	Quignard 1986
$1\text{-}C_{10}$	$1\text{-}C_{15}$	$[Mo]{=}CHCMe_2Ph$[e]/DME	Non-volatile products: 30% yield of $9\text{-}C_{23}$[d] (81% *trans*)	Fox 1994a
$1\text{-}C_{10}$	$1\text{-}C_{15}$	$MoCl_5/SiO_2/Me_4Sn$	79% yield of $9\text{-}C_{23}$[d]	Bykov 1988
$1\text{-}C_{10}$	$2\text{-}C_{16}$	W-based	5% yield of $9\text{-}C_{23}$[d]	Rossi 1975
$1\text{-}C_{14}$	$2\text{-}C_{16}$	Mo-based	10% yield of $13\text{-}C_{27}$[f]	Rossi 1975
$2\text{-}C_{16}$	$9\text{-}C_{18}$	Re-based	13% yield of $9\text{-}C_{23}$[d]	Rossi 1975
$PhCH_2CH{=}CH_2$	$PhCH_2CH{=}CHMe$	Mo- or W-based	16% $PhCH_2CH{=}CHCH_2Ph$ formed at equilibrium; $2\text{-}C_4$ not a primary product	Chevalier 1976

[a] $1\text{-}C_8$ = oct-1-ene; $4\text{-}C_8$ = oct-4-ene; $7\text{-Me-}1\text{-}C_8$ = 7-methyloct-1-ene; $2,5\text{-}Me_2\text{-}2,4\text{-}C_6$ = 2,5-dimethylhexa-2,4-diene; etc. [b] W-based catalysts are generally $WCl_6/EtAlCl_2/EtOH$ or $W(CO)_5(PPh_3)/EtAlCl_2/O_2$; Mo-based catalysts are usually $MoCl_2(PPh_3)_2(NO)_2/EtAlCl_2$, or $[Bu_4N]MoCl(CO)_5/EtAlCl_2$ or MoO_3/Al_2O_3; Re-based catalyst is usually Re_2O_7/Al_2O_3. [c] The *cis* epoxide is a sex pheromone of *Lymantria dispar*. [d] The *cis* isomer is a sex pheromone of the house-fly (*Musca domestica*). [e] $Mo({=}CHCMe_2Ph)({=}NC_6H_3\text{-}i\text{-}Pr_2\text{-}2,6)[OCMe(CF_3)_2]_2$. [f] The *cis* isomer is a sex pheromone of *Musca autumnalis*.

Cross-metathesis of equimolar amounts of styrene with symmetrical olefins (C_4–C_{18}) on Re_2O_7/Al_2O_3 occurs with high conversion at 50°C with 85–90% of the styrene yielding 1-phenylalk-1-enes and only 10% being converted into its self-metathesis product stilbene. In cross-metathesis with terminal alkenes, at least a two-fold excess of the terminal alkene is necessary to obtain comparable results, owing to enhanced self-metathesis of the terminal alkene (Warwel 1985). Selective cross-metathesis between (substituted) styrene and (functionalized) terminal alkenes in the presence of $Mo(=CHCMe_2Ph)(=NC_6H_3\text{-}i\text{-}Pr_2\text{-}2,6)[OCMe(CF_3)_2]_2$ occurs rapidly and selectively at room temperature. High yields of cross-metathesis products with $>95\%$ *trans* selectivity are obtained using 1 mol% catalyst and excess styrene (2 equivalents). Thus, the reaction between styrene and either oct-1-ene, $CH_2{=}CH(CH_2)_3Br$, or $CH_2{=}CH(CH_2)_3OCH_2Ph$ in CH_2Cl_2 as solvent, gives $>85\%$ of the cross-metathesis product (only the *trans* isomer is detected) and $<4\%$ of the self-metathesis product of the terminal alkene (Crowe 1993).

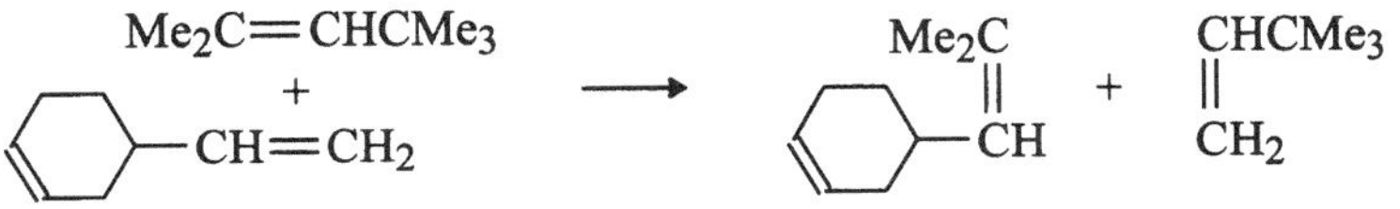

(16)

Cross-metathesis of stilbene with symmetrical olefins is only possible at 100°C with the Re_2O_7/Al_2O_3 catalyst when it is activated with Bu_4Sn. With equimolar amounts of the reactants an equilibrium mixture (50% conversion) is obtained (Warwel 1985).

Some of the reactions listed in Table 9.3 are concerned with the synthesis of sex pheromones of certain insects. The reactions can be conducted either in homogeneous solution or by distillation through MoO_3/Al_2O_3 or Re_2O_7/Al_2O_3 columns at 60–100°C with continuous removal of light olefinic products. Although the physiologically active *cis* isomers can be separated from the *trans* isomers by GC, this is not actually necessary since the *trans* isomers do not interfere with the activity of the *cis* isomers.

9.8 Functionalized olefins

Derivatives of oleic acid that undergo cross-metathesis with simple olefins are listed in Table 9.4, while Table 9.5 gives some examples of cross-metathesis reactions of simple olefins with other functional olefins. Such cross-metathesis reactions may provide useful routes to speciality chemicals such as synthetic perfumes, insect pheromones (Crisp 1988), prostaglandin intermediates (Dalcanale 1985), etc. (see Mol 1982, 1991). Of special interest are ethenolysis reactions, which allow the synthesis of compounds with terminal double bonds. Ethenolysis (and cross-metathesis with lower olefins) has been investigated extensively for

Table 9.4 Examples of cross-metathesis reactions of oleic acid and its derivatives with simple olefins

Derivative[a]	Olefin[b]	Catalyst[c]	Notes	Reference
COOH	C_2	$[Ru]=CHCH=CPh_2$[d]/20°C, 7 bar C_2		Nguyen 1996
COOMe	C_2	WCl_6/Me_4Sn; 70°C, 50 bar C_2	Produces methyl dec-9-enoate and dec-1-ene with high conversion and selectivity	Bosma 1981b; Mol 1983
COOMe	C_2	$Re_2O_7/Al_2O_7/Me_4Sn$; 20°C, 50 bar C_2	As above	Bosma 1981b; Mol 1983
COOMe	C_2	Re_2O_7/SiO_2-Al_2O_3/Bu_4Sn; 40°C, 30 bar C_2	As above	Sibeijn 1992
COOMe	C_2	$[Ru]=CHCH=CPh_2$[d]/20°C, 7 bar C_2		Nguyen 1996
COOMe	1-C_6	WCl_6/Me_4Sn	Formation of 1-C_{10} and methyl dec-9-enoate preferred	Matyska 1989
COOMe	3-C_6	WCl_6/Me_4Sn; 60°C	Convenient synthesis of homologues of methyl oleate	van Dam 1974b
COOMe	*n*-Hexenes	$Re_2O_7/MoO_3/Al_2O_3/Me_4Sn$; 25°C	TON = 80, 3 h reaction time	Warwel 1989b
COOMe	1-C_{10}	WCl_6/R_4Sn		Ichikawa, K. 1976b
COOEt	5-C_{10}	$WCl_4(OC_6H_3$-Cl_2-2,6)/Bu_4Pb; 85°C	80% yield of ethyl tetradec-9-enoate after 2 h (90% selectivity)	Quignard 1986
COOEt	5-C_{10}	$MoCl_5/SiO_2/Me_4Sn$; 90°C	71% yield of ethyl tetradec-9-enoate in 4 h; *cis/trans* = 18/82	Bykov 1988
COOEt	5-C_{10}	MoO_3/SiO_2[e]/cyclopropane; 50°C	58% conversion in 7 min, TON = 145 (92% selectivity)	Berezin 1991
CH_2OCOMe	1-C_6	$WCl_6/Me_3Al_2Cl_3$	15% yield of tetradec-9-enyl acetate; 20% yield of dec-9-enyl acetate	Nakamura, R. 1976b
CH_2OCOMe	5-C_{10}	$WCl_6/Me_3Al_2Cl_3$	26% yield of tetradec-9-enyl acetate	Nakamura, R. 1976b
CH_2OSiMe_3	*n*-Hexenes	$Re_2O_7/Al_2O_3/Me_4Sn$; 65°C	88% conversion in 2 h	Warwel 1988a
CH_2Cl	1-C_6 (excess)	$Re_2O_7/Mt_xO_y/Al_2O_3$; 25°C	The expected four cross-products are formed in comparable amounts	Nakamura, R. 1977a
CH_2Cl	*trans*-5-C_{10} (excess)	$Re_2O_7/Mt_xO_y/Al_2O_3$	70% conversion of the chloride into 14-chlorotetradec-5-ene and tetradec-5-ene	Nakamura, R. 1977a

[a] The group listed replaces COOH in oleic acid, *cis*-$CH_3(CH_2)_7CH=CH(CH_2)_7COOH$. [b] 2-$C_6$ hex-2-ene, 5-C_{10} dec-5-ene, etc. [c] The reactants are used either as neat liquids or in solution. [d] $Ru(=CHCH=CPh_2)(Cl)_2(PCy_3)_2$. [e] MoO_3/SiO_2 photoreduced in CO.

Table 9.5 Typical examples of cross-metathesis reactions of functionalized olefins with simple olefins

Olefin[a]	Functionalized olefin	Catalyst	Notes	Reference
C_2	Methyl linoleate	$Re_2O_7/Al_2O_3/Me_4Sn$; 20°C	High conversion yielding hept-1-ene, methyl dec-9-enoate and penta-1,4-diene	Boelhouwer 1985
C_2	Methyl linoleate	WCl_6/Me_4Sn; 30°C	Main products methyl dec-9-enoate and hept-1-ene	Ahmad 1994
C_2	Methyl linolenate	$Re_2O_7/Al_2O_3/Me_4Sn$; 20°C	High conversion yielding but-1-ene, methyl dec-9-enoate and penta-1,4-diene	Boelhouwer 1985
C_2	Glycerol trioleate	$Re_2O_7/Al_2O_3/Bu_4Sn$; 20°C	Produces dec-1-ene and a triester which can be hydrogenated to tricaprine	Mol 1994b
1–C_4	$CH_2{=}CH(CH_2)_8OCOMe$	$W[{=}C(OMe)Ph](CO)_5/SnCl_4/SiCl_4$; 90°C	2/1 *trans/cis* mixture of dodec-9-enyl acetate[b]	Banasiak 1985
2-C_4	$CH_2{=}CHCH_2SMe$	$[W]{=}CHCMe_3$ (**1** in Ch. 3); 20°C	Conversion 95%, selectivity 100%, product is 3/1 *trans/cis* mixture	Couturier 1993b
1-C_5	$CH_2{=}CHSi(OEt)_3$	$RuCl_3.nH_2O$; 100°C	Conversion 80%, 24% yield of alkenylsilane	Foltynowicz 1991
1-C_6	$CH_2{=}CH(CH_2)_6OCOMe$	$WCl_4(OC_6H_3\text{-}Cl_2\text{-}2,6)_2/Me_4Sn$; 80–90°C	36% yield of dodec-7-enyl acetate, selectivity 76%	Crisp 1988
1-C_6	$CH_2{=}CH(CH_2)_7COOMe$	$[Mo]{=}CHCMe_2Ph^c/DME$	High conversion yielding methyl tetradec-9-enoate	Fox 1994a
1-C_6	$CH_2{=}CH(CH_2)_7CH_2OSiMe_3$	$[Mo]{=}CHCMe_2Ph^c/DME$	30% yield of 1-(trimethylsiloxy)-tetradec-9-ene	Fox 1994a
2-C_6	$CH_2{=}CH(CH_2)_2COOMe$	WCl_6/Me_4Sn; 50°C	Conversion 58%	Otton 1980
3-C_6	$CH_2{=}C(Me)COOMe$	WCl_6/Me_4Sn; 90°C	Conversion 40%	Verkuijlen 1977a
3-C_6	$CH_2{=}CH(CH_2)_3OCOMe$	WCl_6/Me_4Sn; 70°C	Conversion 57%; yield of acetate 37%	Levisalles 1980b
3-C_6	$CH_2{=}CH(CH_2)_3OCOMe$	$WCl_4(OC_6H_3\text{-}Cl_2\text{-}2,6)_2/Me_4Sn$; 85°C	60% yield of hept-4-enyl acetate, selectivity 75%	Quignard 1986
1-C_7	$CH_2{=}CH(CH_2)_2COOEt$	$WCl_6/DSCB^d$; 70°C	Conversion 40%, selectivity 63–73%	Bespalova 1992
1-C_7	$CH_2{=}CHCH_2CN$	$WCl_6/DSCB^d$; 60–80°C	Conversion 20–40%, selectivity 60%	Bespalova 1992
3-C_7	$CH_2{=}CH(CH_2)_nCN$	WCl_6/Me_4Sn; 100°C	$n = 1$–4, maximum reactivity for $n = 2$	Bosma 1981a
4-C_8	$CH_2{=}CH(CH_2)_2CN$	$Re_2O_7/Al_2O_3/Me_4Sn$; 100°C	Conversion 61%	Warwel 1982b
1-C_{10}	$CH_2{=}CHCN$	$[Mo]{=}CHCMe_2Ph^c$; 20°C	72% yield of non-2-enenitrile, *cis/trans* = 9/1	Crowe 1995
1-C_{18}	$CH_2{=}CHSi(OEt)_3$	$RuCl_3.nH_2O$; 100°C	Conversion 95%, 70% yield of alkenylsilane	Foltynowicz 1991
1-C_{18}	$Me(CH_2)_7CH{=}CH(CH_2)_{11}$-COOMe	WCl_6/Me_4Sn; 80°C	The C_{30} ester product can be converted into 1-triacontanol[e]	Penninger 1983

[a] C_2 ethene, 1-C_5 pent-1-ene, 1-C_7 hept-1-ene, etc. [b] An insect pheromone. [c] $Mo({=}CHCMe_2Ph)({=}NC_6H_3\text{-}i\text{-}Pr_2\text{-}2,6)[OCMe(CF_3)_2]_2$. [d] DSCB = 1,1,3,3-tetramethyl-1,3-disilacyclobutane. [e] A plant growth stimulant.

applications in the oleochemical industry in order to convert fatty acids, fatty acid esters, and fatty oils into shorter-chain products (Mol 1994b).

The yield in cross-metathesis between methyl oleate and unsaturated diesters catalyzed by WCl_6/Me_4Sn shows a marked dependence on the number of CH_2 units between the C=C bond and the ester groups; on $Re_2O_7/Al_2O_3/Me_4Sn$ the conversion rates are low, and self-metathesis of methyl oleate predominates (Kohashi 1985).

Unsaturated macrocyclic lactones (macrolides) can be obtained via cross-metathesis of an alkenyl ester with an unsaturated acid ester (Villemin 1980). 1-Triacontanol, a plant growth stimulant, can be produced in a relatively simple process via cross-metathesis of an unsaturated fatty acid ester with either an alk-1-ene (Penninger 1983) or a cyclic alkene (Villemin 1983). It also appears posible to cross-metathesize a conjugated alkadiene with an unsaturated ester on an $Re_2O_7/Al_2O_3/Me_4Sn$ catalyst; reactions (17a) and (17b). All of the expected products are found in the reaction mixture, and 15% of the starting ester is converted into methyl 7-methylocta-4,6-dienoate, one of a class of important aroma ingredients (Woerlee 1984).

$$\begin{array}{c} Me_2C{=}CH{-}CH{=}CMe_2 \\ + \\ CH_2{=}CH(CH_2)_2COOMe \end{array} \rightleftharpoons \begin{array}{c} Me_2C \\ \| \\ CH_2 \end{array} + \begin{array}{l} CH{-}CH{=}CMe_2 \\ \| \\ CH(CH_2)_2COOMe \end{array} \tag{17a}$$

$$\begin{array}{c} Me_2C{=}CH{-}CH{=}CMe_2 \\ + \\ CH_2{=}CH(CH_2)_2COOMe \end{array} \rightleftharpoons \begin{array}{r} Me_2C{=}CH{-}CH \\ \| \\ CH_2 \end{array} + \begin{array}{l} CMe_2 \\ \| \\ CH(CH_2)_2COOMe \end{array} \tag{17b}$$

The perfume constituent non-5-ene-2-one is obtained in 61% yield via cross-metathesis of silylated allyl acetone with 2 equiv of oct-4-ene, reaction (18), using $Re_2O_7/MoO_3/Al_2O_3/Me_4Sn$ as catalyst (Warwel 1989c).

$$\begin{array}{l} CH_2{=}CHCH_2CH{=}C(OSiMe_3)Me \\ + \\ PrCH{=}CHPr \end{array} \longrightarrow \begin{array}{c} CH_2 \\ \| \\ PrCH \end{array} + \begin{array}{l} CHCH_2CH{=}C(OSiMe_3)Me \\ \| \\ CHPr \quad \big\downarrow H_3O^+ \\ PrCH{=}CH(CH_2)_2COMe \end{array} \tag{18}$$

Cross-metathesis of erucic acid, $CH_3(CH_2)_7CH{=}CH(CH_2)_{11}COOH$, with excess oct-4-ene (Hummel 1975a) and of oleic acid with excess hex-2-ene (Hummel 1971) has been observed, although only the hydrocarbon metathesis product has been identified.

With $Ru_3(CO)_{12}/HSiPh_3$ (1/10), cross-metathesis of vinyltrimethylsilane/styrene (2/1) at 80°C gives a yield of 59% $PhCH{=}CHSiMe_3$ (based on styrene), along with a 60% yield of the self-metathesis product 1,2-bis(trimethylsilyl)ethene (Seki, Y. 1989). The complex $Ru(H)(Cl)(CO)(PPh_3)_3$ catalyzes selectively the cross-metathesis of vinylsilanes of general formula $CH_2{=}CHSiMe_2R$ (R = Me, OEt) with monosubstituted alkenes, $CH_2{=}CHR'$ (R' = Me, Ph, CO_2Me, OBu), at 120–140°C (Wakatsuki 1991).

Various compounds of general formula $RCH{=}CHSiMe_{3-n}Ph_n$ ($n = 0$–3, R = alkyl, Ph) can be prepared via cross-metathesis of the corresponding vinylsilanes with alk-1-enes or styrene in the presence of $RuCl_2(PPh_3)_3/C_6H_6$, $[RuCl_2(CO)_3]_2$, $Ru(acac)_3$, or $RhCl(PPh_3)_3$; reaction (19). Good yields are obtained above 100°C when an excess of alk-1-ene is used (Foltynowicz 1993a; Marciniec 1993).

$$\begin{matrix} CH_2{=}CHSiMe_{3-n}Ph_n \\ + \\ CH_2{=}CHR \end{matrix} \longrightarrow \begin{matrix} CH_2 \\ \| \\ CH_2 \end{matrix} + \begin{matrix} CHSiMe_{3-n}Ph_n \\ \| \\ CHR \quad (>90\%\ \textit{trans}) \end{matrix} \qquad (19)$$

Cross-metathesis of equimolar mixtures of vinyltriethoxysilane with various liquid alk-1-enes (C_5–C_{18}), reaction (20), proceeds in the presence of $RuCl_3.nH_2O$ or $RuCl_2(PPh_3)_3$ at 70–120°C in competition with self-metathesis of vinyltriethoxysilane; no self-metathesis of the alk-1-enes is observed. Isomers of the alk-1-ene and oligomers of both olefins are also formed (Marciniec 1988: Foltynowicz 1991, 1993b). An excess of the alk-1-ene gives high yields (78–88%) of alkenyltriethoxysilane (*cis* and *trans*) (Foltynowicz 1989). With propene both cross-metathesis and self-metathesis of vinyltriethoxysilane are catalyzed by $RuCl_2(PPh_3)_3$ in benzene at 100°C. But-1-ene does not give any products of cross-metathesis under similar conditions (Marciniec 1994b). In the presence of these ruthenium complexes, cross-metathesis of vinyltriethoxysilane with alk-2-enes occurs much more slowly than with alk-1-enes, and the predominant pathway seems to be the preliminary isomerization of alk-2-enes to alk-1-enes followed by their cross-metathesis with vinyltriethoxysilane (Foltynowicz 1991).

$$\begin{matrix} CH_2{=}CHSi(OEt)_3 \\ + \\ CH_2{=}CH(CH_2)_nMe \end{matrix} \longrightarrow \begin{matrix} CH_2 \\ \| \\ CH_2 \end{matrix} + \begin{matrix} CHSi(OEt)_3 \\ \| \\ CH(CH_2)_nMe \end{matrix} \qquad (20)$$

At 70–100°C, $RuCl_2(PPh_3)_3$ catalyzes cross-metathesis of vinyltriethoxysilane with styrene, which dominates over the self-metathesis of vinyltriethoxysilane. Similar results are obtained for cross-metathesis with 4-vinylcyclohexene. A mixture of vinyltriethoxysilane and vinyltrimethylsilane gives cross-metathesis and self-metathesis of the former, but no self-metathesis of the latter (Foltynowicz 1991).

Cross-metathesis of trisubstituted vinylsilanes of general formula $CH_2{=}CHSiR^1R^2R^3$ with excess dec-1-ene, catalyzed by $RuCl_2(PPh_3)_3/O_2$ in benzene, gives alkenylsilanes, $CH_3(CH_2)_7CH{=}CHSiR^1R^2R^3$, where R^1, R^2, R^3 can be methyl, alkoxy, phenyl, or trimethylsiloxy (Marciniec 1992). However, when allyltriethoxysilane is reacted with dec-1-ene under similar conditions, the substrate first isomerizes to propenyltriethoxysilane, which then cross-metathesizes with dec-1-ene to give 1-triethoxysilyldec-1-ene. A mixture of allyltrimethylsilane and allyltriethoxysilane gives only products of cross-metathesis of the two propenylsilanes, which are first formed by isomerization (Marciniec 1994a).

WCl_6/Me_4Sn catalyzes the cross-metathesis of alkenylsilanes of the general formula $CH_2{=}CH(CH_2)_nSiMe_3$ ($n = 1, 2$) with *cis*-pent-2-ene at 25°C (Berglund 1985, 1986b). Allyltrichlorosilane and allyltrimethoxysilane also give reasonable conversions. With allyltrimethylsilane ($n = 1$) the reaction also proceeds with WCl_6 alone, but addition of a Lewis acid, e.g. $AlBr_3$, substantially raises the activity (Berglund 1986b). Cross-metathesis of allyltrimethylsilane with alk-1-enes gives an alkenylsilane, which can be desilylated to give a homologue of the original alkene; sequence (21) (Finkel'shtein 1989). For instance, hex-1-ene can be converted into hept-1-ene by this means.

$$\begin{matrix} CH_2{=}CHCH_2SiMe_3 \\ + \\ CH_2{=}CHCH_2R \end{matrix} \xrightarrow[-\,C_2H_4]{Re_2O_7/Al_2O_3/Bu_4Sn} \begin{matrix} CHCH_2SiMe_3 \\ \| \\ CHCH_2R \end{matrix} \xrightarrow[60°C]{TFA/DMSO} \begin{matrix} CH{=}CH_2 \\ | \\ CH_2CH_2R \end{matrix} \qquad (21)$$

Cross-metathesis reactions of allylsilanes with unsaturated esters are useful for producing Si-containing unsaturated esters. Thus, ethylpent-4-enoate can be reacted with allyltrimethylsilane in the presence of WCl_6/1,1,3,3-tetramethyl-1,3-disilacyclobutane at 70°C to give ~40% conversion of the ester and up to 90% selectivity (Bespalova 1990); reaction (22). A series of allylsilanes has been prepared in good yield by cross-metathesis of allyltrimethylsilane with various functionalized terminal olefins, such as substituted styrenes and unsaturated ethers, bromides and nitriles, using 2 mol % $Mo({=}CHCMe_2Ph)({=}NAr)[CMe(CF_3)_2]_2$ as catalyst in DME at 20°C (Crowe 1996).

$$\begin{matrix} CH_2{=}CHCH_2SiMe_3 \\ + \\ CH_2{=}CH(CH_2)_2COOEt \end{matrix} \longrightarrow \begin{matrix} CH_2 \\ \| \\ CH_2 \end{matrix} + \begin{matrix} CHCH_2SiMe_3 \\ \| \\ CH(CH_2)_2COOEt \end{matrix} \qquad (22)$$

Cross-metathesis of ω-unsaturated nitriles, $CH_2{=}CH(CH_2)_nCN$, with *cis*-hept-3-ene in the presence of WCl_6/Me_4Sn at 100°C takes place with a conversion of $\geqslant 42\%$ for $n = 1$, 2, or $\geqslant 5$ (Bosma 1981a, 1985). In the presence of the catalyst system WCl_6/1,1,3,3-tetramethyl-1,3-disilacyclobutane allyl cyanide ($n = 1$) undergoes cross-metathesis with both linear and cyclic olefins, and with allyltrimethylsilane (Bespalova 1992). Acrylonitrile ($n = 0$) cross-metathesizes with a number of terminal alkenes, including functionalized alkenes, in the presence of $Mo({=}CHCMe_2Ph)({=}NC_6H_3\text{-}i\text{-}Pr_2\text{-}2,6)[OCMe(CF_3)_2]_2$ in dichloromethane at 20°C, with a high-*cis* selectivity for the product nitrile (Crowe 1995).

Another example of metathesis of olefins bearing functional groups at the α-position is the cross-metathesis of tetracyanoethene with ethyl α-cyanocinnamate to give 1,1-dicyano-2-phenylethene and ethyl 2,3,3-tricyanopropenoate, eqn. (23), in the presence of certain rhenium(I)enolate complexes at 20°C (Hirano 1994).

$$\begin{matrix} (NC)(EtOOC)C{=}CHPh \\ + \\ (NC)_2C{=}C(CN)_2 \end{matrix} \longrightarrow \begin{matrix} (NC)(EtOOC)C \\ \| \\ (NC)_2C \end{matrix} + \begin{matrix} CHPh \\ \| \\ C(CN)_2 \end{matrix} \qquad (23)$$

10

Acetylenes

10.1 Introduction

Metathesis-type reactions of acetylenes fall into two categories: true metathesis, in which the triple bond is completely broken, and olefin-type metathesis, in which only two of the three bonds are broken. The latter results in polymerization and the formation of cyclic oligomers, especially trimers. The overall reactions are illustrated by eqns. (1)–(3).

$$2\,EtC{\equiv}CMe \rightleftharpoons EtC{\equiv}CEt + MeC{\equiv}CMe \qquad (1)$$

$$n\,PhC{\equiv}CH \longrightarrow \text{+}CPh{-}CH\text{+}_n \qquad (2)$$

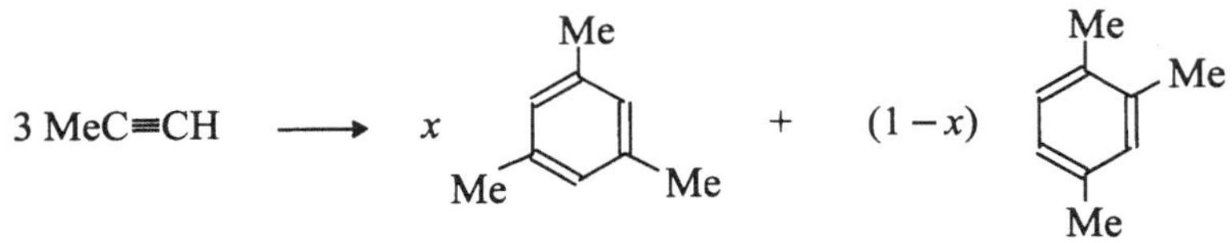

(3)

Reactions of type (1) proceed via metal *carbyne* complexes (Scheme 10.1). The most direct evidence is that such complexes (Mt = Mo, W, Re) can act as initiators for the metathesis of $R^1C{\equiv}CR^2$ (Wengrovius 1981; Weinstock 1991) and of RC≡CH in the initial stages (Mortreux 1995), and for the ROMP of cycloalkynes (Krouse 1989; Zhang, X-P. 1994). Metallacyclobutadienes have been prepared (Pedersen 1982; Latham 1986) and some can act as initiators of acetylene metathesis (Churchill 1984; Schrock 1985); their ready formation as intermediates (Scheme 10.1) can be expected on theoretical grounds (Anslyn 1988; Woo 1993; Lin, Z. 1994). Metal carbyne complexes have to be finely tuned to be capable of bringing about productive metathesis of internal acetylenes, otherwise reaction either stops after the first step, or gives other products (Strutz 1985), or does not proceed at all. A potential side reaction is the formation of a metallatetrahedrane complex by tautomerization of the metallacyclobutadiene intermediate or by its direct formation from the reactants (Schrock 1986b); but, with the right choice of catalyst, the metathesis reaction occurs rapidly and with high selectivity. Examples are given below.

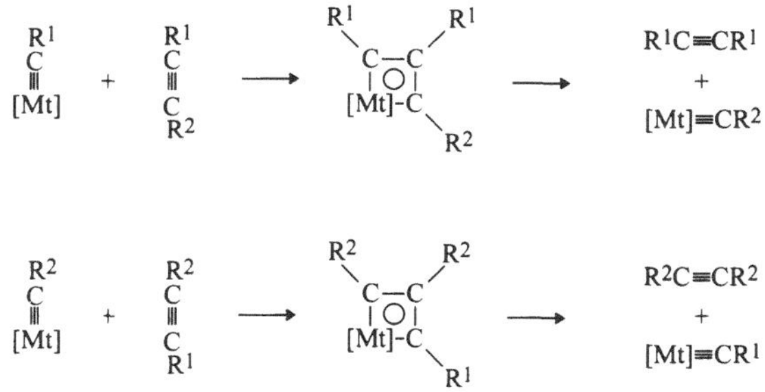

Scheme 10.1 Propagation of acetylene metathesis by metal carbyne and metallacyclobutadiene chain carriers.

In contrast, reactions of type (2) are propagated by metal *carbene* complexes, Scheme 10.2, and may be regarded as olefin metathesis reactions of a particular kind. Again, the most direct evidence is that such complexes (Mt = Ta, Mo) can initiate the living polymerization of RC≡CR (Wallace 1989), RC≡CH (Schrock 1994b), and the cyclopolymerization of 1,6-heptadiynes (Fox 1994b). Metallacyclobutene complexes have also been isolated (Mt = Ti, Ta) and their facile interconversion with substituted vinylalkylidene complexes well established (Tebbe 1980; Wallace 1989; Doxsee 1993). The formation of cyclic trimers (benzene derivatives) during the metathesis polymerization of acetylenes finds a ready explanation in terms of an intramolecular backbiting reaction of the propagating metal carbene complex.

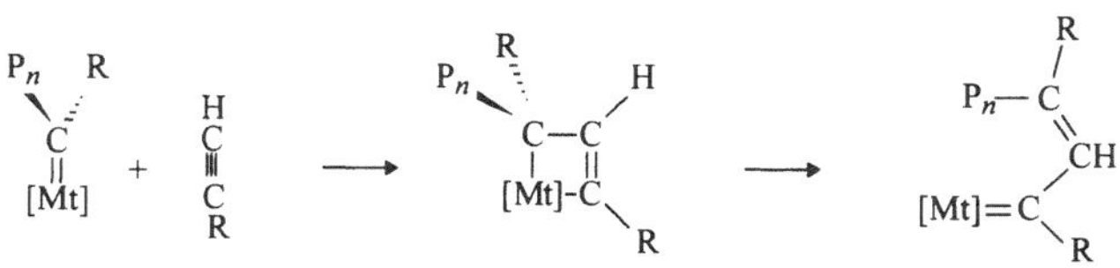

Scheme 10.2 Propagation of polymerization of acetylenes by metal carbene and metallacylobutene carriers.

Some of the first-generation olefin metathesis catalysts, such as WO_3/SiO_2 and MoO_3/SiO_2, are also catalysts for the metathesis of internal acetylenes (Pennella 1968; Mortreux 1980). Such catalysts are therefore assumed to generate metal carbenes by reaction with olefins, but metal carbynes by reaction with acetylenes. There is also a group of catalyst systems, such as $Mo(CO)_6$/ArOH (Mortreux 1974, 1975/76, 1977, 1978; Stockel 1978; Devarajan 1979) and $MoO_2(acac)_2/Et_3Al$/ArOH (Petit 1982; Bages 1990b) that are specific for metathesis of internal acetylenes and do not work for olefins. Such catalysts are evidently capable of generating only metal carbyne initiating species.

Rather surprisingly some metal carbyne complexes bring about the *polymerization* of acetylenes by a reaction of type (2) rather than their metathesis by a reaction of type (1) (Katz 1984; Weiss 1988c). Such complexes are coordinatively saturated and therefore inactive, but can isomerize by ligand migration as in eqn. (4) to yield a coordinatively unsaturated carbene complex which initiates polymerization.

$$PhC{\equiv}W(CO)_4Br \longrightarrow (Ph)(Br)C{=}W(CO)_4 \qquad (4)$$

For the reactions of RC≡CH with $Me_3CC{\equiv}W(OCMe_3)_3$, metathesis products can be detected in the early stages, but the metathesis reaction is rapidly overtaken by polymerization (Bray 1993; Mortreux 1995). This results from elimination of a hydrogen atom from the intermediate metallacyclobutadiene, leading to the formation of a metal carbene complex which then propagates the polymerization reaction; Scheme 10.3 (also see McCullough 1983).

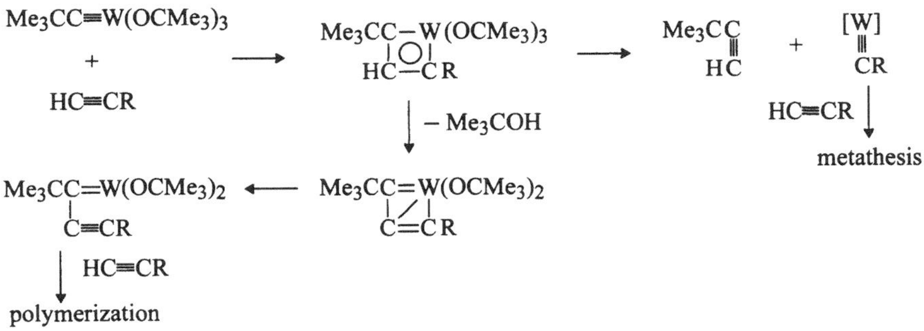

Scheme 10.3 Formation of a metal carbene carrier by reaction of a metal carbyne complex.

10.2 Metathesis reactions involving total cleavage of the C≡C bond

10.2.1 Metathesis of internal acetylenes by ill-defined catalysts

The metathesis of internal acetylenes, eqn. (1), was first observed by Pennella (1968) using WO_3/SiO_2 at 350°C as catalyst. Proof of triple-bond cleavage in such reactions was later obtained by various labelling experiments. For example, the forward reaction (5), catalyzed by MoO_3/SiO_2 at 350°C gives but-2-yne that has twice the specific radioactivity of the reactant hex-2-yne-1-^{14}C (Mortreux 1972). Again, the forward reaction of eqn. (6), which is also catalyzed by MoO_3/SiO_2 at 350°C, gives 1-phenylbut-1-yne in which the C-1 NMR peak (δ 81.04) is enriched to the same extent as *each* acetylenic carbon in the reactant (δ 90.41), while C-2

(δ 92.01) has the same proportion of ^{13}C as normal PhC≡CEt (Mortreux 1980). The back reaction in eqn. (6) is catalyzed by $Mo(CO)_6$/4-bromophenol in heptane at 120°C and gives diphenylacetylene with twice the ^{13}C content in the acetylenic carbons compared with C-1 in the reactant (Mortreux 1978).

$$2\ C_3H_7C{\equiv}C^{14}CH_3 \rightleftharpoons C_3H_7C{\equiv}CC_3H_7 + {}^{14}CH_3C{\equiv}C^{14}CH_3 \qquad (5)$$

$$Ph^{13}C{\equiv}{}^{13}CPh + EtC{\equiv}CEt \rightleftharpoons 2\ Ph\underset{1}{^{13}C}{\equiv}\underset{2}{C}Et \qquad (6)$$

The most active and most selective MoO_3/SiO_2 catalysts are prepared by contacting SiO_2 with (π-allyl)$_4$Mo and then oxidizing the surface. Such catalysts have five times the intrinsic activity of catalysts prepared by the impregnation method, and 95–100% selectivity for the metathesis of hex-2-yne at 0.05 bar/350°C (Mortreux 1980).

The reaction of internal acetylenes $R^1C{\equiv}CR^2$ leads to an equilibrium mixture of $R^1C{\equiv}CR^1$, $R^1C{\equiv}CR^2$, and $R^2C{\equiv}CR^2$, with composition near to 1 : 2 : 1, showing that such reactions are essentially thermoneutral. In the cross-metathesis of PhC≡CEt with PrC≡CMe on MoO_3/SiO_2 at 350°C the equilibrium mixture contains the expected eight compounds PhC≡CPh, EtC≡CEt, PrC≡CPr, MeC≡CMe, PhC≡CPr, EtC≡CMe, PhC≡CMe, and PrC≡CEt, in addition to the reactants. However, the initial rate of production of PrC≡CPr and EtC≡CEt is more than three times that of PrC≡CEt (Mortreux 1980).

The remarkable cocatalytic effect of certain phenols for the metathesis of internal acetylenes by molybdenum compounds such as $Mo(CO)_6$ was discovered by Mortreux (1974). Thus $Mo(CO)_6$/resorcinol (1/6) is a good catalyst for the metathesis of $PhC{\equiv}CC_6H_4$-Me-4 in decalin at 160°C (88% selectivity) and for the metathesis of BuC≡CPr in toluene at 110°C (Mortreux 1974, 1975/76). Cross-metathesis experiments, using $Mo(CO)_6$/3-chlorophenol as catalyst, show that non-4-yne is more reactive than $PhC{\equiv}CC_6H_4$-4-Me (Mortreux 1977). This type of catalyst has also been used to make $PhC{\equiv}C(CH_2)_4OCOMe$ by cross-metathesis of two symmetrical alkynes (Kaneta 1995b). $MoO_2(acac)_2$, mixed with Et_3Al (1/6) for 15 min, and then added to the phenol and substrate in toluene at 110°C is also very effective, not only for the metathesis of BuC≡CPr and $PhC{\equiv}CC_6H_4$-Me-4, but also for $EtC{\equiv}C(CH_2)_3Cl$ (Petit 1982). $BuC{\equiv}C(CH_2)_2OCOMe$ is more readily metathesized by $Mo(CO)_6$/PhOH. Other functionalized acetylenes which have been metathesized in this way include $PhC{\equiv}C(CH_2)_2Y$, where Y = OH, OAc, Br, COOH, CO_2Me, and CN (Villemin 1982).

Replacement of a carbonyl ligand on molybdenum by an electron-donating isonitrile ligand has an adverse effect on the activity (Adams, K.P. 1985) and the presence of an atmosphere of CO can bring the metathesis reaction to a halt (Devarajan 1979). On the other hand, a more acidic phenol, such as 3-chlorophenol, has a beneficial effect on the reactivity. Some systems are even active at room temperature. Thus $MoO_2(acac)_2/Et_3Al/PhOH$ and $MoO(OPh)_4/Et_3Al/PhOH$ cause metathesis of BuC≡CPr at 30°C (Bencheick 1982).

When the substrate itself bears a phenolic group, as in 4-HO-C_6H_4C≡CMe, it acts as its own cocatalyst with $Mo(CO)_6$ and gives a good yield (78%) of metathesis product on heating in toluene. If the 4-HO group is replaced by 4-MeO, metathesis will only occur if a phenol is added. If 4-HO is replaced by 2-HO, then metathesis gives way to the formation of cyclic trimer (Kaneta 1995a).

The mode of action of the phenolic cocatalysts in these systems is not fully understood. One suggestion, based on the identification of the initial products, is that the reaction is propagated by a metal *carbene* complex, the necessary rearrangement of the intermediate metallacyclobutene complex being facilitated by transfer of a proton from the phenol, a different proton being subsequently reclaimed by the phenoxide ion; see Scheme 10.4 (Bages 1990b). It is also possible that the reaction involves the replacement of carbonyl ligands by phenoxo ligands; the fact that metal carbyne complexes that initiate acetylene metathesis usually bear three RO ligands (R = aryl or alkyl) points in this direction; see below.

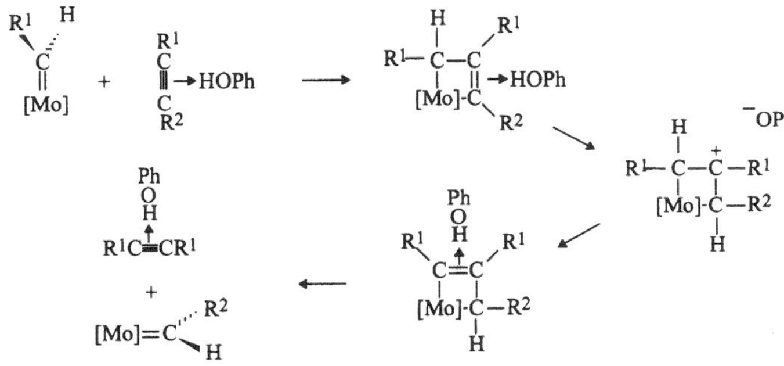

Scheme 10.4 Propagation of acetylene metathesis by a metal carbene chain carrier, facilitated by proton transfer.

Catalysts such as $Mo(CO)_6$/MeCN bring about the metathesis of hept-2-yne, but not of hept-1-yne; instead oligomers are formed from the latter, no doubt by a mechanism analogous to that in Scheme 10.3 (Tsonis 1992).

10.2.2 Metathesis of acetylenes by well-defined metal carbyne initiators

The simple one-step metathesis, reaction (7), between a metal carbyne complex and a symmetrical disubstituted acetylene was first observed by Wengrovius (1981).

The ability of metal carbyne complexes of the type Mt(≡$CCMe_3$)$(OR)_3$ to metathesize internal acetylenes by a chain mechanism depends on a delicate

$$Me_3CC{\equiv}WCl_3(PEt_3) + PhC{\equiv}CPh \xrightarrow{70°C} Me_3CC{\equiv}CPh + WCl_3(PEt_3){\equiv}CPh \qquad (7)$$

balance between electronic and steric factors. The reaction of $W({\equiv}CCMe_3)(OCMe_3)_3$ with $PrC{\equiv}CEt$ produces an equilibrium mixture of $PrC{\equiv}CEt$, $PrC{\equiv}CPr$, and $EtC{\equiv}CEt$ in less than 1 min at 25°C, and the ^{13}C NMR spectrum of the products indicates the presence of the three $W({\equiv}CR)(OCMe_3)_3$ species, where R = Et, Pr, and CMe_3 (Wengrovius 1981). The metathesis of $W({\equiv}CCMe_3)(OCMe_3)_3$ with $PhC{\equiv}CPh$ is first-order with respect to both reactants at 0°C (Sancho 1982). Reaction (8) is an apparent single-step reaction but there is in fact an ongoing degenerate exchange of the reactant with the product metal carbyne complex.

$$Me_3CC{\equiv}W(OCMe_3)_3 + EtC{\equiv}CPh \longrightarrow Me_3CC{\equiv}CEt + W(OCMe_3)_3{\equiv}CPh \qquad (8)$$

In reaction (9) the products are: (i) $EtC{\equiv}CCMe_3$, resulting from the metathesis reaction; and (ii) a metallacyclobutadiene complex, produced by addition of a second molecule of the acetylene to the initially formed carbyne complex. This metallacyclobutadiene complex is able to initiate the metathesis of $EtC{\equiv}CPr$ and $PrC{\equiv}CBu$, and must therefore be in equilibrium with its dissociation products. It exchanges carbyne moieties with $C_2D_5C{\equiv}CC_2D_5$ at a rate that is independent of the substrate concentration, showing that the dissociation of the metallacyclobutadiene complex into the metal carbyne complex is rate-controlling (Churchill 1984). However, when the OR ligands of the complex are $OCH(CF_3)_2$ instead of OC_6H_3-*i*-Pr_2-2,6, the rate of exchange with $C_2D_5C{\equiv}CC_2D_5$ is first order in substrate, indicating an associative mechanism (Freudenberger 1984).

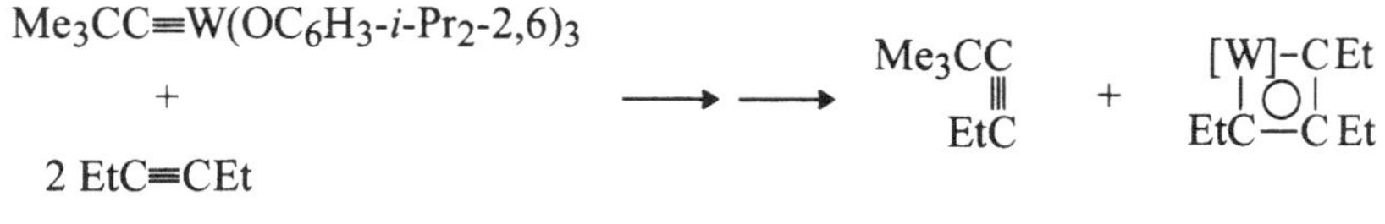

(9)

When OR is $OCMe_3$ metathesis occurs when the metal is tungsten but not when it is the less electrophilic molybdenum (McCullough 1984). However, the molybdenum complexes exhibit metathesis activity when OR is an electron-withdrawing ligand, such as $OCMe_2(CF_3)$, $OCMe(CF_3)_2$ or $OC(CF_3)_3$ (McCullough 1984), but not when OR is $OCHMe_2$ or OCH_2CMe_3. In the latter case

EtC≡CPr is polymerized rather than metathesized, showing that the metal carbyne is then readily converted to a metal carbene complex, which initiates polymerization. Again, although $W(\equiv CCMe_3)(OC_6H_3\text{-}i\text{-}Pr_2\text{-}2,6)_3$ is a successful metathesis catalyst for EtC≡CPr, the analogous $(OC_6H_3\text{-}Me_2\text{-}2,6)_3$ complex is inactive, while the $(OC_6H_3\text{-}t\text{-}Bu_2\text{-}2,6)_3$ complex suffers conversion to a metal carbene complex (Churchill 1984). The role of the alkoxide or phenoxide ligands in these reactions has been reviewed by Schrock (1995a).

The rhenium carbyne complex $Re(\equiv CCMe_3)(=NAr)[OCMe(CF_3)_2]_2$ is active for metathesis of internal acetylenes, but analogous complexes, where OR is $OCMe_2(CF_3)$, $OCMe_3$, or $OC_6H_3\text{-}i\text{-}Pr_2\text{-}2,6$, are not active; the rhenium centre is insufficiently electrophilic to enable the acetylenic substrate to overcome the barrier to reaction (Schrock 1988d; Weinstock 1991).

Some examples of single-step metathesis reactions of acetylenes have already been mentioned. This type of reaction can also occur with binuclear complexes; eqns. (10–12) (Schrock 1982; Chisholm 1984a,b, 1989; Strutz 1984; Listemann 1985; Latham 1986; Koutsantonis 1991). Reaction (10) proceeds via a μ-C_2H_2 complex. Reaction (12), in the absence of quinuclidine (quin), gives only [Mo]≡CPh.

$$(Me_3CO)_3W{\equiv}W(OCMe_3)_3 + HC{\equiv}CH \rightleftharpoons 2\ HC{\equiv}W(OCMe_3)_3 \qquad (10)$$

$$(Me_3CO)_3W{\equiv}W(OCMe_3)_3 + RC{\equiv}CR\ (R = Me, Et, Pr) \rightleftharpoons 2\ RC{\equiv}W(OCMe_3)_3 \qquad (11)$$

$$(Me_3CO)_3Mo{\equiv}Mo(OCMe_3)_3 + HC{\equiv}CPh \xrightarrow{\text{quin}} (Me_3CO)_3Mo(quin){\equiv}CH + (quin)Mo(OCMe_3)_3{\equiv}CPh \qquad (12)$$

The interaction of the carbyne ligand with the substituents on the acetylene is an important factor in determining whether reaction will occur or not. Thus reaction (13) occurs readily, whereas reaction (14) does not, while reactions (15) and (16), leading to complete metathesis of the substrate, proceed rather slowly at 25°C (McCullough 1985; Krouse 1987, 1989). As already discussed (Scheme 10.3), reaction (13) faces strong competition from a reaction leading to a metal carbene complex and subsequent polymerization of the monosubstituted acetylene.

$$\mathrm{Me_3CC{\equiv}Mo(OCMe_3)_3 + HC{\equiv}CPr \xrightarrow{\text{fast}} Me_3CC{\equiv}CH + Mo(OCMe_3)_3{\equiv}CPr} \quad (13)$$

$$\mathrm{Me_3CC{\equiv}Mo(OCMe_3)_3 + EtC{\equiv}CPr \not\longrightarrow Me_3CC{\equiv}CEt + Mo(OCMe_3)_3{\equiv}CPr} \quad (14)$$

$$\mathrm{PrC{\equiv}Mo(OCMe_3)_3 + PrC{\equiv}CEt \xrightarrow{\text{slow}} PrC{\equiv}CPr + Mo(OCMe_3)_3{\equiv}CEt} \quad (15)$$

$$\mathrm{EtC{\equiv}Mo(OCMe_3)_3 + EtC{\equiv}CPr \xrightarrow{\text{slow}} EtC{\equiv}CEt + Mo(OCMe_3)_3{\equiv}CPr} \quad (16)$$

The complex $W(\equiv CMe)(Cl)(PMe_3)_4$ undergoes stoichiometric metathesis with PhC≡CPh but the product PhC≡CMe remains coordinated to the metal centre (Atagi 1992). The stoichiometric metathesis between $W(\equiv CEt)(OCMe_3)_3$ and $MeC{\equiv}CRuCp(CO)_2$ has been reported (Latesky 1987); also the metathesis reaction of EtC≡CPr with a complex of rhenium containing an Re≡Re bond (Diefenbach 1988).

10.2.3 Ring-opening metathesis polymerization of cyclic alkynes

The ROMP of cyclooctyne, eqn. (17), was first observed by Krouse (1989) using $Mo(\equiv CPr)(OCMe_3)_3$ as initiator. Fig. 10.1 shows the ^{1}H NMR spectrum of the initiator before and after addition of 15 equivalents of cyclooctyne. The triplet (δ 0.74) due to the methyl protons in the propyl group of the initiator and the triplet (δ 0.92) due to the same protons in the propagating species are quite distinct, and show that the initiator has been totally converted to propagating species. The triplet (δ 3.02) due to the methylene protons at the other end of the chain is slightly downfield from the triplet (δ 2.95) due to $[Mo]{\equiv}CCH_2CH_2CH_3$ in the initiator, showing that the propagating species is living. The ^{13}C NMR spectrum of the polymer exhibits a sharp resonance at δ 80.4, characteristic of C≡C in a chain and quite distinct from that for C≡C in cyclooctyne (δ 94.5).

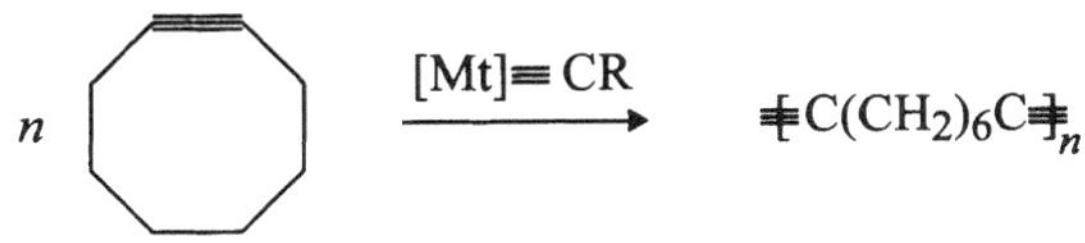

(17)

Another initiator for this polymerization is $(Me_3CO)_3W{\equiv}W(OCMe_3)_3$, but analysis of the products by MS shows that they are macrocyclic oligomers. Their ring-size distribution, as determined by GPC, conforms to the Jacobson–Stockmayer relationship; see Ch. 11. It is evident that with the tungsten catalyst

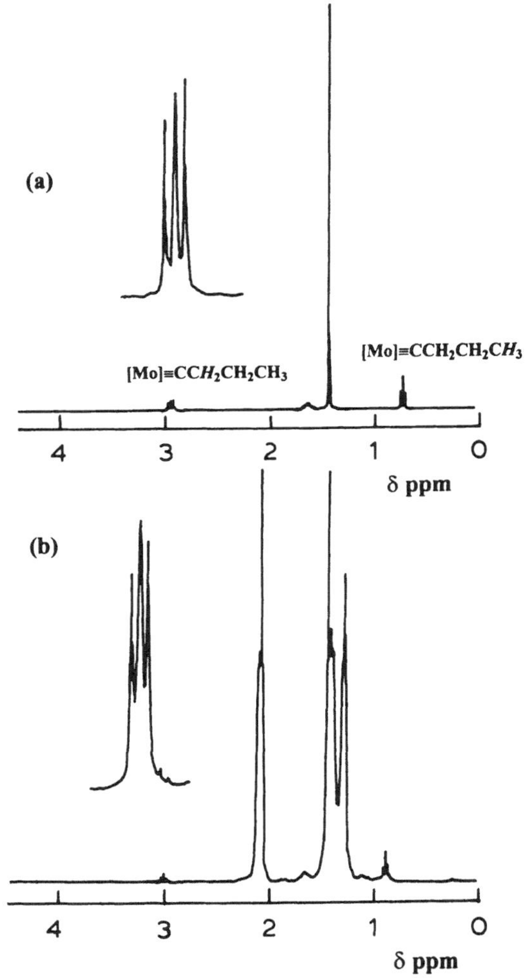

Fig. 10.1 300 MHz ^{1}H NMR spectra in C_6D_6 (a) of $Mo({\equiv}CPr)(OCMe_3)_3$, and (b) after addition of 15 equivalents of cyclooctyne to form $(Me_3CO)_3Mo[{\equiv}C(CH_2)_6C{\equiv}]_n CPr$. Note the slight shifts in the positions of the $[Mo]{\equiv}CCH_2$ and CH_2CH_3 triplets in the polymer compared with the initiator (Krouse 1989).

there is strong competition between the backbiting and propagation reactions. The molybdenum carbyne catalyst is less reactive and gives mainly linear polymer if the reaction is quenched with an excess of phenylacetylene within 2 min of addition of the monomer; but if the quenching is delayed for 15 min, only cyclic species are subsequently isolated. When the backbiting reaction occurs at the end of the chain the initiator is regenerated. A small proportion of initiator is therefore present when the system reaches equilibrium. For example when the initial ratio of cyclooctyne to $Mo(\equiv CPr)(OCMe_3)_3$ is 10, the equilibrium mixture contains 5% of the original initiator; this proportion becomes smaller if the original ratio of monomer to initiator is increased. The rest of the initiator ends up as very short chain living polymer. The equilibrium concentration of cyclooctyne is too small to measure, but a significant proportion of cyclic dimer (cyclohexadeca-1,9-diyne) is formed (Krouse 1989).

The ROMP of tetrasilacycloocta-3,7-diynes can also be initiated by W≡C or W≡W complexes of the above type; eqn. (18).

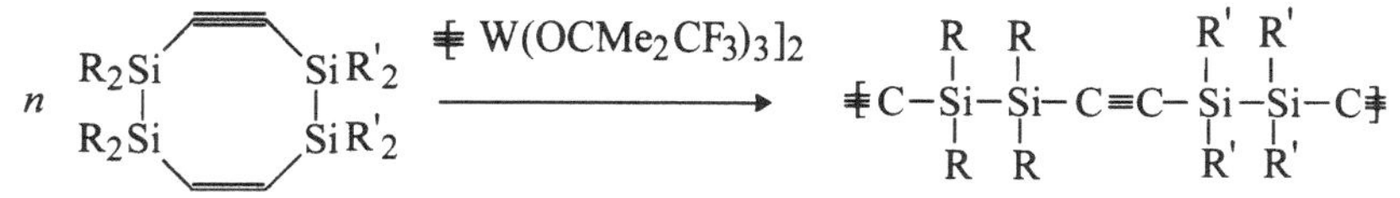

(18)

When $R = R' = Me$, addition of 3 equivalents of monomer to 1 equivalent of $[\equiv W(OCMe_2CF_3)_3]_2$ results in complete consumption of monomer over 3 h at 25°C to yield a partially soluble polymer. Relatively little of the catalyst is used so that propagation must be much faster than initiation. When $R = R' = Et$ there is no reaction with this catalyst. Hence when $R = Me$ and $R' = Et$ one may predict that there will be a strong preference for the reaction to proceed via the metallacyclobutadiene complex in which the bulkier $SiEt_2$ substituent is always placed away from the sterically congested metal. This leads one to expect that the polymer will have an all-HT structure and therefore only two ^{29}Si NMR signals, as indeed is observed. This is in contrast to the anionically produced ring-opened polymer which has additional signals from HH and TT structures. The monomer with $R = R' = Et$ can be polymerized using the more active catalyst $W(\equiv CMe)(OCMe_2CF_3)_3$ (Zhang, X.P. 1994).

10.2.4 Metathesis of diynes

The acyclic diyne metathesis polymerization of dodeca-2,10-diyne is catalyzed by $W(\equiv CEt)(OCMe_3)_3$; eqn. (19). But-2-yne is eliminated and the product is an off-white insoluble powder with the same T_m as the polymer prepared from cyclooctyne (Section 10.2.3) (Krouse 1989).

$$n\ MeC{\equiv}C(CH_2)_6C{\equiv}CMe \xrightarrow{[W]\equiv CEt} MeC\!\!\!+\!\!C(CH_2)_6C\!\!+\!\!\!_n CMe + (n-1)\ MeC{\equiv}CMe \quad (19)$$

10.3 Metathesis reactions involving cleavage of two of the three C≡C bonds

10.3.1 Establishment of the mechanism: a short history

It was first observed by Woon (1974) and Farona (1974) that acetylenes could be polymerized by catalysts of the type $Mo(CO)_3$(toluene). This was followed by the discovery that conventional metathesis catalysts such as $MoCl_5$ (Masuda 1974) and WCl_6 (Navarro 1976; Masuda 1976), with or without a cocatalyst, could also bring about polymerization of acetylenes. At first there was some doubt as to whether these polymerizations were being propagated by the metathesis mechanism (Scheme 10.2) or whether a Ziegler–Natta mechanism was operating. However, the observation that metal carbene complexes could react with acetylenic molecules to form simple adducts as in reaction (20) (Fischer, H. 1980), and the fact that such complexes could initiate the polymerization of acetylenes, albeit somewhat slowly, but cleanly and in fair yield, soon allayed these doubts.

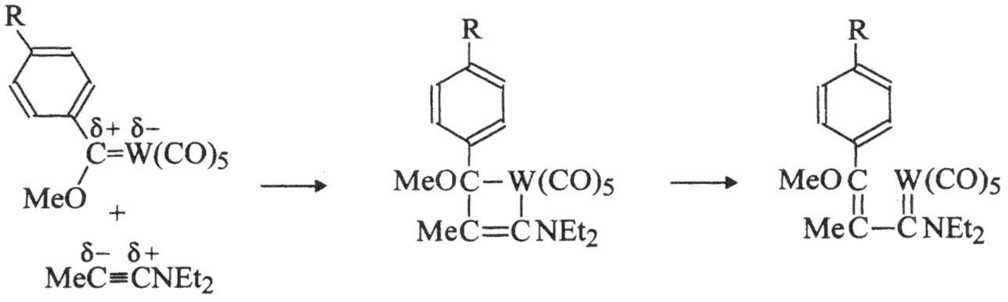

(20)

For example, with a 100 : 1 ratio of $Me_3CC{\equiv}CH$ to $W[{=}C(OMe)Ph](CO)_5$ at 60°C, a 28% yield of polymer ($M_n = 260\ 000$) is obtained in 20 h (Katz 1980a).

Numerous pieces of supporting evidence came from studies of acetylene/olefin systems. Small proportions of phenylacetylene (1–2%) were found to have a strong cocatalytic effect not only on the ROMP of cyclopentene, cycloheptene and cyclooctene, but also on the metathesis of *cis*-pent-2-ene, catalyzed by $W[{=}C(OMe)Ph](CO)_5$ or WCl_6 (Katz 1980c,d, 1982; Makovetskii 1981). Analysis of the resulting polymer of cyclopentene by GPC, using simultaneous refractive index detection (sensitive to the polymer chains) and UV detection (sensitive to phenylacetylene units) showed that phenylacetylene units were present in two forms: first, at the initiating end of long-chain polymer of cyclopentene, and second, as low-molecular-weight homopolymer of phenylacetylene (Han, C-C. 1985; Weiss 1986b). Quantitative studies provided the following explanation. The metal carbene species generated from WCl_6 and $PhC{\equiv}CH$ react 3000 times as fast with phenylacetylene as they do with cyclopentene, giving rise to the low-molecular-weight homopolymer of phenylacetylene. However, if a cyclopentene molecule manages to add to the initiating species, the new metal carbene species favours the addition of cyclopentene by a large factor, particularly when cyclopentene is present in excess. The relative magnitudes of the various rate constants are such that, when the proportion of phenylacetylene is 1–2%, there is virtually no chance of incorporation of another unit of phenylacetylene into the growing chain; but at higher proportions of phenylacetylene this is not so and the net result is that the rate of polymerization passes through a maximum and falls away sharply as the phenylacetylene concentration is increased (Katz 1982). If one starts from a 1 : 1 mixture of cyclopentene and phenylacetylene, the latter is polymerized first, followed by the cyclopentene (Masuda 1991a). The product is then essentially a mixture of the two homopolymers, although the initially formed chains of cyclopentene units will have a very short initial block of phenylacetylene units, reducing to a single unit as the phenylacetylene is used up.

Another approach is to look for the expected end-group in the polymer of an acetylene initiated by an olefin metathesis catalyst. For this purpose, Clark, M. (1982) used $ReCl(CO)_5/EtAlCl_2$ first to initiate the intramolecular metathesis of octa-1,7-diene at 95°C, so generating the intermediate $[Re]{=}CH(CH_2)_4CH{=}CH_2$. If phenylacetylene is then added, allowed to react at 98°C, and the reaction finally quenched with methanol, the polymer formed ($M_n = 2135$) contains aliphatic hydrogens corresponding to one C_7 fragment per polymer chain, as expected. Again, for the polymer of phenylacetylene, made by $W[{=}C(OCH_2CH_3)Ph](CO)_5$ initiation at 50°C, the $={C(OCH_2CH_3)Ph}$ end-group is detectable as the CH_2 quartet at δ 3.85 (Kunzler 1988b).

The most direct proof of the metal carbene mechanism comes from the use of electron-deficient ($< 18e$) metal carbene complexes as initiators. Thus $W({=}CHCMe_3)({=}NAr)(OCMe_3)_2$ initiates the polymerization of acetylene itself and triblock copolymers can be made by successive addition of norbornene, acetylene, and norbornene to the initiator. Provided the centre block contains not more than 10 units of acetylene, such triblock copolymers have narrow MWD ($M_w/M_n < 1.1$), characteristic of a living system (Schlund 1989). Again, $Ta({=}CHCMe_3)(DIPP)_3(THF)$, where DIPP = 2,6-diisopropylphenoxide, reacts with 1 equivalent of $MeC{\equiv}CMe$ to give a THF-free metallacyclobutene complex, which on addition of pyridine yields a vinylalkylidene complex by opening of the metallacyclobutene ring. This can then be used to initiate the polymerization of up to 200 equivalents of $MeC{\equiv}CMe$ to give a living polymer, which, after termination with benzaldehyde, is found to have a very narrow MWD ($M_w/M_n < 1.05$) (Wallace 1989). Likewise $Mo({=}CHCMe_2Ph)({=}N\text{-adamantyl})[OCH(CF_3)_2]_2$(2,4-lutidine) initiates the living polymerization of $HC{\equiv}CC_6H_4\text{-}(SiMe_3)\text{-}2$ to give a polymer with MW proportional to the amount of monomer consumed and with $M_w/M_n = 1.05$ (Schrock 1994b).

Further evidence comes from the structure of statistical copolymers of cycloalkenes (M_1) with acetylenes (M_2) in which the M_1M_2 junctions are present in sufficient proportion to be detectable by NMR. For example, a copolymer of cyclooctene and 2,4-dichlorophenylacetylene, containing 15% of M_1 units, has been prepared using $WCl_4[OCH(CH_2Cl)_2]_2/Et_2AlCl$ as catalyst. The reactivity ratios are such that these units are nearly all flanked by M_2 units and the olefinic protons of the M_1 units give a signal at δ 5.62, somewhat shifted from that for the olefinic protons in the homopolymer of M_1 (δ 5.3) (Tlenkopachev 1989). This is direct proof that acetylenes can add to propagating metal carbene complexes. Another example is the copolymerization of norbornene (M_1) with phenylacetylene (M_2) using WCl_6 as catalyst in toluene at 30°C. The reactivity ratios, $r_1 = 0.20$, $r_2 = 7.0$, show that both types of propagating metal carbene complex P_1 and P_2 (in which the previous units added were M_1 and M_2, respectively) prefer to add M_2 rather than M_1. In the ^1H-^1H COSY NMR spectrum of the copolymer there are two correlations for the M_1M_2 olefinic protons, one at the intersection of δ 5.2 and 6.1, the other at the intersection of δ 5.5 and 6.6. These are assigned to the protons in the *cis* and *trans* M_1M_2 dyads **1** and **2**. The presence of these peaks proves that the product is a statistical copolymer and not a mixture of homopolymers (Masuda 1991a, 1992a).

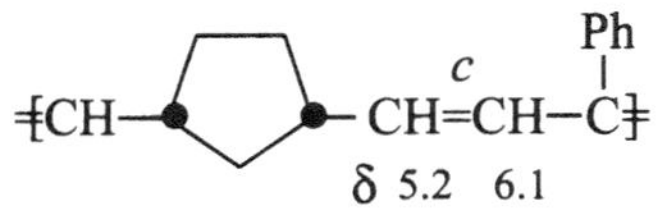

(**1**)

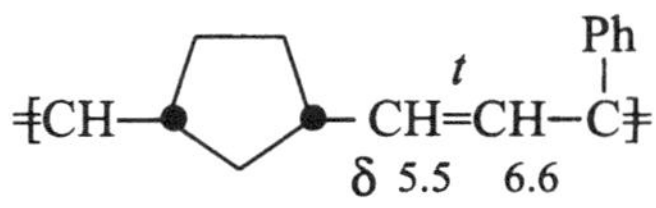

(**2**)

If the polymerization of acetylenes occurs according to Scheme 10.2, the triple bond in the monomer should become a single bond in the polymer. A very elegant experiment to prove that this is so, albeit under restricted circumstances, was performed by Katz (1985b). Phenylacetylene was labelled and polymerized as indicated in eqn. (21). ^{13}C NMR nutation spectra of the polymer at 77 K (Fig. 10.2) showed that the ^{13}C labels in adjacent ^{13}C nuclei were separated by 1.48 Å (single bond). In contrast, polymer made from the same monomer using $Ti(OBu)_4/Et_3Al$ as catalyst had ^{13}C labels separated by 1.36 Å (double bond), indicating polymerization by a Ziegler–Natta mechanism. In these experiments it was essential to hold the temperature below 0°C, otherwise the single and double bonds became scrambled.

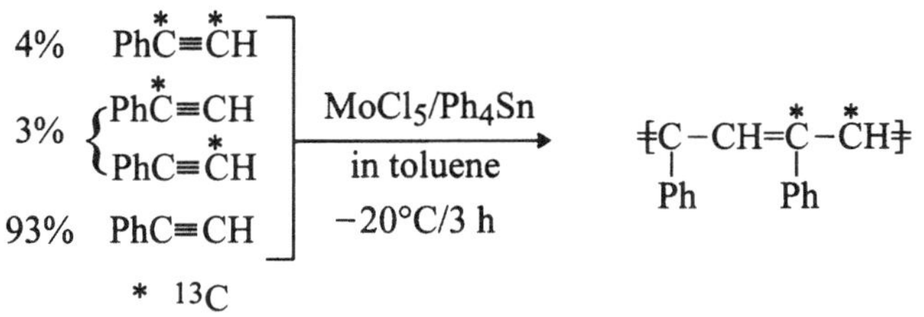

(21)

A further confirmation of the metal carbene mechanism is provided by enyne intramolecular metathesis reactions such as that depicted in eqn. (22). The C≡C bond in the substrate becomes the single bond attaching the alkenyl group to the phenanthrene ring (Katz 1985a, 1989); also see Section 10.4.

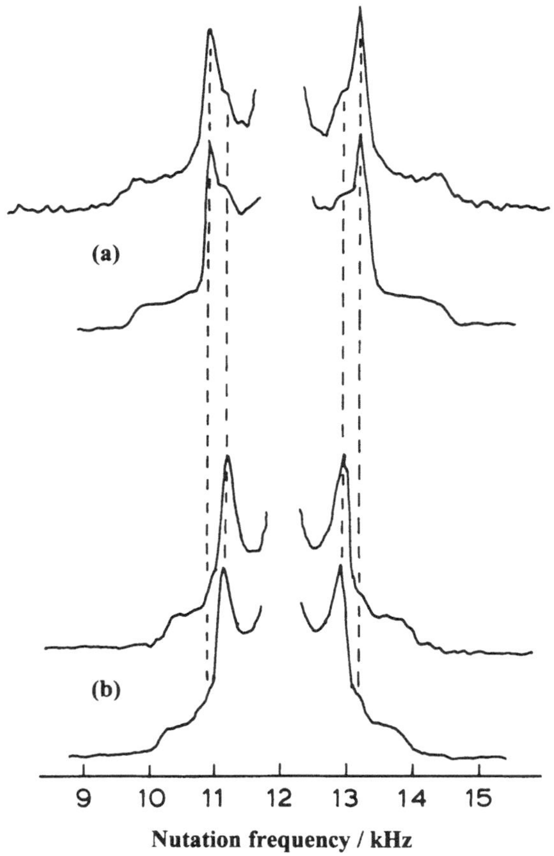

Fig. 10.2 ^{13}C NMR nutation spectra of poly(phenylacetylene) at 77 K. (a) Sample prepared according to eqn. (21) using $MoCl_5/Ph_4Sn$ as catalyst. (b) Sample prepared from the same labelled monomer using $Ti(OBu)_4/Et_3Al$ as catalyst. The lower of each pair of curves is a simulated spectrum. The centre peaks are due to isolated ^{13}C nuclei and have been cut off for clarity. The inner and outer pairs of dashed lines mark, respectively, the peaks of curves arising from pairs of ^{13}C nuclei separated by single and double bonds (Katz 1985b).

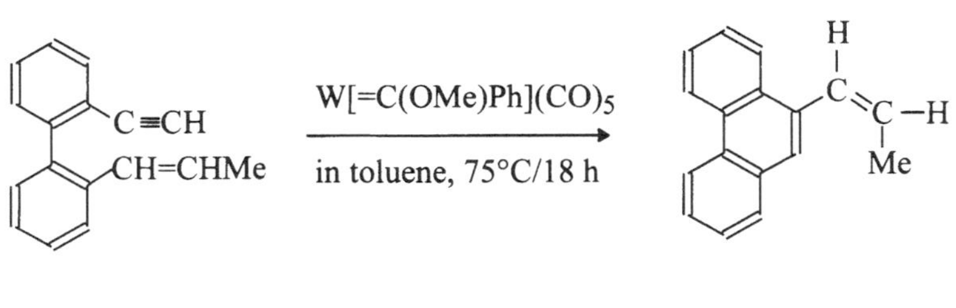

(22)

10.3.2 Metathesis polymerization of acetylene

Since 1985 a number of typical olefin metathesis catalysts have been found to induce the polymerization of acetylene. The complex $W[=C(OMe)CH_2CH_2CH{=}CH_2](CO)_4$ in contact with 1 atm pressure of acetylene in hexane at 20°C gives a 25% yield of polymer in 48 h (Liaw 1985). With Et_3Al as cocatalyst ($Al/W = 20$) the activity is enhanced by a factor of 15, an effect that is attributed to coordination of one of the CO ligands to Et_3Al, thereby assisting the displacement of the coordinated vinyl group by acetylene (Soum 1986; also see Liaw 1993). For the titanium complex $TiCp_2(PMe_3)_2$ used by Alt (1987) there is evidence that acetylene first displaces one of the PMe_3 ligands, forming a titanacyclopropene complex. This may either rearrange to $[Ti]{=}C{=}CH_2$, which then initiates metathesis polymerization of acetylene, or it may add another molecule of acetylene to give a metallacyclopentadiene complex, which does not initiate further polymerization. $WCl_6/BuLi$ is an effective catalyst (Theophilou 1987), but $WCl_6/EtAlCl_2$ and WCl_6/Et_3Al give low yields of polymer (Amass 1987b).

In general the polyacetylene formed with the above catalysts is a black intractable material, insoluble in all solvents, and having a high *trans* content. The initiator $Mt({=}CHCMe_3)({=}NC_6H_3\text{-}i\text{-}Pr_2\text{-}2,6)(OCMe_3)_2$, ($Mt = Mo$, W), in the presence of quinuclidine, provides much better control over the polymerization of acetylene, allowing the preparation of soluble polymers, containing up to nine monomer units. The quinuclidine forms stronger complexes with the propagating species than with the initiating species, thereby slowing down propagation relative to initiation. The living ends may be terminated in a controlled way by cleavage with pivaldehyde. In the HPLC of such polymers (Fig. 10.3) both the all-*trans* oligomers ($n = 3$–9), and those containing one or more *cis* double bonds are resolved (Schlund 1989; Park, L.Y. 1991).

The reaction between acetylene and ethene, catalyzed by WCl_6 in benzene at 20°C, gives polybutadiene (85% 1,2-, 15% 1,4-) of high MW. The initial formation of butadiene probably occurs by alternate addition of acetylene to $[W]{=}CH_2$ to form $[W]{=}CHCH{=}CH_2$, and chain transfer with ethene (Zaliznaya 1990).

10.3.3 Metathesis polymerization of monosubstituted acetylenes

In Table 10.1 are summarized the alkylacetylenes that have been polymerized by metathesis catalysts. The halides $NbCl_5$, $TaCl_5$, $MoCl_5$, and WCl_6 can be used without a cocatalyst because acetylenes themselves react readily with the halide to generate an initiating metal carbene complex, perhaps through a sequence such as eqns. (23a) or (23b) (Weiss 1986b).

$$WCl_6 + HC{\equiv}CR \longrightarrow Cl_5WCH{=}CRCl \xrightarrow{-HCl} Cl_4W{=}C{=}CRCl \qquad (23a)$$

$$Cl_5WCH{=}CRCl \longrightarrow Cl_4W{=}CHCRCl_2 \qquad (23b)$$

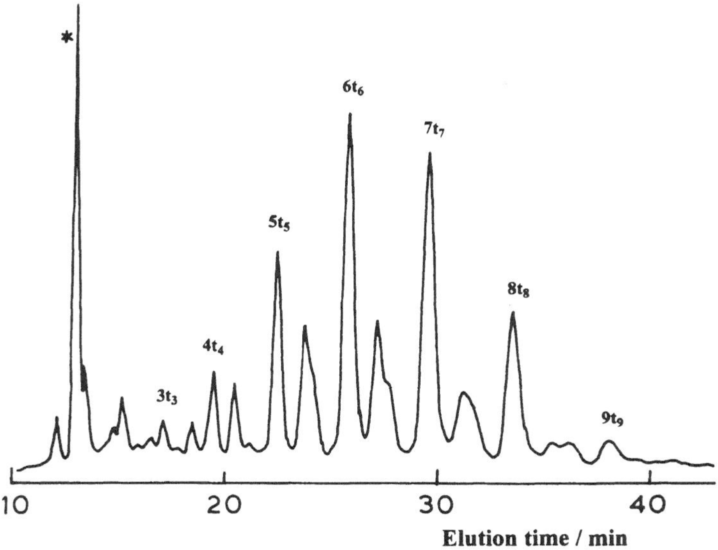

Fig. 10.3 Reversed-phase high-performance liquid chromatography (HPLC) of $Me_3C(CH{=}CH)_xCMe_3$ prepared by the polymerization of acetylene (5 equivalents) in toluene, initiated by $W({=}CHCMe_3)({=}NAr)(OCMe_3)_2$ (1 equivalent) in the presence of quinuclidine (5 equivalents) and terminated by Me_3CCHO. The peak marked $3t_3$ is for the oligomer with $x = 3$ and 3 *trans* double bonds; likewise for the other marked peaks. The unmarked peaks refer to oligomers containing at least one *cis* double bond. The peak marked with an asterisk is the internal standard (guaiazulene) (Schlund 1989).

Most of the catalysts will polymerize most of these acetylenes but, depending on the system, the product may be linear polymer only ($MW = 10^3$–10^6), cyclic oligomers only (see cases mentioned below), or a mixture of linear polymer and cyclic trimers (Yokota 1993). NMR spectra generally indicate that the linear polymers have an all-HT structure (Katz 1980b), so that one might have expected backbiting reactions to yield only the symmetrical 1,3,5-substituted cyclic trimer. In fact the 1,2,4-isomer is generally formed in larger amounts than the 1,3,5-isomer. For a random association of three alkyne molecules, the ratio of 1,3,5 to 1,2,4 structures would be 1 : 3. For the cases where the cyclic trimers are the only products the observed ratios are 1 : 1–3.7 for the WO_3/SiO_2-catalyzed reactions of HC≡CR (R = Me, Et, Pr), 1 : 1.3 for the corresponding MoO_3/SiO_2-catalyzed reactions (Moulijn 1972), 1 : 1.5 for the $TaCl_5$-catalyzed reaction of HC≡CBu, and 1 : 3 for the corresponding $NbCl_5$-catalyzed reaction (Masuda 1983a). It seems therefore that, so long as monomer adds in the same direction to the propagating metal carbene complex, there is a good chance of forming linear polymer; but if a molecule of monomer adds in reverse fashion, there is a strong probability of backbiting to form the 1,2,4-trisubstituted benzene. The tendency towards backbiting is markedly reduced or totally eliminated by the presence of bulky

Table 10.1 References to the metathesis polymerization of HC≡CR (R = alkyl)[a]

R	Catalyst type				
	$MtCl_x$-based[b]	$MtCl_xL_y$-based[c]	$Mt(CO)_xL_y$-based[d]	$[Mt]{=}CR^1R^2$[e]	$[Mt]{\equiv}CR'$[f]
Me [g]				[1]	[2]
Et		[3, 4]			
Pr			[5]		[6]
i-Pr	[4]				
Bu	[7–9]	[10]	[11]	[12] [k]	
i-Bu			[8] [k]		
s-Bu	[8]				
t-Bu	[13,14,30,31] [15–17] [l]	[18] [l]	[19][20] [k]	[21,22][23] [k]	[24]
Pent			[25]	[5][26–28]	
Hex	[9]			[24]	[24,29]
Oct [h]	[9]		[15,25]		
Cy [i]	[30]				
Cxy [j]	[30]				

[a] [1] Liaw 1985, [2] Katz 1984, [3] Nakayama 1992, 1993, [4] Masuda 1985e, [5] Levisalles 1981b, [6] Weiss 1990b, [7] Higashimura 1982, [8] Masuda 1982a, [9] Yokota 1993, [10] Masuda 1983a, [11] du Plessis 1991, [12] Foley 1983, [13] Masuda 1980b, [14] Okano, Y. 1982, [15] Kunzler 1987a, [16] Kunzler 1987b, [17] Kunzler 1988a, [18] Nakano 1994, [19] Tamura 1993, [20] Tamura 1994, 1996, [21] Fischer, H. 1988, [22] Liaw 1993, [23] Liaw 1995, [24] Weiss 1989b, [25] Tsonis 1992, [26] Lucas 1989, 1993, [27] Meziane 1985, 1988, [28] Fontanille 1992, [29] Weiss 1988c, [30] Gal 1995a, [31] Hezhong 1992. [b] e.g. WCl_6, $MoCl_5$, $NbCl_5$, $TaCl_5$, with or without cocatalyst. [c] e.g. $WCl_4(OC_6H_3\text{-}Me_2\text{-}2,6)_2/EtMgBr$. [d] e.g. $Mo(CO)_3(mes)$. [e] e.g. $W(=CHCMe_3)(OCMe_3)_2(OC_6H_5)_2$. [f] e.g. $W(\equiv CCMe_3)(CH_2CMe_3)_3/SiO_2$. [g] Also $WO_3/SiO_2/90°C$ or $MoO_3/SiO_2/350°C$ (Mortreux 1972; Moulijn 1972). [h] Also $HC{\equiv}CC_nH_{2n+1}$, $n = 10, 12, 14, 16$ (Yokota 1993). [i] Cy = cyclohexyl. [j] Cxy = cyclohex-1-enyl. *k* Photochemical initiation. [l] Living polymers formed.

groups either on the metal centre or on the monomer. For example, HC≡CBu/$TaCl_5$ or $NbCl_5$ (Masuda 1983a) and HC≡CCH($SiMe_3$)R/$TaCl_5$ (Masuda 1987a) give only cyclic trimers, whereas HC≡CCMe_3/$TaCl_5$ (Kunzler 1987a), and HC≡CCMe_3/Nb(OC_6H_3-Me_2-2,6)Cl_4THF/*t*-BuMgCl (Nakayama 1993) give high polymer (MW~10^6). HC≡CCH_2CH_2Ph also gives polymer (MW ~ 10^4–10^5) with a number of metathesis catalysts (Oh 1993c).

The stereochemistry of the propagation step can sometimes be modified by changing the solvent. Thus CH≡CCMe_3 with $MoCl_5$ in anisole gives a highly stereoregular all-*cis* polymer, whereas in toluene the polymer formed contains 24% *trans* double bonds (Okano, Y. 1982). The catalyst system $MoOCl_4/Bu_4Sn/EtOH$ (1/1/1) is remarkable in three respects: (i) it is active in toluene at −30°C; (ii) it gives a 97%-*cis* polymer; (iii) the polymer is living, although the catalyst efficiency is only about 2% (Nakano 1994).

Numerous other monosubstituted acetylenes can be polymerized with these catalysts as summarized in Table 10.2 for Si-containing monomers and in Table 10.3 for various non-aromatic monomers containing functional groups.

Much work has also been carried out on phenylacetylene and its derivatives. Phenylacetylene itself can be polymerized by all types of catalyst; for example, by

Table 10.2 Metathesis polymerization of HC≡CR (Si derivatives)

R	Catalyst (examples)	Notes	Reference
$-SiMe_3$	$WCl_6/Bu_4Sn/30°C$ $(OC)_4W{=}C(OMe)$ / $CH_2{=}CH(CH_2)_2$	80% yield; $M_n = 7000$ (soluble part) 90% yield in 5 h/20°C $M_n = 30\,000$ (soluble part)	Okano,Y. 1984 Liaw 1985
$-SiMe_2X$	$WCl_6/Ph_4Sn/80°C$ $WCl_6/Et_3SiH/30\text{-}80°C$	X = Hex, *t*-Bu, $(CH_2)_xPh$ ($x = 0,1,2$); $M_n = 10\,000 - 20\,000$	Tajima 1987 Masuda 1992c
$-CH(CH_2)_xMe$ $SiMe_3$	$MoCl_5/Et_3SiH/30°C$ $Mo(CO)_6/CCl_4/h\nu$	$x = 0,2,4,6$; $M_n = 39\,000 - 94\,000$ for $x = 4$; $TaCl_5$ gives mainly cyclic trimers	Masuda 1987a
$-CH(CH_2)_2Me$ $SiMe_2X$	$MoCl_5/Ph_3Sb/0°C$ $WCl_6/Ph_3Sb/0°C$	X = Hex; $M_n = 200\,000$ X = Ph; $M_n = 8200$	Masuda 1989a
(−)-$CH(CH_2)_4Me$ MeSiPh 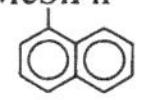	$Mo(CO)_3(MeCN)_3$	64% yield in 24 h/30°C $[\alpha]^{25}_D = -9.5°$ for monomer, −98° for polymer	Tang 1989
Me_3Si 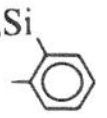	$MoOCl_4/Bu_4Sn/EtOH$ $W(CO)_3(MeCN)_3$ $[Mo]{=}CHCMe_2Ph$ [a] $W(CO)_6/SnCl_4/h\nu$	Living system; $M_w/M_n = 1.11$ Living system; $M_w/M_n = 1.04$ $M_n = 470\,000$	Masuda 1987c, 1993 Tang 1989 Schrock 1994b Tamura 1994[b]
Me_3Si 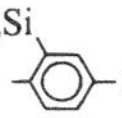 X	W-based	X = Cl, Me, MeO, Me_3SiO, etc.	Yamaguchi 1991
Me_3Si 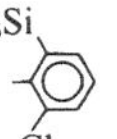 Cl	W-based	$M_n = 180\,000$	Yamaguchi 1991
X 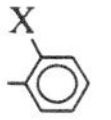	$MoOCl_4/Bu_4Sn/EtOH$ (1/1/10 at −30°C)	X = Me_3SiCH_2, $(Me_3Si)_2CH$, Et_3SiCH_2, *t*-$BuMe_2SiCH_2$; living systems: M_n proportional to % conversion	Seki, H. 1995a

[a] $Mo({=}CHCMe_2Ph)({=}N\text{-}1\text{-adamantyl})[OCH(CF_3)_2]_2$(2,4-lutidine).
[b] See also Masuda 1990a, Tamura 1993; Akiyoshi 1992.

those based on metal halides or metal aryloxo complexes to give linear polymer (Masuda 1982c; Abe 1994; Balcar 1994b, 1995), cyclic trimers (Masuda 1980c), or a mixture of both (Kunzler 1990b); by those based on metal carbonyls (Masuda 1980a; Park, J.S. 1989; du Plessis 1991; Shivasubramaniam 1991; Ganesamoorthy 1992; Tamura 1993, 1996; Vosloo 1993a,b, 1995; Abe 1994; Szymańska-Buzar 1994; Luyt 1995); by thermal, photochemical or chemical activation of Fischer-type metal carbene complexes (Ho, H.T. 1982d; Foley 1983; Liaw 1992); and by other tungsten carbene complexes (Weiss 1989b).

Of particular interest are the *ortho*-substituted phenyl derivatives because, unlike their *meta* and *para* isomers, they generally give living systems when initiated by $MoOCl_4/Bu_4Sn/EtOH$ (1/1/1), allowing the preparation of block copolymers. This is the case for phenylacetylenes with the following *ortho*-substituents: Me, $CHMe_2$,

Table 10.3 Metathesis polymerization of HC≡CR, where R contains a heteroatom other than Si, and no aromatic group

R	Catalyst (examples)	Notes	Reference
-$(CH_2)_x$ OH and other groups containing OH	$MoCl_5/EtAlCl_2$/60°C	x = 1, 2: high yield, insoluble product; x = 3, 4: 55% yield, M_n = 1600	Gal 1995b Gal 1994b,c
-CH_2Hl	$Mo(OEt)_5/EtAlCl_2$	64% yield, insoluble product	Lee, W.C. 1988, Gal 1992a
(Hl = Cl, Br)	$MoCl_5$/N-hexylcarbazole $W(CO)_5(PPh_3)/Et_3Al_2Cl_3$	80% yield, soluble polymer 87% yield, brown powder	Kunzler 1990a Siskos 1994
-$(CH_2)_3Cl$	WCl_6/Me_4Sn/20°C [a]	36% yield, M_n = 7400	Ho, T.H. 1985
-$(CF_2)_xCF_3$	WCl_6/Ph_4Sn/30°C	x = 3, 5; 86% yield, white solid	Tsuchihara 1988
-CR_2NH_2	$Mo(OEt)_5/EtAlCl_2$/90°C $WCl_6/EtAlCl_2$/90°C	R = H, 65% yield, insoluble black powder R = Et, 60% yield, insoluble brown powder	Gal 1992b
-$(CH_2)_3CN$	WCl_6/Me_4Sn/20°C [a]	54% yield, M_n = 5800	Ho, T.H. 1985
-COOH	$MoCl_5$/60°C	low mol. wt. polymer, water-soluble	Masuda 1982d
-COOMe	$MoCl_5$/60°C	low mol. wt. polymer, dioxane-soluble	Masuda 1982d
-$(CH_2)_3COOMe$	WCl_6/Me_4Sn/20°C [a]	65% yield, M_n = 6900	Ho, T.H. 1985
-$CH_2CH(PO(OEt)_2)COOEt$	$WCl_6/EtAlCl_2$/80°C	72% yield, M_n = 8500, yellow polymer	Gal 1995c
-$CH_2PPh_3^+Br^-$	$WCl_6/EtAlCl_2$/90°C		Gal 1994d
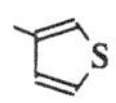	WCl_6/Ph_4Sn/20°C WCl_6/Bu_4Sn/60°C	71% yield 96% yield, M_n = 21600, brown polymer	Tanaka, S. 1986 Gal 1986
-ferrocenyl -ruthenocenyl	Mo=C complex Mo=C complex	soluble head-tail living polymer, see text soluble head-tail living polymer, see text	Buchmeiser 1995 Buchmeiser 1995

[a] WCl_6/Ph_4Sn ineffective.

$GeMe_3$, Cl, Br, CF_3 (Akiyoshi 1992; Masuda 1992b; Mizumoto 1993, 1995; Kaneshiro 1995), $SiMe_3$ (Akiyoshi 1992), and also for (2,3,5,6-tetrafluoro-4-butylphenyl)acetylene (Masuda 1994c). As with HC≡CMe_3 (Nakano 1994), the steric hindrance offered by the substituents close to the metal centre in the propagating species is evidently a decisive factor in giving a living system with this catalyst. Even more instructive is the behaviour of the *o*-$SiMe_3$ monomer with Mo(=$CHCMe_2Ph$)(=N-adamantyl)$(OR)_2$(base) initiators, in which the first step is represented by eqn. (24) (Schrock 1994b, 1996).

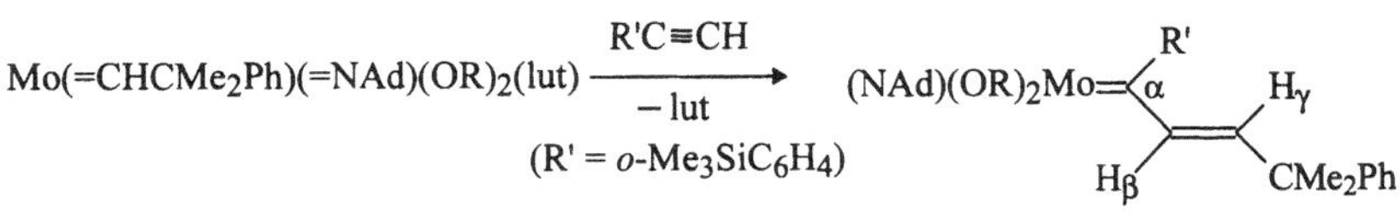

(24)

(3) (4)

With initiator **3** (OR = $OCH(CF_3)_2$, lut = 2,4-lutidine, Ad = adamantyl), the first insertion product, identified by NMR, is **4**, in which the phenyl substituent is attached to the α-carbon (head structure), the double bond is *trans* ($J_{\beta\gamma}$ = 15.6 Hz), and the metal centre is base-free. The monomer is thus able to displace the lutidine ligand in **3** and then react, but the lutidine is unable to coodinate to **4** where less space is available. It is also notable that benzaldehyde reacts less readily with **4** than it does with the initiator **3**, even though it is base-free. When further monomer is added, the MW of the polymer formed is directly proportional to the amount of monomer consumed, and $M_w/M_n = 1.04$ when $M_n = 13\,900$. Similar results are obtained using Mo(=CHCMe$_2$Ph)(=NC$_6$H$_3$-Me$_2$-2,6)(OC$_6$F$_5$)$_2$(quinuclidine) as initiator, although the reaction is not so fast. Initiators with bulkier alkoxy or imido ligands either do not react, or fail to show the characteristics of a living system. The key to success with this monomer is therefore to use a metal carbene complex with relatively small ligands. For the shorter-chain polymers (DP < 25) the absorption spectrum initially shows two maxima separated by about 50 nm but on standing gives largely the red-shifted form; this change is attributed to *cis* → *trans* isomerization. The living polymerization of ethynylferrocene is likewise readily initiated by Mo(=CHCMe$_2$Ph)(=NC$_6$H$_3$-Me$_2$-2,6)[OCMe(CF$_3$)$_2$]$_2$ in toluene or THF. All-HT polymers are formed which remain soluble if DP $\leqslant$ 40. The first addition product is all-*trans* and propagation is *via* the disubstituted carbene complex (α-addition). Ethynylruthenocene behaves similarly (Buchmeiser 1995).

Substituted phenylacetylenes that have been polymerized by metathesis catalysts also include those with the following substituents: 2- and 4-methyl- (Yamaguchi 1991; Mizumoto 1995; Vijayaraj 1995), various 2-alkyls (Abe 1989, 1994), 2,5-di-*t*-butyl- and 2-phenyl- (Masuda 1994a), 2-methyl-5-*t*-butyl- (Kouzai 1994c), 2,6-dimethyl-4-*t*-butyl- (Yoshida, T. 1996), 2-perfluorohexyl- (Seki, H. 1995b), 2-trifluoromethyl- (Muramatsu 1985), 2-fluoro-, 3-trifluoromethyl- and 3-chloro- (Tamura 1993; Mizumoto 1995), 2-chloro- and 4-chloro- (Yamaguchi 1991; Vijayaraj 1995), 2,5-bis(trifluoromethyl)- (Tsuchihara 1990), pentafluoro- (Yoshimura 1991), 2,4-dichloro-, 4-bromo- and 3,4-dimethoxy- (Tlenkopachev 1989, 1995a), 2- and 4-methoxy- (Vijayaraj 1995); see also Table 10.2. Good yields of soluble polymer of high MW are obtained in most cases.

Some α-naphthyl derivatives, with a CF$_3$ group in the 4, 6, or 7 positions, have been polymerized using W(CO)$_6$/*hv*/CCl$_4$ as catalyst (Okano, T. 1988; see also Yamaguchi 1992). Monosubstituted acetylenes containing N-carbazolyl groups have been polymerized by MoCl$_5$/Bu$_4$Sn (Nakano 1995) and WOCl$_4$/Ph$_4$Sn (Kmínek 1989), but the polymers are insoluble and not easily characterized. However, soluble copolymers with other acetylenes can be made, with high MW. Some monosubstituted acetylenes bearing a phenylcyclohexyl mesogenic moiety have been polymerized with MoCl$_5$/Ph$_4$Sn (Oh 1993a,b).

The more highly substituted polymers of the monosubstituted acetylenes can generally be prepared with higher MW than for the less highly substituted polymers. They are also usually colourless or yellow, amorphous, readily soluble, and non-conducting, in sharp contrast to polyacetylene itself, which is black, insoluble, and semi-conducting, showing metallic conduction when doped. This is because the substituents force the main chains to take up twisted, non-conjugated conformations.

10.3.4 Metathesis polymerization of disubstituted acetylenes

Systems that have been studied are summarized in Table 10.4. $W(CO)_6$ in hexane is effective as a photochemical initiator of polymerization of non-functionalized disubstituted acetylenes, such as MeC≡CMe, only if 1% of a monosubstituted acetylene is added. Polymerization then begins after an induction period of 30 min. During this time it is likely that an active metal carbene complex is generated by sequence (25). Dialkylacetylenes are presumably less able to undergo the corresponding alkyl group shift. In contrast, initiation by $Mt(CO)_6/h\nu/CCl_4$ (Mt = Mo, W) is effective for the polymerization of PhC≡CCl without the use of a cocatalyst, and gives high polymer ($M_w \sim 10^6$). Reaction continues in the dark after an initial irradiation period during which time the initiating species $[Mt]{=}CCl_2$ is thought to be generated (Higashimura 1982; Masuda 1982a). $SnCl_4$/toluene can also be used in place of CCl_4 (Tamura 1994).

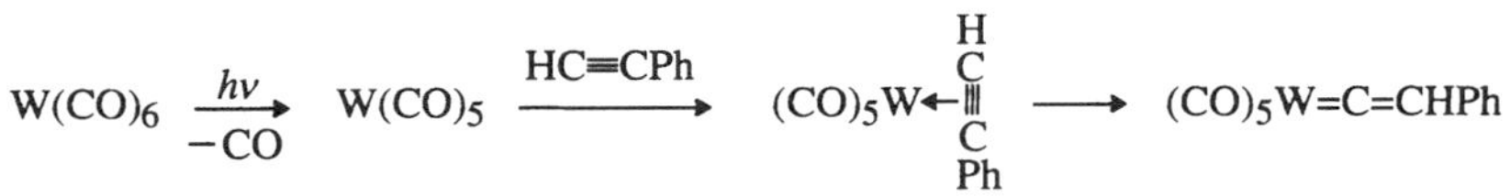
(25)

Disubstituted acetylenes such as PhC≡CPh can severely retard the propagation of the ROMP of cyclopentene (Han, C-C. 1985; Weiss 1986b). Conversely, terminal olefins such as vinyltrimethylsilane can act as chain-transfer agents in the polymerization of MeC≡CPh catalyzed by $NbCl_5/Bu_4Sn$, or of $ClC{\equiv}CC_6H_{13}$ or ClC≡CPh catalyzed by $MoCl_5/Bu_4Sn$. The end-groups derived from the transfer agent can be detected when the MW is sufficiently low (Masuda 1991b; Kouzai 1993).

With WCl_6 or $MoCl_5$ as catalyst it is generally necessary to use a cocatalyst such as Ph_4Sn, unlike the situation with monosubstituted acetylenes, where the substrate acts as its own cocatalyst (Section 10.3.3). With $NbCl_5$ and $TaCl_5$ a cocatalyst is not always necessary, although its use may affect the course of the reaction. For example, for the $TaCl_5$-catalyzed polymerization of MeC≡CPh in toluene at 80°C, the conversion reaches 100% in 6 h, but the polymer then degrades rapidly to oligomer (Masuda 1985c,d). With $TaCl_5/Ph_4Sn$ (1/1) under the same conditions the reaction is complete in 1 h and the MW of the polymer remains stable for at least 24 h; one may suppose that bulky ligands attached to the tantalum centre then prevent secondary metathesis reactions of the double bonds in the polymer chain. Similar observations have been made with $PhC{\equiv}CC_6H_4$-X-4 as monomer, where X = H (Niki 1987), CMe_3, etc. (Kouzai 1994a,b), and $SiMe_3$ (Tsuchihara 1991, 1992).

The structure of the monomer can be a crucial factor with these catalysts. Thus MeC≡CBu is polymerized by both $MoCl_5/Ph_4Sn$ and WCl_6/Ph_4Sn in toluene at 30°C (Higashimura 1982), but the more sterically hindered $MeC{\equiv}CCHMe_2$ is polymerized only by WCl_6/Ph_4Sn at 60°C (Masuda 1982a).

The relative proportions of linear polymer and cyclic trimers produced from disubstituted acetylenes are dependent on the catalyst system and are therefore

Table 10.4 References to the metathesis polymerization of disubstituted acetylenes[a]

Substrate	Catalyst type[b]									
	A	B	C	D	E	F	G	H	I	J
MeC≡CMe	[1–4]	[5]	[6]	[7]			[8]			
MeC≡CR [c]		[5,9]			[10–12,59]		[10,13,14]	[15]	[15]	[16]
$MeC{\equiv}CSiMe_2CH_2X$ [d]					[17]			[18–20]	[17,19,21]	
MeC≡CAr			[22]				[23–25]	[39]	[11,26–28,39]	
EtC≡CR [e]		[29,30]					[10,31–33,59]	[15,26,59]	[15,59]	
EtC≡CPh								[15,27]		
PrC≡CPr		[34]					[32,59]	[15,59]	[15,59]	
$BuC{\equiv}CC_6H_4OC_9F_{17}$								[35]		
PhC≡CX [f]					[36]		[25,33,37,38]		[15,35,40–42]	
HOOCC≡CCOOH					[36]					
$XC{\equiv}CSiMe_3$ [g]					[43]		[43]			
$R^1C{\equiv}CSR^2$ [h]					[44–46]		[46]			
$C_6F_5C{\equiv}CR$ [i]									[47]	
RC≡CHl [j]			[48,49]	[11,50,51]	[12,52–55]	[56–58]				

[a] [1] Katz 1980a,b, [2] Foley 1983, [3] Liaw 1985, [4] Wallace 1989, [5] Katz 1984, [6] Levisalles 1981a,b, [7] Landon 1985, [8] Navarro 1976, [9] du Plessis 1991, [10] Higashimura 1982, [11] Higashimura 1983, [12] Masuda 1991b, [13] Masuda 1982a, [14] Stockel 1978, [15] Niki 1987, [16] Pennella 1968, [17] Kunzler 1987a, [18] Masuda 1983b, [19] Masuda 1985a, [20] Fujimori 1988, [21] Isobe 1986, [22] Woon 1974, [23] Sasaki, N. 1976, [24] Hasegawa, K. 1977a, [25] Masuda 1979, [26] Masuda 1982b, [27] Masuda 1985c,d, [28] Kouzai 1993, [29] McCullough 1984, [30] McCullough 1985, [31] Tlenkopatchev 1995a, [32] Masuda 1981, [33] Tlenkopachev 1989, [34] Weiss 1988c, [35] Yoshimura 1994, [36] Masuda 1982d, [37] Han, C-C. 1985, [38] Weiss 1986b, [39] Kouzai 1995, [40] Kouzai 1994a,b, [41] Tsuchihara 1991, 1992, [42] Tachimori 1994, 1995, [43] Gal 1989, [44] Matsumoto 1991, [45] Masuda 1990b, [46] Watase 1995, [47] Yoshimura 1986, [48] Tamura 1993, [49] Tamura 1994, [50] Masuda 1982e, [51] Yamagata 1984, [52] Yoshimura 1988, [53] Gal 1985, [54] Masuda 1984b, [55] Masuda 1985b, [56] Masuda 1987c, [57] Masuda 1987b, [58] Masuda 1989b, [59] Masuda 1986. [b] A = metal carbene complexes; B = metal carbyne complexes; C = metal carbonyl complexes; D = photochemical systems; E = $MoCl_5$-based catalysts; F = $MoOCl_4$-based catalysts; G = based on tungsten halide complexes; H = $NbHl_5$-based catalysts; I = $TaHl_5$-based catalysts; J = WO_3/SiO_2. [c] R = alkyl. [d] X = H, $SiMe_3$, CH_2SiMe_3. [e] R = Et, Pr, Bu. [f] X = Ph, C_6H_4Y (Y = $SiMe_3$, OC_9F_{17}, carbazolyl, etc.), β-naphthyl, COOH. [g] X = furyl, pyridyl, thienyl. [h] R^1 and R^2 = alkyl or aryl. [i] R = Me, Et, Bu. [j] R = alkyl, 2-thienyl; Hl = Cl, Br.

disubstituted acetylenes are dependent on the catalyst system and are therefore kinetically controlled. This is well exemplified by the data in Table 10.5. For the polymerization of EtC≡CPh, catalyzed by $NbCl_5$, cyclic trimers (< 20%) are formed concurrently with polymer (> 80%) and are not the result of secondary metathesis (Masuda 1985c).

Certain metal alkyne complexes are capable of initiating the polymerization of dialkylacetylenes, no doubt first being converted into metal carbene complexes. For complexes of the type $Mo(\equiv CCMe_3)(OR)_3$ the nature of OR is critical for a substrate such as EtC≡CPr. If OR is $OCMe_3$, there is no reaction; if OR is $OCMe(CF_3)_2$, the acetylene is metathesized; only if OR is $OCHMe_2$ or OCH_2CMe_3 does the acetylene polymerize (McCullough 1984, 1985).

The catalyst system $MoOCl_4/Bu_4Sn/EtOH$ (2/2/1) is remarkable in that it gives living polymers of RC≡CCl (R = Bu, Hex) having narrow MWDs ($M_w/M_n = 1.1$–1.4), just as it does with certain monosubstituted acetylenes (Section 10.3.3). With $MoOCl_4$ or $MoOCl_4/Bu_4Sn$ as catalyst, the MW of the polymer increases with conversion but the polymers have a broader MWD indicating the occurrence of secondary metathesis reactions; Fig. 10.4 (Masuda 1987c, 1989b). It has been suggested that the role of EtOH in the three-component catalyst system is to replace one of the chloride ligands by an ethoxy ligand in the propagating metal carbene species thereby making it less prone to undergo termination or secondary metathesis reactions (Masuda 1992b). The ^{13}C NMR spectra of these polymers are as expected for a regular HT structure, but it is not possible to determine the proportion of *cis* double bonds (Masuda 1987b).

Polymers of other unsymmetrical disubstituted acetylenes, such as MeC≡CPr, have somewhat broad ^{13}C NMR signals, and may well contain HH and TT as well as HT structures (Higashimura 1982). These polymers are generally white and have intrinsic viscosities which are sometimes nearly proportional to the MW, for example, the polymer of EtC≡CPh (Masuda 1985c). This indicates that the polymer molecules have a fairly rigid but twisted backbone with very little

Table 10.5 Products of metathesis polymerization of EtC≡CBu and $MeC\equiv CSiMe_3$ by Nb- and Ta-halide catalysts[a] (Masuda 1982b, 1983b)

	Monomer (M)			
	EtC≡CBu		$MeC\equiv CSiMe_3$	
Catalyst	Polymer[b] (%)	Cyclic trimer (%)	Polymer[c] (%)	Cyclic trimer (%)
$NbCl_5$	100	0	100	0
$NbBr_5$	38	62	100	0
NbI_5	56	15	0	0
$TaCl_5$	82	8	100	0
$TaBr_5$	43	57	95	5

[a] In toluene at 80°C for 24 h; $[M]_o = 1.0$, $[cat]_o = 0.02$. [b] $M_w = (2–32) \times 10^5$. [c] $M_w = (2–9) \times 10^5$.

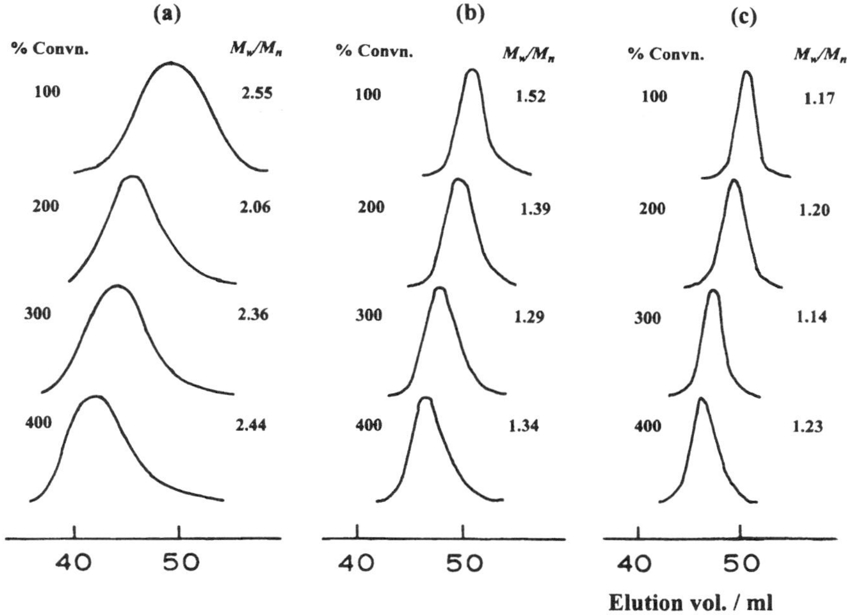

Fig. 10.4 Molecular weight distribution curves, based on polystyrene calibration, of poly(1-chlorooct-1-yne)s obtained with catalysts (a) $MoOCl_4$, (b) $MoOCl_4/Bu_4Sn$, and (c) $MoOCl_4/Bu_4Sn/EtOH$, in toluene at 30°C. Four successive batches of monomer (0.10 M) were added at 5 min intervals to the catalyst mixture ($[MoOCl_4] = 0.02$ M), and completely polymerized at each stage. The curves, from top to bottom, correspond to successive stages of reaction (Masuda 1989b).

conjugation of the double bonds, unlike polyacetylene and polymers of *linear* monosubstituted acetylenes (Masuda 1982a).

Further details, including the physical properties of polyacetylenes, may be found in the reviews of Masuda (1987d, 1994b) and Breslow (1993).

10.3.5 Metathesis polymerization of diynes: cyclopolymerization

The monomer HC≡C−C≡CX ($X = SiMe_3$) is readily polymerized by WCl_6/Ph_4Sn (1/1) in toluene at 40°C to give a soluble polymer ($M_n \sim 10^4$), which, when cast as a film and exposed to a mercury lamp for a few seconds, becomes completely insoluble (Ozaki 1986). The initial reaction is probably metathesis polymerization through the C≡CX bond, while the subsequent cross-linking by irradiation occurs through the other triple bond. Other 1,3-diynes (X = alkyl, hydroxymethyl, phenyl, carbazolyl, etc.) can be polymerized in this way. The polymerization of $PhC{\equiv}C{-}C{\equiv}CCH_2OH$ by $NbCl_5/Bu_4Sn$ in toluene at 80°C gives an 80% yield of soluble polymer ($M_w = 3200$), which on heating under vacuum to 800°C is converted to a graphite-like structure with a high conductivity even in the absence of dopant. The first stage is thought to involve preferential metathesis polymerization through the C≡CCH_2OH bond (Lee, H.J. 1993, 1994a).

The metathesis polymerization of diynes having four single bonds between the triple bonds (dipropargyl compounds) yields cyclopolymers. The structural units may contain a cyclohexene ring, eqn. (26a), or a cyclopentene ring, eqn. (26b), with the possibility of both in the same chain (see below).

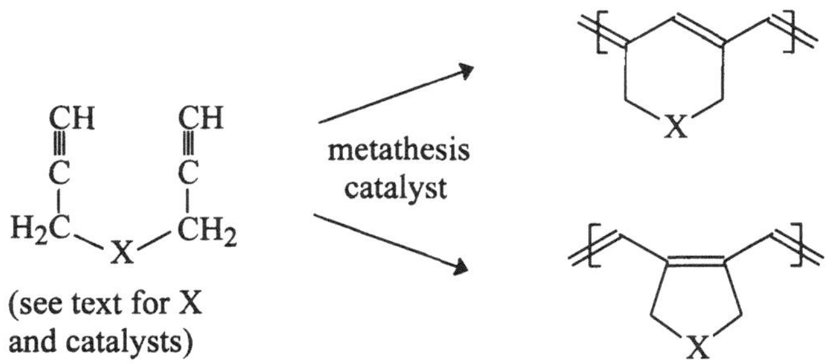

(26a)

(26b)

This reaction occurs with a wide variety of diynes in which X may be: CPh_2 (Jang 1990), $C(CO_2R)_2$ (Ryoo 1990; Han, S.H. 1991; Fox 1992, 1994b), C⟨$(CO.O)_2$⟩CMe_2 (Jin 1993a), $C[CO_2CH_2(CF_2)_5CF_2H]_2$ (Koo 1993), $C[CO_2(CH_2)_6$N-carbazolyl$]_2$ (Park, J-W. 1993), CHOH (Kim, Y-H. 1989), CPh(OH) (Kim, Y-H. 1992), $C(OPh)_2$ (Ahn, H-K. 1992), $CMe(OSiMe_2CMe_3)$ (Koo 1993), CHY or CY_2, where Y is a mesogenic group (Jin 1991, 1993b), $CHCO_2(CH_2)_2O_2CC_6H_3$-$(NO_2)_2$-3,5 (Lee, J-H. 1995), $C[PO(OEt)_2]_2$ and $C(CO_2Et)[PO(OEt)_2]$ (Lee, H-J. 1994b), NY, where Y is a mesogenic group (Choi, S-J. 1994a,b,c,d), $N^+R_2Z^-$, where R = hexyl, Z = Br, BPh_4, tos (Kang 1993; Kim, S-H. 1994a), O (Gal 1987; Lee, H-J. 1995a), S, SO or SO_2 (Gal 1988, 1990, 1993), SiR_2 where R = Me, Ph (Kim, Y-H. 1988; Choi, S-K. 1990), GeR_2, where R = Me, Ph (Cho, O-K. 1990). Where the monomer contains a mesogenic group, both the monomer and the cyclopolymer exhibit liquid crystalline properties. Copolymerization of the diynes with X = $C(CO_2Et)_2$ and X = $C[CO_2(CH_2)_6$N-carbazolyl$]_2$ gives a conductive copolymer (Park, J-W. 1994). The cyclopolymerization of RC≡CCH_2OCH_2C≡CH (R = Me, $SiMe_3$) gives insoluble polymers; but their copolymers with diethyl dipropargyl malonate are soluble (Lee, H-J. 1995b). Tripropargyl ammonium bromide gives a conjugated polymeric salt having two cyclic recurring units per monomer unit (Gal 1994a). Tetrapropargylammonium bromide gives a cross-linked cyclopolymer (Gal 1995d).

Of the first-generation catalysts, those based on $MoCl_5$ are the most effective, either with a cocatalyst such as $EtAlCl_2$ or Bu_4Sn, or alone, at 60–90°C in the usual solvents. Molybdenum carbene complexes give living systems, allowing the preparation of block copolymers of 1,6-diynes with norbornadiene derivatives (Fox 1994b). The polymers are generally soluble and highly coloured, with MW = 10^3 – 10^5, and become conducting when doped with iodine. ^1H and ^{13}C NMR spectra confirm the disappearance of triple bonds and the formation of double bonds during these reactions. Up to 1994 it was generally assumed that the rings formed were mainly six-membered, but there is good evidence that, when $X = C(CO_2R)_2$, five-membered rings are also formed (Fox 1994b). Thus, on addition of the initiator $Mo(=CHCMe_3)(=NAr)[OCMe(CF_3)_2]_2$ to the monomer with R = Et, the carbene proton resonance of the initiator (δ 12.14) is replaced by resonances at 12.41 and 13.03 ppm, assigned to the carbene protons of living **5** and **6** respectively; eqns. (27a) and (27b). Furthermore, in the ^{13}C NMR spectrum of the polymer, there are two distinct carbonyl signals (δ 170.8, 172.0) and two groups of quaternary carbon signals (δ 54–55, 57–58), the latter being assigned to six- and five-membered rings, respectively, by comparison with the spectra of model compounds. For the five-membered rings the 57–58 ppm signal shows fine structure that may be attributed to the influence of the ring structure in the adjacent units. On changing the initiator, the solvent, or the ester group in the monomer, there is some variation in the proportions of five- and six-membered rings formed (80/20–21/79), but the extremes are not reached.

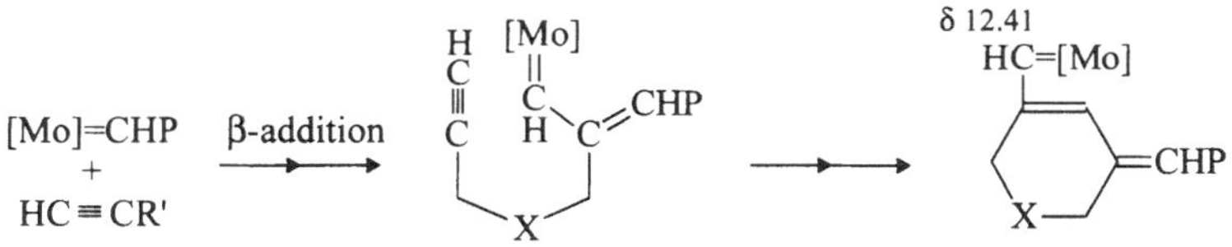

(27a)

(**5**)

P = polymer chain
R' = $CH_2XCH_2C{\equiv}CH$
X = $C(CO_2Et)_2$

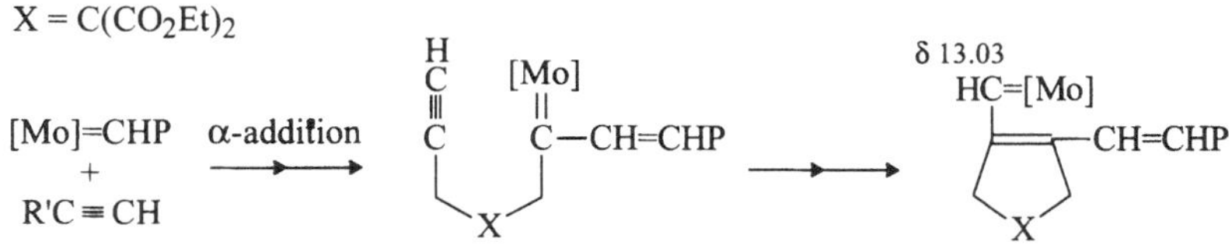

(27b)

(**6**)

Such reactions clearly involve two successive types of propagation step: (i) double-bond exchange between Mt=C and the first C≡C of the monomer; and (ii) rapid intramolecular cyclization by double-bond exchange between Mt=C and the second C≡C of the monomer. If the first step occurs by β-addition (with the substituent β to the metal centre), the subsequent intramolecular reaction is likely to yield a six-membered ring, eqn. (27a), whereas if it occurs by α-addition, the subsequent reaction is more likely to yield a five-membered ring, eqn. (27b). This observation is somewhat surprising because we have already seen that simple monosubstituted acetylenes, containing one C≡C bond, appear to polymerize only by α-addition to give an all-HT structure (Section 10.3.3). It is therefore unexpected that a monosubstituted acetylene in a diyne can add in both directions. Further work is needed to establish whether the result with $X = C(CO_2R)_2$ is the exception rather than the rule. Preliminary reports indicate that when X possesses one extremely bulky substituent ($X = CHCH_2OSiPh_2Bu^t$), the resulting polymer has mostly one type of ring structure (Lee, J-H. 1995). Sterically demanding substituents (preferably two) on X certainly encourage the second, intramolecular, step and reduce the proportion of residual pendant C≡C groups in the polymer, which, if they subsequently react with each other, can give rise to a component with double the MW of the main product. The use of molybdenum carbene initiators, and suitably substituted benzaldehydes as terminating agents, allows the preparation of cyclopolymers having electron-donating or electron-attracting groups at one or both ends of the chains. When the chains are sufficiently short, say DP = 5, and the two end-groups are the same, the absorption maximum for the 'push–push' polymer is at a rather longer wavelength than in the corresponding 'pull–pull' polymer, but when the DP is 20 or more, the chain length exceeds the effective conjugation length and the end groups have comparatively little effect (Fox 1994b).

The diynes $HC{\equiv}C(CH_2)_nC{\equiv}CH$ ($n = 2, 4, 6, 8$) have also been polymerized. For $n = 2$, using $NbCl_5$ as catalyst, the product is mainly trimer, but for $n = 6$ or 8, a polymer with a highly branched structure appears to be formed (Srinivasan 1988). For $n = 4$, it is surprising to find that $Mo(CO)_6/HOC_6H_4$-Cl-3 in toluene at 120°C causes polymerization rather than metathesis (cf. Section 10.2.1); the product is thought to be a cyclopolymer containing enchained cycloheptene rings (Vosloo 1991).

10.3.6 Copolymerization of acetylenes

Acetylenes copolymerize with each other and with cycloalkenes under the influence of olefin metathesis catalysts, as summarized in Tables 10.6 and 10.7, respectively. Each table is divided into two sections according to whether the catalyst system gives long-lived propagating species or not.

With non-living systems, it is possible to make statistical copolymers by adding the monomer mixture to the catalyst system. For copolymers of acetylene with phenylacetylene, made with $MoCl_5$ as catalyst, the UV/visible absorption band moves to longer wavelengths as the proportion of acetylene in the polymer increases, corresponding to an increasing average conjugation length (see Fig.10.5).

Footnotes to Table 10.6

[a] W-1 = WCl_6; W-2 = WCl_6/Ph_4Sn; W-3 = $W[OCH(CH_2Cl)_2]_2Cl_4/Et_2AlCl$; Mo-1 = $MoCl_5$; Mo-2 = $MoCl_5/Bu_4Sn$; Mo-3 = $Mo(CO)_6/HOC_6H_4$-Cl-3; Mo-4 = $MoOCl_4/Bu_4Sn/EtOH$ (2/2/1); Mo-5 = $MoOCl_4/Bu_4Sn/EtOH$ (1/1/1); Ta-1 = $TaCl_5/Ph_3Bi$ at 80°C; Ta-2 = $TaCl_5/Et_3SiH$ at 80°C; Nb-1 = $NbCl_5/Ph_3Bi$ at 80°C. Reaction in benzene or similar solvent except where indicated. [b] $M_2 < M_1$ denotes that M_2 is less reactive than M_1 in copolymerization. [c] ρ^+ is the Hammett reaction parameter obtained from the slope of log $(1/r_1)$ against the Brown σ^+ value characteristic of the substituent X. [d] At 120°C. [e] Block copolymer formed exclusively, regardless of the order of addition of monomers. [f] Block copolymer formed exclusively if the order of addition is M_1, M_2; for the reverse order of addition some homopolymer is also formed. [g] Some homopolymer formed as well as block copolymer for either order of addition of monomers (− 30°C).

Table 10.6 Metathesis copolymerization of acetylenes with each other

M_1	M_2	Catalyst [a]	Notes [b]	Reference
Statistical copolymers (non-living systems)				
HC≡CPh	$HC{\equiv}CC_6H_4$-X-4	W-1	$\rho^+ = -0.47$ [c]	Hasegawa, K. 1975
	X = Cl, Me, MeO			
		W-2	$\rho^+ = -0.47$ [c]	Hasegawa, K. 1977b
	$MeC{\equiv}CC_6H_4$-X-4	W-1	$\rho^+ = -0.96$ [c]	Hasegawa, K. 1977a
	HC≡CH	Mo-1		Weddigen 1983
		Mo-1	Conjugation length = 2–10	Kanischka 1988
	HC≡CR	W-2		Masuda 1984a
	R = Bu, Hex		$M_2 > M_1$ [b]	
	R = CHEtMe		$M_2 \sim M_1$	
	R = CMe_3		$M_2 < M_1$	
	R = CMe_3	W-1,Mo-1	$M_2 \sim M_1$	Masuda 1988
	$HC{\equiv}CC_6H_4$-X-2	W-1,Mo-1		Masuda 1988
	X = CF_3		$M_2 < M_1$	
	X = $SiMe_3$		$M_2 \sim M_1$	
	X = Me	W-1,Mo-1	$M_2 \sim M_1$	Abe 1989
	MeC≡CPent	W-2	$M_2 \sim M_1$	Masuda 1984a
	EtC≡CBu		$M_2 \sim M_1$	
	PrC≡CPr		$M_2 \sim M_1$	
	XC≡CPh	W-2		Masuda 1984a
	X = Me, Ph, Cl		$M_2 < M_1$	
HC≡CH	$HC{\equiv}CCMe_3$	Mo-1	Polymer + cyclic trimers ($3M_2$ and $2M_2 + M_1$)	Kanischka 1988
$HC{\equiv}CCMe_3$	$HC{\equiv}CCH_2X$	Mo-2		Nakano 1995
	X = N-carbazolyl			
$[HC{\equiv}C(CH_2)_2]_2$	HC≡CBu	Mo-3 [d]		Vosloo 1991
$MeC{\equiv}CSiMe_3$	RC≡CPh	Ta-1	$M_2 < M_1$	Hamano 1988
	R = Me, Bu			
	MeC≡CPent	Nb-1	$M_2 > M_1$	Hamano 1988
	PrC≡CPr		$M_2 < M_1$	
EtC≡CPr	$HC{\equiv}CC_6H_3$-Cl_2-2,4	W-3	$M_2 > M_1$	Tlenkopatchev 1995a
PhC≡CPh	$HC{\equiv}CC_6H_3$-Cl_2-2,4	W-3		Tlenkopatchev 1995a
PhC≡CPh	$PhC{\equiv}CC_6H_4$-X-4	Ta-2	$M_2 \sim M_1$	Yoshimura 1994
	X = OC_9F_{17}			
Block copolymers (living systems)				
ClC≡CBu	$ClC{\equiv}CC_{14}H_{29}$	Mo-4	Triblocks ($M_w/M_n = 1.24$)	Masuda 1987c, 1989b
ClC≡CHex	HC≡CX	Mo-5		Akiyoshi 1992
	X = C_6H_4-CF_3-2		[e]	
	X = C_6H_4-$SiMe_3$-2		[f]	
	X = C_6F_4-Bu-4		[f]	
	X = CMe_3		[g]	
	ClC≡CPh	Mo-5	[e]	Akiyoshi 1992

Table 10.7 Metathesis copolymerization of acetylenes with cycloalkenes

M_1	M_2[a]	Catalyst[b]	Notes[c]	Reference
Statistical copolymers (non-living systems)				
HC≡CPh	CPE, CHP, COC	W-1,W-2	M_1 acts as cocatalyst	Katz 1982, Han, C-C. 1985
	CPE	W-1	$M_2 \ll M_1$ [d]	Masuda 1991a
	NBE	W-1	$r_1 = 7.0$, $r_2 = 0.2$[e]	Masuda 1991a, Makio 1993a
	MCPE	W-1	$M_2 < M_1$	Makovetsky 1992a
$HC{\equiv}CC_6H_4$-R [f]	NBE	W-1	$M_2 < M_1$	Makio 1993b
$HC{\equiv}CC_6H_3$-Cl_2-2,4	COC	W-based	$M_2 < M_1$	Tlenkopachev 1989
$HC{\equiv}CC_6H_4$-CF_3-4	NBE	W-1	$M_2 < M_1$ [e]	Makio 1993a
$HC{\equiv}CC_6H_4$-CF_3-2	NBE	W-1	$M_2 < M_1$ [g]	Makio 1993a
$HC{\equiv}CC_6H_3$-$(CF_3)_2$-2,5	NBE	W-1	$r_1 = 1.9$, $r_2 = 17$	Makio 1993b
$HC{\equiv}CC_6F_4$-Bu-4	NBE	W-1	$r_1 = 0.3$, $r_2 = 5.3$	Makio 1993b
$HC{\equiv}CCMe_3$	NBE	W-6	$r_1 = 2.2$, $r_2 = 0.4$	Liaw 1996
ClC≡CHex	NBE	Mo-1	$M_2 < M_1$	Masuda 1991a
ClC≡CHex	NBE	Mo-2	$r_1 = 0.7$, $r_2 = 6.4$	Ohgane 1994
$ClC{\equiv}CC_6H_4$-Cl-4	NBE	Mo-2	$r_1 = 0.6$, $r_2 = 5.4$	
$ClC{\equiv}CC_6H_5$	NBE	Mo-2	$r_1 = 1.0$, $r_2 = 3.1$	
$ClC{\equiv}CC_6H_4$-Me-4	NBE	Mo-2	$r_1 = 3.5$, $r_2 = 3.5$	
Block copolymers (living systems)				
HC≡CH	CPE	W-3[h]	Diblock + homopolymer of M_2	Farren 1987, 1989
	NBE	W-4	$(M_2)_x(M_1)_y(M_2)_z$ triblock [i]	Schlund 1989; Park, L.Y. 1991
HC≡CPh	NBE	W-5	Diblock [j]	Gita 1993
$(HC{\equiv}CCH_2)_2CX_2$ X = CO_2Et	DCMNBD	Mo-3	[k]	Fox 1994b
MeC≡CMe	NBE	Ta-1	Diblock	Wallace 1989

[a] CPE = cyclopentene; MCPE = mixture of 3- and 4-methylcyclopentene; CHP = cycloheptene; COC = cyclooctene; NBE = norbornene; DCMNBD = 2,3-dicarbomethoxynorbornadiene. [b] W-1 = WCl_6; W-2 = W[=C(OMe)Ph]$(CO)_5$ at 50°C; W-3 = $WCl_6/EtAlCl_2$; W-4, = W(=$CHCMe_3$)(=NAr)$(OCMe_3)_2$ (quinuclidine); W-5 = $W(CO)_6/CCl_4/h\nu$; W-6 = W[=C(OMe)$(CH_2)_2$CH=CH_2]$(CO)_4$; Mo-1 = $MoCl_5$; Mo-2 = $MoCl_5/Bu_4Sn$ at −20°C; Mo-3 = Mo(=$CHCMe_3$)(=NAr) [OCMe$(CF_3)_2$]$_2$; Ta-1 = Ta[C(Me)C(Me)$CHCMe_3$](DIPP)$_3$(py), where DIPP = diisopropylphenoxide. Reaction in toluene or similar solvent at ambient temperature unless otherwise stated. [c] $M_2 < M_1$ denotes that M_2 is less reactive than M_1. [d] M_2 reacts only after M_1 has all polymerized. [e] Copolymer and cyclic trimer of M_1 are formed. [f] R = Me, Cl, CF_3 in the 2-, 3-, or 4-positions; see Figs. 10.7 and 10.8. Electron-withdrawing substituents make M_1 less reactive; electron-repelling substituents make M_1 more reactive. The Hammett ρ value, from the slope of log $(1/r_2)$ against σ, is -1.1. [g] Only copolymer is formed; no cyclic trimer of M_1. [h] M_2 is added first, followed by Et_3Al, which transforms the active site into one that is more active for subsequent addition of M_1. The block length of M_2 units can be controlled by the use of a transfer agent (oct-1-ene); average block length of M_1 units is about 10. [i] $M_w/M_n = 1.09$ if $y < 10$. [j] M_2 and M_1 added successively to the previously irradiated catalyst solution. [k] Copolymer containing cyclic units derived from M_1.

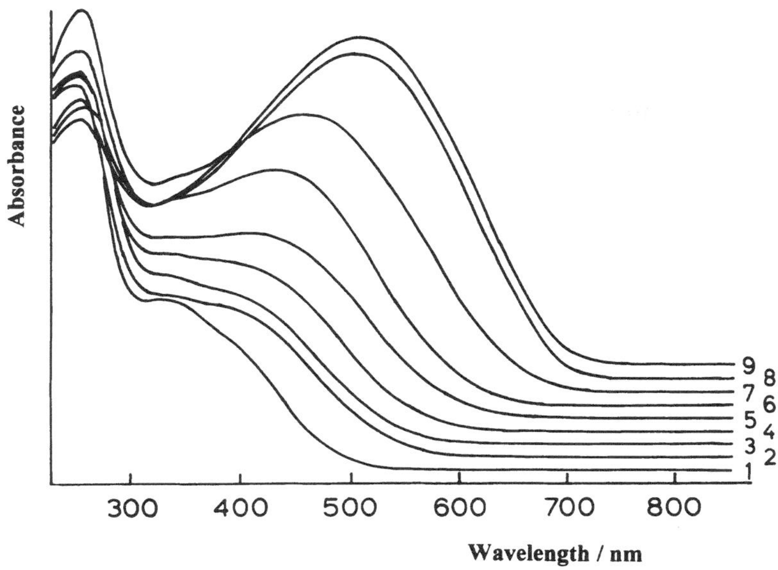

Fig. 10.5 UV/visible absorption spectra of statistical copolymers of phenylacetylene with acetylene, containing from zero to 81.7 mol% acetylene units in going from spectrum 1 to spectrum 9. Solvent, CH_2Cl_2. The absorption maximum at 508 nm corresponds to an average of 9–10 conjugated double bonds (Kanischka 1988).

Some copolymer composition curves for non-living systems are shown in Figs. 10.6–10.8. The Hammett parameters ρ^+ are negative (Table 10.6) indicating that the chain carrier is electrophilic, i.e. the carbene carbon bears a net positive charge, with the counterbalancing negative charge spread to some extent over the chloride ligands. The product of the two reactivity ratios is sometimes considerably larger than 1.0, e.g. 13 in the case of $HC{\equiv}CC_6H_4$-CF_3-2/NBE, and 30 for $HC{\equiv}CC_6H_3$-$(CF_3)_2$-2,5/NBE (Makio 1993a,b). The copolymers in these cases are thus very blocky, which appears to be associated with the presence of the 2-CF_3 group. Occasionally, as in the copolymerization of phenylacetylene and cyclopentene, the product of the reactivity ratios is so great that the cross-propagation reactions scarcely occur and the product is essentially a mixture of the two homopolymers (see Section 10.3.1).

When the propagating species are long-lived, it is possible to make block copolymers of controlled chain length by sequential addition of monomers, for example, with $ClC{\equiv}CC_4H_9$/$ClC{\equiv}CC_{14}H_{29}$ and with acetylene/norbornene (see Tables 10.6 and 10.7).

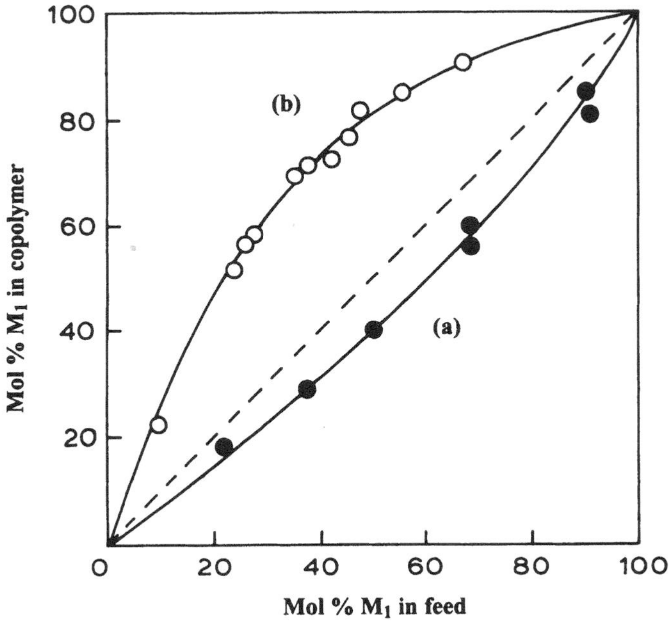

Fig. 10.6 Composition curves for the copolymerizations of 1-chlorodec-1-yne (M_1) with (a) dec-2-yne and (b) 1-chloro-2-phenylacetylene (M_2), catalyzed by $MoCl_5/Bu_4Sn$ (1/1) in toluene at 30°C (Masuda 1987b).

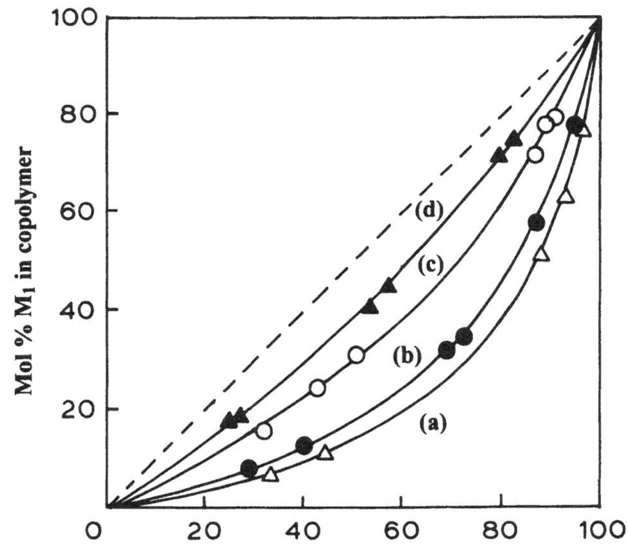

Fig. 10.7 Composition curves for the copolymerizations of norbornene (M_1) with *p*-substituted phenylacetylenes (M_2), catalyzed by WCl_6 in toluene at 30°C: (a) 4-CH_3, (b) 4-H, (c) 4-Cl, (d) 4-CF_3 (Masuda 1992a; Makio 1993b).

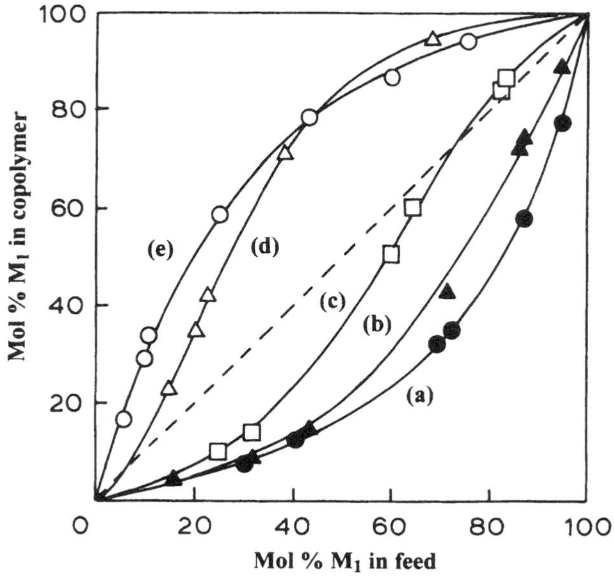

Fig. 10.8 Composition curves for the copolymerizations of norbornene (M_1) with *o*-substituted phenylacetylenes (M_2), catalyzed by WCl_6 in toluene at 30°C: (a) 2-H, (b) 2-Cl, (c) 2-CF_3, (d) 2,5-$(CF_3)_2$, (e) 2,3,5,6-F_4-4-Bu; r_1r_2 values: 1.1, 2.9, 13, 30, and 1.7, respectively (Masuda 1992a; Makio 1993b).

10.4 Metathesis reactions of enynes and dienynes

Enyne intramolecular metathesis reactions, of the type shown in eqn. (22), can be very useful in organic synthesis. Such reactions, catalyzed by tungsten or chromium carbene complexes, have been described by Katz (1985a, 1989), Mori, M. (1992), and Watanuki (1993, 1994, 1995). The catalyst $Ru(=CHCH=CPh_2)(Cl)_2(PCy_3)_2$ is likely to be increasingly used for this purpose because of its greater stability, ease of handling, and good yields. It has been applied to the synthesis of various five-, six-, and seven-membered heterocycles, e.g. eqn. (28) (Kinoshita 1994). Ring closures of enynes affected by $[RuCl_2(CO)_3]_2$ at 80°C are also likely to proceed by a carbene mechanism (Chatani 1994).

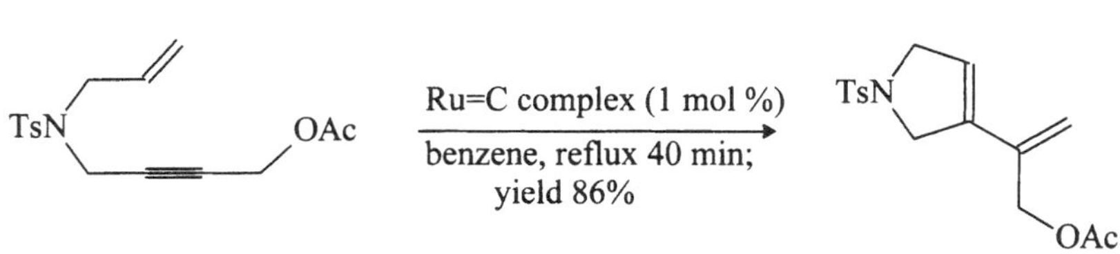

(28)

The ruthenium carbene catalyst also effects the ring-closing metathesis of many acyclic dienynes to form fused bicyclic rings, containing five-, six-, and seven-membered rings, e.g. eqn. (29). The reaction may be assumed to take place in two stages: (i) intramolecular metathesis of the alkyne with the sterically less hindered double bond; followed by (ii) metathesis with the remaining exocyclic double bond (Kim, S-H. 1994b, 1996; Zuercher 1996).

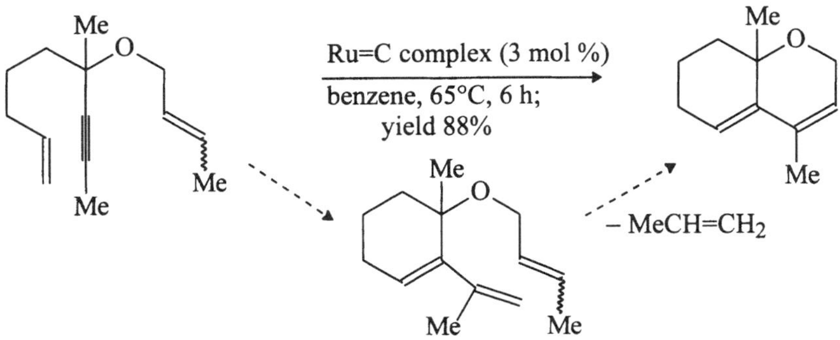

(29)

Many natural products contain fused bicyclic structures, and the dienyne metathesis reaction may well open up new and more efficient routes for their synthesis.

10.5 Other metathesis routes to polyacetylenes

The direct metathesis polymerization of acetylenes is not the only route to polyacetylenes using olefin metathesis chemistry. Below are summarized some of the other methods that have been developed in recent years.

(1) ROMP of cyclooctatetraene and its derivatives including those with chiral substituents (see Section 12.6).

(2) ROMP of fused-ring cyclobutene derivatives followed by a reverse Diels–Alder reaction to eliminate a benzene compound, leaving polyacetylene (see Section 13.2). This method can also be adapted to make the centre block of a triblock copolymer with norbornene, having morphologies ranging from spherical to cylindrical to lamellar (see Section 14.4).

(3) ROMP of benzvalene, followed by an $HgCl_2$-catalyzed isomerization reaction to yield polyacetylene (see Section 13.3.1).

(4) Metathesis reactions with the elimination of benzaldehyde (Schopov 1985, 1989; Jossifov 1991a,b, 1993a,b); eqns. (30) and (31).

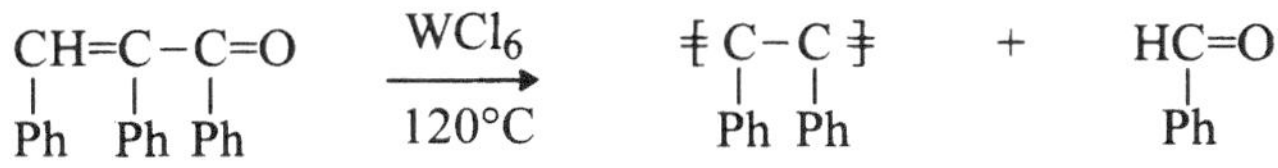

(30)

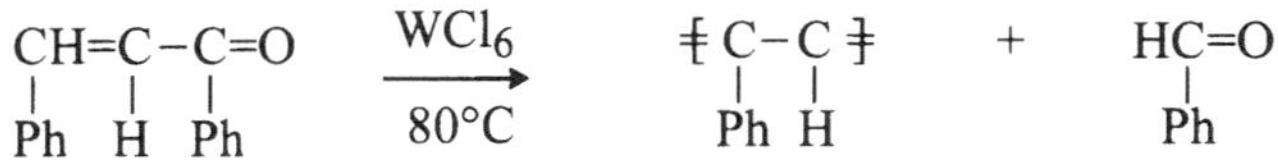

(31)

For further information on polyacetylenes, particularly their physical properties, see the reviews of Masuda (1987d, 1994b) and Breslow (1993).

11

Ring-Opening Metathesis Polymerization: General Aspects

11.1 Introduction

In Section 3.4 we outlined some of the features of ROMP that provide evidence for the metal carbene chain mechanism. In this chapter we consider some more general aspects of this group of metathesis reactions, such as the effect of ring size on polymerizability, the formation of cyclic oligomers, living systems, molecular weight distributions, and stereochemistry. Subsequent chapters will deal with individual monomers grouped according to ring size and type.

11.2 Thermodynamic aspects

11.2.1 Effect of ring size and substitution on thermodynamic polymerizability

It is well known that ring-opening polymerizations of all types are thermodynamically favoured for 3-, 4-, 8-, and larger-membered ring compounds (Ivin 1974, 1991) and will proceed if a mechanism is available, as in the case of the cycloalkenes. For 5-, 6-, and 7-membered rings the situation is more critical and the sign of ΔG for the polymerization process may be sensitive to a number of physical factors, such as monomer concentration, temperature, and pressure, and to chemical factors such as the nature of the substituents and their position in the ring. When bridging groups are present, e.g. when the compound is bicyclic, ΔG for the opening of a particular ring will tend to be more negative as a result of an increased strain energy in the monomer. It must be remembered that for addition polymerization processes of any kind, of which ROMP is a particular case, the boundary between polymerizabilty and non-polymerizabilty, corresponding to ΔG negative and positive respectively, is very distinct. There is, in fact, a sharp temperature, known as the ceiling temperature, above which polymerization will not occur at a given monomer concentration [M]. Conversely, at a given temperature, there is a certain equilibrium concentration $[M]_e$ below which polymerization to high polymer will not occur.

The effect of ring size and substituents on the polymerizability of some cycloalkenes having 5-, 6-, and 7-membered rings is illustrated in Table 11.1. It may be safely assumed that the compounds listed on the right-hand side of Table 11.1 do not undergo ROMP because ΔG is positive under all conditions that have been tried. Such non-polymerizable compounds can, in fact, be prepared from appropriate dienes by RCM and many examples will be found in Ch. 8. Substituents on a given carbon usually have an unfavourable effect on ΔG in any ring-opening polymerization, either making ΔG less negative or changing the sign from negative to positive, as for 1-methylcyclopentene and 3-isopropylcyclopentene. However, when there are multiple subsituents at adjacent ring positions, as in 1,1,2,2-tetramethyl-1,2-disilacyclopent-3-ene, the interaction between the substituents can provide sufficient ring strain to make ROMP possible; see Section 12.3.4.

The failure of cyclohexene to give high polymer is due to its low strain energy, $-\Delta H$ being too small to offset the positive value of $-T\Delta S$. However, at low T and high [M] it is possible to obtain a small proportion of oligomer; see Section 12.4.1. The reverse reaction, namely the elimination of cyclohexene from molecules in which two double bonds are separated by five single bonds, occurs readily, for example, in the intramolecular metathesis of octa-1,7-diene and of deca-2,8-diene.

Table 11.1 Polymerizability, by ring opening, of some cycloalkenes containing 5-, 6-, and 7-membered rings[a]

Polymerizable	Not polymerizable
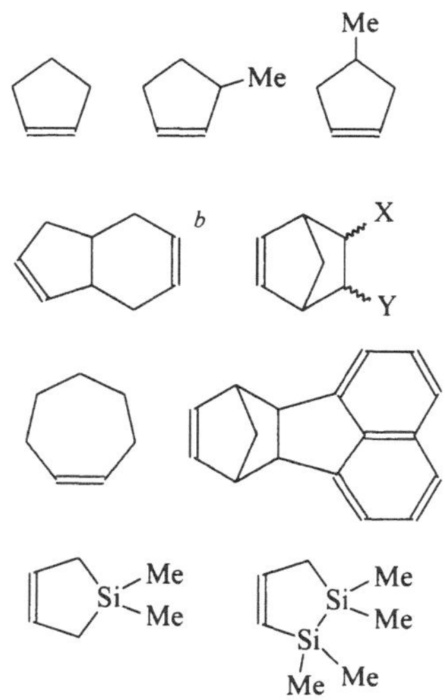	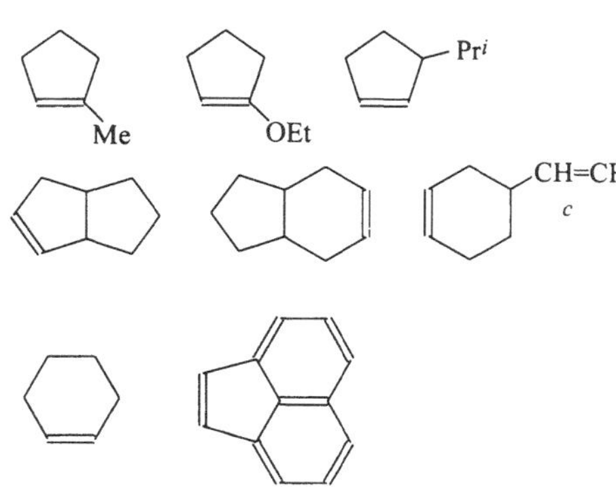

[a]For references see Ch. 12 and 13. [b]Only the 5-membered ring opens. [c]Reacts with oct-4-ene only at the vinyl group.

It is also eliminated from the polymer made by the ROMP of *cis,trans*-cyclodeca-1,5-diene, initiated by $WCl_6/EtAlCl_2$ (Hocks 1975b); eqn. (1). When the monomer adds to the propagating metal carbene, it may form either $[Mt]{=}CH(CH_2)_4CH{=}CH(CH_2)_2CH{=}CHP_n$, which may immediately split out cyclohexene, or $[Mt]{=}CH(CH_2)_2CH{=}CH(CH_2)_4CH{=}CHP_n$, which may split out cyclohexene in a secondary reaction of the polymer.

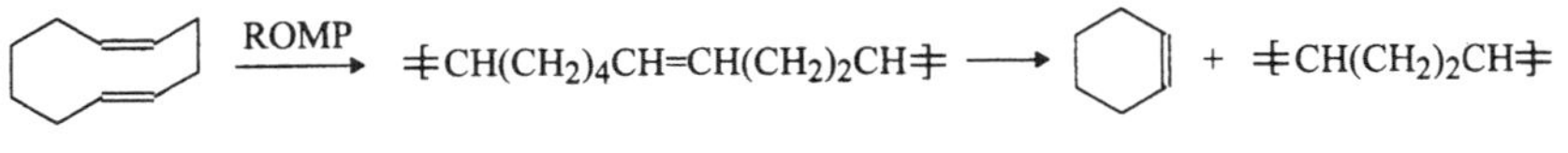

(1)

Some thermodynamic data for the ROMP of cycloalkenes are summarized in Table 11.2. For cyclohexene the thermodynamic functions of the polymer must be estimated from the data for related polymers. Note that ΔG° is negative for the polymerization of all monomers save cyclohexene.

The directly determined values of $[M]_e$ for cyclopentene are plotted as a function of temperature in Fig. 11.1 for catalyst systems giving high-*cis* and high-*trans* polymers, respectively. At a given temperature $[M]_e$ for the formation of high-*cis* polymer is about seven times higher than that for the formation of high-*trans* polymer, showing that the *cis* polymer is thermodynamically less stable than the *trans* polymer to the extent of about $RT \ln 7 = 5$ kJ mol^{-1}, compared with 2.3 kJ mol^{-1} derived from the data in Table 11.2. The agreement is reasonable, bearing in mind the need to make corrections for phase changes.

In the ROMP of norbornadiene and the dicyclopentadienes, the opening of the second double bond is thermodynamically less favourable than the opening of the first double bond, but nevertheless occurs if the concentration of the repeat units is somewhat above 0.2 M (see Sections 13.4.1.1 and 13.4.3).

Table 11.2 Thermodynamics of ROMP of cycloalkenes: ΔH°, ΔS°, and ΔG° for conversion of liquid monomer into solid amorphous polymer at 25°C (Lebedev 1992, 1994; see also Kranz 1972; Lebedev 1977a,b; Cherednichenko 1979; Hocks 1975a)

Monomer	Polymer[a]	$-\Delta H^\circ$ kJ mol^{-1}	$-\Delta S^\circ$ J K^{-1} mol^{-1}	$-\Delta G^\circ$ kJ mol^{-1}
Cyclobutene	*cis*	121	52	105
Cyclopentene	*cis*	15.4	51.8	0.3
	trans	18	52	2.6
Cyclohexene	*cis*	−2	31	−6.2
	trans	2	28	−7.3[b]
Cycloheptene	70% *trans*	18	37	7
Cyclooctene	48% *trans*	13	9	13
Cycloocta-1,5-diene	*cis*	25	5	19
	trans	33	5	24
Norbornene	45% *trans*	62.2	50	47

[a] *Trans* polymers are generally of lower energy than *cis* polymers. [b] Another value, $-\Delta G^\circ = 20$ kJ mol^{-1}, is based on calorimetric measurements on an ethene–butadiene copolymer containing 6% 1,2-butadiene units, and does not provide a proper comparison with the other values in the table (Lebedev 1988).

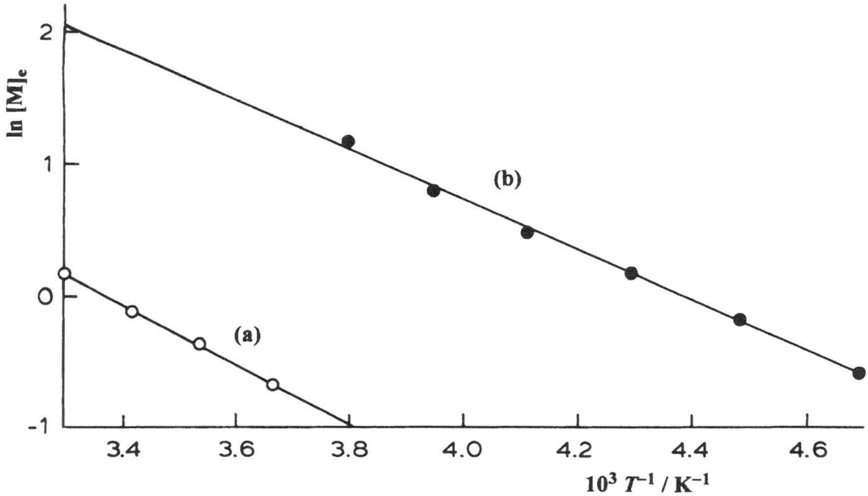

Fig. 11.1 Equilibrium concentration of cyclopentene $[M]_e$, determined using (a) WCl_6/$EtAlCl_2$ as catalyst (high-*trans* polymer) and (b) using $WCl_6/(CH_2{=}CHCH_2)_4Si$ as catalyst (high-*cis* polymer) (Ofstead 1972; Makovetskii 1976).

When attempting the ROMP of any new monomer, it should be remembered that the most favourable thermodynamic conditions are high [M] (which makes ΔS less negative), low T (if ΔH is negative), and high P (if ΔV is negative).

11.2.2 Formation of cyclic oligomers

The main facts concerning the concurrent formation of cyclic oligomers and high polymer have been presented in Section 3.4. Table 11.3 summarizes some of the systems where these cyclic oligomers have been identified. A typical GC trace is shown in Fig. 3.2. The cyclic nature of these oligomers has been demonstrated by mass spectra of both the oligomers and their hydrogenated products (Höcker 1974; Reif 1981a).

The equilibrium between linear polymer M_y and cyclic oligomers c-M_x containing x units may be represented by eqn. (2).

$$M_y \rightleftharpoons M_{y-x} + c\text{-}M_x \tag{2}$$

The equilibrium constant K_x for a given value of x is very closely given by eqn. (3), since, if the polymer chains are long, $[M_y] \approx [M_{y-x}]$.

$$K_x = [c\text{-}M_x] \tag{3}$$

K_x decreases with increasing x as may be seen qualitatively from the GC traces (e.g. Fig 3.2) and quantitatively from the plots of log K_x against log x in Fig 11.2. Above a certain value of x the plots are linear and have a slope of -2.5 as required by the

Table 11.3 Cyclic oligomers, containing *x* monomer units, detected in the ROMP of cycloalkenes

Monomer	Catalyst[a]	*x*	Reference
Cyclopentene	W-1	3–8	Ofstead 1972
	W-2	2–5	Kuteinikov 1975
	W-3	3-11	Witte 1978
Cycloheptene	W-7	2	Kress 1995
Cyclooctene	W-4	2–11	Scott 1969
	W-4	2–12	Höcker 1972, 1974, 1976, 1977, 1991; Reif 1984
Cycloocta-1,5-diene [b]	W-4	(3–20)/2 [c]	Scott 1969, Kelly 1975
	Re-1	(3–14)/2	Kumobayashi 1976
	Re-1	(4–13)/2	Sato 1977b
	Re-1	(3–14)/2	Saito 1979
	W-5	(3–12)/2	Chauvin 1978
Cyclododecene	W-2	2	Kuteinikov 1975
	W-4	2–7	Höcker 1976
Cyclopentadecene	W-4	2–7	Höcker 1980
Cyclododeca-1,5,9-triene	Re-1	(4–8)/3 [c]	Saito 1979
Cyclohexadeca-1,5,9,13-tetraene	Re-1	(3–8)/4 [c]	Saito 1979
Cycloeicosa-1,5,9,13,17-pentaene	Re-1	(3–7)/5 [c]	Saito 1979
Norbornene	W-6	2–14	Reif 1981a,1984

[a] W-1 = $WCl_6/EtAlCl_2$; W-2 = $WCl_6/i\text{-}Bu_2AlCl$; W-3 = $WCl_{4.2}(OCH_2CH_2Cl)_{1.8}/Et_2AlCl$; W-4 = $WCl_6/EtAlCl_2/EtOH$; W-5 = $W[{=}C(OMe)Ph](CO)_4(PPh_3)/TiCl_4$; W-6 = WCl_6/Me_4Sn (1/2); W-7 = $W[{=}\overline{C(CH_2)_3C}H_2](Br)_2(OCH_2CMe_3)_2.GaBr_3$; Re-1 = Re_2O_7/Al_2O_3. [b] See Thorn-Csányi (1995a) concerning the kinetics of formation of the oligomers, especially *ttt*-cyclododeca-1,5,9-triene. [c] Cyclic oligomers are formed in steps of C_4H_6 units.

theory of ring-chain equilibria (Jacobson and Stockmayer 1950). Although the original theory gives the correct form of dependence of K_x on x, the calculated values of K_x are too high; but when the theory is refined by the use of the rotational–isomeric states model (with no adjustable parameters) the calculated values are in satisfactory agreement with experiment for rings containing more than about 30 atoms (Suter 1988). The discrepancy between theory and experiment for smaller rings is the result of the failure of the assumption that the rings are unstrained. Their strain energy has been estimated by means of a Monte Carlo configurational search using molecular mechanics (MM3), leading to much better agreement between the calculated and experimental values of K_x for rings of all sizes (Chen, Z-R. 1995).

Strictly speaking, the value of K_x depends not only on x but also on the *cis/trans* ratio in the polymer and in the cyclic oligomer. However, the theory predicts that, in the range 20–90% *trans*, the effect will be very small (Suter 1988). Provided that the catalyst is not too stereoselective, *cis/trans* equilibrium can be readily attained in both polymer and cyclic oligomer. The observed value of K_x for a given x then represents the sum total of the concentrations of the various oligomeric *cis/trans*

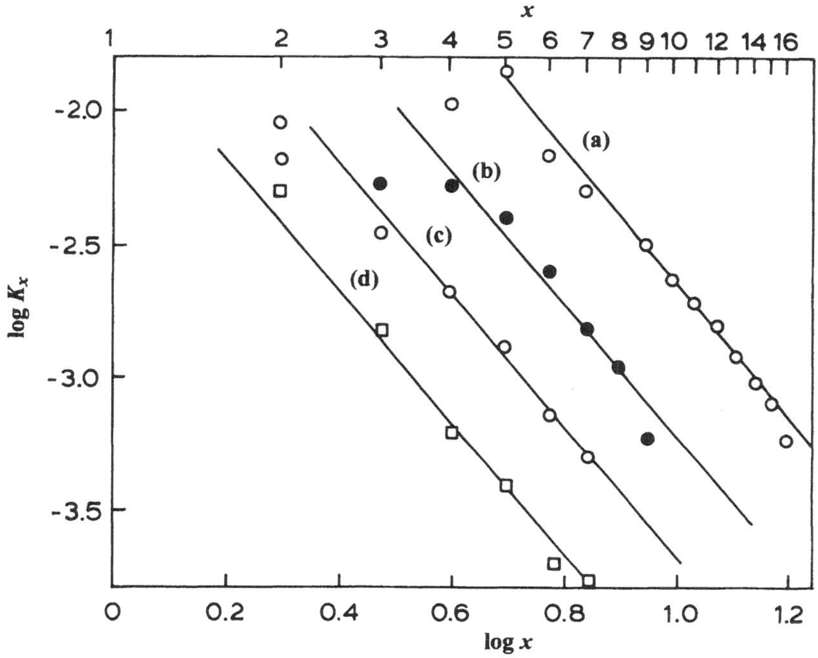

Fig. 11.2 Variation of log K_x with the number of units x in the cyclic oligomer formed from (a) cyclooctadiene (expressed as C_4H_6 units), (b) cyclooctene, (c) cyclododecene, and (d) cyclopentadecene (Höcker 1980).

isomers having this value of x, in equilibrium with high polymer of a particular *cis/trans* ratio (about 20/80); (Drapeau 1985). The proportions of the various possible C_{16} cyclic oligomers derived from cycloocta-1,5-diene at an early stage of reaction may be seen from Fig. 3.2(b). The average *cis* content falls from 70% for the C_{16} fraction to about 40% for C_{44} and higher fractions. For the smaller rings the probability of ring closure is evidently higher when there is a greater proportion of *cis* double bonds in the ring to be formed. Similarly, in the cyclic oligomers derived from *cis*-cyclooctene, the C_{16} components consist of mainly the *cis,cis* and *cis,trans* isomers (Höcker 1977, 1980; Wideman 1968) and the *cis* content decreases from 80% for C_{16} to 52% for the C_{104} oligomer (Höcker 1974).

The overall cyclic oligomer concentration in equilibrium with linear high polymer, expressed in terms of repeat units, is 0.125 M for norbornene in chlorobenzene at 20°C (Höcker 1984; Reif 1984) and somewhat higher (0.3–1.0 M) for the monocyclic monomers. For example, in the case of cyclooctadiene, when the starting concentration is 2.7 M, the equilibrium mixture contains 0.95 M cyclic oligomers, 1.75 M linear high polymer (all molarities expressed in terms of C_4H_6 units), and a negligible concentration of monomer (Fig. 3.1). If the starting concentration is less than 0.95 M, no linear polymer is formed and the product consists entirely of cyclic oligomers. Advantage can be taken of this dilution effect if it is desired to make cyclic oligomers for synthetic work; in some cases the product then consists mainly of cyclic dimers (Wideman 1968; Ofstead 1977; Warwel 1987c; Anhaus 1993; Kress 1995). For *cis*-cyclooctene the equilibrium distribution of cyclic oligomers is approximately independent of temperature between − 15° and 90°C as expected for a near-athermal process (Höcker 1977).

During the approach to equilibrium two extremes of behaviour can be observed. In the first case the kinetics may be such as to favour the initial formation of high polymer. At a later stage the polymer re-enters the metathesis reaction and strips out cyclic oligomers until final equilibrium is reached. This tends to happen if one or more of the following conditions is satisfied: (i) if the monomer is highly strained; (ii) the catalyst activity is not too great; and (iii) $[M]_o$ is high. The propagation reaction is then favoured relative to the first-order backbiting reaction of the metal carbene. If no backbiting occurs, equilibrium with cyclic oligomers is never achieved, living polymers are obtained, and a narrow (Poisson) MWD is observed. The backbiting reaction can sometimes be slowed down by the use of an additive which coordinates to the metal centre; for example, the addition of catalytic amounts of isoprene to WCl_6/Me_4Sn (1/2) prevents the formation of oligomers from norbornene (Reif 1983). When backbiting does occur as a slow secondary reaction, the viscosity of the reaction medium, and of the polymer formed, passes through a maximum as the reaction proceeds. Examples are shown in Fig. 11.3; see also Amass (1982).

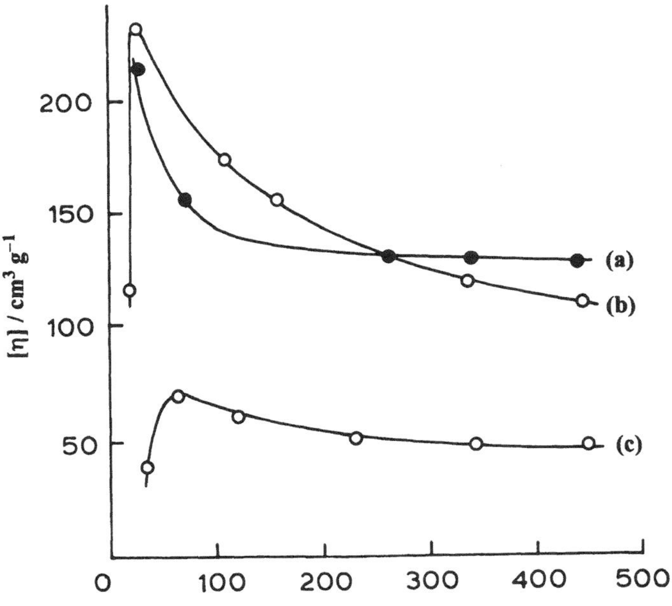

Fig. 11.3 Plots of limiting viscosity number [η] against reaction time for the polymers formed from (a) cyclopentene, (b) cyclooctene, and (c) cyclododecene. Catalyst: WCl_6/ *i*-Bu_2AlCl (1/5) in benzene at 25°C (Dolgoplosk 1977a).

The other extreme case is where the kinetics are such as to favour the initial formation of cyclic oligomers, which then convert slowly into an equilibrium mixture of high polymer and cyclic oligomers. This tends to happen if one or more of the following conditions is satisfied: (i) the monomer has low ring strain; (ii) the catalyst is of high activity, and (iii) $[M]_o$ is on the low side (but still large enough for the equilibrium oligomer concentration to be exceeded). In this situation the oligomer concentration will build up to a maximum and then fall back to its equilibrium value as the high polymer is formed. A prime example is the ROMP of *endo*-dicyclopentadiene catalyzed by $ReCl_5/Me_4Sn$ (Fig. 11.4). Within a few minutes most of the monomer is transformed into oligomers, while the formation of high polymer begins slowly and approaches an asymptotic value only after 4 h.

Another example is the ROMP of *cis*-cyclooctene (0.25 M) catalyzed by $WCl_6/EtAlCl_2/EtOH$ in chlorobenzene at 0°C, where transitory excess concentrations of cyclic dimer and trimer are observed. The *ct* dimer rather than the *cc* dimer tends to disappear after the maximum concentration of dimers has been reached (Höcker 1980). Some initial bias towards the formation of cyclic oligomers is also sometimes seen in the ROMP of cyclopentene (Witte 1978), and of cyclooctadiene (see Fig. 3.2). Even when the polymer/oligomer ratio is changing with time, the relative proportions of the various oligomer components are often close to their ultimate equilibrium values (Chauvin 1978; Saito 1979). However, with cyclooctadiene the most abundant cyclic oligomeric species initially formed is $(C_4H_6)_4$, but this eventually gives way to $(C_4H_6)_3$ as a result of secondary metathesis reactions (Thorn-Csányi 1995a). The equilibrium concentration of the

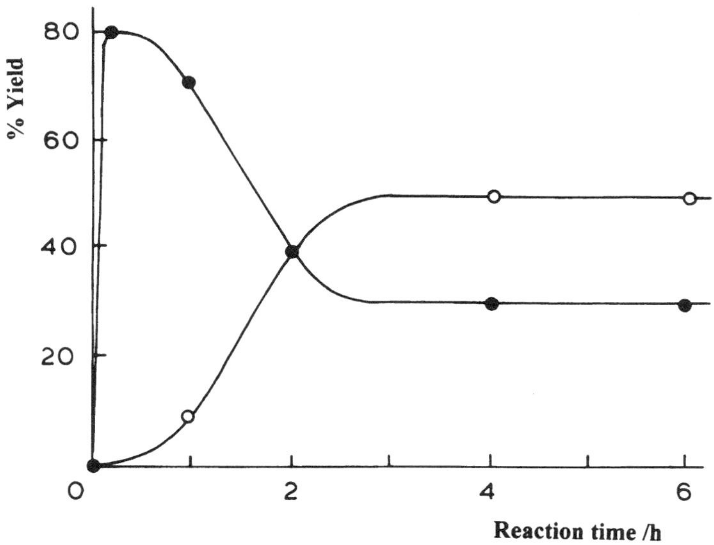

Fig. 11.4 Variation of yield of oligomers (●) and polymers (O) with time in the ROMP of *endo*-dicyclopentadiene catalyzed by $ReCl_5/Me_4Sn$ (1/1) in CCl_4 at 50°C; $[M]_o = 3$ M (Pacreau 1987).

monomer $(C_4H_6)_2$ is extremely small (Fig. 3.1) and there is no detectable formation of cyclobutene. Mathematical analyses of the various possibilities have been given by Matyjaszewski (1980), Slomkowski (1985) and Martl (1988).

11.3 Efficiency of initiation

Metal carbene complexes used as initiators (I) are not always fully consumed before the monomer (M) has been completely polymerized. This can happen when the propagation rate constant k_p is somewhat greater than the initiation rate constant k_i.

$$\mathrm{I} + \mathrm{M} \rightarrow \mathrm{P}_1 \qquad k_i$$

$$\mathrm{P}_n + \mathrm{M} \rightarrow \mathrm{P}_{n+1} \qquad k_p$$

An expression for k_p/k_i in terms of the fraction of initiator remaining can be obtained by dividing the rate of reaction of the monomer by that of the initiator, substituting $[\mathrm{P}] = [\mathrm{I}]_o - [\mathrm{I}]$, and integrating between the limits $[\mathrm{M}] = [\mathrm{M}]_o$ and 0, and $[\mathrm{I}] = [\mathrm{I}]_o$ and $[\mathrm{I}]_\infty$, where $[\mathrm{I}]_\infty$ is the final concentration of initiator. This leads first to eqn. (4) and, after integration, to eqn. (5). In order to determine values of k_p/k_i experimentally from this relationship, presented graphically in Fig. 11.5, it is best to adjust $[\mathrm{M}]_o/[\mathrm{I}]_o$ so that the measured value of $[\mathrm{I}]_\infty/[\mathrm{I}]_o$ is in the middle range, preferably near 0.5. When P_1 and P_n ($n > 1$) give separate signals in the NMR spectrum, as is often the case, k_p/k_i can be determined directly from the value of $[\mathrm{I}]/[\mathrm{P}_1]$ at the maximum concentration of $[\mathrm{P}_1]$, since at this point the rate of formation of P_1 is equal to its rate of disappearance. Some values obtained in these two ways are collected in Table 11.4.

$$d[\mathrm{M}]/d[\mathrm{I}] = 1 + (k_p/k_i)([\mathrm{P}]/[\mathrm{I}]) = [1 - (k_p/k_i)] + (k_p/k_i)([\mathrm{I}]_o/[\mathrm{I}]) \qquad (4)$$

$$k_p/k_i = \{([\mathrm{M}]_o/[\mathrm{I}]_o) + ([\mathrm{I}]_\infty/[\mathrm{I}]_o) - 1\}/\{\ln([\mathrm{I}]_o/[\mathrm{I}]_\infty) + ([\mathrm{I}]_\infty/[\mathrm{I}]_o) - 1\} \qquad (5)$$

It is frequently found that the initiation reaction is more stereoselective than the propagation reaction, with the preferential formation of *trans* double bonds in the first step; for examples, see Ch. 13. For the ROMP of norbornene initiated by Mo-1, Mo-3, and Mo-5, the propagation reaction is the same for all three reactions. The increasing value of k_p/k_i therefore results from a decreasing value of k_i, to be attributed to the increasing bulk of the alkylidene group in the initiator. Usually $k_p > k_i$, but for the first two entries in Table 11.4, where the monomers are substituted in both the 5- and 6-positions, $k_p < k_i$. Thus steric hindrance to the propagation reaction can sometimes be greater than for the initiation reaction.

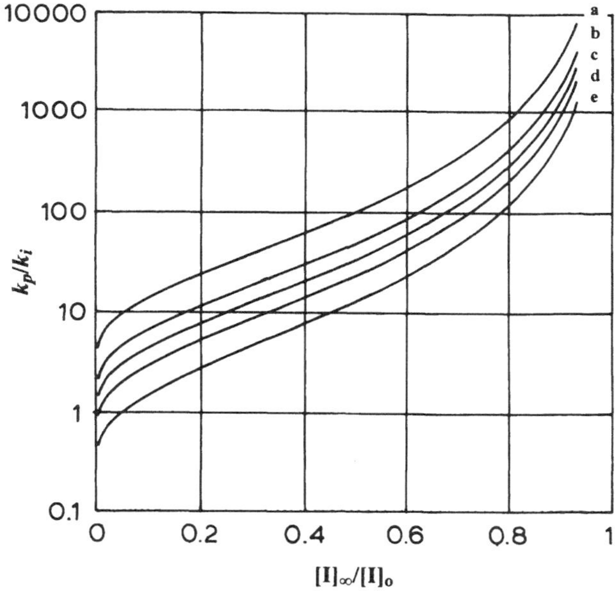

Fig. 11.5 Relationship between k_p/k_i and $[I]_\infty/[I]_o$ when $[M]_o/[I]_o$ is (a) 20, (b) 10, (c) 7, (d) 5, or (e) 3; see eqn. (5) (Benedicto 1995).

consumed is quite small (Schrock 1988b). When the aim is to produce a living polymer with a narrow MWD, it is best to choose an initiator which is consumed before too much monomer is used up. Provided that $[M]_o/[I]_o > 100$, a value of k_p/k_i up to about 10 can be tolerated. Living polymers derived from Mo and W carbenes are best terminated by reaction with an aldehyde; but those derived from Ru carbenes must be terminated by use of a transfer agent such as an unsaturated ether. Termination with di- and tri-aldehydes can, in principle, lead to polymer molecules with double or treble the MW of the living polymer (Dounis 1996a).

With non-carbene catalysts there is comparatively little information about the efficiency of conversion of the transition metal compound to the metal carbene initiating species. The activity of the catalyst system generally rises to a maximum and then dies away as side reactions lead to the destruction of both the initiating and propagating species. It is therefore not easy to obtain quantitative kinetic data with such catalyst systems. For WCl_6/Me_4Sn in CD_2Cl_2 the overall efficiency of initiation is estimated to be less than 0.7%; see Section 3.7 (Ivin 1987). Another example is the ROMP of norbornene catalyzed by $[Ru(H_2O)_6](OTs)_2$. This reacts with the substrate to form an intermediate ruthenium alkyl complex but only a small proportion of this undergoes the α-hydride elimination reaction to give the initiating ruthenium carbene complex (Karlen 1994; Mühlebach 1994). Non-carbene catalysts do occasionally give polymers of narrow MWD and this may be taken as evidence of a high initiation efficiency. For example, tungsten

Table 11.4 k_p/k_i values mostly determined from eqn. (5)[a]

Monomer	Catalyst [b]	Solvent	k_p/k_i	Reference
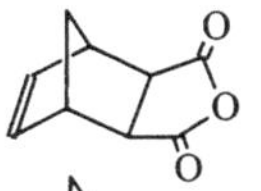	W-1	CD_2Cl_2	0.26	Ivin 1988
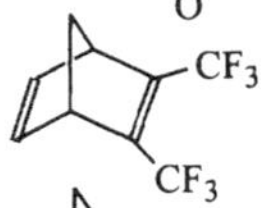	Mo-1	C_6D_6	0.72	Bazan 1991a
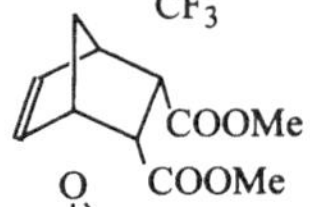	W-1	CD_2Cl_2	1.0	Ivin 1992
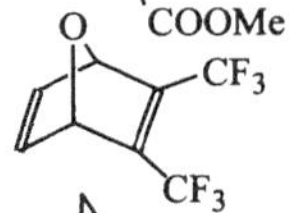	Mo-2	$C_6D_5CD_3$	2.4	Bazan 1991c
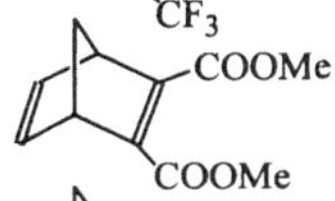	Mo-1	C_6D_6	3	Bazan 1991a
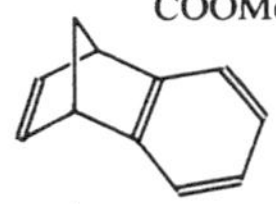	Mo-1	$C_6D_5CD_3$	7	Bazan 1990
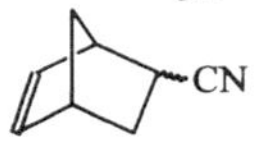	Mo-1	$C_6D_5CD_3$	7	Bazan 1991a
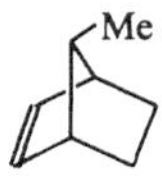	Mo-1	C_6D_6	9	Feast 1992c
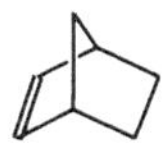	Mo-1	C_6D_6	12	Bazan 1991a
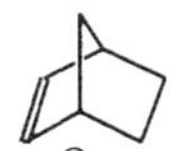	Mo-3	$C_6D_5CD_3$	30	Benedicto 1995
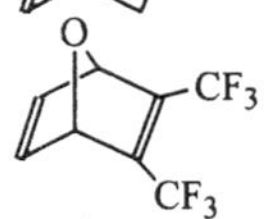	Mo-4	$C_6D_5CD_3$	40	Bazan 1991c
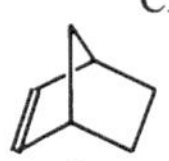	Mo-5	C_6D_6	270	Bazan 1990
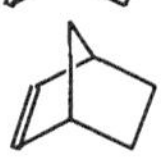	Ru-1	CD_2Cl_2/C_6D_6	Very large	Nguyen 1993
	Ru-2	CH_2Cl_2	170	Schwab 1996
	Ru-3	CH_2Cl_2	0.1–0.8[c]	Schwab 1995,1996

initiating ruthenium carbene complex (Karlen 1994; Mühlebach 1994). Non-carbene catalysts do occasionally give polymers of narrow MWD and this may be taken as evidence of a high initiation efficiency. For example, tungsten tetraphenylporphyrinate in toluene gives polymers of cyclopentene and cyclooctene with an initial M_w/M_n of 1.1 (Coca 1994).

11.4 The use of chain-transfer agents

Cross-metathesis reactions between cyclic and acyclic compounds are dealt with in detail in Ch. 15. Alk-1-enes are the preferred transfer agents when it is desired to reduce the MW of polymers produced by ROMP in non-living systems. The transfer constants are rather small and it is usually necessary to employ a fairly high concentration of transfer agent if a substantial reduction of MW is required to ease solution of the polymer or to improve the resolution of its NMR spectrum. End-groups derived from the transfer agent can be detected and identified (Ivin 1978b; Bell B. 1992; France 1993b). Polymers with functional end-groups at both ends (telechelic polymers) can be prepared by the use of appropriate unsaturated transfer agents, preferably the more reactive *cis* isomers; for example, diesters (Cramail 1991a), bis(silyl) ethers (Hillmyer 1993), borane derivatives (Chung 1992a), or protected diols (Hillmyer 1995c); see Ch. 15.

With living systems the MW of the polymer can be reduced by increasing $[I]_o$, but this requires the use of large amounts of initiator. The same result can be achieved at low $[I]_o$ by inclusion of a good transfer agent (Crowe 1990). The determination of transfer constants (k_{tr}/k_p) in living systems is not as straightforward as in non-living systems. A detailed consideration of the problem has been

---

Footnotes to Table 11.4

[a] Ambient temperature.

[b] Mo-1= Mo(=CHCMe$_3$)(=NC$_6$H$_3$-*i*-Pr$_2$-2,6)(OCMe$_3$)$_2$
Mo-2= Mo(=CHCMe$_2$Ph)(=NC$_6$H$_3$-*i*-Pr$_2$-2,6)[OCMe$_2$CF$_3$]$_2$
Mo-3= Mo(=CHCMe$_2$Ph)(=NC$_6$H$_3$-*i*-Pr$_2$-2,6)(OCMe$_3$)$_2$
Mo-4= Mo(=CHCMe$_2$Ph)(=NC$_6$H$_3$-*i*-Pr$_2$-2,6)[OCMe(CF$_3$)$_2$]$_2$

Mo-5= 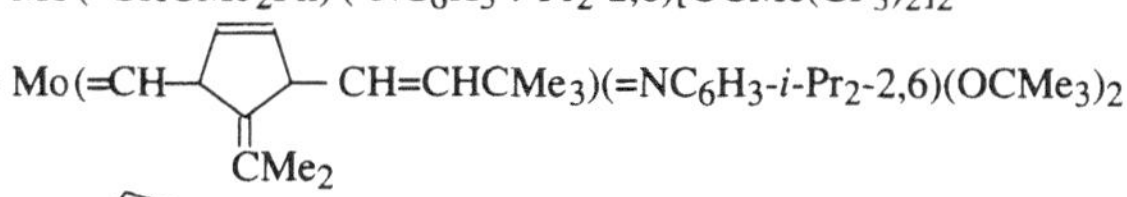

W-1= 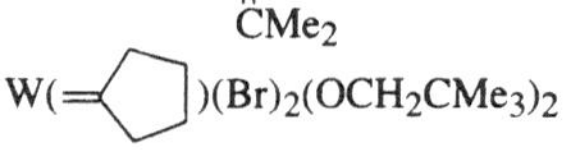

Ru-1= Ru(=CHCH=CPh$_2$)(Cl)$_2$(PCy$_3$)$_2$; (Cy = cyclohexyl)
Ru-2= Ru(=CHCH=CPh$_2$)(Cl)$_2$(PPh$_3$)$_2$
Ru-3= Ru(=CHC$_6$H$_4$X-*p*)(Cl)$_2$(PPh$_3$)$_2$

[c] *p*-X in Ru-3 (k_p/k_i values): H (0.11), F (0.21), Me (0.34), NMe$_2$ (0.38), OMe (0.38), NO$_2$ (0.43), Cl (

as though the transfer agent were not present and a living polymer will be formed with a number average degree of polymerization $DP_n = [M]_o/[I]_o$. In time the transfer agent will react with the living polymer to give dead polymer of the same DP, and a new carbene species. Addition of a fresh batch of monomer would allow the process to be repeated and, provided that the initiation efficiency for the new carbene species was high, the second batch of polymer would again have a narrow MWD. Such conditions are realized in practice with $Mo(=CHCMe_3)(=NAr)(OCMe_3)_2$ as initiator, styrene (30 equivalents) as transfer agent, and norbornene added in eight successive batches of 80 equiv at 40 min intervals. The GPC of the total product shows a main peak (95% of the whole) with $M_w/M_n = 1.07$. When the monomer is 2,3-bis(trifluoromethyl)norbornadiene the product consists only of low-molecular-weight oligomers. This is because k_p is much smaller and the transfer constant therefore much larger than for norbornene, so that the chains have no chance to grow to their full length before suffering reaction with the transfer agent (Crowe 1990).

A different situation exists when the transfer constant is high and the amount of transfer agent used is such that it is all consumed before polymerization is complete. The remaining monomer will then polymerize in the absence of transfer agent and give rise to a fraction with a much higher MW than that of the initially formed polymer. The product will thus have a bimodal MWD. There may also be some cyclic oligomers formed by the backbiting reaction. The most reliable method of measuring the transfer rate constant k_{tr} in such a system is to observe directly by NMR the rate of removal of chain transfer agent (CTA) by reaction with the propagating species. In this way the reaction of $CH_2{=}CHCMe_3$ with the propagating species in the ROMP of norbornene initiated by $Mo(=CHCMe_2Ph)(=NAr)(OCMe_3)_2$ is found to have $k_{tr} = 3 \times 10^{-5}\ \text{M}^{-1}\ \text{s}^{-1}$ at 22°C compared with $k_p = 17\ \text{M}^{-1}\ \text{s}^{-1}$. Once the rate constants are known, the relative amounts of reagents in the reaction can be optimized to secure the required MW and MWD. It will be evident that the slopes of (Mayo) plots of $1/(DP_n)$ against $[CTA]/[M]_o$ do not provide a reliable method of determining k_{tr}/k_p in living systems unless care is taken to obviate the aforementioned pitfalls (Benedicto 1995).

For systems giving high-*cis* polymers it is not always easy to find a good transfer agent. For example, pent-1-ene is a very poor transfer agent for the ROMP of cyclopentene initiated by $WF_6/EtAlCl_2$ at 0°C in chlorobenzene, which gives polymer with 97% *cis* content. However, hepta-1,6-diene is quite effective, and much more so than either hexa-1,5-diene or octa-1,7-diene. Two factors are thought to be involved. First, in the polymer chain the *cis* double bond closest to the metal carbene is likely to remain near the metal centre until displaced by another olefin molecule. Second, the internal structure of hepta-1,6-diene is identical with the repeat unit of the polymer so that, after coordination of its first double bond to the metal centre, the second double bond can readily displace the coordinated double bond of the polymer chain, leading ultimately to its total detachment from the metal centre; see Scheme 11.1 (Ofstead 1980).

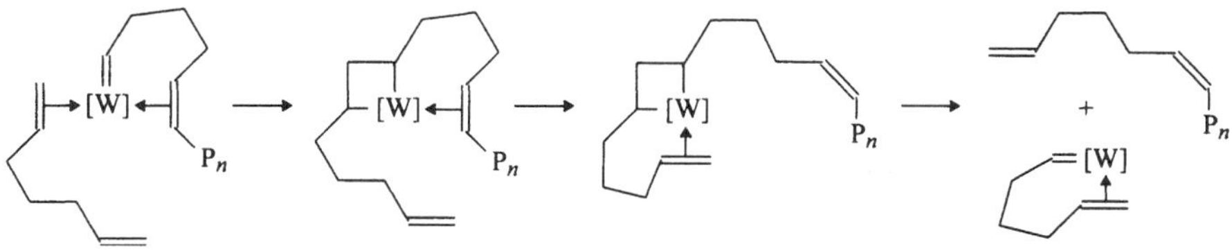

Scheme 11.1 Mechanism accounting for chain transfer with hepta-1,6-diene in the ROMP of cyclopentene catalyzed by $WF_6/EtAlCl_2$ to give high-*cis* polymer (Ofstead 1980).

The possibility of using alkenylcyclopentenes as transfer agents has been examined, with the idea that the cyclopentene ring will first be opened in the normal way, and that this will be followed by an internal metathesis reaction involving the alkenyl group, so as to split out a cyclohexenyl unit forming the end-group of the polymer chain. A combination of unfavourable circumstances makes this difficult to achieve in practice (Schrock 1989).

11.5 Molecular weight distributions

In non-living systems the initiating species may be generated over a period of time, and the termination reactions are generally not well defined. In some cases there will also be backbiting reactions leading to the formation of cyclic oligomers. For such systems it is not surprising to find that M_w/M_n of the linear polymer increases with time (Vardanyan 1973) and sometimes exceeds the thermodynamic value of 2.0. In living systems, on the other hand, provided that initiation takes place over a short period of time during which not too much monomer is consumed, all the chains grow to much the same length and an initial Poisson distribution results ($M_w/M_n < 1.1$). This again may broaden with time as a result of secondary metathesis reactions.

Bimodal distributions can arise for various reasons.

(i) Cyclic oligomers may be formed either before, during, or after the formation of linear high polymer, as discussed in Section 11.2.2. Depending on the packing material of the GPC column, the cyclic oligomers may give a single broad peak in the chromatogram, or a number of peaks from the individual oligomers (see Fig. 3.2).

(ii) A small, variable, proportion of polymer having double the MW of the main fraction is sometimes observed. This results from the reaction of living polymer with adventitious oxygen to give a terminal aldehyde, which then couples with a second chain by a Wittig-like reaction; see Section 13.3.3.7 and Perrott (1995).

(iii) When living ROMP is conducted in the presence of an inadequate supply of an efficient transfer agent, the initially formed polymer will have a much lower MW than that formed after the transfer agent has all been consumed (see Section 11.4).
(iv) When an initiator consisting of a racemic mixture of non-interconverting enantiomers is used to initiate the ROMP of a chiral monomer, two types of propagating species are formed, which may have sufficiently different propagation constants to give rise to two distinct peaks in the GPC of the polymer (see Fig. 13.7).

11.6 Polymer microstructure

Cycloalkenes are of three broad types:

(i) Monomers such as cyclopentene give ring-opened polymers in which only *cis/trans* double-bond isomerism is possible. Each methylene group in the polymer is then situated between a *cc*, or a *ct*, or a *tt* pair of double bonds as shown in **1**, where the first letter denotes the configuration of the nearest double bond.

c *c* *t* *t*

$-CH{=}CHCH_2CH_2CH_2CH{=}CHCH_2CH_2CH_2CH{=}CHCH_2CH_2CH_2CH{=}CH-$ **(1)**

α-C *cc* *cc* *ct* *tc* *tt* *tt*

β-C *cc* *ct*(≡*tc*) *tt*

(ii) Prochiral monomers such as 4-methylcyclopentene or norbornene exist in only one form but give polymers containing chiral centres. Such polymers may contain *m* and *r* dyads as well as *cis* or *trans* double bonds. The tactic dyads may be represented in projection form as in **2** and **3**.

Me Me H

$=CHCH_2\overset{|}{C}CH_2CH{=}CHCH_2\overset{|}{C}CH_2CH{=}CHCH_2\overset{|}{C}CH_2CH{=}$ **(2)**

H H Me

|—— *m* ——+—— *r* ——|

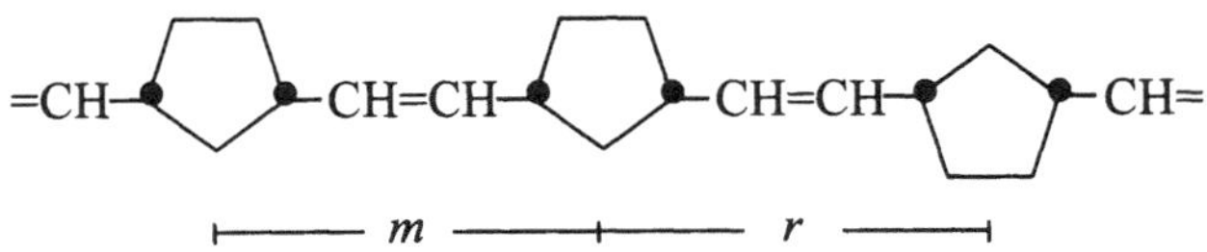

(3)

(iii) Monomers such as 3-methylcyclobutene and 5,5-dimethylnorbornene may give rise to polymers with head–head (HH), head–tail (HT), and tail–tail (TT) dyads as shown in **4** and **5** respectively. Each of these can have *m* or *r* forms and may contain a *cis* or *trans* double bond. If the polymer is made from a single enantiomer, as represented in **5**, the *m* dyads are necessarily HT, and the *r* dyads HH or TT. This is the basis of one method for the determination of tacticities in such polymers; see Section 11.6.2.1.

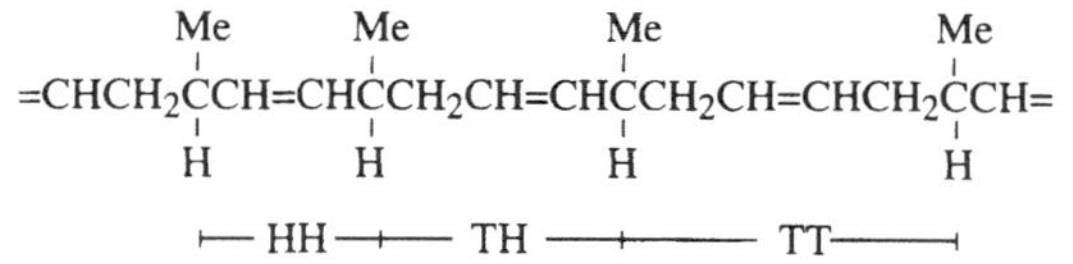

(4)

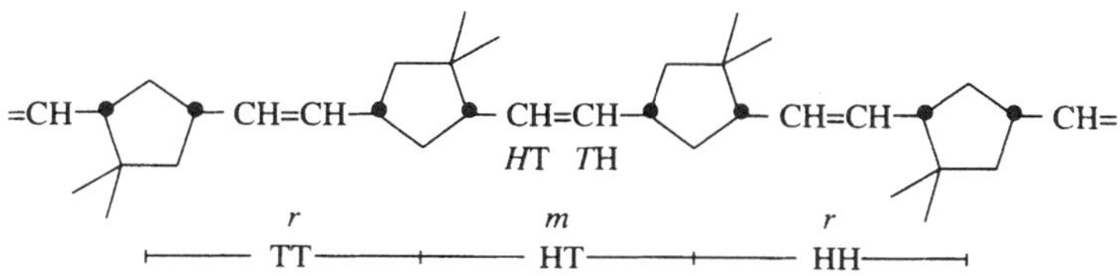

(5)

^{13}C NMR spectroscopy can provide a great deal of information about these three different types of isomerism, which will now be considered in detail.

11.6.1 Double bonds and their distribution

The ^{13}C NMR spectrum of poly(1-pentenylene) is shown in Fig. 11.6. The α-carbons give four lines and the β-carbons three lines corresponding to the dyad structures in **1**. The positions of these lines are listed in Table 11.5 along with those for other ring-opened polymers of cycloalkenes. α-*Cis* carbons always appear about 5 ppm upfield from α-*trans* carbons as in simple olefins; this rule provides the starting point for making assignments in the spectra of these polymers. In poly(1-pentenylene) the chemical shifts of the α-carbons (C-3,5) are about 0.5 ppm upfield from those in the polymers of the other cycloalkenes, an effect which is attributed to a higher proportion of gauche conformations about the CH_2-CH_2 bonds arising from the influence of the γ-olefinic carbons (Dounis 1995).

In the spectrum of poly(1-butenylene), the α-carbons are sensitive to the configuration of the nearest double bond, but almost insensitive to that of the next nearest double bond, even though it is only three bonds away; two opposing effects must cancel in this case. The olefinic carbons on the other hand are sensitive to the configuration of the next nearest double bond, which is four bonds away. When the polymer is made from *cis,cis*-cycloocta-1,5-diene, every other double bond in the chain is preformed and must be *cis*. The spectrum of freshly formed polymer therefore shows only *tc*, *cc*, and *ct* olefinic peaks at 130.15, 129.63, and 129.46

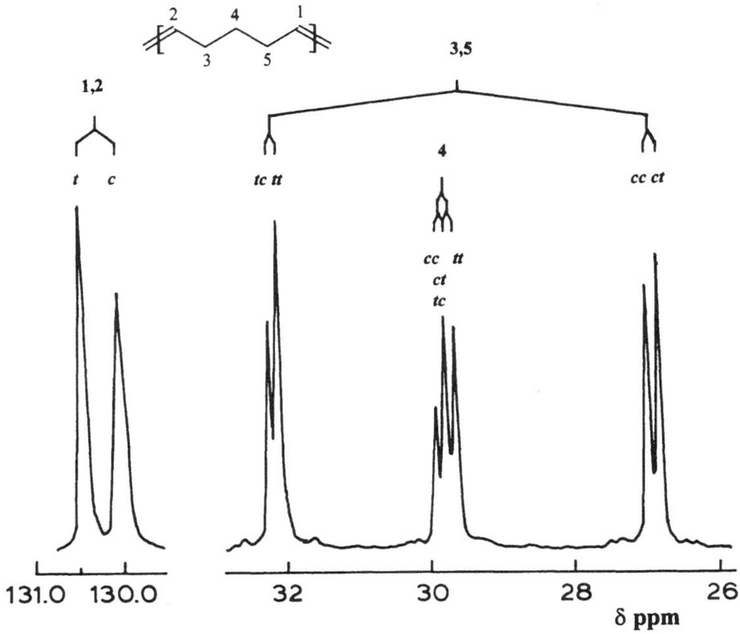

Fig. 11.6 22.63 MHz ^{13}C NMR spectrum of a sample of poly(1-pentenylene) containing 47% *cis* double bonds (Ivin 1979a).

ppm, respectively, and no *tt* peak. However, with some catalysts the double bonds in the polymer undergo rapid secondary metathesis, leading in a very short time to the formation of *tt* dyads (peak at 130.04 ppm) (Syatkovskii 1981; Ivin 1982a). The *cis* content eventually falls below 50% (Tanaka Y. 1977). This is a general phenomenon for all ring-opened polymers of monocyclic alkenes (Syatkovskii 1981). These secondary reactions also cause an adjustment of M_w/M_n towards the most probable value of 2 (Calderon 1981).

In the ^{13}C NMR spectra of polymers of norbornene, all the ring carbons are sensitive to the configuration of the two nearest double bonds, while the olefinic carbons are sensitive to the three nearest double bonds, even though the third is five bonds away. The line positions are listed in Table 11.5. All these splittings disappear in the spectrum of the hydrogenated polymers so that none appear to be the result of *m/r* isomerism (Ivin 1977b; Hamilton 1984a).

The *cis* content in all these polymers can be readily determined from either their ^{13}C or ^{1}H NMR spectra. The ^{1}H NMR spectrum of polynorbornene gives well-resolved olefinic proton signals (multiplets) at 5.35 ppm (*t*) and 5.21 ppm (*c*), also α-proton signals (fine structure unresolved) at 2.8 ppm (*c*) and 2.45 ppm (*t*) (Larroche 1982). The ^{13}C NMR spectra give, in addition, information on the proportions of double-bond dyads and triads, and hence on the distribution of *cis* and *trans* double bonds. There are two main questions to be answered: (i) what

Table 11.5 ^{13}C NMR line positions for some ring-opened polymers of cycloalkenes;[a] solvent $CDCl_3$; positions in ppm downfield from tetramethylsilane (Greene 1986, Dounis 1995)

Repeat unit	Carbon	Position and assignment			
2 3 4 1 =CHCH$_2$CH$_2$CH=	C-1,2	130.06 (*tc*),	129.94 (*tt*),	129.54 (*cc*),	129.38 (*ct*)
	C-3,4 (α)	32.71 (*tt*),	32.66 (*tc*),	27.38 (*cc*/*ct*)	
2 3 4 5 1 =CHCH$_2$CH$_2$CH$_2$CH=	C-1,2 [b]	130.30 (*ttt*)		129.81 (*ccc*)	
	C-3,5 (α)	32.19 (*tc*),	32.05 (*tt*),	26.88 (*cc*),	26.73 (*ct*)
	C-4 (β)	29.82 (*cc*),	29.66 (*ct*),	29.53 (*tt*)	
2 3 4 =CHCH$_2$CH$_2$	C-1,2 [b]	130.31 (*ttt*)		129.84 (*ccc*)	
5 CH$_2$	C-3,7 (α)	32.58 (*t*),	27.19 (*c*)		
=CHCH$_2$CH$_2$ 1 7 6	C-4,6 (β)	29.67 (*cc*),	29.64(*ct*),	29.56 (*tc*),	29.53(*tt*)
	C-5 (γ)	28.98 (*cc*),	28.84 (*ct*),	28.71 (*tt*)	
2 3 4 5 =CHCH$_2$CH$_2$CH$_2$	C-1,2 [b]	130.33 (*ttt*)		129.86 (*ccc*)	
|	C-3,8 (α)	32.62 (*t*),	27.22 (*c*)		
=CHCH$_2$CH$_2$CH$_2$ 1 8 7 6	C-4,7 (β)	29.75 (*c*),	29.64 (*t*)		
	C-5,6 (γ)	29.23 (*cc*),	29.19(*ct*),	29.09 (*tc*),	29.05(*tt*)
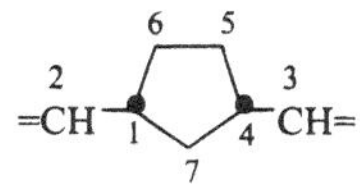	C-2,3 [b,c]	133.43 (*tcc*),	134.35 (*tct*),	134.27 (*ccc*),	134.18 (*cct*)
		133.54 (*ttc*),	133.42 (*ttt*),	133.42 (*ctc*),	133.25 (*ctt*)
	C-1,4 (α)	43.50 (*tc*),	43.24 (*tt*),	38.84 (*cc*),	38.58 (*tc*)
	C-5,6 (β)	33.27 (*cc*),	33.06 (*ct*),	32.54 (*tc*),	32.37 (*tt*)
	C-7 (γ)	42.89 (*cc*),	42.20 (*ct*),	41.47 (*tt*)	

[a] Data for other polymers may generally be found in the appropriate reference; see Ch. 12. and 13. [b] The signals for C-1,2 double-bond triads are resolved but cannot be fully assigned in 100 MHz spectra; they are not resolved in the spectra of polymers of cyclodecene or cyclododecene (Dounis 1995). [c] *tcc*, etc. denote double-bond triads. The central letter refers to the double bond in which the carbon is situated, the first letter refers to the next nearest double bond, and the third letter to the double bond on the other side of the central double bond. The sequence *tcc* thus corresppnds to the actual sequence in the chain (Greene 1986).

determines whether *cis* or *trans* double-bond formation is preferred; and (ii) how can blocky *cis/trans* distributions be accounted for.

The relative ease of formation of the intermediate *cis* and *trans* metallacycles determines whether *cis* or *trans* double bonds are eventually formed. A number of factors appear to be involved. When the number of ligands around the metal, other than the carbene, is no more than three, the main controlling factor is the relative extent of 1,2- and 1,3-interactions between the substituents in the metallacyclobutane ring (C-1 is the carbon derived from the reacting carbene). When 1,3-interactions are dominant a *cis* reactant will tend to give a *cis* product, but when 1,2-interactions are dominant a *cis* reactant may give an all-*trans* product, as discussed in Section 3.3. Examples are given in Ch. 13. When the metal site is more crowded, the steric effects of the ligands appear to favour the formation of *cis* double bonds. For example, $W(=CPh_2)(CO)_4$ (derived from the pentacarbonyl by the loss of CO) generally gives high-*cis* polymers. Chelating diolefins in particular can have a dramatic effect on the *cis* content of the polymer. Thus in the ROMP of norbornene catalyzed by $RuCl_3$ the inclusion of a small amount of *endo*-dicyclopentadiene (DCPD) raises the *cis* content of the polymer from 5% to 95%. The ROMP of *endo*-DCPD itself by $RuCl_3$ also gives high-*cis* polymer, as expected if the monomer provides a bidentate chelating ligand as well as being the reactant. In contrast, *exo*-DCPD, which cannot act as a bidentate ligand, gives high-*trans* polymer (Hamilton 1986). Maleates and their derivatives also have the ability to raise the *cis* content to quite high levels with $MoCl_5$-based catalysts, again probably through a chelation mechanism (Ivin 1979c; Amir-Ebrahimi 1995).

The distribution of the *cis* and *trans* double bonds in a given polymer chain may be expressed in terms of the ratios $r_t = (tt)/(tc)$ and $r_c = (cc)/(ct)$. If the probability of formation of a *cis* double bond is independent of the configuration of the previous double bond, the distribution will be random (Bernoullian) and characterized by a single parameter $r_t = 1/r_c$. Figure 11.7 shows that this is the case for polymers of norbornene with less than 35% *cis* content, but for polymers with more than 50% *cis* content the distribution is generally somewhat blocky, with $r_t r_c$ reaching values of 8 or more in some cases.

This pattern of behaviour is found not only for the ROMP of norbornene (Ivin 1979c; Greene 1986; Bencze 1996) but also for its derivatives, and for cyclopentene (Ceausescu 1986, 1988a). A very significant observation is that r_t seldom falls below unity even when the *cis* content is high. This means that at high

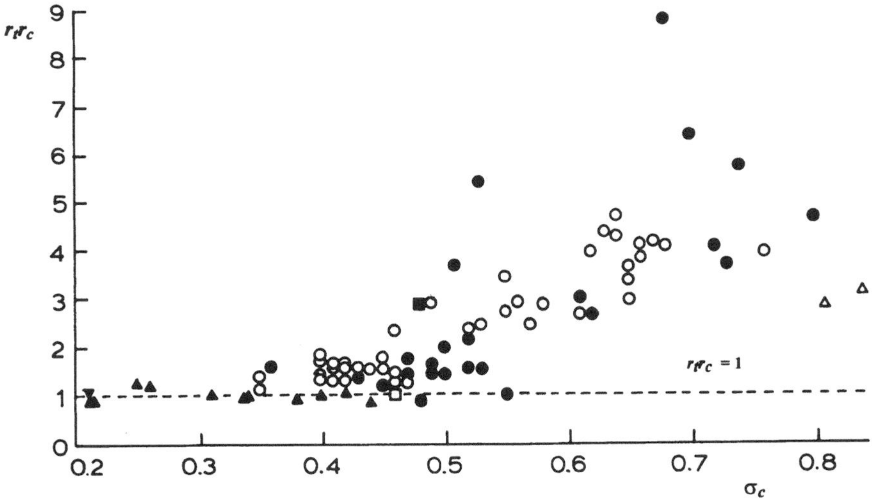

Fig. 11.7 Plot of $r_t r_c$ against σ_c for polymers made by ROMP of norbornene using various catalysts: ○ = Mo-based; ● = W-based; △ = Re-based; ▲ = Ir-based; □ = Os-based; ■ = V-based. Dashed line corresponds to a random distribution of *cis* and *trans* double bonds (Ivin 1979c).

cis content the *trans* double bonds tend to occur in pairs and are not to be found singly in the chain. To explain both the blockiness and the *tt* pair phenomenon, it is necessary to postulate the presence of three kinetically distinct propagating species when the *cis* content is high. Detailed analysis of the proportions of double-bond triads in polynorbornene samples shows that the propagating species which have just formed a *cis* double bond (P_c) all have the same selectivity but those which have just formed a *trans* double bond have different selectivity according to whether the last but one double bond formed was *cis* (P_{tc}) or *trans* (P_{tt}). The particular case of $W(CO)_6/h\nu$ as catalyst is illustrated in Scheme 11.2.

$$P_c \xrightarrow{+M} P_c \xrightarrow{+M} P_{tc} \xrightarrow{+M} P_{tt} \xrightarrow{+M} P_c$$

Bond formed:	*c*	*t*	*t*	*c*
Probability:	0.86	0.14	0.81	0.49

Scheme 11.2 Probabilities of forming *cis* or *trans* double bonds in the $W(CO)_6/h\nu$ catalyzed ROMP of norbornene.

The probabilities of forming a *trans* double bond for the three distinct steps are 0.14, 0.81, and 0.51, respectively. The problem is thus to explain how a *cis*-directing species (P_c) is followed by one which is strongly *trans*-directing (P_{tc}), and then by a third which is non-selective (P_{tt}), giving an overall *cis* content of 74% (Greene 1986). It has been argued that P_{tc} and P_{tt} differ essentially in their ligand geometry as a result of the disrotatory modes of cleavage of the metallacyclobutane precursors. P_{tt} and P_c may differ more in the extent of crowding of the metal site by the polymer chain itself, which may or may not involve coordination of double bonds to the metal centre (Gillan 1988). The *cis* content tends to fall and the blockiness to disappear as the monomer concentration is reduced (Ardill 1985) or as the temperature is raised (see Fig. 11.8). This would suggest that, given time, the two species P_{tc} and P_{tt} become kinetically identical through relaxation to a species with common ligand geometry represented by P_t. Polymers of cyclooctene do not exhibit *cis/trans* blockiness, showing that lengthening of the repeat unit tends to eliminate the kinetic distinction between P_t and P_c (Babitskii 1981; Denisova 1983a,b).

11.6.2 Tacticity

11.6.2.1 Determination of tacticity

Tacticities have been determined for polymers of various derivatives of norbornene, norbornadiene and 7-oxanorbornene. Three methods have been used: (i) by polymerizing single enantiomers of 5-, or 5,5-, or *endo,exo*-5,6-disubstituted derivatives, and determining whether the polymers have an HT (*m*) or HH/TT (*r*) structure, or both (*m* and *r*); (ii) by polymerizing 5,6-disubstituted derivatives in which the substituents contain a chiral centre of a single handedness, and

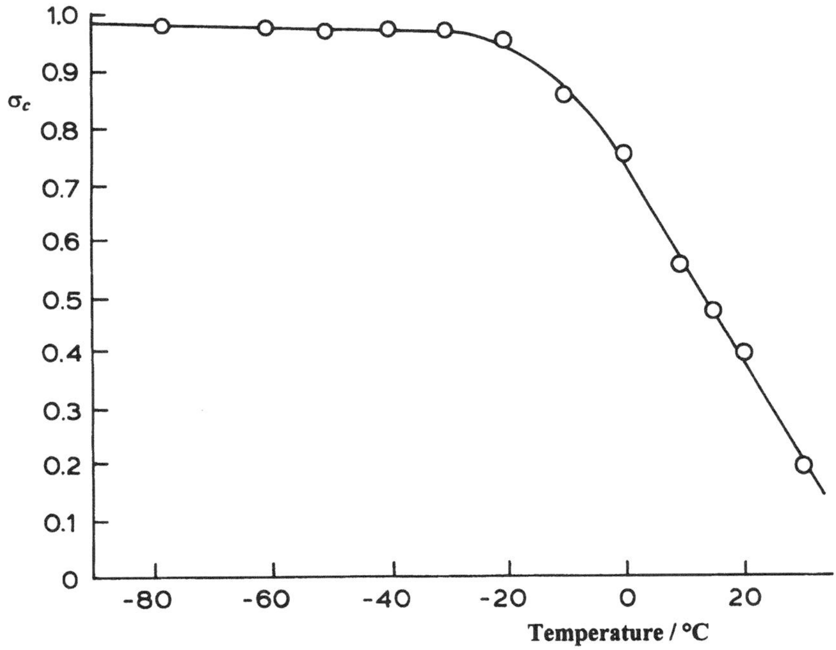

Fig. 11.8 Variation of the fraction of *cis* double bonds, σ_c, with preparation temperature for the ROMP of cyclopentene catalyzed by $WCl_6/(CH_2{=}CHCH_2)_4Si$ (Oreshkin 1977).

determining whether the olefinic protons in the polymer are coupled (*m*) or not (*r*); (iii) by polymerizing prochiral monomers and arguing by analogy with the results on closely related chiral monomers. The first method requires some assumptions about the magnitude of ^{13}C NMR substitution parameters, but the second method is absolute. The third method should be safe if applied with caution.

The first method may be illustrated by reference to the ^{13}C NMR spectrum of an all-*trans* polymer made from (+)-5,5-dimethylnorbornene (Fig. 11.9a). The four olefinic lines are assigned to the four types of carbon shown in **5** (*T*H*m*, TT*r*, HH*r*, *H*T*m*), on the following basis. When 5,5-dimethyl substituents are placed on polynorbornene, the olefinic carbons in **5** will experience the following substitution effects: for *T*H, $(\delta_1 + \delta_2)$, where δ_1 is the effect of the two methyl groups on the tail side, *via* three single bonds, and δ_2 is the effect of the two methyl groups on the head side, *via* two single bonds and a double bond; likewise for TT, $(\delta_1 + \varepsilon)$; HH, $(\delta_2 + \gamma)$; and *H*T, $(\varepsilon + \gamma)$. γ values are always negative (usually -1 to -3 ppm for a single methyl substituent), while δ values are generally small and positive (0–2 ppm), and ε values are very small (< 0.5 ppm). The predicted line order, from downfield to upfield, is therefore *T*H, TT, HH, *H*T. The spacings of the downfield pair of lines, $(\delta_1 + \delta_2) - (\delta_1 + \varepsilon)$, should be the same as for the upfield pair, $(\delta_2 + \gamma) - (\varepsilon + \gamma)$, as indeed is observed. Also, a long chain must contain equal numbers of TT and HH carbons, and an equal number of *H*T and *T*H carbons. The

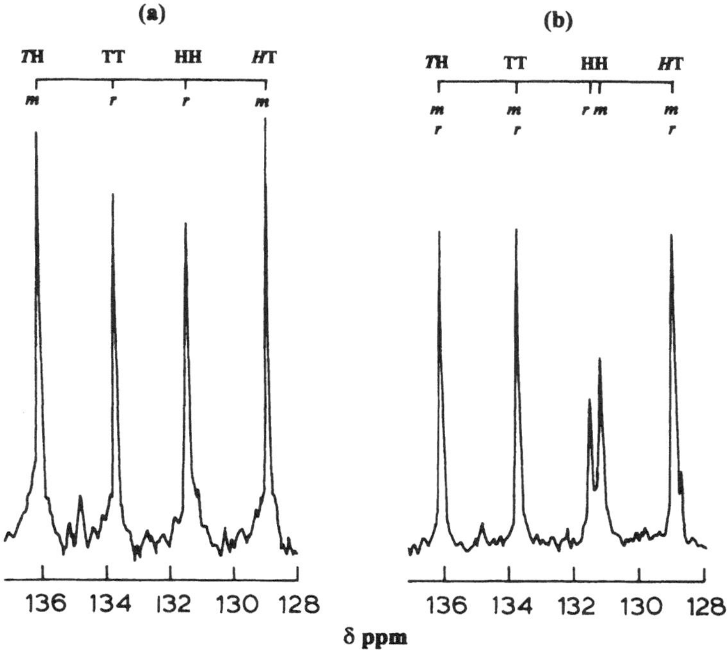

Fig. 11.9 ^{13}C NMR spectrum (olefinic region) of an all-*trans* polymer made from (a) (+)- and (b) (±)-5,5-dimethylnorbornene. Catalyst: $RuCl_3$ in EtOH/PhCl at 75°C. Optical purity of (+)-monomer 83% (Ivin 1982a).

intensities of the outer pair of lines (*T*H, *H*T) should therefore be the same, as should the intensities of the inner pair (TT, HH), again as observed. These assignments lead to the conclusion that, for this polymer, the fraction of *m* dyads, $(\sigma_m)_t$, is 0.57.

The substitution parameters, along with those of the singly substituted methyl compounds, are summarized in Table 11.6. The *cis* olefinic lines are generally in a different position from the *trans* olefinic lines but the line order remains the same.

Table 11.6 ^{13}C NMR substitution parameters (ppm): chemical shifts of olefinic carbons in polymers of *exo*-5-methyl-, *endo*-5-methyl-, and 5,5-dimethylnorbornene, less those in polynorbornene, for all-*cis* and all-*trans* polymers.

	Parameters for *cis* C=C			Parameters for *trans* C=C			
Substituents	$\delta_1+\varepsilon$	$\delta_2-\varepsilon^a$	$\gamma+\varepsilon$	$\delta_1+\varepsilon$	$\delta_2-\varepsilon^a$	$\gamma+\varepsilon$	Reference
exo-5-methyl	0.54	1.81	−1.87	0.33	1.77	−1.61	Ivin 1980b
endo-5-methyl	0.55	0.75	−3.28	0.39	1.42	−2.79	Ivin 1977b
5,5-dimethyl[b]	0.80	1.96	−4.83	0.57	2.31	−4.26	Ho, H.T. 1982b

[a] Average spacing of upfield pair of lines and downfield pair of lines; see text. [b] The effect of two methyl substituents on C-5 is somewhat less than the sum of the effects of the individual substituents.

Table 11.7 Polymers of norbornene derivatives showing *m*/*r* splitting of the *trans* HH olefinic resonance in the ^{13}C NMR spectrum; X = *exo*-5-substituent, Y = *endo*-5-substituent. Catalyst: $RuCl_3.3H_2O$ in EtOH/PhCl at 50–70°C, except where indicated otherwise (Ivin 1982a)

X	Y	HH line positions/ ppm		Splitting/ ppm	Relative intensities/ (downfield/upfield)[a]
CH_3	CH_3	131.43	131.14	0.29	1.3
CH_3	CH_2OH	130.58	129.98	0.60	2
CH_2OH	CH_3	131.10	130.80	0.30	[b]
CH_3	COOH	131.32	131.02	0.30	1.0 (0.47)[c]
COOH	CH_3	130.70	130.56	0.14	
H	$COOCH_3$	131.36	130.97	0.39	0.8
H	$COOC_2H_5$	131.61	131.02	0.59	
CH_3	$COOC_2H_5$	131.29	130.96	0.33	1.1
$COOCH_3$	CH_3	130.82	130.63	0.19	1.0

[a] For X = Y = CH_3 the upfield peak is *m* and the downfield peak is *r*. It is likely that the same line order applies in most other cases. Where no entry is given the *trans* peaks are not sufficiently well resolved from overlapping *cis* peaks. [b] $IrCl_3$ catalyst. [c] $OsCl_3$ catalyst (all-*trans* polymer); note variation of tacticity with catalyst.

The substitution parameters, though different from those for *trans* dyads, remain within the expected ranges.

Turning now to the spectrum of the polymer made from racemic monomer (Fig. 11.9b), we see that the HH *trans* resonance is split into two peaks, one in the same position as in Fig. 11.9a, already assigned to HH,*r*; the other, 0.29 ppm upfield, can therefore be assigned to HH,*m*. The relative intensities of these two peaks provide confirmation of the value of $(\sigma_m)_t = 0.57$. The other three peaks, *T*H, TT, and *H*T, do not show detectable *m*/*r* splitting.

The sensitivity of the HH *trans* olefinic carbons to *m*/*r* dyads is a general phenomenon for polymers of 5,5-disubstituted norbornenes and is sometimes observed for polymers of 5-substituted norbornenes (see Table 11.7). However, to determine $(\sigma_m)_t$ in such cases requires the assumption that the line order is the same as in the polymer of 5,5-dimethylnorbornene, which may not always be true. In the HH dyad the substituents in adjacent units will interact to different extents in the *m* and *r* dyads. This will affect the relative conformational populations about the bonds between the olefinic carbons and the rings, and hence the chemical shifts. In the HT and TT dyads the substituents in adjacent units will interact to a lesser extent, and in these cases the differential effect in the two types of dyad is evidently not enough to be detectable in the chemical shift. An *m*/*r* splitting is also sometimes observed in the HH (*trans*) component of the signal from the ring carbon C-4 lying between the C-3 olefinic carbon and the C-5 carbon bearing the substituents (Ivin 1982a,b).

Tacticity determinations by the first method have been carried out using enantiomers of the following derivatives of norborn-2-ene: 1-methyl- (Hamilton 1983, 1984a), *exo*-5-methyl- (Ivin 1979b, 1980b), 5,5-dimethyl- (Ho, H.T. 1982a,b), *endo,exo*-5,6-dimethyl- (O'Dell 1994; Sunaga 1994), 1,7,7-trimethyl-

(Feast 1994d), *endo*-5-methoxymethyl- (Ivin 1994a), *endo*-5-acetate (Steinhäusler 1994a,b), *endo,exo*-5,6-dimethoxymethyl- (O'Dell 1994), and *endo,exo*-5,6-dicarbomethoxy- (O'Dell 1994); and the following derivatives of 7-oxanorborn-2-ene: *exo*-5-methoxymethyl- (Novak 1988a), and *endo*-5-methoxymethyl- (Benedicto 1992).

Some monomer/catalyst systems give highly tactic polymers and we shall consider later in more detail the various extreme cases. Figure 11.10a shows an example of the spectrum of an all-*cis* syndiotactic polymer, which has an HH, TT structure when made from a single enantiomer. (The small HT peaks arise from the presence in the monomer of a small amount of the minor enantiomer.) When the polymer is made from racemic monomer, Figure 11.10b, the polymer may still be

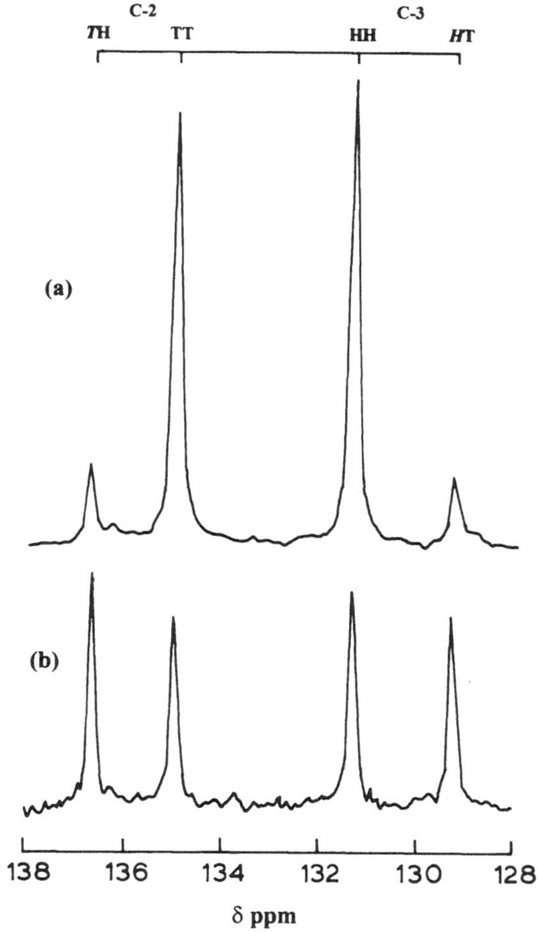

Fig. 11.10 ^{13}C NMR spectrum (olefinic region) of all-*cis* syndiotactic polymer made from (a) (+)- and (b) (±)-5,5-dimethylnorbornene. Catalyst: $ReCl_5$ in PhCl at 20°C. Optical purity of (+)-monomer, 86% (Ho, H.T. 1982b).

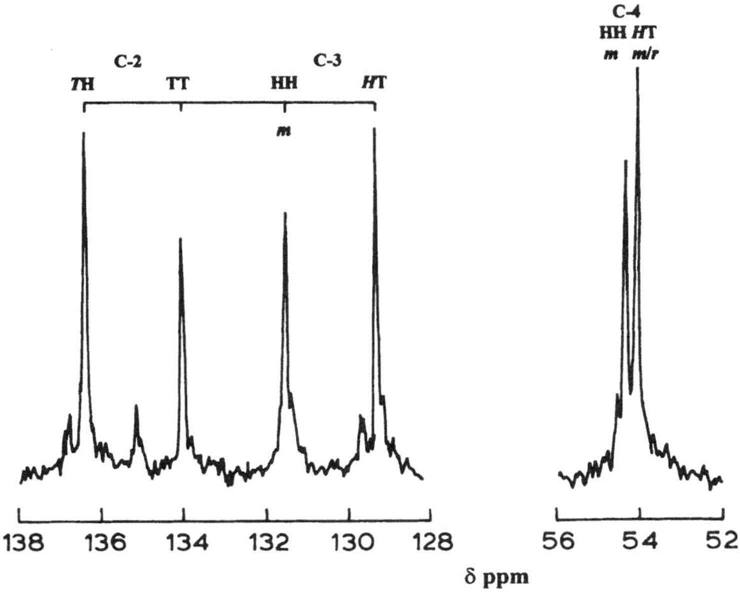

Fig. 11.11 ^{13}C NMR spectrum (olefinic region) of 85% *trans* isotactic polymer made from (±)-5,5-dimethylnorbornene. Catalyst: $W(CO)_3$(mesitylene)/$EtAlCl_2$/*exo*-2,3-epoxynorbornane in PhCl at 20°C. [M] = 0.4 M. Note the absence of the HH,*r* peak; cf. Fig.11.9b (Devine 1982).

presumed to be syndiotactic, but the substituents show no HT bias, which means that the two enantiomers are randomly incorporated. An example of the spectrum of a high-*trans* isotactic polymer is shown in Fig. 11.11. In this case one can work with racemic monomer, the HH,*m* peak having been assigned on the basis of the result shown in Fig. 11.9.

All resonances in the ^{13}C NMR spectra of all-HT polymers of racemic 1-methylnorbornene are insensitive to tacticity, but C-6 in the spectrum of the *hydrogenated* polymer does show *m*/*r* splitting (0.05 ppm), when the precursor is made with $OsCl_3$ as catalyst. The downfield component (39.12 ppm) is dominant when the experiment is repeated with partially resolved monomer and is therefore assigned to *m* dyads. The upfield component (39.05 ppm) is the only peak observed in the spectrum of the hydrogenated all-*cis*, all-HT polymer made from *racemic* monomer with $ReCl_5$ as catalyst, and it may be concluded that this is a syndiotactic alternating copolymer of the two enantiomers. This structure is forced on the polymer by the steric exclusion of *cis* HH structures, which are never observed in any polymers of this monomer (Hamilton 1983, 1984a; also see Couturier 1992).

The ROMP of 1,7,7-trimethylnorbornene initiated by a molybdenum carbene complex gives an all-*trans*, all-HT polymer, which is necessarily isotactic when made from a single enantiomer. It is atactic when made from racemic monomer showing that the two enantiomers add randomly to the growing chain (Feast 1994d).

The second method of determining tacticities is essentially an extension of the first, the only difference being that the chiral centre is placed in the side chain(s) of a monomer that would otherwise be prochiral. The adjacent olefinic protons in the polymer may be equivalent and uncoupled (HH or TT), or non-equivalent and coupled (HT). The application of this method to the case of 5,6-dicarboxylic esters of norbornadienes is described in Section 13.4.1.4. It turns out that an all-*trans* polymer is syndiotactic and an all-*cis* polymer is isotactic, which is the opposite way round to what was found for 5,5-dimethylnorbornene (see above).

The third method of determining tacticities may be illustrated by reference to the polymers of *anti*-7-methylnorbornene. The ^{13}C NMR spectra of all-*trans* and high-*cis* polymers are shown in Figs. 11.12 and 11.13. The first is clearly atactic, all carbons except C-7 being sensitive to tacticity. The assignments are based on a comparison with the spectrum of a polymer made with $W(CO)_3$(mesitylene)/ $EtAlCl_2$/*exo*-2,3-epoxynorbornane as catalyst and the assumption that, as in the polymer of 5,5-dimethylnorbornene made with this catalyst, the *trans* double bonds are always associated with *m* dyads. Likewise the high-*cis* polymer made with $ReCl_5$ as catalyst is assumed to be syndiotactic as for the polymer of 5,5-dimethylnorbornene made with this catalyst. The spectrum of the hydrogenated

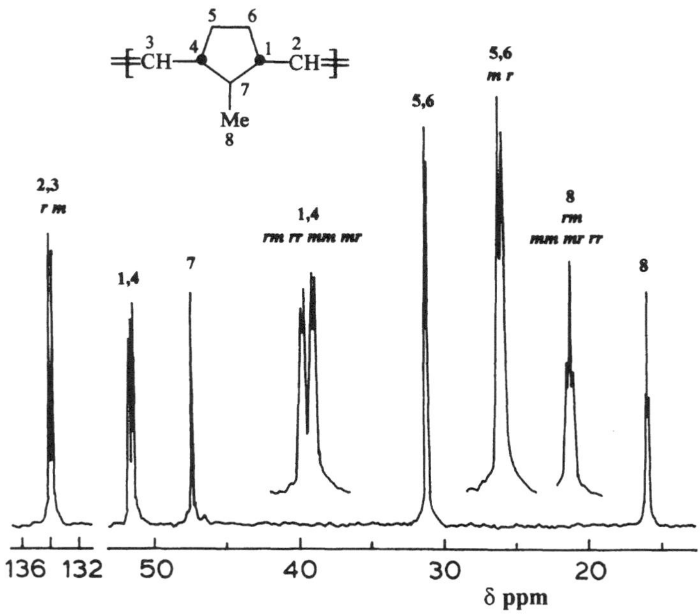

Fig. 11.12 ^{13}C NMR spectrum of an all-*trans* polymer of *anti*-7-methylnorbornene. Catalyst: $RuCl_3$ at 60°C (Hamilton 1985b).

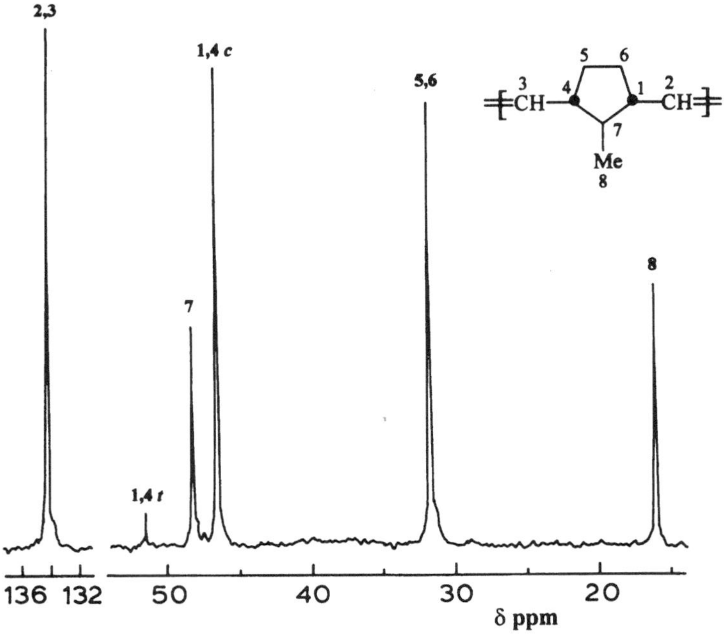

Fig. 11.13 ^{13}C NMR spectrum of a 90%-*cis* polymer of *anti*-7-methylnorbornene. Catalyst: $ReCl_5$ (Hamilton 1985b).

atactic polymer shows larger *m*/*r* splittings than in the spectrum of its precursor, although C-7 remains insensitive to tacticity. The hydrogenated syndiotactic polymer gives essentially a single set of lines, as expected (Hamilton 1985b).

Tacticities in polymers of prochiral monomers made with various catalysts have been determined in this way for the following: *anti*-7-methylnorbornene (Hamilton 1985b; Gilliom 1988; Kress 1990; Feast 1992c), *syn*-7-methylnorbornene (Kress 1990; Feast 1992c), *endo,endo*-5,6-dimethylnorbornene (Greene 1987a), hydrogenated polymers of *exo,exo*-5,6-dimethylnorbornene (Greene 1987a), spiro(norbornene-7,1′-cyclopropane) (Seehof 1993a), and 5,6-bis(trifluoromethyl)-norbornadiene (Bazan 1989).

Tacticities in hydrogenated polymers, expressed as the fraction of *m* dyads $(\sigma_m)_H$, are a weighted mean of those for the *cis* and *trans* dyads in the precursor, $(\sigma_m)_c$ and $(\sigma_m)_t$, respectively. Equation (6) is sometimes useful for determining the third σ_m value when only two are readily determinable, or as a means of checking internal consistency. For an example, see Ivin (1990).

$$(\sigma_m)_H = \sigma_c(\sigma_m)_c + (1 - \sigma_c)(\sigma_m)_t \tag{6}$$

Table 11.8 Some limiting cases of tactic polymers made by ROMP

			Predominant structure				
Monomer [a]	Catalyst [b]	σ_c [c]	*c/r*	*c/m*	*t/m*	*t/r*	Reference
5,5-DMNBE	$ReCl_5$	~1.0	✓				Ho, H.T. 1982b
	W-4	~1.0	✓				Eilerts 1995
anti-7-MNBE	$ReCl_5$	~1.0	✓				Hamilton 1985b
(±)-1-MNBE	$ReCl_5$	1.0	✓				Hamilton 1984a
7-MNBD	$OsCl_3$	0.97 [d]	✓				Hamilton 1993
MOMONBE	W-1	0.97	✓				Benedicto 1992
endo,exo-							
5,6-DMNBE	Mo-1	0.85		✓			Sunaga 1994
5,6-$(CO_2R^*)_2$NBD	Mo-2	0.99		✓			O'Dell 1994
5,6-$(CO_2Me)_2$NBD	W-5	>0.95		✓			O'Donoghue 1996
anti-7-MNBE	Ta-1	0.99		✓			Mashima 1996
5,5-DMNBE[e]	W-2	0.15			✓		Devine 1982
MOMONBE	Ru-1	0.04			✓		Benedicto 1992
(±)-1-MNBE	$OsCl_3$	0.0			✓		Hamilton 1984a
(−)-1,7,7-TMNBE	Mo-3	0.0			✓		Feast 1994d
5,6-$(CO_2R^*)_2$NBD	Mo-4	0.06				✓	O'Dell 1994
anti-7-MNBE	W-2	0.45	✓		✓		Hamilton 1985b
5,5-DMNBE	W-3	0.61	✓		✓		Ho, H.T. 1982a

[a] MNBE = methylnorbornene; DMNBE = dimethylnorbornene; TMNBE = trimethylnorbornene; MNBD = methylnorbornadiene; MOMONBE = *endo*-5-methoxymethyl-7-oxanorborn-2-ene; 5,6-$(CO_2R^*)_2$NBD = 5,6-dicarboalkoxynorbornadiene, (R^* = (−)-menthyl). [b] W-1 = W (=$CHCMe_3$)(=NAr)[$OCMe(CF_3)_2$]$_2$, (Ar = 2,6-diisopropylphenyl); W-2 = $W(CO)_3$(mesitylene)/$EtAlCl_2$/*exo*-2,3-epoxynorbornane; W-3 = WCl_6/Bu_4Sn/20°C; W-4 = W(=NC_6H_3-Me_2-2,6)$(Cl)_3$(OArO)(OEt_2)/Et_2AlCl; W-5 = *syn*-W(=$CHCMe_3$)(O)(OC_6H_3-Ph_2-2,6)$_2$(PMe_3); Mo-1 = Mo(=$CHCMe_3$)(=NAr)[OCMe$(CF_3)_2$]$_2$; Mo-2 = Mo(=$CHCMe_2Ph$)(=NAr)[$OC(CF_3)_3$]$_2$; Mo-3 = Mo(=$CHCMe_2Ph$)(=NAr)[OCMe-$(CF_3)_2$]$_2$; Mo-4 = Mo(=$CHCMe_2Ph$)(=NAr)[$OCMe_3$]$_2$; Ru-1 = [RuCl(μ-Cl)η^3:η^3-$C_{10}H_{16}$)]$_2$, ($C_{10}H_{16}$ = 2,7-dimethyloctadienediyl); (;Ta-1 = Ta(=CHPh)Cp* (η^4-C_4H_6) generated from Ta$(CH_2Ph)_2$Cp*(η^4-C_4H_6) at 65°C (Cp* = pentamethylcyclopentadienyl). [c] Fraction of double bonds with *cis* configuration. [d] Mainly *anti* repeating units. [e] [M] = 0.4 M.

11.6.2.2 Interpretation of tacticity

Three broad types of tacticity may be distinguished in polymers made by ROMP: (i) fully tactic polymers, which may be divided into four subgroups *c/r*, *c/m*, *t/m*, *t/r*, as listed in Table 11.8, and a fifth group in which the polymer has intermediate *cis* content but in which only *c/r* and *t/m* structures are found; (ii) completely atactic polymers, which may be of any *cis* content; and (iii) polymers of intermediate tacticity, where $(\sigma_r)_c$ is sometimes but not always equal to $(\sigma_m)_t$.

A given catalyst does not always result in polymer of a given type of tacticity, it can vary with the monomer, temperature and dilution. This is to be expected if second-order propagation processes are competing with first-order epimerization or relaxation processes involving a reorganization of the ligand geometry around the

Table 11.9 Tacticities of some polymers made using WCl_6/R_4Sn (1/2) as catalyst/cocatalyst system; solvent PhCl. Symbols defined in text.

Cocatalyst	Temperature/ (°C)	Monomer [a]	σ_c	$(\sigma_r)_c$	$(\sigma_m)_t$	$(\sigma_m)_H$	Reference
Bu_4Sn	20	5,5-DMNBE	0.61	0.98	0.97	0.39	Ho, H.T. 1982a
Bu_4Sn	100	5,5-DMNBE	0.71	0.52	0.44	0.47	Ho, H.T. 1982a
Ph_4Sn	20	5,5-DMNBE	0.69	0.63	0.55	0.43	Ho, H.T. 1982a
Bu_4Sn	20	5,6-DMNBE	0.94	0.6			Greene 1987a
Bu_4Sn	20	*anti*-7-MNBE	0.52	0.53	0.59	0.53	Hamilton 1985b

[a] See footnote to Table 11.8; 5,6-DMNBE is the *endo,endo* isomer.

metal centre of the propagating species. The catalyst systems WCl_6/R_4Sn provide good examples of the way in which a change of conditions can give rise to a substantial change in tacticity (see Table 11.9). Thus the polymer of 5,5-DMNBE is fully tactic (*c*/*r*, *t*/*m*) when the cocatalyst is Bu_4Sn at 20°C, but if it is changed to Ph_4Sn, or the temperature is raised to 100°C, or the monomer changed to *anti*-7-MNBE, the polymer formed is atactic or nearly so. Dilution effects are less well documented. For the ROMP of 5,5-DMNBE catalyzed by $RuCl_3$, $(\sigma_m)_t$ falls towards 0.5 with increasing dilution, but when the initiator is $W[=C(OMe)Ph](CO)_5$, no effect of dilution is observed at 100°C even though $(\sigma_m)_t$ and $(\sigma_r)_c$ have intermediate values of about 0.75 (Ho, H.T. 1982a). It appears that, in the latter case, the tacticity is governed by competing second-order reactions involving metal carbene complex and monomer.

As in Ziegler–Natta polymerization, steric control of the propagation step may either involve the interaction of the monomer with a chiral metal centre (enantiomorphic sites model), or it may involve the interaction of the monomer with the chiral centres in the repeating unit(s) adjacent to the metal centre (chain-end model).

Consider first the enantiomorphic sites model. Scheme 11.3 shows a basis for the prediction of tacticity in the ROMP of norbornene and related compounds (Ivin 1977a). This scheme involves a number of assumptions, which will be listed and then discussed. First, it is assumed that the propagating species may have left- or right-handed forms (P_l or P_r), represented here in terms of octahedral symmetry about Mt, with one position vacant for the acceptance of monomer. Second, it is assumed that norbornene presents its less hindered *exo* face (the five-membered unsaturated ring) to the metal centre. Third, it is assumed that the Mt=C and C=C double bonds approach each other in parallel alignment in forming the transition state leading to the metallacyclobutane intermediate X, and likewise in reverse in the subsequent rearrangement of X. Fourth, it is assumed that the configuration P_l or P_r retains its integrity between successive propagation steps, i.e. there is no ligand migration, rotation about the Mt=C double bond, or other significant change of geometry that would result in epimerization or relaxation to an achiral form.

If the propagating metal carbene complex does not have a predetermined vacant position, but is instead trigonal–bipyramidal or tetrahedral, it may still behave as in

Scheme 11.3 provided that the ligands other than the carbene offer an asymmetric environment, which controls the direction of approach of the monomer. If this is not the case, there will not be a favoured direction of approach unless the chain-end effect comes into play. Barriers to rotation about Mt=C have been measured by observation of NMR coalescence temperatures (Guggenberger 1975; Schrock 1978; Kress 1987a). In some cases, these are high enough that epimerization by rotation about Mt=C is unlikely to be important, but in others such a process may be as fast as or faster than the propagation step.

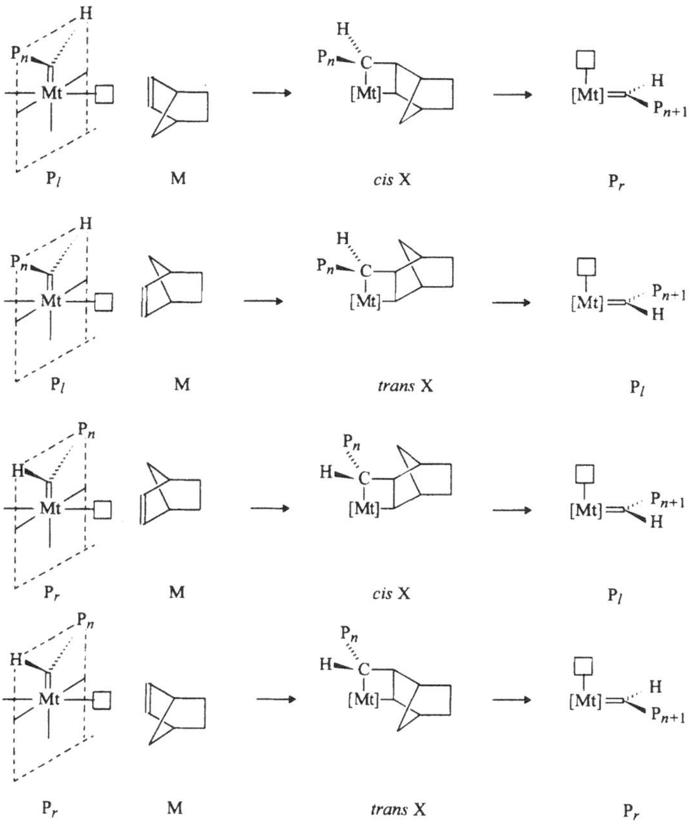

(7)

(8)

(9)

(10)

Scheme 11.3 The four possible modes of addition of norbornene (M) to a chiral propagating species (P_l or P_r). P_n denotes a chain containing n monomer units. The four unspecified ligands may be the same or different.

The second assumption may be regarded as proved beyond reasonable doubt. It is known that the additions of most electrophiles to norbornene occur at the *exo* face (March 1977); also, the structure of the metallacyclobutane formed by the addition of norbornene to the Tebbe reagent, 'Ti(=CH_2)Cp_2', does correspond to reaction at the *exo* face (Gilliom 1986b). Furthermore, the fact that *syn*-7-methylnorbornene is much less reactive than its *anti*-isomer towards ROMP using most initiators is also in keeping with reaction at the *exo*-face (Hamilton 1985b).

The third assumption will be valid if the filled and unfilled orbitals on the Mt=C and C=C bonds match correctly as discussed in Section 3.8. These conditions are more likely to be fulfilled in cases where the metal centre is in a high oxidation state, e.g. W(IV), Mo(IV), and Re(V) (d^2 configuration, regarding the carbene as a neutral ligand). For metals in a lower oxidation state, e.g. Os(II), Ir(I), and Ru(II) (d^4 configuration), the initial approach is more likely to be with the Mt=C and C=C bonds orthogonal (Taube 1983; Kress 1987a; Novak 1988b; Youinou 1988).

From Scheme 11.3 it may be seen that reactions (7) and (9) result in the formation of *cis* double bonds in the polymer chain, and a metal carbene complex in which the chirality of the metal centre is *opposite* to that in the reacting complex. This means that, in an all-*cis* polymer, the enchained rings will have alternating configurations, i.e. the polymer will be syndiotactic (*c/r*). At the other extreme, an all-*trans* isotactic polymer (*t/m*) will result from the repeated occurrence of either reaction (8) or reaction (10), and, if the P_l and P_r species propagate independently, the result will be a racemic mixture of polymer molecules. More detailed considerations show that when both *cis* and *trans* double bonds are formed in accordance with Scheme 11.3, then the *cis* junctions will always be associated with *r* dyads, and *trans* junctions with *m* dyads (Ivin 1977a, 1983). Scheme 11.3 thus correctly predicts the observed tacticities in the first, third, and fifth groups of results listed in Table 11.8. Cases of intermediate tacticity, where $(\sigma_r)_c = (\sigma_m)_t = 0.5–1.0$, can also be interpreted in terms of this scheme if it is modified to include partial epimerization of P_l and P_r between propagation steps. Some examples approximating to this type of behaviour are shown in Table 11.9.

This leaves the second (*c/m*) and fourth (*t/r*) groups in Table 11.8 to be explained in a different way. In these cases it must be the stereochemistry of the polymer chain itself that is controlling the mode of addition of the next monomer unit, either before, during, or after the conversion of the metallacyclobutane complex into the propagating metal carbene complex. We have already seen that, in the formation of high-*cis* polymers of norbornene, the *c/t* selectivity depends not only on the configuration of the previously formed double bond but, when this is *trans*, also on the one before that (Scheme 11.2). It is therefore very reasonable to expect that, in high-*cis* polymers made using carbene initiators with symmetrical tetrahedral ligand geometry, any *m* selectivity must arise through the influence of the configuration of the previously added monomer unit or units. While we cannot argue in the same way for high-*trans* polymers, where the *c/t* selectivity is generally independent of the configuration of the previously formed double bond, nevertheless, the *m/r* selectivity with such initiators may still be sensitive to the configuration of the previously added unit, so accounting for the comparatively rare case of predominantly *t/r* structure.

11.6.3 Head–tail bias

The extent of HT bias in the polymers of substituted cycloalkenes is very dependent on: (i) the location and nature of the substituent(s); and (ii) the catalyst. Monomers with a substituent at the double bond generally give strongly biased polymers with most catalysts. For example, 1-methylcyclobutene with $W(=CPh_2)(CO)_5$ gives an 85% *cis* polymer with HH : HT : TT = 1 : 8 : 1 (Katz 1976a). The presence of one or two substituents at the α-position results in fully biased polymers with some but not all catalysts. When the substituent(s) are further removed from the double bond there is usually very little HT bias in the polymer (see, for example, Figs.11.9b and 11.10b).

Some examples of monomers giving polymers with a completely regular HT structure are shown in Table 11.10. In most cases the polymers are either high-*cis* or high-*trans* and then have exceptionally simple ^{13}C NMR spectra (see Fig. 11.14). A few, however, are of intermediate *cis* content.

With the polymers of **13** there is a remarkable variation of HT bias with catalyst. Thus $RuCl_3$ at 60°C gives an unbiased all-*trans* polymer (HH : HT : TT = 1 : 2 : 1), whereas $IrCl_3$ at 75°C gives a strongly biased high-*trans* polymer, the extent of bias increasing with dilution of the monomer. $OsCl_3$ at 60°C gives a polymer containing 16% *cis* double bonds, which occur only in HT dyads, and 84% *trans* double bonds, which occur mainly in HT dyads (HH : HT : TT = 1 : 11.3 : 1). Catalysts such as WCl_6/R_4Sn (R = Me, Bu) at 20°C give polymers with 40–70% *cis* double bonds, but containing absolutely no *cis* HH structures. Fig. 11.15 shows the olefinic region of the ^{13}C NMR spectrum for a polymer of this kind, having a modest overall bias. Note that (*trans* HH) = (*trans* TT) + (*cis* TT), and that (*cis* HH) is not present. Models indicate that the formation of such a structure would be extremely difficult (Hamilton 1985a).

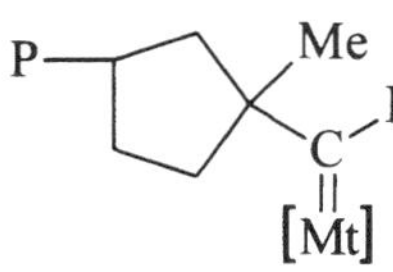

(**17**) P_H

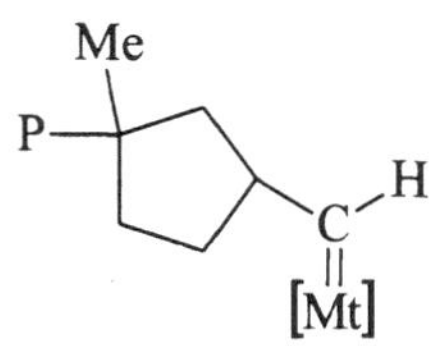

(**18**) P_T

The magnitude of the HT bias is clearly determined by both polar and steric effects, and may also be governed to some extent by relaxation processes between propagation steps. In living systems, the head species P_H (**17**) is generally present in higher concentration than the tail species P_T (**18**) (see e.g. Fig. 3.4). It is likely that in the extreme cases of total HT bias the propagation proceeds generally through the head species. In the more general case it is necessary to consider a total of eight propagation steps as shown in Scheme 11.4. This scheme gives eqn. (11) for an all-*trans* polymer and eqn. (12) for an all-*cis* polymer. In these equations, HT represents the sum of *HT* and *TH* in the ^{13}C NMR spectra. No distinction

Table 11.10 Systems giving fully HT-biased polymers[a]

Monomer		Catalyst	Temperature °C	σ_c	Reference
6	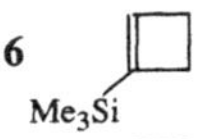	$W(=CPh_2)(CO)_5$	40	1.0	Katz 1980b
7	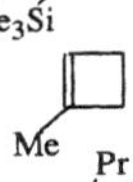	$Mo(=CHCMe_2Ph)(=NAr)(OCMe_2CF_3)_2$	20	1.0	Wu 1994a
8a	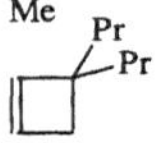	$Mo(=CHCMe_2Ph)(=NAr)(OCMe_3)_2$	20	0.0	Alder 1994
8b	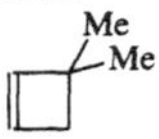	$Mo(=CHCMe_3)(=NAr)(OCMe_3)_2$	20	0.0	Wu 1995a
9	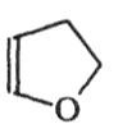	$Cr(=CPh_2)(CO)_5$	40	0.5	Thu 1981b
10	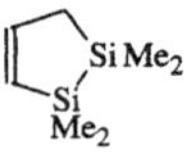	$Mo(=CHCMe_2Ph)(=NAr)[OCMe(CF_3)_2]_2$	25	0.03	Sita 1993
11	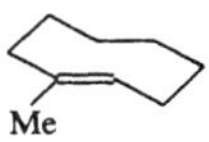	$W(=CPh_2)(CO)_5$	50	0.24	Lee, S.J. 1976
12	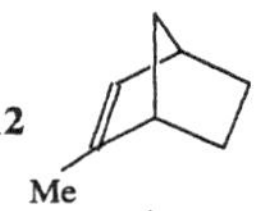	$W(=CPh_2)(CO)_5$	40	~0.4	Katz 1980b
13	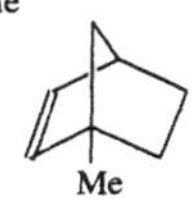	$ReCl_5$	20	1.0	Hamilton 1984a, 1985a
14	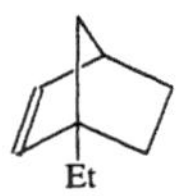	$IrCl_3$	75	0.09	Hamilton 1988
15	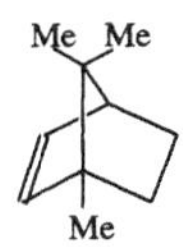	$Mo(=CHCMe_2Ph)(=NAr)[OCMe(CF_3)_2]_2$	20	0.00	Feast 1994d
16	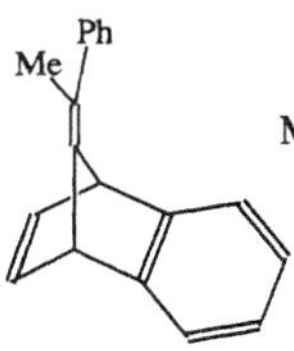	$Mo(=CHCMe_3)(=NAr)(OCMe_3)_2$	20	0.00	Schimetta 1994b

[a] σ_c is the fraction of *cis* double bonds in the polymer.

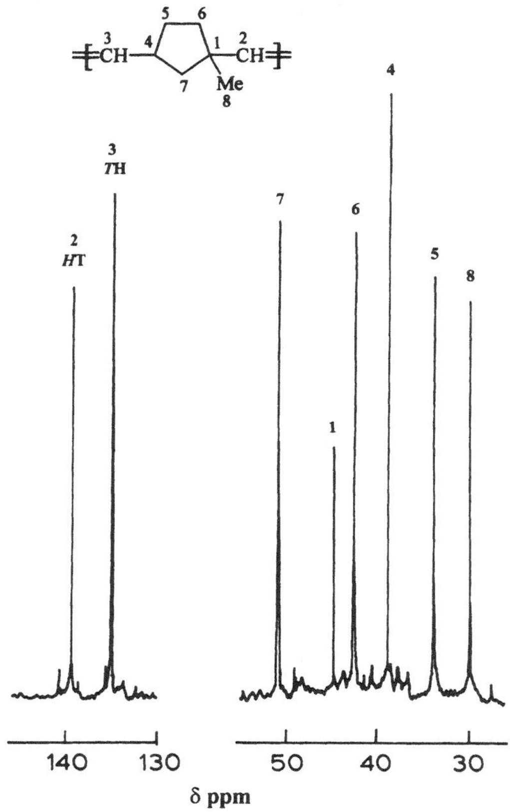

Fig. 11.14 ^{13}C NMR spectrum of all-*cis*, all-HT polymer of 1-methylnorbornene. Catalyst: $ReCl_5$ (Hamilton 1984a, 1985a).

		C=C formed	Dyad	Rate constant
$P_H + M$	→ P_H	*cis*	HT	k_{1c}
	→ P_T	*cis*	HH	k_{2c}
	→ P_H	*trans*	HT	k_{1t}
	→ P_T	*trans*	HH	k_{2t}
$P_T + M$	→ P_T	*cis*	HT	k_{3c}
	→ P_H	*cis*	TT	k_{4c}
	→ P_T	*trans*	HT	k_{3t}
	→ P_H	*trans*	TT	k_{4t}

Scheme 11.4 Mechanism of formation of HT-biased polymers.

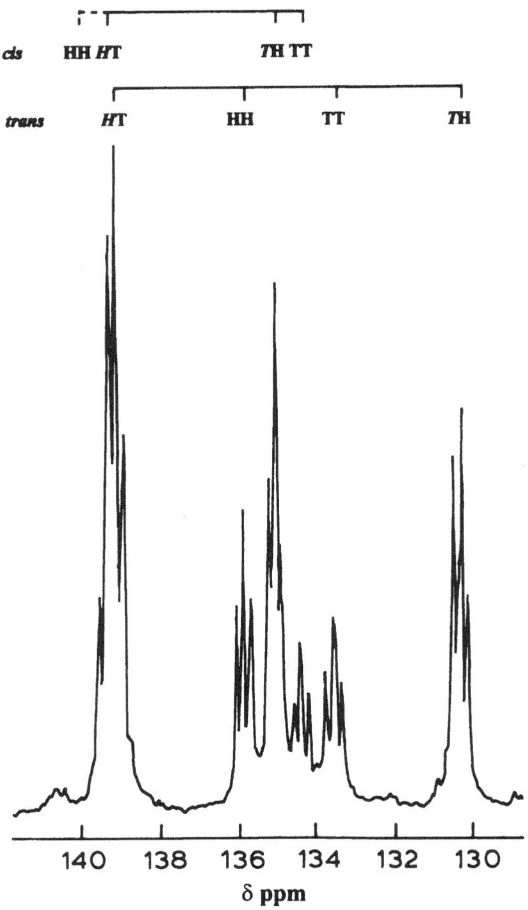

Fig. 11.15 ^{13}C NMR spectrum (olefinic region) of a 48% *cis* polymer of 1-methylnorbornene. Catalyst: $Mo_2(OAc)_4/EtAlCl_2$. (HT+TH)/(HH+TT) = 2.8. The fine structure arises from double-bond triads, e.g. *cct*, *ccc*, *tct*, *tcc* (Hamilton 1984a).

can be made experimentally between HT structures formed by $P_H + M \rightarrow P_H$ and $P_T + M \rightarrow P_T$.

$$(HT/HH)_t = (k_{1t}/k_{2t}) + (k_{3t}/k_{4t}) \tag{11}$$

$$(HT/HH)_c = (k_{1c}/k_{2c}) + (k_{3c}/k_{4c}) \tag{12}$$

For an all-*trans* polymer without HT bias, $k_{1t} = k_{2t}$, and $k_{3t} = k_{4t}$, so that $(HT/HH)_t = 2$. When the polymer is biased, it is more likely that P_H is dominant, as discussed above, with $k_{1t} \gg k_{2t}$, and $k_{3t} \ll k_{4t}$, rather than P_T dominant and reversed relationships. The absence of bias when **13** is polymerized with $RuCl_3$

shows that, in this system, the electronic and steric factors determining the rate constant ratios in eqn. (7) are rather delicately balanced, with probably a rather low polarity in the Ru=C bond of the propagating species. The increase in HT bias with increasing dilution observed in some cases may be interpreted in terms of two distinct P_H (or P_T) species: an unrelaxed form of higher energy, which is less H/T discriminating, and a relaxed form of lower energy, which is more H/T discriminating.

12

Monocyclic Alkenes and Polyenes

12.1 Introduction

The general features of ROMP of cycloalkenes have been presented in Ch. 11. Here we shall describe the behaviour of the monocyclic alkenes and polyenes, including heterocyclic rings, taking them in order of ring size.

Acetylenes may be regarded as the first members of the series; their polymerization using olefin metathesis catalysts is described in Ch. 10. There is no recorded attempt to polymerize cyclopropene with metathesis catalysts; the product would probably be cyclohexa-1,4-diene rather than polymer.

12.2 Four-membered rings

12.2.1 Cyclobutene

The ROMP of cyclobutene was first observed by Dall'Asta (1962) using $TiCl_4$/Et_3Al (1/3) in heptane at $-50°C$ as catalyst. The IR spectrum of the polymer indicated that it contained mainly *cis* units (**1**) and a small proportion of *trans* units (**2**). Cyclic units (**3**), characterized by a weak band at 2645 cm^{-1}, were also detected. Clean ROMP can be obtained with a variety of catalysts as listed in Table 12.1. Polymers formed with other catalysts and containing a proportion of units of type **3** are likely to be mixtures of poly-**3** and poly-(**1**+**2**), produced by independent mechanisms. However, if the propagating species for the two mechanisms are capable of interconversion, it is possible that the two types of unit may be present in the same chain, as in the polymerization of norbornene with certain catalysts (see Ch. 4). In this connection it may be significant that dramatic changes in the proportion of units of type **3** can be brought about by changing the catalyst (based on a given transition metal), the cocatalyst, their ratio, the solvent, and temperature; see Ivin (1983) for details.

Table 12.1 *Cis* content σ_c of ring-opened polymers of cyclobutene obtained with various initiators

Initiator system	Solvent/temperature (°C)	σ_c	Reference
$W(=CPh_2)(CO)_5$	Toluene/20°	0.93	Katz 1976c
$W[=C(OMe)Ph](CO)_5$	Toluene/50°	0.90	Katz 1976d
$Mo(\pi\text{-}C_4H_7)_4$	Benzene/30°	0.91	Kormer 1972
$W(\pi\text{-}C_4H_7)_4$	Benzene/30°	0.90	Kormer 1972
$W(=CHCMe_3)(=NAr)(OCMe_3)_2(PMe_3)$[a]	Toluene/20°	0.78[b]	Wu 1992
$RuCl_3$	Water [c]/50°	0.5	Natta 1965a
VCl_4/BuLi (1/3.5)	Toluene/−20°	0.5	Dall'Asta 1968a
$Cr(acac)_3/Et_3Al$ (1/2.5)	Toluene/−20°	0.3	Natta 1963
$V(acac)_3/Et_3Al$ (1/10)	Toluene/−10°	0.2	Natta 1963
$RuCl_3$	Ethanol/20°	0.0	Natta 1965a

[a] Ar = 2,6-diisopropylphenyl. [b] Blocky *c/t* distribution. [c] With emulsifying agent

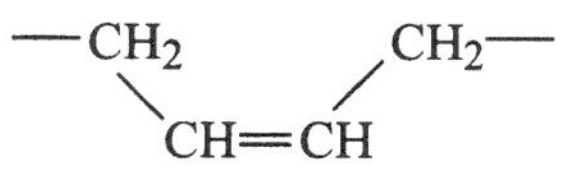

(1)

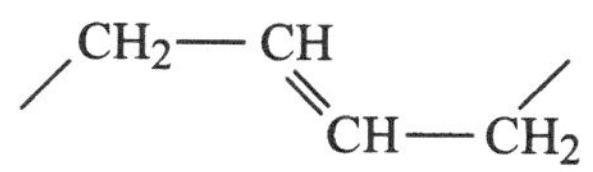

(2)

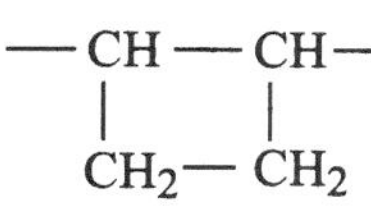

(3)

The initiator $W(=CHCMe_3)(=NAr)(OCMe_3)_2$, in the absence of PMe_3, gives a polymer with a relatively broad MWD ($M_w/M_n > 2$). This is because propagation is much faster than initiation and only a very small fraction of initiator is consumed ($k_p/k_i \sim 1000$ at -60°C). In the presence of PMe_3, an equilibrium is established between free PMe_3 and PMe_3 bound to the complex for both initiating and propagating species. However, the binding to the propagating species at 25°C ($K \sim 10^5\ \text{M}^{-1}$) is much stronger than to the initiating species ($K \sim 500\ \text{M}^{-1}$) because it is sterically less hindered; and in the presence of sufficient PMe_3 (10 equivalents) the propagation reaction is slowed down relative to initiation to such an extent that all the initiator is consumed and the MWD becomes very narrow ($M_w/M_n = 1.03$). The reaction then shows well-defined kinetics and all the characteristics of an ideal living system. Similar results are obtained when the reaction is initiated with the corresponding molybdenum carbene complex. The polymers may also be hydrogenated to give very well-defined samples of polyethene (Wu 1992, 1994b).

12.2.2 Monocyclic cyclobutene derivatives

Of the compounds **4–15**, all except **15** undergo clean ROMP. The polymer of **15**, obtained in low yield using WCl_6/Me_4Sn as catalyst, is a black insoluble powder of reduced chlorine content, caused by loss of HCl and development of conjugation (Brunthaler 1985).

Table 12.2 Proportions of *cis* double bonds σ_c and HT dyads in ring-opened polymers of 1-methylcyclobutene (**4**) made with different initiators

Initiator	Solvent/ temperature °C	σ_c	HH : HT : TT	Reference
$W(=CHCMe_3)(=NAr)[OCMe(CF_3)_2]_2$	Toluene/20°	0.95	20 : 60 : 20	Wu 1994a
WCl_6/tetraisobutylaluminoxane [a]		0.78	15 : 70 : 15[b]	Kormer 1980
$W(=CHCMe_3)(=NAr)(OCMe_3)_2$	Toluene/35°	0.80	0 : 100 : 0	Wu 1994a
$Mo(=CHCMe_2Ph)(=NAr)(OCMe_2CF_3)_2$	Toluene/20°	1.00	0 : 100 : 0	Wu 1994a

[a] See Syatkowsky (1979). [b] The HT dyads have a higher *cis* content (83%) than HH and TT (67%).

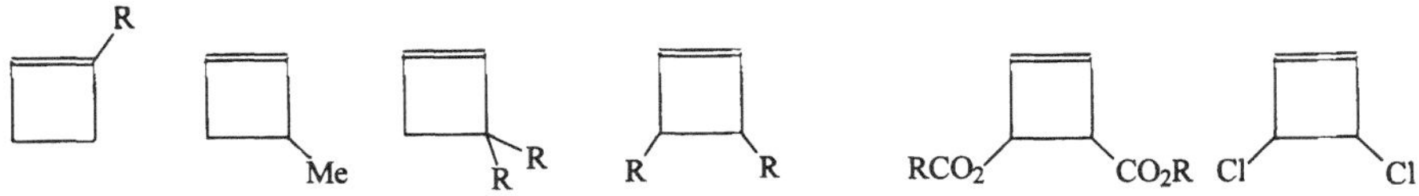

(**4**) R = Me (**5**) R = $SiMe_3$ (**6**) (**7**) R = Me (**8**) R = Pr (**9**) R = $CHMe_2$ (**10**) R = CH_2OMe (**11**) R = CH_2OCH_2Ph (**12**) R = Me (**13**) R = Et (**14**) R = CH_2Ph (**15**)

Polymers of **4** generally have a high proportion of *cis* double bonds but with varying degrees of HT bias (Table 12.2). The molybdenum carbene initiator is the only one to give an all-*cis*, all-HT, polymer, identical with natural rubber, *cis*-1,4-polyisoprene. This remarkable result clearly stems from a fine balance between the electrophilicity of the metal centre, and the steric interactions during the approach of the monomer to the initiator. Very little of the initiator is consumed and the propagating species cannot be detected by ^1H NMR, indicating a high value of k_p/k_i. The yield of polymer is 78% (MW ~ 20 000), but the MWD is inevitably broad ($M_w/M_n = 2.5$). Propagation can be slowed down relative to initiation by addition of 10 equivalents of PPh_2Me to the initiator. About half the initiator is then consumed during polymerization and the carbene proton in the propagating species can be detected as a triplet, corresponding to the tail species $[Mo]{=}CHCH_2CH_2$-$C(Me){=}CHP_n$ (δ 13.2, $J = 6.3$ Hz) (Wu 1994a). The ROMP of 1-trimethylsilyl-cyclobutene (**5**), initiated by $W(=CPh_2)(CO)_5$ at 40°C, also gives an all-*cis* all-HT polymer and the Me_3Si groups in the polymer may be replaced by PhS groups by reaction with PhSCl (Katz 1980b, 1983).

The behaviour of 3-methylcyclobutene (**6**) with various catalysts resembles that of cyclobutene. Early work with $RuCl_3$ and other catalysts gave ring-opened polymers having 15–65% *cis* content, but of unknown tacticity or HT bias (Natta 1965a; Dall'Asta 1968a). With $Mo(=CHCMe_3)(=NAr)(OCMe_3)_2$ as initiator, the propagating species gives a ^1H NMR spectrum containing two carbene proton doublets (δ 11.95, $J = 6.6$ Hz; δ 11.50, $J = 9.5$ Hz), assigned to the *anti* and *syn* rotamers of the head species $[Mo]{=}CHCHMeCH_2CH{=}CHP$ in which the polymer chain P′ in $[Mo]{=}CHP'$ points away from (*anti*) or towards (*syn*) the (=NAr) ligand. This polymer contains 84% *cis* double bonds and the methyl substituents are randomly oriented with respect to both *cis* and *trans* double bonds (Wu 1995a). In contrast 3,3-dimethylcyclobutene (**7**) with the same initiator gives the 'tail' propagating species [Mo]=$CHCH_2CMe_2CH$=CHP, characterized by a carbene proton triplet (δ 8.53, $J = 5.9$ Hz), and an all-*trans*, all-HT, polymer (Wu 1995a). Likewise the ROMP of 3,3-dipropylcyclobutene (**8**) initiated by Mo(=CH-CMe_2Ph)(=NAr)$(OCMe_3)_2$, also gives an all-*trans*, all-HT polymer **16**. The ^{13}C NMR spectrum of the hydrogenated polymer contains four signals from methylene carbons confirming that it has the all-HT structure **17**. On the other hand, the initiator Mo(=$CHCMe_2Ph$)(=NAr)$[OCMe(CF_3)_2]_2$ gives a largely *cis* HT polymer, and catalysts such as WCl_6/Et_3Al give polymers, which are neither stereospecific nor regiospecific. The polymers **16** and **17** are crystalline when freshly prepared, with melting endotherms at 102°C and 136°C, respectively, but crystallize rather slowly when cooled from the melt (Alder 1994).

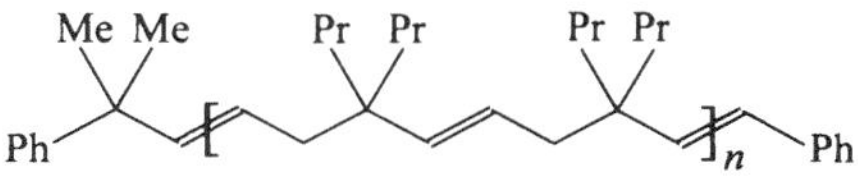

(**16**)

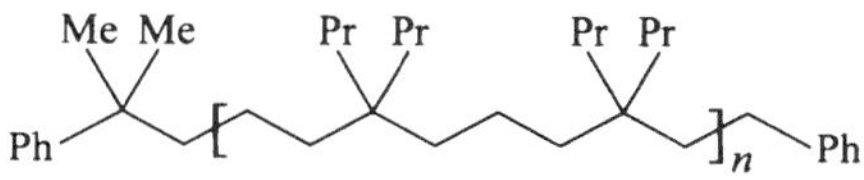

(**17**)

The ROMP of **9**, initiated by WCl_6/Me_4Sn, gives a polymer with a simple four-line ^{13}C NMR spectrum, but it has not been established whether it is all-*cis* or all-*trans*. The former is more likely (Brunthaler 1985). The monomers **10–14** all give living polymers when initiated by Mo(=$CHCMe_3$)(=NAr)$(OCMe_3)_2$ and related complexes. The ROMP of **13** shows well-behaved second-order kinetics in C_6D_6 at 25°C. The rate constant is about 10 times lower than for the reaction of 5,6-dicarbomethoxynorbornadiene, no doubt owing to the greater deactivating effect of the COOR groups in **13**. Polymers of **12** and **13** have 45–93% *cis* content depending on the initiator (Perrott 1995, 1996).

The ROMP of **18** can be effected using [Ti(=CH_2)Cp_2] sources as catalyst (Swager 1987), also with $WOCl_4/EtAlCl_2$ in chlorobenzene at −78°C, admitting the monomer as a gas (Grubbs 1987a); eqn. (1). The polymer forms as a transparent highly soluble material that becomes conductive (10^{-3} $ohm^{-1}cm^{-1}$) when doped with iodine. The undoped polymer shows a single absorption peak at 278 nm ($\varepsilon = 20\,000$ $M^{-1}cm^{-1}$) (per triene unit), the position of which shows that the π-system of the polymer chain is segregated into triene segments. This is the result of steric interactions that force the polymer backbone into a non-planar conformation.

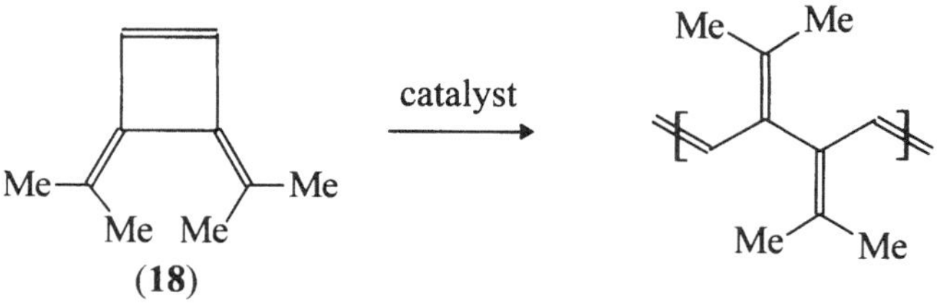

(**18**)

(1)

For bicyclic derivatives of cyclobutene see Section 13.2.

12.3 Five-membered rings

12.3.1 Cyclopentene

Ever since the discovery of the ring-opening polymerization of cyclopentene (Eleuterio 1957) the potential importance of the elastomeric product has stimulated a large amount of work on this monomer. We have already touched on various aspects: the equilibrium monomer concentration (Fig. 11.1); the relative proportions of high polymer and cyclic oligomer (Figs. 3.1, 11.3 and 11.4), and the detection of various oligomers (Table 11.3); the variation of *cis* content with preparation temperature (Fig. 11.8); and the effect of chain transfer agents (Section 11.4; see also Ch. 15).

Given time and an active catalyst, the polymerization reaches equilibrium between monomer, cyclic oligomers, and linear polymer. However, with many catalyst systems, the rate constants are such that the polymer tends to be formed first (Fig. 11.3), and under some conditions the cyclic oligomers can easily escape detection (Arlie 1974). It is curious that *cis,cis*-cyclodeca-1,6-diene, the cyclic dimer, is not normally found in the cyclic oligomers nor will it undergo ROMP in the presence of $WCl_6/EtAlCl_2/EtOH$ at $-30°C$, although it degrades to cyclopentene with this catalyst at 50–80°C (Küpper 1973, 1974). These observations presumably reflect the relative free energies of the species concerned in this system.

Metastable equilibrium between monomer and high-*cis* polymer can be achieved using $WCl_6/(CH_2{=}CHCH_2)_4Si$ as catalyst (see Fig. 11.1). $WF_6/EtAlCl_2$ also gives high-*cis* polymer, the *cis* content remaining high throughout the reaction. With less stereospecific catalysts, secondary *cis/trans* isomerization occurs as the reaction proceeds, leading ultimately to an equilibrium *cis* content of about 16% (De Figueiredo 1979; Ofstead 1980; Syatkovskii 1981) and a much lower equilibrium monomer concentration (see Figs. 12.1 and 11.1, respectively).

Catalysts that give high-*cis* polymers include $WCl_6/Et_3Al/(PhCOO)_2$ at $-40°C$ (Minchak 1972), $W({=}CPh_2)(CO)_5$ at 40°C (Katz 1976c), $WCl_6/(CH_2{=}CHCH_2)_4Si$ at $-30°C$ (Oreshkin 1977; Pakuro 1982; Makovetskii 1985), fresh WCl_6 solution exposed to UV radiation (Yakovleva 1977b), $WF_6/Et_3Al_2Cl_3$ (2/1) at $-30°C$ (Günther 1970), $MoCl_5/Et_3Al$ at $-30°C$ (Natta 1964a), $W({\equiv}CPh)Br(CO)_4$ at 22°C under nitrogen (Katz 1984), WCl_6/Ph_4Sn at $-20°C$ (Ceausescu 1986, 1988b), $Ti({=}CHCMe_2CH{=}CH_2)Cp_2$ at 40°C (Grubbs 1987b), generally in aromatic solvents; also see Dimonie (1992), Wei (1994) and Dragutan (1994). Without exception the *cis* content falls as the temperature is raised for a given catalyst system (see Figs. 11.8 and 12.2). Other important factors are dilution and ageing time (Fig. 12.2), and catalyst/cocatalyst ratio. The effect of increasing $Et_3Al_2Cl_3/WF_6$ ratio (Fig. 12.3) may be the result of a decreasing oxidation state of the tungsten site and/or replacement of the fluoride by the less electronegative chloride ligands. When the electron density at the metal centre is low, a reduction in temperature will allow increased coordination of double bonds in the polymer chain

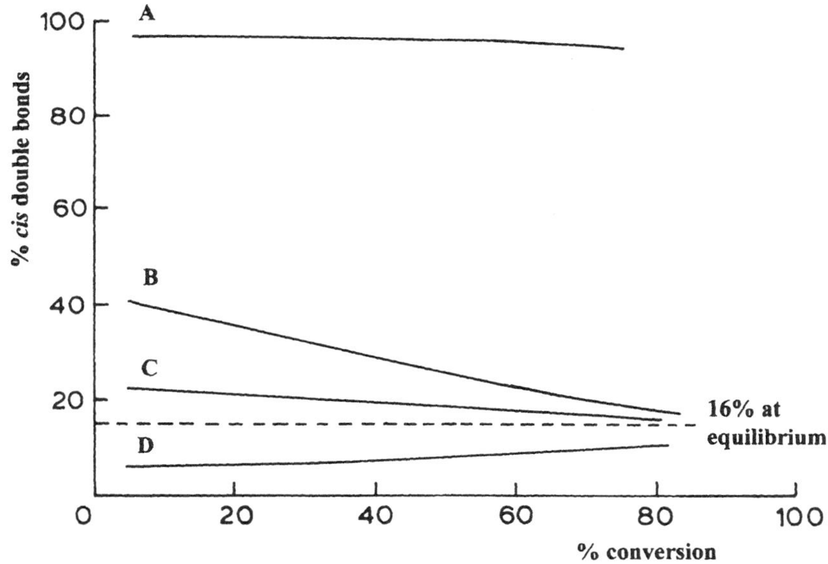

Fig. 12.1 Effect of catalyst and conversion on *cis* content of polymer formed from cyclopentene at 0°C: (A) $WF_6/EtAlCl_2$; (B) $WCl_6/EtAlCl_2/EtOH$; (C) $WCl_6/EtAlCl_2/HOCH_2CH_2Cl$; (D) $WCl_6/EtAlCl_2/HOCH_2CH_2CN$ (Ofstead 1980).

to the metal site. The consequent crowding appears to favour the direction of approach of the monomer which leads to the formation of a *cis* double bond.

Many catalyst systems give polymers with a *trans* content in the region of the equilibrium value (80–90%) (Drapeau 1985). This is sometimes a consequence of secondary metathesis reactions resulting in isomerization of the originally formed *cis* double bonds (see Fig. 12.1). High-*trans* polymer is formed with $TiCl_4/Et_3Al$ or WCl_6/Et_3Al at −30°C using bulk monomer (Natta 1964a, 1966a), and with WCl_6/*i*-Bu_3Al (Haas 1970). The kinetic behaviour of the WCl_6/*i*-Bu_3Al-initiated polymerization is complex and the mechanism not well understood (Amass 1980, 1990; Ceausescu 1985).

A third component in the catalyst system sometimes has a remarkable effect on the rate and/or stereochemistry of the ROMP of cyclopentene. Such effects must arise from a modification of the ligands in the initiating and propagating species as a result of reaction with the third component. For example, the activity of WCl_6/*i*-Bu_3Al is greatly enhanced by the presence of phenols, which convert chloro ligands into aryloxo ligands (Chen, W. 1981). Again, the addition of benzoyl peroxide to WCl_6/Et_3Al at −30°C causes the product to be high-*cis* instead of high-*trans*. Another example is WCl_6/Et_4Sn (1/4), which, if mixed at −30°C in toluene and added to cyclopentene, gives a high-*cis* polymer. The same mixture is ineffective if prepared at 20°C and cooled to −30°C, but if made at 20°C in the presence of diethyl ether it gives a high-*trans* polymer (Pampus 1972, 1974).

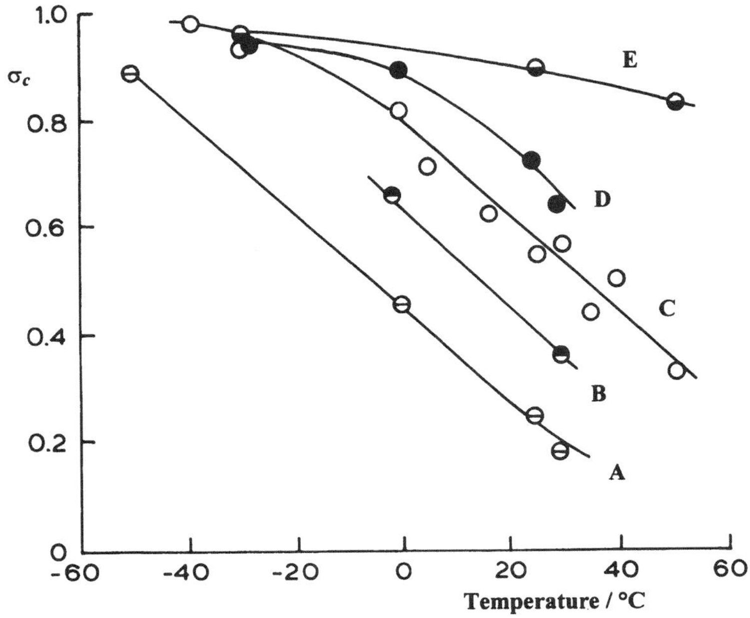

Fig. 12.2 Effect of temperature on the fraction of *cis* double bonds, σ_c, in polymer made from cyclopentene using γ-irradiated WCl_6 as catalyst: (A, B, C) ethyl chloride solution; (D, E) no added solvent; (C, E) freshly prepared solutions; (A, D) WCl_6 aged in monomer before irradiation; (B) aged in ethyl chloride before irradiation (Sheinker 1979).

Polymers of intermediate *cis* content may be made with a great variety of catalyst systems and it is not difficult to control the *cis* content by minor adjustments of the conditons; see Ivin (1983). Among the more recently used catalyst systems the following are of particular interest.

(i) $W(=CHCMe_3)(=NAr)(OCMe_3)_2$ (0.04 M in benzene) will polymerize 50 equivalents of cyclopentene to give an equilibrium mixture containing about 5% monomer (0.1 M) at -60°C and about 95% monomer (1.9 M) at 60°C. The monomer may be stripped completely from the living polymer by continuous evacuation, reforming the original initiator (Schrock 1988c, 1989). In order to make a polymer of narrow MWD ($M_w/M_n = 1.08$) with this system, it is necessary to work at -40°C and to terminate the reaction after 1 h so as to forestall the tendency towards a thermodynamic distribution (Schrock 1989). Star polymers can be made by terminating such a living polymer with a polyfunctional aldehyde (Dounis 1996a).

(ii) Tetraphenylporphyrinatotungsten tetrachloride/tetraisobutylaluminooxane (1/2) polymerizes cyclopentene (4.8 M) in toluene to give a 20% *cis* polymer with a surprisingly narrow MWD ($M_w/M_n = 1.2$) (Coca 1994). The polymer

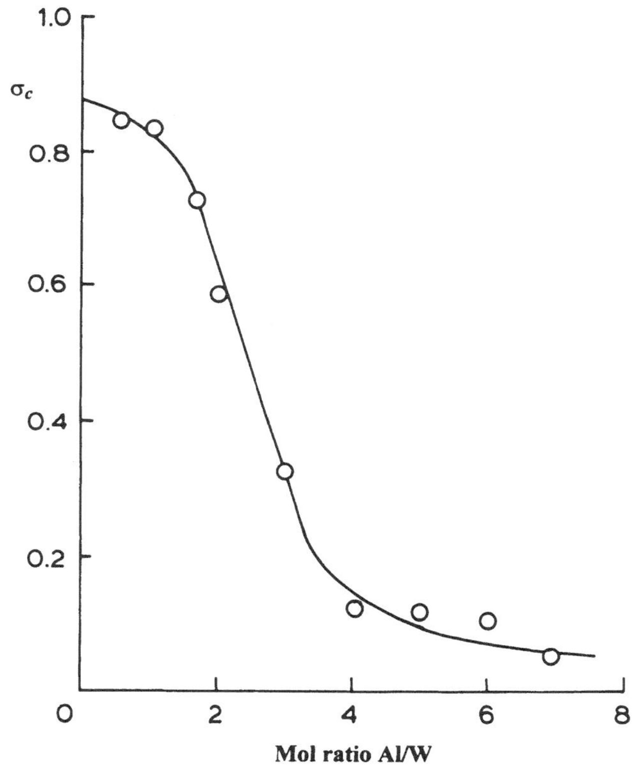

Fig. 12.3 Fraction of *cis* double bonds, σ_c, in polymer made from cyclopentene in toluene at −30°C as a function of Al/W ratio. Catalyst: $WF_6/Et_3Al_2Cl_3$ (Günther 1970).

formed initially using $(C_{17}H_{35}COO)_2MoCl_3/Et_2AlCl$ as catalyst also has a narrow MWD (Pakuro 1976). These may also be living systems.

(iii) $Ru(=CHCH=CPh_2)(Cl)_2(PCy_3)_2$, where Cy = cyclohexyl, brings about the ROMP of cyclopentene, the propagating species being quite stable and detectable by ^1H NMR (Nguyen 1993). Certain other ruthenium complexes, such as diruthenium(II,II) tetrakis carboxylates (Demonceau 1992) and $Ru(H_2O)_6(OTs)_2$ (France 1993a) are also effective in the presence of diazoesters. Previously it had been thought essential to have a strong electronegative ligand attached to ruthenium, as in the system $RuH_2(PPh_3)_4$/TFA (Porri 1975), but now it appears that electron-rich ligands such as PCy_3 are required (Nguyen 1993).

(iv) Tungsten-based systems in which phenylacetylene is used as cocatalyst for the ROMP of cyclopentene are described in Section 10.3.1.

12.3.2 Monocyclic cyclopentene derivatives

A 2 : 3 mixture of compounds **20** and **21** readily undergoes ROMP in the presence of $WCl_6/PhC{\equiv}CH$ in toluene, giving a 90% *trans* polymer and attaining an equilibrium monomer concentration of 2.7 M at 20°C in a few hours (Makovetsky 1987, 1992a). Individual enantiomers of **20** have been prepared by RCM (Section 8.2.1) and polymerized by $Mo({=}CHMe_2Ph)({=}NAr)[OCMe(CF_3)_2]_2$ at − 30°C to give a 52% yield of a 74% *trans* polymer (Sita 1995). Likewise **21**, with the same initiator at − 55°C, gives a 51% yield of 60% *cis* polymer, with a blocky *cis/trans* distribution ($r_t r_c = 6.3$); the hydrogenated polymer is atactic as shown by its ^{13}C NMR spectrum (Feast 1996).

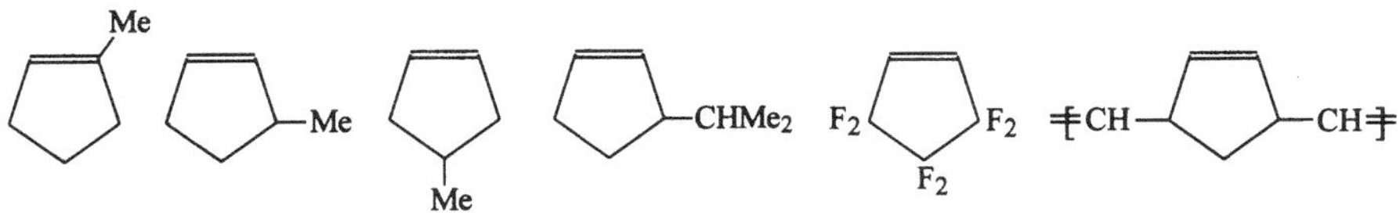

(19) (20) (21) (22) (23) (24)

Compounds **19** (Makovetsky 1992a), **22** (Günther 1970) and **23** (Feast 1980) do not polymerize, probably because their free energy of polymerization is positive. However, the fact that 1% of **19** can completely inhibit the polymerization of **20** and **21** indicates that it is likely to add preferentially to the active site forming the head carbene complex, $[W]({=}CMeCH_2CH_2CH_2CH{=}CHR)$, which is then unable to add any of these three monomers. It should be capable of copolymerization with norbornene. 3- and 4-Alkenylcyclopentenes have been tested as chain transfer agents for the ROMP of norbornene, but without much success (Schrock 1989).

The ring-opened polymer of norbornadiene consists of 3,5-disubstituted cyclopentene units (**24**). When the concentration of these units is kept below 0.2 M the polymer remains soluble, but above this concentration, in the presence of WCl_6/Me_4Sn (1/2), it gels. This is caused by cross-linking, brought about by the ROMP of the enchained cyclopentene rings (see Section 13.4.1.1).

12.3.3 Cyclopentadiene

Cyclopentadiene with $WCl_6/AlBr_3$ as catalyst gives a high yield of powdery amorphous polymer (Marshall 1969); with WCl_6 as catalyst in toluene, a soluble polymer of low MW is formed (Sumitomo 1980). It also copolymerizes with cyclopentene (Ofstead 1979) and norbornene (Minchak 1979), but the products have not been closely characterized.

12.3.4 Heterocyclic rings

The ROMP of 2,3-dihydrofuran (**25**) has been achieved using a chromium carbene complex as initiator; eqn. (2). The repeat units are all-HT, with 50% *cis* double

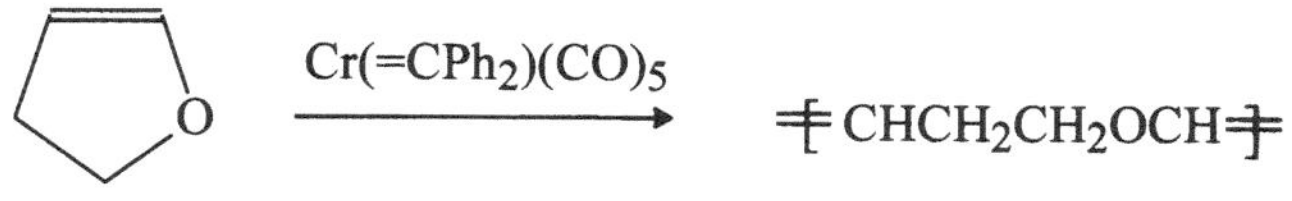

(2)

(**25**)

bonds. The reaction is remarkable in two respects: first, it is a very rare case of metathesis initiated by a chromium compound and, second, it fails with the corresponding tungsten complex. A homologous series of cyclic oligomers is also formed. Various *cis/trans* isomers are present as indicated by multiple GC peaks; after hydrogenation only one GC peak is seen for each ring size (Thu 1981b, 1985; Höcker 1984).

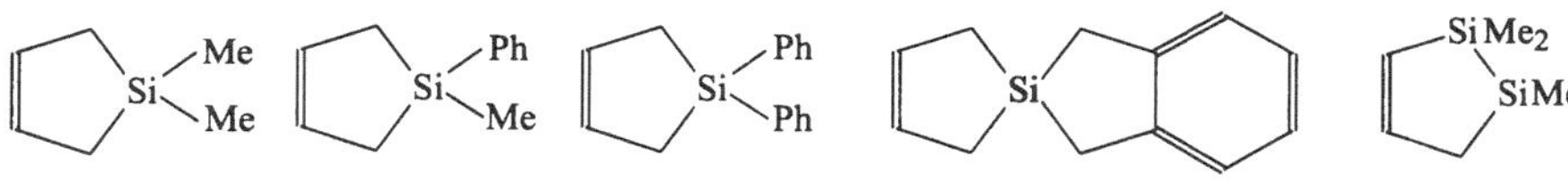

(**26**) (**27**) (**28**) (**29**) (**30**)

Four disubstituted 1-silacyclopent-3-enes (**26**–**29**) have been studied; also (**30**). When polymerized in bulk, using $Mo(=CHCMe_2Ph)(=NAr)[OCMe(CF_3)_2]_2$ as initiator, **27** and **28** give high polymer, ($M_n \approx 40\,000$) (Anhaus 1991, 1993), as does **30** ($M_n \approx 10\,100$) (Sita 1993), but **26** gives low polymer ($M_n \approx 4000$) not only with this initiator but also with the catalyst systems WCl_6/*i*-Bu_3Al/Na_2O_2 (Lammens 1971), $Re_2O_7/Al_2O_3/Bu_4Sn$ (Finkel'shtein 1981), and WCl_6/Me_4Sn (Anhaus 1993). WCl_6/Ph_4Sn is also an effective catalyst for the ROMP of **28** and **29** provided that a small amount of cyclopentene or cyclohexene is present to assist initiation (Stonich 1991b; Wang 1992). WCl_6/Me_4Sn (and WCl_6/Ph_4Sn) gives a high-*cis* polymer of **28**, whereas the Mo carbene complex gives a polymer with only 45% *cis* double bonds. The corresponding polymers of **27** contain 80% and 25% *cis* double bonds, respectively. The former has a ^{29}Si NMR spectrum (in $CDCl_3$) with three signals at δ $-4.31(cc)$, $-4.44(ct)$ and $-5.18(tt)$ in the ratio 72 : 20 : 8, corresponding to a blocky *cis/trans* distribution ($r_t r_c \approx 5.5$); the latter has a random distribution ($r_t r_c \approx 1$). These polymers all have a broad MWD ($M_w/M_n \geq 2$). The polymer of **29** initiated by WCl_6/Ph_4Sn has 84% *cis* double bonds and both the ^{29}Si and benzylic ^{13}C NMR signals are sensitive to *cc*, *ct*, *tt* dyads (Wang 1992).

When **28** is sufficiently dilute, its polymerization is no longer thermodynamically allowed but, in the presence of the Mo carbene initiator, a mixture of cyclic dimers, *cc* (84%), *ct* (14%) and *tt* (2%), is readily formed, in equilibrium with the monomer. The *cc* dimer has been isolated and its crystal structure determined

(Anhaus 1993). These cyclic dimers can also be made by metathesis degradation of the polymers (Section 16.3.4) or by ADMET of diallyldiphenylsilane (Section 8.3).

The ROMP of neat **30** is remarkable in that it proceeds to 96% conversion in 18 h at 25°C to yield an all-HT, 97% *trans* polymer. Dilution of the living polymer with benzene causes neither reversion to monomer nor backbiting to form cyclic oligomers. The driving force for polymerization in this case derives from the relief of torsional strain in the monomer caused by interaction between eclipsed methyl groups on the adjacent silicon atoms (Sita 1993).

12.4 Six-membered rings

12.4.1 Cyclohexene

The failure of cyclohexene to form long-chain polymer by ROMP has already been discussed in Section 11.2.1. However, if a 5 M solution of cyclohexene in toluene is added to a WCl_6/Me_4Sn solution at 25°C and then cooled to −77°C, 12% of the cyclohexene is consumed, and is regenerated on warming to 25°C. If cold wet acetone is added to the reaction mixture at −77°C, the products are found by GC to consist of oligomers containing 2–6 monomer units. The number of peaks indicates that several *cis/trans* isomers of the various oligomers are present, with *cis* double bonds preferred. Exposure of the mixture of oligomers to fresh catalyst brings about reversion to cyclohexene.

If norbornene is added to a mixture of WCl_6/Me_4Sn/cyclohexene (1/2.4/5) in toluene at 25°C, 5% of the cyclohexene is consumed and incorporated into the polymer of norbornene, as shown by the presence of 7.5% 1,6-hexanediol in the products obtained by ozonolysis of this polymer followed by reduction with $LiAlH_4$. If the copolymer is allowed to stand overnight in the presence of the catalyst at 25°C the cyclohexene units are split out, with 99% recovery of the original cyclohexene.

These two pieces of evidence show that cyclohexene can add to metal carbene complexes to some extent, but that, at low temperature, backbiting to give oligomers is preferred to propagation, while at room temperature, the product of addition can be trapped, at least for a time, by reaction with norbornene (Patton, P.A. 1986).

There is also some indirect evidence for the interaction of cyclohexene with catalyst systems. First, the presence of cyclohexene assists the formation of the initiating species in the WCl_6/Ph_4Sn-catalyzed ROMP of **28** (Stonich 1991b). Second, the presence of cyclohexene increases by 30% the rate of ROMP of cycloocta-1,5-diene catalyzed by WCl_6/Me_4Sn at 25°C, without itself being consumed (Patton, P.A. 1987). However, there is no NMR spectroscopic evidence that $W[\overline{=C(CH_2)_3CH_2}](Br)_2(OCH_2CMe_3)_2/GaBr_3$ can interact with cyclohexene, although it does so with cycloheptene and cyclooctene. Ring-strain relief is therefore of some importance in the formation of these cycloalkene adducts (Kress 1992).

The formation of polymers containing $=CH(CH_2)_4CH=$ units is possible through the ROMP of an appropriate cyclic diene, such as cycloocta-1,3-diene, or by a double-bond shift reaction of a polymer such as poly(1-pentenylene). Such units can be eliminated as cyclohexene so long as metathesis activity is present in the system (Klavetter 1988).

12.4.2 Cyclohexadienes

Cyclohexa-1,3-diene is polymerized by WCl_6 but the structure of the polymer is unknown (Laverty 1976a). It is eliminated by backbiting during the ROMP of cycloocta-1,3,5-triene initiated by $W(=CHCMe_3)(=NAr)[OCMe(CF_3)_2]_2$ (Klavetter 1988). Cyclohexa-1,4-diene does not undergo ROMP because it is thermodynamically more stable than its polymer, as shown by its production from methyl linoleate by intramolecular metathesis using WCl_6/Me_4Sn at 80°C as catalyst (Verkuijlen 1974).

12.4.3 Heterocyclic rings

The ROMP of 2,3-dihydropyran, initiated by $Mo(CO)_6/CBr_4/h\nu$, has been reported (Liu 1986). The formation of six-membered silacyclohex-3-ene derivatives by metathesis reactions of appropriate dienes is also known (Ushakov 1989).

12.5 Seven-membered rings

The ROMP of cycloheptene can lead mainly to dimer or to high polymer depending on the conditions, with a temperature-dependent equilibrium monomer concentration (Kress 1995). When a 30% solution in hexane is refluxed in a Soxhlet apparatus through a fixed bed of Re_2O_7/Al_2O_3, preferably pretreated with Me_4Sn, a 68% yield of cyclic dimers is obtained, the proportions being close to the equilibrium values: *tt* 88%, *tc* 9%, *cc* 3% (Warwel 1987b,c). A 95% *cis* polymer can be obtained using $W(=CPh_2)(CO)_5$ as initiator in toluene at 40°C (Katz 1976c), 80% *cis* polymer using $WCl_6/PhC{\equiv}CH$ (Katz 1982) or $W({\equiv}CPh)Cl(CO)_4/O_2$ (Katz 1984), 20–50% *cis* polymer using tungsten or molybdenum carbene complexes (Kress 1992, 1995; Dounis 1995), 35% *cis* polymer using an Ir-TFA complex (Porri 1975), and ~90% *trans* polymer using $MoCl_5/Et_3Al$, etc. (Natta 1966a). WCl_6/Bu_4Sn gives a white tacky polymer (Hein 1973). The ^{13}C NMR line positions for the polymer are given in Table 11.5.

2,3,4,5-Tetrahydrooxepin-2-yl acetate is a seven-membered unsaturated cyclic ether, which undergoes ROMP to give an 80% *cis* polymer (Höcker 1984).

12.6 Eight-membered rings

12.6.1 The cyclooctenes and their derivatives

Cis- and *trans*-cyclooctene, **31** and **33**, respectively, and their derivatives **34–38**, all undergo ROMP (Katz 1980b); also **39** (Hillmyer 1995b, Stumpf 1995), **40** and **41** (Stumpf 1995), **42–44** (Hillmyer 1995b), **45** (Reddy 1983), **46** (Couturier 1993a), **47** (Chung 1992b), **49** (Hillmyer 1995b), **50** and **51** (Cho 1993a,b). Only **32** (Katz 1980b) and **48** (Hillmyer 1995b) fail to polymerize, perhaps owing to unfavourable choice of catalyst and conditions.

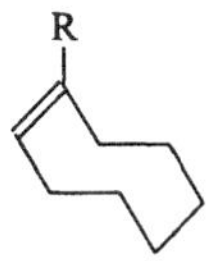

(**31**) R = H
(**32**) R = Me

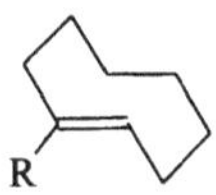

(**33**) R = H
(**34**) R = Me

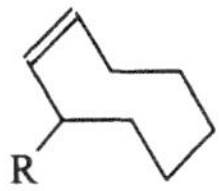

(**35**) R = Me
(**36**) R = Ph

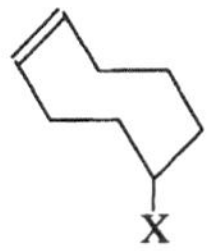

(**37**) X = Me, (**38**) X = Cl, (**39**) X = Br,
(**40**) X = CO_2Me, (**41**) X = CO_2SiMe_3,
(**42**) X = OH, (**43**) X = OAc,
(**44**) X = $O.CO.CMe(=CH_2)$, (**45**) X = I,
(**46**) X = SR', (**47**) X = BEt_2, (**48**) X = CN

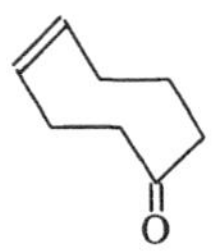

(**49**)

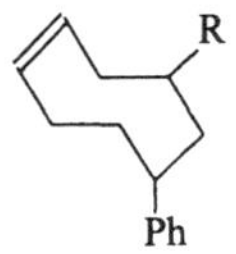

(**50**) R = Me
(**51**) R = Ph

The ROMP of **31** was first observed by Natta (1966a,b). At 25°C the equilibrium concentration of monomer is about 0.002 M, while that of the cyclic oligomers is about 0.25 M (Wasserman 1968; Scott 1969; Höcker 1980). No high polymer is therefore formed if $[M]_o$ is 0.25 M but, starting from 5 M, the product is 96% high polymer and about 4% cyclic oligomers. The proportion of oligomers as a function of $[M]_o$ is shown in Fig. 12.4. High conversion to cyclic oligomers containing a substantial proportion of dimers can be achieved by refluxing pure **31** through alumina-supported Mo, W, or Re catalysts at 140°C (Alkema 1968, 1971; Turner 1968). A 30% yield of cyclic dimers, in their equilibrium proportions (*tt* 16%, *tc* 58%, *cc* 26%), may be obtained by passing a 0.02 M solution of **31** in pentane through Re_2O_7/Al_2O_3 at 15–50°C (Eberle 1986; Warwel 1987b). Under certain conditions, the proportion of dimer initially formed can exceed the ultimate equilibrium value (Höcker 1980). In the early stages of reaction catalyzed by WCl_6/Me_4Sn, the concentration of oligomers is kinetically controlled and is proportional to $(DP)^{-3/2}$, in accordance with the backbiting mechanism. Their final concentration is governed by the ring-chain equilibrium; see Section 11.2.2 (Reif 1984).

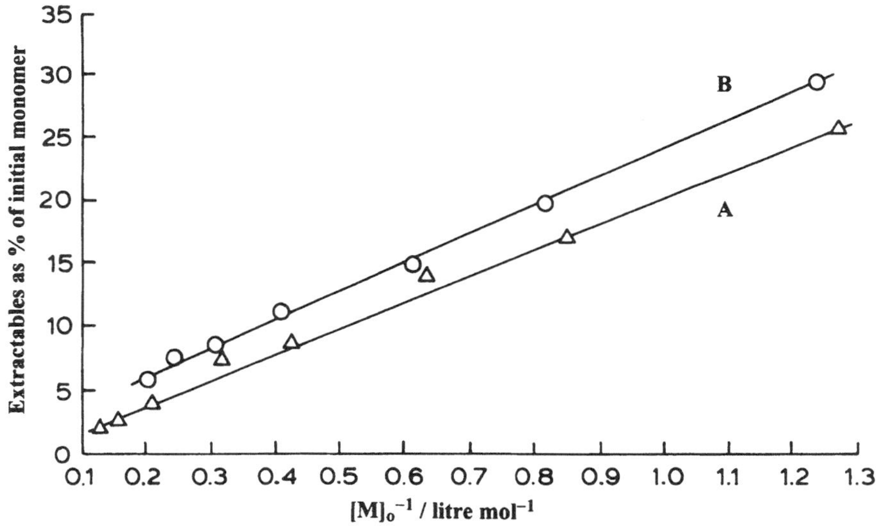

Fig. 12.4 The proportion of extractable cyclic oligomers as a function of the reciprocal initial monomer concentration $[M]_o$ for the ROMP of (A) *cis*-cyclooctene, and (B) *cis,cis*-cycloocta-1,5-diene. Catalyst: $WCl_6/EtAlCl_2/EtOH$ (1/4/1) in benzene (Scott 1969)

The *trans* monomer **33** gives a 43% *cis* polymer very rapidly in the presence of $MoCl_2(PPh_3)_2(NO)_2/EtAlCl_2$ (Larroche 1983) and is polymerizable by $Ru(=CHCH=CPh_2)(Cl)_2(PPh_3)_2$ (Nguyen 1993).

The fractions of *cis* double bonds in the polymers of **31** and **33** are readily determined from their ^{13}C NMR spectra (see Table 11.5). High-*cis* polymers of **31** are formed using the following catalysts: $W(=CPh_2)(CO)_5$ (Katz 1976c); $WF_6/Et_3Al_2Cl_3/CCl_3CH_2OH$ (1/2/1) in cyclohexane at −30°C (Zimmermann 1976a); and $WCl_6/PhC{\equiv}CH$ (Katz 1982). Catalysts based on $WCl_6/i\text{-}Bu_3Al$ or $WCl_6/i\text{-}Bu_2AlCl$, to which a third component such as chloranil or water has been added, all give 75–90% *cis* polymers in the early stages of reaction. These have a random *cis/trans* distribution in contrast to the blocky distributions generally found with high-*cis* polymers of other monomers. Secondary metathesis reactions of the double bonds in the polymer chain cause the *cis* content to fall as the reaction proceeds (Kress 1995), at a rate that is markedly dependent on the proportions and mode of mixing of the catalyst components (Syatkowsky 1979; Denisova 1983a,b). High-*trans* polymers are formed using WCl_6/Et_3Al at 20°C (Natta 1966a). The commercial applications of the high-*trans* polymer are covered in Ch. 17. A great many catalyst systems give polymers of intermediate *cis* content; for example, $MoCl_2(PPh_3)_2(NO)_2/EtAlCl_2$ gives a 67% *cis* polymer (Larroche 1983). Ru-based systems (Demonceau 1992; France 1993a; Stumpf 1995; Herrmann 1996) and those based on carbene complexes of tungsten (Blosch 1991; Kress 1992, 1995; Gamble 1993), molybdenum (Vaughan 1995), and ruthenium (Nguyen 1993; Schwab 1996) can also be used.

The ^{13}C NMR spectrum of the 24% *cis* polymer of **34**, initiated by $W(=CPh_2)(CO)_5$, indicates a perfect HT structure (Lee, S.J. 1976). When the monomer attaches itself to the propagating metal carbene complex, one orientation of the methyl group must be strongly preferred, just as it is in the cross-metathesis or degenerate metathesis of alk-1-enes. In the case of **35** both high polymer and cyclic oligomers (20%) are formed (Calderon 1967c). The IR spectrum of the hydrogenated polymer shows a weak band at 755 cm^{-1} characteristic of the structure $-CH(CH_3)CH_2CH_2CH(CH_3)-$, with an intensity corresponding to 5% HH and, therefore, to 5% TT and 90% HT structures (Dall'Asta 1972a). The methyl group in the 3-position thus still has a significant effect in directing the orientation of the monomer during the propagation reaction, though not as strong as for **34**. The hydrogenated polymer of **35** has quite a high rate of crystallization in spite of the HH,TT irregularity and its atacticity (Gianotti 1971; Dall'Asta 1972b).

The 5-substituted cyclooctenes (**37–47**, **49**) generally give unbiased polymers, the substituent being too far away from the reaction site to influence the direction of addition of monomer. This is particularly clearly seen in the ^{13}C NMR spectrum of the polymer of **49**, made using $Ru(=CHCH=CPh_2)(Cl)_2(PCy_3)_2$ as initiator, the olefinic region consisting of two well-defined symmetrical quartets (1 : 1 : 1 : 1) attributable to *cis* and *trans* olefinic carbons within HH, HT, TH, TT structures (Hillmyer 1995b).

The 5-methyl compound (**37**) polymerizes at much the same rate as **31** in the presence of $WCl_6/Et_3Al_2Cl_3$ in toluene at 20°C, to give a 30% *cis* polymer. The *cis* content is increased to 68% if the reaction is conducted at − 10°C and at low Al/W ratio (Sato 1977a). The polymer of **36** has the expected ^1H NMR spectrum for a ring-opened polymer, with no sign of migration of the phenyl group from the α-position (Calderon 1967c). The 5-chloro compound (**38**) with WCl_6/Et_2AlCl as initiator gives an oily product ($M_n = 800$) with a ring-opened structure, but $WCl_6/EtAlCl_2$ gives a white powder lacking in unsaturation (Sato 1977a).

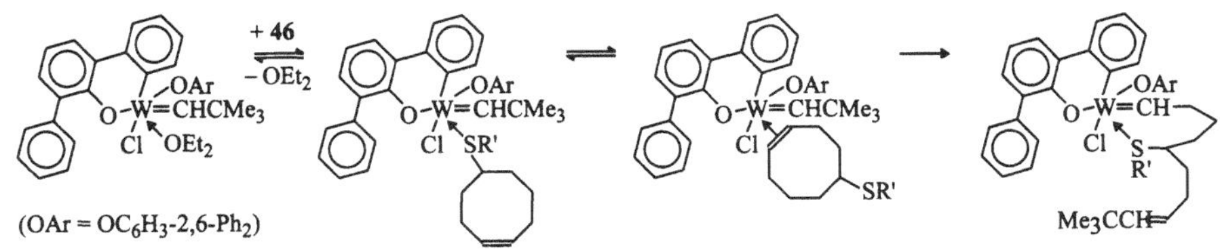

(**52**) (**53**) (**54**) (**55**)

The conversion–time curves for the ROMP of the 5-alkylthiocyclooctenes (**46**) with R′ = Et, Bu, Hex, *c*-Hex, *t*-Bu) initiated by **52** are shown in Fig. 12.5. For R′ = Bu the rate of polymerization is proportional to both monomer and initiator concentrations. The most reactive monomers are those with branched alkyl substituents on the sulfur atom. Thus for R′ = *t*-Bu, reaction is 95% complete in about 10 min; the ROMP of cyclooctene itself is at least five times faster under

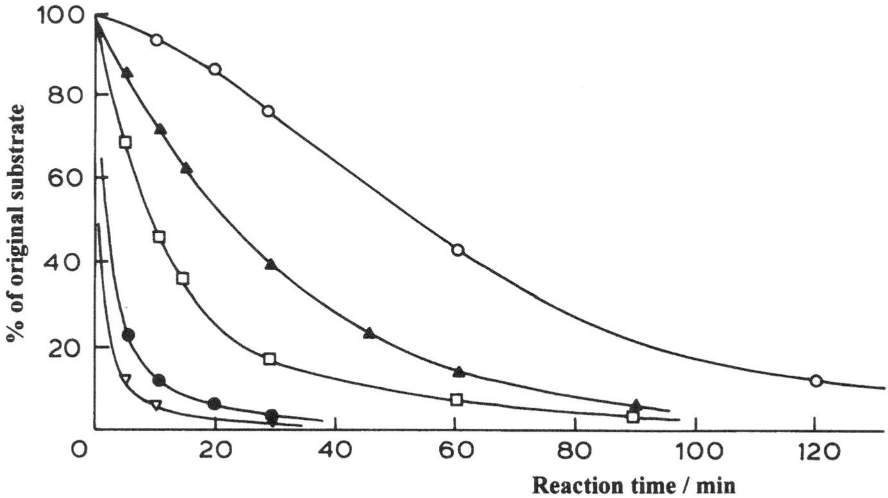

Fig. 12.5 ROMP of 5-alkylthiocyclooctenes **46** (0.5 M) initiated by **52** (0.005 M) in chlorobenzene at 20°C. Curves from top (slowest) to bottom (fastest): alkyl = ethyl, butyl, hexyl, cyclohexyl, *t*-butyl (Couturier 1993a).

the same conditions. These variations in rate are likely to be connected with the strength of coordination of the sulfur atoms in the monomer and/or the propagating species to the tungsten centre, as in **53** and **55**, respectively. Such coordination will be impeded when R′ is *t*-Bu or *c*-Hex, allowing a higher equilibrium concentration of **54**, the precursor to the transition state that leads to addition of monomer (Couturier 1993a,b).

Monomer **47** undergoes very rapid ROMP in the presence of $W(=CHCMe_3)(=NAr)[OCMe(CF_3)_2]_2$ to yield an 80% *trans* polymer **56**, which can then be converted to **57**; or to **58** if the polymer is first reacted with 9-borabicyclononane (9-BBN). The ^{13}C NMR spectra of these polymers show that HH and TT structures are present as well as the HT structures **56–58**. There are two modes of reaction, both for the formation of **56** and for its subsequent reaction with 9-BBN. In the saturated polymer, therefore, the CH(OH) groups may be separated by two or four, as well as three, CH_2 groups, and may have an *m* or *r* relationship with each of the adjacent CH(OH) groups. After acetylation of **58** the polymer has an M_n of 122 000 ($M_w/M_n = 2.8$) (Ramakrishnan 1990; Chung 1991b, 1992b).

The ROMP of **50** and **51**, catalyzed by $WCl_4(OC_6H_3\text{-}Ph_2\text{-}2,6)_2/Et_4Pb$ (1/2) at 15°C, gives high yields of polymers having intermediate *cis* content (Cho, I. 1993a,b).

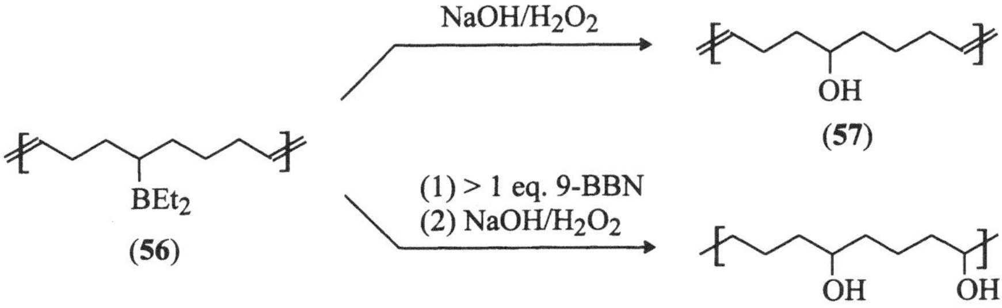

(56) **(57)** **(58)**

12.6.2 Unsubstituted cyclooctapolyenes

All the compounds **59–63** undergo ROMP with the usual catalysts but, if the reactions of **59**, **62** and **63** are carried out in dilute solution, C_6 ring compounds are eliminated by backbiting reactions, while **60** and **61** form cyclic oligomers (see Fig. 12.4). Thus with $W(=CHCMe_3)(=NAr)[OCMe(CF_3)_2]_2$ in dilute solution **59** yields cyclohexene. This suggests that the preferred mode of addition is that to give $[W]=CH(CH_2)_4CH=CHCH=CHCMe_3$ rather than $[W]=CHCH=CH(CH_2)_4CH=CHCMe_3$, since the former can eliminate cyclohexene immediately. With the same catalyst in dilute solution, **62** yields some cyclohexa-1,3-diene and **63** gives a 75% yield of benzene (Klavetter 1988).

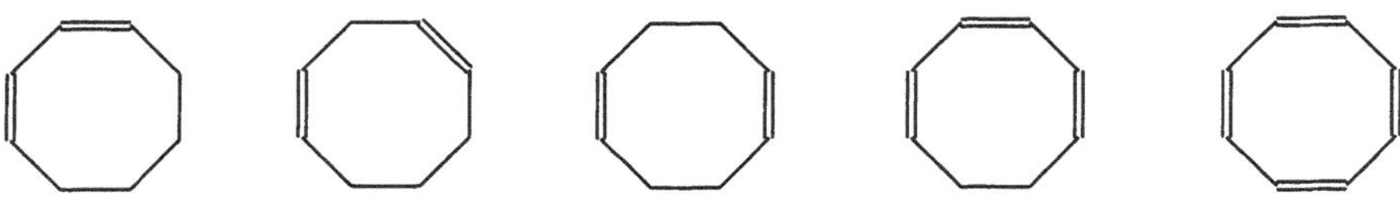

(59) **(60)** **(61)** **(62)** **(63)**

The ROMP of cycloocta-1,3-diene (**59**) was first observed by Korshak (1982) and Tlenkopachev (1983) using $WCl_4[OCH(CH_2Cl)_2]_2/Et_2AlCl$ as catalyst, with or without an ether additive. The polymer chains contain mainly 1,3-diene structures corresponding to HT addition, but 1,3,5-trienes and isolated C=C bonds are also present, indicating some HH and TT addition, i.e. opening of the double bonds in both directions. The polymer is rubbery when first formed but rapidly becomes brittle on exposure to air (Tlenkopachev 1983). **59** can have a strong inhibitory effect on some metathesis reactions but not on others, depending on its ability to compete with the substrate for the vacant site at the metal centre (Zerpner 1986). Sometimes it can influence a reaction without inhibiting it. For example, its presence can increase the *cis* content of the polymer formed by the ROMP of **61** (Zerpner 1987).

The ROMP of cycloocta-1,4-diene (**60**) is catalyzed by $WCl_6/EtAlCl_2/EtOH$ and gives a 45% *cis*, rubbery, polymer in 67% yield. Virtually no cyclohexa-1,4-diene is found among the products. This would have been expected to be formed by elimination from HH structures and the polymer therefore probably has a regular HT structure. Initially the double bonds must be alternately *cis* and *trans* because only one of the two double bonds of the monomer is involved in the propagation reaction, the other retaining its original *cis* configuration. The presence of slightly less than 50% *cis* double bonds indicates that some secondary *cis/trans* isomerization has occurred (Holtrup 1985b, 1986).

We have already made frequent reference to the ROMP of *cis,cis*-cycloocta-1,5-diene (**61**), for example, its cross-metathesis reactions (Section 3.2), the equilibrium between monomer, cyclic oligomers, and polymer (Fig. 3.1 and Table 11.3), the distribution of cyclic oligomers (Figs. 3.2 and 11.2), and the proportion of cyclic oligomers as a function of dilution (Fig. 12.4). Most catalyst systems have been based on compounds of tungsten; those based on compounds of molybdenum (Minchak 1982; Dounis 1995), rhenium (Kumobayashi 1976; Sato 1977b; Saito 1979), and ruthenium (Nguyen 1993; Schwab 1996) are also effective. $MoCl_5$/Et_2AlI in benzene at 25°C gives a 52% yield of 99% *cis* polymer after 1 h. Most catalysts give initially a polymer of at least 80% *cis* content, since one of the *cis* double bonds is preformed. The *cis* content can be raised to 90% or more by: (i) working at low temperature, for example, $H_2WO_4/AlCl_3$ at −20°C gives a polymer with 93% *cis* double bonds (De Figueiredo 1979); and (ii) the use of phenolic or silane additives (Castner 1977b; Bell, A.J. 1976), or particular cocatalysts such as $PhCHN_2$ (Dolgoplosk 1974), polygermanes (Kolesnikov 1982), cyclodienes (Küpper 1979; Zerpner 1987), or acetylenes (Makovetsky 1992a). With some catalysts, such as WCl_6/R_4Sn (R = Ph, Bu), the initial *cis* content is high (85%) and remains so, with no sign of the development of *tt* dyads with increasing reaction time. With others, secondary *cis/trans* isomerization leading to *tt* dyads may occur, also acid-catalyzed double-bond shift reactions. Thus for catalysis by $WCl_6/EtAlCl_2$, three different régimes may be distinguished: (i) with Al/W = 4, and adding the monomer last, the reaction is moderately fast (34% yield in 40 s) giving polymer with 80% *cis* content and no sign of isomerization reactions; (ii) with Al/W = 10, and again adding the monomer last, the reaction is faster (82% yield in 5 s), but the distribution of *cis* and *trans* double bonds has already become random; (iii) with Al/W = 20, and $EtAlCl_2$ added last, the reaction is fast (85% yield in 30 s) giving polymer in which both *cis/trans* isomerization and double-bond shift reactions have occurred, with [$=CH(CH_2)_nCH=$] units, $n = 1$, 2, 3, formed in the ratio 23 : 54 : 23 (Ivin 1982a). With $WCl_6/(i\text{-}Bu_2Al)_2O$ as catalyst the newly formed *cis* and *trans* double bonds are randomly disposed; and the activation energy for the formation of a *trans* double bond is 31 kJ mol^{-1} higher than for the formation of a *cis* double bond (Syatkowsky 1981).

The ROMP of cyclooctatetraene (**63**) to give polyacetylene was first reported by Korshak (1985) and Tlenkopachev (1986) using $W[OCH(CH_2Cl)_2]_nCl_{6-n}$/$Et_2AlCl$ ($n = 2$ or 3) as catalyst at 20°C. When the reaction is conducted in toluene the yield of black insoluble polymer is only 6%, but if the monomer is condensed on to a solid layer of catalyst a yield of up to 40% polymer is obtained. In the former case there is a much greater tendency towards formation of oligomers, amongst which the cyclic products of backbiting reactions, $(CH{=}CH)_n$, $n=5$–8, can be identified by MS. The nature of the polyacetylene formed by the second method depends on the initial Al/W ratio. When Al/W = 1 the polymer is formed as a blue–black film containing 84% *cis* double bonds, but when Al/W = 2 the film is golden and contains only 39% *cis* double bonds. A 50% yield of polymer can also be obtained from neat monomer with WCl_6/BuC≡CH as catalyst (Makovetsky 1992a). Better control over this reaction can be achieved using $W(=CHCMe_3)(=NAr)[OCMe(CF_3)_2]_2$ as catalyst (Klavetter 1988). Dissolution of catalyst in neat **63** produces, within a few seconds at room temperature, a high-quality lustrous silver film with smooth surface morphology. When first prepared, its CP-MAS ^{13}C NMR spectrum shows two olefinic peaks: a stronger peak at 126.4 ppm (*cis*) and a weaker peak at 132.2 ppm (*trans*). On heating the sample, the spectrum changes, giving a main peak at 135.9 ppm (*trans*) and an upfield shoulder (*cis*). The chemical shift of the *trans* olefinic carbons are known to be sensitive to the configuration of the surrounding double bonds. Heating thus induces *cis/trans* isomerization to produce long segments of *trans-transoid* structure within the polymer chains. When doped by exposure to iodine the polymer acquires a conductivity greater than 100 ohm^{-1} cm^{-1}. Linear copolymers of varying conjugation length can be produced by the inclusion of a second monomer such as **61** during the preparation of the film, allowing a wide range of conductivities in the doped copolymers to be obtained (Klavetter 1989).

12.6.3 Substituted cycloocta-1,5-dienes

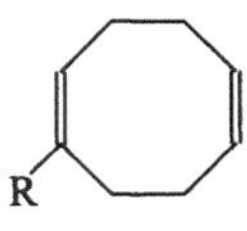

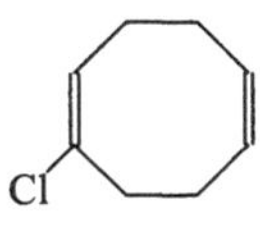

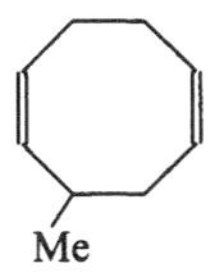

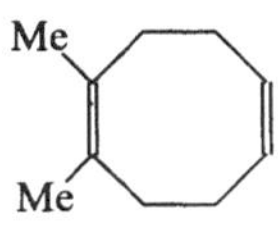

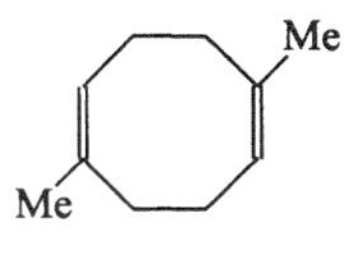

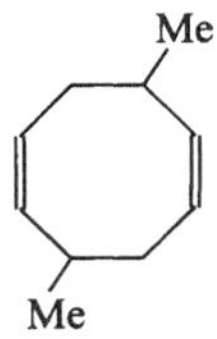

(**64**) R = Me (**66**) (**67**) (**68**) (**69**) (**70**)
(**65**) R = Et

Of the compounds **64–70**, all except **69** are polymerizable by ring-opening (Küpper 1974; Ofstead 1969; Syatkowsky 1981; Red'kina 1981; Khachaturov 1982). Ring-opening by metathesis occurs only at the unsubstituted double bonds in **64**, **65**, **66** and **68**. A good example is the polymer of **64**, made using $WCl_6/(i\text{-}Bu_2Al)_2O$ as catalyst. It has a structure corresponding to an alternating polymer of 1,4-isoprene units, containing *cis* double bonds as in the monomer, and 1,4-butadiene units in which the double bonds (60–75% *cis*) are formed in the propagation reaction. The upfield part of the ^{13}C NMR spectrum of this polymer is shown in Fig. 12.6. The signal for C-7 adjacent to a *cis* double bond is 5.25 ppm upfield from that for C-7 adjacent to a *trans* double bond, as expected; likewise for the two C-4 signals. C-7 is on the head side of the isoprene unit and is subject to an upfield γ-substitution

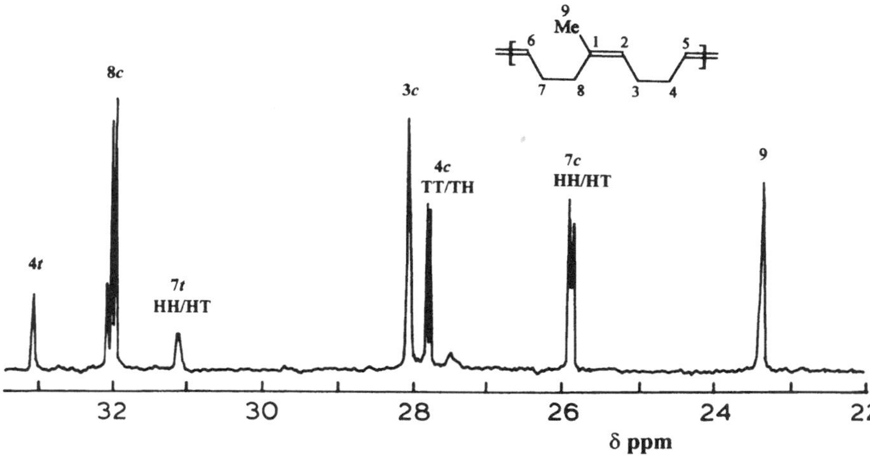

Fig. 12.6 Upfield region of the ^{13}C NMR spectrum of a ring-opened polymer of **64**; the butadiene units contain approximately 75% *cis* double bonds (Khachaturov 1982).

effect from the methyl group attached to C-1, whereas C-4 is on the tail side of the isoprene unit and suffers only a small δ-substitution effect. This is the basis for deciding which pair of signals belong to C-7 and which to C-4; the observed value of $(\gamma - \delta)$ is -1.95 ppm. The C-7 *cis*, C-4 *cis*, C-7 *trans*, and C-8 *cis* signals show further fine structure, each being split into two equal peaks owing to the influence of the next nearest isoprene unit, which may have either the same orientation as its neighbour (HT or TH) or the opposite orientation (HH or TT). The equal intensities show that the isoprene units are randomly oriented, which means that the methyl group in **64** has no influence on the direction of opening of the unsubstituted double bond during the propagation reaction.

The polymer of **68** exhibits none of the IR bands characteristic of 1,4-isoprene units, which would have been formed had both double bonds taken part in ROMP. In the ROMP of **65**, cyclic oligomers are formed, which are made up of whole units of monomer with little sign of the sesqui-oligomers such as are formed from **61**. Likewise in the ROMP of **64**, only traces of the sesqui-oligomers are formed. However, with increasing conversion the proportion of sesqui-oligomers gradually increases, indicating that the substituted double bonds are not entirely unreactive (Ofstead 1969).

12.6.4 Monosubstituted cyclooctatetraenes (RCOT)

Ring-opened polymers have been made from monomers having various substituents R: Br (Klavetter 1988), Me, *i*-Pr, Bu, *s*-Bu, *t*-Bu, neopentyl, 2-ethylhexyl, octyl, octadecyl, cyclopropyl, cyclopentyl, phenyl, methoxy, and *t*-butoxy (Gorman 1989, 1990, 1991a,b, 1993), $SiMe_3$ (Ginsburg 1989; Sailor 1990), and the chiral substituents CHMeCHMe(OMe), CHMeCHMe($OSiMe_2CMe_3$), and

$CH_2CHMe(OSiMe_2CMe_3)$ (Moore 1991). The initiators $W(=CHCMe_3)(=NC_6H_3$-*i*-Pr_2-2,6)$[OCMe(CF_3)_2]_2$ and $W[=CHCC_6H_4$-(OMe)-2]$(=NC_6H_5)[OCMe(CF_3)_2]_2$ (THF) are both very effective, but even with pure monomer some backbiting reaction occurs in competition with propagation, giving rise to 7–16% C_6H_5R. No benzene is detected, showing that ring-opening does not occur at the substituted double bond. Ring-opening may, however, occur at any of the other three double bonds, so that the product is a polyacetylene having substituents placed on average on every fourth or fifth double bond (elimination of C_6H_5R will tend to decrease the frequency), but never on adjacent double bonds, unlike the polymers of substituted acetylenes (see Ch. 10).

The initially formed polymers have a high *cis* content and, except when R is Me, are generally soluble in tetrahydrofuran, chloroform and benzene ($M_n \sim 10^4$–10^6). On standing at room temperature, the solutions slowly change colour, and a new absorption maximum appears at longer wavelengths. This change is fastest for the polymers with straight-chain substituents and slowest for those with bulky substituents. It is accelerated by exposure to light (Fig. 12.7) and is the result of *cis/trans* isomerization. Long sequences of *trans* double bonds allow a much greater degree of conjugation, giving rise to a low-energy $\pi \rightarrow \pi^*$ electronic absorption. For monomers with straight-chain or alkoxide substituents, the predominantly *trans* polymer comes out of solution as it is formed. In contrast, polymers containing a secondary or tertiary substituent adjacent to the backbone remain soluble in the mainly-*trans* form. Effective conjugation lengths of up to 30 double bonds have been observed for these soluble polymers.

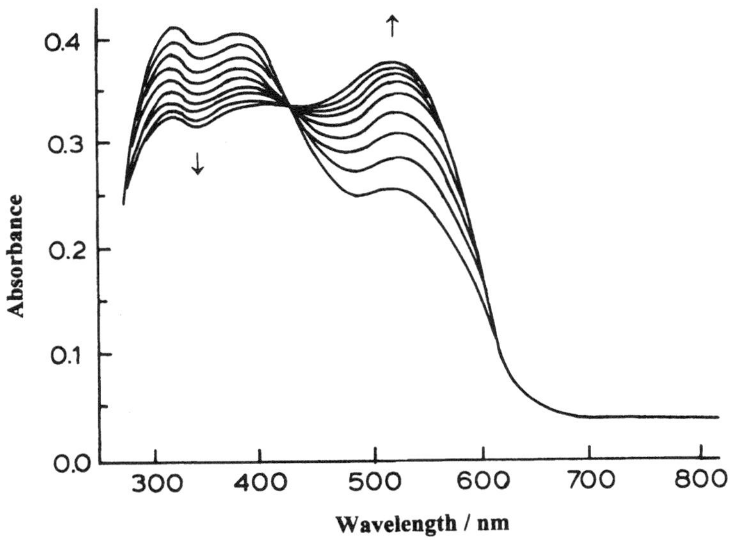

Fig. 12.7 Absorption spectra of polymer of trimethylsilylcyclooctatetraene in CCl_4 (10^{-6} M) obtained between eight periods of photolysis (10 s each) (Ginsburg 1989).

For the polymer of *s*-BuCOT in benzene, *cis/trans* isomerization can be monitored by the change of absorbance at 560 nm. The reaction is first order, with a half-life of about 27 min at 65°C and an activation energy of 89 kJ mol^{-1}. However, the isomerization of the trisubstituted *cis* double bond proceeds only part way. Its reaction can be followed separately by means of ^{1}H NMR spectra, the methine proton signal of the side groups being sensitive not only to the configuration of the nearest double bond but also that of the next nearest double bond: *tt* 2.75, *tc* 2.51, *cc* 2.12 ppm. At equilibrium the preference for the *cis* configuration means that the proportion of long *trans* sequences is small. During the isomerization process, the absorption spectra exhibit an isosbestic point at 400 nm, which is consistent with a mechanism involving multiple isomerization of the *cis* double bonds in one chain by a cooperative motion.

When R = *t*-Bu, the polymer is freely soluble and yellow–orange in colour ($\lambda_{max} = 432$ nm after isomerization); it also remains an insulator in the presence of iodine. This indicates a very low effective conjugation length even after isomerization. This is attributed to a twist in the polymer chains, caused by the bulky *t*-Bu groups. This effect is much less for the polymers with other R groups and their films can be made conducting by doping with iodine (Gorman 1989). For the polymers bearing chiral substituents, the backbone $\pi \rightarrow \pi^*$ transition shows substantial circular dichroism, the magnitude of which is characteristic of a disymmetric chromophore. The chiral side groups thus twist the main chain predominantly in one sense rather than just perturbing that chromophore electronically (Moore 1991).

The possible application of these materials to form surface barrier solar cells and to make conductor/insulator/conductor sandwiches by sequential polymerization of different monomers has been explored (Sailor 1990; Jozefiak 1991; Gorman 1991b).

12.7 Nine-membered rings

Apart from generalized patent claims (e.g. Natta 1966b), there is one brief reference to the ROMP of *cis*-cyclononene using $WCl_6/LiAlH_4$ as catalyst, but the reaction is not clean (Rossi 1976a). *Cis,cis*-cyclonona-1,5-diene also polymerizes with this catalyst to give a highly viscous solution, which later becomes more fluid (Rossi 1976b,c). The ^{1}H NMR spectrum of the polymer ($M_n = 2200$) shows that it contains olefinic : allylic : other protons in the ratio 4 : 8.4 : 3.5 instead of the theoretical ratio 4 : 8 : 2 for a ring-opened polymer. This indicates the formation of a small proportion of cyclic saturated structures by polymerization through the double bond. One of the by-products is cyclopentene, formed by a backbiting reaction of the propagating species $[W]{=}CH(CH_2)_3CH{=}CH(CH_2)_2CH{=}CHR$.

12.8 Ten-membered rings

cis-Cyclodecene, with $WOCl_4/Et_2AlCl$ or WCl_6/Et_2AlCl (1/2.5–5) as catalyst at 0–20°C, gives a 75–85% *trans* polymer which melts around 62°C. Its highly crystalline monoclinic form is converted into a triclinic form on stretching (Dall'Asta 1968b). Metal carbene complexes also give high-*trans* polymers (Dounis 1995).

cis,trans-Cyclodeca-1,5-diene is polymerized by WCl_6 in conjunction with $EtAlCl_2$, Et_2AlCl, or Et_3Al, the activity decreasing in that order. $MoCl_5$-based catalysts are much less effective. The product is an oil or elastomer. The ^1H NMR spectrum of the polymer formed after a short time shows olefinic : allylic : other protons in the correct proportions of 4 : 8 : 4 (Furukawa 1973). However, after a longer time, substantial amounts of cyclohexene appear in the products and the *trans* content of the polymer increases from 66% to 90% (Hocks 1975b). At the same time the spectrum of the polymer tends towards that of 1,4-polybutadiene. As discussed in Section 11.2.1, the ready elimination of cyclohexene is to be expected. *cis,trans*-Cyclodeca-1,5-diene also acts as a cocatalyst with WCl_6 for the polymerization of various cycloalkenes (Küpper 1979). Methyl (*cis,trans* or *trans,cis*)-3,7-cyclodecadienecarboxylate has been successfully polymerized by ring-opening using $WCl_4(OC_6H_3\text{-}Ph_2\text{-}2,6)_2/Et_4Pb$ as catalyst at 60°C. ^{13}C NMR spectra show that the monomer consists of only one of these isomers and that the polymer has a very regular structure (Cho, I. 1991).

The failure of *cis,cis*-cyclodeca-1,6-diene to polymerize has been discussed earlier (Section 12.3.1). In conjunction with WCl_6 it is said to catalyze the polymerization of *cis,cis*-cycloocta-1,5-diene and of cyclododecene at 20°C (Küpper 1979).

12.9 Twelve-membered and other rings

12.9.1 Cyclododecene and derivatives: catenanes

The larger the ring size in a cyclic monoalkene the greater is the proportion of *trans* isomer in the *cis/trans* equilibrium monomer mixture. When cyclododecene containing 32% of *trans* isomer is treated with $WCl_6/EtAlCl_2/EtOH$ the *cis* isomer undergoes ROMP proportionately faster than the *trans* isomer, but when treated with WCl_6/Me_4Sn the reverse is the case (Höcker 1980). When $WCl_6/i\text{-}Bu_2AlCl$ is used as catalyst, the residual monomer retains its original composition (32% *trans*), while the polymer formed contains 70% *trans* double bonds (Vardanyan 1972); also see Ceausescu (1988b), Coca (1994). Metal carbene complexes give high-*trans* polymers (Dounis 1995).

Both cyclic oligomers ($n = 2$–14) and high polymers are produced from cyclododecene (Fig. 11.2). A rather remarkable phenomenon is observed when

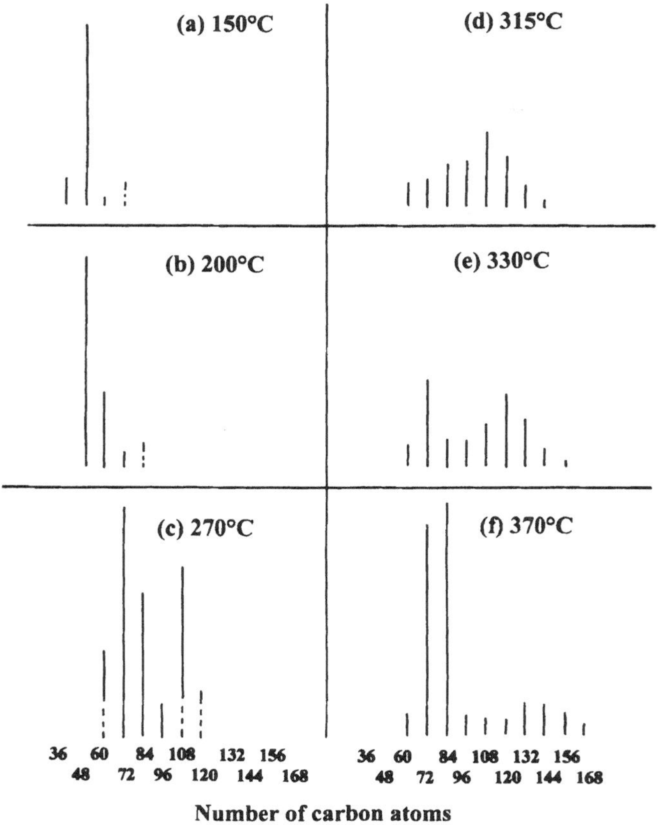

Fig. 12.8 Relative intensities of molecular ions from mass spectra of cyclic oligomers ($n = 3$–14) of cyclododecene, using sample temperatures from 150° to 370°C. Peaks with broken lines in (a), (b), and (c) have been amplified by a factor of 30. Allowance has been made for satellite peaks. The relative intensities at different temperatures are not correlated. Some of the lower members ($n = 2$, 3) of the series were removed before analysis (Ben-Efraim 1970).

the cyclic oligomers are submitted to mass spectral analysis (see Fig. 12.8). If the sample is held at 150°C the main component reaching the mass spectrometer is the tetramer (C_{48}). As the temperature is raised to 315°C progressively heavier molecules (C_{60}–C_{144}) reach the ionization chamber and the maximum intensity moves towards higher mass numbers. However, at 330° and 370°C there is a marked jump in the relative intensity of the hexamer (C_{72}) and heptamer (C_{84}) peaks (Ben-Efraim 1970). Similar observations have been made on the fully hydrogenated cyclic oligomers (Wolovsky 1970). It is especially significant that at the higher temperatures no monomeric species (C_{12}) is detected.

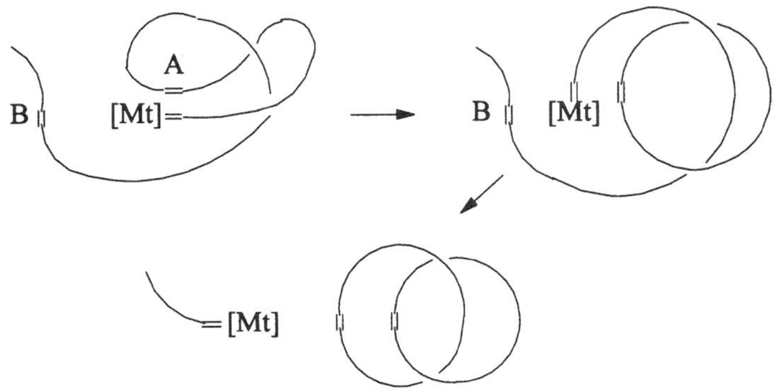

(3)

These observations may be interpreted in terms of the formation of catenanes during the original polymerization reaction. Such molecules consist of interlocking but chemically unconnected rings. Catenation may occur by successive intramolecular metathesis reactions at bonds A and B in sequence (3); see also Gruter (1996). At 330–370°C one of a pair of interlocked rings may break, allowing the two parts to enter the ionization chamber separately, so accounting for the increased yield of C_{72} and C_{84} peaks at these temperatures.

Apart from the catalyst systems already mentioned, the following have also been used: WCl_6/Et_2AlCl and the less active $MoCl_5/Et_3Al$, both of which give high-*trans* polymers (Natta 1966a). WCl_6/cyclic dienes (Küpper 1979) and $(\pi\text{-}C_4H_7)_4W/AlBr_3$ (Kormer 1972) give polymers of intermediate *cis* content.

3-Methyl-*trans*-cyclododecene has also been polymerized; the IR spectrum of the hydrogenated polymer indicates that it contains about 5% each of HH and TT structures (Dall'Asta 1972a).

12.9.2 Cyclopentadecene

A *cis/trans* (18 : 82) mixture of cyclopentadecenes undergoes ROMP. As with cyclododecene the fractional conversion of the *cis* isomer is faster than that of the *trans* isomer when $WCl_6/EtAlCl_2/EtOH$ is used as catalyst, and the reverse with WCl_6/Me_4Sn. The proportions of cyclic oligomers ($n = 2$–7) fall off with n in the expected manner (Fig. 11.2) (Höcker 1980).

12.9.3 Paracyclophane and ferrocenophane derivatives

[2.2]Paracyclophan-1-ene (**71**) can be regarded as a 12-membered ring but its rigid structure is akin to that of cyclobutene. ROMP of **71** is initiated by $Mo(=CHCMe_2Ph)(=NC_6H_3\text{-}i\text{-}Pr_2\text{-}2,6)[OCMe(CF_3)_2]_2$ in toluene, giving a living polymer with a carbene proton singlet at 12.84 ppm, the intensity of which increases as the reaction proceeds slowly to completion (18 h). The polymer (**72**) has 98% *cis* double bonds, but on irradiation or exposure to catalytic amounts of iodine, the double bonds undergo *cis/trans* isomerization and the polymer comes out of solution. Many other Mo- and W-based catalysts are also effective but give insoluble polymer, presumably because of a high *trans* content.

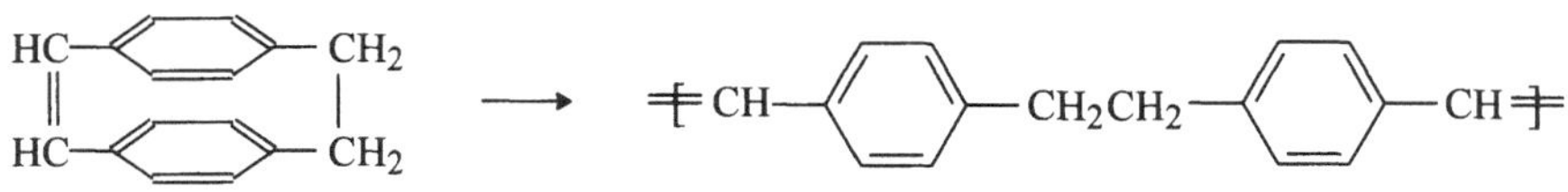

(**71**) (**72**)

A solution of the high-*cis* polymer fluoresces when excited by irradiation at 330 nm. The initial spectrum displays a weak emission at about 370 nm attributed to 2% of *trans*-stilbene segments originally present in the chains. With continued irradiation (2 min) there is an increase in the emission from this band as more *trans* double bonds are generated, and a new emission appears at 445–500 nm. After further irradiation, this intense red-shifted luminescence becomes the predominant feature before the polymer precipitates. In a statistical copolymer of 9% of **71** and 91% of norbornene, the units of **72** are isolated between norbornene units, and the fluorescence spectrum is confined to the shorter wavelength region with a maximum around 360 nm (Miao 1994a).

Similar results have been obtained for a derivative of **71** in which one of the CH_2 hydrogens is substituted by $OSiMe_2CMe_3$. The polymer is unbiased but differs from that made from **71** in that, after isomerization to a high-*trans* polymer, it remains soluble in organic solvents. The silyl group in the polymer can be removed by treatment with Bu_4NF to give the hydroxy analogue, which can then be dehydrated thermally at 105°C or catalytically (HCl) at 25°C to give poly(*p*-phenylenevinylene) (Miao 1994b, 1995; Bazan 1996a).

The ROMP of [2.2]paracyclophane-1,9-diene (**73**) yields poly(*p*-phenylene-vinylene) (**74**) as an insoluble yellow fluorescent powder. Soluble copolymers can be made with an excess of cyclopentene (Thorn-Csányi 1992b), cycloocta-1,5-diene (Thorn-Csányi 1993a), or cyclooctene (Thorn-Csányi 1994c). The UV–visible absorption spectra of the copolymers with cyclooctene show separate peaks for sequences of one, two, and three *p*-phenylene-vinylene units at 290, 345, and about 390 nm, respectively, with a Bernoullian distribution. The formation of the odd members of this series evidently involves dissection of the two halves of the original monomer units by secondary metathesis reactions.

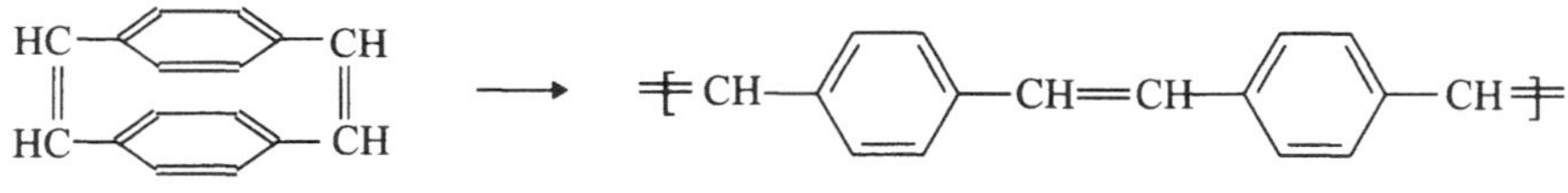

(**73**) (**74**)

Attempts to bring about the ROMP of ferrocenophanes, in which the two cyclopentadienyl rings are linked by a divinylene group, have met with limited success. Soluble oligomers ($M_n = 1700$) are obtained using a tungsten carbene complex as catalyst. Soluble polymers, probably of higher MW, can be obtained by placing a methoxy group on the carbon adjacent to a Cp ring or by copolymerizing with *s*-butylcyclooctatetraene (Stanton 1995).

12.9.4 Higher cyclic polyenes

Table 12.3 summarizes some of the higher polyenes that will undergo ROMP. NMR spectra show that the reaction may be carried out cleanly without the intervention of transannular reactions or double-bond shift reactions. The proportions of linear polymer and extractable cyclic oligomers depend on the initial concentration of monomer. The distribution of cyclic oligomers is essentially the same as that obtained by the polymerization of the parent cyclic mono-ene. The ring strain in monomers containing 20 or more carbon atoms is low; the free energy change for the ROMP of such monomers is therefore close to zero. There is some evidence that once large rings are formed they remain stable (Holtrup 1985a).

In competition experiments, using Re_2O_7 as catalyst, CDT is considerably less reactive than CHT, CEP or cycloocta-1,5-diene (COD) (Saito 1979). It appears that, like *cis,cis*-cyclodeca-1,6-diene (see Section 12.8), this particular ring is more stable than the others. Yet it can act as a cocatalyst, when used in small quantities with WCl_6, for the ROMP of COD (Küpper 1979).

Table 12.3 ROMP of cyclic trienes, tetraenes, and pentaenes (obtained as cyclic oligomers by the ROMP of cyclooctadiene or cyclooctene)

Ring monomer	Catalyst system	Solvent [a]/ temperature (°C)	Reference
$[CH{=}CH(CH_2)_2]_3$ [b]	$WCl_6/EtAlCl_2$	B/50°	Calderon 1967c
[b]	$WCl_6/EtAlCl_2/EtOH$	B/20°	Scott 1969
[c]	$WCl_6/AlBr_3$	N/20°	Marshall 1969
[c]	$(\pi\text{-}C_3H_5)_4W/Al_2O_3\text{-}SiO_2$	T/50°	Oreshkin 1971
	$(\pi\text{-}C_3H_5)_4W/CCl_3COOH$	T/50°	Oreshkin 1971
[c]	$WCl_6/EtAlCl_2/PhOH$		Khodzhemirov 1976b
	Re_2O_7/Al_2O_3	V/40°	Saito 1979
$[CH{=}CH(CH_2)_2]_n$	$W[{=}C(OMe)Ph](CO)_4/PPh_3/TiCl_4$	C/30°	Chauvin 1977, 1978
$[CH{=}CH(CH_2)_2]_4$ [d]	Re_2O_7/Al_2O_3	V/40°	Saito 1979
$[CH{=}CH(CH_2)_2]_5$ [e]	Re_2O_7/Al_2O_3	V/40°	Saito 1979
$[CH{=}CH(CH_2)_6]_3$ [f]	$WCl_6/EtAlCl_2$	B/25°	Ofstead 1972; Höcker 1977
$[CH{=}CH(CH_2)_6]_n$ [g]	$Re_2O_7/Al_2O_3/Me_4Sn$	P/20°	Eberle 1990

[a] B = benzene; T = toluene; C = chlorobenzene; P = pentane; N = none; V = various. [b] 40/60 (*cis, trans,trans*)/(*trans,trans,trans*) cyclododecatriene (CDT). [c] *cis,trans,trans*-CDT. [d] Cyclohexadecatetraene (CHT). [e] Cycloeicosapentaene (CEP). [f] Cyclotetraeicosatriene (CTT). [g] $n \geq 3$. Results in redistribution of cyclic oligomers to give 39.5 wt % for $n = 2$ and 60.5 wt % for $n \geq 3$.

9-Phenyl-1,5-cyclododecadiene undergoes ROMP in the presence of $WCl_4(OC_6H_3\text{-}Cl_2\text{-}2,6)_2/Et_4Pb$ in PhCl at 80°C; 98% yield in 2 h (Cho, I. 1995).

12.9.5 7-Hexadecen-16-olide (ambrettolide)

Ambrettolide is an unsaturated cyclic lactone, which readily undergoes ROMP to an unsaturated linear polyester ($M_n = 95\,000$) in the presence of WCl_6/Me_4Sn (Ast 1976b, 1977a).

13

Polycyclic Alkenes

13.1 Introduction

Various aspects of the ROMP of norbornene and its derivatives have been covered in Ch. 1, 3, 4, 10, 11, and 12. Here we shall give a general survey of the types of polycyclic alkene that undergo ROMP, taken in order of ring size of the unsaturated ring; see also Feast (1995).

13.2 Monomers containing a fused cyclobutene ring

The ROMP of the following compounds has been reported: **1** (Ivin 1982a; Wu 1993); **2** (Dall'Asta 1968d); **3** (Dall'Asta 1968d; Thu 1981a); **4** (Coleman 1988); **5** (Edwards 1984; Park, L.Y. 1991); **6** (Feast 1987; Park, L.Y. 1991); **7** (Edwards 1984; Leising 1984; Knoll 1988, 1989; Schrock 1988a,c; Allen 1989; Park, L.Y. 1991, 1992; Dounis 1996b); **8** (Feast 1987); **9** (Feast 1985c; Jones 1991); **10** (Edwards 1984; Fischer, W. 1993); **11** (Edwards 1984); **11a,b,c** (Perrott 1995). In all cases the reactive double bond is that in the C_4 ring. The repeat units in the polymer have an *erythro* structure corresponding to the *cis* relationship of the bonds which attach the cyclobutene ring to the rest of the ring system in the monomer (M).

When $Ru(=CHCH=CPh_2)(Cl)_2(PPh_3)_2$ in CH_2Cl_2 is added to 20 equivalents of **1** at 40°C, all the initiator I is consumed and ROMP proceeds at a rate proportional to both [I] and [M], with an apparent k_p of 0.183 M^{-1} min^{-1}. The ^{1}H NMR spectrum shows the presence of three propagating species in equilibrium, one of which is probably the main propagating species. One has two PPh_3 ligands, while the other two are presumed to have one such ligand; the ^{31}P NMR spectrum shows two signals (Wu 1993). This system exhibits all the characteristics of a living polymerization and the polymer has 58% *cis* double bonds; cf. 62% with WCl_6/$EtAlCl_2$/EtOH as catalyst (Ivin 1982a). On adding the ruthenium carbene initiator to a mixture of norbornene and **1**, all the **1** reacts first and then the norbornene so that a block copolymer is formed. The same result may be achieved by adding norbornene to the system after all the **1** has been consumed (Wu 1993). In the ROMP of **2**, the C_4 ring opens in preference to the C_5 ring, though little is known about the detailed structure of the polymer (Dall'Asta 1968d). Clean ROMP of **3** can be achieved using the catalysts WCl_6/Me_4Sn at 20°C or $W(=CPh_2)(CO)_5$ at 50°C; no oligomers are formed even at low [M] (Thu 1981a).

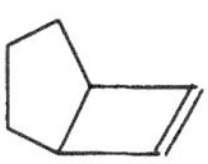

(**1**)

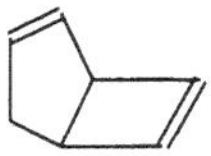

(**2**)

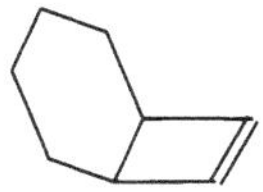

(**3**)

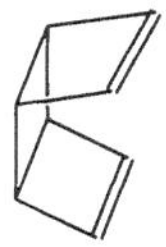

(**4**)

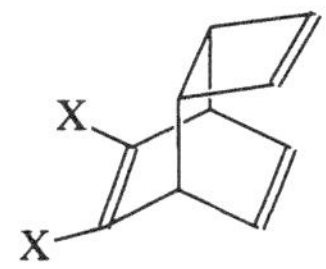

(**5**) X = H
(**6**) X = COOMe
(**7**) X = CF_3

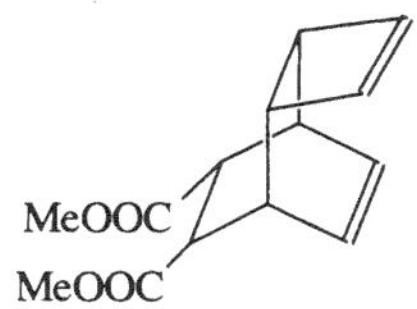

(**8**)

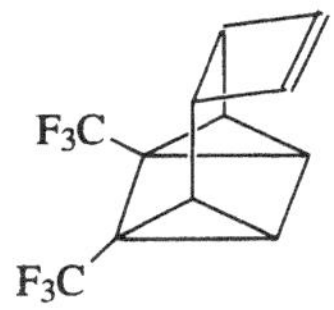

(**9**)

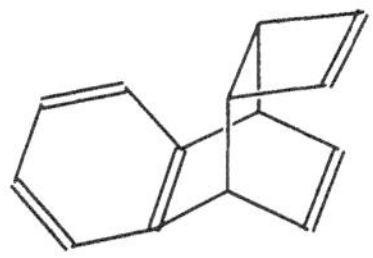

(**10**)

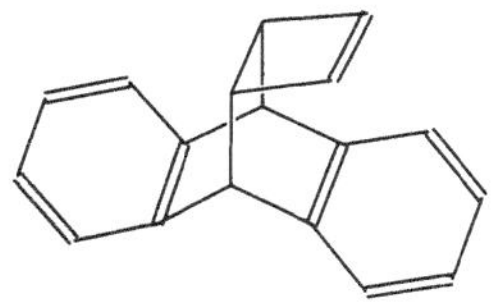

(**11**)

(**11a**)

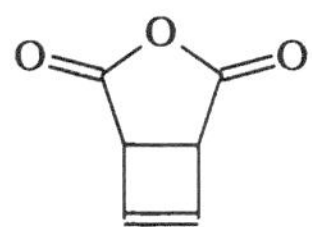

(**11b**)

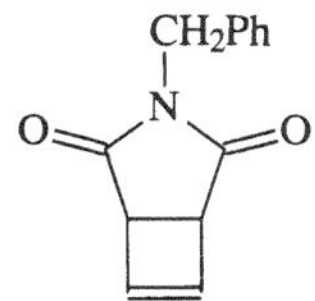

(**11c**)

Monomers **5–7**, **9–11** give ring-opened polymers, which undergo a retro-Diels–Alder reaction, either at room temperature or on heating to 120°C, with the elimination of benzene, naphthalene, anthracene, dimethyl phthalate, and 1,2-bis(trifluoromethyl)-benzene from the polymers of **5**, **10**, **11**, **6**, **7** (and **9**), respectively, and the formation of polyacetylene, as illustrated for **7** in sequence (1). The elimination reaction is accompanied by a change of colour from yellow to deep red as longer polyenes are formed. The polymer **7P** is moderately stable at 20°C with a half-life of 20 h (Edwards 1984). This is long enough for films to be made and stressed uniaxially, and then converted to highly oriented non-fibrous crystalline films of polyacetylene (Leising 1984). The monomer **9** is readily made by photoisomerization of **7**. The polymer of **9** is more stable than that of **7**, but its conversion to polyacetylene is much more exothermic and not so easily controlled (Feast 1985c; Jones 1991). The double bonds B formed in the retro-Diels–Alder reaction are initially *cis*, as may be some of the double bonds A formed in the metathesis reaction, but these quickly isomerize to *trans* above 100°C. In a differential scanning calorimetry (DSC) trace of the polymer of **7**, the exotherm associated with this isomerization process can be detected at a higher temperature (110°C) than that of the retro-Diels–Alder reaction (70°C), but not in the case of the polymer of **9**, where the former is merged with the latter (110°C).

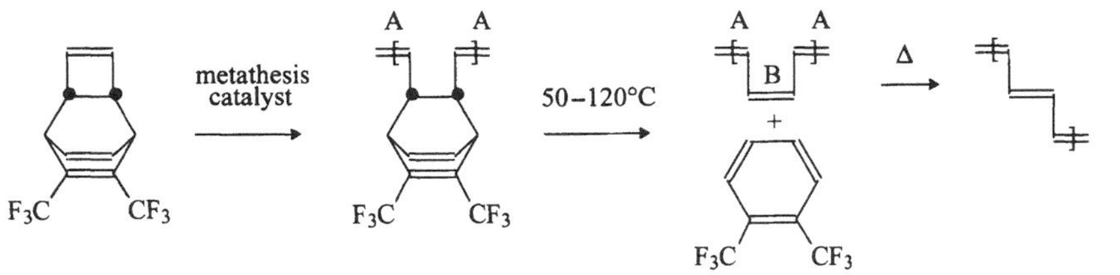

(7) (7P) (1)

The order of stability of the polymers is **5P** < **7P** < **6P** < **9P** ≈ **10P** < **11P**. The last two are quite stable at room temperature and must be heated to 100–200°C to effect their conversion to polyacetylene. Polymers made using WCl_6/Me_4Sn as catalyst (MW ~ 40 000) contain comparable proportions of *cis* and *trans* double bonds (Edwards 1984). Polymer **8P**, unlike **6P**, is quite stable because the retro-Diels–Alder reaction is less favoured (it would give a cyclohexadiene, not a benzene derivative). Polymers of **8** having various *cis* contents (6–77%) have been prepared, but no information is available concerning their tacticity (Feast 1987).

A closer insight into these reactions, particularly with monomer **7**, has been gained through the use of $Mt(=CHCMe_3)(=NC_6H_3\text{-}i\text{-}Pr_2\text{-}2,6)(OCMe_3)_2$ as initiator, where Mt is W (Knoll 1989) or Mo (Park, L.Y. 1991), and using Me_3CCHO, $Me_3CCH{=}CHCHO$, or $C_7H_{15}CHO$ to terminate the chains in a Wittig-like reaction, eqn. (2), either before or after the retro-Diels–Alder reaction has occurred.

$$\begin{matrix}[Mt]{=}CHP \\ + \\ O{=}CHCMe_3\end{matrix} \longrightarrow \begin{matrix}[Mt] \\ \| \\ O\end{matrix} + \begin{matrix}CHP \\ \| \\ CHCMe_3\end{matrix} \qquad (2)$$

When polyacetylene made in this way is passed through a column of silica gel under nitrogen at − 40°C, one can isolate homologues containing up to 13 double bonds. Reaction of the tungsten carbene initiator with **7** gives a *trans* double bond, as also does the reaction of Me_3CCHO with the living end, while the propagation reaction gives 75% *trans* double bonds. On the other hand, the initially formed double bond from the retro-Diels–Alder reaction is always *cis*. Such a procedure therefore yields a series of polyenes with $(2n+1)$ double bonds in which initially the most abundant components are those having alternate *trans* and *cis* double bonds: $(tc)_n t$. Termination of the living ends with $Me_3CCH{=}CHCHO$ gives a series of polyenes with an even number of double bonds, but the termination reaction is not stereospecific. Termination with 0.5 equivalent of an unsaturated dialdehyde such as OHCCH=CHCH=CHCH=CHCHO can be used as a means of joining two chains together and extending the polyene sequence, but solubility and stability problems limit the usefulness of this procedure (Park, L.Y. 1991).

With the molybdenum carbene initiator the various stages of reaction may be followed very closely by ^1H NMR as shown in Fig. 13.1. Spectrum (a), taken 20 min after mixing equivalent proportions of monomer and initiator, shows two main carbene proton doublets ($J_{\alpha\beta}=6$ Hz) at 11.21 and 11.09 ppm as well as the singlet from residual initiator at 11.24 ppm. The doublets are assigned to the protons α_n in the products (**12**) of addition of one and two molecules of **7**, respectively. Already in this spectrum may be seen small doublets ($J_{\alpha\beta}=11$ Hz) at 12.63 and 12.48 ppm assigned to the corresponding protons α'_n in the products of the retro-Diels–Alder reaction, which occurs most readily at the unit adjacent to molybdenum in **13**. In spectrum (b) these signals have strengthened at the expense of the α_n protons, as also have those around 8 ppm (β'_n in **13**). In **13** the newly formed double bond is *cis*. On heating to 50°C for 90 min the *cis* double bonds are largely converted to *trans* giving rise to new carbene proton doublets at 11.96 and 11.85 ppm (α''_n in **14**) and a new signal for the β''_n protons; see spectrum (d). In **12-14** the alkylidene ligand may be oriented either towards the nitrogen atom in the imido ligand (*syn* rotamer) or away from the nitrogen atom in the imido ligand (*anti* rotamer). The *syn* rotamer predominates in each case, but the *anti* rotamers are also detectable for **13** and **14** as indicated in Fig. 13.1. The carbene proton signals from the species corresponding to $n \geq 3$ have also been resolved (Park, L.Y. 1991). The rate constant for **12** $\rightarrow$ **13** at 25°C is 1.06×10^{-2} min^{-1}, while that for **13** $\rightarrow$ **14** is 1.80×10^{-3} min^{-1}, some six times smaller. These rate constants are independent of solvent (C_6D_6 and THF-d_8). When reaction is initiated by the tungsten complex, the rate constants are somewhat smaller; also the rate of the retro-Diels–Alder reaction for the unit adjacent to the tungsten centre in the analogue of **12** is about ten times faster than for the unit that is further away (Knoll 1989).

The ROMP of **7** may be used as the first stage in the preparation of polyacetylene molecules with mesogenic (liquid-crystalline) functional groups at the chain ends. The ROMP of **7** is initiated by a molybdenum carbene complex and the living ends terminated by reaction with a substituted benzaldehyde bearing a mesogenic group, followed by heating to effect the retro-Diels–Alder reaction as in eqn. (1) (Widawski 1995). In a similar manner block copolymers, containing ring-opened units of **10**, can be prepared and then heated to 220°C to eliminate naphthalene, so yielding copolymers containing polyacetylene blocks (Fischer, W. 1993). If it is desired to introduce a controlled proportion of cross-links during the ROMP, say, of

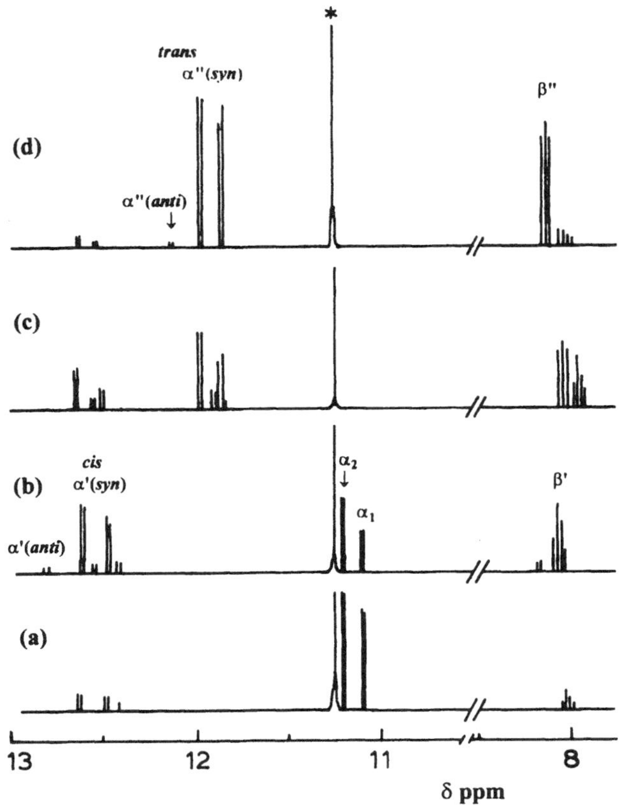

Fig. 13.1 500 MHz ^1H NMR spectra for the reaction of Mo(=$CHCMe_3$)(=NC_6H_3-*i*-Pr_2-2,6)($(OCMe_3)_2$) with 1.0 equivalent of **7** in C_6D_6, after (a) 20 min at 25°C, (b) 60 min at 25°C, (c) heating to 50°C for 30 min, and (d) heating to 50°C for 90 min. The asterisk marks the singlet from the initiator. For assignments, see **12–14** and text (Park, L.Y. 1991).

cyclooctene, one may use a small proportion of *syn*-tricyclo[4.2.0.0^{2,5}]octa-3,7-diene, **4**, as cross-linking agent. With WCl_6/Me_4Sn as catalyst, both double bonds in **4** react completely and, if sufficient **4** has been used, the cyclobutane rings which form the cross-links give rise to observable signals at 41.2 ppm in the ^{13}C NMR spectrum and 3.5 ppm in the ^1H NMR spectrum (Coleman 1988). The physical properties of polyacetylenes prepared in these various ways have been closely studied (Allen 1989; Jones 1991; Park, L.Y. 1992).

It should be noted that monomers that contain (i) a cyclobutene ring substituted at the olefinic carbons by COOMe or CF_3, and (ii) a norbornene ring system, undergo ROMP by preferential opening of the norbornene ring (see Section 13.3.5) (Saunders 1995).

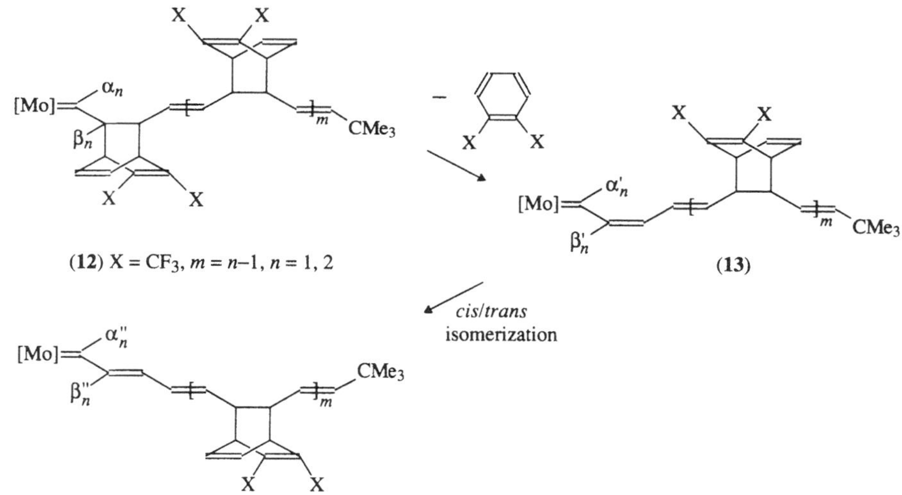

(**12**) X = CF_3, $m = n-1$, $n = 1, 2$

(**13**)

(**14**)

13.3 Monomers containing a fused cyclopentene ring and one double bond

13.3.1 Fused ring monomers other than norbornene derivatives

Compounds in this category are bicyclo[3.3.0]octene (**15**), tricyclo[5.2.1.$0^{2,6}$]dec-3-ene (**16**), *exo*-tricyclo[5.2.1.$0^{2,6}$]dec-8-ene (**17**), benzvalene (**18**), and delta-cyclene (**19**). Of these, **15** and **16** do not undergo ROMP with conventional catalysts (Ofstead 1972), but **17** does so (Oshika 1968). The ROMP of **18** proceeds smoothly using tungsten carbene initiators, and films of the polymer **18P** can be cast directly from the reaction mixture (Swager 1988, 1989). The polymer has a tendency to cross-link and to decompose spontaneously once isolated in dry form, so is best handled in solution, especially as the decomposition can be explosive. The DSC thermogram of the polymer shows an exotherm at 153°C, attributed mainly to isomerization to polyacetylene **20**, eqn. (3), and a second exotherm at

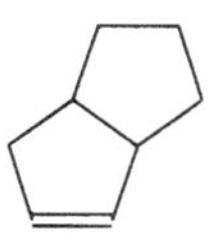
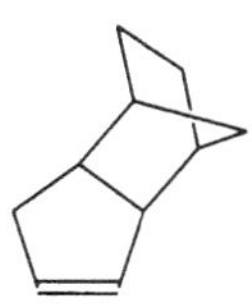
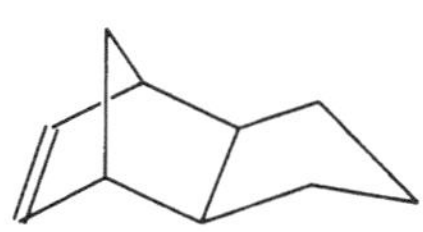
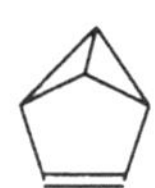
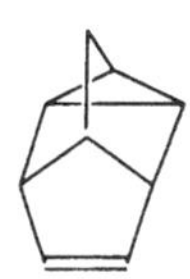

(**15**) (**16**) (**17**) (**18**) (**19**)

308°C, of uncertain origin. The polymer **18P** made using $W(=CHCMe_3)(=NAr)(OCMe_3)_2$ as initiator has a very simple three-line ^{13}C NMR spectrum but it is uncertain whether the structure is all-*cis* or all-*trans*. Polymers **18P** made with other initiators give spectra of greater complexity, probably due to partial isomerization to **20**. For clean conversion of **18P** to **20** it is best to treat freshly cast films with a 5% solution of $HgCl_2$ in THF. The films turn red within seconds, to blue–green over the next 30 s, and finally to a black, silvery, shiny film within 2–3 min.

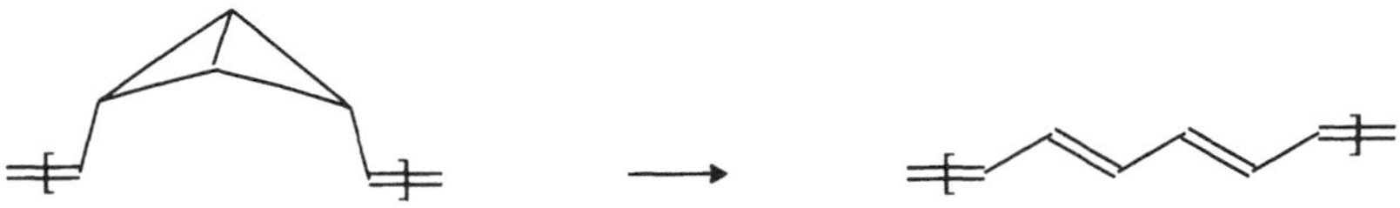

(3)

(18P) **(20)**

The ROMP of **19** leads to polymers of high MW, with a range of *cis* content depending on the catalyst: 27–35% for $Mo(=CHCMe_2Ph)(=NAr)(OCMe_3)_2$, 60–70% for $RuCl_3$/60°C, 70% for WCl_6/Ph_4Sn, and 100% for $ReCl_5$ (Lautens 1989, 1990, 1991). With $ReCl_5$ as catalyst a small amount of saturated low-molecular-weight polymer is also produced, probably by a cationic mechanism since the yield depends on the amount of residual moisture in the system. The all-*cis* polymer can be fully epoxidized using dimethyldioxirane, but in *cis/trans* polymers only the *cis* double bonds are epoxidized (Lautens 1990).

13.3.2 Norbornene (bicyclo[2.2.1]hept-2-ene)

By virtue of its cheapness and ready polymerizability by all metathesis catalysts norbornene (**21**) has received a great deal of attention. Its ROMP according to the first step in sequence (4) was first reported in the open literature by Truett (1960), using $TiCl_4/LiAl(C_7H_{15})_4$ as catalyst. Evidence for structure **22** came from its ozonolysis to form *cis*-cyclopentane-1,3-dicarboxylic acid (**23**), later borne out by the ^{13}C NMR spectrum (see Table 11.5).

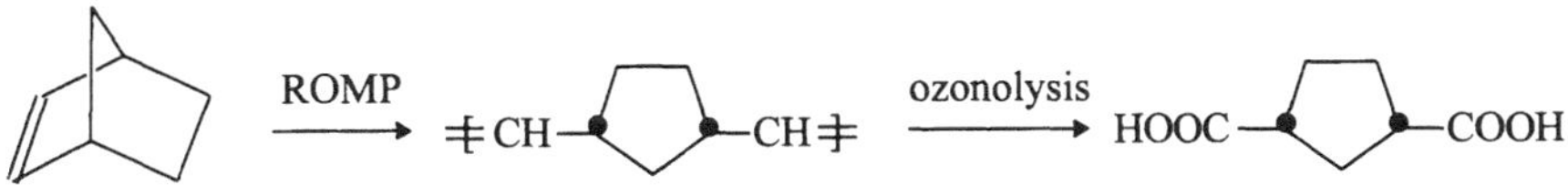

(4)

(21) **(22)** **(23)**

With $TiCl_4/i\text{-}Bu_3Al$ (2/1) as catalyst system the polymer formed shows little unsaturation, but a 1/2 catalyst/cocatalyst ratio gives ring-opened polymer. With Et_3Al as cocatalyst the cationic side reactions can be suppressed by the inclusion of a tertiary amine in the reaction mixture (Saegusa 1964; Tsujino 1964, 1965; Winstein 1977). More recently, a number of titanacyclobutane complexes have been prepared which can act as initiators of living ROMP of norbornene (Gilliom 1986b, 1988; Petasis 1993). Such initiators can be used to make block copolymers (Cannizzo 1988) and adapted to the production of star-shaped polymers (Risse 1989c). The living ends can be terminated with benzophenone to yield a $Ph_2C=$ end-group (Cannizzo 1987) or with terephthalaldehyde to yield a $=CHC_6H_4$-(CHO)-4 end-group (Risse 1989d).

Tantalum carbene complexes such as $Ta(=CHCMe_3)(OC_6H_3$-*i*-Pr_2-2,6$)_3$(thf) and $Ta(=CHCMe_3)(SC_6H_2$-*i*-Pr_3-2,4,6$)_3$(py) are effective, provided the conditions are such as to allow the coordinated base (thf or py) to give way to monomer. In the first example, the initially formed tantalacyclobutane complex has been isolated and shown to have a trigonal–bipyramidal structure, and to polymerize norbornene at a rate that is independent of [M]. The rearrangement of the intermediate tantalacyclobutane complex to form the tantalum carbene complex thus controls the rate of polymerization. In the second example the rate is first order in both initiator and monomer, and the tantalum carbene complex is thus the more stable chain carrier. In both cases the polymer has been cleaved from the metal with benzaldehyde and shown to be nearly monodisperse (Wallace 1987, 1988). The ditantalacyclobutadiene complex, $(Me_3SiCH_2)_4Ta_2(\mu\text{-}CSiMe_3)_2$ also induces the ROMP of norbornene, but only in the presence of an equivalent of oxygen. Here again the rate is zero order in monomer (Eilerts 1992). An effective $[Ta](CH_2Ph)_2$ precursor complex has also been reported; see Table 13.1.

The earliest molybdenum-based catalyst to be used was $MoO_3/\gamma\text{-}Al_2O_3$ (Eleuterio 1957), but the yield of polymer was low. Non-carbene catalysts have been based on compounds such as $MoCl_5$ (Sartori 1963; Ivin 1979c; Minchak 1981), $(\pi\text{-}C_4H_7)_4Mo$ (Kormer 1972), $MoCl_2(PPh_3)_2(NO)_2$ (Larroche 1982, 1983), $Mo(CO)_6$ (Greene 1986, Tamura 1996), $Mo(CO)_5$(py) (Johnston 1991b, 1992), $(Bu_4N)_2(Mo_6O_{19})$ (McCann 1993, 1995), $[Mo_2(MeCN)_8](BF_4)_2$ and related complexes (McCann 1991, 1994), usually with a cocatalyst; (see also McCann 1996).

Molybdenum carbene complexes of the type $Mo(=CHR)(=NAr)(OCMe_3)_2$ are very effective for the ROMP of norbornene (Schrock 1988c) and can be manipulated: (i) with chain transfer agents such as penta-1,3-diene and styrene to reduce the MW without sacrificing the narrow distribution (Crowe 1990); (ii) with norbornadiene dimer to produce star polymers and star block copolymers (Bazan 1991a); and (iii) with substituted benzaldehydes as terminating agents, to give polymers with a variety of functionalized end-groups (Mitchell 1991); see also Fischer, W. (1993) for another example of a block copolymer made with this initiator. The ROMP of norbornene in THF (with $R = CMe_2Ph$, $Ar = C_6H_3$-*i*-Pr_2-2,6) is first-order in both monomer and initiator; $k_p = 4.3$ M^{-1} s^{-1}at 20°C, $\Delta H_p^{\ddagger} = 43$ kJ mol^{-1}, $\Delta S_p^{\ddagger} = -84$ J $K^{-1}mol^{-1}$ (Heroguez 1994). With the same initiator in toluene at 22°C, and using $Me_3CCH{=}CH_2$ as transfer agent, $k_p = 17$ M^{-1} s^{-1} and $k_{tr} = 3 \times 10^{-5}$ M^{-1} s^{-1} (Benedicto 1995). A molybdenum carbene complex with a tridentate ligand [tris(pyrazolyl)borate] is effective in the presence of $AlCl_3$ (Vaughan 1995).

An interesting heterogeneous catalyst has been obtained by adding one drop of 1.8 M $EtAlCl_2$ in toluene to a crystal (20 mg) of $(Bu_4N)_2(Mo_6O_{19})$; its surface colour changes from yellow to dark brown, corresponding to Mo(VI) → Mo(V) → Mo(IV). If a solution of norbornene in toluene is added after 30 s, there is instant polymerization to a gel, which is partially soluble in chloroform and contains 33% *cis* double bonds. On removing the crystal from the polymer with forceps, its surface colour is restored by atmospheric oxidation and it can be used to repeat the process without loss of activity. Similar results are obtained with related tungstates (McCann 1993, 1995).

A great number of tungsten compounds have been used to catalyze the ROMP of norbornene; see Ivin (1983). They can be divided into three main groups: (i) WCl_6, with various cocatalysts (Ho, H.T. 1982c; Reif 1983; Ivin 1987; Cramail 1990, 1991b; Bespalova 1994a); (ii) other non-carbene catalysts, particularly those with aryloxy ligands which are sometimes bidentate (Barnes 1994) and may need a cocatalyst (Dietz 1993), also tungsten carbonyl complexes (Bencze 1984, 1985a,c, 1996); and (iii) tungsten carbene complexes with monodentate ligands, of which two may be alkoxy or aryloxy (Kress 1985, 1987b,c, 1988, 1989; Schrock 1987b, 1988c; Boutarfa 1988; Greene 1989; Johnson 1993), or with one bidentate ligand (Couturier 1992, van der Schaaf 1992, 1993, 1994; VanderLende 1994), or one tridentate ligand (Blosch 1991); and 1,3-ditungstacyclobutadiene complexes (Eilerts 1992). Also in the third group are complexes such as $W[=C(OMe)Ph](CO)_5$ and $W(=CPh_2)(CO)_5$, which, surprisingly, give polymers of different *cis* content (and tacticity in substituted norbornenes) (Thoi 1982), indicating that the initiating species is not simply formed by loss of CO, but must contain at least one different permanent ligand. In the former case, there is evidence that one permanent ligand may be derived from the original carbene ligand (Bechera 1986).

Rhenium catalyst systems are mostly based on $ReCl_5$ and give polymers of high *cis* content (Oshika 1968; Ivin 1977a, 1979c; Greene 1986); likewise Re_2O_7/Al_2O_3 gives an all-*cis* polymer, but when pretreated with Me_4Sn gives a polymer with comparable proportions of *cis* and *trans* double bonds showing that a different active species must be involved (Moloy 1994). One rhenium carbene complex is reported to be active (Toreki 1993).

Catalyst systems based on Ru, Os, and Ir compounds differ from all others in that they will initiate ROMP of norbornene and its derivatives not only in the usual organic non-polar and weakly polar solvents, but also in ethanolic solution and in aqueous emulsion (Rinehart 1965, 1968). Michelotti (1965b) was the first to show that these metals were active in the form of their hydrated trichlorides in ethanol: $IrCl_3$ shows some activity at 25°C but $RuCl_3$ must be heated to 80°C. In all cases the polymer precipitates from the reaction medium as it is formed. The hydrate of $Ru(OTs)_2$ is effective at 50°C both in protic solvents giving high-*trans* polymers (France 1993a; Mühlebach 1994), and in supercritical carbon dioxide to give a high-*cis* polymer (Mistele 1995, 1996). Arene complexes of this salt are very active when exposed to UV radiation (Karlen 1995). $RuCl_2(PPh_3)_2(py)_2$ and related complexes are active even at 20°C in the presence of oxygen (Ivin 1981b). The monomer first undergoes catalytic epoxidation, followed by formation of an oxaruthenacyclobutane complex, which can then generate the propagating carbene complex. Catalysis by OsO_4 at 60°C probably operates in a similar fashion (Hamilton 1990a). Ethyl diazoacetate enhances the activity of ruthenium compounds for the ROMP of norbornene (Demonceau 1992). The complex $Ru(=CHCH=CPh_2)(Cl)_2(PR_3)_2$, (R = cyclohexyl), polymerizes 142 equivalents of norbornene in CD_2Cl_2/C_6D_6 (1/4) at room temperature in less than a minute, giving a polymer with 86% *trans* double bonds, but very little of the initiator is used and the polymer has a broad MWD (Nguyen 1993). The corresponding complex with R = Ph is somewhat less active but gives a living system; the carbene proton in the propagating species appears at 17.79 ppm in the ^1H NMR spectrum (Nguyen 1992). These complexes are also active when supported on polystyrene (Nguyen 1995). The complexes $Ru(=CHR)(Cl)_2(PPh_3)_2$, (R = Me, Et, Ph), are much more efficient [$k_i/k_p = 9$ when R = Ph, compared with 0.006 for $Ru(=CHCH=CPh_2)(Cl)_2(PPh_3)_2$] and give polymers with very narrow MWD ($M_w/M_n = 1.04$) (Schwab 1995, 1996). The binuclear complex $(RuClCp)_2(=CHCH=CPh_2)$ is not very active (Gagné 1992).

Dienes can have a significant effect on the course of reaction by coordinating to the metal centre. Thus isoprene (6×10^{-4} M) can completely suppress the formation of cyclic oligomers during the ROMP of norbornene catalyzed by WCl_6/Me_4Sn (Reif 1983). In the absence of isoprene 40% of the product consists of oligomers containing from 2 to 14 monomer units as detected by GPC (Reif 1981a). Again, the activity of $RuCl_3$ is much reduced in the presence of *endo*-dicyclopentadiene (DCPD) and the stereoselectivity of polymerization is greatly altered. Thus $RuCl_3$ alone gives a 5% *cis* polymer of norbornene, whereas $RuCl_3$ which has been pretreated with DCPD gives a 95% *cis* polymer (Gillan 1988). For $OsCl_3$, the effect is somewhat smaller, 29% *cis* and 85% *cis*, respectively, while with $IrCl_3$ the difference is very small, 36% and 43% *cis* respectively (Greene 1986; Gillan 1988). The ability of the doubly coordinated diene molecule to influence the relative ease of approach of the monomer to form *cis* and *trans* double bonds is thus a sensitive function of the nature of the metal. A similar effect is observed with certain bis(allyl)ruthenium(IV) complexes as catalysts, which give polymers of norbornene containing 30–90% *cis* double bonds (Wache 1995a,b; Herrmann 1996).

Some examples of the *cis* content of the polymers formed using various catalysts are given in Table 13.1. The NMR spectra of these polymers are insensitive to tacticity but the spectra of their fully hydrogenated products are sensitive to tacticity when run under the most favourable conditions (Al-Samak 1996).

13.3.3 Monosubstituted norbornenes

Here we shall adopt a uniform numbering system for norbornene derivatives, with the double bond always in the 2-position, even when it should strictly be numbered as the 5-position. This will make it easier to list and compare such compounds.

Hydrogenation of polymers of substituted norbornenes is frequently a valuable aid to the determination of structural detail and is best carried out using diimide (NH=NH) generated *in situ* by the decomposition of *p*-toluenesulfonohydrazide in xylene at 120°C. This procedure works well and selectively even in the presence of groups such as COOMe and PPh_2 (Sohn 1995).

13.3.3.1 Norbornenes with an alkyl, alkenyl, or aryl substituent

The methyl and ethyl norbornenes **24–30** all readily undergo ROMP. As normally made, **24-28** consist of racemic mixtures of the two enantiomers. The racemic mixture of **24** gives an all-HT polymer (Katz 1980b). The polymers of **25** and **26** usually have a strong or complete HT bias, except when made with certain catalysts (Hamilton 1983, 1984a, 1985a,d, 1988; Boutarfa 1988; Gilliom 1988; Greene 1989;

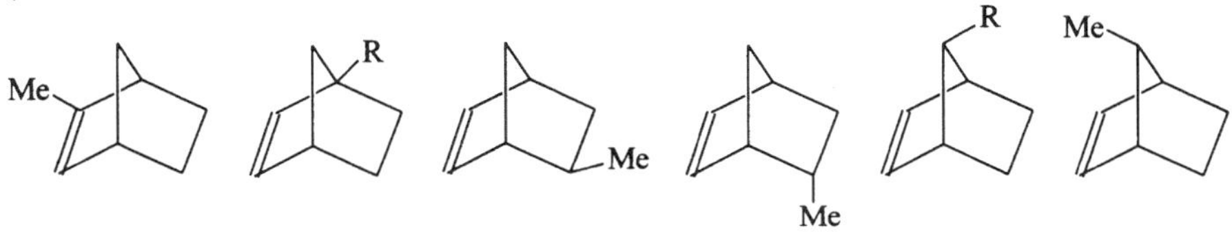

(24) **(25)** R = Me **(26)** R = Et **(27)** **(28)** **(29)** R = Me **(30)** R = Et **(31)**

Table 13.1 Fraction of double bonds σ_c having *cis* configuration in some polymers of norbornene formed by ROMP [a]

Catalyst system	Solvent[b]/ temperature (°C)	σ_c	Reference
Re_2O_7/Al_2O_3 (slurry)	H/20°	1.00	Moloy 1994
[W](=$CHSiMe_3$) complex [c]	B/20°	1.00	van der Schaaf 1994
[W](=$CHCMe_3$) complex [d]	T/ − 30°	> 0.95	Schrock 1987b
$RuCl_3$/*endo*-DCPD [e]	C/E/60°	0.95	Gillan 1988
[Ta](=CHPh) complex [g]	T/65°	0.97	Mashima 1996
$W(\pi\text{-}C_4H_7)_4$	B/30°	0.85	Kormer 1972
$MeReO_3/MeAlCl_2$	C/25°	0.84	Herrmann 1991
WCl_6/BuLi (1/6)	H/20°	0.74	Ivin 1979c
WCl_6/Ph_4Sn (1/1)	C/20°	0.43	Ivin 1979c
$IrCl_3$	C/E/20°	0.36	Greene1986
WCl_6/Me_4Sn (1/2)	D/20°	0.24	Ivin 1987
[Ru](=CHCH=CPh_2) complex [f]	B/D/20°	0.14	Nguyen 1993
$RuCl_3$	C/E/60°	0.05	Gillan 1988

[a] Secondary metathesis sometimes causes σ_c to fall as the reaction proceeds; the values therefore vary according to the reaction conditions.
[b] B = benzene; C = chlorobenzene; D = dichloromethane; E = ethanol; H = hexane; T = toluene.
[c] W(=$CHSiMe_3$)(=NPh)(CH_2SiMe_3)(8-quinolinolate).
[d] W(=$CHCMe_3$)(=NC_6H_3-*i*-Pr_2-2,6)[$OCMe(CF_3)_2]_2$.
[e] DCPD = dicyclopentadiene.
[f] Ru(=CHCH=CPh_2)$(Cl)_2(PCy_3)_2$; Cy = cyclohexyl.
[g] Produced by heating $Ta(CH_2Ph)_2Cp^*(\eta^4\text{-}C_4H_6)$ at 65°C.

Couturier 1992). With **27** and **28** most catalysts give polymers with randomly oriented substituents (Ivin 1980b, 1981a). Thus the further away the substituent is from the double bond, the less effect it has on the head or tail orientation of the incoming monomer. In polymers of single enantiomers the orientation of the substituents in the polymer is correlated with the ring tacticity. Details are given in Ch. 11.

Monomers **29–31** are prochiral and give polymers of varying tacticity, depending on the catalyst (Hamilton 1985b; Greene 1987b; Mashima 1996). As normally made 7-methylnorbornene is a 50 : 50 mixture of the *syn*- and *anti*- isomers, **31** and **29**, respectively. When this mixture is treated with WCl_6/Me_4Sn the *anti*-isomer is selectively polymerized and the *syn*-isomer can be recovered from the final reaction mixture in good yield (Hamilton 1985b, 1994). The *syn*-methyl group, being close to the double bond, must offer considerable steric hindrance as the monomer approaches with its *exo* face towards the propagating metal carbene complex derived from this catalyst. More active initiators, such as $W(CO)_3$(mesitylene)/$EtAlCl_2$/epoxide, $W(=CHCMe_3)(Br)_2(OCH_2CMe_3)_2/GaBr_3$ or $Mo(=CHCMe_3)$-$(=NAr)(OCMe_3)_2$, are less discriminating, and although the *anti*-isomer still reacts preferentially at the beginning of reaction, the *syn*-isomer reacts later to give its own distinctive propagating species, detectable by ^1H NMR; the product is then a block or tapered-block copolymer of the two isomers (Hamilton 1985b; Kress 1985, 1990; Feast 1992c). With some metal carbene initiators, it is possible to observe not only the propagating metal carbene complexes by NMR (Kress 1985), but also the various isomeric metallacyclobutane intermediates (Kress 1988, 1989) (see Ch. 3).

Other norbornene derivatives having the following hydrocarbon substituents in the 5-position (*endo/exo* mixture) also undergo ROMP: Me_2CH (Tenney 1979b), $CH_2{=}CHCH_2CH_2$ (Dekking 1961), C_nH_{2n+1} ($n = 8$–12) (Tenney 1979a), and C_6H_5 (Rinehart 1965, 1968; Tanaka, Y. 1976; Komatsu 1977).

13.3.3.2 Norbornenes with a silicon-containing substituent

The ROMP of norbornenes with the following substituents in the 5-position (*endo/exo* mixtures) has been reported: $SiMe_3$, **32**; $SiMe_2(CH_2SiMe_3)$, **33**; $SiMe_2[(CH_2)_3$-9-carbazolyl], **34**; $CH_2Si(Me)(CH_2CH_2CH_2)$, **35**; $SiMeCl_2$, **36**; $SiCl_3$, **37**; $Si(OMe)_3$, **38**; and $Si(OEt)_3$, **39** (Zimmermann 1976b; Streck 1982; Schrock 1990a; Finkel'shtein 1991, 1994, Makovetsky 1992b). Catalyst systems such as WCl_6/Et_3Al or $WCl_6/i\text{-}Bu_2AlCl$ can be used, except for the ROMP of **38** and **39**, where it is better to use Lewis-acid-free catalysts such as $RuCl_2(PPh_3)_3$ at 60°C (Finkel'shtein 1994) or a molybdenum carbene complex (Schrock 1990a). The most interesting of the polymers are **32P** and **34P**, the former because of its considerably enhanced permeability and diffusion coefficients for light gases, compared with polynorbornene (Finkel'shtein 1991), and the latter because of its ability to form charge transfer complexes with acceptors such as 2,4,7-trinitro-9-fluorenone (Finkel'shtein 1994).

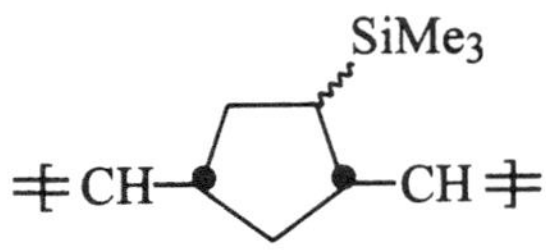

(32P)

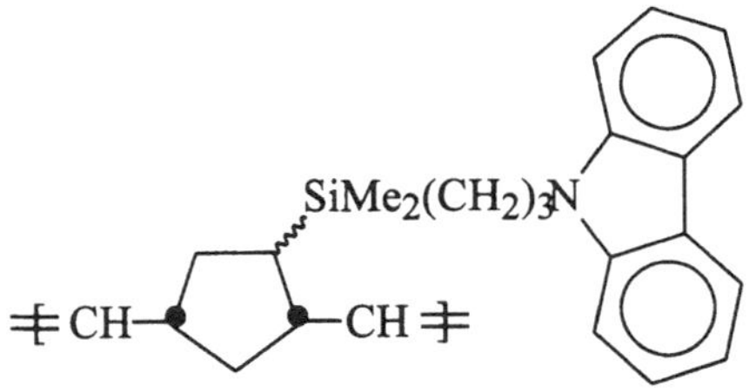

(34P)

13.3.3.3 Norbornenes with a COOR substituent

Monomers in this category fall into three groups: (i) those with small substituents, giving polymers, which may be thermoplastic (Ueshima 1974); (ii) those with substituents of intermediate size and of such a character that the polymers can have liquid-crystalline or other interesting properties; and (iii) macromonomers in which a polymer is terminated by a norbornenyl group and which can be used to make graft copolymers by ROMP. Except where stated the monomers are *endo*/*exo* mixtures.

The first group of monomers includes those with the following substituents: *endo*-COOMe, **40**; *endo*-COOEt, **41**; *exo*-COOMe, **42**; *exo*-COOEt, **43**; and $COOC_6H_2Br_3$, **44**. All types of metathesis catalyst are effective (Rinehart 1965, Hepworth 1973; Ho, H.T. 1989) including metal carbene complexes (Ivin 1992) and $RuCl_2(PR_3)$(*p*-cymene)/$Me_3SiCH_2N_2$ (Stumpf 1995). The ethyl ester polymers can also be produced by heating the carboxylic acid monomer with $IrCl_3$ in ethanol (Michelotti 1965a; Ho, H.T. 1989). Monomer **44** gives a fire-resistant polymer (Imaizumi 1977). Polymers with the following *cis* contents have been prepared from **40** (10–85%), **41** (15–45%), **42** (10–60%), and **43** (5–30%). The ^{13}C NMR spectra indicate that the substituents are randomly oriented in the polymers (Ho, H.T. 1989).

The reaction of **40** with a tungsten cyclopentylidene initiator has been followed in CD_2Cl_2 at 27°C by ^1H NMR. The initiator is all consumed in 15 min and the monomer (3 equivalents) in about an hour. The head and tail propagating species give separate carbene proton signals, and the head species formed by addition of one unit of monomer may be distinguished from the head species containing more than one monomer unit. The final concentration of head species is nearly twice that of the tail species, indicating a somewhat lower overall reactivity of the head species in the propagation steps. With **42** as monomer, the spectra of the head and tail propagating species are indistiguishable, although the first addition product gives a distinct spectrum (Ivin 1992).

(45)

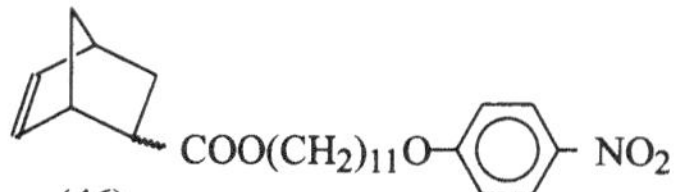

(46)

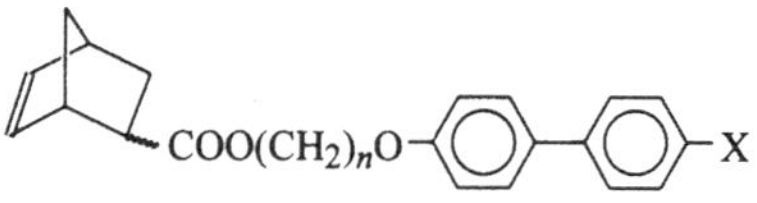

(47) $X = NO_2$, $n = 11$
(48) $X = OMe$, $n = 2–12$
(49) $X = CN$, $n = 3–12$

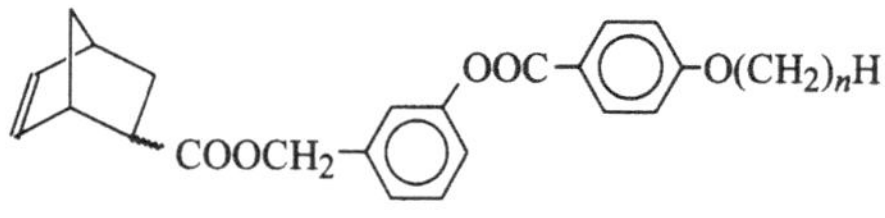

(50) $n = 1–6$

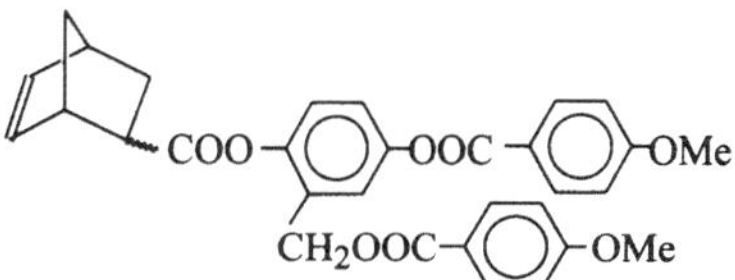

(51)

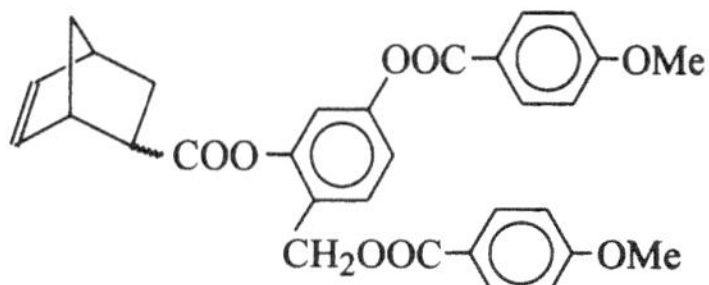

(52)

In the second group, some monomers (**45–47**) have been polymerized by $RuCl_3$ (Laschewsky 1993), and others in a more controlled fashion by $Mo(=CHCMe_3)(=NAr)(OCMe_3)_2$ or related complexes: **48** (Komiya 1992a, 1992b), **49** (Komiya 1993b), **50–52** (Pugh 1992, 1994); see also Weck (1996b). Both **45** and its polymer form monolayers, but the polymer exhibits a higher collapse pressure and reduced collapse area compared with the monomer. The hexadecyl chains must evidently be able to pack more closely in the monolayer of polymer. The polymers of **46** and **47** do not form well-defined monolayers.

With **48** and **49** not only has the spacer length n been varied but a series of polymers of narrow MWD ($M_w/M_n = 1.05–1.28$, DP = 5–100) has been prepared in high yield through the use of living systems. The polymers of **48** ($n = 2–8$) exhibit an enantiotropic nematic mesophase. The glass transition and isotropization temperatures increase with increasing MW and reach a limit at about 30–50 repeat units (30°C and 90°C, respectively, for $n = 8$). With longer spacers ($n = 9–12$) some side-chain crystallization is observed in the lower-molecular-weight materials (DP = 10–20), along with a nematic or smectic mesophase, but this is suppressed at higher MW (DP = 50–100). The polymer of **49** ($n = 3$) is amorphous whereas with longer spacers ($n = 4–12$) the polymers display enantiotropic nematic mesophases that are independent of MW, and no side-chain crystallization is observed. Similar results have been obtained with the polymers of **50–52**. The phase behaviour becomes independent of MW at about 25 repeat units and the transition temperatures decrease with increasing n in the polymer of **50**. The polymer backbone in fact has little effect on the transition temperatures of side-chain liquid-crystalline polymers displaying nematic mesophases, even when the chemical structures of the backbones are substantially different, as with polynorbornenes *vs* polyacrylates. This result is consistent with the proposal that mesogens jacket the extended polymer chain (Pugh 1992). Monomers containing a long 5-substituent and bearing a methacrylate end-group have also been prepared and polymerized to give products with potential electro-optic applications (Betterton 1992).

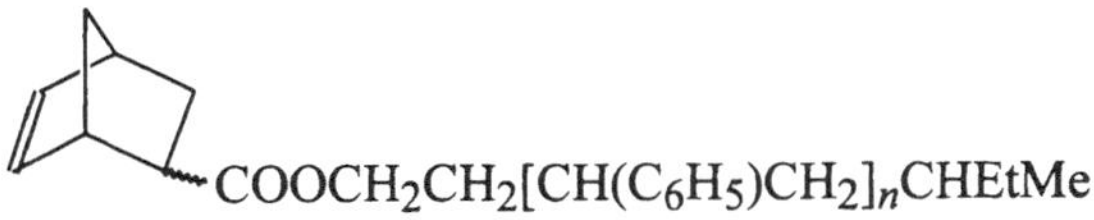

(53)

The third group in this category is exemplified by the macromonomer **53**, which can be made by first initiating the anionic polymerization of styrene with *s*-BuLi, end-capping with ethylene oxide, and then reacting with norborn-2-ene-5-carbonyl chloride (Norton 1989). It can then be copolymerized with norbornene using WCl_6/Me_4Sn as catalyst to give poly(norbornene-*graft*-styrene) copolymers. The ROMP of the macromonomer itself ($M_n = 2700–11\,000$) proceeds to high conversion only when initiated by a molybdenum carbene complex, yielding comb-like polymers of high MW. The solution behaviour of these polymacromonomers is very dependent on both the MW of the monomer, which governs the length of the polystyrene side chains and that of the final polymer (Breunig 1995). For an alternative way of producing this macromonomer, see Herogues (1996).

13.3.3.4 Norbornenes with an OCOR substituent

Monomers **54–57** all undergo ROMP. Michelotti (1965a) first reported the reaction of **57** using $IrCl_3$, and Hepworth (1973) that of *endo/exo* mixtures of **54** using $WCl_6/Et_3Al_2Cl_3$; see also Ueshima (1974), Kobayashi, Y. (1977), Thorn-Csányi (1992a) and Balcar (1994a).

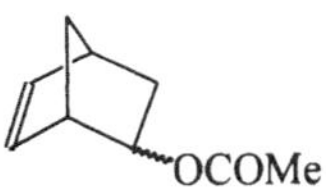

(54)

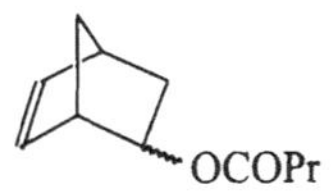

(55)

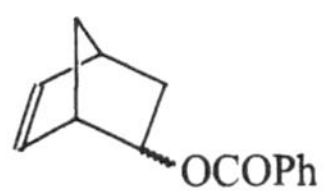

(56)

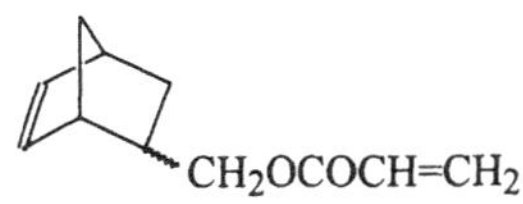

(57)

Conventional catalysts may be used to give unbiased polymers of (±)-*endo*-**54** with 16–75% *cis* content (Ivin 1993a). Tungsten and molybdenum carbene complexes are much more reactive. Thus $W(=CHCMe_3)Cl(CH_2CMe_3)(OAr)_2$-$[O(CHMe_2)_2]$ induces complete polymerization of (±)-*endo*-**54** in 10 s at 25°C (Quignard 1985b). With (−)-*endo*-**54** initiated by $Mo(=CHCMe_2Ph)$-$(=NAr)(OCMe_3)_2$ a 30% *cis* polymer is formed, which is nearly atactic with respect to both *cis*- and *trans*-centred ring dyads, but when initiated by $Mo(=CHCMe_2Ph)(=NAr)[OCMe(CF_3)_2]_2$ the polymer has a much higher *cis* content (87%) and the *cis*-centred dyads are biased towards isotactic, as shown by the inequality, TH > TT, of the olefinic signals in the ^{13}C NMR spectrum (Stein-häusler 1994a). The 87% *cis* polymer has nearly twice the specific rotation of the 30% *cis* polymer. The enantiomers of *endo*-**55** and *endo*-**56** likewise give optically active polymers (Steinhäusler 1994b).

13.3.3.5 Norbornenes with a hydroxy or alkoxy substituent

Monomer *exo*-**58** polymerizes at 20°C in the presence of $IrCl_3$ whereas *endo*-**58** requires heating to 60°C (Michelotti 1965a). *Endo*-**59** will also polymerize with this catalyst at 60°C and with various others to yield polymers having 10–48% *cis* double bonds and random orientation of substituents (Ivin 1993b).

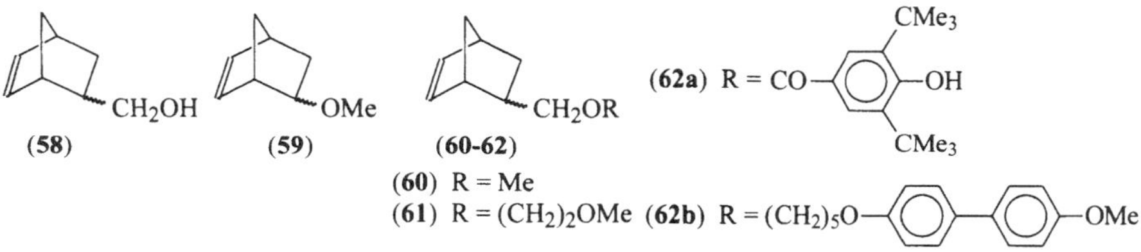

(**58**) (**59**) (**60-62**)

(**60**) R = Me
(**61**) R = $(CH_2)_2OMe$
(**62a**) R = CO–(3,5-di-CMe3-4-OH-phenyl)
(**62b**) R = $(CH_2)_5O$–(biphenyl)–OMe

Racemic *endo*-**60** and *exo*-**60** also undergo ROMP to give unbiased polymers with a range of *cis* contents. With optically active *endo*-**60** $RuCl_3$ gives an atactic polymer (10% *cis*); WCl_6/Me_4Sn also gives an atactic polymer (88% *cis*). On the other hand, $ReCl_5$ gives an 87% *cis* polymer that shows a syndiotactic bias (Ivin 1994a). A derivative of **60**, in which the methyl group is replaced by 6-*N*-carbazoylhexyl, undergoes ROMP with various catalysts to yield high-*trans* polymers that have been characterized by cyclic voltammetry. The carbazole units in the side chains can be oxidatively coupled to form dicarbazyls (Lee, J-H. 1993). Monomer *endo*-**61** has been polymerized in a living system and used to make block copolymers which offer potential for binding zinc and cadmium compounds through the oxygen donors (Cummins 1992). Monomer **62a** undergoes ROMP when catalyzed by $RuCl_3$ at 70°C, giving high-*trans* unbiased polymer ($M_n = 2 \times 10^5$) (Tlenkopatchev 1995b). Monomer **62b** undergoes ROMP with Mo- and W-based catalysts to yield polymers ($M_n = 31\text{–}130 \times 10^3$) that show an isotropic/smectic transition at about 70°C on cooling (Kim, S-H. 1993).

13.3.3.6 Norbornenes with a cyano substituent

The ROMP of an *endo*/*exo* mixture of 5-cyanonorborn-2-enes (**63**) can be achieved with various catalysts: WCl_6/Me_4Sn at 70°C (Matsumoto 1976), WCl_6/Et_2AlCl at 60°C (Asrar 1992a), and most effectively with WCl_6/Et_2AlOEt at 20°C to form a 92% *cis* polymer (Thorn-Csányi 1991a, 1992a). Metal carbene complexes give living systems. With $Mo(=CHCMe_3)(=NAr)(OCMe_3)_2$ in THF the propagating species give rise to two pairs of carbene proton doublets in the ^1H NMR spectrum: two at about 11.53 ppm attributed to the tail adducts of the *endo*- and *exo*-isomers, and two at 11.29 and 11.37 ppm attributed to the head adducts; the head and tail adducts are present in roughly equal proportions. The reaction apparently does not proceed in toluene (Bazan 1991b). This initiator can also be used to make star-block copolymers (Bazan 1991a). A tungsten carbene complex gives complete reaction in 10 s (Quignard 1985b).

The relative reactivity of *endo*-**63** and *exo*-**63** depends on the catalyst system: with the highly reactive WCl_6/Et_2AlOEt the residual monomer has the same composition as the starting mixture (*endo/exo* = 57/43), but with the much less active WCl_6/Et_2AlCl, *exo*-**63** reacts more rapidly than *endo*-**63**. This is also the case when the isomers are reacted separately. For *exo*-**63** 90% of monomer reacts in 40 min at 60°C whereas with *endo*-**63** only 25% of monomer reacts in this time. The two polymers have 58% and 25% *cis* double bonds, respectively. There is also a difference in the degree of HT bias. The polymer of *exo*-**63** shows only a slight overall bias (HT/HH = 1.2) whereas the polymer of *endo*-**63** exhibits a moderate overall bias (HT/HH = 2.8). In each case the bias in the *trans*-centred dyads is greater than that in the *cis*-centred dyads (Asrar 1992a).

The polymer of **63** is a potential thermoplastic for which there are numerous patents; various additives are claimed to have beneficial effects on the yield when using WCl_6/Et_2AlCl as catalyst (Kobayashi, Y. 1979). Optical discs can be made having a heat distortion temperature of 145°C (Ueshima 1974; Hirata 1989; Washiyama 1989).

13.3.3.7 Norbornenes with a halogenated substituent

Proof that 5-trifluoromethylnorborn-2-ene (**64**) can undergo ROMP catalyzed by WCl_6/Ph_4Sn was first obtained by Feast (1979, 1980). With *endo*-**64**, both $OsCl_3$ and $MoCl_5/Me_4Sn$ give 90% *trans* atactic polymers with no HT bias. $ReCl_5$ gives a 92% *cis* polymer, also unbiased, but probably tactic. With *exo*-**64**, $OsCl_3$ gives an all-*trans* polymer that is probably atactic, although the spectrum does not show any tactic splitting of the olefinic HH carbons (Blackmore 1986a,b). Living ROMP of **64** is initiated by $Mo(=CHCMe_3)(=NAr)(OCMe_3)_2$ to give a polymer with $M_w/M_n = 1.09$ (Feast 1994b).

In ROMP it is sometimes found that the GPC of the product shows two peaks: a main peak and a much smaller peak at twice the MW of the main peak; for example, in the ROMP of *exo*-dicyclopentadiene (Cannizzo 1988). The reason for this was first elucidated in the living polymerization of *endo/exo*-**64** initiated by $Mo(=CHCMe_3)(=NAr)(OCMe_3)_2$. If a trace of oxygen is present after the monomer has polymerized there is premature termination of some of the living ends to form an aldeyde end-group, which in turn terminates a second chain to which it becomes attached. If benzaldehyde is then added, the main portion of the living ends react to give a benzylidene end-group. The GPC of the product then shows the aforementioned two peaks if an RI detector is used, but only the main peak if a UV detector is used, set to detect absorption by the aromatic ring (Feast 1992b). 5-(1′-Bromotetrafluoroethyl)-norborn-2-ene also polymerizes smoothly with this initiator (Bazan 1991b). Good yields of polymer can be obtained in the

ROMP of norbornenes substituted with various fully fluorinated alkyl substiuents (Perez 1988).

The ROMP of 5-chloronorborn-2-ene (**65**) and 5-bromonorborn-2-ene (**66**) was first reported in various patents (Hepworth 1973; Kobayashi, M. 1978). A 50% *cis* polymer of **65** is formed when $WCl_6/(CH_2{=}CHCH_2)_4Sn$ is employed as catalyst (Makovetsky 1992b). The ROMP of 5-chloromethylnorborn-2-ene (**67**) proceeds readily with Ru-, Ir-, Mo-, and W-based catalysts to give unbiased polymers with a range of *cis* contents (Makovetsky 1992b; Ivin 1994b). With $ReCl_5$, *exo*-**67** gives an all-*cis* polymer but *endo*-**67** does not react cleanly.

13.3.3.8 Norbornenes with an amine or amide substituent

The ROMP of norbornenes bearing various 5-substituents of the type CH_2NHR has been reported. Thus, when R = H, Me or $CHMe_2$, effective catalysts are $W({=}CPh_2)(CO)_5/EtAlCl_2/O_2$ or $W(CO)_3(\text{mesitylene})/EtAlCl_2/O_2$ in chlorobenzene at 25°C, but the cocatalyst must be used in unusually large excess (W/Al = 1/40). When R = H, the *exo* isomer is more reactive than the *endo* isomer, but both are fully consumed. All the polymers have 50% *cis* double bonds and can be rendered soluble in water by quaternization of the amine groups. They can also be made more readily soluble through the use of pent-1-ene as a chain transfer agent (Larroche 1984). The compound with $R = C_6H_4NHC_6H_5$ has been used as a comonomer with dicyclopentadiene to act as a built-in antioxidant (Matlack 1990).

Various norborn-2-ene-5-amides and N-substituted amides are also polymerizable (Matsumoto 1977a; Komatsu 1977; Kobayashi, M. 1978). In particular a series of glycomonomers represented as NBE-*exo*-5-CONH(glu)R_4 have been prepared by the reaction of norborn-2-ene-*exo*-5-carbonyl chloride with glucosamine hydrochloride; R = H or a protecting group (COMe, COPh, or $SiEt_3$). These monomers all undergo ROMP when initiated by $Ru({=}CHCH{=}CPh_2)(Cl)_2(PCy_3)_2$ in benzene at 50°C. The reaction of the unprotected monomer proceeds quantitatively when carried out in aqueous emulsion but the polymer is rather insoluble in all solvents, unlike the polymers of the protected monomers. In principle this range of monomers can be extended to include those attached to biologically relevant carbohydrates via flexible extenders (Fraser 1995a, Nomura 1996).

13.3.3.9 Miscellaneous

Exo-**68** can be made by hydroboration of norborna-2,5-diene with 9-borabicyclononane (9-BBN) using excess of diene. It undergoes ROMP to yield a polymer **69**, which can be readily oxidized to the hydroxy derivative **70**. This is insoluble in water and common organic solvents, but soluble in mixed solvents such as $CDCl_3/CD_3OD$. The *cis* content varies with the catalyst: 93% with $W({=}CHCMe_3)({=}NAr)[OCMe(CF_3)_2]_2$, 48% with WCl_6/Me_4Sn, and 26% with $W({=}CHCMe_3)({=}NAr)(OCMe_3)_2$. The high-*cis* polymer has a relatively simple

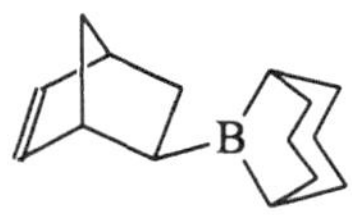

(68)

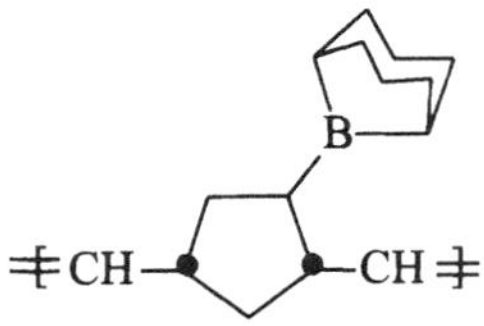

(69)

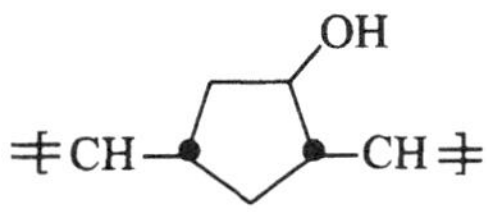

(70)

^{13}C NMR spectrum, with two main pairs of equally intense olefinic lines: 135.36 (TH), 134.00 (TT), 133.24 (HH), 131.88 ppm (HT), indicating a random orientation of the substituents. The corresponding signals for the *trans*-centred dyads may be seen in the spectra of the polymers obtained with the other two catalysts, but with fine structure caused by *tt/tc* splitting: 134.42 (TH), 132.90 (TT), 132.05 (HH), 130.78 ppm (HT) (Chung 1991a,c).

The hydrophilic (OH) groups in these otherwise hydrophobic polymers allow the formation of stable monolayers at an air–water interface. The high-*cis* polymer occupies 38 $Å^2$/monomer unit which is much larger than the 9 $Å^2$/monomer unit occupied by the 74% *trans* polymer, suggesting that the *cis* polymer is rather rigid and lies stretched on the water surface with most of its OH groups at the interface. The T_g values for these polymers are in the range 118–140°C, much higher than those for polymers of norbornene (5°C) (Ramakrishnan 1989a; Chung 1991a,c).

Norbornenes with 5-substituents containing various metallic elements have been prepared and polymerized. The polymers containing tin have biocidal properties (Streck 1982, Lequan 1984). Block copolymers containing lead, palladium, platinum and other metals are described in Ch. 14.

13.3.4 5,5-Disubstituted norbornenes

Compounds in this category that undergo ROMP are listed in Table 13.2. Polymers of the racemic monomers show little sign of HT bias for any value of the *cis* double-bond content. Detailed tacticity studies have been made on the 5,5-dimethyl compound as described in Ch. 11. The high-*cis* syndiotactic polymer has a higher T_g (106°C) than the atactic polymer (47°C) (Barnes 1994).

The 5-methylene compound ($XY = CH_2$) undergoes clean ROMP only with certain metathesis catalysts such as WCl_6/Et_3Al (1/2), giving polymer **71**, which contains 66% *cis* double bonds in the main chain and methylene substituents that are randomly oriented (Ivin 1980a; Brigodiot 1981). Catalysts such as WCl_6/ $EtAlCl_2$ contain sufficient acidic species to favour the exclusive formation of polymer **72** by a cationic mechanism.

Table 13.2 ROMP of 5,5-disubstituted norbornenes (substituents *exo*-X, *endo*-Y)

X	Y	Catalyst system	Reference
$=CH_2$		$IrCl_3$	Rinehart 1968
		WCl_6/Et_3Al, $MoCl_5/Et_3Al$	Ivin 1980a
CH_3	CH_3 [a]	Various	Ho, H.T. 1982a,b
		$[W](=CHCMe_3)$ [b]	Kress 1985
		$[W](=CHCMe_3)$ [c]	Greene 1989
		$W(=Z)(Cl)_2(OArO)(THF)$ [d]	Barnes 1994
		$W(=Z)(Cl)_3(OArO)(OEt_2)$ [d]	Eilerts 1995
CH_3	CN	Mo- and W-based	Ueshima 1974
CH_3	CH_2OH	$RuCl_3$	Ivin 1982a
CH_2OH	CH_3	$RuCl_3$	Ivin 1982a
CH_3	COOH	$RuCl_3$, $OsCl_3$	Ivin 1982a
COOH	CH_3	$RuCl_3$, $OsCl_3$	Ivin 1982a
CH_3	COOR [e]	Various	Rinehart 1968; Hepworth 1973; Ueshima 1974
$COOCH_3$	CH_3	Various	Ivin 1982; Ho, H.T. 1989
CN	CN	$WCl_6/Et_2AlCl/(CH_2OEt)_2$	Nakamura, J. 1979
CH_2Cl	CH_2Cl	Various	Ivin 1994b
Spiro compounds [f]		W-based	Watkins 1994

[a] Racemic and enantiomeric. [b] $W(=CHCMe_3)(Br)_2(OCH_2CMe_3)_2/GaBr_3$; the head propagating species is present in higher concentration than the tail species and is thus the less reactive. [c] Seven different complexes. [d] $Z = O$ or NC_6H_3-Me_2-2,6; cocatalyst Et_2AlCl. The *cis* content of the polymer varies from 41% to 100% depending on the nature of the chelating diolate ligand. [e] R = Me or Et. [f] Norborn-2-ene-5-spiro-3′-*exo*-succinic anhydride, and norborn-2-ene-5-spiro-3′-*exo*-N-phenylsuccinimide and derivatives.

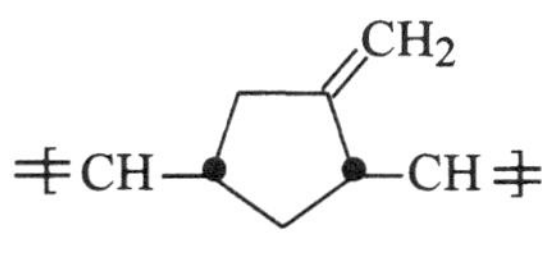

(71)

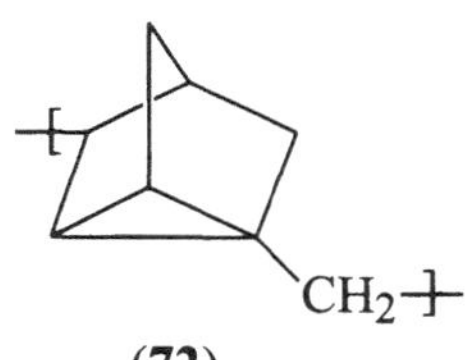

(72)

13.3.5 5,6-Disubstituted norbornenes

Monomers in this category can be divided into three groups: (i) those in which the two substituents are the same (**73–93** in Table 13.3); (ii) those in which the substituents are different; and (iii) those in which the substituents form part of a ring system.

Some selected features of the systems in Table 13.3 are the following.

(i) The metal carbene complex initiators generally give living polymerizations. The carbene proton doublet from the first propagating carbene complex is often resolved (about 0.02 ppm downfield) from that of the longer chain species, e.g. for **75**.

Table 13.3 ROMP of 5,6-disubstituted norbornenes (both substituents R the same)

R	Isomers [a]			Catalyst type [b]	Reference
73 CH_3	*xx*		*nn*	WC	Kress 1985
	xx	*xn*	*nn*	A,B	Greene 1987a
			nn	WC	Kress 1988
			nn	WC	Kress 1989
	xx			WC	Greene 1989
		(+)*xn*		MoC	O'Dell 1994; Sunaga 1994
74 COOH	*xx*			A	Kurosawa 1976a
75 COOMe		*xn*		B	Rinehart 1968
		xn		A	Ueshima 1974
			nn	MoC	Murdzek 1987b
			nn	MoC	Schrock 1988c
			nn	WC	Ho, H.T. 1989
	xx	*xn*		MoC	Bazan 1991b
	xx	*xn*	*nn*	WC	Ivin 1992
		(+)*xn*		MoC	O'Dell 1994
76 COOEt		*xn*		B	Laschewsky 1993
77 $COO(CH_2)_{12}H$	*xx*			B	Bell, B. 1994
78 $COOCH_2(CF_2)_6F$	*xx*			B	Bell, B. 1994
79 $COOSiMe_3$		*xn*		MoC	Saunders 1992
80 $COOCHMeCH_2(sty)_nBu$ [c]		*xn*		MoC	Feast 1994a
81 COO(pte) [d]		*xn*		MoC	Albagli 1993
82 OCOMe	*xx*			MoC	Bazan 1991b, Zenkl 1992b
			nn	MoC	Fischer, W. 1993
83 O(CO)OMe	*xx*			MoC	Schimetta 1994a, 1995
84 O(CS)SMe	*xx*			MoC	Schimetta 1994a, 1995
85 CH_2OMe		*xn*		MoC	Cummins 1992
		xn		MoC	Saunders 1992
		(+)*xn*		MoC	O'Dell 1994
CH_2OMe [f]		*nn*		RuC	Lynn 1996
86 CH_2SMe		*xn*		MoC	Cummins 1992
87 CN				A	Ueshima 1974
88 Cl		*xn*	*nn*	B	Rinehart 1968
89 CH_2Cl				A	Kurosawa 1976b
	xx [e]		*nn*	A	Shahada 1990
90 $(CF_2)_nF$ ($n = 4,6,8$)					Perez 1988
91 CH_2NHCMe_3		*xn*		MoC	Cummins 1991
92 $CH_2NHSiMe_3$		*xn*		MoC	Cummins 1991
93 PPh_3		*xn*		MoC	Chan 1992a

[a] *xx*, *xn*, *nn* denote *exo,exo-*, *exo,endo-*, *endo,endo-*, respectively; (+) denotes enantiomer. [b] A = $TiCl_4$-, $MoCl_5$-, WCl_6-, or $ReCl_5$-based, or similar; B = Ru-, Os-, or Ir-based; C = metal carbene complexes (Mo=C, or W=C or Ru=C). [c] sty = styrene; n = 4–9. [d] pte = phenothiazin-10-ylethyl. [e] Monomer assumed to be mainly *xx* but line positions in the ^{13}C NMR spectra of the polymer are more consistent with the monomer being mainly *nn*. [f] Also $CH_2OSiMe_2CMe_3$.

(ii) The intermediate *transoid* metallacyclobutane species can sometimes be detected at low temperature, but not the corresponding *cisoid* species. With *endo,endo*-**73** the first *transoid* metallacyclobutane complex formed from $W(=CHCMe_3)(Br)_2(OCH_2CMe_3)_2/GaBr_3$ is particularly stable at −38°C. It can be produced in 70% yield and its rearrangement followed at higher temperatures (see Fig. 3.7). Diastereoisomers (precursors of *m* and *r* dyads) in the subsequently formed metallacyclobutanes can also be distinguished.

(iii) Absolute tacticities have been determined for polymers of *exo,endo*-**73**, -**75**, and -**85** using single enantiomers. Depending on the initiator one can obtain a high-*trans* atactic polymer or a high-*cis* isotactic polymer. Polymers of racemic *exo,endo*-monomers and of the prochiral *exo,exo*- and *endo,endo*-monomers sometimes show *m/r* splittings in the ^{13}C NMR spectra of both their high-*trans* and high-*cis* polymers and more clearly in the spectra of their fully hydrogenated products (see e.g. Fig. 13.2).

(iv) The polymer of **74** cannot be made directly but can be produced indirectly by hydrolysis of the polymer made from the *exo* adduct of cyclopentadiene and maleic anhydride. Copolymers containing these units are useful as mouldings, flocculants, adhesives and ion exchangers.

(v) The reactions of the isomers of **75** with a tungsten cyclopentylidene complex proceed with measurable speed in CD_2Cl_2 at 25°C. For the *endo,endo*-isomer the initiation and propagation rate constants are about the same: 5.0×10^{-3} $\text{M}^{-1}\ \text{s}^{-1}$ (Ivin 1992).

(vi) Monomer **79** has been used to make amphiphilic star-block copolymers in which the trimethylsilyl esters have been converted to carboxylic acids.

(vii) Well-defined comb graft copolymers can be produced from **80**, containing an average of 4, 7 or 9 styrene units in the side chains.

(viii) Monomer **81** has been used to make block copolymers with norbornene, end-capped by reaction with 1-pyrenecarboxaldehyde. Fluorescence emission from the pyrene end-groups is quenched by electron transfer from the phenothiazine group to the excited singlet state of the pyrene group, to an extent which depends on the structure of the copolymer, in particular on the closeness of the two groups in the chain.

(ix) The polymer of *exo,exo*-**82** is a white powder that can be cast from toluene as a flexible transparent film. On heating at 300°C the film becomes red–black and insoluble, 2 equivalents of acetic acid being lost; eqn. (5). The all-*trans* polymer of **84** breaks down at 200°C to give a similar product as does the 93% *trans* polymer of **83** at 270°C. Attempts to catalyze this type of reaction at lower temperature have not been very successful (Zenkl 1992b; Fischer 1993).

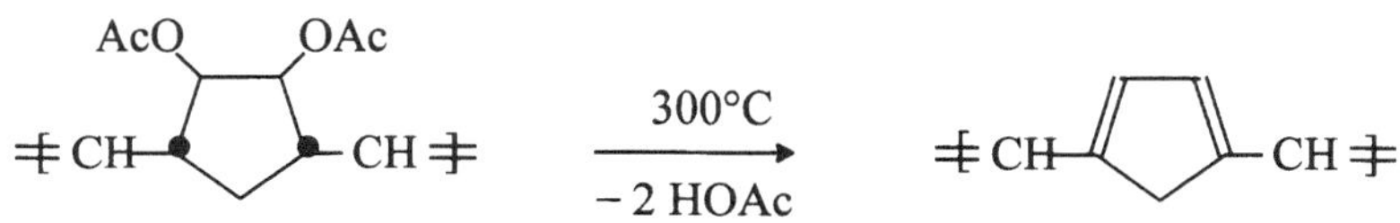

(5)

(x) The polymers of **85**, **86**, **91**, **92**, and **93** are notable in being able to form complexes with metal compounds as O and S donors (to Zn or Cd compounds), N donors (to Sn or Pb compounds), or P donors (to Au or Ag compounds). Diblock copolymers of these monomers with methyltetra-cyclododecene are described in Ch. 14.

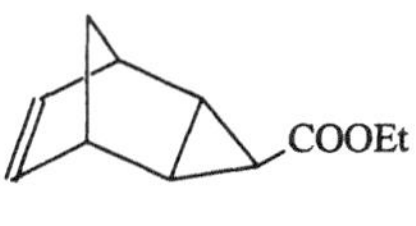

(94)

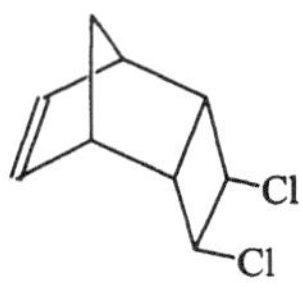

(95)

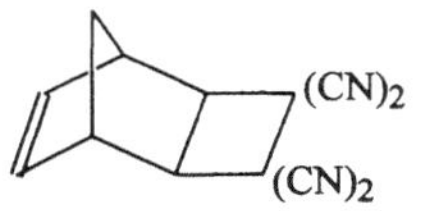

(95a)

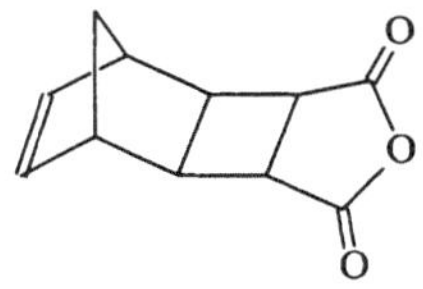

(95b)

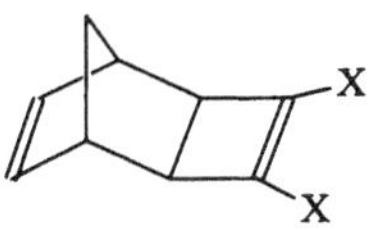

(95c) X = CF_3
(95d) X = CO_2Me
(95e) X = CN

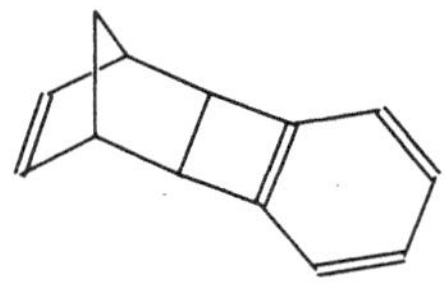

(96) (*endo* and *exo*)

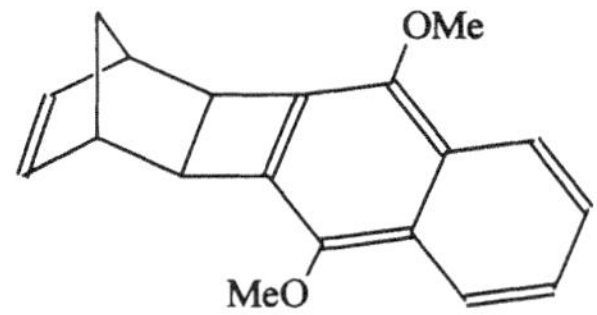

(96a)

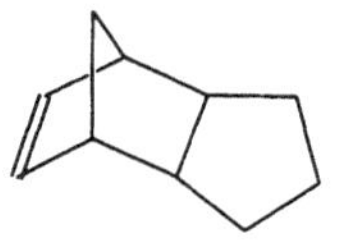

(97) (*endo* and *exo*)

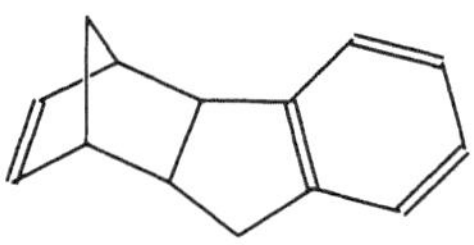

(98)

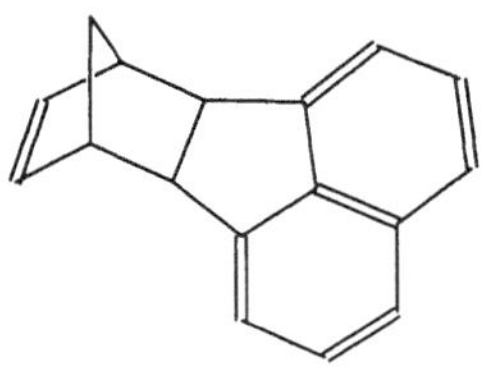

(99)

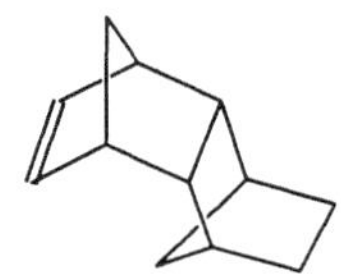

(100)

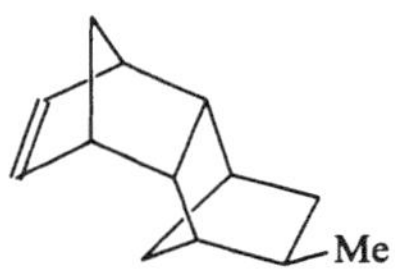

(101) (main isomer)

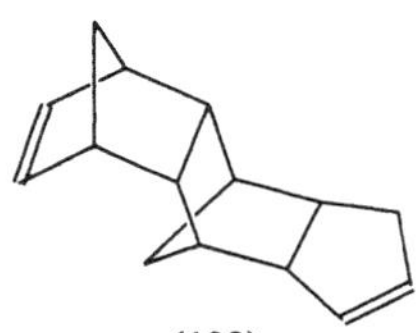

(102)

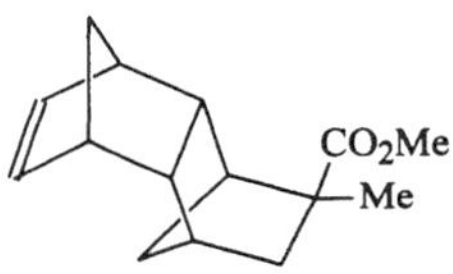

(103)

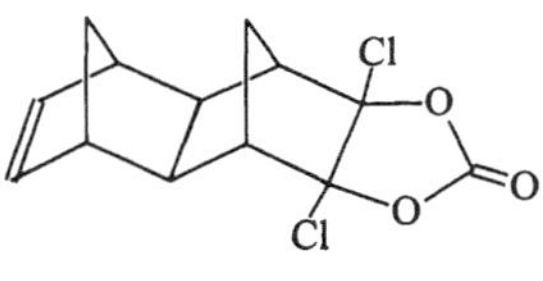

(104)

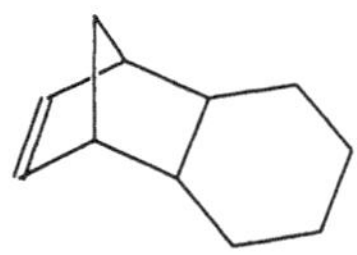

(105)

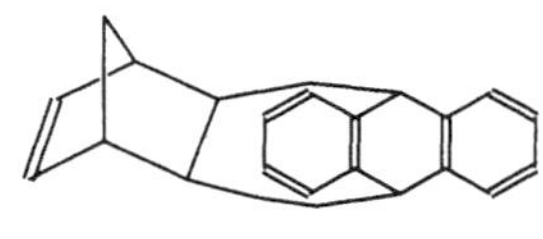

(106)

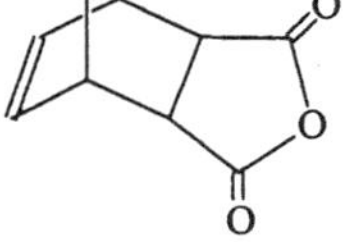

(107) (*endo* and *exo*)

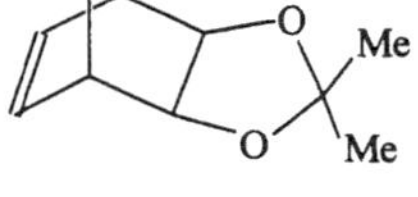

(108) (*exo*)

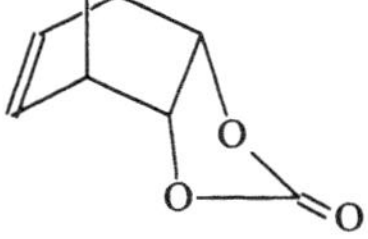

(109) (*endo*)

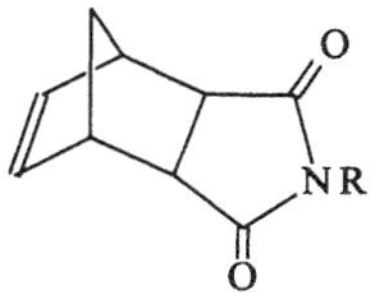

(110) R = H, **(111)** R = Me, **(112)** R = Pr, **(113)** R = Ph,
(114) R = C_6H_4X (*o*, *m*, *p*; X = F, Cl, Br, I), **(115)** R = C_6F_5,
(116) R = *CHMeCO_2Me (optically active and racemic monomers)

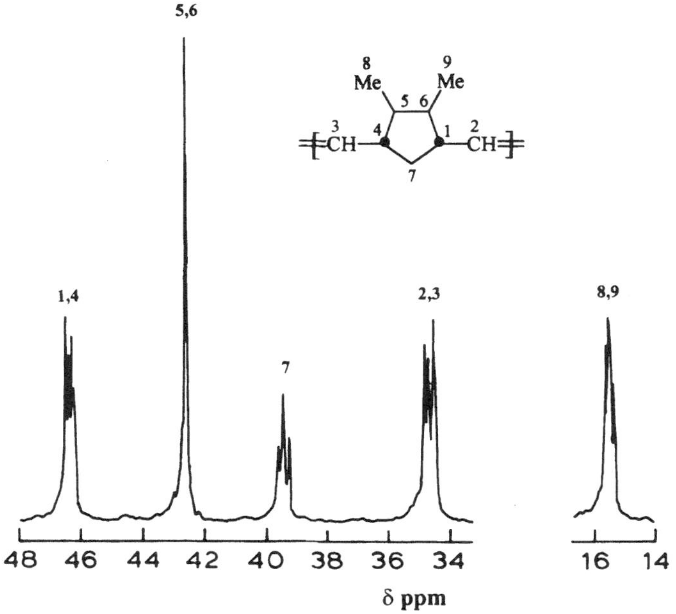

Fig. 13.2 62.8 MHz ^{13}C NMR spectrum of the polymer obtained by hydrogenating a 53% *cis* polymer of *exo,exo*-5,6-dimethylnorbornene made using WCl_6/Me_4Sn as catalyst. All carbons except C-5,6 are sensitive to tactic triads (Greene 1987a).

(xi) Bis[(4′-cyanobiphenyl-4-yl)oxy-*n*-alkyl]norborn-2-ene-5,6-carboxylates, with alkyl chain lengths 2-12, undergo ROMP. Both monomers and polymers exhibit liquid-crystalline phases (Ungerank 1995).

In the second group of monomers are 5-cyano-6-phenylnorborn-2-ene (Ueshima 1974) and *trans*-5-carbomethoxy-6-ferrocenylnorborn-2-ene. In the latter case fluorescence quenching studies have been made as for the polymers and block copolymers derived from **81** (Albagli 1992b, 1993).

The third group of monomers include compounds **94** (Ast 1976c), **95** (Alimuniar 1985), **95a** (Feast 1995; Saunders 1995), **95b** (Saunders 1995, ROMP unsuccessful), **95c,d** (Saunders 1995), **95e** (Feast 1995), **96** (El-Saafin 1982), **96a** (Saunders 1995), **97** (Kobayashi, S. 1967; Oshika 1968; Rinehart 1968; Hamilton 1986), **98** (Sjardijn 1990a), **99** (El-Saafin 1982; Sjardijn 1990a), **100** (Nishi 1990b), **101** (Asrar 1991), **102** (Nishi 1989, 1990a), **103** (Goto 1990a,b; Komiya 1991; Iio 1991), **104** (Feast 1985a), **105** (Reif 1983), **106** (Sjardijn 1990a), **107** (Ueshima 1975; Kurosawa 1976a; Matsumoto 1977a,b; Castner 1982; Ivin 1988; Bell, B. 1994), **108** (Bazan 1991b), **109** (Feast 1985a), **110** (Matsumoto 1977a,b; Asrar 1989), **111** (Hardiman 1992), **112** (Matsumoto 1977a,b), **113** (Asrar 1992b, 1993; Bazan 1991b), **114** (Asrar 1994a), **115** (Blackmore 1988; Bazan 1991b), **116** (Coles 1994).

The ROMP of **95c** and **95d** proceed through the opening of the norbornene ring and not the sterically hindered cyclobutene ring, and the same is probably true of **95e**, which gives an insoluble polymer, and also of **96a**.

In much of the early work the monomers used were mixtures of isomers. For example, *endo,anti*-**95**, as prepared, contains minor amounts of the *exo,syn*-isomer, a smaller amount of the *endo,syn*-isomer, but no *exo,anti*-isomer. Again, **101** is the main isomer (66%) in a mixture with seven others of which only one is present in significant amount (33%). In general the *endo*-isomers are less reactive than the *exo*-isomers, no doubt because of the greater degree of steric hindrance when the *endo*-isomer approaches the propagating complex. However, *endo*-isomers that are unreactive with A- or B-type catalysts, as defined in Table 13.3, have sometimes been found to react slowly when placed in contact with a metal carbene initiator. For example, when 3.2 equivalents of *exo*-**107**, containing some of the *endo*- isomer, are mixed with $W[=\overline{C(CH_2)_3C}H_2](OCH_2CMe_3)_2Br_2$ in CD_2Cl_2 at 25°C, the *exo*-monomer reacts first to give the propagating species P_1, which then adds further *exo*-monomer to give P_n ($n > 1$). P_1 and P_n ($n > 1$) are distinguished by their $(OCH_2CMe_3)_2$ NMR signals at 4.44, 4.39, and 4.46, 4.41 ppm, respectively, the two neopentoxy ligands being non-equivalent in each case. The concentration of P_1 passes through a maximum after about 10 min and then declines as it is replaced by P_n. The ratio of concentrations of P_1 and initiator at this maximum gives the ratio of initiation to propagation rate constants, k_i/k_p, as 3.8 (see Fig. 3.5). Once the initiator has all been consumed, the concentration of propagating species is constant and the remaining monomer disappears with a half-life of 74 min, corresponding to $k_p = 3.15 \times 10^{-3}$ M^{-1} s^{-1}. The carbene proton signals for P_1 and P_n are not resolved; both give a doublet at 11.245 ppm. Towards the end of the reaction, when the *exo*-monomer peaks have practically disappeared, a second carbene proton doublet appears in very low intensity at 11.59 ppm and other weak peaks appear in the upfield part of the spectrum. On addition of pure *endo*-isomer to the reaction mixture, in amount similar to that of the original *exo*-isomer, the weak doublet at 11.59 ppm grows at the expense of the *exo*-P_n doublet at 11.245 ppm, until after 3 h it represents 70% of the tungsten carbene proton signal. The fall in concentration of the *endo*-monomer is also several times that of the new carbene proton species, indicating that more than one molecule of *endo*-monomer has added to the chain (Ivin 1988). Likewise *endo*-**109** cannot be homopolymerized by the relatively short-lived WCl_6/Me_4Sn catalyst but can be incorporated into a copolymer with norbornene. *Exo*-**107** and *exo*-**108**, initiated with a molybdenum carbene complex, give polymers containing 30% and 40% *cis* double bonds, respectively. Polymers of *exo*-**107** are readily converted to diester or half-ester derivatives by reaction with alcohols. It is also possible to carry out ROMP of *exo*-**107** and esterification in one operation using $RuCl_3$ or $OsCl_3$ at 70°C as catalyst (Bell, B. 1994). The polymer of *exo*-**108** is readily converted to the diol derivative by hydrolysis with aqueous trifluoroacetic acid (Bazan 1991b).

The ROMP of *endo*-**113** gives a 15% yield of a low-molecular-weight product. In contrast, the ROMP of *exo*-**113** proceeds to high conversion with the formation of high-molecular-weight polymers. The *cis* contents range from 0% ($RuCl_3$ catalyzed) to 48%. The bridgehead protons at C-7 in the all-*trans* polymer are non-equivalent (1.74 and 2.23 ppm) (Asrar 1992b). The 7-isopropylidene derivative of **113** also gives a high-*trans* polymer with $RuCl_3$ (Shahada 1994). The T_gs of the polymers of *exo*-**114** vary with the substituent X: from 199°C (X = *m*-I) to 270°C (X = *o*-Br) (Asrar 1994a,b). The ROMP of *exo*-**115** catalyzed by $MoCl_5/Me_4Sn$ at 60°C gives a high yield of high-*trans* polymer. *Endo*-**115** fails to homopolymerize but can be copolymerized to some extent with its *exo*-isomer (Blackmore 1988).

Potentially the most interesting polymers in the third group are those of **116**, made by reacting **107** with (+)- or (−)-alanine methyl ester, and thus containing a chiral substituent. The chirality and molecular recognition capacity of the resulting polymers might ultimately be useful as a template for controlling the architecture of other polymers formed in their presence. Both *endo*-**116** and *exo*-**116** readily undergo ROMP under the influence of $Mo(=CHCMe_2Ph)(=NC_6H_3\text{-}i\text{-}Pr_2\text{-}2,6)(OCMe_3)_2$ to give 78–91% *trans*, optically active polymers of narrow MWD (Coles 1994). The synthesis of numerous monomers related to **116** has been reported (Biagini 1995).

For the ROMP of norbornene attached to C_{60} at the 5,6-positions, see Ch.14.

13.3.6 Other polysubstituted norbornenes

Various other polymerizable di-, tri-, and tetra-substituted norbornenes are listed in Table 13.4. The polymer of **117** is insoluble but its solid-state ^{13}C NMR spectrum indicates that the polymerization has occurred mainly by ring opening.

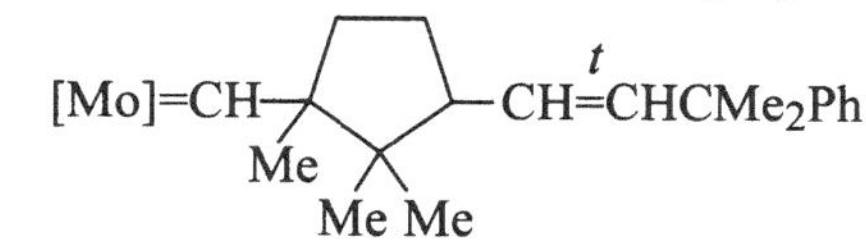

(130)

1,7,7-Trimethylnorbornene, **118**, was originally thought to be unpolymerizable (Hamilton 1985b), but it does in fact yield to the very active and long-lived molybdenum carbene initiator shown in Table 13.4. In CD_2Cl_2 at 20°C the head alkylidene adduct P_{1H} (**130**), in which the C=C bond is *trans*, is formed after a few hours, reaching a maximum of more than half the original initiator concentration, and then declines very slowly as more monomer adds to P_{1H} to give an all-*trans*, all-HT polymer. The first-order rate constant for the decay of initiator in the presence of excess monomer M shows a complex dependence on the monomer concentration, tending towards first-order at low [M] and zero-order at high [M]. This is interpreted in terms of a mechanism involving an equilibrium between the *syn* and *anti* rotamers of the initiator, in which the conversion of the dominant *syn* rotamer into the minor *anti* rotamer is rate-determining at high [M], with $k = 6.1 \times 10^{-5}\ s^{-1}$, while at low [M] the addition of M to the *anti* rotamer becomes rate-determining (Feast 1994d). The value of k agrees well with the value determined directly in toluene by photochemical displacement of the equilibrium between the rotamers (Oskam 1992, 1993a). Polymer made from a single enantiomer, being all-HT, is necessarily isotactic, while polymer made from racemic monomer is found to be atactic; every carbon, except that corresponding to the *anti*-7-methyl carbon, is sensitive to tacticity.

The spiro compound, **119**, is more readily polymerizable than *syn*-7-methylnorbornene (**31**); the cyclopropyl group evidently offers less steric hindrance than the *syn*-methyl group in the propagation reaction. The pattern of the fine structure of the ^{13}C NMR spectrum of the high-*trans* polymers of **119** is similar to that of the high-*trans* polymers of *anti*-7-methylnorbornene (**29**) and *m*/*r* assignments can be made on this basis (Seehof 1993a). The high-*trans* polymer of **29** is completely atactic ($m/r = 50/50$), but the high-*trans* polymers of **119** are somewhat biased towards syndiotactic dyads ($m/r = 38/62$).

For monomers **124** (62% *exo*-*i*-C_3F_7-*endo*-CF_3, 38% *endo*-*i*-C_3F_7-*exo*-CF_3), **125** (55% *exo*-CF_3, 45% *endo*-CF_3), and **126** (48% *exo*-$(CF_2)_3$, 52% *endo*-$(CF_2)_3$), the first-named *exo*-isomer is in each case the more reactive so that if polymerization stops short of 100% conversion, the residual monomer is enriched in the other isomer (Seehof 1993b).

Films of the polymer of **127** exhibit high permeability to gases because of their high fractional free volume. However, this is offset by a somewhat lower discrimination between different gases, as compared with films of other polymers (Teplyakov 1992).

Polymers of **128** have a narrow MWD if prepared in THF but not if prepared in chlorobenzene. The polymer has a T_g of 110°C and decomposes at 300°C, losing two molecules of acetic acid per repeat unit. It is also readily hydrolyzed to the polydiol, which is soluble in $CF_3COOH/CHCl_3$ and degrades with loss of water at 300°C (Bazan 1991b).

Monomer **129** is not readily polymerized by the molybdenum carbene complex shown in Table 13.4. Instead there is reversible formation of a molybdacyclobutane complex, the rearrangement of which to a metal carbene complex and subsequent propagation has to compete with side reactions which drain away the carbene complexes (Bazan 1991b). However, ROMP does occur with WCl_6/Me_4Sn as catalyst at 70°C (Feast 1985a) and a film of the polymer **129P** can be readily hydrolyzed by 0.5 M NaOH to give a black product containing the conjugated units **131**; eqn. (6). This product can also be generated as a surface layer on its precursor polymer, but unfortunately is vulnerable to atmospheric oxidation (Feast 1986b).

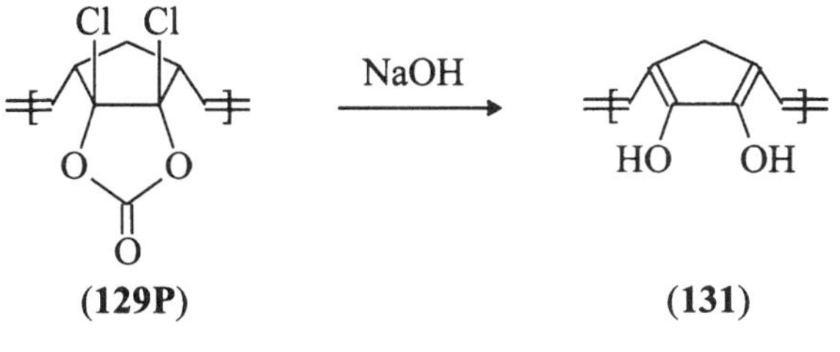

(129P) **(131)** (6)

Table 13.4 ROMP of other polysubstituted norbornenes

Monomer	Substituent position[a]							Catalyst	σ_c [b]	Reference
	1	5 *endo*	5 *exo*	6 *endo*	6 *exo*	7 *syn*	7 *anti*			
117		=CH_2		=CH_2				WCl_6/Me_4Sn		Shahada 1992
118	CH_3					CH_3	CH_3	Mo-1[c]	0.0	Feast 1994d
119						CH_2–CH_2		$RuCl_3$	0.05 [d]	Seehof 1993a
								WCl_6/Ph_4Sn	0.15 [e]	
								WCl_6/PhC≡CH	0.75	Finkel'shtein 1993
120		CF_3	F	F	F			WCl_6/Ph_4Sn		Feast 1980
121		CF_3 /	F	F	F			WCl_6/Ph_4Sn		Feast 1980
								Mo-1[c], Mo-2 [g]		Feast 1994b
122		CF_3	F	F	CF_3			WCl_6/Ph_4Sn		Feast 1980
123		CF_3 /	Cl	Cl /	CF_3			WCl_6/Ph_4Sn		Feast 1980
124		CF_3 /	F	F /	*i*-C_3F_7			WCl_6/Ph_4Sn		Seehof 1992, 1993b
125		CF_3 /	F	F	F	CH_2–CH_2		WCl_6/Ph_4Sn	0.08	Seehof 1992, 1993b
126		X [f] /	F	Y [f] /	F			WCl_6/Ph_4Sn	0.20	Seehof 1992, 1993b
127		CF_3	CF_3	F	F					Teplyakov 1992
127a		CF_3	CF_3	CN	CN			Mo-2 [g,h]		Feast 1995
128			OAc				OAc	Mo-2 [g]		Bazan 1991b
129		X [i]	Cl	Y [i]	Cl			WCl_6/Me_4Sn	0.45	Feast 1985a
										Feast 1986b
								Mo-2 [g]		Bazan 1991b
129a		CN	CN	CN	CN			Mo-2 [g,h]		Feast 1995

[a] A solidus in these columns indicates that the monomer used was a mixture of *endo* and *exo* isomers. [b] σ_c = fraction of *cis* double bonds in polymer. [c] Mo(=$CHCMe_2Ph$)(=NC_6H_3-*i*-Pr_2-2,6)[OCMe$(CF_3)_2$]$_2$. [d] *m*/*r* = 38/62 for *trans*-centred dyads. [e] *m*/*r* = 44/56 for *trans*-centred dyads. [f] XY = $CF_2CF_2CF_2$. [g] Mo(=$CHCMe_3$)(=NC_6H_3-*i*-Pr_2-2,6)$(OCMe_3)_2$. [h] No reaction. [i] XY = O(CO)O.

Norbornene derivatives with a keto group in the 5-position do not undergo ROMP (Feast 1985a).

13.4 Monomers containing a fused cyclopentene ring and more than one double bond

13.4.1 Norbornadiene and its derivatives without fused aromatic rings

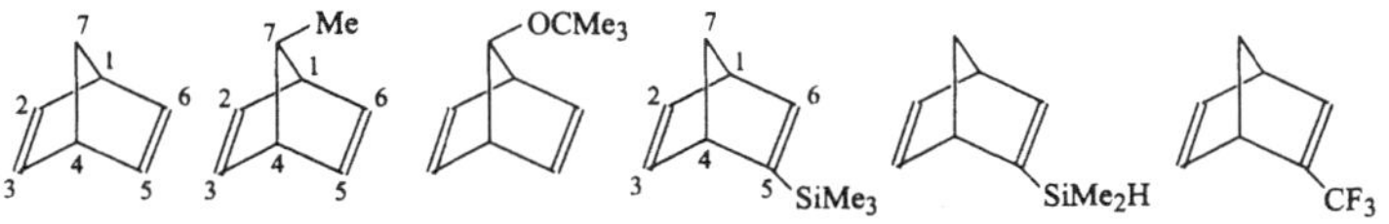

(132) (133) (134) (135) (136) (137)

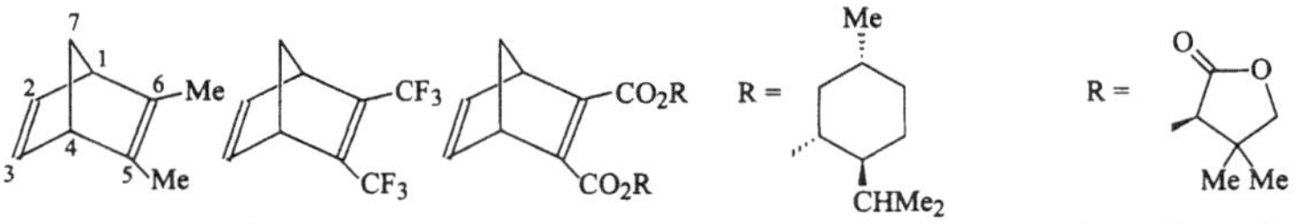

(138) (139) (140a) R = Me, (140b) R = (–)-menthyl, (140c) R = (–)-pantalactonyl

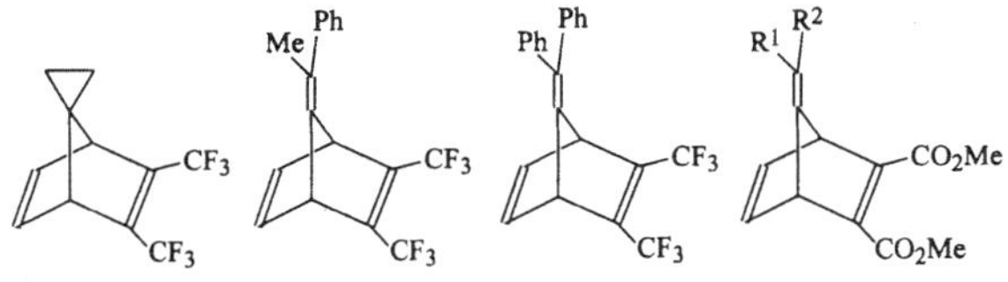

(141) (142) (143) (144–146)

	R^1	R^2
(144)	Me	Me
(145)	Me	Ph
(146)	Ph	Ph

Monomers of this kind that undergo ROMP are **132** (Ivin 1978a; Reif 1981b; Graimann 1985; Gillan 1989; Bell, B. 1992; Hamilton 1993; van der Schaaf 1993), **133** (Hamilton 1993), **134** (Hamilton 1992), **135** (Stonich 1991a), **136** (Stonich 1991c), **137** (Feast 1979, 1980; Blackmore 1987), **138** (Graimann 1985), **139** (Feast 1979, 1980; Stelzer 1982; Alimuniar 1986; Bazan 1989, 1990; Feast 1990, 1992a; Oskam 1992; McConville 1993), **140a** (Bazan 1989, 1990, 1991a; McConville 1993), **140b,c** (O'Dell 1994), **141** (Seehof 1993b), **142** (Feast 1985b), **143** (Feast 1985b), **145** (Feast 1985b), **146** (Feast 1985b), but not **144** (Bazan 1990). For ease of comparison between these monomers and those derived from norbornene, we shall adopt a uniform numbering system as shown for **132**, **133**, **135**, and **138**, with $C^2{=}C^3$ always unsubstituted. This sometimes differs from the IUPAC numbering system.

Where one or both of the 5,6-positions are substituted (**135–143**, **145–146**), ROMP invariably occurs only by cleavage of the unsubstituted double bond. For monomers substituted at the 7-position only (**133–134**), less active catalysts favour cleavage of the less hindered $C^2{=}C^3$ double bond, but more active catalysts show little discrimination between $C^2{=}C^3$ and $C^5{=}C^6$.

13.4.1.1 Norbornadiene (bicyclo[2.2.1]hepta-2,5-diene)

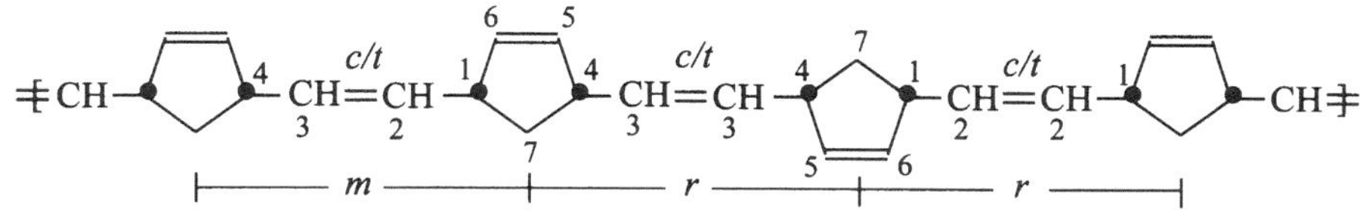

(147)

With norbornadiene (**132**) the first double bond to react can open in one of four ways, as for norbornene, to give either a *cis* or *trans* double bond, located within an *m* or *r* ring dyad, **147**. Polymers have been produced with *cis* contents ($C^2{=}C^3$) ranging from 90% ($OsCl_3$ catalyst) to 37% ($MoCl_5/Bu_4Sn$ catalyst). Provided that hex-1-ene is used as chain-transfer agent (40% of monomer concentration), the polymers are soluble, allowing well-resolved ^{13}C and ^1H NMR spectra to be obtained. These spectra can be fully assigned in terms of the structures represented in **147** and in no case is there any sign of structures arising from the opening of the $C^5{=}C^6$ bond (Fig. 13.3). However, if the concentration of $C^5{=}C^6$ bonds in the polymer solution exceeds 0.2 M in the presence of WCl_6/Me_4Sn, there is immediate formation of a gel which cannot be stirred. This process is reversed by dilution (Reif 1981b). The opening of the second double bond, leading to cross-linking of the polymer, is thus thermodynamically allowed above a critical concentration, as for the ROMP of cyclopentene itself. From the slight variation of this critical concentration with temperature (−40° to 18°C) one obtains $\Delta H^\circ = -4.6$ kJ mol^{-1} and $\Delta S^\circ = -2.9$ J K^{-1} mol^{-1} (standard state 1 M) for the opening of the second double bond (Höcker 1984).

Several other features are worthy of note in the ROMP of norbornadiene. First, $OsCl_3$ gives a polymer with a much higher *cis* content (90%) than for the ROMP of norbornene (40%); and $RuCl_3$ fails to give any polymer at all, yet with norbornene it gives a high-*trans* polymer. These two differences are ascribed to the di-*endo* chelation of one molecule of norbornadiene to the metal carbene centre, acting as a spectator ligand. For $OsCl_3$ the resultant crowding of the reaction site favours the approach of the monomer leading to the formation of a *cis* double bond, while for $RuCl_3$ the crowding is such as to prevent reaction altogether (Bell, B. 1992). Secondly, like the polymers of norbornene, the high-*cis* polymers of norbornadiene tend to have a blocky *cis/trans* distribution; and the chemical shifts in both polymers are insensitive to tacticity (Ivin 1978a, Bell, B. 1992). Thirdly, a

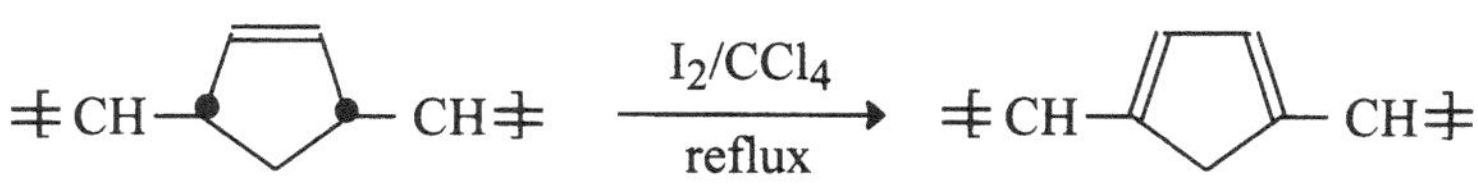

(7)

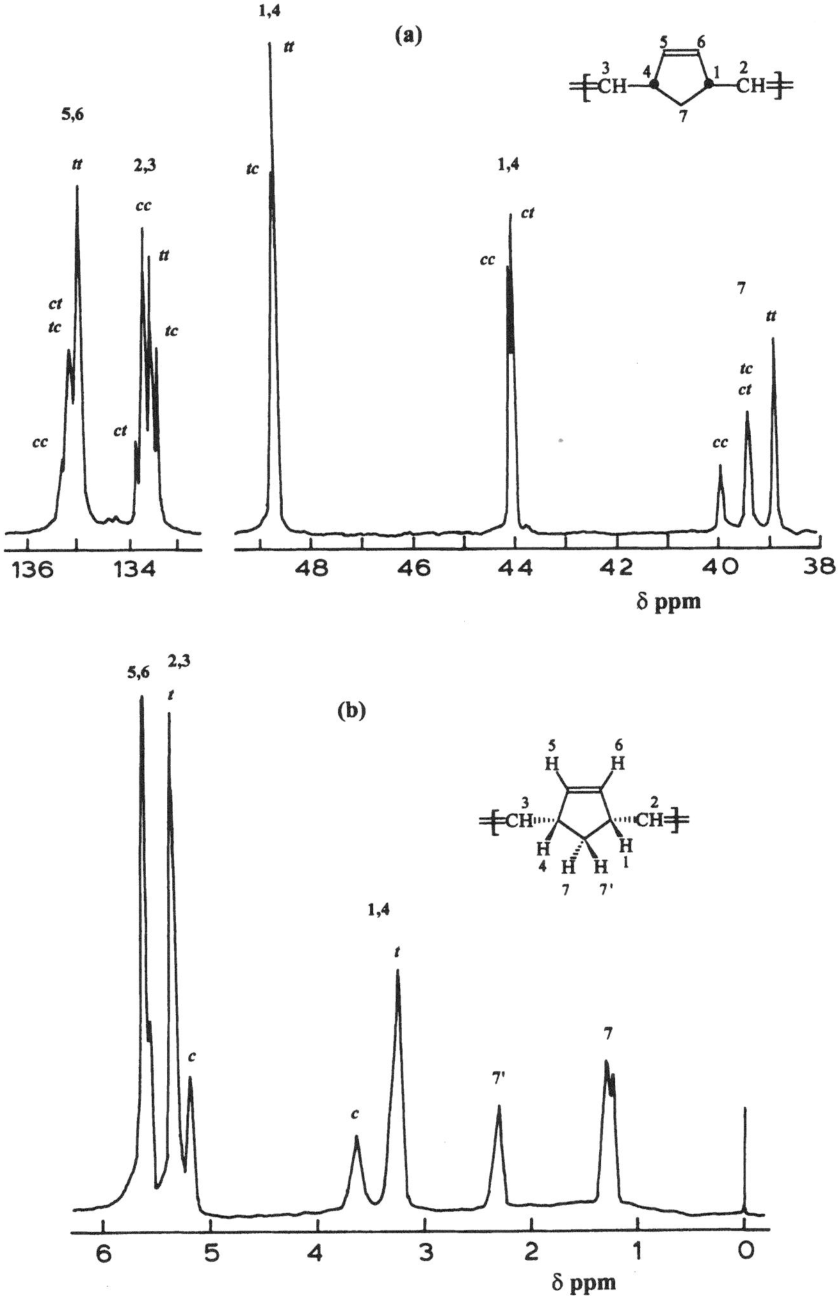

Fig. 13.3 (a) 62.5 MHz ^{13}C NMR spectrum and (b) 300 MHz ^{1}H NMR spectrum of the 37% *cis* polymer of norbornadiene prepared using $MoCl_5/Bu_4Sn$ as catalyst. *c* and *t* refer to the adjacent $C^2{=}C^3$ double bond (Bell, B. 1992).

[W](=$CHSiMe_3$) complex causes very rapid formation of an insoluble polymer (van der Schaaf 1993). Fourthly, the polymer may be dehydrogenated to give a black lustrous rigid solid, which is strongly paramagnetic ($g = 2.0027$) and may be presumed to contain some units of the type shown in eqn. (7) (Gillan 1989).

13.4.1.2 7-Substituted norbornadienes

Like **132**, the monomers **133**, **133a**, and **134** give high-*cis* polymers with $OsCl_3$, containing 7%, 11%, and 30% *syn* units, respectively (Hamilton 1992, 1993, 1995). With Mo(=$CHCMe_2Ph$)(=NC_6H_3-*i*-Pr_2-2,6)[$OCMe(CF_3)_2]_2$ as initiator, both **133** and **134** give high-*cis* polymers containing approximately 50% *syn* units (Ivin 1995). Some of the possible dyad sequences are illustrated in **148**. The reasons for the variations in the *syn/anti* unit ratio are not clear. A ratio of 7/93 corresponds to an energy difference of 7 kJ mol^{-1}, which may be interpreted in terms of repulsion of the methyl group to approach of the monomer to the C^5=C^6 double bond on the *exo* face. When the ratio is near to 50/50, the question arises as to whether the monomer now reacts on the *endo* face where the substituent will have no effect, or whether it still approaches on the *exo* face but with such a high degree of attraction to the electron-deficient metal centre that the repulsive effect of the substituent is brushed aside. There is less discrimination when the substituent is *t*-butoxy than when it is methyl, and the role of the lone pairs of electrons on the oxygen atom must then be considered, in particular whether they can assist the approach of the monomer to the C^5=C^6 double bond on the *exo* face. This might happen through an interaction of a lone pair with a π^* orbital on this double bond. In this way the metallacyclobutanation step becomes a pseudo [3 + 2] reaction thereby facilitating both formation and cleavage of the metallacycle. Another possibility is that the reaction may be assisted by simultaneous coordination of the metal centre to the oxygen atom and the *syn* double bond (*syn–exo* chelation) (Hamilton 1993).

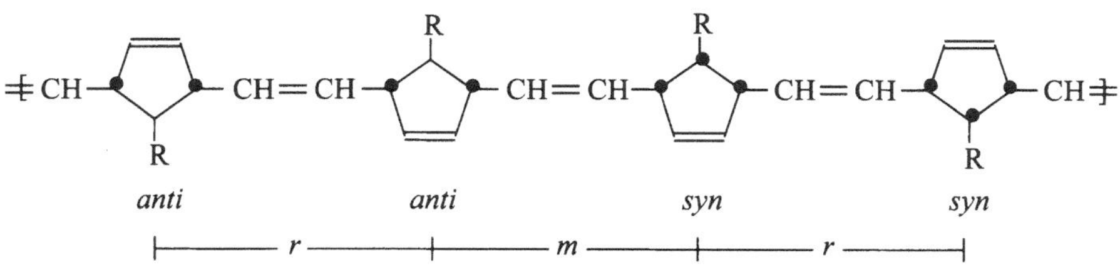

(148)

Polymers of **133** can be made with *cis* contents ranging from 97% ($OsCl_3$ catalyst) to 20% ($MoCl_5/Me_4Sn/Et_2O$ catalyst); a somewhat narrower range (80–33%) is observed for polymers of **133a**. The ^{13}C NMR spectra of the polymers of **133**, containing mainly *anti* units, show a similar *c/t* and *m/r* splitting pattern to that observed in the spectra of polymers of **29**. Tacticity assignments have been made by comparing the spectra of the fully hydrogenated polymers of **133** and **133a** with those of **29**. $ReCl_5$ gives an 80% *cis* polymer of **133** in which the *cis*-centred dyads are *r* and the *trans*-centred dyads are *m*. Very high syndiotacticity is also observed in the 97% *cis* polymer made with $OsCl_3$ as catalyst. In contrast, the 42% *cis* polymer made with WCl_6/Me_4Sn as catalyst is essentially atactic. $RuCl_3$ is ineffective for **133** and **133a**, as for **132** (Hamilton 1993).

13.4.1.3 5-Substituted norbornadienes

Monomers **135-137** are all readily polymerized by WCl_6/Me_4Sn. The NMR spectra of the polymers show that: (i) only $C^2{=}C^3$ is opened; (ii) the substituents are randomly oriented; (iii) the *c/t* ratio is approximately 50/50 in each case; and (iv) C-1 and C-4 each show the expected *c/t* pair of signals separated by about 5 ppm (*cis* upfield). For **137** the *cis* content obtained with other catalysts is: $Mo(=CHCMe_2Ph)(=NC_6H_3\text{-}i\text{-}Pr_2\text{-}2,6)(OCMe_3)_2$, 10%; $MoCl_5/Me_4Sn$, 30%; $OsCl_3$, 45%; $RuCl_3$, 44%; $IrCl_3$/TFA, 46%; $ReCl_5$, 86%; $Mo(=CHCMe_2Ph)(=NC_6H_3\text{-}i\text{-}Pr_2\text{-}2,6)[OCMe(CF_3)_2]_2$, 95%; a pattern somewhat different from that observed with **132–134**. The high-*cis* polymer gives a simple 1 : 2 : 1 triplet for the C-2,3 olefinic carbons, consistent with an unbiased polymer, with the TT and HH signals overlapping to give the central line. The 45% *cis* polymer gives a five-line spectrum for these carbons, resulting from overlap of the *cis* triplet with the corresponding *trans* signal. In the high-*cis* polymer the C-1, C-4, and C-7 signals also show HH/HT and TT/TH splittings (Blackmore 1987; Feast 1994b).

13.4.1.4 5,6-Disubstituted norbornadienes

The ^{13}C NMR spectrum of the polymer of **139** made with WCl_6/Ph_4Sn as catalyst is shown in Fig. 13.4. The *c/t* fine structure is assigned by comparison with spectra of polymers made with other catalysts such as $RuCl_3/Me_4Sn$ (30% *cis*) and $MoCl_5/Me_4Sn$ (13% *cis*) (Alimuniar 1986; corrected from the earlier assignment). The range of *cis* content has been extended to the two extremes through the use of molybdenum carbene complexes as initiators. Thus **149a** and **149b** give all-*trans* polymers having ^{13}C NMR spectra with single lines for C-2,3 (133.50), C-1,4 (49.80), and C-7 (37.20 ppm), while **149c** and **149g** give all-*cis* polymers (Bazan 1990; McConville 1993). However, the all-*cis* polymer formed with **149g** gives a single C-7 line at 38.38 ppm, but that formed with **149c** gives three lines: at 38.38, 37.61, and 36.44 ppm, in order of diminishing intensity. This fine structure in an all-*cis* polymer can only be the result of *m/r* splitting. One can therefore say immediately that the former polymer is fully tactic (100/0), and the latter partially tactic in the same direction (74/26), but without knowing whether the line order is *mm, mr* (≡ *rm*), *rr*, or the reverse. Nor can one be sure that the all-*trans* polymers made using **149a** and **149b** are tactic since no all-*trans* polymers have been made that show fine structure for C-7 (or any other carbon), and it is not uncommon for

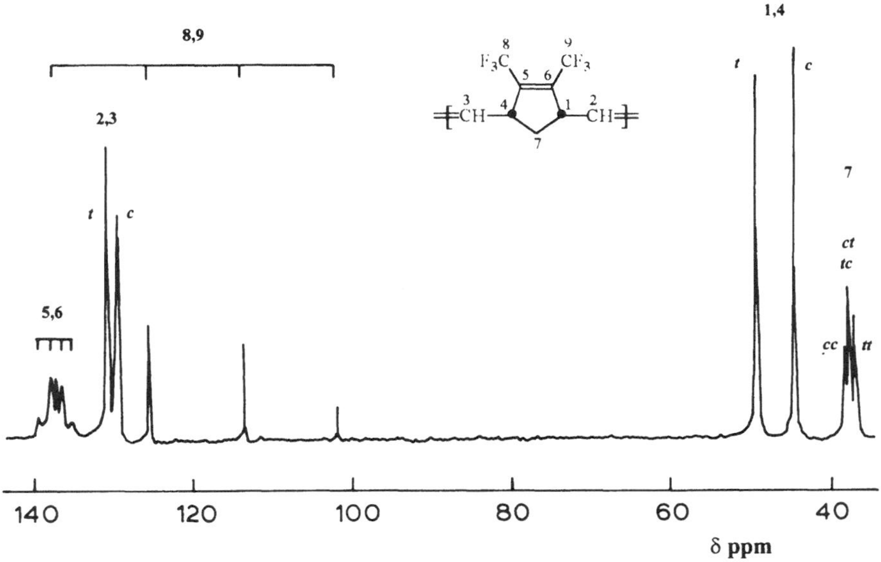

Fig. 13.4 22.63 MHz ^{13}C NMR spectrum of ring-opened polymer of **139** (46% *cis*); solvent $(CD_3)_2CO$. C-8,9 and C-5,6 show fluorine splittings, $J(C-F)=273$ Hz and $J(C-C-F)=31.8$ Hz, respectively; *t/c* and *tt/tc/cc* assignments refer to the nearest $C^2{=}C^3$ bonds (Alimuniar 1986; Feast 1980; corrected assignments). The C-7 signal also contains *m/r* fine structure (see text).

m/r splittings to be unresolved in atactic polymers. The fact that the *trans* polymer is semi-crystalline with a melting point at 200°C, and can be fibred and cold-drawn, lends support to the belief that it is highly tactic (Bazan 1989; Feast 1990). On the basis of the known tacticities of polymers of chiral ester monomers **140b**,**c** made with the same initiators (see below), one may speculate that the polymer made using **149g** is isotactic (all-*cis*) and that made using **149a** is syndiotactic (all-*trans*) (O'Dell 1994). The *trans* polymer of **139** has an unusually high relaxed dielectric constant, which accords with a predominantly syndiotactic structure, but the relatively low value for the *cis* polymer can be interpreted in terms of either a syndiotactic structure (Davies 1993, 1995, 1996) or an isotactic structure (O'Dell 1994).

An all-*cis* polymer of **139** made with initiator **149f** in toluene has the opposite tacticity ($\sigma_m \sim 0.12$ with the above assumption) to that made using **149c** in THF ($\sigma_m \sim 0.87$). A marked temperature dependence of the *cis* content of the polymer formed using **149h** in THF (100% at −35°C, 24% at 60°C) has been interpreted in terms of an equilibrium between *syn* and *anti* rotamers, with *cis* C=C formed mainly by addition to the THF-free *syn* rotamer, and *trans* C=C formed mainly by addition to the THF-free *anti* rotamer (Schrock 1995b). In Ch. 3 we have seen that the reaction of **139** with **149a** proceeds mainly *via* the less abundant, but much more reactive *anti* rotamer of the initiating and propagating species.

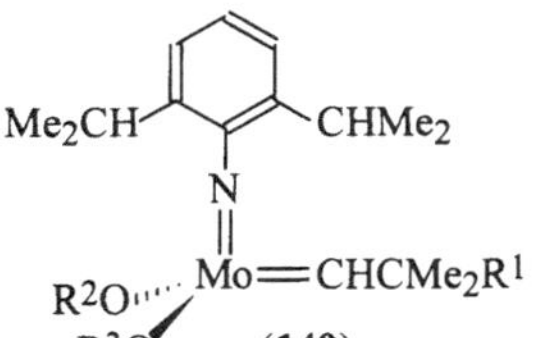

(**149**)

(**149a**) $R^1 = Me$, $R^2 = R^3 = CMe_3$
(**149b**) $R^1 = Ph$, $R^2 = R^3 = CMe_3$
(**149c**) $R^1 = Ph$, $R^2 = R^3 = CMe(CF_3)_2$
(**149d**) $R^1 = Ph$, $R^2 = CMe_3$, $R^3 = CMe(CF_3)_2$
(**149e**) $R^1 = Me$, $R^2 = R^3 = CMe(CF_3)_2$
(**149f**) $R^1 = Ph$, $R^2 = R^3 = C(CF_3)_2(CF_2CF_2CF_3)$

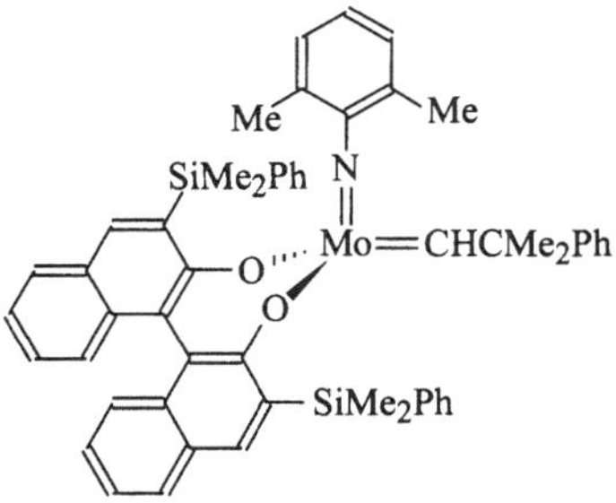

(**149g**) (racemic)

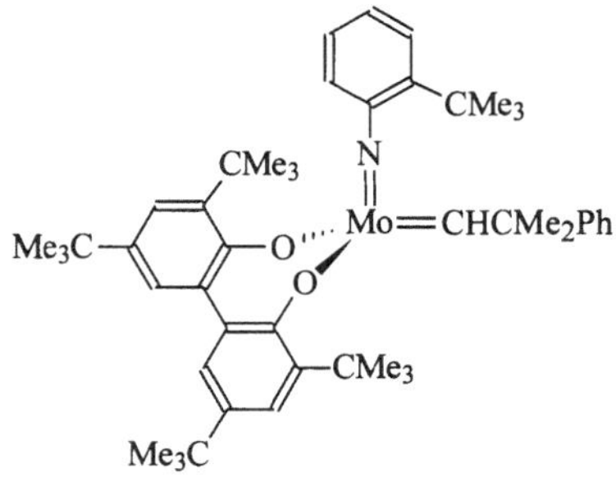

(**149h**)

Fig. 13.5 shows that the first four propagating species may be distiguished in the ^{1}H NMR spectrum; for the longer chains the *anti* and *syn* rotamers are just resolved, in the ratio of 1 : 6. The corresponding ^{13}C NMR signals are also resolved: 253.1 and 252.6 ppm. In benzene at 22°C, $k_p/k_i = 0.72$ and $k_p = 0.057$ M^{-1} s^{-1}. The reactivity of **139** is some 30 times less than that of **140a**. This may be ascribed mainly to a lower electron density at the unsubstituted double bond in **139** (Bazan 1990; Fox 1993).

The ROMP of **139** initiated by a mixture of **149b** and **149c** in trifluorotoluene might have been expected to give a mixture of all-*trans* and all-*cis* polymer. Instead the *cis* and *trans* double bonds are distributed throughout the chains, as shown by the C-7 fine structure (Davies 1993). This is because the alkoxy ligands in **149b** and **149c** undergo rapid exchange to form an equilibrium mixture with **149d**, and because the rates of exchange of alkoxy ligands between the corresponding propagating species are much faster than the rates of addition of monomer. A polymer of any desired *cis* content can thus be prepared from **139** by mixing with appropriate proportions of the two initiators (Feast 1992a). One can also make a *trans/cis* stereoblock polymer by using **149a** as initiator of the first batch of monomer and, before the addition of a second batch of monomer, converting the OR ligands to $OCMe(CF_3)_2$ by reaction with $HOCMe(CF_3)_2$ (Broeders 1996). A similar strategy has been used to make block copolymers in which short emissive poly(*p*-phenylenevinylene) sequences are encased in a fluorinated polymer matrix (Bazan 1996b).

The tungsten analogue of **149a** adds to **139** to form a *transoid* square–pyramidal metallacyclobutane complex, which can be isolated and characterized at low temperature. Its rearrangement to the propagating metal carbene complex can also

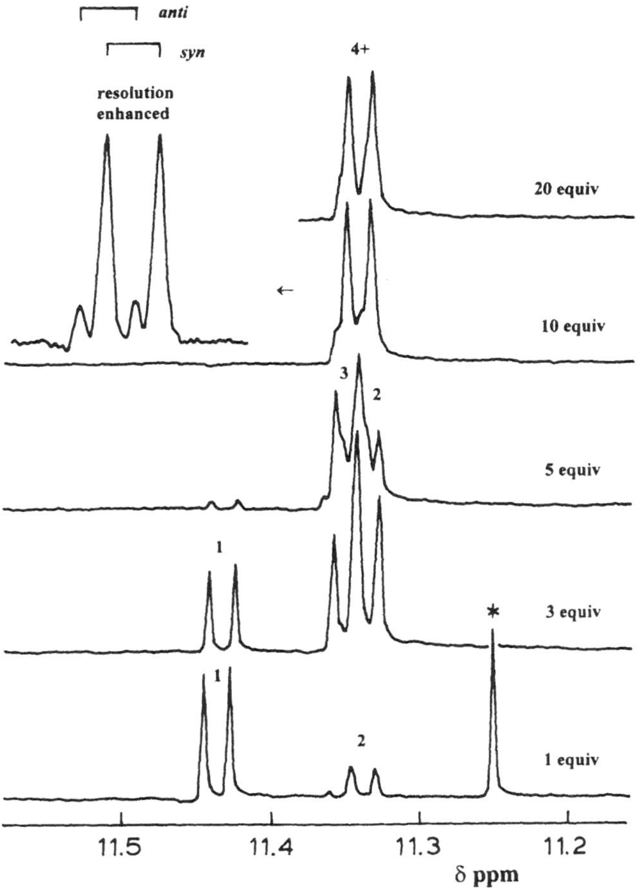

Fig. 13.5 500 MHz ^{1}H NMR spectra in the carbene proton region of samples containing $Mo(=CHCMe_3)(=NAr)(OCMe_3)_2$ and 1, 3, 5, 10, and 20 equivalents of **139**. Numbers above the resonances refer to the number of insertions of monomer. The asterisk marks the singlet from the initiator (Bazan 1989).

be followed but the product is not stable, and the system is therefore not satisfactory for making a well-defined polymer (Bazan 1990).

The ROMP of the diester **140a** is initiated by **149a** in C_6D_6. The carbene proton signal of the living polymer consists of two doublets in the ratio 5 : 1 at δ 11.694 and 11.652, respectively, attributed to the *syn* and *anti* rotamers, though which is which is uncertain. The signals from the first insertion product are well resolved from those of the longer chains. The double bonds formed are mainly *trans* ($\sim$95%) and the ^{13}C NMR spectrum shows a single set of lines, apart from the very small peaks arising from the presence of about 5% *cis* double bonds. This strongly suggests but does not prove that the polymer is tactic. The value of k_p/k_i is 3 and the initiator is fairly quickly consumed when there is an excess of monomer. The polymer has $M_w/M_n = 1.06$, and decomposes above 200°C (Bazan 1989, 1990; Feast 1994b).

In order to throw light on the vexed question of the tacticity of the polymer of **140a**, the chiral diesters **140b** and **140c** have been prepared and polymerized using several molybdenum carbene initiators. The protons H-2 and H-3 attached to C-2 and C-3 are now non-equivalent both in the monomer and polymer. In an *m* dyad the double bonds (*cis* or *trans*) will all be of the type $-C^2H^2=C^3H^3-$, so that, if the chemical shift difference is not too small, the signal for H-2,3 will be an AB quartet, split further by coupling to the adjacent ring protons (cf. **147**). On the other hand, in a fully syndiotactic polymer the double bonds in the *r* dyads will be alternately of the types $-C^2H^2=C^2H^2-$ and $-C^3H^3=C^3H^3-$, and there will be no coupling between H^2 and H^3. The ^1H–^1H correlation spectrum (COSY) of the all-*trans* polymer of **140b** made with **149b** as initiator, Fig. 13.6a, shows clearly that the olefinic protons of the *trans* double bond are not coupled (δ 5.568 and 5.534), and that the polymer is therefore fully syndiotactic. Note also that the couplings to the adjacent allylic protons, H-1 and H-4, are not resolved. In the ^{13}C NMR spectrum the C-1 and C-4 signals are just resolved (δ 46.7 and 46.8), while C-7, being always situated within an *rr* triad, gives a single peak at δ 37.5. For the all-*cis* polymer of **140b** made with **149g** as initiator the ^{13}C NMR spectrum gives a single set of peaks as expected for a tactic polymer. Its COSY spectrum, Fig. 13.6b, shows that the H-2 and H-3 protons (δ 5.51 and 5.37) *are* coupled and that the dyads are therefore isotactic (*m*). In this case there is some observable coupling to H-1 and

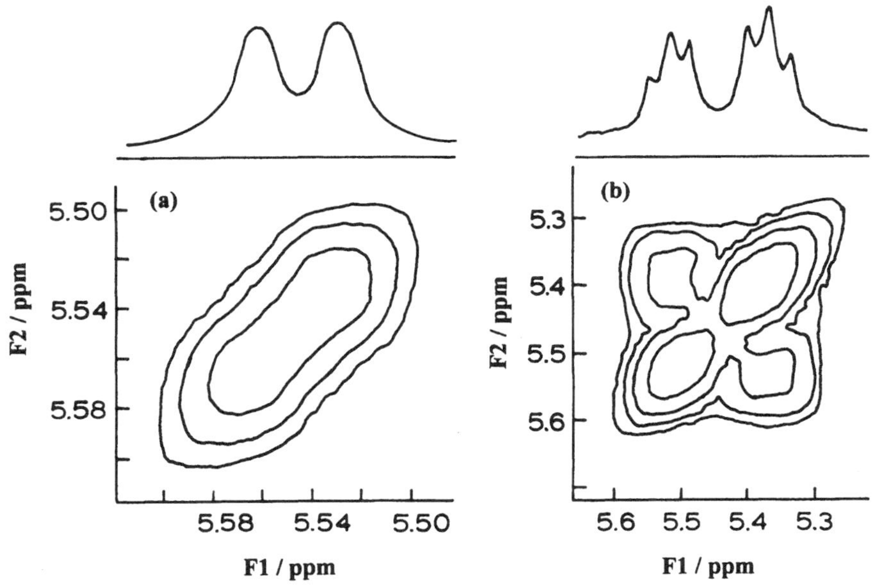

Fig. 13.6 ^1H–^1H COSY spectra of the olefinic protons H-2,3 for (a) *trans* polymer of **140b** initiated by **149b**, and (b) *cis* polymer of **140b** initiated by **149g** (O'Dell 1994).

H-4, and on irradiation of these protons, the signal collapses to the expected AB quartet with a coupling constant of 10 Hz characteristic of *cis* C=C. It seems safe enough to carry over these conclusions to the polymers of **140a** but to do so for the polymers of **139** is less secure (McConville 1993, O'Dell 1994).

The all-*cis* polymers in the series **140** (R = Me, Et, *i*-Pr, *t*-Bu), initiated by **149c** in toluene, have an isotactic bias ($\sigma_m \sim 0.78$, 0.84, 0.81, 0.97, respectively), as judged by the C-7 fine structure in the ^{13}C NMR spectra, assuming that the line order is the same in each case (Schrock 1995b). A remarkable observation has been made for the system **140c**/**149g**. The GPC of the polymer ($M_n = 28\,200$) shows two sharp peaks (Fig. 13.7), each of which has $M_w/M_n \sim 1.06$, and taken together, 1.13. This has been attributed to the fact that the initiator is a 1 : 1 mixture of non-interconvertible enantiomers and that highly regular all-*cis* isotactic chains grow separately from each enantiomeric metal site. The chirality at the metal site, interacting with the chiral monomer, controls the rate of propagation, while the chirality of the chain 'end', i.e. the previously added monomer unit, independently controls the stereochemistry of the next monomer addition. This effect is not observed with **140b** as monomer, where the rates of propagation at the two types of site are presumably not sufficiently different (O'Dell 1994).

The ROMP of the dicyano analogue of **139**, also the tricyclic monomer having [$CH_2C(CN)_2C(CN)_2CH_2$] attached to the 5,6-positions, can be initiated by **149a**, the latter giving a 97% *trans* polymer (Feast 1995).

The ROMP of **141** proceeds readily in the presence of WCl_6/Ph_4Sn at 70°C to give an 80% *trans* polymer (Seehof 1993b). The ROMP of the bicyclofulvene derivatives **142**, **143**, **145**, and **146** is catalyzed by $MoCl_5/Ph_4Sn$ at 70°C to give reasonable yields of high-molecular-weight products, probably containing mainly *cis* double bonds. WCl_6/Ph_4Sn, however, is ineffective (Feast 1985b). Surprisingly, **144** is not polymerized by **149a**, although it adds one molecule slowly at 25°C to give the molybdenum carbene adduct. According to both ^1H and ^{13}C NMR spectra,

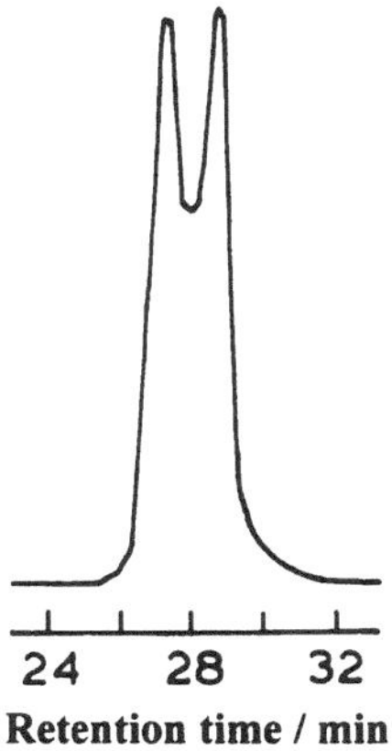

Fig. 13.7 GPC trace of all-*cis* isotactic polymer of **140c** initiated by **149g** (O'Dell 1994).

only one rotamer is present in solution. An X-ray study shows that, in the crystal, the molecule is in the form of the *syn* rotamer and this is presumably also the dominant form in solution. The coupling constant for the carbene CH (in C_6D_6) is J_{CH} =128 Hz. This matches closely the value (130 Hz) for the major isomer of the product of reaction of **140a** with **149a**, which is therefore probably also the *syn* rotamer (see above). Both **149a** and its adduct with **144** are pseudo-tetrahedral around the molybdenum centre. The rate of reaction of **149a** with **144** ($k = 1.3 \times 10^{-3}$ $\text{M}^{-1}\text{s}^{-1}$ at 22°C) is some 500 times smaller than its rate of reaction with **140a** as to be expected for reaction at the *exo* face. Although the adduct of **149a** with **144** will not add more **144**, it will add norbornene, with a k_p/k_i ratio of 270 (Bazan 1990).

13.4.2 Norbornadiene derivatives with fused aromatic rings

The monomers **150–159** all undergo ROMP readily at 20°C in chlorobenzene by opening of the unsubstituted double bond of the norbornadiene ring system: **150** (El-Saafin 1982; Cannizzo 1988; Bazan 1990; Hamilton 1995), **151** (Cannizzo 1988), **152** and **153** (Feast 1991c), **154–156** (Feast 1986a), **157–159** (Schimetta 1994b, 1995). This was first demonstrated for **150** using WCl_6/Ph_4Sn (1/2) as catalyst and later for **152–156** using WCl_6/Me_4Sn. The *cis* contents are mostly 40–50% and are somewhat lower with $MoCl_5/Me_4Sn$ as catalyst, becoming as low as 4% for **156**. The polymer of **153** is insoluble and has not been characterized. The titanacyclobutane complex $Cp_2\overline{TiCH_2CMe_2CH_2}$ has also been tested as initiator for the ROMP of **150** and **151** in toluene but is of limited value since the polymer

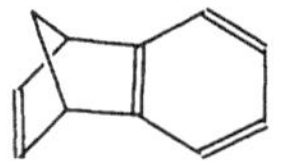

(150)

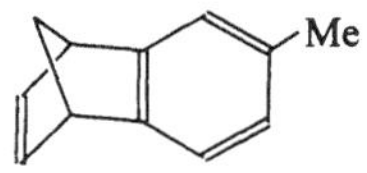

(151)

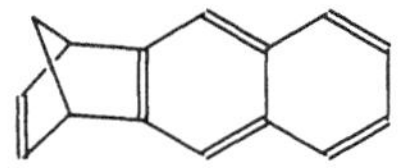

(152)

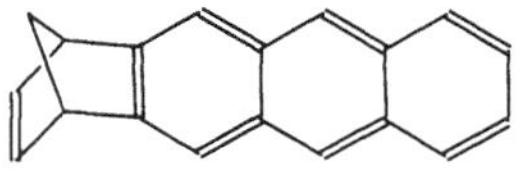

(153)

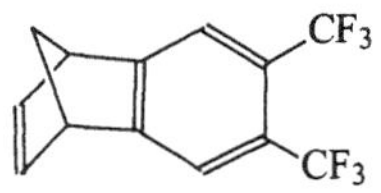

(154)

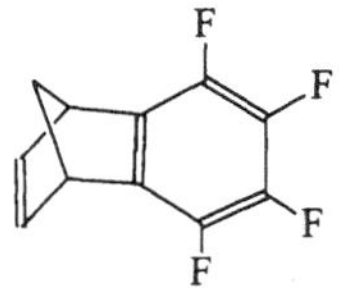

(155)

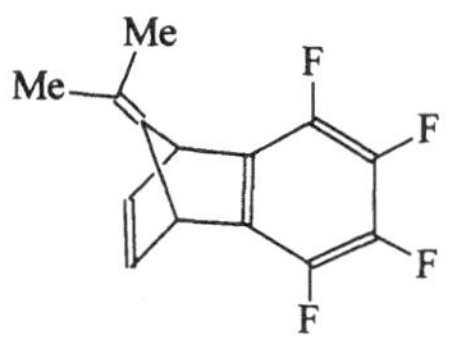

(156)

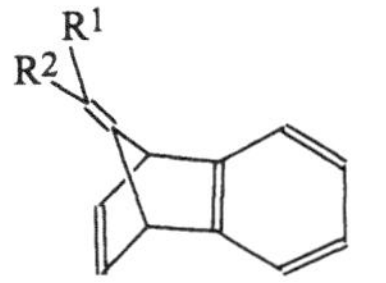

(157) $R^1 = R^2$ =Me
(158) R^1 = Me, R^2 = Ph
(159) $R^1 = R^2$ = Ph

comes out of solution after the addition of only 9 units of monomer (Cannizzo 1988). The reason for this is not entirely clear because Mo(=CHC-Me_3)(=NAr)($OCMe_3$)$_2$ effects smooth and rapid ROMP of 100 equivalents of **150** in toluene ($k_p/k_i = 7$) to give a 24% *cis* polymer with $M_w/M_n = 1.05$. The ^{1}H NMR spectrum of the propagating species shows a complex carbene proton signal consisting of at least four overlapping doublets in the 11.73–11.66 ppm region. This fine structure may arise from: (i) a sensitivity to the *cis* or *trans* nature of the nearest C=C bond, induced by the presence of the benzene ring in the repeat unit; and (ii) the occurrence of *syn* and *anti* rotamers (Bazan 1990). The preparation of various other monomers of this kind has been reported (Petasis 1996).

The most remarkable results for this group of monomers are those obtained with the fulvene derivatives **157–159**. Their ROMP is initiated by both **149a** and **149e** in toluene at 20°C, the latter giving the faster reaction. For each initiator the rate decreases in going from **157** to **158** to **159**, i.e. as the bulk of the substituents becomes larger. Surprisingly the *cis* content of each polymer is independent of the catalyst: 20% for **157**, 0% for **158**, and 100% for **159**. The NMR spectra of the polymers of **158** are in keeping with an all-HT structure, **160**: the olefinic protons are non-equivalent (δ 5.4 and 5.0), and couple with each other ($^3J = 14.7$ Hz) and with the allylic protons at δ 4.3 ($^3J = 8.7$ and 8.3 Hz); and the ^{13}C NMR spectrum consists of a single line for each carbon, suggestive of a highly tactic polymer. If isotactic, each enantiomer will have polymerized separately to yield a racemic

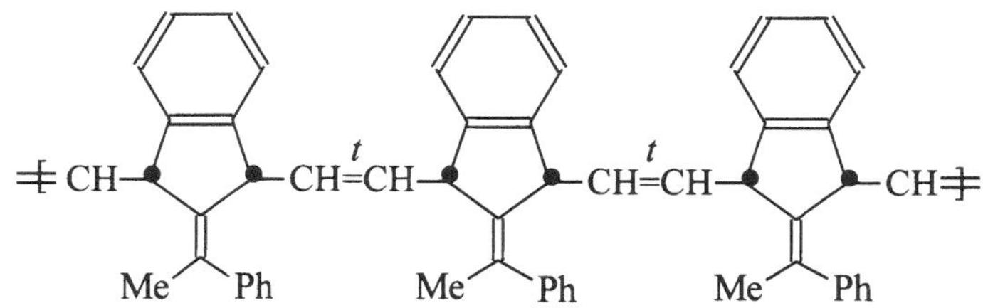

(160)

mixture of polymer chains; if syndiotactic, the polymer will consist of an alternating copolymer of the two enantiomers (Schimetta 1994b).

All the above polymers containing fused aromatic rings are susceptible to photoxidation, but remain stable at low temperature in the absence of air and light. They can also be dehydrogenated to yield conjugated polymers.

13.4.3 Dicyclopentadienes

The ROMP of the common *endo* isomer **161**, catalyzed by systems such as $MoCl_5$/Et_3Al and WCl_6/Et_2AlCl (Oshika 1968; Dall'Asta 1969), was first reported in the late 1960s. Since then numerous other catalysts have been shown to be effective, including those based on $TiCl_4$ (Winstein 1977), titanacyclobutane complexes (Cannizzo 1988; Risse 1989d; Fisher 1992), $IrCl_3$ and $RuCl_3$ (Hamilton 1986), $ReCl_5$ (Hamilton 1986; Pacreau 1987), various $WCl_n(OR)_{6-n}$ compounds (Basset 1988; Boutarfa 1991; Heroguez 1992; Balcar 1992a; Bell, A. 1992), the robust complex $Mo(=NC_6H_3\text{-}i\text{-}Pr_2\text{-}2,6)(=NCMe_3)(CH_2CMe_3)_2$ in conjunction with a phenolic activator (Bell, A. 1994a), and various polymetallates (Goodall 1993, 1995). Since the 1980s the ROMP of **161** has been developed into a commercial process for the production of quite large objects by reaction injection moulding (see Ch. 17). The *exo*-isomer **163** also readily undergoes ROMP.

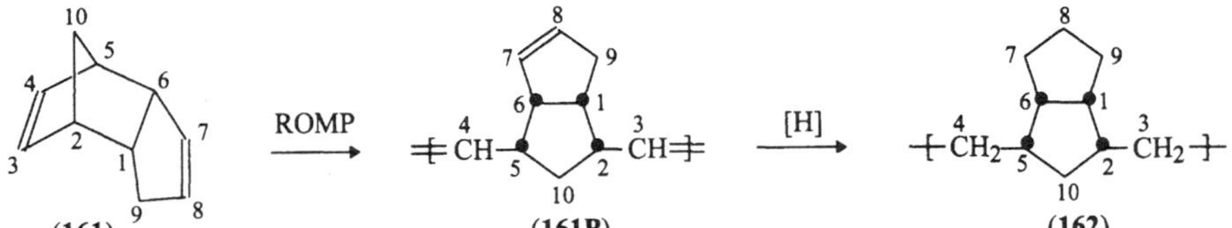

(161) **(161P)** **(162)**

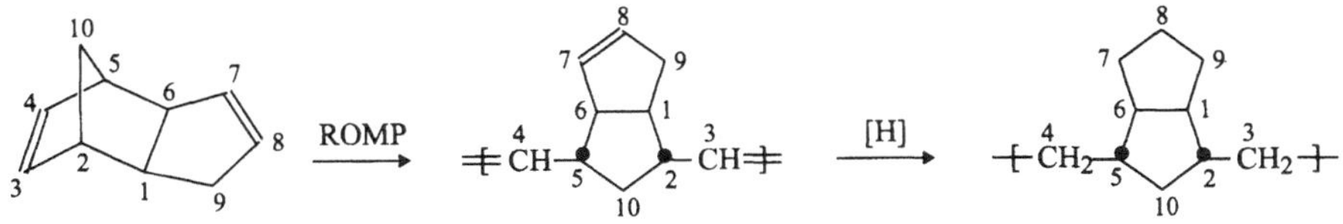

(163) **(163P)** **(164)**

In many cases the polymer formed is only partially soluble in organic solvents, but sometimes it is fully soluble, particularly if the initial monomer concentration is not too high, or if a chain-transfer agent has been used to restrict the MW. Insolubility of the polymer indicates a certain degree of cross-linking and the question arises as to how this comes about, in particular whether the double bonds in the rings of **161P** and **163P** can also open by metathesis. The opening of these double bonds, situated in a disubstituted cyclopentene ring, is thermodynamically much less favourable than the opening of the double bond in the norbornene ring system. However, there is now good evidence that there is a critical concentration above which these double bonds do in fact undergo the metathesis reaction thereby giving rise to cross-links. The best documented case is that of the ROMP of **163** initiated by various tungsten or molybdenum carbene complexes, I. For $[I]_o = 0.005$ M in toluene at 70°C and $[M]_o$ up to 1.0 M, the polymer is completely soluble and the polymerization living ($M_w/M_n = 1.14$); but with neat monomer (7 M), the product consists of about 50% insoluble polymer, 25% soluble polymer, and 25% unreacted monomer. If the soluble polymer is isolated, dissolved in fresh initiator solution, and gradually concentrated by evaporation, up to 35% insoluble material is obtained. On the other hand, if the insoluble polymer is treated with fresh initiator solution for 36 h, 70% goes into solution and this soluble product is identical with the original soluble fraction (Fisher 1992). The opening and closing of the second double bond is thus reversible, and the behaviour is parallel to that observed for the polymer of norbornadiene (see Section 13.4.1.1). For the ROMP of the *endo*-isomer **161** catalyzed by $ReCl_5/Me_4Sn$ in CCl_4 at 50°C, the critical concentration for the opening of the second double bond appears to be close to that of the neat monomer as judged by the change of viscosity as the reaction proceeds; but when catalyzed by $W(OC_6H_3\text{-}Cl_2\text{-}2,6)_2Cl_4/Me_4Sn$ in toluene at 60°C, the critical concentration appears to be not much greater than 0.3 M (Pacreau 1987). Thus there may be a solvent effect but more work is needed to define these critical concentrations more precisely. Cross-linking may also occur by cationic reactions through the second double bond, especially for those catalyst systems known to generate acidic species. Such cross-linking would not be subject to a critical concentration effect and would not be reversible; see also Davidson, T. A. (1996).

The aforesaid $ReCl_5/Me_4Sn$-catalyzed ROMP of the *endo*-isomer (3 M in CCl_4) is notable in another respect, namely that within a few minutes the main part of the monomer is transformed into oligomers (presumably cyclic), while the formation of high polymer begins slowly and reaches an asymptotic value corresponding to 50% conversion of monomer after 4 h. The concentration of oligomers passes through a maximum and reaches an ultimate equilibrium value corresponding to 30% of the original monomer (see Fig. 11.4).

With a titanacyclobutane complex as initiator, the relative rates of ROMP of **21** (norbornene), **163**, and **161** are about 20 : 4 : 1. For norbornene the rate is independent of [M]. In all three cases the rate of ROMP is likely to be governed by the rate of opening of the titanacyclobutane chain carrier to form the intermediate metal carbene complex (Cannizzo 1988). The increase in stability of the metallacyclobutane bearing an *endo* substituent is a feature which is also found in the ROMP of other substituted norbornenes (see Section 3.5.2). In contrast the rate of ROMP of **161** induced by $ReCl_5/Me_4Sn$ (1/1.5) is first order in both catalyst and monomer, and here the addition of monomer to the metal carbene complex must be rate-determining (Pacreau 1987).

The *cis* content of the $C^3{=}C^4$ bonds in **161P** can be varied from 19% using $IrCl_3$ as catalyst to 100%, using $RuCl_3$ or $ReCl_5$ as catalyst. The result for $RuCl_3$ is quite remarkable since with most norbornene derivatives the polymer obtained using $RuCl_3$ as catalyst is high *trans*. The high-*cis* content for **161P** made with this catalyst is attributed to the ability of the monomer to provide a chelated spectator ligand in which both double bonds are coordinated to the metal centre. The resultant crowding favours the direction of approach of the reacting monomer molecule that leads first to the formation of a *cisoid* metallacyclobutane complex and then to a *cis* double bond (see Section 11.6.1). This effect is not found with the *exo*-isomer because the two double bonds cannot coordinate simultaneously to the metal centre. In this case $RuCl_3$ gives a 10% *cis* polymer **163P**, while $ReCl_5$ still gives a 100% *cis* polymer. The ^{13}C NMR spectra of **161P**, **163P**, and their fully hydrogenated products **162** and **164** show that the repeating units are randomly oriented in the chains. Tacticity splittings (*m/r*) can be observed for C-3 and C-4 in the high-*trans* polymer of **161**; for C-2,5, C-3,4, and C-10 in **164**; and for C-2,5 in **162**. Most of the polymers appear to be atactic except for those made with $ReCl_5$ which are probably tactic (Hamilton 1986). The properties of **162** and **164** have been described (Kodemura 1995).

13.4.4 Miscellaneous

The ROMP of **165** proceeds readily at $-30°C$ to give a high-*trans* polymer when catalyzed by WCl_6/Et_2AlCl and a 40% *cis* polymer when catalyzed by $MoCl_5/$

Et_3Al. Only the C_5 ring is opened. As expected, the analogous compound in which the double bond in the C_5 ring has been hydrogenated, cannot be polymerized by metathesis (Dall'Asta 1968c). The reaction does not proceed to completion but only to an equilibrium monomer concentration (Ofstead 1972).

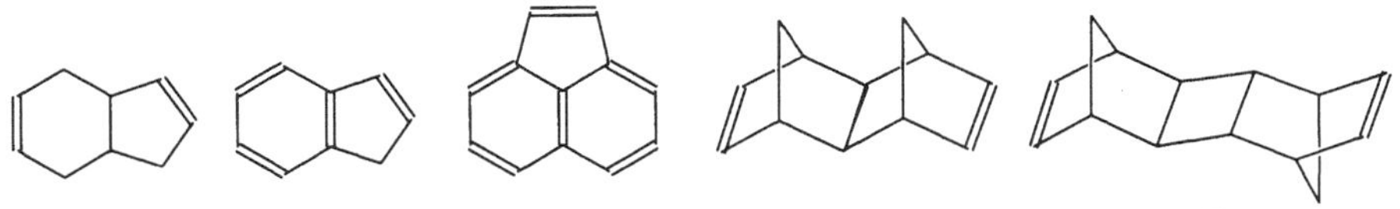

(165) **(166)** **(167)** **(168)** **(169)**

The ROMP of indene **166** has been attempted using WCl_6/Ph_4Sn (El-Saafin 1982) and $Mo(CO)_5(py)/EtAlCl_2$ (Johnston 1991b), but with only limited success. The uptake of bromine by the product and the presence of olefinic resonances in the ^1H NMR spectrum give hope that a cleaner reaction might be found with another catalyst. Acenaphthylene, **167**, on the other hand, shows little prospect of being susceptible to ROMP (El-Saafin 1982).

Compounds of the type **168** and **169**, having two double bonds of essentially equal reactivity to ROMP are ideal as cross-linking agents. They have been used, for example, to aid cross-linking during the ROMP of *endo*-dicyclopentadiene (Minchak 1987). Star-block copolymers can be made through the controlled use of **169** (Bazan 1991a). For other polycyclic compounds containing two double bonds (but only one reactive) see Nishi (1989, 1990a).

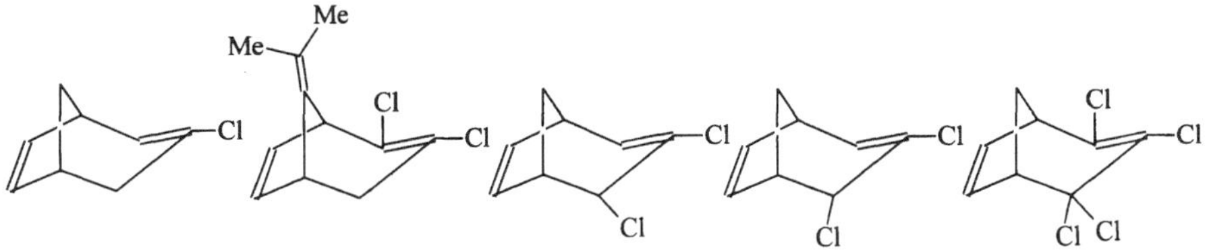

(170) **(171)** **(172)** (*exo*) **(173)** (*endo*) **(174)**

The bicyclo[3.2.1] compounds **170–174** have a structure derived from cyclopentene and contain a second double bond in the three-atom bridge. Only **170–172** undergo ROMP, catalyzed by WCl_6/Me_4Sn or $MoCl_5/Me_4Sn$. The failure of **173** and **174** to polymerize is not the result of any poisoning of the catalyst, since it remains capable of polymerizing norbornene. Rather it is to be attributed to the adverse effect of the *endo* chlorine substituent (Alimuniar 1985).

13.5 Bicyclo[2.2.1] compounds containing heteroatoms in the ring system

13.5.1 7-Oxanorbornene and its derivatives

Monomers **175–199** are all capable of ROMP if the catalyst is properly chosen: **175** (Benedicto 1992), **176, 184** (Novak 1988a), **177** (Novak 1988a; Ellsworth 1991), **178** (Benedicto 1992; Novak 1988a), **179, 192b** (Bell, B. 1994), **180** (Novak 1988b; Lu 1994), **181** (Novak 1987, 1988a,b; Feast 1991a,b; Benedicto 1992; France 1993b; Lu 1992, 1993a,b, 1994), **182, 185–188** (Lu 1994), **183** (Novak 1987), **189** (Feast 1991d), **190a** (Drent 1991; Feast 1991d; Zenkl 1992a; Gilbert 1993; McArdle 1995; Karlen 1995; Stumpf 1995), **190b** (Karlen 1995); **191** (Lu 1993a, 1994), **192a** (Karlen 1995), **193, 195, 196** (Bazan 1991c), **194** (Grubbs 1990), **197** (Lu 1993a, 1994; Viswanathan 1993a), **198** (Viswanathan 1993b), and **199** (Hillmyer 1991, 1992; Lynn 1996). Most of the work has been done with $RuCl_3$, $OsCl_3$, or $IrCl_3$ as catalysts at 50–80°C in water, aqueous emulsion, or mixtures of an alcohol with water or an aromatic solvent. Various other ruthenium compounds have also been used as noted below. Tungsten or molybdenum carbene complexes in toluene are effective at 20°C with monomers that do not contain hydroxyl groups. Thus $W(=CHCMe_3)(=NAr)[OCMe(CF_3)_2]_2$ gives polymers of very high *cis* content with **175**, **178**, and **181**, while $Mo(=CHCMe_3)(=NAr)$-$(OCMe_3)_2$ gives a living polymer (60% *cis*) of **195**. However, $Mo(=CHCMe_2Ph)$-$(=NAr)(OCMe_3)_2$ with **193** gives only a 2 : 1 mixture of the two initial *transoid* metallacyclobutane square–pyramidal complexes (in which the cyano group nearest to the metal is in either the *endo* or *exo* position). The more powerful initiator $Mo(=CHCMe_2Ph)(=NAr)(OCMe_2CF_3)_2$ brings about the formation of a living polymer of **193**, having $T_g = 193$°C and decomposing above 240°C. Ruthenium carbene complexes initiate the living ROMP of **199** in aqueous emulsion. This is the first example of a living polymerization occurring in the presence of water by any mechanism (Lynn 1996).

The high-*cis* polymer of **178** (see above), when made from enantiomeric monomer, has a mainly HH, TT structure and is therefore largely syndiotactic. On the other hand, the 96% *trans* polymer made from enantiomeric monomer with $[RuCl(\mu\text{-}Cl)(\eta^3{:}\eta^3\text{-}C_{10}H_{16})]_2$ as catalyst, ($C_{10}H_{16}$ = 2,7-dimethyloctadienediyl), has an HT structure and is therefore essentially isotactic. These tacticities are as predicted from the pseudo-octahedral model, if the ligands are not labile and one site is available for coordination of monomer (see Ch.11). Polymer of **178** made with $[Ru(H_2O)_6](OTs)_2$ as catalyst has a *cis* content of about 45%, but the tacticities with respect to each type of double bond cannot be determined because of overlap of the *cis* and *trans* olefinic lines in the ^{13}C NMR spectrum. However, the hydrogenated polymers of **181** made with the same catalysts give ^{13}C NMR spectra with fine structure for the ring carbons, that can only be the result of *m/r* splitting. Assignments based on the known tacticities of the high-*cis* and high-*trans* polymers of **178** indicate that the *m/r* ratios for polymers of **181** having

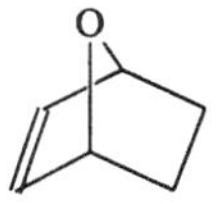

(175)

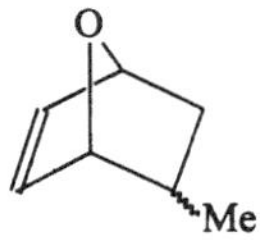

(176)

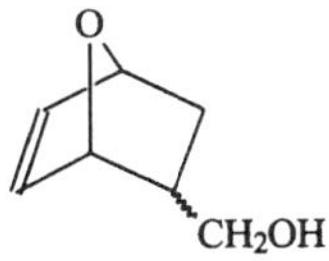

(177)

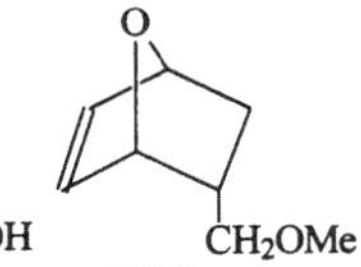

(178)

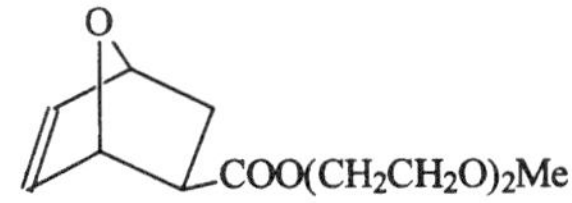

(179)

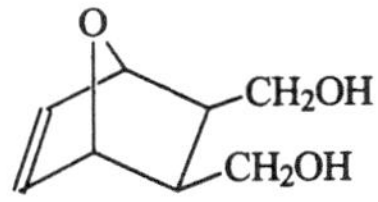

(180)

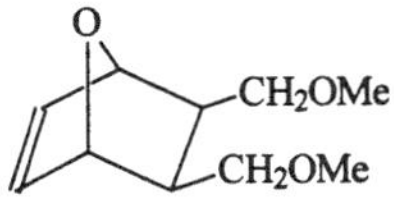

(181)

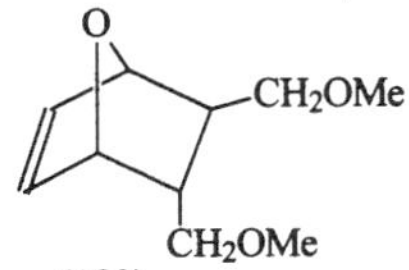

(182)

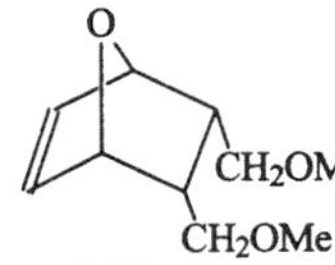

(183)

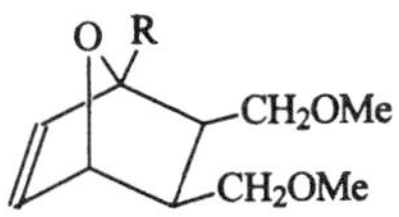

(184) R = Me, Et

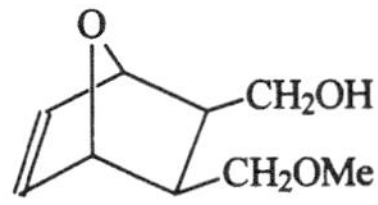

(185)

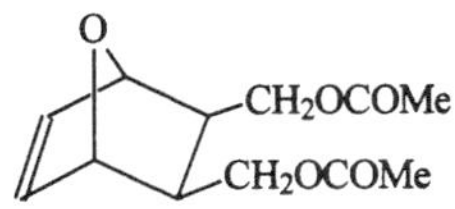

(186)

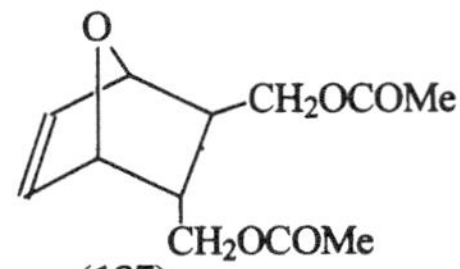

(187)

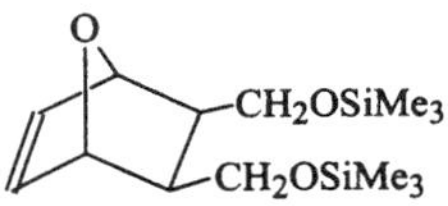

(188)

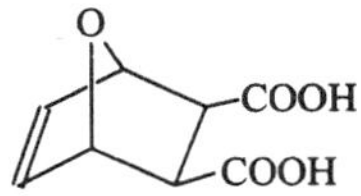

(189)

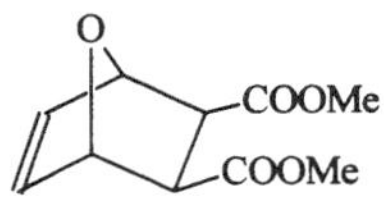

(190a)

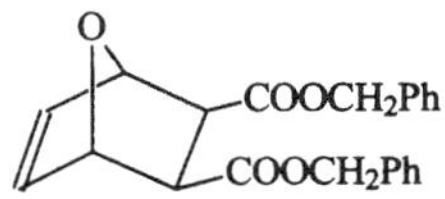

(190b)

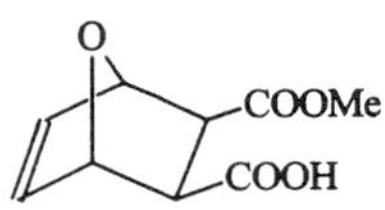

(191)

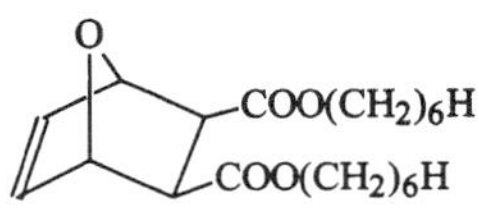

(192a)

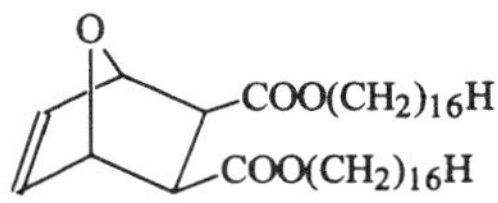

(192b)

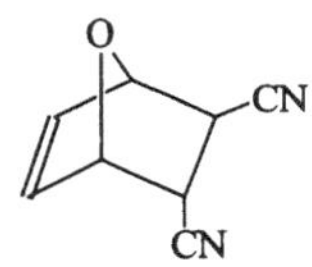

(193)

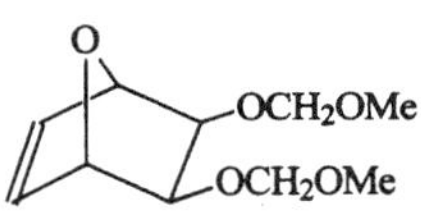

(194)

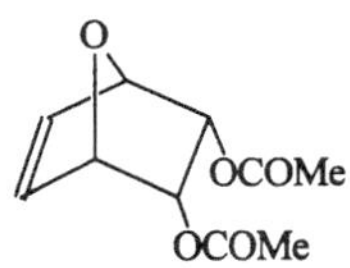

(195)

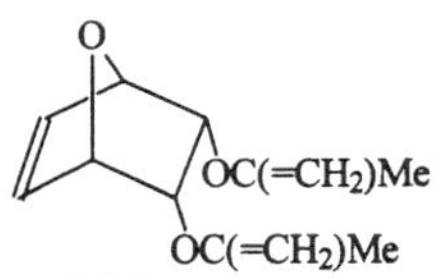

(196)

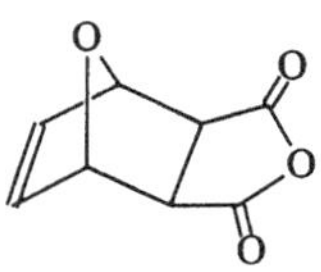

(197)

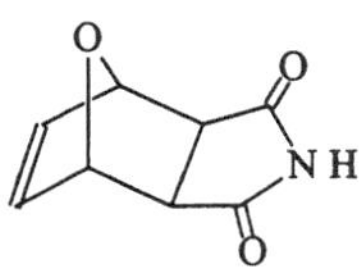

(198)

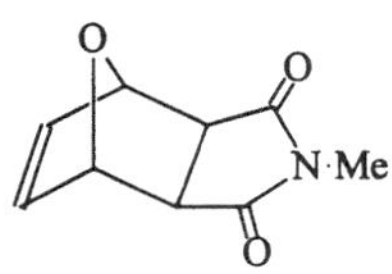

(199)

intermediate *cis* content are similar to their *t/c* ratios: $RuCl_3$ (85/15), $[Ru(H_2O)_6](OTs)_2$ (58/42); but this does not prove that *trans* and *cis* double bonds are associated only with *m* and *r* dyads, respectively (Benedicto 1992).

With $RuCl_3$ as catalyst, the *cis* content of the polymer is often markedly dependent on the solvent and on the monomer/catalyst ratio. For example, the polymer formed from **181** contains 7% *cis* double bonds when the solvent is a 5/1 mixture of benzene/ethanol but 34% when the solvent is ethanol (Novak 1988a); and with **190a** the polymer contains 18% *cis* double bonds when the solvent is THF but about 50% when the solvent is a 1/1 mixture of ethanol/water (Gilbert 1993). For the ROMP of monomer **182** in water/ethanol (10/90), the *cis* content of the polymer rises from 5% to 30% when the monomer/catalyst ratio is raised from 30 to 500. Increasing the proportion of water in the solvent also increases the *cis* content as well as the ultimate rate of polymerization (Lu 1994). Another feature of catalysis by $RuCl_3$ is that polymerization is generally preceded by an induction period (IP) during which the concentration of propagating ruthenium carbene species is building up to a stationary state. The length of the IP is very dependent on the solvent. Thus with **180** in chlorobenzene/ethanol at 55°C, it is 2–3 days, but in water it is only about 30 min; and if the same solution is used to initiate the ROMP of further batches of monomer, the IP drops to 10 s (Novak 1988b). Similar observations have been made for the ROMP of **181**. The subsequent kinetics are quite remarkable: not only is the rate of polymerization proportional to the *square* of the monomer concentration but it *decreases* with increasing catalyst concentration. All these observations point to a mechanism in which: (i) the solvent can influence the course of the reaction through coordination to the metal centre of the propagating species; (ii) the steady-state concentration of propagating species is proportional to the monomer concentration; and (iii) the catalyst is involved in termination as well as initiation (Lu 1992). There is good evidence that initiation proceeds through an Ru(II) π-complex with the monomer (Novak 1988b, 1992).

Derivatives of acyclic olefins can be used as chain-transfer agents in these polymerizations. The most effective are those with a terminal double bond. For example, in the ROMP of **181** catalyzed by $[Ru(H_2O)_6](OTs)_2$, the transfer constant (k_{tr}/k_p) for $CH_2{=}CHCH_2CH_2OH$ is 0.21 (see Ch. 15). The size of the polymer particles produced by emulsion polymerization of **181**, using $RuCl_3$ with a non-ionic surfactant is of the order of 0.03 μm (Lu 1993b).

In many cases, side reactions are liable to occur during the ROMP of these monomers, for example, esterification and trans-esterification when the ROMP of carboxylic acids, anhydrides, or esters are carried out in solvents containing alcohols, or hydrolysis when carried out in water (Drent 1991; Lu 1993a, 1994). Retro-Diels–Alder reactions can also be a problem. Thus, although **199** is cleanly polymerized to high conversion by $[Ru(H_2O)_6](OTs)_2$ under mild conditions (55°C), its *endo*-isomer fails to polymerize because the retro-Diels–Alder reaction produces N-methylmaleimide, which complexes with the catalyst and puts it out of action. The *exo*-N-phenyl analogue is likewise unable to polymerize (Hillmyer 1992).

Much interest centres on whether these polyfuran derivatives can behave like crown ethers. The polymer of **181** does in fact coordinate alkali metal ions. The flexible binding cavities formed by this polymer also allow it to complex preferentially large polyaromatic cationic dyes such as methylene blue and rhodamine 6G (Novak 1988a). Analogues of **181**, in which the methyl groups are replaced by $(CH_2)_mCH_3$ ($m = 9$, 13, 15, 17, 19, 21), also undergo ROMP when heated with $RuCl_3$ in ethanol. For $m = 15$ the polymer has much the same value of $[\eta]$ in toluene as in THF, indicating similar hydrodynamic behaviour of the polymer molecules in the two solvents, and therefore most probably a coil rather than helical conformation, contrary to previous suggestions (Kapellen 1994).

The synthesis of an agglutination inhibitor by aqueous ROMP has been successfully accomplished. Cell-surface oligosaccharides have been implicated as key participants in many intercellular recognition events, and the ROMP of suitable monomers with carbohydrate substituents offers a means of obtaining polymers of potential recognition capacity. A diester of **189** has been prepared in which the ester groups are $COO(CH_2)_2$(glu), where (glu) is an α-glucose substituent attached via a C-glycoside linkage. The polymer, made using $RuCl_3$ in water as catalyst, is soluble in water and is 2000 times as effective as the monomer in inhibiting erythrocyte agglutination by concanavalin A (Con A, a carbohydrate-binding protein). The application of ROMP to the synthesis of these polyglycomers offers new opportunities for the design of materials for modulation of cell adhesion, immobilization of particular cell types, and study of multivalency in extracellular interactions (Mortell 1994, 1996).

13.5.2 7-Oxanorbornadiene derivatives

The ROMP of **200-204a** has been reported: **200** (Hamilton 1990b; Bazan 1991c; Monakov 1992; Viswanathan 1993a; Amir-Ebrahimi 1995), **201**, **202** (Monakov 1992), **203**, **204** (Bazan 1991c; Karlen 1995), **204a** (Karlen 1995). $RuCl_3$ in PhCl/EtOH at 100°C gives a 93% *trans* polymer of **200**, $OsCl_3$ in PhCl/EtOH at 70°C gives a 50% *trans* polymer, while $MoCl_5/Me_4Sn/Et_2O$ gives a 90% *cis* polymer.

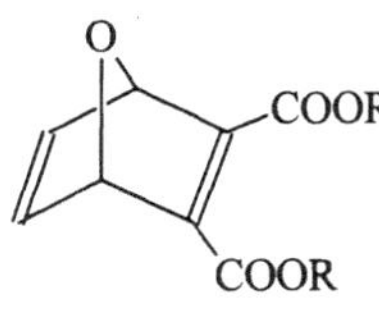

(**200**) R = Me
(**201**) R = Et
(**202**) R = Pr

(**200–202**)

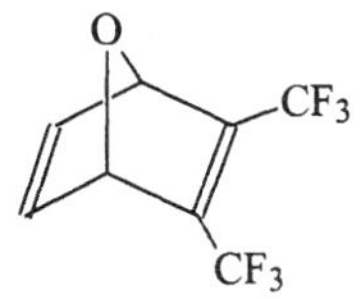

(**203**)

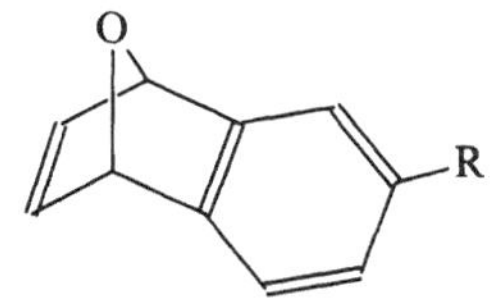

(**204**) R = H; (**204a**) R = Me

The ^{13}C NMR spectra show clearly that the double bond and not the 1,4-epoxide ring has opened. The high-*trans* polymer of **200** is readily dehydrogenated by refluxing in benzene with a stochiometric amount of dichlorodicyanobenzoquinone (DDQ) to afford the soluble, purple–red ($\lambda_{max} = 460$ nm), conjugated polyene **205**, having an estimated conjugation length of 10 double bonds; eqn. (8). This material is moderately paramagnetic and gives a symmetrical ESR signal with $g = 2.0027$. The high-*cis* polymer of **200** is less readily dehydrogenated and gives a red product solution, but UV irradiation of this solution causes the colour to change quickly to purple–red as a result of *cis* → *trans* isomerization, analogous to the behaviour of high-*cis* polyacetylene (see Ch. 10 and 12). These polymers are readily hydrolyzed to the corresponding polymers of the disodium carboxylates, the UV-visible spectra of which are dependent on the *cis* content, pH, state (solution or film), and sample history. This is to be attributed to the dependence of the conjugation length on these variables (Amir-Ebrahimi 1995). The dehydrogenation reaction succeeds only for polymers in which the enchained five-membered rings are already unsaturated; it does not work for the analogous polymer of **190a**. Likewise the polymers of norbornadiene derivatives can be dehydrogenated in this way, but not those of norbornene.

(8)

(205)

Unlike the ROMP of **190a**, that of **200** catalyzed by $RuCl_3$ in PhCl proceeds without an induction period. In the series of esters, **200–202**, the rate declines with increasing bulk of the ester group while the *cis* content of the polymer goes up from 15% to 22% to 26%. The presence of the 7-oxa group enhances the reactivity (Monakov 1992).

With $Mo(=CHCMe_2R)(=NC_6H_3\text{-}i\text{-}Pr_2\text{-}2,6)(OCMe_3)_2$ as initiator, the monomers **200**, **203**, and **204** give initially the relatively stable *transoid* metallacycles **206–210**. An X-ray study of **209** confirms that the geometry is square–pyramidal and shows that the oxygen atom in the 7-position of the norbornadiene system is 3.33Å from the metal centre, too great a distance for a bond to the metal to be considered. After 24 h in C_6D_6 at room temperature, about 75% of **208** is still present. As the metallacycle opens, eqn. (9), carbene proton resonances appear at 11.233 ppm (major) and 11.072 ppm (minor) in a constant 9 : 1 ratio assigned to the *syn* and *anti*

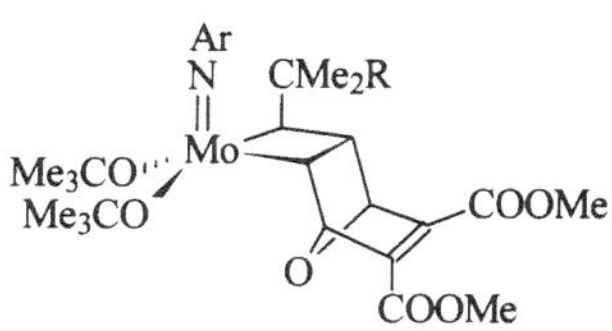

(206) R = Me
(207) R = Ph

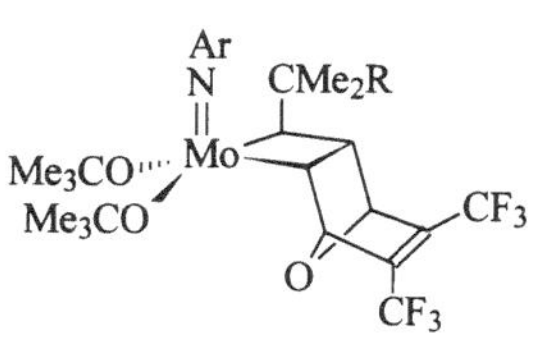

(208) R = Me
(209) R = Ph

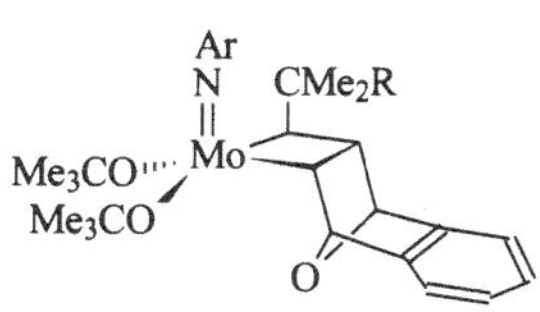

(210) R = Me

rotamers of **211**, respectively. Equilibrium between the rotamers is thus quickly established but at a rate less than 80 Hz (the chemical shift difference).

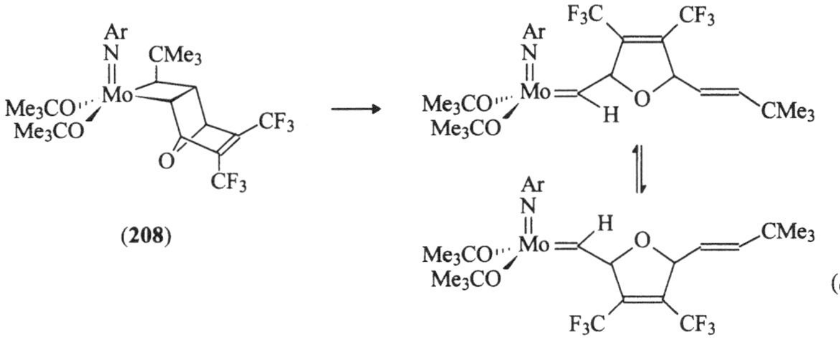

(208)

(*syn*-**211**)

(*anti*-**211**)

(9)

The former signal is a doublet, $J = 8$ Hz, being split by the adjacent proton; but the latter is a sextet as a result of additional fluorine splitting. The conversion of **208** to **211** in C_6D_6 is strictly first order, $k = 4.2 \times 10^{-5}\ s^{-1}$ at 35°C. The corresponding reactions of **206**–**210** have rate constants at 35°C that increase in the order: **209** ($8.6 \times 10^{-6}\ s^{-1}$) < **208** ($4.2 \times 10^{-5}\ s^{-1}$) < **207** ($2.7 \times 10^{-4}\ s^{-1}$) < **206** ($4.5 \times 10^{-4}\ s^{-1}$) < **210** (60% conversion after 1 h at 0°C). The phenyl group in **207** and **209** thus has a stabilizing effect relative to the methyl group, while the benzo group in **210** has a destabilizing effect. The molybdenum carbenes formed from **206** and **207** show distinct carbene proton doublets from the *syn* and *anti* rotamers (9 : 1, $J \sim 8$ and 4.5 Hz, respectively) but polymerization of further monomer is quickly forestalled by destruction of the living ends. However, polymerization of **200** can be achieved by switching to the more active Mo(=$CHCMe_2Ph$)(=NC_6H_3-*i*-Pr_2-2,6)($OCMe_2CF_3$)$_2$ in CH_2Cl_2. This initiator is also successful for the ROMP of **203** ($k_p/k_i = 2.4$); so too is Mo(=$CHCMe_3$)(=NC_6H_3-*i*-Pr_2-2,6)[OCMe(CF_3)$_2$]$_2$ ($k_p/k_i = 30$–50), but no intermediate metallacycle can be observed at 20°C. For the reaction of **204** initiated by Mo(=$CHCMe_3$)(=NC_6H_3-*i*-Pr_2-2,6)($OCMe_3$)$_2$, the molybdenum carbene complex formed by the rapid rearrangement of **210** is capable of adding further monomer to yield long-chain living polymer ($M_w/M_n = 1.06$). The initial carbene proton signal consists of two doublets (*syn*/*anti* = 9/1) but the longer chains give a more complex pattern consisting of two sets of up to four doublets, probably arising from *c*/*t* and/or *m*/*r* splitting (Bazan 1991c).

Competition experiments have shown that **203** reacts more than 50 times faster with Mo(=$CHCMe_2Ph$)(=NAr)($OCMe_3$)$_2$ than does its norbornadiene analogue (**139**). The oxygen in the 7-position enhances the activity, presumably by assisting the initial coordination of the monomer to the molybdenum centre, although it is released once the metallacycle has formed. The comparative stability of the metallacycles formed from the 7-oxanorbornadiene derivatives is not the result of any direct interaction between the oxygen and the metal, and must therefore be due to inductive effects. Further competition experiments have shown that **203** also reacts about 30 times faster than norbornene with this initiator, but that when norbornene does react, the resulting metal carbene adds norbornene much more readily than **203**. When a molecule of **203** does manage to add to the growing chain to form the metallacycle, further addition of monomer is effectively halted (Bazan 1991c).

13.5.3 Compounds containing N atoms in the ring system

The ROMP of **212** proceeds best with WCl_6/Et_3Al (1/4) as catalyst in PhCl at 60°C (34% yield). The ^{13}C NMR spectrum of the polymer in $CDCl_3$ shows only one line for each carbon, which indicates that the polymer is certainly all-HT with one dominant double-bond configuration (Cho 1985). The ROMP of **213** using a W-based catalyst at 70°C gives an 11% yield of soluble polymer, $M_n = 5240$. If the benzyl group in the monomer is replaced by methyl it fails to polymerize (Watkins 1994).

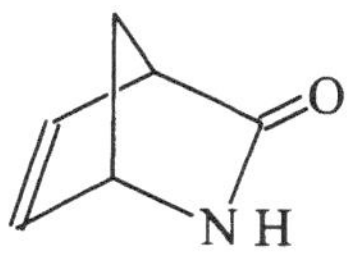

(212)

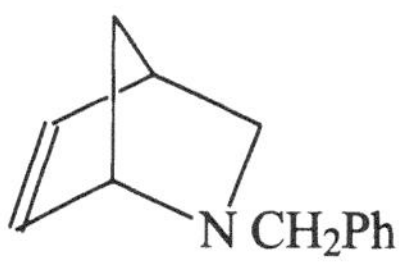

(213)

13.6 Other bicyclic compounds

13.6.1 Bicyclo[2.2.2] compounds

The ROMP of bicyclo[2.2.2]oct-2-ene (**214**), eqn.(10), requires attack on the steric equivalent of the *endo* face of norbornene, although the electron density distribution is not the same. ROMP can in fact be brought about in chlorobenzene at 20°C by $W(CO)_3$(mesitylene)/$EtAlCl_2$/2,3-epoxynorbornane containing a trace of norbornene and Me_4Sn. A 50% yield of gel is thereby obtained. The ^{13}C NMR spectrum of the polymer is entirely consistent with the ring-opened structure **214P**. The double bonds are 34% *cis* and the *c/t* distribution is somewhat blocky. The main chain bonds attached to the rings must necessarily have a formal *cis* relationship, as in the monomer. For a chair configuration, this corresponds to an axial/equatorial relationship. The absence of *a/e* fine structure in the ^{13}C NMR spectrum shows that either there is rapid $ae \leftrightarrow ea$ interconversion or that the rings exist in an intermediate twist-boat conformation (Hamilton 1985c).

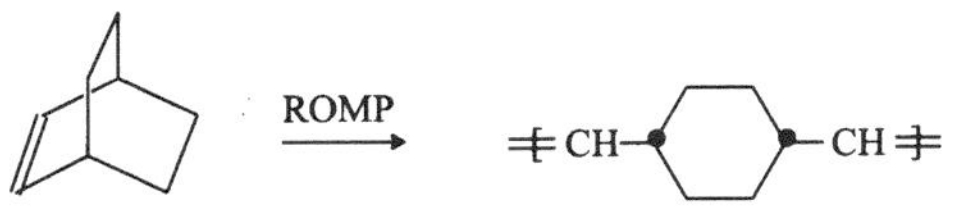

(10)

(214) **(214P)**

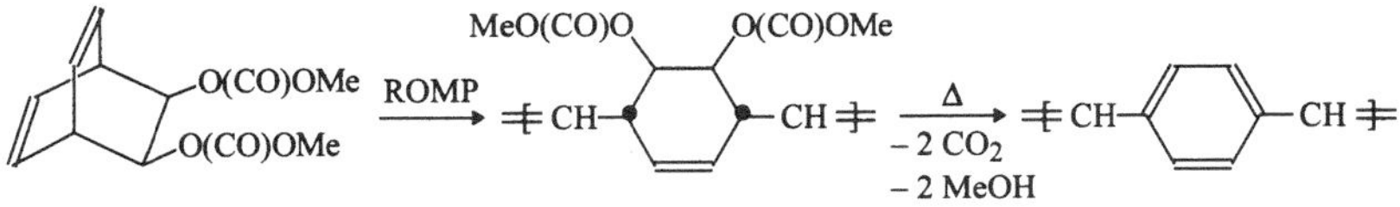

(11)

(215) **(215P)** **(216)**

The ROMP of **215** proceeds readily at 25°C in CD_2Cl_2 when initiated by Mo(=$CHCMe_2Ph$)(=NC_6H_3-*i*-Pr_2-2,6)($OCMe_2CF_3$)$_2$, giving a living polymer which can be terminated by capping with benzaldehyde. The carbene proton of the propagating species exhibits a doublet at 12.69 ppm. The reaction proceeds much more slowly in the coordinating solvent THF. In both solvents, propagation is somewhat faster than initiation. The ^{13}C NMR spectrum indicates approximately equal proportions of *cis* and *trans* bonds between the rings in **215P**; there is also secondary splitting of the vinylene carbons (*m/r* or secondary *c/t* splitting). Reaction is presumed to occur on the face of the monomer that does not bear substituents, but whether the opening of one double bond is preferred over the other is not clear. On heating a film of **215P** to 280°C under a flow of argon, it loses carbon dioxide and methanol to afford a yellow film of poly(1,4-phenylenevinylene) (**216**). The pyrolysis temperature can be reduced to 80°C by the addition of catalytic amounts of tri-*n*-octylamine (Conticello 1992).

The synthesis of poly(1,4-naphthylenevinylenes) (PNV) can be effected by the ROMP of benzobarrelenes. The introduction of C_6 or C_{11} alkyl substituents to the backbone of the rigid, planar PNV structure renders the polymer soluble in organic solvents. Doped films of these polymers show excellent conductivity (Pu 1996).

13.6.2 Bicyclo[n.1.0] compounds

Bicyclo[5.1.0]oct-2-ene (**217**) polymerizes according to eqn. (12) in the presence of W[=C(OMe)Ph](CO)$_5$/$TiCl_4$ at 50°C to give **217P**. The ^{13}C NMR spectrum shows eight lines in the olefinic region (129.35–130.45 ppm), probably stemming from *c/t* splitting of the TH, TT, HH, HT resonances (Kolesnikov 1979).

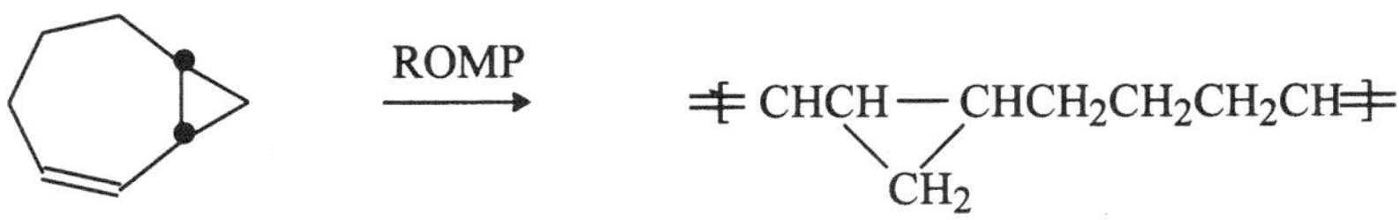

(12)

(217) **(217P)**

A number of bicyclo[6.1.0]non-4-enes (**218**) are polymerized by the ususal metathesis catalysts, eqn. (13). The polymer **218P**, with X = Y = H, made with $WCl_6/(i\text{-}Bu_2Al)_2O$ as catalyst at 63°C, contains 62% *cis* double bonds, and gives a well-defined ^{13}C NMR spectrum with olefinic carbons at 130.36 (*t*) and 129.85 (*c*), α-CH_2 at 33.25 (*t*) and 27.90 (*c*), β-CH_2 at 29.15, γ-CH at 15.77, and CH_2 in the cyclopropyl ring at 11.14 ppm (Syatkowsky 1980). Other catalyst systems give polymers with *cis* contents between 19 and 75% (Pinazzi 1973a; Kolesnikov 1979). Polymers with X = Y = Cl or Br (Pinazzi 1974), and with X = H, Y = COOEt or $COO(CH_2)_2OMe$ (Ast 1976c, 1977a) have also been prepared.

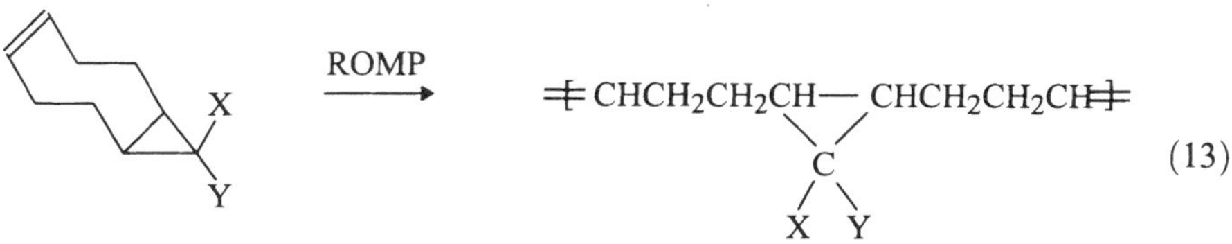

(**218**) (**218P**)

The bicyclo[10.1.0] compound **219** is polymerized by WCl_6/Me_4Sn, eqn. (14) (Ast 1976c).

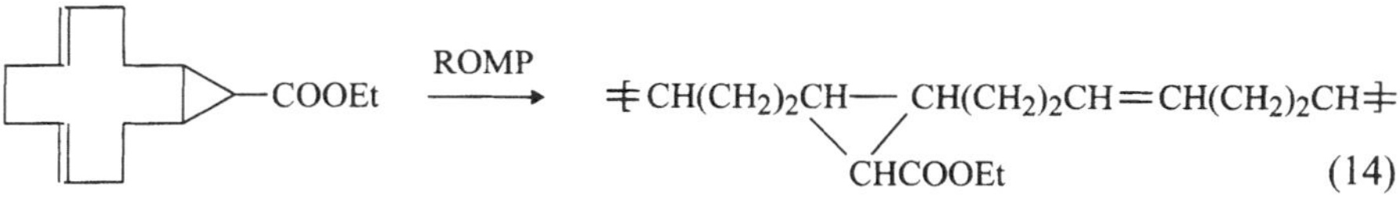

(**219**)

14

Copolymers of Cycloalkenes

14.1 Introduction

There are several ways of using the olefin metathesis reaction to generate copolymers:

(i) By metathesis copolymerization of a mixture of monomers, one of which can be an acetylene (see Section 10.3.6).
(ii) By sequential addition of two or more monomers to a living metal carbene system so as to yield a block copolymer; star-block copolymers can also be produced by the use of suitable diene monomers to provide branch points.
(iii) By making a homopolymer by ROMP and then causing it to react either at its end-group(s) or at functional groups along the chains, so as to produce block, graft, comb, or even statistical copolymers.
(iv) By ROMP of a mixture of a normal cycloalkene with a macromonomer containing a cycloalkenyl end-group; this will give a type of graft copolymer.
(v) By transformation of the propagating species to or from a metal carbene complex, allowing the formation of block copolymers in which one block is produced by ROMP and the other by an entirely different mechanism. Another possibility is to change the nature of the metal carbene complex so as to allow a second monomer to add to a living polymer formed from the first monomer.

Occasional reference has been made in earlier chapters to the formation of copolymers. Here we give a more systematic account to illustrate methods (i)–(v). For the ADMET copolymerization of linear dienes see Section 8.11.

14.2 Direct metathesis copolymerization

The formation of a true copolymer from cyclopentene (M_1) and norbornene (M_2), using WCl_6/Ph_4Sn as initiator, is very clearly demonstrated by the ^{13}C NMR spectrum of the copolymer (see Fig. 14.1). Four groups of peaks may be seen, corresponding to the olefinic carbons in M_1M_1 and M_2M_2 dyads, and the two types of olefinic carbon in an M_1M_2 dyad, designated M_1M_2 and M_2M_1 in **1**, and

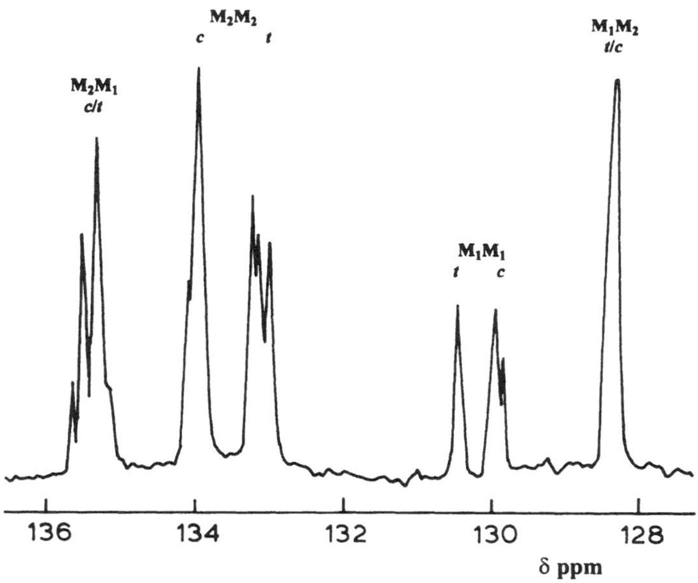

Fig. 14.1 ^{13}C NMR spectrum (olefinic region) of a copolymer of cyclopentene (M_1) and norbornene (M_2) containing 33% M_1 units. Peaks labelled M_1M_1 and M_2M_2 are found in the spectra of the respective homopolymers. Those labelled M_1M_2 and M_2M_1, see **1**, are characteristic of M_1M_2 dyads (Ivin 1979d).

necessarily equal in number. The fine structure within each group derives from the *cis/trans* isomerism of the two nearest double bonds. Provided that the copolymer is made at low conversion, the reactivity ratios can be determined directly from such a spectrum, knowing the feed composition. Thus r_1 and r_2 are given by eqns. (1) and (2), where the round brackets embrace the proportions of dyads. Experimentally no distinction can be made between M_1M_2 and M_2M_1 dyads formed in the cross-propagation reactions $P_1 + M_2 \rightarrow P_2$ and $P_2 + M_1 \rightarrow P_1$, respectively. M_1M_2 and M_2M_1 written above the peaks in Fig. 14.1 correspond to the indicated carbon atoms in **1**, but the resonances in each case come equally from M_1M_2 and M_2M_1 sequences in the actual polymer chain.

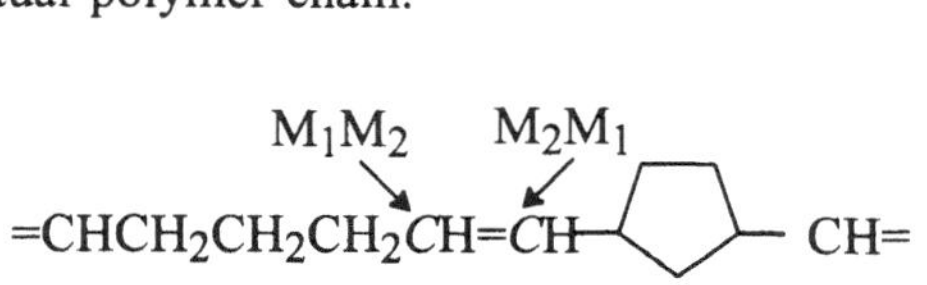

(1)

$$r_1 = \frac{(M_1M_1)}{(M_1M_2)} \bigg/ \frac{[M_1]}{[M_2]} \quad (1); \qquad r_2 = \frac{(M_2M_2)}{(M_2M_1)} \bigg/ \frac{[M_2]}{[M_1]} \quad (2)$$

Detailed studies of this and other monomer systems have shown that the copolymer composition formed from a given feed, and hence r_1 and r_2 for a given pair of monomers, are markedly dependent not only on the catalyst system, but sometimes on the method of mixing. Thus, for cyclopentene (M_1)/norbornene (M_2) with various catalysts, r_1 values vary from 0.07 to 0.62 and r_2 values from 2.6 to 70. The copolymers are often compositionally blocky (e.g. $r_1r_2=3$ for WCl_6/Bu_4Sn catalysis), but sometimes the distribution of monomer units is close to ideal (Bernoullian), with $(M_1M_1):(M_1M_2):(M_2M_2)=x^2:2x(1-x):(1-x)^2$, where x is the fraction of M_1 units in the copolymer. This is the case for WCl_6/Ph_4Sn catalysis, where $r_1r_2=1.4$ (Ivin 1979d, 1982a,b). Statistical copolymers of cyclopentene and norbornene can also be obtained using $Ru(OTs)_2(OH_2)_6$ as catalyst but only homopolymer of M_2 is obtained with $Ru(=CHCH=CPh_2)(Cl)_2(PPh_3)_2$ as catalyst. However, on changing the phosphine ligands to PCy_3, a copolymer is again obtained (Hillmyer 1995a).

For catalysis by WCl_6/Ph_4Sn, the composition of the copolymer is not sensitive to the catalyst/cocatalyst ratio or to the method of mixing, suggesting that there is only one initiating species in this system. This is not so for $WCl_6/EtAlCl_2$, which is strongly discriminating in favour of norbornene to an extent that depends sharply on the catalyst/cocatalyst ratio and on whether the cocatalyst or the monomer mixture is added last. Similar variations are observed in the cyclooctene (M_1)/norbornene (M_2) system. Thus, using $WCl_6/EtAlCl_2$ (1/4), with $EtAlCl_2$ added last, the product approximates to a homopolymer of M_2, while for $WCl_6/EtAlCl_2$ (1/10), with the monomer mixture added last, the spectrum of the product resembles that of a mixture of the two homopolymers; but for $WCl_6/EtAlCl_2$ (1/4), with monomer mixture added last, significant proportions of M_1M_2 dyads are observed. Such results suggest the presence of at least two types of initiating species, one of which may dominate under some conditions. Another important observation is that the *cis* content is often considerably different for the double bonds within the three types of dyad (Ivin 1982b).

There are evidently many factors determining the overall composition of copolymers made in this way and the experimental reactivity ratios can only be regarded as overall average values. To represent the propagating species as P_1 and P_2, according to whether the last unit added was M_1 or M_2, may clearly be an oversimplification. Apart from the fact that any given catalyst system may generate more than one type of initiating species, we have the possibility that each of these may exist in more than one conformation and/or configuration in which the previously formed double bond may or may not be coordinated to the metal site at the moment of reaction with the next monomer molecule. Penultimate unit effects, including the configuration of the previous double bond, may also influence the reactivity, and the reversible nature of the propagation reaction also may sometimes have to be taken into account, particularly for the addition of cyclopentene (M_1) to the propagating species P_1 above 0°C. Some of these factors have been discussed by Efimov (1994, 1995).

Some apparent reactivity ratios for various pairs of monomers have been summarized (Ivin 1983). Broadly speaking, reactivities of monomers towards a given propagating species lie in the order: cyclobutene derivatives > norbornene and its 5-substituted derivatives > cyclopentene and larger rings. Thus, for the reaction of 5-norbornen-2-yl acetate (M_1) with cyclooctene (M_2), catalyzed by WCl_6/Me_4Sn, $r_1 = 1/r_2 = 132$, whereas for its reaction with norbornene (M_2) or with dicyclopentadiene (M_2), $r_1 \approx r_2 \approx 1$ (Balcar 1992b, 1994a). Likewise, the reactivity ratios for the copolymerization of norbornadiene and norbornene, catalyzed by $OsCl_3$, are close to unity (Bell, B. 1992), as also are those for the copolymerization of cyclooctatetraene and cycloocta-1,5-diene, catalyzed by $W(=CHCMe_3)(=NAr)[OCMe(CF_3)_2]_2$ (Klavetter 1988). Deltacyclene (**19** in Ch. 13) is 2.4 times as reactive as norbornene when catalyzed by $RuCl_3$ (Lautens 1993). Alternating copolymers, corresponding to $r_1 = r_2 = 0$, have not been found in ROMP except in the special case of the alternating copolymerization of the enantiomers of 1-methylnorbornene, catalyzed by $ReCl_5$ to give an all-*cis*, all-HT polymer (see Section 11.6.2.1).

In a living system, if M_1 is much more reactive than M_2 and polymerization is allowed to proceed to completion, the end-product is a tapered block copolymer, in which only the middle section contains units of both monomers, e.g. with *anti*-7-methylnorbornene (M_1)/*syn*-7-methylnorbornene (M_2), catalyzed by a tungsten cyclopentylidene complex (Kress 1990), or norbornene (M_1)/cyclooctatetraene (M_2), catalyzed by $W(=CHCMe_3)(=NAr)[OCMe(CF_3)_2]_2$ (Klavetter 1988). In the extreme case, the cross-propagation reactions may be so slight that the product is indistinguishable from a perfect block copolymer, e.g. with bicyclo[3.2.0]hept-2-ene (M_1)/norbornene (M_2) catalyzed by $Ru(=CHCH=CPh_2)(Cl)_2(PPh_3)_2$ (Wu 1993; Hillmyer 1995a), or with *anti*-7-methylnorbornene (M_1)/*syn*-7-methylnorbornene (M_2), catalyzed by $Mo(=CHCMe_3)(=NAr)(OCMe_3)_2$ (Feast 1992c). The successive polymerization of the two monomers can be readily followed by ^1H NMR (see Fig. 14.2).

For 5-substituted norbornenes the *exo*-isomer is usually rather more reactive than the *endo*-isomer. An extreme case is provided by the *exo*- and *endo*-isomers obtained by the Diels–Alder reaction of cyclopentadiene and maleic anhydride. The aforementioned tungsten cyclopentylidene complex at first brings about the selective ROMP of the *exo*-isomer, but when this is all consumed, the carbene proton signal corresponding to the addition of the *endo*-isomer begins to appear and this isomer then slowly adds to the chains (Ivin 1988). Likewise the *endo*-isomer of *N*-pentafluorophenylnorborn-5-ene-2,3-dicarboximide (**115** in Ch. 13) will copolymerize to some extent with its *exo*-isomer in the presence of $MoCl_5/Me_4Sn$ but does not undergo ROMP by itself (Blackmore 1988). Other examples of this kind, where M_2 will not homopolymerize, are the copolymerization of norbornene (M_1): (i) with cyclopentene (M_2), catalyzed by $Ru(OTs)_2(OH_2)_6$ (Hillmyer 1995a); (ii) with *endo*-3a,4,7,7a-tetrahydro-4,7-methano-1,3-benzodioxol-2-one (M_2) (**109** in Ch. 13), catalyzed by WCl_6/Me_4Sn (Feast 1985a); and (iii) with cyclohexene (M_2), catalyzed by WCl_6/Me_4Sn (Patton, P.A. 1986, 1987). Cyclopentene (M_2) can also

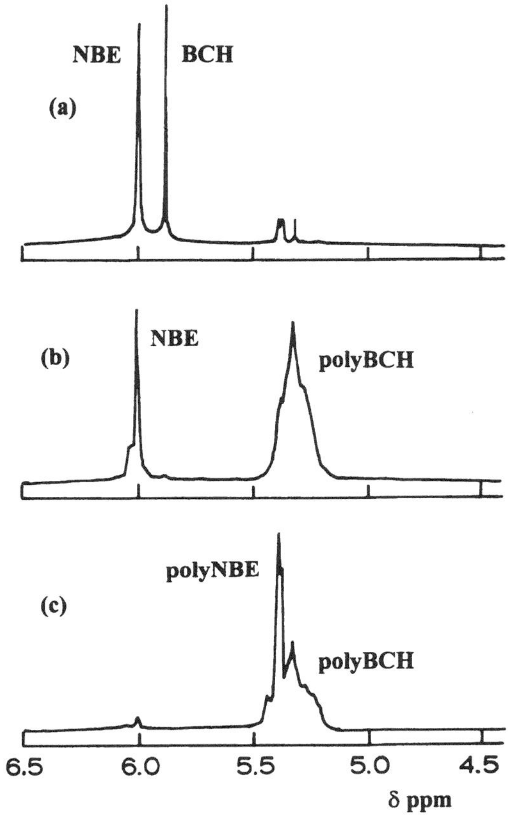

Fig. 14.2 ^{1}H NMR spectra (olefinic region) for the copolymerization of bicyclo[3.2.0]hept-2-ene (BCH) and norbornene (NBE) initiated by $Ru(=CHCH=CPh_2)(Cl)_2(PPh_3)_2$ in dichloromethane at 40°C: (a) initial, (b) after 31 min, (c) final (Hillmyer 1995a).

be copolymerized with the Diels–Alder *exo* adduct of cyclopentadiene and maleic anhydride at 52°C under conditions that are thermodynamically unfavourable for the ROMP of cyclopentene itself (Castner 1982). This means that the propagation reaction $P_1 + M_2 \rightarrow P_2$, followed by $P_2 + M_1 \rightarrow P_1$ will sometimes occur even when $P_2 + M_2 \rightarrow P_2$ is impossible or improbable.

Norbornene (M_1) has been copolymerized with a number of its derivatives and with 7-oxanorbornene derivatives (M_2), using $RuCl_3$ and other catalysts. The monomers are of comparable reactivity ($r_1 = 0.5$–2.6) and the behaviour close to ideal ($r_1r_2 = 0.9$–2.2) (Monakov 1995; Tlenkopatchiev 1995c).

The fullerene monomer (**3**), made by the cycloaddition of quadricyclane (**2**) to C_{60}, eqn. (3), can be copolymerized with an excess of norbornene in the presence of $Mo(=CHCMe_3)(=NAr)[OCMe(CF_3)_2]_2$ to yield a high-molecular-weight, soluble, film-forming copolymer (86% *cis*), containing 1% of C_{60} derivative, and exhibiting electronic and electrochemical properties, which are typical of the carbon cluster (Zhang, N. 1995).

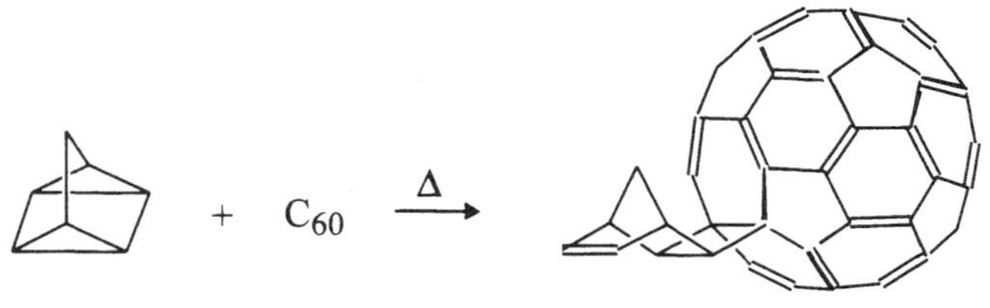

(3)

(2) **(3)**

In the ROMP of cyclooctadiene (M_1) with 4,7-dihydro-2-phenyl-1,3-dioxepin (M_2) catalyzed by $Ru(=CHR)(Cl)_2(PCy_3)_2$, where $R = CH{=}CPh_2$ or Ph, M_2 is about half as reactive as M_1. By using a small proportion of M_2, one can produce a polymer of M_1 containing occasional M_2 units, which may be broken by hydrolysis to yield 1,4-hydroxytelechelic polybutadiene ($M_w/M_n \sim 1.2$) (Fraser 1995b).

The metathesis copolymerization of the cyclobutene derivative, 7,8-bis(trifluoromethyl)tricyclo[4.2.2.0^{2,5}]deca-3,7,9-triene (M_1) (**7** in Ch. 13), with norbornene or cyclopentene (M_2) gives copolymers that are readily converted into acetylene copolymers by elimination of 1,2-bis(trifluoro)benzene from the M_1 units, but the compositional sequence distribution in these copolymers is difficult to establish (Ramakrishnan 1989b).

Secondary metathesis reactions are sometimes encountered during metathesis copolymerization, leading to a reshuffling of the units in the chain and resulting eventually in a random distribution. This occurs, for example, in the copolymerization of [2.2]paracyclophane-1,9-diene (M_1) with cyclopentene, cyclooctene, and cycloocta-1,5-diene, respectively, using a tungsten carbene complex as initiator. Spectral data show that the units of $=CHC_6H_4CH{=}CHC_6H_4CH=$ initially derived from the ROMP of M_1 are almost completely dissected into $=CHC_6H_4CH=$ units distributed at random in the chain; see Section 12.9.3 (Thorn-Csányi 1993a,c, 1994c). Another case is the copolymerization of certain 7-oxanorbornene derivatives (**181** and **191** in Ch. 13) using $RuCl_3$ as catalyst; statistical copolymers are produced no matter whether the monomers are mixed initially or added sequentially (Lu 1993).

The use of comonomers, including cross-linking agents, in the ROMP of dicyclopentadiene is described in Section 17.11. Numerous references to other metathesis copolymerization systems can be found in the patent literature.

14.3 Cyclic co-oligomers

The formation of cyclic co-oligomers by backbiting reactions is to be expected when one or both monomers are monocyclic and the concentration of monomers is not very high. Such co-oligomers were first observed in the reaction of cyclooctene (M_1) with cyclopentene (M_2) catalyzed by $WCl_6/EtAlCl_2$ in chlorobenzene (1.0 M in each monomer). GC/MS analysis shows the presence of cyclic oligomers having mass numbers corresponding to $(M_1)_x(M_2)_y$ with $(x, y) = (0, 1)$, (1, 0–2), (2, 0–3), (3, 0–4), (4, 0–4), (5, 0–5), (6, 0–3), (7, 0–2), and (8, 0). Under the conditions used for the preparation of this copolymer, the addition of M_2 to P_2 is reversible, whereas the addition of M_1 to P_1 or P_2 is essentially irreversible. It is, therefore, not surprising to find that y never exceeds x by more than one. The chances are that, when a cyclic oligomer is split off from the metal carbene complex by intramolecular metathesis, there will not be more than one M_2 unit adjacent to the metal atom and the chain itself will contain only isolated M_2 units, sandwiched between M_1 units. It also follows that for oligomers with $x = y$ the isomers with alternating M_1 and M_2 units are likely to predominate (Arlie 1974).

Cyclic oligomers have also been extracted from the products of metathesis copolymerization of norbornene (M_1) and cycloocta-1,5-diene (M_2) catalyzed by WCl_6/Me_4Sn. The main oligomers are derived from M_2 alone, in units of C_4H_6 ($x = 0$, $2y = 2$–6), but some co-oligomers are also formed, containing one unit of M_1 ($x = 1$, $2y = 2$–4) (Stelzer 1982). Similarly for the reaction between cycloocta-1,3-diene (M_1) and cycloocta-1,3,5,7-tetraene (M_2), with a W-based catalyst, the cyclic co-oligomers having $x = 1$, $y = 0.5$, 1, 2 and $x = 2$, 3, $y = 2$ have been identified (Korshak 1985).

14.4 Block copolymers by sequential addition of monomers to living systems

Early attempts at making block copolymers using first-generation catalysts met with limited success. Thus, when cyclooctatetraene (M_2) is added to a freshly polymerized solution of cyclooctene (M_1), made with a W-based catalyst, the UV spectrum of the product indicates that at least two or three molecules of M_2 have added to the chains (Korshak 1985).

The great advance in this area came with the development of the well-defined metal carbene initiators. The first indications that these would provide a practical method of making long-chain block copolymers came with the observation that the propagating tungsten carbene species P_1 derived from monomer M_1 (norbornene) could be readily converted, by dosing with a second monomer M_2 (*endo*-5-methylnorbornene), to the propagating species P_2, and then back again to P_1 by a further addition of M_1 (Kress 1985). This was followed by qualitative observations of the appropriate increases in MW after each addition of monomer (Greene 1988) and by more quantitative studies that showed that block copolymers could be made with a narrow MWD ($M_w/M_n < 1.1$), even from norbornene derivatives containing ester groups (Murdzek 1987b). These observations have since been extended to a host of other systems, mostly involving norbornene (Table 14.1) or methyltetracyclododecene (Table 14.2) as one of the monomers. The latter has the advantage of producing block copolymers that are more easily microtomed to a thickness of 300–400 Å for study by transmission electron microscopy (TEM) (Chan 1993).

In the formation of block copolymers by sequential addition of monomers, it generally does not matter which monomer is polymerized first, and diblock or multiblock copolymers of narrow MWD and of any desired sequence length are readily prepared. Occasionally one may be troubled by slow initiation relative to propagation at the second stage, leading to a broadened MWD, but such difficulties can usually be overcome by careful attention to ligands, for example, in system 1 by adding PMe_3 after the polymerization of M_1 in order to slow down the rate of propagation of M_2 relative to its rate of addition to P_1. Secondary metathesis reactions of the copolymer can also broaden the MWD. Termination is usually effected by reaction of the living ends with aldehydes; ketones can be used for terminating titanacyclobutane ends, while unsaturated ethers are used for terminating ruthenium carbene complexes.

If the two component monomers are of similar polarity, e.g. system 3, there is no observable phase separation within the block copolymer, and DSC shows a single glass transition temperature T_g. When there is a larger difference in polarity, microphase separation can generally be detected by one or more of a variety of techniques. The simplest test is to look for the two T_g peaks of the separate amorphous phases. These are seen in the products from systems 11 and 12, and in a block copolymer made from two fluorinated norbornene derivatives (Feast 1994b). Small-angle X-ray scattering (SAXS) can also be used, but most impressive is TEM. Some examples for metal-containing block copolymers are shown in Fig. 14.3. The morphology of Si-containing block copolymers can also be observed by TEM. As the proportion of M_2 units in the block copolymer decreases, the morphology changes in the usual way from lamellar to cylindrical to spherical and all three can be observed in the same system (Saunders 1991).

In the metal-containing block copolymers the metal component can be reacted in various ways without too much effect on the morphology. Thus, Pb and Zn components can be converted to the sulfides by exposure to H_2S, while Pd components can be reduced to the metal by exposure to H_2 at 100°C. Ag and Au components can be decomposed to the metal at 150°C resulting in small clusters of metal atoms, which reside largely within the original microdomains (20–100 Å for Ag, 15–40 Å for Au) (see Fig. 14.3f). Further development along these lines can be expected to produce interesting new materials, consisting as they do of conducting or semi-conducting nanoclusters within a non-conducting matrix.

The M_2 units in the block copolymers from systems 14–16 readily eliminate an aromatic ring compound on heating, giving $M_{2'}$ units, =CHCH=CHCH=, equivalent to two acetylene units. $M_{2'}M_1M_{2'}$ triblocks made in this way can be doped with WF_6 and their morphology studied by scanning electron microscopy (SEM). They can also be surface etched by metathesis degradation so as to enhance the spherical domains of M_1 units (Stelzer 1992). The reverse type of triblock, $M_1M_{2'}M_1$, can also be made, in which the average number of double bonds in the centre block can be closely controlled (see Fig. 14.4). Comparison with $M_1M_{2'}$

Table 14.1 Block copolymers made from norbornene (M_1) by sequential addition of monomers

System	Monomer M_2	Catalyst [a]	Reference [b]
1	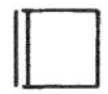	W-1	[1]
2	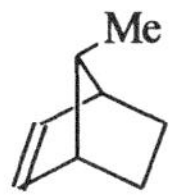	W-2	[2]
3	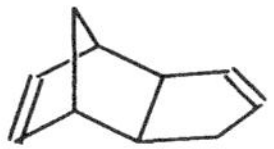	Ti-1	[3]
4	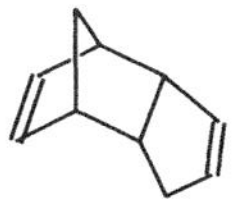	Ti-1	[3]
5	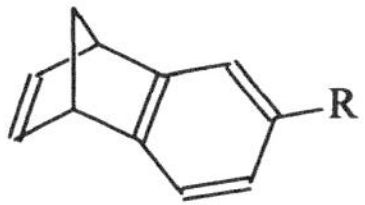 R = H, Me	Ti-1	[3]
6	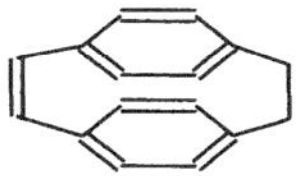	Mo-1	[4]
7	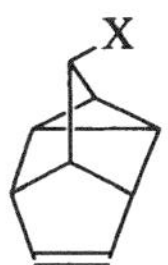 X = H, Me	Mo-2	[5]
13 [d]	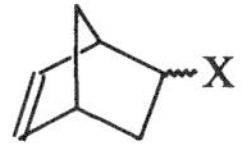 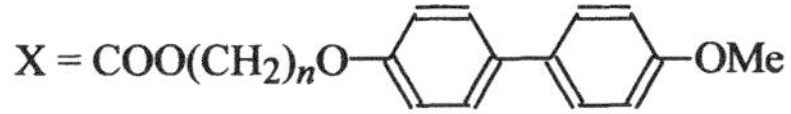 (n = 3, 6)	Mo-2	[10]
14	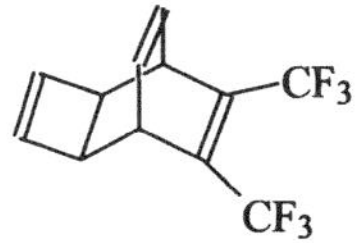	Mo-2 W-1	[11] [12]
15	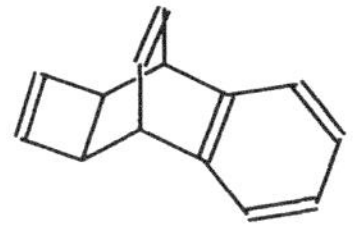	W-1 Mo-2 Ti -1	[12] [13] [14]
16	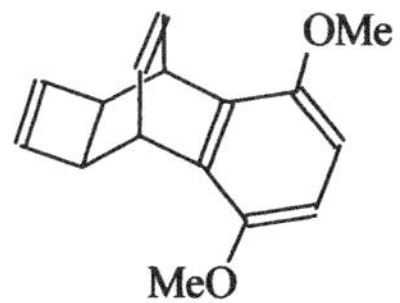	W-1	[12]
17	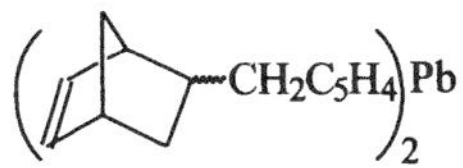	Mo-2	[15]

8–10	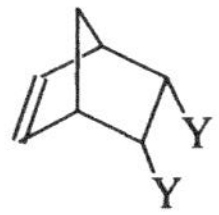		
8	Y = OCOMe	Mo-2	[6]
9	Y = COOMe	Mo-2	[7]
10	Y = $CH_2OSiBu^tMe_2$	Ru-1	[8]
11/12	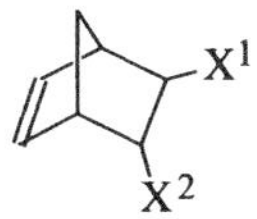	Mo-2	[9]
11 [c]	$X^1 = X^2 = COO(CH_2)_2$(phenothiazine)		
12 [c]	X^1 = COOMe, X^2 = ferrocenyl		
18	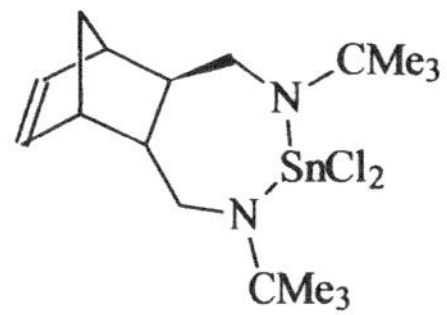	Mo-2	[16]
19	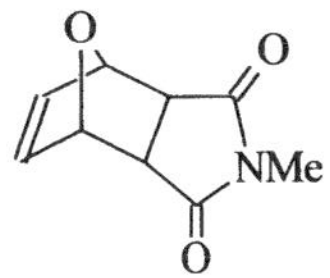 and other 7-oxa-norbornene derivatives	Ru-2 Ru-3	[8] [17]

[a] W-1 = W(=$CHCMe_3$)(=NAr)$(OCMe_3)_2$; W-2 = W(=$\overline{C(CH_2)_3CH_2}$)$(Br)_2(OCH_2CMe_3)_2$;

Ti-1 = Cp_2Ti 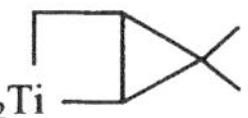 ; Mo-1 = Mo(=$CHCMe_2R$)(=NAr)$[OMe(CF_3)_2]_2$, (R = Me or Ph); Mo-2 = Mo(=$CHCMe_2R$)(=NAr)$(OCMe_3)_2$, (R = Me or Ph); Ru-1 = Ru(=CHCH=CPh_2)$(Cl)_2(PPh_3)_2$; Ru-2 = Ru(=CHCH=CPh_2)$(Cl)_2(PCy_3)_2$; Ru-3 = $(PCy_3)_2(Cl)_2$Ru=CH-p-C_6H_4CH=Ru$(Cl)_2(PCy_3)_2$.

[b] [1] Wu 1994b, [2] Greene 1988, [3] Cannizzo 1988, [4] Miao 1994a, [5] Lautens 1993, [6] Fischer, W. 1993, [7] Murdzek 1987b, [8] Kanaoka 1995, [9] Albagli 1992a, 1992b, [10] Komiya 1993a, [11] Krouse 1988, [12] Saunders 1991, [13] Stelzer 1992, [14] Stelzer 1991, [15] Sankaran 1990, [16] Cummins 1991, [17] Weck 1996a.

[c] Redox-active block copolymer.

[d] T_g of each block and the isotropization temperature of the mesophase are independent of the composition of the block copolymer and the same as those of the respective homopolymers. 5-Cyanonorborn-2-ene and methyltetracyclododecene have also been used as M_1.

Table 14.2 Block copolymers made from methyltetracyclodoecene[a](M_1) by sequential addition of monomers

System	Monomer M_2	Catalyst [b]	Reference [c]
20	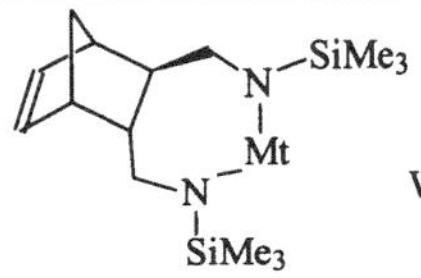 Mt: (a) Sn, (b) Pb	W-1	[1]
21	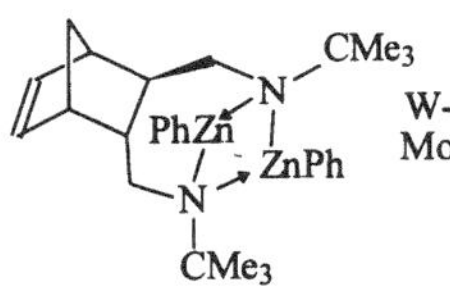	W-1, Mo-2	[1]
22-24	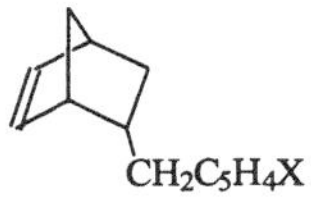	Mo-2	
22	X = Pd(η^3-1-phenylallyl)		[2, 3]
23	X = $PtMe_3$		[2]
24	X = $PbOSO_2CF_3$		[4]
	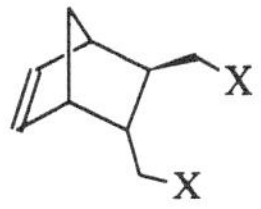		
25 [d]	X = OMe	Mo-2	[5]
26 [d,e]	X = SMe	Mo-2	[5]
27 [d,e]	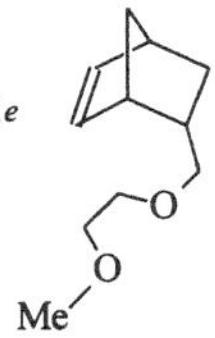	Mo-2	[5]
28 [f]	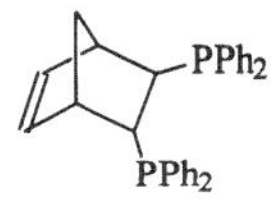	Mo-2	[6]

[a] 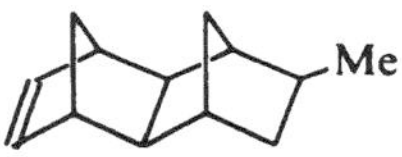

[b] As in Table 14.1. [c] [1] Cummins 1991, [2] Chan 1992b, [3] Chan 1993, [4] Tassoni 1994, [5] Cummins 1992, [6] Chan 1992a.
[d] The block copolymer reacts with $ZnPh_2$ to yield a block copolymer in which the M_2 units are bound to Zn.
[e] The block copolymer reacts with $Cd[C_6H_3(CF_3)_2\text{-}3,5]_2$ to yield a block copolymer in which the M_2 units are bound to Cd.
[f] The block copolymer reacts with Ag(COD)(Hfacac) or $Au(PMe_3)Me$, with release of COD or PMe_3, respectively, to yield a block copolymer in which the M_2 units are bound to Ag or Au.

diblocks having the same number of $M_{2'}$ units shows that the triblocks must be restricted in the extent to which their double bonds can be brought into conjugation by twisting of the polyene chain, whereas in the diblocks the polyene chains are more mobile and more readily brought into conjugation (Krouse 1988). In system 8, acetic acid can be eliminated from the pendant ester groups again giving rise to polyene block sequences, which can be doped to give semi-conducting domains within an insulating matrix (Fischer, W. 1993).

Amphiphilic star-block copolymers can be readily prepared by the use of a suitable polycyclic diene such as **4** or the more readily available **5**. For example, monomer **6** is first added to the initiator Mo-2 (Table 14.1), followed by norbornene

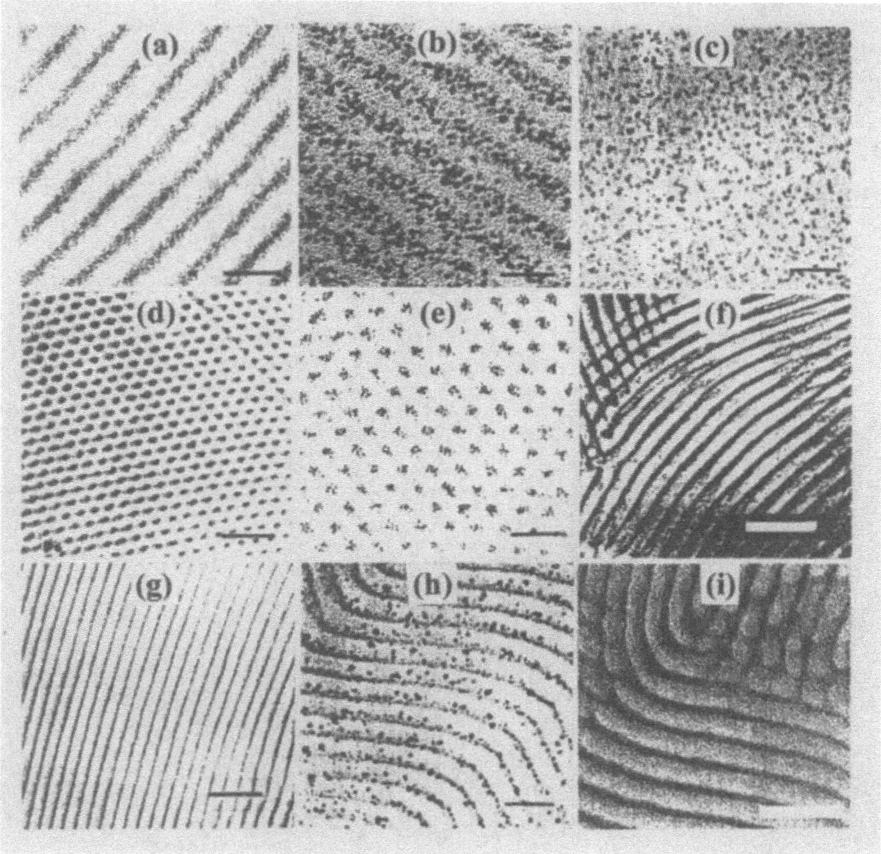

Fig. 14.3 Electron micrographs of some metal-containing block copolymers produced by ROMP; units of $M_1 : M_2$ in the copolymer shown in brackets. For systems, see Tables 14.1 and 14.2. (a) Pd, system 22 (113 : 50), lamellar, before reduction, bar = 250 Å; (b) same as (a) but after reduction to metal, bar = 250 Å; (c) Pd, system 22 (163 : 10), spherical, after reduction to the metal, bar = 500 Å; (d) Au, system 28 (25 wt% M_2 block), cylindrical (hexagonal packing), before heating, bar = 1000 Å; (e) same as (d) but after heating to form metallic Au particles, bar = 500 Å; (f) Sn, system 18 (120 : 25), lamellar, bar = 1000 Å; (g) Ag, system 28 (28 wt% M_2 block), lamellar, before heating, bar = 1000 Å; (h) same as (g) but after heating to form metallic Ag particles, bar = 1000 Å; (i) Zn, system 21 (250 : 80), lamellar, bar = 1000 Å. (a), (b), (c) Chan 1992b; (d), (e), (g), (h) Chan 1992a; (f),(i) Cummins 1991.

to make the hydrophobic block. The living diblock is then treated with 5–6 equivalents of **5** to form a star polymer. The trimethylsilyl esters are converted to carboxylic acids by soaking the cast film of the polymer in water for 2–3 days. The end-product then has a hydrophobic core of polynorbornene and a hydrophilic outer layer (Saunders 1992; see also Bazan 1991a).

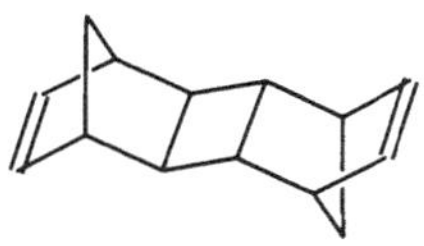

(4)

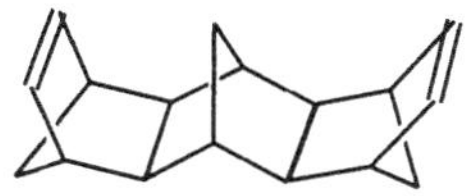

(5)

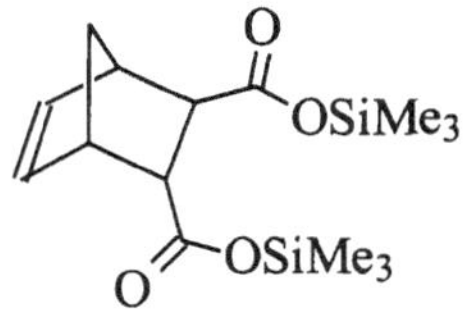

(6)

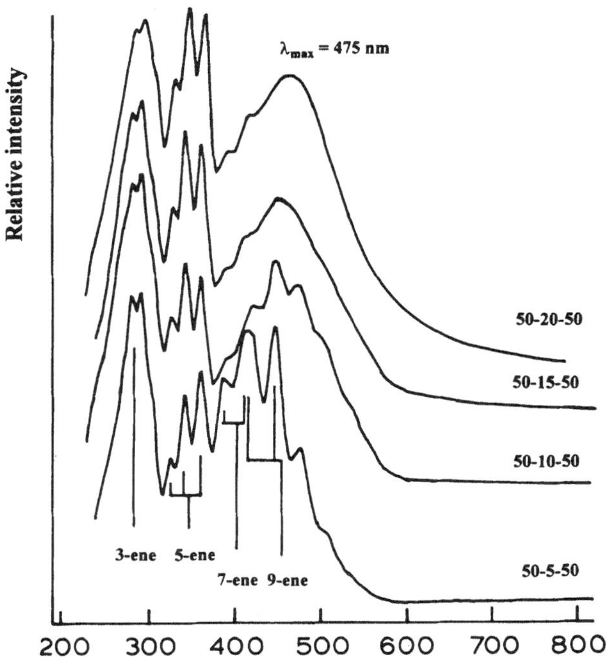

Fig. 14.4 UV–visible spectra of polynorbornene–polyene–polynorbornene triblock copolymers. The average numbers of units in each block are indicated on the right, e.g. the '50–10–50' triblock has a centre block of 10 units containing an average of $(2 \times 10) + 1 = 21$ double bonds (Krouse 1988).

14.5 Block copolymers by modification of homopolymers

14.5.1 By coupling of living polymers

If the ROMP of norbornene is initiated with a titanacyclobutane complex, the resulting living polymer chains may be coupled so as to give an ABA triblock copolymer by reaction with half an equivalent of a polymer containing keto end-groups. This has been done using poly(oxy-2,6-dimethyl-1,4-phenylene) to provide block B. The original phenolic hydroxyl end-groups in this polymer must first be reacted with 4-fluorobenzophenone in the presence of K_2CO_3 to give the required keto end-groups. A copolymer formed in this way with three equal blocks shows only a single T_g (70°C) compared with 36° and 90°C for A and B, respectively; thus there is no phase separation in this triblock material, even though A and B are not miscible. However, with *exo*-dicyclopentadiene in place of norbornene, the triblock does show microphase separation (Risse 1989b).

A variation on this principle has been used to make well-defined ABA triblock copolymers, where A is polymethyltetracyclododecene (see Table 14.2) and B is a predetermined number (9, 10, 11, or 12) of =CHCH= units. This is an improvement over the method of producing block copolymers using systems 14–16 (Table 14.1) where only the *average* number of such units can be controlled. Initiator Mo-2 is first used to make the A blocks and the living polymer is then reacted with an *excess* of hexa-2,4-diene-1,6-dial to give **7**; eqn. (4). The terminal aldehyde groups are finally coupled by reaction with half an equivalent of unsaturated bisphosphonate ester (**8**) to give the triblock (**9**). Alternatively coupling can be effected by reaction with a difunctional molybdenum alkylidene complex in which the two molybdenum atoms are linked by [=CHCH=]$_3$. Octa-2,4,6-triene-1,8-dial can also be used as terminator thereby introducing an extra double bond. In all cases the end-result is a continuous polyene sequence in block B (Park, L.Y. 1991; Craig 1993, 1994). If the excess dialdehyde reagent were removed from a dead polymer such as **7**, the aldehyde end-group could be used to terminate a different living polymer made by ROMP, so as to generate a diblock copolymer. Anthracene dialdehyde has also been used as a coupling agent, though not for making block copolymers (Wu 1994b).

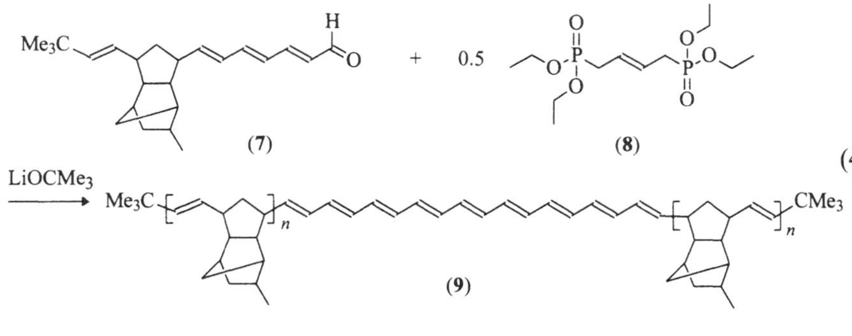

(4)

14.5.2 By transformation of propagating species

14.5.2.1 Metal carbene to metal carbene

Two examples may be cited. First, if a mixture of *anti-* and *syn-*7-methylnorbornene (M_1 and M_2, respectively) is treated with $W[=\overline{C(CH_2)_3C}H_2]$-$(Br)_2(OCH_2CMe_3)_2$, M_1 is selectively polymerized. If an equivalent of $GaBr_3$ is now added, converting [W]−Br into $[W]^+GaBr_4^-$, the metal carbene becomes much more active allowing M_2 to add to the living ends. However, propagation is much faster than initiation at this second stage and the product is a mixture of block copolymer and homopolymer of M_1 (Kress 1990).

Secondly, when a mixture of norbornene (M_1) and 5-acetoxy-1-cyclooctene (M_2) is treated with $Ru(=CHCH=CPh_2)(Cl)_2(PPh_3)_2$ in CH_2Cl_2, the M_1 is selectively polymerized in less than 3 h. If 4 equivalents of PCy_3 are now added, the PPh_3 ligands are displaced, giving a more active system and allowing M_2 to add to the chains, resulting in an increase in MW. However, if the reaction is not terminated quickly, secondary metathesis reactions are likely to occur, converting the initial block copolymer into a random copolymer (Hillmyer 1995a). See also Section 13.4.1.3 (Bazan 1996b).

14.5.2.2 Metal carbene to metal alkyl

The living chain carrier **10** for the ROMP of norbornene can be converted into the alkyl titanocene methoxide complex **11** by reaction with methanol; eqn. (5). This can then be used in conjunction with $EtAlCl_2$ to propagate the ZNP of ethene, so forming a block copolymer of norbornene and ethene (Tritto 1993).

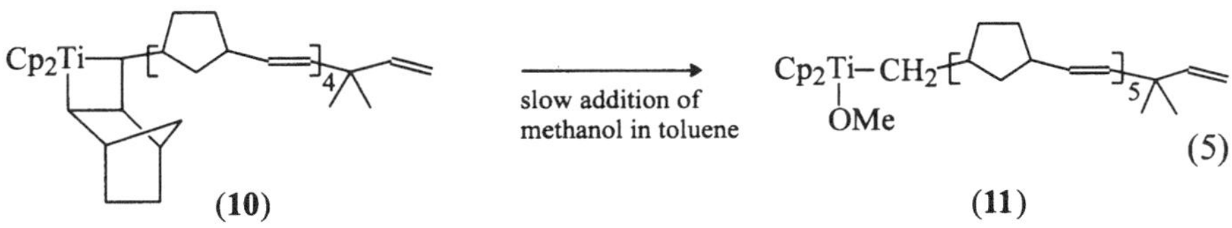

(10) (11) (5)

14.5.2.3 Metal carbene to aldehyde

If the living ROMP of norbornene is terminated with a nine-fold excess of terephthalaldehyde, the chains formed carry an aldehyde end-group, which, when activated by $ZnCl_2$, can be used to initiate the aldol-group-transfer polymerization of *tert*-butyldimethylsilyl vinyl ether (Risse 1989a, 1991).

14.5.2.4 Anionic to metal carbene

Living anionic polymerization of styrene can be initiated by butyllithium in cyclohexane. If a solution of WCl_6 in cyclohexane is added to such a living polymer solution (MW = 61 000, W/Li = 1/4) and then exposed to gaseous acetylene, the latter reacts to give a soluble product for which the GPC gives two peaks, one corresponding to the original polystyrene and the other, MW > 200 000, consisting of a block copolymer of styrene and acetylene. It seems likely that the reaction of the living anionic polystyrene with WCl_6 converts it partially into a living tungsten carbene complex, which propagates the metathesis polymerization of acetylene (Stowell 1987). Similar results have been obtained using cyclopentene in place of acetylene (Amass 1987a).

14.5.3 By cross-metathesis of unsaturated homopolymers

When *cis*-1,4-polybutadiene is added to a solution of cyclododecene that has been polymerized using $WCl_6/Et_3Al_2Cl_3$ as catalyst, a block copolymer is obtained, but at the same time the polybutadiene chains undergo considerable intramolecular degradation. The amount of degradation can be reduced to acceptable levels by careful control of the reaction conditions. However, the technological properties of such block copolymers are not superior to those of blends (Streck 1982).

The ADMET polymerization of 1,4-divinylbenzene gives rise to linear oligomers containing up to 20 units from which the dimer, trimer, and tetramer can be readily isolated and then cross-metathesized with polybutadiene using a tungsten carbene initiator. The UV–visible spectrum of the product shows that the oligomer sequence remains intact in the resulting copolymer (Thorn-Csányi 1994a).

14.5.4 By cross-metathesis of unsaturated polymers with cycloalkenes

Work on such systems has been mainly confined to establishing that the reaction will occur, and to demonstrating the superiority of the mechanical properties of the products compared with random copolymers or blends of homopolymers of the same overall composition. For example, a copolymer made from *cis*-1,4-polybutadiene and the methyl ester of norborn-2-ene-5-carboxylic acid has four times the strength of the original polymer (Ikeda 1977b). Polybutadiene and styrene/butadiene copolymers, both block and random, will undergo metathesis with a variety of cycloalkenes and norbornene derivatives using tungsten-based catalysts; polyisoprene fails to react with cyclopentene but will do so with norbornene derivatives (Dolgoplosk 1977b; Pampus 1970; Streck 1973; Matsumoto 1977a,b; Ikeda 1977a,b; Ast 1977c). Graft copolymer may be formed if the polymer carries double bonds in the side chains, as in 1,2-polybutadiene. Polymers of norbornene and its derivatives are resistant to secondary metathesis, but isoprene/isobutene copolymers, polychloroprene and poly(1-pentenylene) will react with norbornene derivatives (Ikeda 1977a,b).

14.6 Comb and graft copolymers

14.6.1 Comb polymers

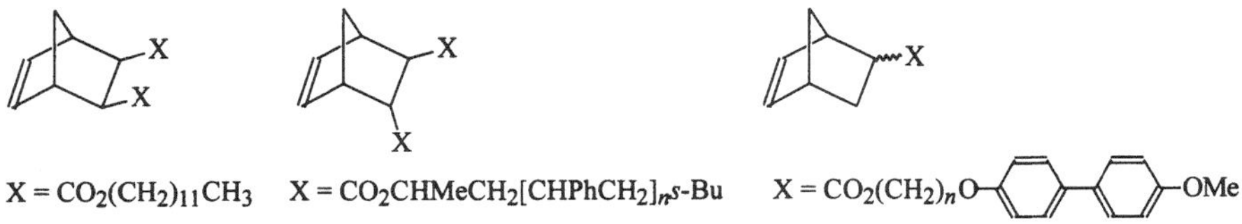

(12) **(13)** ($n = 4$–9) **(14)** ($n = 2$–8; *exo/endo* = 1/3)

These are homopolymers that have short regular side chains and are akin to graft copolymers. They can be produced by the ROMP of monomers such as **12** (Bell, B. 1994), **13** (Feast 1994a), and **14** (Komiya 1992b, 1993b). The polymers of **12** are hydrogels and can take up a moderate amount of water. The ROMP of **13** only proceeds to completion if the polystyrene side-chains are kept reasonably short ($n = 4$, 7, or 9). Polymers of monomers such as **14** exhibit a nematic or smectic mesophase resulting from side-chain crystallization. Isotropization temperatures increase with increasing MW, becoming constant at about 30–50 repeat units. Side-chain crystallization tends to be suppressed as the MW increases.

14.6.2 Copolymers with short grafts

These can be made by copolymerizing two monomers, one of which contains a side-chain such as C_{10} or C_{12}, e.g. the copolymerization of norbornene dicarboxylic esters with 5-decylnorborn-2-ene. Alternatively, the homopolymer of the diester can first be hydrolyzed to the acid form and then partially reacted with 1-dodecylamine. The viscosity of aqueous solutions of such hydrophobically modified polymers increases sharply with increase in concentration as a result of intermolecular association (McArdle 1995).

14.6.3 Copolymers with long grafts

Three examples may be cited. First, the macromonomer ω-norbornenylstyrene has been made by reacting a living anionic polystyrene with ethylene oxide to give a hydroxyl end-group, which is then reacted with norborn-5-ene-2-carbonyl chloride. The macromonomer so formed can be copolymerized with norbornene using WCl_6/Me_4Sn as catalyst. The resulting copolymer contains 3–16 polystyrene grafts per chain of 500 units and gives clear films showing that the microdomains are smaller than the wavelength of visible light (Norton 1989).

Another macromonomer has been made by reacting the terminal hydroxyl groups of poly(oxy-2,6-dimethyl-1,4-phenylene) (MW = 2000–7000) with norborn-5-ene-2-carbonyl chloride. When this macromonomer is reacted with the

5-$COO(CH_2CH_2O)_3CH_3$ derivative of norbornene, using $RuCl_3$ as catalyst, copolymers of MW up to 250 000 are obtained which contain about 2% of the macromonomer units. If the macromonomer contains only *exo*-substituted end-groups, it is more reactive than when it contains 55% *exo* and 45% *endo* end-groups (Grutke 1994).

Thirdly, if living polynorbornene (A) having titanacyclobutane ends is terminated with a short-chain polyether ketone (B) containing about 3 keto groups per chain, a triblock copolymer ABA is formed, with a single graft of A emanating from the B block (Risse 1989b).

14.7 Copolymers by ROMP in conjunction with radical reactions

The ROMP of cyclooctene-5-methacrylate and its copolymerization with cyclooctadiene is catalyzed by $Ru(=CHCH=CPh_2)(Cl)_2(PCy_3)_2$ in the presence of *p*-methoxyphenol as radical inhibitor. The double bonds in the methacrylate groups are inert towards metathesis. After chain transfer with ethyl vinyl ether to release the polymer from the ruthenium centre, it can be cross-linked by radical polymerization through the methacrylate side-chains (Maughon 1995).

The simultaneous ROMP of norbornene or dicyclopentadiene on the one hand, and the radical polymerization of styrene or methyl methacrylate on the other, gives a polymer blend (Sjardijn 1990b).

15

Cross-metathesis between Cyclic and Acyclic Olefins

15.1 Introduction

Studies of this kind fall broadly into two groups as indicated in Table 15.1. Those labelled P are mainly concerned with the effect of small quantities of acyclic olefin M_2 on the MW, yield, and *cis* content of the polymer produced from the cyclic olefin M_1; in this case the ratio $[M_2]/[M_1]$ is usually 0–0.05. Those labelled T involve the use of much higher proportions of M_2 (with $[M_2]/[M_1] \approx 0.5–2$), with the object of producing telomers with well-defined end-groups, in some cases for synthetic purposes. For example, highly substituted cyclopentane and tetrahydrofuran derivatives can be readily made by the cross-metathesis of substituted norbornenes or 7-oxanorbornenes with hex-3-ene or 1,4-dimethoxybut-2-ene catalyzed by ruthenium carbene complexes (Schneider, M.F. 1996). If the acyclic olefin is unsymmetrical and represented by Q^1Q^2, where Q^1 and Q^2 are the alkylidene moieties, three series of telomers may be produced, $Q^1(M_1)_nQ^1$, $Q^1(M_1)_nQ^2$, $Q^2(M_1)_nQ^2$, where n is the number of M_1 units. The lowest members of the series ($n = 1$) are dienes, and it is sometimes possible to detect, separate, and identify the *cc*, *ct*, and *tt* isomers.

We shall consider the evidence for the presence of Q^1 and Q^2 end-groups in the polymer and telomer molecules, the dependence of the MW of the polymer on $[M_2]/[M_1]$, the effect on the *cis* content of the polymer, and the kinetic information that can be derived from quantitative studies.

15.2 End-groups and telomers

For polymers of modest MW (< 5000), the most direct evidence concerning end-groups is provided by ^{13}C NMR spectra such as that shown in Fig. 15.1 for a polymer made from cyclopentene in the presence of pent-1-ene. Such spectra show the presence of equal proportions of the two end-groups derived from M_2 but do not give any information about the relative proportions of $Q^1(M_1)_nQ^1$, $Q^1(M_1)_nQ^2$, and $Q^2(M_1)_nQ^2$. Evidence for this comes from GC/MS investigation of the telomers ($n = 1–5$). An example is shown in Fig. 15.2 for the same system. All three groups of telomers are detected, in the ratio 1 : 40 : 1. The dominance of the unsymmetrical series of telomers is characteristic of cross-metathesis between cyclic olefins and any acyclic olefin of structure $RCH{=}CH_2$ or $R^1R^2C{=}CH_2$. This is observed for example, in the reactions of cyclooctene with propene or pent-1-ene (Hérisson 1971); of norbornene with pent-1-ene or isobutene (Rossi 1974), or with hex-1-ene or octa-1,7-diene (Bencze 1984), or with styrene (Noels 1988); and of 5-cyanonorbornene with pent-1-ene (Thorn-Csányi 1992a). However, the magnitude of the effect for a given pair of olefins depends to some extent on the catalyst (Kelly 1975) and diminishes with time (Lal 1975), indicating that secondary metathesis reactions can cause a reshuffling of end-groups amongst the telomers. In the reaction of cyclooctene with hex-1-ene, catalyzed by $WCl_6/EtAlCl_2/EtOH$, the ratio $Q^1M_nQ^2/Q^1M_nQ^1$ in the products is 5.5 for $n = 1$ but only 2.7 for $n = 2$, showing that the secondary reactions occur more easily at the internal double bond. This is to be expected because of its lower polarity. The ready occurrence of secondary reactions shows that no proper kinetic data can be obtained unless the product ratios are extrapolated back to zero time.

The fact that the unsymmetrical product dominates in the cross-metathesis of a cyclic olefin and a terminal olefin is an advantage for the efficient synthesis of the brown algae pheromones multifidene (yield 31%) and viridiene (yield 30%) via the cross-metathesis between bicyclo[3.2.0]hepta-2,6-diene and but-1-ene or butadiene, respectively (molar ratio 1/2), in the presence of 1 mol% of $Ru({=}CHCH{=}CPh_2)(Cl)_2(PCy_3)_2$; Scheme 15.1 (Randall 1995).

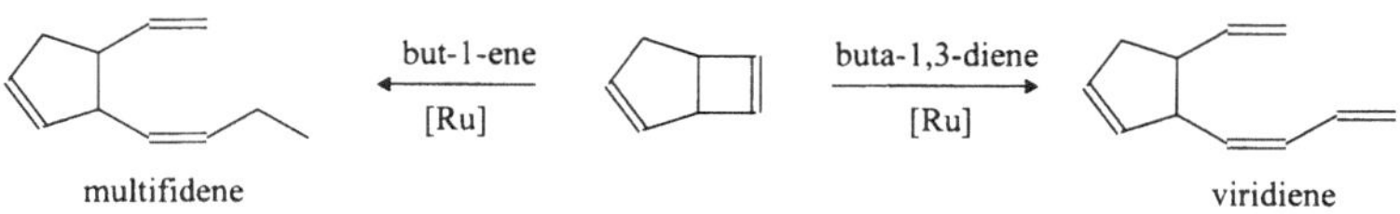

Scheme 15.1 Selective cross-metathesis reactions of bicyclo[3.2.0]hepta-2,6-diene.

By employing hex-1-ene as a chain transfer agent low-molecular-weight homopolymers of norbornadiene and copolymers with norbornene can be obtained. The products are freely soluble and afford high-quality NMR spectra, which are truly representative of higher-molecular-weight polymers and which have been fully assigned to establish the detailed microstructure (Bell, B. 1992).

When M_2 is an unsymmetrical internal olefin, such as pent-2-ene, the initial proportions of the three series of telomers are usually much closer to the statistical ratio 1 : 2 : 1. However, even here, careful studies show that there is an initial bias towards the unsymmetrical species. For example, in the cross-metathesis of cyclooctene with hex-2-ene, catalyzed by $MoCl_2(PPh_3)_2(NO)_2/Me_3Al_2Cl_3$ in chlorobenzene, the initial ratio for $n = 1$ is $1 : 3.25 \pm 0.17 : 1$, moving towards 1 : 2 : 1 with increasing time (Katz 1977a). Cases where statistical ratios have been observed, after periods ranging from 14 min to 25 h, include cyclopentene/pent-2-ene (catalyzed by $WOCl_4/Bu_4Sn$) for $n = 1$–4; also for cyclooctene/pent-2-ene with the same catalyst ($n = 1$, 2) and for cycloocta-1,5-diene/pent-2-ene ($n = 1$–4 units

Table 15.1 References[a] to cross-metathesis between cyclic (M_1) and acyclic (M_2) olefins. P denotes study of polymer; T denotes study of telomers

	M_1[c]						
M_2[b]	CBE	CPE	CHX	CHP	COC	COD	NBE
C_2	T(74)	T(1)	T(2)	T(65)	T(3,62)	T[j](3,62)	
C_3					T(4)	P(61) T(5,56)	
1-C_4		P(6–9)			T[j](80)		
◇=◇	T(74)	T(84)					
◇=CH_2		T(84)					T(86)
1-C_5		P(6,10,11,47) T(12,13)		T(80)	T(4)	P(61) T(5,55)	P(14) T(14,15)
1-C_6		P(61) T(80,84)	T(80)		T(16,80)	P(61) T(5,55,56)	P(17,18,85) T(57) P[h](18) T[h](19,82)
1-C_7						P(61) T(5)	
1-C_8		T(80)			T(5,80)		P[d](20,21)
1-C_9/1-C_{12}						T(56)	
2-C_4		P(8,14,22)			P(48,49) T(4,5,23,24)	T(25)	T(14,15,26)
2-C_5		P(10,14,22) T(4)	T(4)		P(22) T(4)	P(26) T(4)	P(14,46) T(14,15)
2-C_6					T(24)		
3-C_6	T(83)				P(27)		T(90)
4-C_8	T(58)				P(48,49) T(5,23,24)	T(28,29)	
5-C_{10}							T[d](78)
2-Me-C_3		P(8,30) T(67)				T(62)	T(14,15)
2-Me-2-C_4		P(8,10)					T(15)
4-Me-4-C_8						T(31)	
Styrene							P(68) T(64,81[g])
Dienes		P(11,32,33,87) T(19,72)				P(75) T(34)	P(50,54,68) T(57)

Cyanides[e]	T(79)		T(79)		
Ethers[e]	P(35,36)		T(37)		T(90)
Esters[e]	P(38,39,63)		T(40,41)	PT(51,52,88)	P(14,43)
	T(59,60)			T(42,59,66)	T(69,70,77[d],79)
Alcohols[e]					T[d](77)
Diols[e]					P[d](71)
Silylethers[e]		P(76,89)			
Silanes[e]			T(79)		
Boranes[e]				P(73)	
Halides[e]	P(44,45)			T(53)	
Sulfides[e]	T(86)				T(86)

[a] References: (1) Ray 1966, (2) Crain 1972, (3) Zuech 1970, (4) Hérisson 1971, (5) Günther 1970, (6) Minchak 1972, (7) Pampus 1970, (8) Witte 1978, (9) Grahlert 1975, (10) Ofstead 1980, (11) Kelly 1975, (12) Lal 1972, (13) Korshak 1980, (14) Porri 1974, (15) Rossi 1974, (16) Lal 1975, (17) Tanaka, Y. 1976, (18) Ivin 1982a, (19) Bencze 1980, (20) Hepworth 1973, (21) Winstein 1977, (22) Porri 1975, (23) Katz 1975, (24) Katz 1977a, (25) Sherman 1976, (26) Singleton 1970, (27) Scott 1971, (28) Pinazzi 1973b, (29) Pinazzi 1977a, (30) Streck 1971, (31) Pinazzi 1976, (32) Montecatini 1968, (33) Streck 1972d, (34) Pinazzi 1977b, (35) Streck 1972c, (36) Streck 1976, (37) Ast 1977a, (38) Streck 1972a, (39) Furukawa, J. 1978, (40) Otton 1979, (41) Otton 1980, (42) Pinazzi 1980, (43) Ivin 1977b, (44) Streck 1972b, (45) Imaizumi 1979, (46) Howard 1980b, (47) Ivin 1978b, (48) Badamshina 1981, (49) Korshak 1982, (50) Minchak 1981, (51) Seyferth 1981a, (52) Reyx 1982a, (53) Reyx 1982b, (54) Reif 1983, (55) Turov 1983, (56) Shapiro 1983, (57) Bencze 1984, (58) Brunthaler 1985, (59) Reyx 1986, (60) Reyx 1987, (61) Yefimov 1988, (62) Chaumont 1988, (63) Amass 1988, (64) Noels 1988, (65) Warwel 1989a, (66) Reyx 1990, (67) Schaper 1990, (68) Crowe 1990, (69) Cramail 1991a, (70) Cramail 1991b, (71) Feast 1991a, (72) Bencze 1991, (73) Chung 1992a, (74) Finkel'shtein 1992a, (75) Thorn-Csányi 1993b, (76) Hillmyer 1993, (77) France 1993b, (78) Finkel'shtein 1993, (79) Bespalova 1994a, (80) Bykov 1994, (81) Balcar 1994a, (82) Bell, B. 1992, (83) Wilson 1976, (84) Vdovin 1988, (85) McCann 1995, (86) Couturier 1993b, (87) Makovetskii 1985, (88) Nubel 1996, (89) Hillmyer 1995c, (90) Schneider, M.F. 1996. [b] C_2 ethene, 1-C_4 but-1-ene, 2-Me-C_3 2-methylpropene, etc. [c] CBE = cyclobutene derivatives; CPE = cyclopentene; CHX = cyclohexene; CHP = cycloheptene; COC = cyclooctene; COD = cycloocta-1,5-diene; NBE = norbornene. [d] Norbornene derivatives. [e] Unsaturated. [f] Also cyclodecene. [g] Dicyclopentadiene. [h] Norborna-2,5-diene. [j] Also cyclododecene.

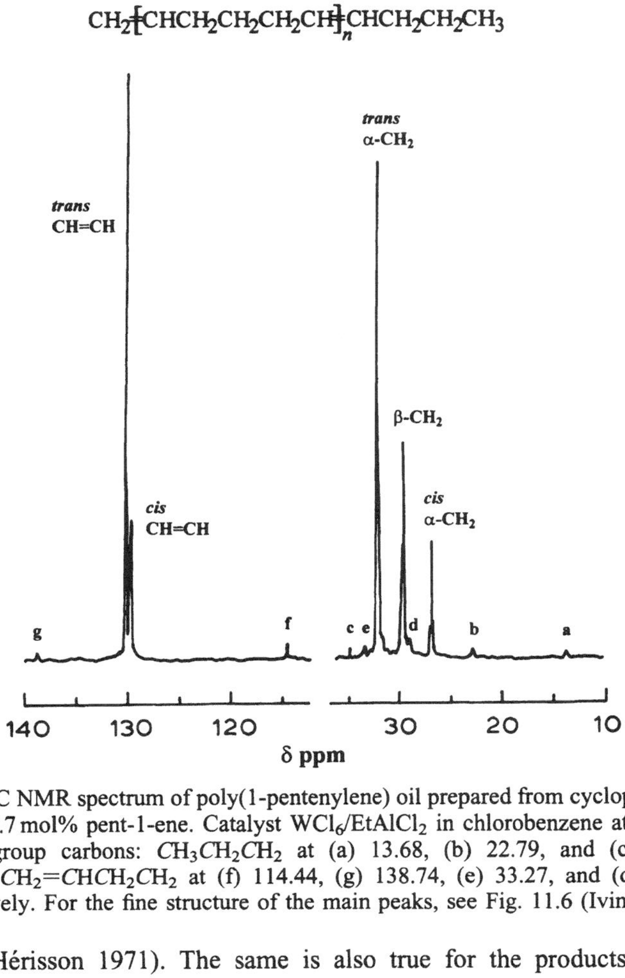

Fig. 15.1 ^{13}C NMR spectrum of poly(1-pentenylene) oil prepared from cyclopentene in the presence of 6.7 mol% pent-1-ene. Catalyst $WCl_6/EtAlCl_2$ in chlorobenzene at 20°C. Peaks due to end-group carbons: $CH_3CH_2CH_2$ at (a) 13.68, (b) 22.79, and (c) 34.78 ppm, respectively; $CH_2{=}CHCH_2CH_2$ at (f) 114.44, (g) 138.74, (e) 33.27, and (d) 28.91 ppm, respectively. For the fine structure of the main peaks, see Fig. 11.6 (Ivin 1978b).

of C_4H_6) (Hérisson 1971). The same is also true for the products with $n = 1$ obtained from norbornene/pent-2-ene using an Ir-based catalyst (Porri 1974; Rossi 1974), and likewise for the system cycloocta-1,5-diene/4-methyloct-4-ene with $WCl_6/EtAlCl_2$ as catalyst (Pinazzi 1976). The reaction of ambrettolide with 4-pentenyl butyl ether, catalyzed by WCl_6/Me_4Sn at 90°C, also gives the three telomers ($n = 1$) in a ratio close to 1 : 2 : 1 (Ast 1977a).

In the reaction of cycloocta-1,5-diene (M_1) with oct-4-ene (M_2), the proportion of any given linear polyene formed at equilibrium increases to a maximum and then declines as the ratio $[M_1]/[M_2]$ is increased. If the double bonds were distributed at random between M_1, M_2, and the linear polyenes, the optimum content of double bonds in linear dienes should occur at $[M_2]/[M_1] = 2$, and for trienes at $[M_2]/[M_1] = 1$. Experimentally, the maximum proportion of double bonds in the linear diene and triene species is found to occur at the somewhat higher values of $[M_2]/[M_1] = 2.5–3$ and 2–2.5, respectively (Pinazzi 1977a). This is a consequence of the

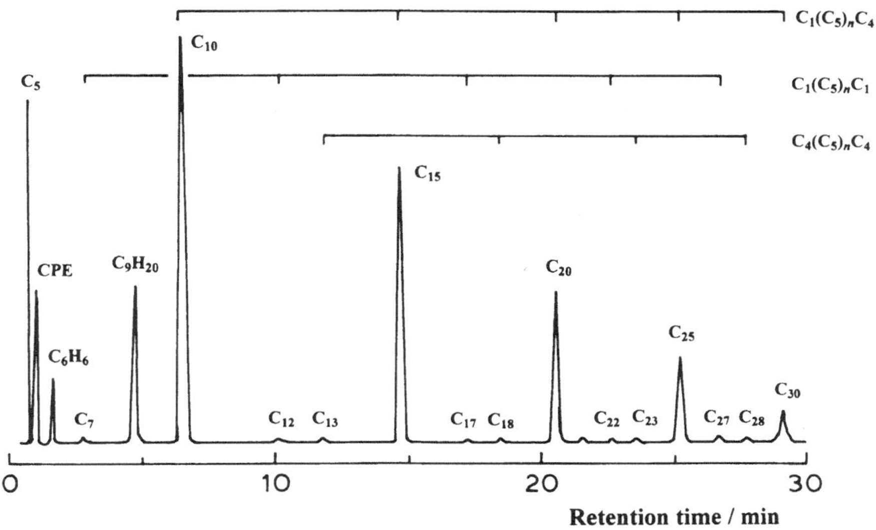

Fig. 15.2 GC of the polyenes obtained by cross-metathesis of cyclopentene (CPE) and pent-1-ene (C_5). Catalyst system: $WCl_6/EtAlCl_2/CCl_3CH_2OH$ in benzene at 25°C. Internal standard: C_9H_{20} (Kelly 1975).

fact that the ring-opening process is slightly exothermic, and so the double bonds are not randomly distributed.

In the reaction of cycloocta-1,5-diene (M_1) with ethene, useful α,ω-olefins are produced (see also Ch. 17). Ethenolysis of M_1 produces deca-1,5,9-triene, which, on further ethenolysis, produces two molecules of hexa-1,5-diene. As may be seen from Fig. 15.3, excess ethene favours the formation of hexa-1,5-diene, whereas with small amounts of ethene a series of longer chain oligomers are formed (Chaumont 1988). Cross-metathesis of 1-methylcyclobutene with methylenecyclobutane (1 : 2) over $Re_2O_7/Al_2O_3/Bu_4Sn$ or $MoO_3/Al_2O_3/Et_4Pb$ gives three products, reaction (1), of which **3** is the result of cross-metathesis of methylenecyclobutane and **2** (Finkel'shtein 1992a).

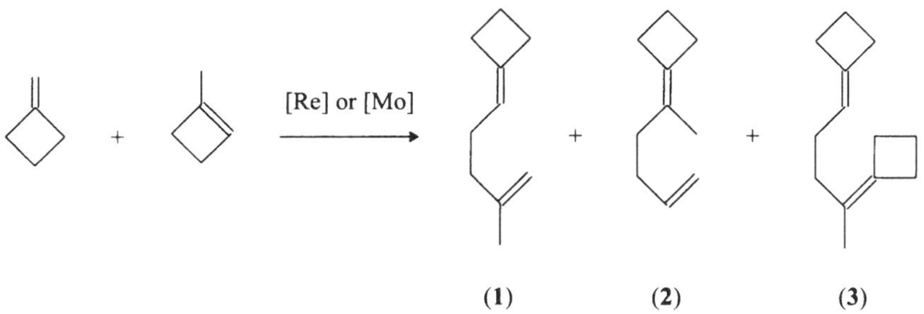

(1)

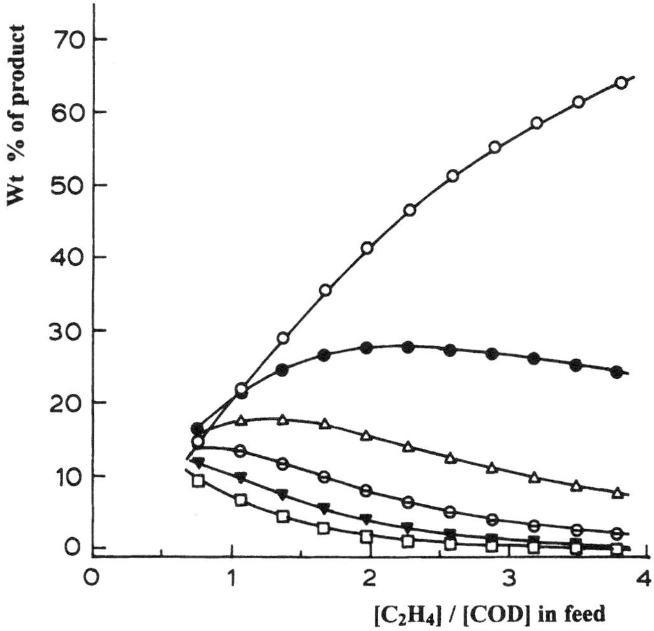

Fig. 15.3 Equilibrium product distribution of the reaction of ethene with cycloocta-1,5-diene (COD) at room temperature. ○ = C_6; ● = C_{10}; △ = C_{14}; ⊖ = C_{18}; ▼ = C_{22}; □ = C_{26} (Chaumont 1988).

The synthesis of α,ω-difunctional telomers by cross-metathesis between cyclic olefins and acyclic functionalized olefins is possible in the presence of functional-group-tolerant catalysts. Thus, cross-metathesis of dimethyl hex-3-enedioate with cycloocta-1,5-diene (Pinazzi 1980; Reyx 1982b, 1990), cyclopentene (Reyx 1986, 1987), or norbornene (Cramail 1991a) in the presence of the catalyst system WCl_6/Me_4Sn allows the synthesis of α,ω-difunctional oligomers, e.g. reaction (2).

 + $MeOOCCH_2CH{=}CHCH_2COOMe$ ⟶

$MeOOCCH_2CH{\neq}\!\left[CH(CH_2)_3CH{\neq}\right]_n CHCH_2COOMe$ ($n = 1, 2, \ldots$) (2)

We have seen that end-groups may be detected by ^{13}C NMR spectra and inferred from mass spectra. Other methods for the detection of end-groups in the products of cross-metathesis of cyclic and acyclic olefins include IR and UV spectra, and elementary analysis. For example, polymers of norbornene made in the presence of pent-1-ene show a band at 1375 cm^{-1} attributable to methyl groups, and bands at 905 and 990 cm^{-1} due to vinyl groups (Porri 1974). Such bands become stronger relative to the main band as $[M_2]/[M_1]$ is increased, thereby reducing the MW. Polymers of cyclopentene made in the presence of 5-(β-naphthoxypent-1-ene) have demonstrated by the presence of one chlorine atom per chain in the polymer (Günther 1970).

As discussed in Section 11.2.1, the polymerization of cyclohexene does not occur because of its very low strain energy. For the same reason, cross-metathesis of cyclohexene with acyclic olefins is strongly disfavoured. Nevertheless, using a high pressure of ethene, it is possible to displace the equilibrium sufficiently to be able to detect a trace of octa-1,7-diene product (Crain 1972). Likewise a small yield of dodeca-1,7-diene may be detected in its reaction with hex-1-ene (Bykov 1994) and of undeca-2,8-diene in its reaction with pent-2-ene (Hérisson 1971).

15.3 Dependence of molecular weight on $[M_2]/[M_1]$

The importance of control of MW of potential commercial polymers has led to numerous studies of the effect of various unsaturated additives on the ROMP of cycloalkenes, especially cyclopentene and norbornene (see Table 15.1).

The effectiveness of acyclic olefins in reducing the MW depends very much on their structure and on the catalyst system, especially on the *cis* content of the polymer formed. The order of effectiveness of M_2 in reducing the MW is generally: $RCH{=}CH_2 > R^1CH{=}CHR^2(cis) > R^1CH{=}CHR^2(trans) \gg R^1R^2C{=}CH_2$. An example of the comparative effectiveness of terminal and internal olefins is shown in Fig. 15.4.

It was first noted by Minchak (1972) that added terminal olefins had much less effect on the MW of the polymer made from cyclopentene when the conditions were such as to produce a high-*cis* polymer. Thus, with $WCl_6/Et_3Al/(PhCOO)_2$ as catalyst at $-50°C$, an all-*cis* polymer is produced and it is necessary to add 10% but-1-ene to reduce the MW to a certain level. At 0°C, where the *cis* content is much less, the same MW can be achieved with the addition of only 0.01% but-1-ene to the cyclopentene. Similarly pent-1-ene is a relatively poor transfer agent for the $WF_6/EtAlCl_2$-initiated polymerization of cyclopentene at 0°C, a system that also gives very high-*cis* polymer. The interpretation of this observation, discussed in detail in Section 11.4, is that chain transfer is much more difficult under *cis*-selective conditions, except with certain specific dienes, because of the intramolecular coordination of the previously formed *cis* double bond at the catalyst site (Ofstead 1980).

In principle, for non-living systems, the slopes of the plots of the reciprocal number average degree of polymerization, $1/DP_n$, against $[M_2]/[M_1]$ give the transfer constants (k_{tr}/k_p). The order of effectiveness of different acyclic olefins M_2 in reducing the MW undoubtedly indicates the relative values of the transfer constants but there are too many uncertainties for absolute values to be derived. Thus, it is generally not known to what extent secondary reactions, especially those leading to cyclic oligomer, have modified the original DP_n, or whether cyclic oligomer has been removed before MW determination; also many results are

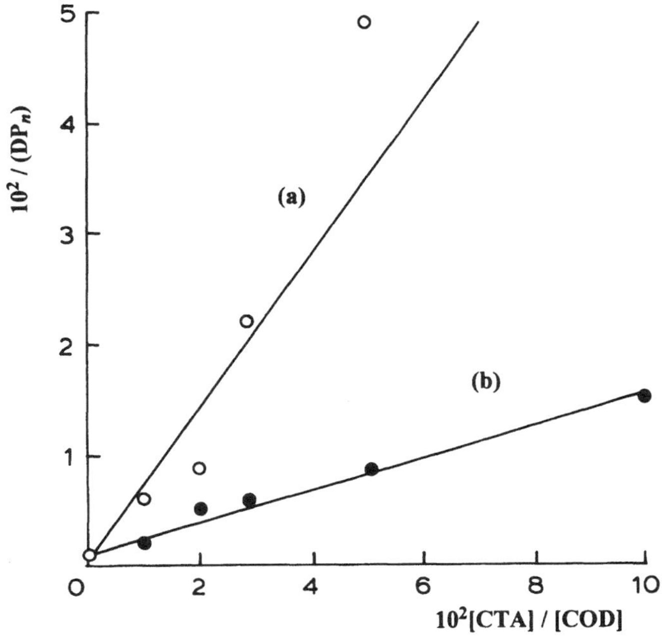

Fig. 15.4 Plot of $1/DP_n$ against the ratio of concentrations of chain-transfer agent (CTA) and cycloocta-1,5-diene (COD). (a) CTA = 6-(9-BBN)-hex-1-ene; (b) CTA = 6-(9-BBN)-dec-5-ene (9-BBN = 9-borabicyclononane) (Chung 1992a).

leading to cyclic oligomer, have modified the original DP_n, or whether cyclic oligomer has been removed before MW determination; also many results are reported in terms of intrinsic viscosity or inherent viscosity, which cannot be directly related to the DP_n. Data such as those in Fig. 15.4 give apparent transfer constants of about 0.5–1 for alk-1-enes and about 0.05 for alk-2-enes, but the results from telomer ratios suggest that such values may be too low (see Section 15.4.3). One case where detailed allowance has been made for ring-chain equilibria is the system cyclooctene/hex-3-ene (Scott 1971).

Added acyclic olefins generally have little effect on the *cis* content of the polymer formed. However, certain unsaturated compounds carrying functional groups can either raise or lower the *cis* content. Thus, catalytic amounts of allyl 2,4,6-tribromophenyl ether added to a $WCl_6/EtAlCl_2/EtOH$ catalyst in hexane can raise the *cis* content of the polymer of cyclopentene from 73% to near 100%, and at the same time improve the yield (Streck 1976). On the other hand, ethyl acrylate, diethyl maleate, and diethyl fumarate can increase the *cis* content in the polymerization of norbornene, while reducing the yield (Ivin 1977b, 1979d). Such additives presumably act by increasing or decreasing the electron density at the metal site in some way, thereby affecting both the rate and probability of *cis* double bond formation.

Unsaturated compounds containing functional groups may be divided into four broad groups according to their effect on the ROMP of cycloalkenes: (i) inhibitors; (ii) inert solvents; (iii) active participants (chain transfer agents); and (iv) activating participants. Over 150 unsaturated compounds containing functional groups have been classified in this way by Streck (1982), taking the ROMP of cyclopentene catalyzed by $WCl_6/EtAlCl_2/EtOH$ as the test reaction. Examples of compounds in these four categories for this reaction are: (i) acrylonitrile; (ii) 1,1-dichloroethene; (iii) unsaturated alcohols and ethers; and (iv) vinyl chloride. For example, in the ROMP of **181** (see Ch. 13) catalyzed by $[Ru(H_2O)_6](OTs)_2$, the transfer constant for $CH_2{=}CHCH_2CH_2OH$ is 0.21. When sufficient transfer agent is used, it is possible: (i) to observe individual oligomers by high-resolution GPC (Fig. 15.5); (ii) to observe by GC the *cis* and *trans* isomers of $Q^1M_1Q^2$; and (iii) to observe by ^1H–^1H COSY NMR the end-groups in the oligomers (France 1993b). Unsaturated esters are less effective as transfer agents, especially when the ester group is close to the double bond. Thus the activity diminishes in the order C=C−C−C−COOMe > C=C−COOMe > C−C=C−C−COOMe > C−C−C=C−COOMe (France 1993b). An example of the use of terminal acyclic functionalized olefins for the control of the polymerization of a 7-oxanorbornene

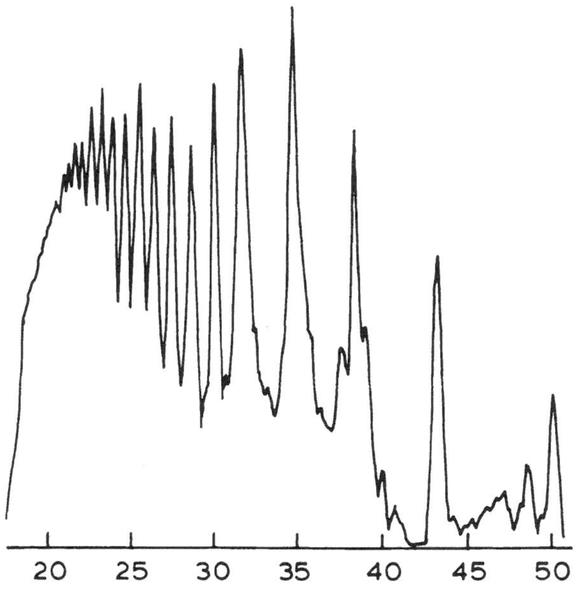

Fig. 15.5 High-resolution GPC trace of polymer of *exo*, *exo*-5,6-bis(methoxymethyl)-7-oxanorbornene (M_1) regulated with but-3-en-1-ol as chain-transfer agent (CTA); $[CTA]/[M_1] = 0.89$, catalyst $Ru(H_2O)_6(OTs)_2$. The shoulders on the sides of the peaks at 38 ml ($n = 1$) and 35 ml ($n = 2$) may indicate the presence of three series of oligomers: $Q^1M_nQ^1$, $Q^1M_nQ^2$, $Q^2M_nQ^2$ (France 1993b).

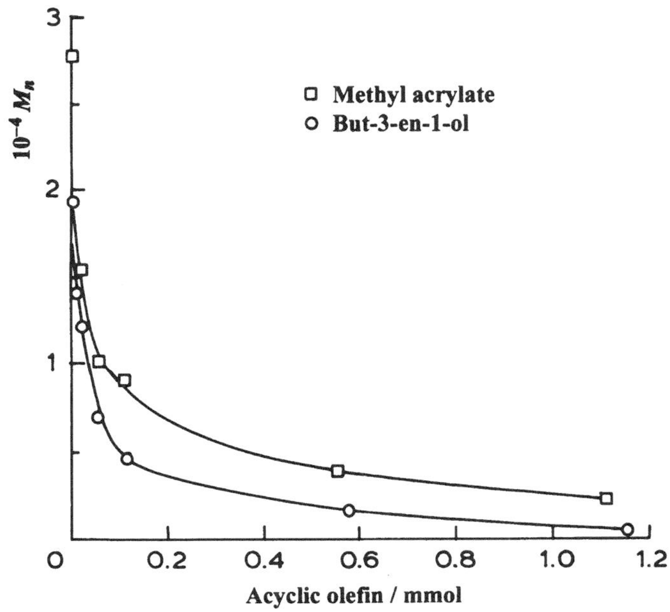

Fig. 15.6 Plots of M_n of polymer formed from *exo,exo*-5,6-bis(methoxymethyl)-7-oxanorbornene (0.62 mmol) as a function of added but-3-en-1-ol and methyl acrylate (0–1.2 mmol) in water (1.1 ml). Catalyst: $Ru(H_2O)_6(OTs)_2$, 55°C, 1 h. The value of M_n in the absence of transfer agent is 60 000 (not shown on the plot) (France 1993b).

derivative is shown in Fig. 15.6. At first the MW drops off rapidly with increasing concentration of transfer agent and then falls more slowly. For the ROMP of this monomer catalyzed by $RuCl_3$ in aqueous emulsion, the transfer constant for *cis*-but-2-ene-1,4-diol is very small (7.3×10^{-5}) and a concentration of transfer agent equal to that of the monomer must be used if it is required to reduce M_n from 122 000 to around 5000. The dimethyl ether of this diol is about 3.7 times less effective than the diol itself (Feast 1991a,b).

The third group also includes unsaturated boranes, which have the advantage of: (i) the stability of borane groups towards tungsten catalysts; (ii) the stability of borane monomers and polymers in hydrocarbon solvents; and (iii) the versatility of borane groups, which can be transformed to a variety of functionalities under mild reaction conditions (Chung 1992a).

The particular problems of finding suitable transfer agents for use in living systems are discussed in Section 11.4.

15.4 Kinetic data

Cross-metathesis between cyclic and acyclic olefins may involve some or all of the propagation and chain-transfer steps shown in Scheme 15.2 for the particular case of telomers $Q^1(M_1)_nQ^1$, $Q^1(M_1)_nQ^2$, $Q^2(M_1)_nQ^2$ results from the sequences $(5b)(3)^{n-1}(4b)$, either $(5b)(3)^{n-1}(4a)$ or $(5a)(3)^{n-1}(4b)$, and $(5a)(3)^{n-1}(4a)$, respectively. Whenever the unsymmetrical series is dominant, as in Fig. 15.2, it means that one of the two transfer reactions, (4a) and (4b), must be much preferred to the other. Now it is already known that in the metathesis of terminal alkenes the degenerate exchange is likely to proceed through a substituted metal carbene, by a reaction analogous to (4b), and that this proceeds much more readily than productive metathesis, analogous to (4a) (see Ch. 5). Hence we may conclude that, in the reaction between cyclopentene and pent-1-ene, catalyzed by $WCl_6/EtAlCl_2/CCl_3CH_2OH$, nearly every chain *begins* with PrCH= and *ends* with $=CH_2$ rather than the other way round. That reaction (4b) does occasionally happen is shown by the presence of about 2% of each series of symmetrical telomers in the products.

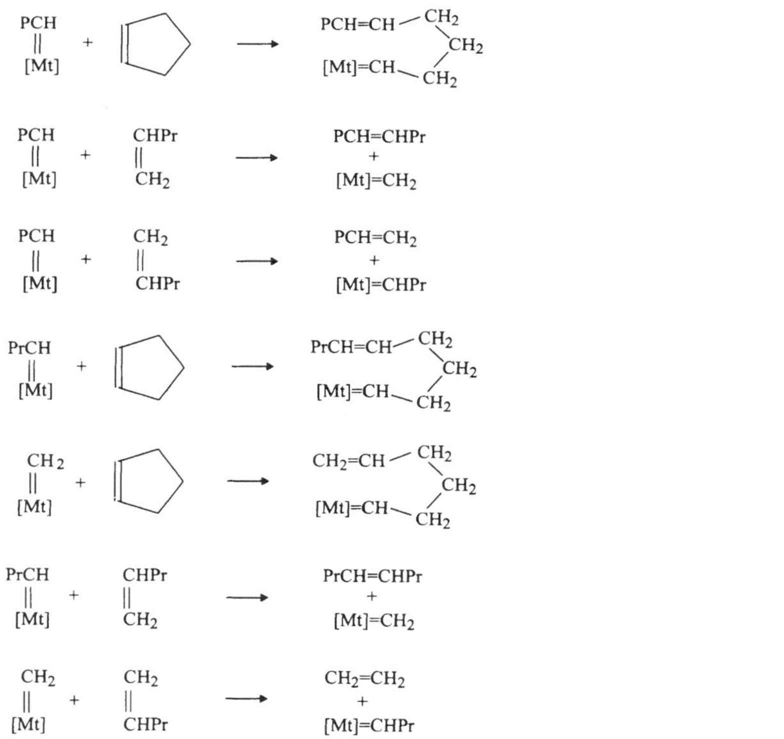

(3)

(4a)

(4b)

(5a)

(5b)

(6a)

(6b)

Scheme 15.2 Mechanism of cross-metathesis between cyclopentene and pent-1-ene; P = polymer chain.

Scheme 15.2 Mechanism of cross-metathesis between cyclopentene and pent-1-ene; P = polymer chain.

Treatment of rate and MW data in terms of Scheme 15.2 requires that due attention be given to the following possible complications: (i) secondary metathesis reactions – only data extrapolated to zero time should be used; (ii) 'pinching off' of cyclic oligomers – this will generally only be serious at low concentrations of the cyclic species M_1; (iii) *cis* and *trans* double-bond formation – each of (3), (4a), (5a), and (6a) represents two reactions, giving *cis* or *trans* double bonds in the products according to the geometry of the intermediate metallacycle. The overall rate constant could, in some instances, be dependent on the concentrations of M_1 and M_2 in so far as these may affect the proportion of *cis* and *trans* double bonds formed.

Three types of rate constant ratios may be derived from experimental data in cross-metathesis systems: (i) k_{4a}/k_{4b} in the case where M_2 is unsymmetrical (Scheme 15.2); (ii) k_{14}/k_{15} in the case where two symmetrical acyclic olefins, Q^1Q^1 and Q^2Q^2 are used (Scheme 15.3); and (iii) $(k_{4a}+k_{4b})/k_3$, which may be found either from telomer ratios or, in principle, from the effect of M_2 on DP_n. Let us consider each of these in turn.

15.4.1 Simple cross-metathesis (k_{4a}/k_{4b})

Let the series of telomers be formed initially in the ratio $1:r:1$, with r representing the unsymmetrical series. In order to solve the stationary-state equations, it is necessary to make the assumption expressed by eqn. (7) (Katz 1975). The rationale of this asumption is as follows. k_{5a}/k_{5b} expresses the relative reactivity of [Mt]=CHPr and [Mt]=CH_2 for the common olefin M_1, while k_{4a}/k_{4b} denotes the relative reactivity of a given metal carbene with the olefin M_2 in its two possible orientations. It is therefore reasonable to suppose that the product of these two ratios will be equal to the relative reactivity of [Mt]=CHPr and [Mt]=CH_2 with the olefin M_2 in the two reactions (6a) and (6b). With this assumption a steady-state treatment of Scheme 15.2 leads to eqn. (8). According to this equation r will be equal to 2 only if $k_{4a}=k_{4b}$, that is, if there is no preferred orientation of M_2 when it reacts with [Mt]=CHP; otherwise it must be greater than 2. [The resemblance of eqn. (8) dealing with bias in end-groups, and eqns. (11) and (12) of Ch. 11, dealing with HT bias in repeat units, should be noted.]

$$(k_{6a}/k_{6b}) = (k_{5a}/k_{5b})(k_{4a}/k_{4b}) \tag{7}$$

$$r = (k_{4a}/k_{4b}) + (k_{4b}/k_{4a}) \tag{8}$$

$$r = x + 1/x \tag{9}$$

Eqn. (8) may be written in the form of eqn. (9) and a value of x, greater than 1, derived from the experimental value of r. The problem then is to decide whether x represents k_{4b}/k_{4a} or its reciprocal. We have argued earlier that, for the particular case shown in Scheme 15.2, we can expect k_{4b}/k_{4a} to be greater than 1. The main factor here is probably the polarity of the olefin with respect to the metal carbene bonds, as indicated in eqn. (4b′). In the case of unsymmetrical internal olefins, a similar but weaker effect may be expected to control the preferred mode of addition, as in eqn. (10).

$$\begin{matrix} \mathrm{PCH}^{\delta+} & & {}^{\delta-}\mathrm{CH_2} \\ \| & + & \| \\ [\mathrm{Mt}]^{\delta-} & & {}^{\delta+}\mathrm{CHPr} \end{matrix} \longrightarrow \begin{matrix} \mathrm{PCH{=}CH_2} \\ + \\ [\mathrm{Mt}]\mathrm{{=}CHPr} \end{matrix} \qquad (4b')$$

$$\begin{matrix} \mathrm{PCH}^{\delta+} & & {}^{\delta-}\mathrm{CHMe} \\ \| & + & \| \\ [\mathrm{Mt}]^{\delta-} & & {}^{\delta+}\mathrm{CHEt} \end{matrix} \longrightarrow \begin{matrix} \mathrm{PCH{=}CHMe} \\ + \\ [\mathrm{Mt}]\mathrm{{=}CHEt} \end{matrix} \qquad (10)$$

Most of the experimental values of r and hence of k_{4a}/k_{4b} must be regarded as lower limits because of the uncertain extent of scrambling by secondary reactions (see Table 15.2). The reactions of cyclooctadiene have not been included in this table because the products usually contain $Q^1{=}CHCH_2CH_2CH{=}Q^2$, which can only be produced by secondary reactions not included in Scheme 15.1 (Pinazzi 1976).

In cross-metathesis with symmetrical olefins only a single series of telomers can be formed. The transfer constants can then be determined directly from the relative rates of consumption of cyclic and linear olefins, making allowance for back reactions. For the cross-metathesis of cyclooctene with internal olefins, catalyzed by $WCl_6/Me_2Si(CH_2SiMe_2CH_2)$, Korshak (1982) obtained the following values for the transfer constants: *cis*-but-2-ene 0.9, *trans*-but-2-ene 0.6, *trans*-oct-4-ene 0.2.

15.4.2 Double cross-metathesis (k_{14}/k_{15})

Double cross-metathesis describes the simultaneous reaction of a cycloalkene M with two symmetrical internal olefins represented as Q^1Q^1 and Q^2Q^2. Fig. 15.7 shows the evolution of the telomer ratios Q^1MQ^2/Q^1MQ^1 and Q^1MQ^2/Q^2MQ^2 with the conversion for the case where M = cyclooctene, Q^1Q^1 = *cis*-but-2-ene, Q^2Q^2 = *cis*-oct-4-ene, and the catalyst as indicated. The initial proportions of the three telomers are $C_{12}:C_{14}:C_{16} = 2.5 \pm 0.25:1:0.090 \pm 0.007$.

The first important conclusion from these experiments is that the unsymmetrical telomer (C_{14}) is formed from the start, which would not be possible on the pairwise mechanism. The results may be interpreted in terms of the metal carbene mechanism shown in Scheme 15.3. In order to obtain useful expressions from this scheme, it is necessary to make an assumption similar to that of eqn. (7) for Scheme 15.2, namely that the rate constant ratios are related by eqn. (19). One then obtains eqns. (20) and (21) from a steady-state treatment of Scheme 15.3. Multiplying the two together gives eqn. (22). The product of the two intercepts in Fig. 15.7 is 4.5 ± 0.6. A more accurate value of 4.1 ± 0.1 is obtained by extrapolating the quantity $[C_{14}]^2/[C_{12}][C_{16}]$ to zero conversion, confirming the prediction of eqn. (22) and justifying eqn. (19) *a posteriori*.

Table 15.2 Experimental values of r and derived values of k_{4b}/k_{4a} [a]

M_1 [b]	M_2 [c]	Catalyst systems	Solvent[d]/ temperature (°C)	r	k_{4b}/k_{4a}	Reference
CPE	1-C_5	$WCl_6/EtAlCl_2/CCl_3CH_2OH$	B/25°	$\geqslant 40$	$\geqslant 40$	Kelly 1975
CPE	2-C_5	$WOCl_4/Et_2AlCl$	C/20°	$\geqslant 2.1 \pm 0.1$	$\geqslant 1 \pm 0.6$	Hérisson 1971
COC	C_3	$WOCl_4/Bu_4Sn$	C/20°	$\geqslant 7 \pm 2$	$\geqslant 7 \pm 2$	Hérisson 1971
COC	1-C_5	$WOCl_4/Bu_4Sn$	C/20°	$\geqslant 21 \pm 2$	$\geqslant 21 \pm 2$	Hérisson 1971
COD	1-C_6	$WCl_6/EtAlCl_2/EtOH$	B/25°	$\geqslant 4 \pm 1$	$\geqslant 4 \pm 1$	Lal 1975
COC	2-C_5	$WOCl_4/Bu_4Sn$	C/20°	$\geqslant 2.1 \pm 0.1$	$\geqslant 1 \pm 0.6$	Hérisson 1971
COC	2-C_6	$MoCl_2(NO)_2(PPh_3)_2/Me_3Al_2Cl_3$	C/20°	3.15 ± 0.17	2.9 ± 0.2	Katz 1977a
NBE	1-C_5	$[(cyclooctene)_2IrCl]_2$	N/20°	Large	Large	Rossi 1975
NBE	2-C_5	$[(cyclooctene)_2IrCl]_2$	N/20°	$\geqslant 2$	$\geqslant 1$	Rossi 1975
NBE	2-Me-C_3	$[(cyclooctene)_2IrCl]_2$	P/20°	Large	Large	Rossi 1975

[a] k_{4b} is assumed to refer to the reaction $PCH{=}[Mt] + R^1CH{=}CHR^2 \rightarrow PCH{=}CHR^1 + [Mt]{=}CHR^2$, where R^2 is more electron-repelling than R^1. For k_{4a}, R^1 and R^2 are reversed. [b] CPE = cyclopentene; COC = cyclooctene; NBE = norbornene. [c] 1-C_5 = pent-1-ene; 2-Me-C_3 = 2-methylpropene; etc. [d] B = benzene; C = chlorobenzene; P = pentane; N = neat.

$$[Mt]{=}Q^1 + M \longrightarrow [Mt]{=}MQ^1 \tag{11a}$$

$$[Mt]{=}Q^2 + M \longrightarrow [Mt]{=}MQ^2 \tag{11b}$$

$$[Mt]{=}Q^1 + Q^2Q^2 \longrightarrow [Mt]{=}Q^2 + Q^1Q^2 \tag{12}$$

$$[Mt]{=}Q^2 + Q^1Q^1 \longrightarrow [Mt]{=}Q^1 + Q^1Q^2 \tag{13}$$

$$[Mt]{=}MQ^1 + Q^1Q^1 \longrightarrow [Mt]{=}Q^1 + Q^1MQ^1 \tag{14}$$

$$[Mt]{=}MQ^1 + Q^2Q^2 \longrightarrow [Mt]{=}Q^2 + Q^1MQ^2 \tag{15}$$

$$[Mt]{=}MQ^2 + Q^1Q^1 \longrightarrow [Mt]{=}Q^1 + Q^1MQ^2 \tag{16}$$

$$[Mt]{=}MQ^2 + Q^2Q^2 \longrightarrow [Mt]{=}Q^2 + Q^2MQ^2 \tag{17}$$

$$[Mt]{=}MQ^1 + M \longrightarrow [Mt]{=}(M)_2Q^1 \tag{18a}$$

$$[Mt]{=}MQ^2 + M \longrightarrow [Mt]{=}(M)_2Q^2 \tag{18b}$$

Scheme 15.3 Mechanism of double cross-metathesis between cyclooctene (M), *cis*-but-2-ene (Q^1Q^1) and *cis*-cot-4-ene (Q^2Q^2).

$$(k_{12}/k_{13}) = (k_{15}/k_{16})(k_{11a}/k_{11b}) \tag{19}$$

$$\frac{d[Q^1MQ^2]}{d[Q^1MQ^1]} = \left(\frac{d[C_{14}]}{d[C_{12}]}\right)_o = \frac{2\,k_{15}\,[Q^2Q^2]}{k_{14}\,[Q^1Q^1]} \tag{20}$$

$$\frac{d[Q^1MQ^2]}{d[Q^2MQ^2]} = \left(\frac{d[C_{14}]}{d[C_{16}]}\right)_o = \frac{2\,k_{14}\,[Q^1Q^1]}{k_{15}\,[Q^2Q^2]} \tag{21}$$

$$\left(\frac{d[C_{14}]}{d[C_{12}]}\right)_o \left(\frac{d[C_{14}]}{d[C_{16}]}\right)_o = 4 \tag{22}$$

The individual intercepts in Fig. 15.7 give $k_{14}/k_{15} = 6.5 \pm 0.9$ for the relative reactivity of the two *cis* olefins at 0°C. The corresponding ratio for the *trans* olefins at 20°C is 3.08 ± 0.16. Thus in each case oct-4-ene is substantially less reactive than but-2-ene, presumably for steric reasons.

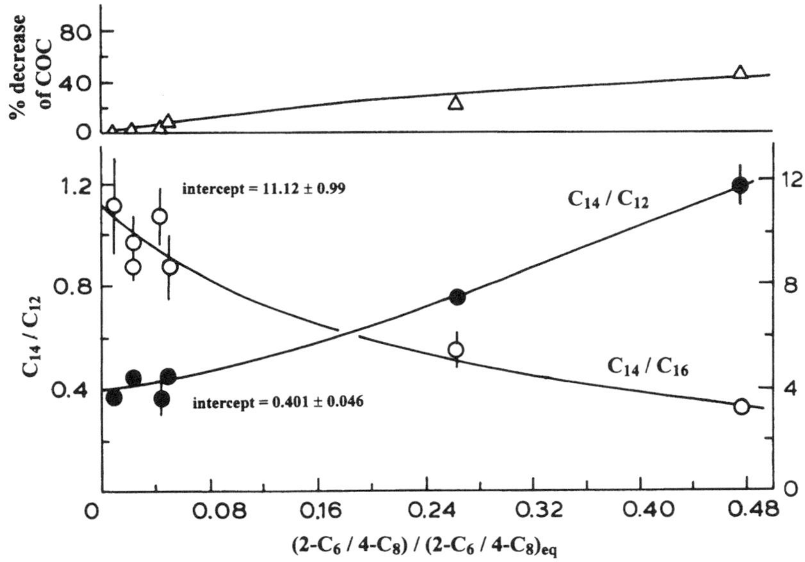

Fig. 15.7 Relative rates of telomer formation in the metathesis reactions between cyclooctene (COC) and a mixture of *cis*-but-2-ene (2-C_4) and *cis*-oct-4-ene (4-C_8) in the ratio $[C_4]/[C_8] = 0.75 \pm 0.05$. Catalyst: $MoCl_2(NO)_2(PPh_3)_2/Me_3Al_2Cl_3$ in chlorobenzene at 0°C. The abscissa provides a measure of the extent of reaction and correlates with the consumption of COC as shown in the upper graph. The telomers produced are represented by $C_{12}(C_2C_8C_2)$, $C_{14}(C_2C_8C_4)$, and $C_{16}(C_4C_8C_4)$ (Katz 1977a).

15.4.3 Transfer constants from telomer ratios

Transfer constants can, in principle, be derived from the relative rates of production of telomers of successively higher n values. Thus for Scheme 15.2 one can derive eqn. (23), where $m = 7$, 10, or 13 depending on which series of telomers is being considered. Likewise for Scheme 15.3, eqn. (24) can be obtained, where $m = 12$, 14, or 16, again depending on the series of telomers; see Katz (1977a) for derivations. Published data on telomer ratios that have been extrapolated to zero time indicate $(k_{4a} + k_{4b})/k_3$ values of about 1 for pent-2-ene/cyclopentene (Hérisson 1971) and about 16 for pent-1-ene/cyclopentene (Kelly 1975), but one must emphasize that they are likely to be dependent on catalyst conditions and especially on the *cis* content.

$$\frac{d[C_{m+5n}]}{d[C_{m+5(n+1)}]} = 1 + \left(\frac{k_{4a} + k_{4b}}{k_3}\right) \frac{[\text{1-C}_5]}{[\text{CPE}]} \qquad (23)$$

$$\frac{d[C_{m+8n}]}{d[C_{m+8(n+1)}]} = 1 + \frac{k_{14}\,[\text{2-C}_4]}{k_{18}\,[\text{COC}]} + \frac{k_{15}\,[\text{4-C}_8]}{k_{18}\,[\text{COC}]} \qquad (24)$$

16

Degradation of Unsaturated Polymers by Metathesis

16.1 Introduction

Unsaturated polymers may undergo two types of olefin metathesis reaction leading to degradation, namely intramolecular and intermolecular. As already discussed in Section 11.2.2, intramolecular reactions may occur during the polymerization of cycloalkenes, leading to the concurrent formation of a series of cyclic oligomers. If the initial rate of propagation is fast compared with the intramolecular degradation reaction represented by eqn. (1), the latter may manifest itself only as a secondary reaction, taking over once the monomer concentration [M] has fallen below a certain level. This accounts for the observation of a maximum viscosity during the polymerization of a number of cycloalkenes (Fig. 11.3). If the linear high polymer is separated from the cyclic oligomers and placed again in contact with the metathesis catalyst, further degradation to cyclic oligomers may occur, showing that the double bonds in the polymer are capable of undergoing secondary metathesis reactions. This is also proved by the fact that many unsaturated polymers can be completely degraded by intermolecular metathesis with acyclic olefins, so as to yield species of the type Q^1MQ^2, where Q^1 and Q^2 are alkylidene moieties derived from the acyclic olefin. Both polymers and copolymers, whether linear, cross-linked, or filled, may be degraded by metathesis. The proportion of fillers may be readily determined in cross-linked polymers by degrading the polymer to soluble products (Hummel 1982b, 1993a; Stelzer 1987a). The nature of the cross-links can be elucidated by GC/MS analysis of the ultimate degradation products. For this purpose it is essential to use a metathesis catalyst that does not give rise to side reactions. $WCl_6/EtAlCl_2$ is therefore to be avoided. WCl_6/R_4Sn is usually satisfactory but the metal carbene complexes are best (Wagener 1991a; Marmo 1993). Surfaces of blended or filled polymers having an unsaturated component can be etched by exposure to a metathesis catalyst, and examined by electron microscopy (Hummel 1982a; Stelzer 1992).

$$[\mathrm{Mt}]\mathrm{M}_n\mathrm{M}_m\mathrm{Q} \longrightarrow [\mathrm{Mt}]\mathrm{M}_m\mathrm{Q} + c\text{-}\mathrm{M}_n \qquad (1)$$

Much of the work in this area relates to the polymers or copolymers of butadiene. The latter may have been made directly either as statistical or block copolymers, or they may have been made by partial reaction of polybutadiene at the double bond or at the allylic position. The distribution of degradation products provides valuable information concerning the mechanism of formation of the copolymer. Specific examples are given below. The possibility of using metathesis degradation for recycling of unsaturated polymers has been discussed by Höcker (1993). For further detail, see the reviews of Hummel (1985, 1990c, 1993b) and Thorn-Csányi (1994d).

16.2 Degradation by intramolecular metathesis

16.2.1 Poly(1-butenylene)

This polymer can be made by ROMP of cycloocta-1,5-diene (COD) or cyclobutene, or by the polymerization of butadiene with appropriate catalysts. If the linear high polymer made from COD is freed from cyclic oligomer by extraction with 50/50 hexane/propan-2-ol and placed again in contact with the same catalyst (WCl_6/$EtAlCl_2$/EtOH), cyclic oligomers are re-formed, in the same proportions as before, as first demonstrated by Scott (1969) (Fig. 16.1). The ability of dead polymer to undergo reactivation, leading to intramolecular degradative metathesis, has since been shown to occur with a variety of other catalysts (Kropocheva 1972; Tlenkopachev 1976; Korshak 1977; Zemtsov 1977; Chauvin 1978). If the initial concentration of high polymer is below the equilibrium concentration of cyclic oligomer that would be attained at higher initial polymer or monomer concentration, the high polymer is completely degraded to cyclic oligomers. The equilibrium concentration is 0.65 M (C_4H_6 units) at 30°C in chlorobenzene. If the initial concentration of polymer is higher than this, it only undergoes partial degradation and the MW of the residual polymer remains remarkably high (Korshak 1976). This indicates that once a dead polymer molecule is converted into an active metal carbene, the latter may strip off cyclic oligomeric species very rapidly until completely broken down. In other words the equilibrium concentration of cyclic oligomer can be achieved through the total degradation of a small number of molecules rather than the partial degradation of all the molecules. The activity of a catalyst system does not always last long enough for degradation to proceed to its maximum possible extent. A fresh dose of catalyst will then induce further degradation (Scott 1969; Pampus 1970; Tlenkopachev 1976; Korshak 1977).

If a high-*cis* polymer is exposed to WCl_6/Me_4Sn, the polymer degrades and about 75% of the *cis* double bonds isomerize to *trans* (Kumar 1983). No cyclic trimers are formed during the degradation of an all-*cis* polymer if a *cis*-selective catalyst such as $(\pi\text{-}C_4H_7)_4Mo$/$EtAlCl_2$ is used. This is because the *ccc* isomer is highly strained and not readily formed by intramolecular metathesis. However, with a non-selective catalyst, cyclic trimers (*ccc*, *cct*, *ctt*, and *ttt*) are formed and the overall *trans* content

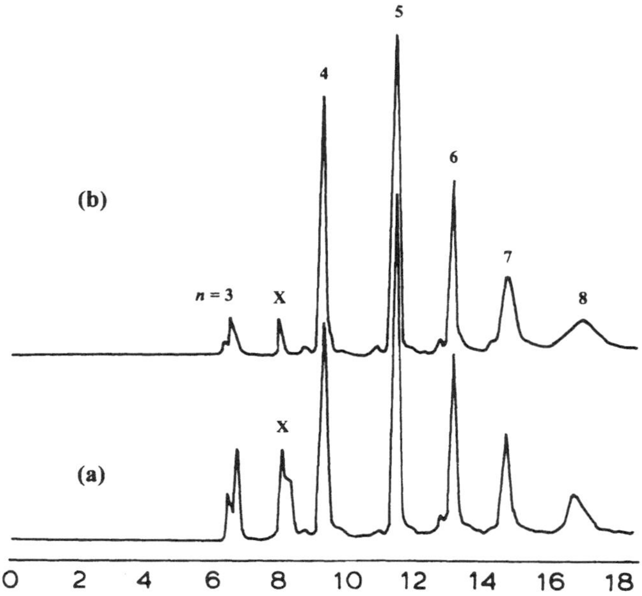

Fig. 16.1 Comparison of GC of macrocyclic species $(C_4H_6)_n$ extracted from (a) the ROMP of COD and (b) the degradation of high-molecular-weight poly(1-butenylene) prepared from COD. Catalyst: $WCl_6/EtAlCl_2/EtOH$ in both cases. X = di-*t*-butyl-*p*-cresol present as stabilizer. The two peaks at $n = 3$ are due to *ctt* and *ttt* isomers (Scott 1969).

increases with time (Kropacheva 1972; Thorn-Csányi 1995a). No cyclobutene and only traces of COD are ever detected in the degradation products.

16.2.2 Poly(1-pentenylene)

The limiting viscosity number of the polymer formed during the ROMP of cyclopentene at 25°C catalyzed by $WCl_6/i\text{-}Bu_2AlCl$ declines with increasing reaction time indicating the occurrence of secondary metathesis degradation reactions of the polymer (Dolgoplosk 1977a). At −20°C with a catalyst system based on $WCl_{4.2}(OCH_2CH_2Cl)_{1.8}$, the high polymer initially formed from low $[M]_o$ (3.3 wt%) is almost completely converted into cyclic oligomers after 5 h, whereas at high $[M]_o$ (15 wt%) there is comparatively little formation of cyclic oligomers even after 22 h (Witte 1978). Propagation is thus favoured relative to intramolecular degradation at lower temperatures as well as at higher [M]. For the ROMP of cyclopentene at 20°C catalyzed by $WCl_6/\overline{Me_2SiCH_2SiMe_2CH_2}$, the rate of fall of MW during the secondary stage of reaction is markedly dependent on the polymer concentration, being four times faster at 5 wt% than at 1 wt%. This shows that, in this system, intermolecular metathesis reactions are occurring in competition with the intramolecular reaction (Korshak 1979; Badamshina 1982).

16.2.3 Miscellaneous

The viscosities of the products of ROMP of cyclooctene and cyclododecene also pass through a maximum with time. The decrease in MW is accompanied by the formation of increasing amounts of cyclic oligomers until an equilibrium position is reached (Kuteinikov 1976). The high-molecular-weight fraction of poly(1-octenylene) can be separated and degraded by metathesis catalysts to cyclic oligomers, $(C_8H_{14})_n$, $n=2$–4, which are formed in the same proportions as in the original preparation.

1-Methylcycloocta-1,5-diene undergoes ROMP in the presence of WCl_6/$EtAlCl_2$/EtOH to yield a polymer having alternating butadiene (Bd) and isoprene (Ip) units (Ofstead 1969). At the same time cyclic oligomeric species $(\mathrm{Bd\ Ip})_n$ ($n=2, 3, 4$) are formed. These two facts indicate that only the unsubstituted double bond takes part in the ROMP of this monomer. After longer reaction times, additional oligomers are formed, the GC peak positions of which suggest that they correspond to the cyclic sesqui-oligomers Bd_3Ip_2, Bd_2Ip_3, Bd_4Ip_3, Bd_3Ip_4. The substituted double bonds in the polymer are thus capable of slow degradative reaction, like those in polyisoprene (Korshak 1982).

Early attempts to bring about the degradation of *cis*-polyisoprene by metathesis were unsuccessful, probably because of the short life of the catalyst (Ast 1970; Pampus 1970). The system $WCl_4[OCH(CH_2Cl)_2]_2$/Et_2AlCl/anisole (1/5/5) is longer-lived and under favourable conditions can reduce the MW of polyisoprene dissolved in benzene at 25°C from 10^6 to 2×10^4 in 10 days, without any marked loss of unsaturation or *cis* double-bond content (Korshak 1982).

The 1 : 1 alternating copolymer (**1**) of butadiene and propene has a structure corresponding to the polymer that would be produced by the hypothetical ROMP of 4-methylcyclohexene (**2**). When a 1–2% solution of **1** in toluene is placed in contact with $WCl_4[OCH(CH_2Cl)_2]_2$/$EtAlCl_2$ the polymer degrades to **2** with a 20% yield; some cyclic oligomers $(C_7H_{12})_n$, $n=3$–7, are also formed (Korshak 1977). The elimination of cyclohexene from the polymer of *cis,trans*-cyclodeca-1,5-diene has been discussed in Section 12.8.

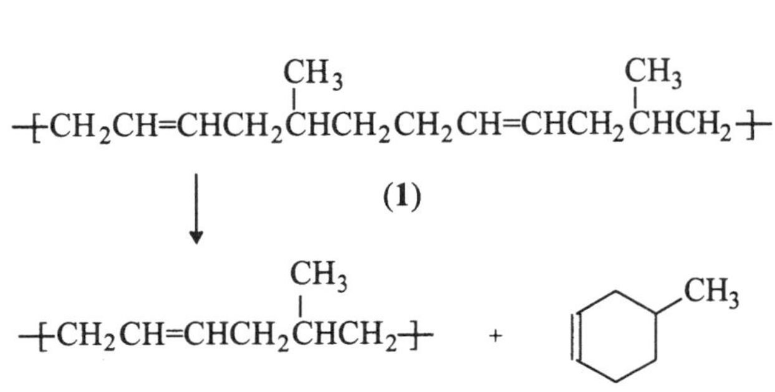

(**1**)

(**2**)

16.3 Degradation by intermolecular metathesis

16.3.1 Linear 1,4-polybutadiene

The first experiments on the cross-metathesis of polybutadiene with internal acyclic olefins were reported by Ast (1970) using hex-2-ene as degrading agent. A total of 0.5 g of polymer and 5 g of hex-2-ene were reacted twice with 0.01 mmol of WCl_6 mixed with 0.06 mmol of $EtAlCl_2$ for 60 min at 20°C. Products containing both one and two units of butadiene were detected, with either two ethylidene or two butylidene end-groups, or one of each. In order to minimize the number of possible products, most of the later work was carried out using symmetrical acyclic olefins, ranging from ethene to tetradec-7-ene. After 1975, WCl_6/R_4Sn was generally used as initiator because of its longer life and relative freedom from side reactions. More recently, the metal carbene initiators have been found to be better still. In what follows, Q denotes the alkylidene unit, which is half the symmetrical internal olefin Q_2, and M_x denotes x 1,4-butadiene units $[=CHCH_2CH_2CH=]_x$.

If the ratio of alkylidene units Q to monomer units M in a system is greater than 10, the main species present at equilibrium are, in order of decreasing concentration, Q_2, QMQ, QM_2Q, and QM_3Q. For example, with Q = butylidene and [Q]/[M] = 10, the relative proportions of these species at equilibrium are 82.9%, 14.1%, 2.4%, and 0.4%, respectively (Hummel 1973a). In these circumstances, the concentrations of cyclic oligomeric species in equilibrium with QM_xQ, $x \geqslant 2$, are negligible; but this is not the case when [Q]/[M] is less than 10. Fig. 16.2 shows the GC of the reaction mixture after 2 min degradation of *cis*-1,4-polybutadiene with tetradec-7-ene, using a ratio [Q]/[M] = 3. The cyclic oligomeric species are formed initially in greater proportion than the linear species but decline in relative concentration as the system approaches equilibrium.

In the degradation of both *cis*- and *trans*-1,4-polybutadiene by hex-3-ene, using $WCl_6/Et_4Sn/Et_2O$ as catalyst, it is necessary to use a 200-fold excess of hex-3-ene in order to ensure the conversion of more than 90% of the polymer into a mixture of the *cc*, *ct*, and *tt* isomers of deca-3,7-diene, QM_1Q (Abendroth 1975; Canji 1978).

Ethene degrades polybutadiene (92% 1,4 linkages) in the presence of $W(=CHCMe_3)(=NAr)[OCMe(CF_3)_2]_2$. Bubbling ethene into a solution of the polymer and initiator in toluene is sufficient to reduce the MW by a factor of 100 and to generate some hexa-1,5-diene (Wagener 1991a). $Ru(=CHPh)(Cl)_2(PCy_3)_2$ is an even better initiator (Watson 1996). $Me_3SiCH_2CH=CHCH_2SiMe_3$ is also very effective as degradation agent with the tungsten initiator. It can be generated *in situ* from $Me_3SiCH_2CH=CH_2$. After an initial rapid reaction, in which ethene is evolved and the internal olefin is formed, the degradation of the polymer continues over a period of 48 h. With $[Q]_o/[M]_o = 10$, the GC/MS of the volatile products shows the presence of QM_xQ, $x = 3$, 4, 5, and no by-products (see Fig. 16.3). If $Mo(=CHCMe_2Ph)(=NAr)[OCMe(CF_3)_2]_2$ is used to catalyze the degradation by $SiClMe_2(CH_2CH=CH_2)$, the sole product is QM_1Q, i.e. α,ω-bis(chlorodimethylsilyl)octa-2,6-diene (Marmo 1993).

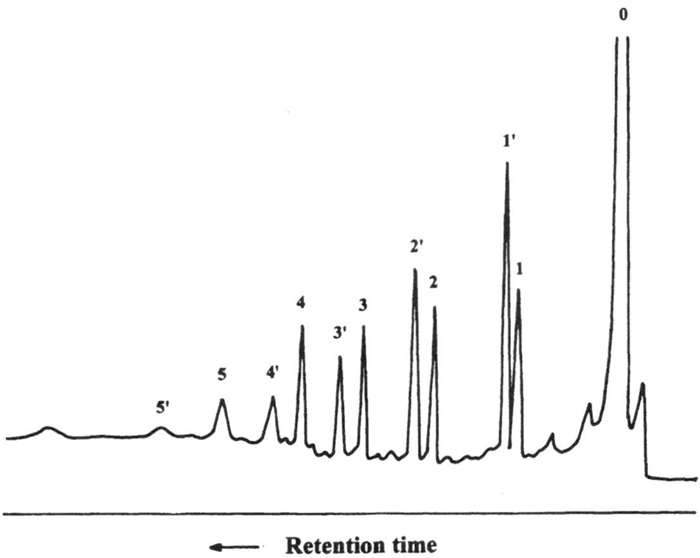

Fig. 16.2 GC of initial metathesis products from *cis*-1,4-polybutadiene (0.33 mol M units dm^{-3}) and tetradec-7-ene (0.5 M) in hexane at 20°C. Catalyst: $WCl_6/EtAlCl_2$. Reaction time 2 min. Cyclic oligomers: $1' = C_{16}$; $2' = C_{20}$; $3' = C_{24}$; $4' = C_{28}$; $5' = C_{32}$. Linear species QM_xQ ($x = 0$–5): $0 = C_{14}$; $1 = C_{18}$; $2 = C_{22}$; $3 = C_{26}$; $4 = C_{30}$; $5 = C_{34}$ (Lorber 1973).

The equilibria set up between Q_2 and QM_xQ, $x = 1, 2, 3...$, may be treated on a statistical basis on the assumptions (i) that the various metathesis reactions are thermoneutral, and (ii) that equilibria involving cyclic species can be neglected. Good agreement between the calculated and observed distribution of products has been found for the reaction of polybutadiene with dodec-6-ene (Hummel 1973b) and with oct-4-ene (Hummel 1973a), for $[Q]_o/[M]_o \geqslant 9$. As degradation proceeds the MWD moves towards the most probable distribution ($M_w/M_n = 2$) (Calderon 1981; Chung 1992a).

The evolution of the distribution function with time can also be treated on a statistical basis assuming that all double bonds are equally reactive, independent of their position in the chain (Stelzer 1977; Martl 1991). Back reactions and cyclization reactions are neglected, which is permissible when Q_2 is in excess and the reaction has not proceeded too far. The fraction of residual double bonds in the polymer declines exponentially with time, as expected (Fig. 16.4). There is a short induction period which decreases with increasing temperature and is presumably caused by the build-up to a steady concentration of active species.

Polybutadiene (MW 3000), capped with ester groups, can be made by metathesis degradation of polybutadiene with dimethyl hex-3-enedioate, catalyzed by WCl_6/Me_4Sn at 100°C (Seyferth 1981a). Telechelic polymers of MW 1000 have also been made by degradation with 6-(9-borobicyclononane)-hex-1-ene, followed by reaction with $NaOH/H_2O_2$ and selective extraction, using hexane/ether (70/30), of

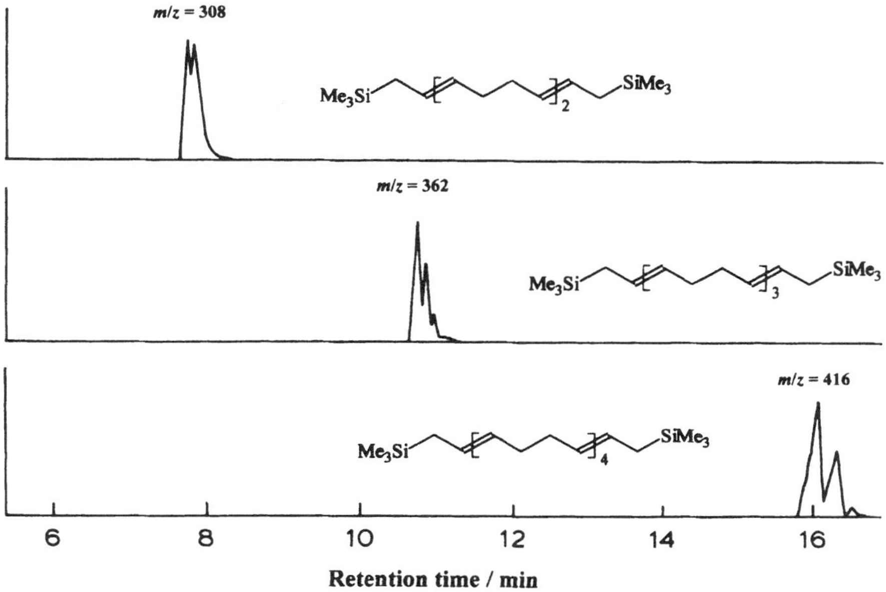

Fig. 16.3 GC with mass-exact labels obtained from the MS of α,ω-bis(trimethylsilyl)-1,4-polybutadiene containing 3, 4, and 5 units of M. For each value of *m/z* there are two or more GC peaks corresponding to the various *cis/trans* isomers that are formed during degradation (Marmo 1993).

the molecules containing two hydroxyl end-groups (Chung 1991e, 1992a,b). Perfectly difunctional telechelic polybutadiene oligomers, containing from one to five butadiene units, have been identified by GC/MS in the products of cross-metathesis with $RCH_2CH{=}CHCH_2R$, where $R = (CH_2)_2COOEt$, $(CH_2)_3OCOMe$, $OSiMe_2$ (CMe_3), or N-phthalimide, using a molybdenum carbene initiator (Marmo 1994, 1995).

16.3.2 Polybutadienes containing substantial proportions of 1,2 units

Cross-metathesis of but-2-ene with polybutadiene containing 10% 1,2 units and 90% 1,4 units gives the expected 81% octa-2,6-diene derived from (1,4) to (1,4) sequences but products derived from (1,2) to (1,4) sequences have not been detected (Michajlov 1970, 1971). A better understanding of the degradation of such polymers has been obtained by cross-metathesis with hex-3-ene, using $WCl_6/Et_4Sn/Et_2O$ as initiator in toluene (Abendroth 1975; Thorn-Csányi 1979). With this system the pendant vinyl groups in 1,2 units give but-1-ene and the triene **3**, but most of the latter suffers metathetical ring closure to give either **4** or **5** (Scheme 16.1). From a polymer containing 19% 1,2 units (as determined both from the IR spectrum of the original polymer or from the amount of but-1-ene produced by

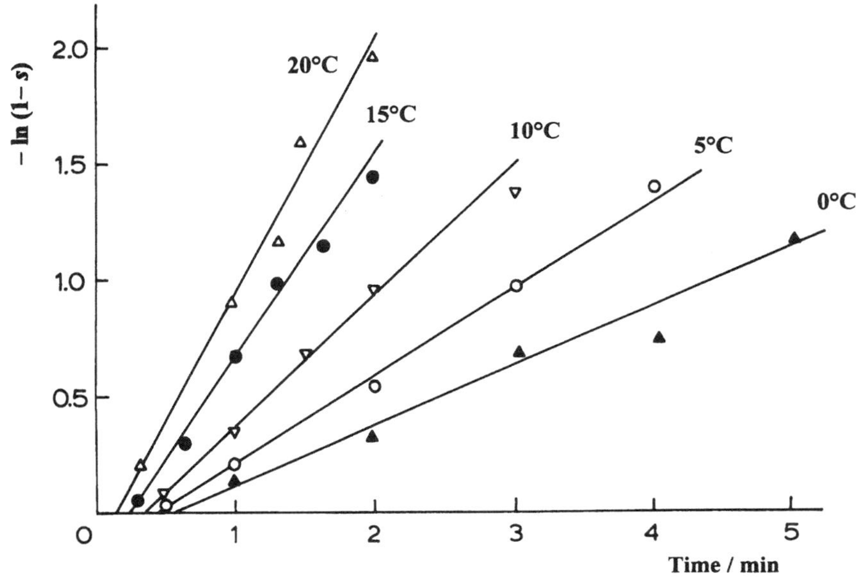

Fig. 16.4 Degradation of 1,4-polybutadiene by metathesis with oct-4-ene. Decline of the fraction of unreacted double bonds in the polymer, $(1-s)$, with time. Catalyst: $WCl_6/EtAlCl_2$ in tetrachloroethene (Stelzer 1977).

degradation) the relative proportions of **3**, **4**, and **5** formed are 0.4 : 3.2 : 6.2. A polymer containing 40% 1,2 units can also be degraded completely by hex-3-ene; the amounts of but-1-ene and deca-3,7-diene formed correspond to a random distribution of 1,2 units in the original polymer (Canji 1977, 1978). However, the complete degradation of a polymer containing 80% 1,2 units is more difficult to achieve and requires repeated dosage with catalyst. The double bonds in the main chain are especially resistant to metathesis. This is attributed to the fact that they are partially shielded by pairs of neighbouring vinyl groups in 1,2 units; not only this, but these vinyl groups react preferentially to give crotyl groups ($CH_3CH{=}CH{-}$), which afford even greater protection to the main-chain double bonds (Canji 1978).

Scheme 16.1 shows that monosubstituted cyclohexenes and cyclopentenes are obtainable as degradation products when the original polymer contains isolated 1,2 units. If the polymer contains adjacent 1,2 units, the intramolecular metathesis reaction can lead to 3,5-disubstituted cyclohexenes and cyclopentenes of the types **6**, **7**, and **8**. Each of these has *tt*, *tc*, and *cc* isomers, of which *tt* is usually dominant (Thorn-Csányi 1979).

Cross-metathesis of dimethyl hex-3-enedioate with polybutadiene containing 20–30% 1,2 units has also been studied, with similar results (Campistron 1986).

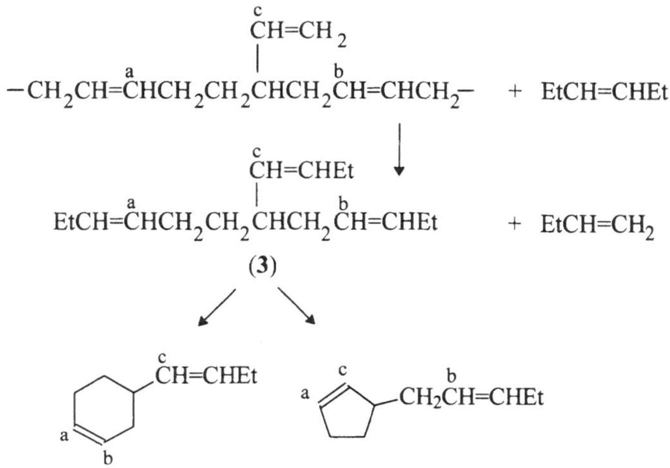

(4) (5)

Scheme 16.1 Products of metathesis degradation of polybutadiene containing 1,2 units.

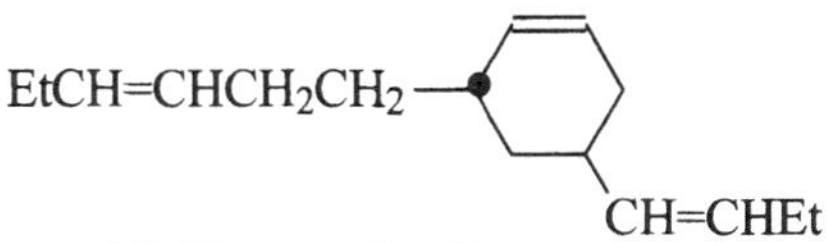

(6) (from *m* dyad)

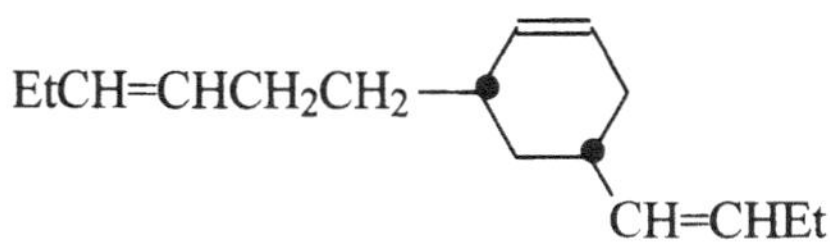

(7) (from *r* dyad)

(8)

16.3.3 Cross-linked polybutadienes

Polybutadiene is readily cross-linked by heating with 1% dicumyl peroxide for 10 min at 145°C. If the product is first extracted with C_2Cl_4 to remove any non-cross-linked material and then placed in contact with $WCl_6/EtAlCl_2/EtOH$ in C_2Cl_4 at 45°C, it rapidly goes into solution (Hummel 1970; Kumar 1981). An added acyclic olefin assists this process but is not essential. It is clear that the network can be readily broken down by the occurrence of either intramolecular or intermolecular metathesis reactions.

With a more highly cross-linked polymer, it is possible to follow the progress of degradation by measuring the loss of weight of rubber balls with time. The rate of degradation by oct-1-ene, initiated by $WCl_6/EtAlCl_2$, increases more than 50-fold when the temperature is raised from $-15°C$ to $60°C$; it also increases when the Al/W ratio is raised from 1.5 to 10.8. Similar results are obtained by degradation with oct-4-ene. Maximum activity is observed for an Al/W ratio of 4, and the activation energy is 34 ± 7 kJ mol^{-1} (Dürr 1971; Hummel 1974).

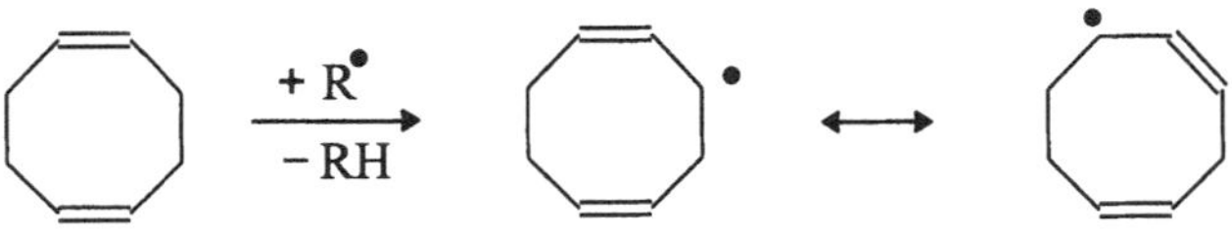

(9) **(10)** **(11)**

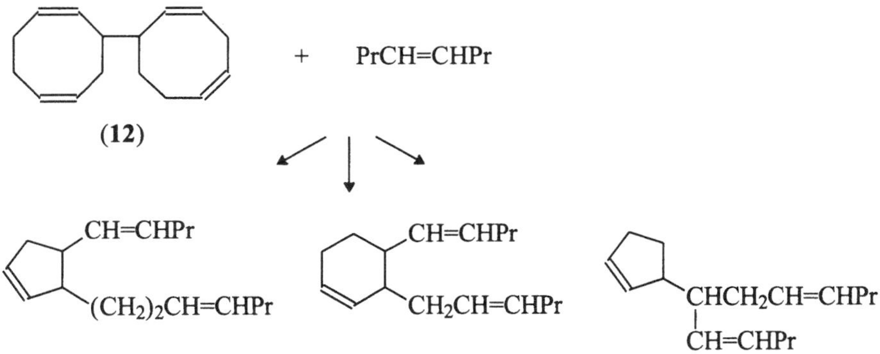

(12)

(13) **(14)** **(15)**

The extent of cross-linking is measured by the swelling constant Q, which is defined as the ratio of the weight of liquid absorbed to the weight of polymer when equilibrium has been reached with the swelling liquid; this is an inverse measure of the proportion of cross-links. The volatile products of degradation of moderately cross-linked polybutadienes ($Q = 1.5$–2.3) by hex-2-ene give GC peaks at C_8, C_{10}, C_{12}, C_{14}, C_{16}, C_{18}, and C_{20}, and are indistiguishable from the products of degradation of non-cross-linked polybutadiene (Ast 1971). More sensitive experiments, in which *cis*-1,4-polybutadiene is cross-linked by heating with dicumyl peroxide in chlorobenzene at 12°C and then degraded with *trans*-oct-4-ene at 20°C using WCl_6/Me_4Sn as catalyst, have shown that double-bond shift reactions of the intermediate allylic radicals do occur during the cross-linking process (Hummel 1983b). This also happens when cycloocta-1,5-diene (**9**) is used as a model compound. Radical **10**, in mesomeric equilibrium with **11**, can combine with

11 to yield **12**, which on metathesis with oct-4-ene should result in a mixture of **13**, **14**, and **15**, all of formula $C_{17}H_{28}$. A small peak of this mass (232) is indeed observed in the MS of the degradation products. The same parent peak is observed in the degradation products from cross-linked polybutadiene but the MS cracking pattern is somewhat different, suggesting that the isomers are formed in different proportions (Hummel 1983b). Use of cyclododeca-1,5,9-triene as a model compound gives similar results to those obtained with **9** (Hummel 1986).

For heavily cross-linked polymers ($Q < 0.3$), there is always an insoluble residue after metathetical degradation. As the degradation proceeds, volatiles analyzable by GC are produced, together with a certain amount of low-molecular-weight extractable polymer, the proportion of which rises to a maximum and then declines as metathesis proceeds towards completion (Ast 1977b; Kumar 1982).

The degradations of vulcanized polybutadienes (Bozkurt 1989; Zümreoglu 1992; Hummel 1993a) and of cross-linked blends with ethene/propene copolymer (Hummel 1987b) have also been studied.

16.3.4 Other homopolymers

Trans-1,4-polyisoprene can be degraded by metathesis with ethene at 5 bar in toluene initiated by a tungsten carbene catalyst. The expected 2-methylhexa-1,5-diene is detected and the MW of the polymer is reduced from 2.5×10^4 to 10^3. The model compound squalene (**16**) is also degraded to the expected products; eqn. (2) (Wagener 1991a).

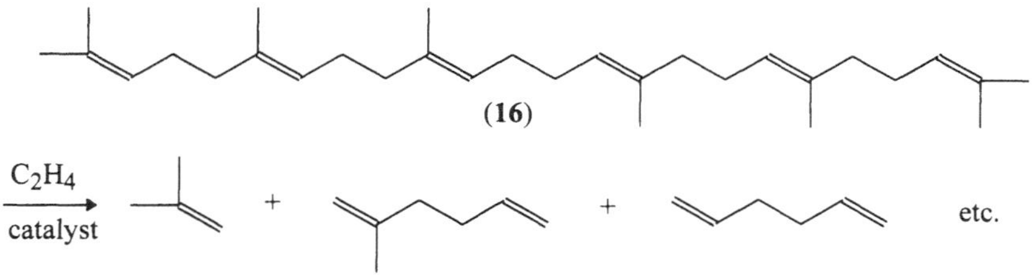

(2)

Cis-1,4-polyisoprene (natural rubber) has been degraded by metathesis with oct-1-ene using both WCl_6/Me_4Sn (Alimuniar 1990) and $WCl_6/Et_3Al_2Cl_3$ (Hummel 1993a) as catalysts. With the former system, there are side reactions leading to some loss of unsaturation, and with the latter there are double-bond shift reactions, which actually assist degradation through the formation of $-CH{=}CH-$ linkages; these are more reactive in metathesis than $-CH{=}CMe-$ linkages. This is an advantage if the purpose is to determine the proportion of filler in a rubber that has been cross-linked by heating with peroxides or by vulcanization (Hummel 1993b). Experiments on the metathesis reaction of the model compound 2-methylpent-2-ene with tetradec-7-ene have also demonstrated the importance of double-bond shift reactions in the metathesis degradation of natural rubber (Hummel 1993a).

Poly(1-pentenylene) can be readily degraded with hex-2-ene to yield the expected C_9, C_{11}, and C_{13} products derived from single monomer units, together with C_{14}, C_{16}, and C_{18} products derived from pairs of monomer units (Ast 1970).

Poly(1-octenylene) having 88% *trans* double bonds is readily degraded by ethene in the presence of a tungsten carbene initiator. The ^{13}C NMR spectrum of the degraded polymer (MW 1000) shows the expected signals from the carbons in the allyl end-groups, while the *trans* double-bond content of the polymer rises to 96% indicating that the *cis* double bonds react more readily than the *trans* double bonds during the metathesis degradation. Some of the ultimate degradation product, deca-1,9-diene, is also detected. The reaction is very clean, with no sign of the side reactions sometimes encountered with WCl_6-based catalysts (Wagener 1991a).

Polynorbornene is more difficult to degrade. Ethene in the presence of a tungsten carbene catalyst does bring about some reduction in MW (factor of five) but no 1,3-divinylcyclopentane is detected (Wagener 1991a). This is another example of the protection against metathesis offered by substituents on the carbons adjacent to the double bonds, in this case in the form of cyclopentane rings.

The polymer **17**, prepared by the ADMET polymerization of neat diallyldiphenylsilane using $Mo(=CHCMe_2Ph)(=NAr)[OCMe(CF_3)_2]_2$ (Anhaus 1991), can be degraded by diluting the living polymer solution with sufficient toluene. This gives a 30% yield of the *tt* cyclic dimer **18**, which separates as a white precipitate. When this is isolated and placed in contact with further initiator, an equilibrium is established between the cyclic monomer **19** and the three cyclic dimers (*cc* : *ct* : *tt* = 73 : 21 : 6), eqn. (3) (Anhaus 1993). The formation of the *tt* cyclic dimer in the initial degradation of **17** is thus kinetically favoured.

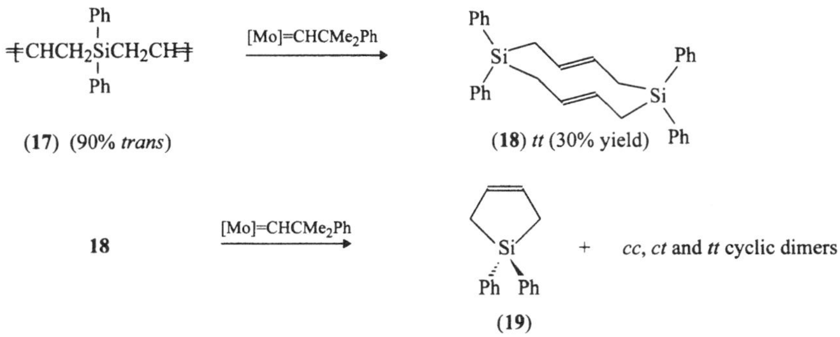

(3)

Polyphenylacetylene can be degraded by reaction with hept-1-ene in the presence of WCl_6 (Severina 1987). Unsaturated polymers containing side-chain ester groups, and polymers of ambrettolide containing ester groups in the main chain have been successfully degraded by metathesis (Ast 1976b).

Polymers of 7-oxanorbornene and 7-oxanorbornadiene derivatives can be degraded to telechelic oligomers using $(Me_3SiCH_2CH=)_2$ with $Mo(=CHCMe_2Ph)(=NAr)[OCMe_2CF_3]_2$ as initiator (Viswanathan 1994).

16.3.5 Polybutadienes modified by partial addition reactions

The main question is whether such reactions yield random or block copolymers. Random addition of a limited amount of reagent XY to the double bonds in the polymer will result in structures of the type $-M_1M_2M_1-$ (**20**), which can then be degraded by metathesis with the symmetrical olefin Q_2 to give $Q(M_1/2)(M_2)$-$(M_1/2)Q$ as one of the products. For example, if $XY = H_2$ and $Q_2 =$ oct-4-ene, this product is $PrCH{=}CH(CH_2)_6CH{=}CHPr$.

$$-CH_2CH{=}CHCH_2 \vdots CH_2CHXCHYCH_2 \vdots CH_2CH{=}CHCH_2-$$
$$(M_1) \qquad\qquad (M_2) \qquad\qquad (M_1)$$

(20)

If the addition reaction is taken to higher conversion, the products contain sequences of M_2 units, and in principle the sequence distribution may be found by GC analysis of the degradation products, provided they are sufficiently volatile. This is the case for $XY = H_2$ or CH_2; products containing up to at least five M units can be detected (see Fig. 16.5). For $XY = Cl_2$, Br_2, HCl, or HBr, only the first and sometimes the second member of the series ($m = 1, 2$) are sufficiently volatile to be detected by GC.

The results of this type of investigation are summarized in Table 16.1. On the heterogeneous Pd/C catalyst, there is a marked variation in hydrogenation behaviour with temperature. At 20°C no double-bond shifts are observed but at

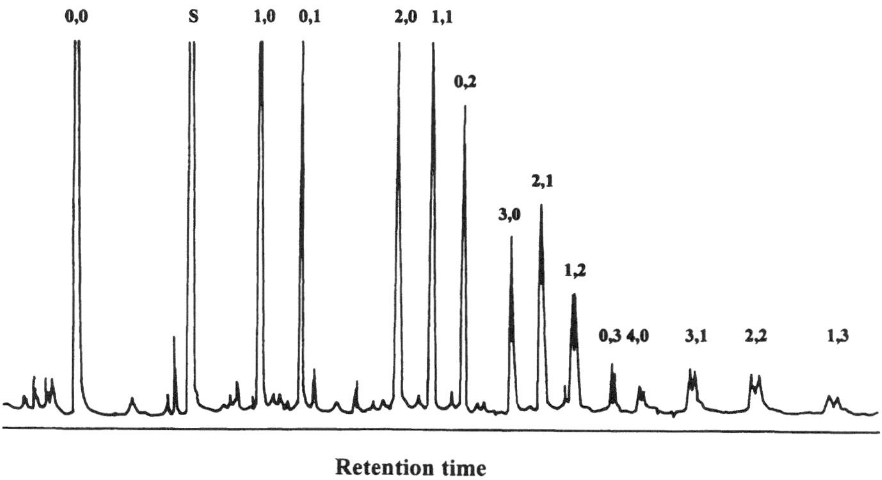

Fig. 16.5 Degradation of partially (25%) cyclopropanated *cis*-1,4-polybutadiene by metathesis with *trans*-oct-4-ene. GC of a reaction mixture in the early stages of reaction. n, m above the peaks correspond to the number of M_1 and M_2 units, respectively, in $Q(M_1/2)\,(M_1)_n(M_2)_m(M_1/2)Q$ (see text). S = internal standard, tetradecane (Hummel 1984).

130°C the shift reaction is so extensive that the GC of the degradation products shows a peak for every carbon number from C_{12} to C_{36}, declining steadily in peak height for peaks beyond C_{15}. If poly(1-octenylene) is partially hydrogenated under similar conditions at 130°C, the GC of the degradation products (using oct-4-ene) again shows a peak at every carbon number, but this time the peak heights show periodic maxima at multiples of C_8; in this case the separation of the original double bonds is evidently too great for complete randomization to take place during hydrogenation (Ast 1977b).

For the homogeneous partial hydrogenation of polybutadiene the proportions of M_1M_1, M_1M_2, and M_2M_2 dyads can be readily calculated from the GC of the degradation products. As may be seen from Fig. 16.6, the distribution is close to random. A more severe test is to compare the experimental and theoretical proportions of each member of the $Q(M_1/2)(M_2)_m(M_1/2)Q$ series. This comparison is shown in Fig. 16.7. Again there is good agreement, regardless of the extent of hydrogenation.

In the heterogeneous catalysis of hydrogenation of polybutadiene, both random and blockwise processes occur simultaneously (Ast 1977b). For the addition of HCl, HBr, Cl_2, and Br_2 (Table 16.1), products corresponding to degradation of $-M_1M_2M_1-$ structures are detected, although one cannot exclude the possibility that blocks of M_2 units are also present. From Fig. 16.8 it may be seen that the metathesis of a double bond in an M_1M_2 dyad (conversion of C into B) occurs more slowly than the metathesis of a double bond in an M_1M_1 dyad.

16.3.6 Polybutadienes modified by partial substitution reactions

Two methods have been used to introduce substituents at the α-methylene position in polybutadienes. The first is by radical-induced bromination using *N*-bromosuccinimide (NBS) with 2,2′-azoisobutyronitrile (AIBN) at 80°C followed by replacement of the bromine substituent with an aryl or alkyl group by means of the Grignard–Wurtz reaction. At least 18 substituents have been introduced in this way (Hummel 1977b, 1978, 1980a,b, 1983a; Sentürk 1979; Stelzer 1987b); sequence (4). The second method is by direct radical aralkylation initiated by the decomposition of dicumyl peroxide at 120°C; eqn. (5) (Hummel 1977a, 1978, 1990a; Demel 1977; Sentürk 1979). By these two methods copolymers containing up to 35% of the substituted units (M_2) have been prepared.

$$\underset{(M_1)}{\text{+CHCH}_2\text{CH}_2\text{CH+}} \xrightarrow[\text{AIBN}]{\text{NBS}} \text{+CHCH}_2\text{CHBrCH+} \xrightarrow{\text{R}^1\text{MgBr}} \underset{(M_2)}{\text{+CHCH}_2\overset{\overset{\text{R}^1}{|}}{\text{C}}\text{HCH+}} \qquad (4)$$

R^1 = isopropyl, butyl, pentyl, cyclohexyl, phenyl, 4-methylphenyl, 4-ethylphenyl, 3,4-dimethylphenyl, 4-isopropylphenyl, 4-fluorophenyl, 3-trifluoromethylphenyl, 4-chlorophenyl, benzyl, 2-methylbenzyl, 4-methylbenzyl, 4-methoxyphenyl, 3,4-dimethoxyphenyl, etc.

Table 16.1 Products of metathesis degradation of partial addition products of *cis*-1,4-polybutadiene

XY	Fraction of C=C reacted	Catalyst for addition reaction	Q_2	Metathesis catalyst	Observed degradation products $Q(M_1/2)(M_1)_n(M_2)_m(M_1/2)Q$ *n*	*m*		Reference
H_2	0.3	Pd/C/cyclohexane/20°C (heterogeneous)	*cis*-PrCH=CHPr	$WCl_6/EtAlCl_2$	0	1		Wewerka 1976
					n	0		
H_2	0.08–0.85	Pd/C/decalin/20°C (heterogeneous)	*cis*-PrCH=CHPr	$WCl_6/Me_4Sn/Bu_2O$			C_{12} to C_{24} (double-bond shift during hydrogenation)	Ast 1977b
H_2	0.08–0.85	4-methylphenylsulfonyl hydrazide/xylene/130°C (homogeneous)	*cis*-PrCH=CHPr	$WCl_6/Me_4Sn/Bu_2O$	0	0–9	Distribution corresponding to random addition of H_2	Ast 1977b, 1979
H_2	0.84–1.0	$RhCl(PPh_3)_3/PPh_3$/toluene/80°C (homogeneous)	$H_2C=CH_2$ (50 bar) *trans*-PrCH=CHPr	WCl_6/Me_4Sn/MeCOOPr/135°C	0	5 (av.)		Shiono 1993
CH_2	0.25	$Zn(CH_2I)_2$ (reagent)		WCl_6/Me_4Sn	0–4	0	See Fig.16.5	Hummel 1984
					0–3	1		
					0–2	2		
					0–1	3		
HCl	0.27–0.60	$SnCl_4$/toluene/15°C	*cis*-PrCH=CHPr	WCl_6/Me_4Sn	0–2	0		Heiling 1976
					0	1		
HBr	0.11–0.50	$AlBr_3/CCl_4$/10°C [a]	*cis*-PrCH=CHPr	WCl_6/Me_4Sn	0–2	0		Kathan 1977
					0–1	1		Hummel 1981
Cl_2	0.3	None/CH_2Cl_2/20°C	*cis*-PrCH=CHPr	$WCl_6/EtAlCl_2$	1	0–1[b]		Thummer 1975
					0	0–2		
Br_2	0.1	None/CH_2Cl_2/20°C	MeCH=CHMe	$WCl_6/EtAlCl_2$	0–3	0		Stelzer 1973
					0–2	1		

[a] The bromine atoms in the addition product can be completely replaced by phenyl groups by reaction with PhMgBr (Kathan 1977), as indicated by the subsequent metathetical degradation products. [b] Kinetic studies show that double bonds adjacent to M_2 groups react rather more slowly than those adjacent to M_1 groups (Hummel 1977c).

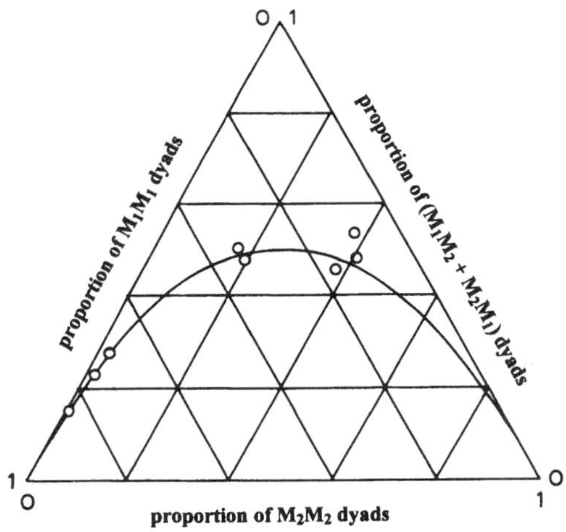

Fig. 16.6 Proportions of the three types of dyad (see **20**) in partially hydrogenated polybutadiene, using homogeneous catalysis. The curve corresponds to a random distribution (Ast 1977b).

The proof of the occurrence of reactions (4) and (5) comes best from an examination of the products of degradation by metathesis with a 10–30-fold excess of *trans*-oct-4-ene (Q_2) initiated by WCl_6/Me_4Sn. In all cases the main products

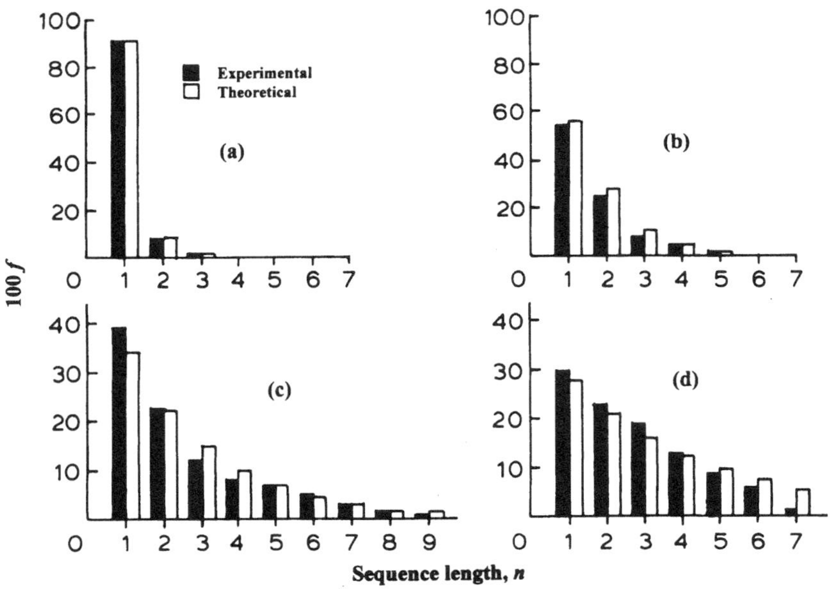

Fig. 16.7 Experimental and theoretical sequence length distribution in homogeneously hydrogenated polybutadienes. Fractions of units hydrogenated: (a) 0.088; (b) 0.442; (c) 0.674; (d) 0.778; *n* = number of units hydrogenated; *f* = fraction of hydrogenated units (Ast 1977b).

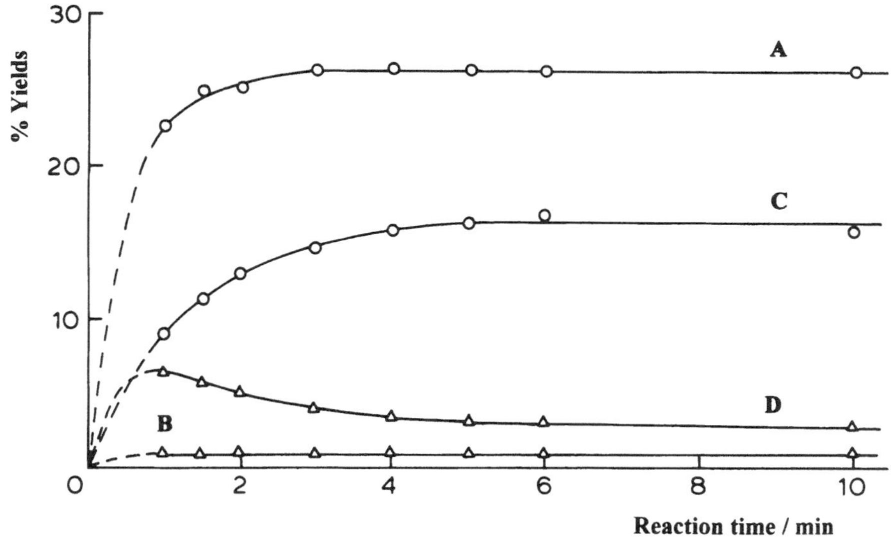

Fig. 16.8 Formation of degradation products from partially (31%) brominated 1,4-polybutadiene by metathesis with oct-4-ene at 10°C: (A) dodeca-4,8-diene,$Q(M_1/2)(M_1/2)Q$; (B) hexadeca-4,8,12-triene, $Q(M_1/2)(M_1)(M_1/2)Q$; (C) 8,9-dibromohexadeca-4,12-diene, $Q(M_1/2)(M_2)(M_1/2)Q$; (D) 8,9-dibromoeicosa-4,12,16-triene, $Q(M_1/2)(M_2)(M_1)(M_1/2)Q$ (Hummel 1977c).

QM_2Q and QM_1Q have been identified by GC/MS. The brominated units themselves do not yield any $QCHCH_2CHBrCHQ$ by metathesis, presumably because HBr is readily lost to give a conjugated polyene. QM_2M_1Q and QM_1M_1Q are also observed in the GC, as well as minor products resulting from double-bond shift reactions, or from intramolecular metathesis and addition of HCl generated by the metathesis catalyst. In favourable cases, the isomers of QM_2M_1Q, with R^1 in the 6- or 7- positions, are resolved in the GC and can be identified by MS, for example, in the case of $R^1 = Ph$, where the 7-substituted product is formed in greater amount. It is also observed that double bonds in the neighbourhood of phenyl substituents are less reactive towards metathesis than double bonds with two adjacent methylene groups (Stelzer 1987b).

$$\underset{(M_1)}{\ddagger CHCH_2CH_2CH \ddagger} \xrightarrow[DCP]{R^2H} \underset{(M_2)}{\ddagger CHCH_2\overset{\overset{\displaystyle R^2}{|}}{C}HCH \ddagger} \qquad (5)$$

R^2 = benzyl, 3,5-dimethylbenzyl, 4-fluorobenzyl, 2,4-dichlorobenzyl, 3,4-dichlorobenzyl, 2-chloro-4-methylbenzyl, 3-chloro-4-methylbenzyl, 4-bromobenzyl, 4-carbomethoxybenzyl, 4-methoxybenzyl, etc.

Table 16.2 Metathesis degradation of copolymers of butadiene (M_1) and monomer M_2

M_2	Type of copolymer [a]	Degradation conditions: Olefin [b]	Degradation conditions: Catalyst	Notes on degradation products analyzed by GC/MS	Reference
Styrene	Statistical (A, R, E)	3-C_6	$WCl_6/Et_4Sn/Et_2O$	33 products containing 0–4 M_2 units. Distinct products from *m* and *r*(1,2 M_1)(1,2 M_1) dyads, cf. **6** and **7**; also from *m* and *r* M_2M_2 dyads. Approximately equal amounts of *m* and *r* in each case	Canji 1977, Thorn-Csányi 1980, 1982, 1986b
	Diblock	C_2	[W]=$CHCMe_2Ph$ [c]	Hexa-1,5-diene + oligomers of M_1 + polystyrene	Wagener 1991b
Propene	Alternating	3-C_6 4-C_8	WCl_6/Me_4Sn WCl_6/Me_4Sn	Various 'defect' structures detected including those containing sequences of 2 and 3 M_2 units. Kinetics studied	Hummel 1987a
	Alternating, aralkylated [d]	4-C_8	WCl_6/Me_4Sn		Hummel 1992
	Alternating, cross-linked [e]	4-C_8	WCl_6/Me_4Sn	Products give evidence of direct cross-links between chains	Hummel 1988 Hummel 1990b
Isobutene	Statistical	DBD [f]	WCl_6/Me_4Sn	Result is telechelic polyisobutene with alkyl borane end-groups	Chasmawala 1995
Acrylonitrile	Statistical	3-C_6	WCl_6/Me_4Sn $WCl_6/PhC{\equiv}CH$	23 products containing 0–2 M_2 units	Stelzer 1988
Trimethylvinylsilane	Statistical (A)	4-C_8 2-C_4	WCl_6/Me_4Sn WCl_6/Me_4Sn	Products containing sequences of up to 6 M_2 units; also a product containing one M_2 unit between 1,4-M_1 and 1,2-M_1 units	Hubmann 1990
Dimethyldivinylsilane	Statistical cross-linked	4-C_8 3-C_6	WCl_6/Me_4Sn WCl_6/Me_4Sn	Products give direct evidence for the nature of the cross-links	Hummel 1991
Ethylvinylbenzene + divinylbenzene	Statistical cross-linked	2-C_4	WCl_6/Me_4Sn	16 products including two containing the cross-linking units (see Table 16.3)	Hubmann 1987
		3-C_6	WCl_6/Me_4Sn	17 products including two containing the cross-linking units	

Ethene	Cross-linked [g] polymers of M_1 and M_2	2-C_8	WCl_6/$EtAlCl_2$	M_1 units removed by degradation leaving polyethene	Hummel 1976
Cyclopentene	Cross-linked [g] polymers of M_1 and M_2	4-C_8	WCl_6/$EtAlCl_2$	Cross-products (C_{17}) formed in addition to products of homodegradation (C_{12}, C_{13}, C_{16}, C_{18})	Hummel 1975b

[a] A = anionic; R = radical; E = emulsion polymerization. [b] 3-C_6 denotes hex-3-ene, etc. [c] $W(=CHCMe_2Ph)(=NAr)[OCMe(CF_3)_2]_2$. [d] Partially aralkylated by reaction with 4-chlorotoluene/dicumyl peroxide. [e] Cross-linked by partial bromination with NBS/AIBN followed by reaction with $BrMg(CH_2)_nMgBr$ ($n = 4$, 6). [f] 1,10-di(9-borabicyclononane)-dec-5-ene. [g] Cross-linked by heating with dicumyl peroxide.

In the reaction of 2-chloro-4-methyltoluene with polybutadiene, initiated by dicumyl peroxide, there is evidence for the formation of some disubstituted units by sequence (6) (Demel 1977).

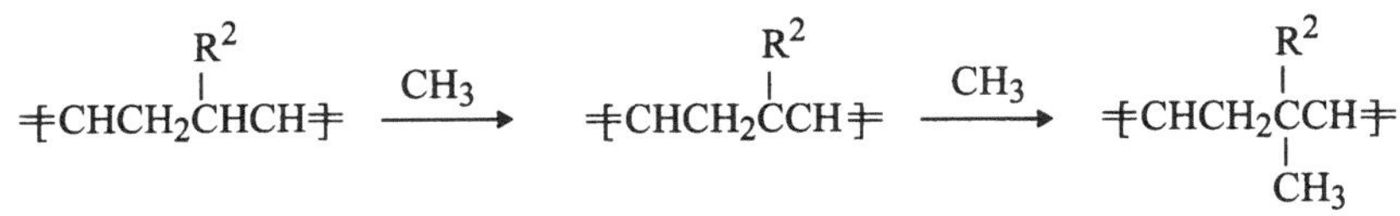

(6)

Cyclooctadienyl substituents at the methylene groups in *cis*-1,4-polybutadiene can be introduced by the radical-induced reaction with cycloocta-1,5-diene. Metathesis degradation of the product with *trans*-oct-4-ene shows that the substitution reaction is accompanied by double-bond shift reactions (Hummel 1990a).

16.3.7 Copolymers of butadiene

Metathesis degradation of a number of copolymers of butadiene (M_1) has been studied (Table 16.2). They include statistical, alternating and block copolymers, sometimes cross-linked either during or after polymerization. In statistical copolymers there will be molecules containing runs of 1, 2, 3, 4 ... units of M_2 in diminishing proportions. These runs remain intact in the degradation products, allowing their distribution to be determined. This distribution depends on the reactivity ratios for the copolymerization and therefore on the nature of the propagating species (Thorn-Csányi 1982, 1986b). 'Alternating' copolymers, such as those with propene, are not always perfectly alternating and may have occasional runs of two or three M_2 units; these also show up readily in the degradation products even when such defects are undetectable ($< 1\%$) in the NMR spectrum of the copolymer (Hummel 1987a). In the case of block copolymers (Wagener 1991b), or two homopolymers that have been subsequently cross-linked (Hummel 1976), degradation leads to the recovery of the polymer of M_2 when this is saturated, e.g. polystyrene or polyethene, though it will of course have alkylidene end- or side-groups attached.

As an example, consider the copolymer of butadiene with 1,4-divinylbenzene and 1,4-ethylvinylbenzene in which the butadiene units are enchained in both 1,2 and 1,4 positions. Chlorobenzene is added to the copolymer under nitrogen until swelling equilibrium is reached. The nitrogen is then replaced by *cis*-but-2-ene (1.1 bar) and the catalyst components added ($WCl_6/Me_4Sn = 1/2$). After 24 h at 35°C methanol is added to destroy the catalyst and the insoluble part removed by centrifugation. The liquid is then concentrated by partial evaporation and analyzed by GC/MS. Sixteen compounds have been identified, made up of units **A–G**, as listed in Table 16.3. Eight of these contain substituted cyclopentene or cyclohexene rings (Hubmann 1987).

Table 16.3 Products resulting from the degradation of a butadiene–ethylvinylbenzene–divinylbenzene copolymer with but-2-ene, in order of increasing GC retention time (Hubmann 1987)

Formula	Compound	Units
C_8H_{14}	Octa-2,6-diene	**A** + 2**G**
C_9H_{14}	3-(But-2-enyl)cyclopentene	**B** + **G**
C_9H_{14}	4-(Prop-1-enyl)cyclohexene	**B** + **G**
$C_{12}H_{20}$	Dodeca-2,6,10-triene	2**A** + 2**G**
$C_{13}H_{22}$	5-(Prop-1-enyl)deca-2,8-diene	**B** + 3**G**
$C_{13}H_{20}$	1-(Cyclopent-2-enyl)octa-2,6-diene	**A** + **B** + **G**
$C_{14}H_{22}$	3-(But-2-enyl)-5-(pent-3-enyl)cyclopentene	**C** + 2**G**
$C_{14}H_{18}$	1-(Cyclohex-3-enyl)-4-ethylbenzene	**D**
$C_{16}H_{26}$	Hexadeca-2,6,10,14-tetraene	3**A** + 2**G**
$C_{18}H_{26}$	5-(4-Ethylphenyl)deca-2,8-diene	**D** + 2**G**
$C_{19}H_{26}$	6-(Cyclopent-2-enyl)-5-(4-ethylphenyl)hex-2-ene	**E** + **G**
$C_{18}H_{22}$	1,4-bis(Cyclohex-3-enyl)benzene	**F**
$C_{20}H_{32}$	Eicosa-2,6,10,14,18-pentaene	4**A** + 2**G**
$C_{22}H_{30}$	1-[1-(But-2-enyl)hex-4-enyl]-4-(cyclohex-3-enyl)benzene	**F** + 2**G**
$C_{22}H_{32}$	10-(4-Ethylphenyl)tetradeca-2,6,12-triene	**A** + **D** + 2**G**
$C_{24}H_{38}$	Tetracosa-2,6,10,14,18,22-hexaene	5**A** + 2**G**

=CH–CH$_2$–CH$_2$–CH=

A

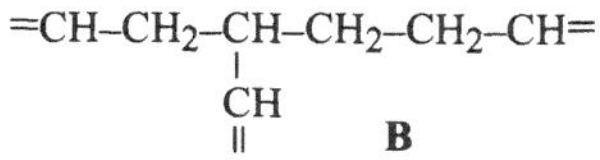

B

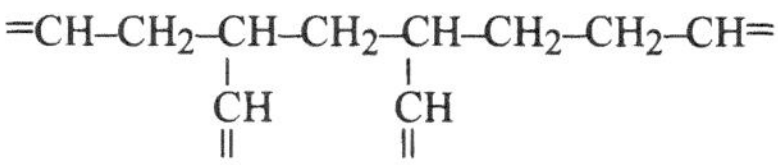

C

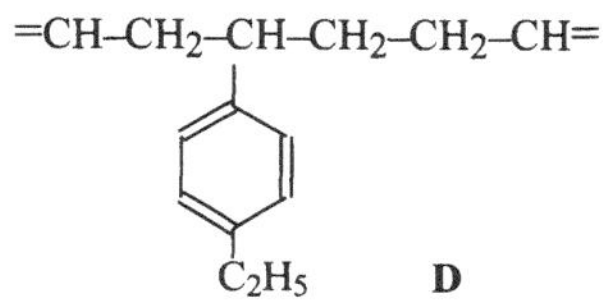

D

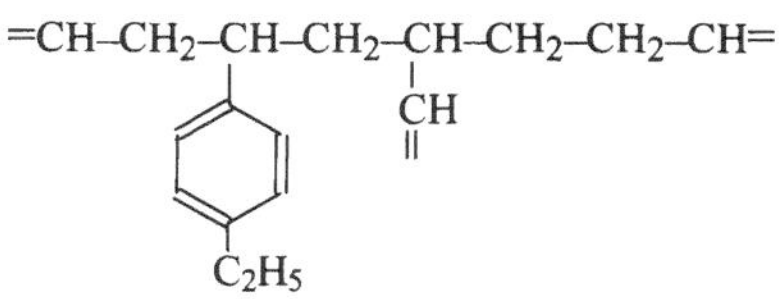

E

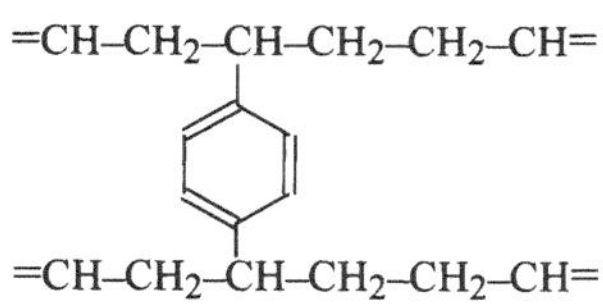

F

=CH–CH$_3$

G

16.3.8 Other copolymers

Copolymers of norbornene and cycloocta-1,5-diene can be degraded with oct-4-ene using WCl_6/Me_4Sn as catalyst to yield (**H** + 2**K**), (**H** + **J** + 2**K**), (2**H** + 2**K**) and (2**H** + **J** + 2**K**) as volatile products. Likewise copolymers of 2,3-bis(trifluoromethyl)-norborna-2,5-diene and cyclopentene can be degraded with oct-4-ene to yield (**L** + 2**K**), (**L** + **P** + 2**K**), (**K** + **P** + **L** + **P** + **K**), (2**L** + 2**K**) and the cyclic (**L** + 2**P**) (Stelzer 1982).

H

$=CH-CH_2-CH_2-CH=$

J

$=CH-CH_2-CH_2-CH_3$

K

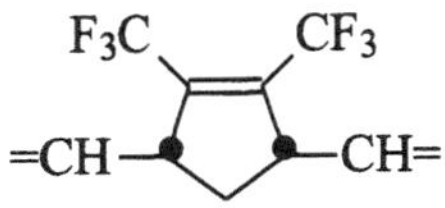

L

$=CH-CH_2-CH_2-CH_2-CH=$

P

This is in contrast to the homopolymer of **L**, which is impossible to degrade and the homopolymer of **H**, which only degrades with difficulty (Stelzer 1982; Wagener 1991a). It is clear that, in the degradation of the copolymers, it is the double bonds between **H** and **J**, and between **L** and **P** that yield to attack by metathesis.

17

Applications of the Olefin Metathesis Reaction

17.1 Introduction

Applications of the olefin metathesis reaction can be divided into three broad groups: (i) those in the field of synthetic organic chemistry, giving new routes to biologically active compounds etc. (see Ch. 4 and 8); (ii) those that have been developed for larger scale production, but not currently in use; and (iii) those that are currently in commercial use, with potential for expansion. Here we give a short summary of the last two categories. For an excellent review of the economic and ecological aspects in applied olefin metathesis, see Streck (1992).

17.2 The Phillips triolefin process

Historically this was the first industrial application of the olefin metathesis reaction and involved the production of high-purity ethene and but-2-ene from propene; reaction (1). The process, based on technology developed by the Phillips Petroleum Co., USA, was operated for 6 years from 1966 by Shawinigan Chemicals near Montreal in Canada. The plant had a capacity of 50 000 tons of feedstock per year. The propene was obtained from a local naphtha cracker. The triolefin process (Fig. 17.1) was operated at near equilibrium conversion, 40–43%, and at high selectivity ($>95\%$); unconverted propene was recycled to the reactor. Besides pure propene, it was also possible to operate the process using a propene/propane cracker fraction. The catalyst used was WO_3/SiO_2, doped with sodium to prevent double-bond shift reactions, at 350–425°C. A change in the economic climate and an increase in demand for propene led to the closure of the plant in 1972. However, the plant is relatively simple and could easily be restarted. Since the but-2-ene can be treated with an isomerization catalyst to yield but-1-ene, the triolefin process provides a route to high-purity but-1-ene, which can be used as a comonomer in polyethene production (Banks 1982).

$$2\,MeCH{=}CH_2 \rightleftharpoons CH_2{=}CH_2 + MeCH{=}CHMe \qquad (1)$$

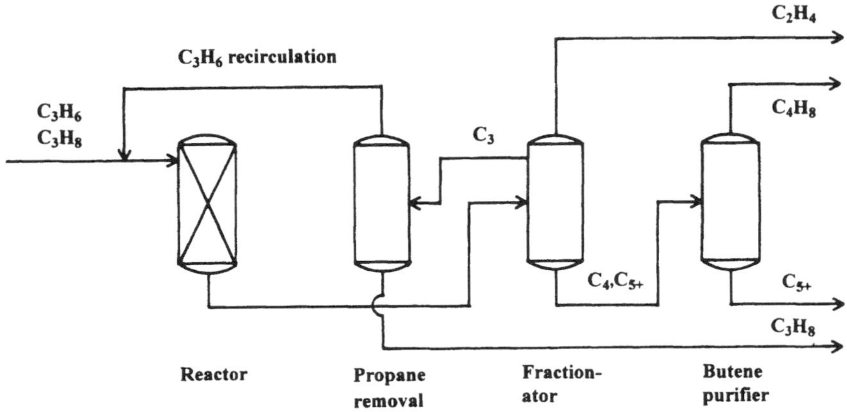

Fig. 17.1 The Phillips Triolefin Process.

A different market situation can make it attractive to use the triolefin process in the reverse direction to produce polymerization-grade propene from ethene and but-2-ene. For this purpose there are two ways of obtaining but-2-ene: via dimerization of ethene, or directly from a naphtha cracker. In 1985 the Lyondell Petrochemical Co. started to operate a plant in Channelview, Texas (USA) for the production of 136 000 tons per year of propene. In this process, part of the ethene from ethane-cracking units is dimerized to but-2-ene using a homogeneous Ni-containing catalyst developed by Phillips, which reacts via the reverse Phillips triolefin process with the rest of the ethene to produce propene. The Institut Français du Pétrole (IFP) and the Chinese Petroleum Corporation (Taiwan) have jointly developed a commercial process, called Meta-4, in which ethene and but-2-ene react with each other in the liquid phase in the presence of an Re_2O_7/Al_2O_3 catalyst at 35°C and 60 bar (Amigues 1990). The process features semi-continuous countercurrent contact between catalyst and the reactant mixture (see Fig. 17.2). As the catalyst descends through the reactor, it becomes less active and must be regenerated. A small percentage is withdrawn periodically from the bottom of the reactor and transferred to the regenerator. The fully regenerated catalyst is returned to the reactor. Ethene is recovered from the reactor effluent and returned to the reactor; propene is produced in the C_3 column and most of the C_4 hydrocarbons are recycled.

17.3 The neohexene process

3,3-Dimethylbut-1-ene (neohexene) is an important intermediate in the synthesis of musk perfume. Its production was begun by the Phillips Petroleum Co. in 1969. Demand has steadily grown (Banks 1982). The process is based on the dimer of

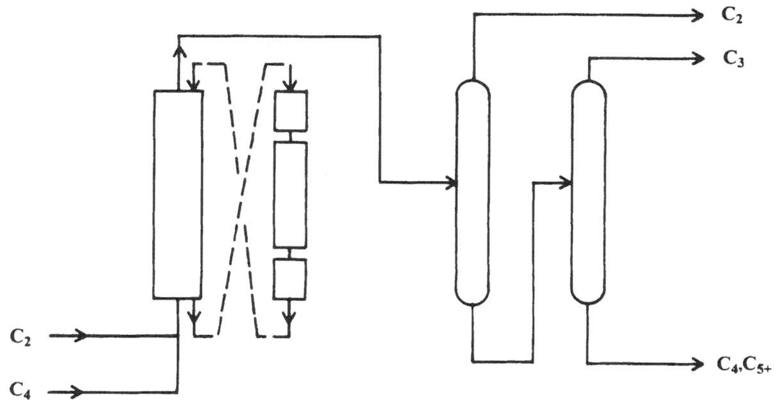

Fig. 17.2 The IFP Meta-4 Process for the production of propene from ethene and but-2-ene (adapted from the IFP brochure).

isobutene, which consists of a mixture of 2,4,4-trimethylpent-2-ene and 2,4,4-trimethylpent-1-ene. Cross-metathesis of the former with ethene yields the desired product; reaction (2). The latter is not wasted since a dual catalyst can be used to ensure that it is converted into 2,4,4-trimethylpent-2-ene as this gets used up by metathesis. With a 1 : 3 catalyst mixture of WO_3/SiO_2 and MgO (370°C, 30 bar, molar ratio ethene/diisobutene = 2/1), an average conversion of the diisobutene of 65–70% and a selectivity to neohexene of approx. 85% is achieved. The coproduct isobutene is recycled to an isobutene dimerization reactor (Banks 1980).

$$Me_3CCH{=}CMe_2 + CH_2{=}CH_2 \rightleftharpoons Me_3CCH{=}CH_2 + CH_2{=}CMe_2 \quad (2)$$

A simplified scheme of the plant is shown in Fig. 17.3. Commercial diisobutene is first fractionated to remove an oxidation inhibitor, which would otherwise poison the dual catalyst system. The fractionated diisobutene along with the ethene stream enters the top of the reactor containing the catalyst. The ethene, consisting of 'make-up' and recycled ethene, is compressed to the required pressure before it enters the reactor. The separation of reactants and products is achieved by stripping and fractionation. The catalyst is reactivated from time to time, using a mixture of air and inert gas to control the temperature of the coke burn-off. In this way the conversion can be maintained between 75% and 50%. The final product has a purity of 96%.

Neohexene is employed to make the class of synthetic musks designated as bicyclic musks. The reaction with *p*-cymene, followed by acetylation, gives one such compound, 7-acetyl-1,1,3,4,4,6-hexamethyltetralin (**1**); sequence (3). Bicyclic musks, with their excellent odour and fixative properties, simulate macrocyclic musks, and have excellent stability towards alkali and light, as well as being colourless and inexpensive to produce.

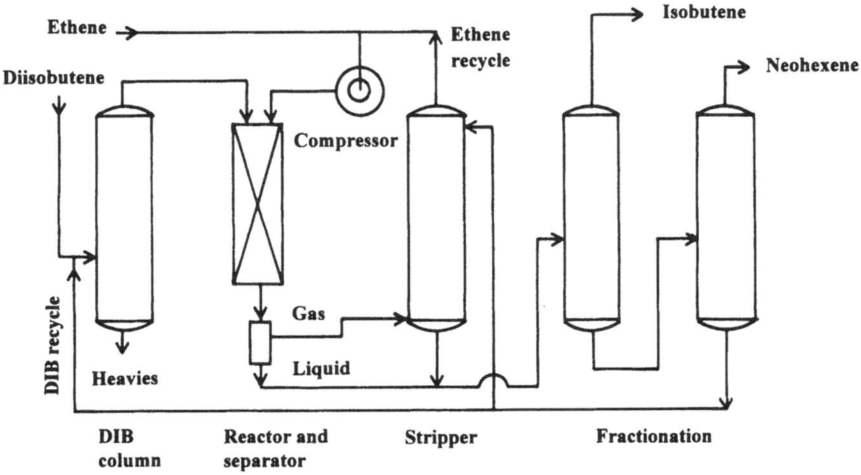

Fig. 17.3 Neohexene plant. DIB = diisobutene (Banks 1982).

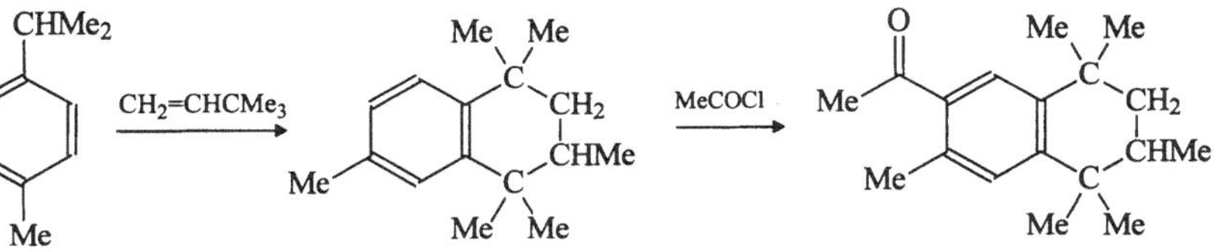

(1) (3)

17.4 The Shell Higher Olefins Process

A large-scale industrial process incorporating olefin metathesis is the Shell Higher Olefins Process (SHOP) for converting ethene to detergent-range alkenes (Fig. 17.4). In the first step of this three-stage process, ethene is oligomerized in the presence of a homogeneous Ni-phosphine catalyst (at 90–100°C and 100–110 bar) in a polar solvent to give a mixture of linear even-numbered alk-1-enes ranging from C_4 to C_{40} with a Schulz–Flory distribution. The olefins formed are immiscible with the solvent; product and catalyst phases are thereby readily separated so that the Ni catalyst can be recycled repeatedly. The C_6–C_{18} alk-1-enes, containing 94–97 wt% linear alk-1-enes, are separated from the product mixture by distillation. The C_6–C_{18} fraction can be further fractionated into individual compounds, which can be used as comonomers in linear-low-density polyethene production or converted into products such as synthetic lubricants, plasticizer alcohols, detergent

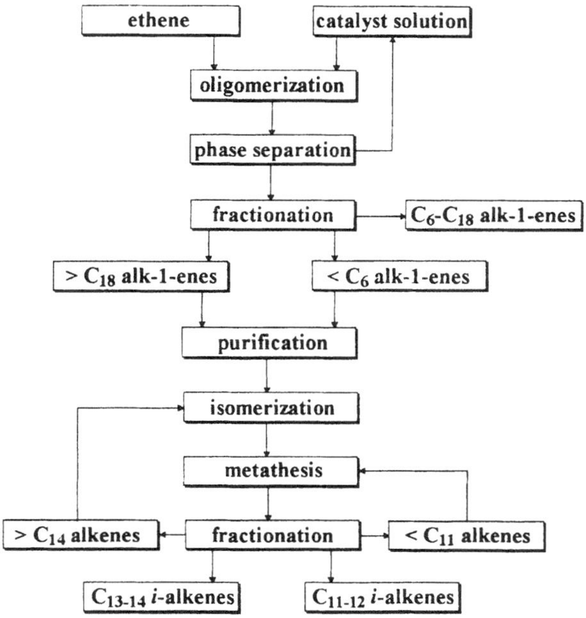

Fig. 17.4 Block diagram of the Shell Higher Olefins Process.

alcohols, synthetic fatty acids, etc. The remaining lighter ($<C_6$) and heavier ($>C_{18}$) alkenes go to purification beds, which remove catalyst and solvent residues that would otherwise deactivate the isomerization catalyst. In the second step, these alkenes undergo double-bond isomerization over a solid alkali–metal-based catalyst to an equilibrium mixture of internal alkenes. This mixture is then passed over a solid metathesis catalyst in the third step, resulting in a statistical distribution of linear internal alkenes with both odd and even numbers of carbon atoms. This yields about 10–15 wt% of the desired detergent-range internal alkenes (C_{11}–C_{14}) per pass, which are subsequently separated by normal distillation. The isomerization and metathesis catalysts operate at 100–125°C and 10 bar; an alumina-supported cobalt molybdate metathesis catalyst is probably used. The remaining lower ($<C_{11}$) and higher ($>C_{14}$) alkenes are recycled. The product consists of $>96\%$ of linear internal C_{11}–C_{14} alkenes (3% branched alkenes, $<0.5\%$ paraffins, and $<0.1\%$ aromatics and conjugated dienes). These are then converted into detergent alcohols via a hydroformylation process or into detergent alkylates (Freitas 1979; Sherwood 1982). Shell Chemicals began large-scale production (200 000 tons per year) in 1979 in Geismar, Louisiana (USA), the capacity being increased in 1989 to 590 000 tons per year. A second plant came on stream in 1982 in Stanlow (UK) to produce, after expansion in 1987, 270 000 tons per year of higher alkenes.

17.5 Other multistage processes involving metathesis

In the commercial manipulation of hydrocarbon feedstocks, the metathesis reaction clearly provides a valuable unit process, which may be combined with other processes either in the same reactor using successive layers of the requisite catalysts, or in consecutive reactors. The neohexene and SHOP processes have already been discussed and three other examples are shown in Scheme 17.1.

$$\text{(1)}\quad \mathrm{MeCH{=}CH_2} \xrightarrow[\text{OM}]{} \mathrm{MeCH{=}CHMe} \xrightarrow[\text{AN}]{\mathrm{Me_3CH}} \mathrm{MeCH_2CHMeCMe_3}$$

$$\text{(2)}\quad \mathrm{MeCH{=}CH_2} \xrightarrow[\text{OM}]{} \mathrm{MeCH{=}CHMe} \xrightarrow[\text{IS}]{} \mathrm{EtCH{=}CH_2} \xrightarrow[\text{OM}]{\mathrm{MeCH{=}CH_2}} \mathrm{EtCH{=}CHMe}$$

$$\mathrm{EtCH{=}CHMe} \xrightarrow[]{\text{IS}} \mathrm{PrCH{=}CH_2}$$

$$\text{etc.} \xleftarrow[\text{OM}]{\mathrm{MeCH{=}CH_2}} \mathrm{BuCH{=}CH_2} \xleftarrow[\text{IS}]{} \mathrm{PrCH{=}CHMe} \xleftarrow[\text{OM}]{\mathrm{MeCH{=}CH_2}} \mathrm{PrCH{=}CH_2}$$

$$\text{(3)}\quad \text{Alkane} \xrightarrow[\text{DH}]{-\mathrm{H_2}} \text{Alkene} \xrightarrow[\text{OM}]{} \text{Alkenes} \xrightarrow[\text{HY}]{\mathrm{H_2}} \text{Alkanes}$$

Scheme 17.1 Multistage processes involving olefin metathesis (OM), alkylation (AN), isomerization (IS), dehydrogenation (DH), and hydrogenation (HY) stages.

The product from the first sequence is useful as an additive to gasoline, where it raises the octane number (Logan 1968a,b). The second sequence, using MgO as catalyst for the double-bond shift reactions, is reported to be capable of giving a wide range of linear olefins, which can first be subjected to ethenolysis so as to yield a mixture of α-olefins, and then fractionated (Crain 1972). The net overall reaction for providing 1 mol of long-chain linear olefins of average carbon number n is given in eqn. (4). Of the reacted propene, 50–63% is converted into the major coproduct ethene. The C_6–C_8 range can be hydroformylated to yield alcohols for use as plasticizers (Anderson 1976), while the C_{12}–C_{16} range can be used for producing synthetic lubricants and a variety of surfactants (Hughes 1975). Technology for this application has been developed in a pilot plant (Banks 1984a).

A redistribution of chain lengths can also be achieved by the third multistage process in Scheme 17.1. Starting from n-butane at 62 bar, and passing it over successive layers of Pt/Al_2O_3 (dehydrogenation), WO_3/SiO_2 (metathesis), and Pt/Al_2O_3 (hydrogenation) at 400°C, the following weight percentages of linear alkanes can be obtained: C_2 3%, C_3 25%, C_4 38%, C_4 38%, C_5 16%, C_6 9%, C_7 4%, C_8 and higher 5%. The rearrangement process is much less effective with branched alkanes (Burnett 1973).

$$(n-2)\,\mathrm{CH_3CH{=}CH_2} \rightleftharpoons (n-3)\,\mathrm{CH_2{=}CH_2} + \mathrm{C}_n\mathrm{H}_{2n} \qquad (4)$$

Phillips have developed a two-step process in which a mixture of oct-1-ene and dec-1-ene is first metathesized to a mixture of C_{14}–C_{18} internal alkenes over a rhenium catalyst, and then dimerized (DM) to lube-oil-range hydrocarbons; reaction (5) (Nelson, W.T. 1983).

$$\begin{matrix} CH_2{=}CH(CH_2)_5CH_3 \\ + \\ CH_2{=}CH(CH_2)_7CH_3 \end{matrix} \xrightarrow{OM} \left\{\begin{matrix} CH_3(CH_2)_5CH{=}CH(CH_2)_5CH_3 \\ CH_3(CH_2)_5CH{=}CH(CH_2)_7CH_3 \\ CH_3(CH_2)_7CH{=}CH(CH_2)_7CH_3 \end{matrix}\right\} \xrightarrow{DM} \begin{matrix} \text{Star-} \\ \text{branched} \\ C_{28}\text{-}C_{36} \\ \text{olefins} \end{matrix} \qquad (5)$$

An interesting alternative route for the production of styrene (normally made from benzene) in a two-step process was studied by Monsanto (Montgomery 1976) and Gulf (Innes 1979, 1981). First, toluene is converted into stilbene (1,2-diphenylethene) by oxidative coupling (dehydrodimerization) at 600°C over alumina-supported PbO, which serves as a catalyst as well as oxygen source; eqn. (6). The PbO-oxidant is periodically regenerated with air. In the second stage, ethenolysis of stilbene at 500°C is conducted over a base-treated WO_3/SiO_2 metathesis catalyst; eqn. (7). The advantage of this process is that toluene is used, which is cheaper than benzene, and only half the amount of ethene is required as compared to the conventional process. However, in practice, the process has its problems. To be useful, this method of making stilbene would have to be greatly improved, largely because of the tremendous amount of oxidant that is required.

$$2\,PhCH_3 + 2\,PbO \longrightarrow PhCH{=}CHPh + 2\,H_2O + 2\,Pb \qquad (6)$$

$$PhCH{=}CHPh + CH_2{=}CH_2 \longrightarrow 2\,PhCH{=}CH_2 \qquad (7)$$

Styrene can also be obtained in a three-step process from toluene, butadiene and ethene. First, buta-1,3-diene is reacted with toluene in the presence of a CaO-supported Na metal catalyst to give 5-phenylpent-2-ene. Next, this is isomerized to 1-phenylpent-1-ene on K_2CO_3/Al_2O_3. Finally, the 1-phenylpent-1-ene is cross-metathesized with ethene on $Re_2O_7/Al_2O_3/Bu_4Sn$ to give styrene and pent-1-ene (Slaugh 1991).

A two-step synthesis of primary alkyl benzenes (precursors for the production of detergents) has been proposed, in which the first step consists of cross-metathesis of terminal or symmetric internal linear olefins with 4-vinylcyclohexene, styrene, or stilbene. Thus, metathesis of linear olefins with 4-vinylcyclohexene in the presence of $WCl_6/R_4Sn/Et_2O$ or Re_2O_7/Al_2O_3 leads to 1-cyclohexenylalk-1-enes. Metathesis of linear olefins with styrene (on Re_2O_7/Al_2O_3) or stilbene (on $Re_2O_7/Al_2O_3/R_4Sn$) yields 1-phenylalkenes (see Ch. 9). The metathesis products are readily converted into primary alkyl benzenes during the second step (Warwel 1985).

17.6 The isoamylene process

Various industrial companies have developed a metathesis process for the production of 2-methylbut-2-ene (isoamylene), which can be used to make isoprene via oxidative dehydrogenation. It is produced via cross-metathesis of isobutene and but-2-ene; reaction (8). Cross-metathesis can also be carried out with propene instead of but-2-ene, or a mixture of the two.

$$Me_2C{=}CH_2 + MeCH{=}CHMe \rightleftharpoons Me_2C{=}CHMe + CH_2{=}CHMe \quad (8)$$

This process has been developed at Phillips as far as the pilot plant stage (Banks 1984a). The feed entering the reactor contained 50 mol % isobutene, the balance being essentially propene and but-2-ene; the catalyst was WO_3/SiO_2. However, this process has never been commercialized because isobutene is being increasingly used to manufacture methyl *t*-butyl ether, a high-octane gasoline blending compound. Moreover, it is cheaper to extract 2-methylbut-2-ene from the C_5-fraction of a naphtha cracker. Shell also developed a process for the production of 2-methylbut-2-ene by the cross-metathesis of isobutene with propene (Shell Tertiary Amylene Process; STAP), but with the advent of polypropene the need for the development of alternative uses for propene has disappeared.

17.7 α,ω-Diolefins

Metathesis technology enables the synthesis of many other high-purity olefins for the speciality chemicals market. Phillips have produced multi-ton quantities of hexa-1,5-diene via ethenolysis of cyclooctadiene and cyclododecatriene over a WO_3/SiO_2 catalyst; reaction (9). High yields of deca-1,9-diene can be obtained from ethenolysis of cyclooctene (obtained by partial hydrogenation of readily available cyclooctadiene), reaction (10). Similarly, tetradeca-1,13-diene can be obtained via ethenolysis of cyclododecene (obtained from cyclododecatriene); (Banks 1982, 1984a).

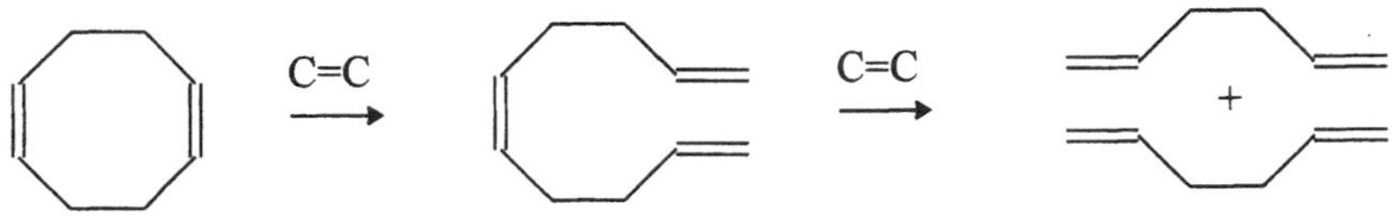

(9)

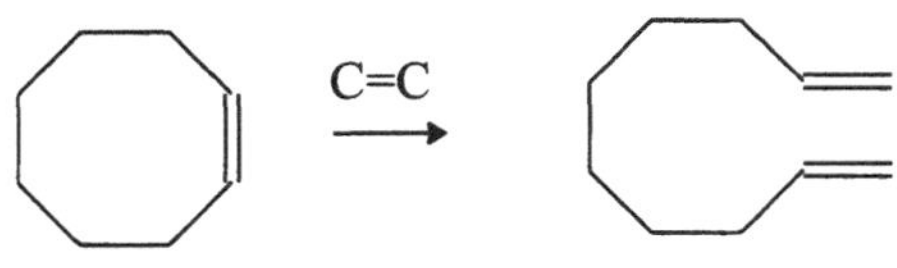

(10)

Shell also developed a process for the manufacture of these α,ω-dienes by ethenolysis of cycloalkenes using a promoted Re_2O_7/Al_2O_3 catalyst (Chaumont

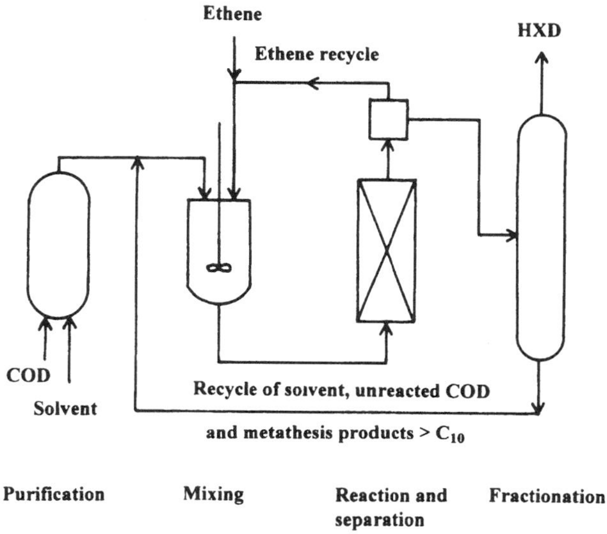

Fig. 17.5 Simplified flow scheme for the production of hexa-1,5-diene (HXD) from cycloocta-1,5-diene (COD) (Chaumont 1988).

1988). The reaction takes place in the liquid phase under extremely mild conditions (0–20°C, 1–2 bar). A simplified flow scheme of the process for the production of hexa-1,5-diene is given in Fig. 17.5.

There are many potential outlets for these kinds of product, viz. cross-linking agents, speciality (co)monomers, and starting materials in the preparation of various α,ω-disubstituted intermediates for the production of aroma chemicals, pharmaceuticals, and agricultural chemicals. The first commercial plant, operating under the name FEAST (Further Exploitation of Advanced Shell Technology), was opened in 1987 in Berre l'Etang (France) and has a capacity of 3000 tons per year (Short 1987).

Shell is also able to produce a branched diolefin, 7-methylocta-1,6-diene, with acceptable yields by reaction (11) (Schaper 1990).

 $+ \quad CH_2{=}CMe_2 \longrightarrow CH_2{=}CH(CH_2)_3CH{=}CMe_2$ (11)

17.8 *trans*-Poly(1-pentenylene)

In 1974 it seemed likely that *trans*-poly(1-pentenylene), sometimes termed *trans*-polypentenamer, was about to make its début as the new all-purpose synthetic elastomer, which could be blended with, or take the place of, natural rubber (Calderon 1974). The properties of the polymer appeared very promising (Graulich 1972). The 'green', i.e. unvulcanized, polymer has high tensile strength and good building tack, both before and after addition of considerable amounts of extending oil and carbon black (Günther 1970; Graulich 1972). Even with a low dosage of cross-linking agents, highly elastic vulcanized rubbers of good resistance to aging and abrasion can be obtained. The melting and glass transition temperatures of the pure polymer are close to those of natural rubber (shown in brackets): $T_m = 20°C$ (30°C); $T_g = -90°C$ (−70°C). Unfortunately, the detailed property profile of the product still leaves something to be desired, at least for applications in the tyre sector. Plans for the large-scale production of this elastomer have therefore been shelved. However, world demand for rubber is still increasing and there is a continuing incentive to produce a synthetic material comparable in properties to those of natural rubber.

17.9 *trans*-Poly(1-octenylene)

The ring-opened polymer of cyclooctene was first marketed in 1980 by Chemische Werke Hüls, Marl, Germany, under the trade name Vestenamer 8012. The number 8012 denotes that the product contains 80% *trans* double bonds and has a viscosity number of 120 $cm^3 g^{-1}$ when measured in 0.1% solution in toluene at 25°C. This polymer also goes under the name *trans*-polyoctenamer (TOR) (Streck 1982). In 1989 an expansion of the Hüls plant brought its capacity up to 12 000 tons per year (Streck 1990a). The polymerization process uses a homogeneous WCl_6-based catalyst.

Vestenamer 8012 possesses extremely unusual properties for an elastomer. At 20°C it is hard and has an exceptionally high viscosity; above 60°C it becomes a fluid with a honey-like consistency. The hardness of the polymer at low temperature is the result of its crystallinity (33%); its fluidity at high temperature is the result of its relatively low MW (60 000). The melting point of the unvulcanized polymer is 55°C and the glass transition temperature of the filler-free vulcanized polymer is −65°C.

Vestenamer 8012 is especially suitable for use in blends with other rubbers since the aforementioned properties are carried over to some extent into the blends, which become stiffer at room temperature and flow more easily at 60°C. Generally 10–30 parts of Vestenamer per 100 parts of blend are sufficient to confer these technical advantages. In this way the following effects can be achieved: (i) easier filler incorporation and distribution; (ii) considerably higher green strength and rigidity of the uncured blend at room temperature; (iii) lower Mooney viscosity at the processing temperature, better mouldability, lower energy uptake of extruder, greatly improved flow behaviour during calendering, extrusion, injection moulding, and pressing; (iv) excellent covulcanization with other elastomers; (v) higher hardness of vulcanizate (controllable by recipe); (vi) higher elasticity of vulcanizate at same hardness level; (vii) very high stability against thermal, oxidative, and photochemical degradation (Streck 1982; Dräxler 1983a,b).

Thus Vestenamer 8012, used as a blending material, seems to offer considerable possibilities for the improvement of properties of the more common elastomers and can be expected to be used in various specialist applications wherever such improvements are critical.

An additional type of Vestenamer has been developed with a lower *trans* content (62%) and therefore lower crystallinity (Vestenamer 6213) to provide for low-temperature applications, where the admixture of the standard type would lead to excessive stiffening (Schneider, W.A. 1988; Diedrich 1989; Streck 1990b).

17.10 Polymers of norbornene

The 90% *trans* polymer of norbornene was put on the market in 1976 by CdF Chimie in France, and in 1978 in the USA and Japan (Marbach 1989). It is now produced by Elf Atochem in Carling (France). The monomer is made by the Diels–Alder reaction of cyclopentadiene and ethene. The polymerization process uses an $RuCl_3/HCl$ catalyst in butanol. The polymer is sold as a moulding powder under the trade name Norsorex and contains a small amount of non-staining antioxidant. It has a high MW (2×10^6), and can absorb up to seven parts of extending oil and ester plasticizers. The plasticized material has good green strength and is highly compatible with fillers. It can be moulded and cured in the usual way at temperatures up to 185°C. The glass transition temperature of 35°C is depressed to −60°C when sufficient plasticizer is present. The vulcanized product has important specialist applications, particularly for engine mountings, anti-vibration mounts, shock-proof bumpers, and flexible couplings. Unusually soft rubbers can be produced (15–30 Shore A hardness) offering advantages over the cellular materials normally used for arm rests, sealing rings, printing rollers, etc.

17.11 Polymers of norbornene derivatives

17.11.1 Polymers of *endo*-dicyclopentadiene (DCPD)

The cheapness of DCPD, obtained as a by-product from the cracking of oil, makes it an attractive candidate for the production of objects by metathesis polymerization, using the reaction injection moulding (RIM) technique.

Following the observations by Oshika (1968) and Dall'Asta (1969) that DCPD could be polymerized by metathesis catalysts, it was shown by Devlin (1970) that it could be polymerized by $MoOCl_3$ to a gel at 35°C and then cured to a hard brown resin at 140°C. Initially attention was concentrated on finding conditions for producing substantially gel-free polymers and copolymers of DCPD through careful control of the order of mixing, the use of chain-transfer agents, e.g. but-1-

ene (Minchak 1977a), and the use of ester solvents for the catalyst (Minchak 1977b, 1978). Copolymers of DCPD with norbornene and its derivatives are easier to process than the homopolymers, and can be compounded with fillers etc. to form rigid articles, such as hub caps, by injection moulding (Tenney 1979a). Cyclopentadiene itself can also be used as comonomer with norbornene derivatives and yields products of high impact strength (Minchak 1979). Gel-free copolymers of cyclopentene and DCPD, having a wide range of composition, can also be prepared (Minchak 1983a).

An important advance was the use of organoammonium molybdates or tungstates of the type $(R'_4N)_{2y-6x}Mo_xO_y$ or $(R'_3NH)_{2y-6x}Mo_xO_y$ as catalyst in conjunction with an aluminium compound such as Et_2AlCl or (RO)RAlCl as cocatalyst (Minchak 1983b, 1984, 1987). These systems have two advantages: first, when R′ is a long chain (e.g. C_{12}) the catalyst is soluble in the monomer and no solvent is required; secondly, the introduction of an alkoxy (or aryloxy) group into the cocatalyst reduces the reactivity to the point where the complete reaction mixture can be made up at room temperature, has a pot life of 1–8 h, and can be injected into the mould at 60–200°C under conditions such that reaction is complete in 2 min. Cross-linking bifunctional metathesis monomers such as **2** can be included so as to produce a thermoset product (Minchak 1987). It should be noted that in DCPD, **3**, it is mainly the norbornene-type double bond that undegoes ROMP to give **4**; the opening of the other double bond by metathesis is thermodynamically less favourable, eqn. (12), but may occur to some extent towards the end of the reaction so producing the cross-links, which cause the product to be largely insoluble in organic solvents (see Ch. 13). With some metathesis catalysts cross-linking may also occur by cationic as well as metathesis reactions. Deliberate inclusion of a cationic initiator, e.g. $SnCl_4$, can assist in this respect (Khasat 1989).

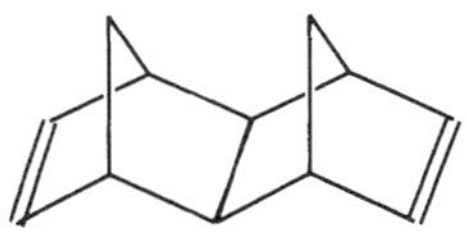

(**2**) 1,4,5,8-dimethano-1,4,4a,5,8,8a-hexahydronaphthalene

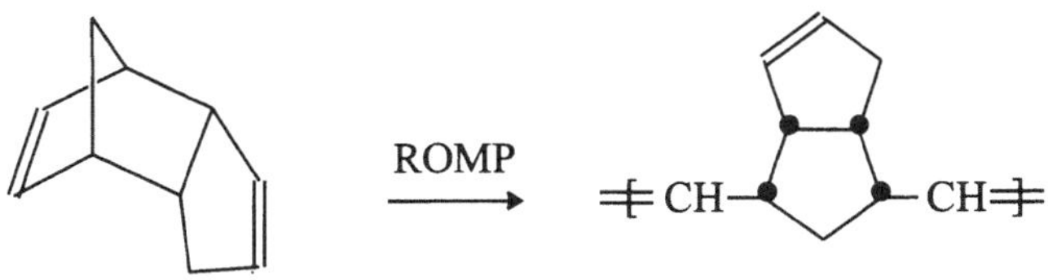

(12)

(**3**) (**4**)

RIM of DCPD by ROMP using two streams rather than one has been extensively tested. A typical arrangement is to place the catalyst, e.g. WCl_6/ROH, in one stream and the cocatalyst (activator), e.g. Et_2AlCl, in the other, together with the monomer in one or other stream. The two streams pass first into a mixing chamber and then into the mould where the heat of reaction raises the temperature within a minute to about 150°C (Matejka 1985). The product is a tough, rigid, thermoset polymer of high modulus and excellent impact strength. Fillers such as 3 mm milled glass or Kraton (a hydrogenated block copolymer of styrene and butadiene) and antioxidants can be included in the injection streams (Klosiewicz 1983, 1984, 1985, 1987; Tom 1985, 1987). Another variation is to include a blowing agent (e.g. $CFCl_3$) in one of the streams so as to produce a cellular product (Leach 1984; Janda 1990).

The catalyst system developed at the Koninklijke/Shell Laboratorium in Amsterdam, The Netherlands, for the RIM polymerization of DCPD is WCl_6/2,6-diisopropylphenol (1/2) with R_3SnH as cocatalyst. Both components are soluble in DCPD and have a long shelf-life. This catalyst system has the advantage of being able to polymerize DCPD of technical quality (van Deursen 1989).

One of the chief problems of the two-stream method is the prevention of premature reaction leading to clogging in the region between the mixing chamber and the mould. This problem can be overcome either by the use of a less active cocatalyst (or a mixture of cocatalysts) or by the inclusion of a Lewis base (moderator), such as benzonitrile, dibutyl ether, tetrahydrofuran, acetylacetone or an alkylacetonacetate, or by use of a comonomer having an ester substituent. In all cases the initial reaction is slowed down to the extent that there may be a short induction period allowing the reaction mixture to reach the mould before the acceleratory period (Klosiewicz 1985; Leach 1986; Nahm 1987; Nelson 1988, 1989; Boutarfa 1991). The function of the Lewis base is to coordinate or chelate to the metal centre, so modifying the initiation and propagation rates.

For objects produced by RIM, it is desirable to have as high a conversion of monomer as possible, since this will affect both the mechanical properties of the product and the extent of residual odour. With many catalyst systems the conversion is often only 90–95%, but with the use of certain additives such as CCl_4 and $C_6H_5CCl_3$, conversions as high as 99.9% have been claimed (Newburg 1984). It is not clear how such additives work but presumably the propagating species is protected in such a way as to make it less vulnerable to destruction.

Another minor problem is that DCPD has a melting point of 32°C, which necessitates heating the input stream if the monomer is very pure. This can be avoided by using a mixture of cyclopentadiene oligomers, which can be readily produced by heating DCPD to 125–250°C in a closed system. A monomer feed of DCPD containing 5–30% of higher oligomers undergoes metathesis polymerization under RIM conditions in a matter of 30 s or less (Matlack 1987; Espy 1988).

The production of large moulded objects (up to 300 kg) from DCPD-based feeds using RIM technology has been developed mainly in the USA by BFGoodrich Co. (under the trade name Telene) and Hercules Inc. (Kirkland 1990). In 1988 Hercules inaugurated a 13 600 ton capacity plant at Deer Park, Texas, to produce poly(dicyclopentadiene) under the trade name Metton. The catalyst used is a combination of WCl_6/$WOCl_4$ and nonylphenol with Et_2AlCl (Breslow 1990). In the

Telene process the preferred catalyst is a trialkylammonium molybdate. Up to 10% trimer of cyclopentadiene is added to the monomer, not only to decrease the melting point of DCPD, but also to increase cross-linking in the polymer. The T_g of the product is typically 150°C. BFGoodrich have licensed their process to the Japanese company Nippon Zeon, which manufactures the product under the name Pentem. There is considerable potential for outdoor applications and perhaps later for objects to be used in more confined spaces.

17.11.2 Polymers of other derivatives

Polymers made from nitrile, ester and halogenated derivatives of norbornene have been the subject of numerous patents, particularly by Japanese companies, but they have not become commercially viable materials. For example, a polymer containing mainly 5-cyanonorbornene units is reported by Showa Denka to have high impact and tensile strength, a wide temperature range for useful properties (−50° to 120°C), good creep resistance, transparency, and good blending properties and workability for extrusion, moulding, or film. Unfortunately, the polymers tend to discolour at high temperature (Ueshima 1974). Such monomers have, however, found application as comonomers with DCPD, etc. (Takahashi 1987; Hara 1989a,b,c,d,e,f, 1990, 1991).

17.11.3 Hydrogenated polymers

Since 1991 Nippon Zeon has produced the polymer Zeonex, obtained by ROMP of norbornene and related (multi-ring) monomers, followed by hydrogenation. The products are amorphous, colourless, transparent polymers with high T_g (140–160°C) and low moisture absorption. These properties make them very suitable for optical applications (e.g. disks, lenses, prisms); (Nishi, Y. 1990c; Tsunokai 1995; Suzuki, T. 1996; see also Aine 1992; Komiya, T. 1992b).

17.12 Miscellaneous

The metathesis of functionalized olefins such as unsaturated fatty acid esters can yield intermediates to a variety of useful products, for example, pheromones, fragrances etc. (Ch. 7). Unfortunately, owing to the relatively large catalyst/substrate ratio required to give good yields such syntheses are not yet commercially viable. A pilot plant at Wacker Chemie, Germany, is now producing the cyclic dimer of cyclooctene by metathesis (see Ch. 12) and thence a cyclic C_{16} ketone, on offer as a synthetic musk perfume (Streck 1990a).

Bibliography

Most of the references to *Chemical Abstracts* refer to patents for which the abstract gives a specific example.

Abbenhuis, H.C.L., Rietveld, M.H.P., Haarman, H.F., Hogerheide, M.P., Spek, A.L., van Koten, G. (1994). *Organometallics*, **13**, 3259.

Abe, Y., Masuda, T., Higashimura, T. (1989). *J. Polymer Sci.*, A, *Polymer Chem.*, **27**, 4267.

Abe, Y., Kouzai, H., Mizumoto, T., Masuda, T., Higashimura, T. (1994). *Polymer J.*, **26**, 207.

Abendroth, H., Canji, E. (1975). *Makromol. Chem.*, **176**, 775.

Adam, G.J.A., Davies, S.G., Ford, K.A., Ephritikine, M., Todd, P.F., Green, M.L.H. (1980). *J. Mol. Catal.*, **8**, 15.

Adams, K.P., Joyce, J.A., Nile, T.A., Patel, A.I., Reid, C.D., Walters, J.M. (1985). *J. Mol. Catal.*, **29**, 201.

Agapiou, A., McNelis, E. (1975). *J. Organomet. Chem.*, **99**, C47.

Aguero, A., Osborn, J.A. (1988). *New J. Chem.*, **12**, 111.

Ahmad, F.B.H., Yarmo, M.A., Alimuniar, A. (1994). *J. Mol. Catal.*, **89**, 357.

Ahn, H-G., Yamamoto, K., Nakamura, R., Niiyama, H. (1992). *Chem. Lett.*, 503.

Ahn, H-G., Yamamoto, K., Nakamura, R., Niiyama, H. (1993). *Chem. Lett.*, 141.

Ahn, H-K., Kim, Y-H., Jin, S-H., Choi, S-K. (1992). *Polymer Bull.*, **29**, 625.

Aine, T., Kawasaki, M., Minami, S. (1992). *Chem. Abstr.*, **116**, 195152.

Akiyoshi, K., Masuda, T., Higashimura, T. (1992). *Makromol. Chem.*, **193**, 755.

Albagli, D., Bazan, G., Schrock, R.R., Wrighton, M.S. (1992a). *Mol. Cryst. Liq. Cryst.*, **216**, 123.

Albagli, D., Bazan, G., Wrighton, M.S., Schrock, R.R. (1992b). *J. Am. Chem. Soc.*, **114**, 4150.

Albagli, D., Bazan, G.C., Schrock, R.R., Wrighton, M.S. (1993). *J. Phys. Chem.*, **97**, 10211.

Aldag, A.W., Lin, C.J., Clark, A. (1977). *Rec. Trav. Chim. Pays-Bas*, **96**, M27.

Aldag, A.W., Lin, C.J., Clark, A. (1978). *J. Catal.*, **51**, 278.

Alder, R.W., Allen, P.R., Khosravi, E. (1994). *J. Chem. Soc., Chem. Commun.*, 1235.

Aliev, R. K., Kadushin, A.A., Tsitovskaya, I.L., Krylov, O.V. (1977). *Izv. Akad. Nauk SSSR, Ser. Khim.*, **26**, 1004.

Aliev, R.K., Tsitovskaya, I.L., Kadushin, A.A., Spiridinov, K.N., Krylov, O.V. (1978). *React. Kinet. Catal. Lett.*, **8**, 257.

Alimuniar, A.B., Edwards, J.H., Feast, W.J. (1985). *J. Mol. Catal.*, **28**, 313.

Alimuniar, A.B., Blackmore, P.M., Edwards, J.H., Feast, W.J., Wilson, B. (1986). *Polymer*, **27**, 1281.

Alimuniar, A., Yarmo, M.A., Rahman, M.Z.A., Kohjiya, S., Ikeda, Y., Yamashita, S. (1990). *Polymer Bull.*, **23**, 119.

Alkema, H.J., van Helden, R. (1968). *Chem. Abstr.*, **69**, 95906.

Alkema, H.J., Medema, D., Wattimena, F. (1970). *Chem. Abstr.*, **72**, 81244.

Alkema, H.J., van Helden, R. (1971). *Chem. Abstr.*, **75**, 140361.

Allen, P.C., Bott, D.C., Brown, C.S., Connors, L.M., Gray, S., Walker, N.S., Clemenson, P.I., Feast, W.J. (1989). In '*Electronic Properties of Conjugated Polymers*' (eds H. Kuzmany, M. Mehring, S. Roth), p.456. Springer, Berlin.

Al-Mashta, F., Davanzo, C.U., Sheppard, N. (1983). *J. Chem. Soc., Chem. Commun.*, 1258.

Al-Samak, B., Amir-Ebrahimi, V., Carvill, A.G., Hamilton, J.G., Rooney, J.J. (1996). *Polymer Int.*, **41**, 85.

Alonso, M.A., Bower, K.E., Johnston, J.A., Farona, M.F. (1988). *Polymer Bull.*, **19**, 211.

Alt, H.G., Engelhardt, H.E., Rausch, M.D., Kool, L.B. (1987). *J. Organomet. Chem.*, **329**, 61.

Amass, A.J., Zurimendi, J.A. (1980). *J. Mol. Catal.*, **8**, 243.

Amass, A.J., Zurimendi, J.A. (1982). *Polymer*, **23**, 211.

Amass, A.J., Bas, S., Gregory, D., Mathew, M.C. (1985). *Makromol. Chem.*, **186**, 325.

Amass, A.J., Gregory, D. (1987a). *Brit. Polymer J.*, **19**, 263.

Amass, A.J., Beevers, M.S., Farren,T.R., Stowell, J.A. (1987b). *Makromol. Chem., Rapid Commun.*, **8**, 119.

Amass, A.J., Dale, A.L., Tighe, B.J. (1988). *Makromol. Chem.*, **189**, 515.

Amass, A.J. (1989). In '*Comprehensive Polymer Science*' (eds G.C. Eastmond, A. Ledwith, S. Russo, P. Sigwalt), **4**, p.109. Pergamon, Oxford.

Amass, A.J., Lotfipour, M., Tighe, B.J., Tuck, C.N., Zurimendi, J.A. (1990). *NATO ASI Ser.*, **C326**, 167.

Amigues, P., Chauvin, Y., Commereuc, D., Lai, C.C., Liu, Y.H., Pan, J.M. (1990). *Hydrocarbon Process.*, **83**, 79.

Amigues, P., Chauvin, Y., Commereuc, D., Hong, C.T., Lai, C.C., Liu, Y.H. (1991). *J. Mol. Catal.*, **65**, 39.

Amir-Ebrahimi, V., Byrne, D., Hamilton, J.G., Rooney, J.J. (1995). *Macromol. Chem. Phys.*, **196**, 327.

Anderson, K.L., Brown, T.D. (1976). *Hydrocarbon Process.*, 119.

Andreini, A., Mol, J.C. (1981). *J. Coll. Interface Sci.*, **84**, 57.

Andreini, A., Mol, J.C. (1985). *J. Chem. Soc., Faraday Trans.,1*, **81**, 1705.

Andreini, A., Xu, X., Mol, J.C. (1986). *Appl. Catal.*, **27**, 31.

Angermund, K., Grevels, F.W., Kruger, C., Skibbe, V. (1984). *Angew. Chem., Int. Ed. Engl.*, **23**, 904.

Anhaus, J.T., Clegg, W., Collingwood, S.P., Gibson, V.C. (1991). *J. Chem. Soc., Chem. Commun.*, 1720.

Anhaus, J.T., Gibson, V.C., Clegg, W., Collingwood, S.P. (1993). *Organometallics*, **12**, 1780.

Anisimov, A.V., Grishkyan, A.A., Ryabov, A.B., Tarakanova, A.V. (1991). *Petrol. Chem.*, **31**, 47.

Anpo, M., Tanahashi, I., Kubokawa, Y. (1982a). *J. Chem. Soc., Faraday Trans.,1*, **78**, 2121.

Anpo, M., Kubokawa, Y. (1982b). *J. Catal.*, **75**, 204.

Anpo, M., Suzuki, T., Kubokawa, Y., Tanaka, F., Yamashita, S. (1984). *J. Phys. Chem.*, **88**, 5778.

Anpo, M., Suzuki, T., Kubokawa, Y., Fujii, T., Kuno, K., Suzuki, S. (1986). *Chem. Express*, **1**, 41.

Anpo, M., Kondo, M., Kubokawa, Y., Louis, C., Che, M. (1987). *Chem. Express*, **2**, 65.

Anpo, M., Kondo, M., Kubokawa, Y., Louis, C., Che, M. (1988). *J. Chem. Soc., Faraday Trans.,1*, **84**, 2771.

Anslyn, E.V., Grubbs, R.H. (1987). *J. Am. Chem. Soc.*, **109**, 4880.
Anslyn, E.V., Brusich, M.J., Goddard, W.A. (1988). *Organometallics*, **7**, 98.
Ardill, H.E., Greene, R.M.E., Hamilton, J.G., Ho, H.T., Ivin, K.J., Lapienis, G., McCann, G.M., Rooney, J.J. (1985). *Am. Chem. Soc. Symp. Ser.*, **286**, 275.
Arlie, J.P., Chauvin, Y., Commereuc, D., Soufflet, J.P. (1974). *Makromol. Chem.*, **175**, 861.
Arnoldy, P., Bruinsma, O.S.L., Moulijn, J.A. (1985). *J. Mol. Catal.*, **30**, 111.
Asakura, K., Nishimura, M., Iwasawa, Y. (1989). *J. Mol. Catal.*, **55**, 159.
Asrar, J. (1989). *Chem. Abstr.*, **111**, 24151.
Asrar, J., Curran, S.A. (1991). *J. Mol. Catal.*, **65**, 1.
Asrar, J., Curran, S.A. (1992a). *Makromol. Chem.*, **193**, 2961.
Asrar, J. (1992b). *Macromolecules*, **25**, 5150.
Asrar, J., Hurlbut, J.B. (1993). *J. Appl. Polymer Chem.*, **50**, 1727.
Asrar, J. (1994a). *Macromolecules*, **27**, 4036.
Asrar, J. (1994b). *Macromol. Reports*, **A31**, Suppl. 6/7, 883.
Ast, W., Hummel, K. (1970). *Naturwiss.*, **57**, 545.
Ast, W., Hummel, K. (1971). *Kautsch. Gummi, Kunstst.*, **24**, 220.
Ast, W., Rheinwald, G., Kerber, R. (1976a). *Makromol. Chem.*, **177**, 39.
Ast, W., Rheinwald, G., Kerber, R. (1976b). *Makromol. Chem.*, **177**, 1341.
Ast, W., Rheinwald, G., Kerber, R. (1976c). *Makromol. Chem.*, **177**, 1349.
Ast, W., Rheinwald, G., Kerber, R. (1977a). *Rec. Trav. Chim. Pays-Bas*, **96**, M127.
Ast, W., Zott, C., Bosch, H., Kerber, R. (1977b). *Rec. Trav. Chim. Pays-Bas*, **96**, M81.
Ast, W., Rheinwald, G., Kerber, R. (1977c). *Chem. Abstr.*, **86**, 141395.
Ast, W., Zott, C., Kerber, R. (1979). *Makromol. Chem.*, **180**, 315.
Atagi, L.M., Critchlow, S.C., Mayer, J.M. (1992). *J. Am. Chem. Soc.*, **114**, 9223.
Babich, E.D., Bespalova, N.B., Vdovin, V.M., Mironov, V.F., Shiryaev, V.I., Kochergin, V.P. (1976). *Izv. Akad. Nauk SSSR, Ser. Khim.*, **25**, 2609.
Babich, E.D., Bespalova, N.B., Vdovin, V.M., Gar, T.K., Mironov, V.F., (1977). *Izv. Akad. Nauk SSSR, Ser. Khim.*, **26**, 960.
Babitskii, B.D., Gavrilova, L.V., Denisova, T.T., Syatkowskii, A.I. (1981). *Dokl. Akad. Nauk SSSR*, **257**, 378.
Badamshina, E.R., Tlenkopachev, M.A. (1981). *Vysokomol. Soedin.*, **B23**, 828.
Badamshina, E.R., Timofeeva, G.I., Korshak, Y.V., Berlin, A.A., Vdovin, V.M., Kutepov, D.F., Pavlova, S.A. (1982). *Polymer Sci. USSR*, **24**, 164.
Bages, S., Petit, M., Mortreux, A., Petit, F. (1990a). *NATO ASI Ser.*, **C326**, 89.
Bages, S., Petit, M., Mortreux, A., Petit, F. (1990b). *J. Mol. Catal.*, **59**, L25.
Baker, R., Crimmin, M.J. (1977). *Tetrahedron Lett.*, 441.
Balcar, H., Matyska, B., Svestka, M. (1983). *J. Mol. Catal.*, **20**, 159.
Balcar, H., Dosedlová, A., Petrusová, L., Matyska, B. (1984). *Coll. Czech. Chem. Commun.*, **49**, 1736.
Balcar, H., Dosedlová, A., Petrusová, L., Matyska, B. (1985). *Coll. Czech. Chem. Commun.*, **50**, 2665.
Balcar, H., Dybal, J. (1986a). *React. Kinet. Catal. Lett.*, **30**, 17.
Balcar, H., Dosedlová, A., Matyska, B. (1986b). *Coll. Czech. Chem. Commun.*, **51**, 753.
Balcar, H., Dosedlová, A., Matyska, B. (1987). *J. Mol. Catal.*, **41**, 367.
Balcar, H., Dosedlová, A. (1989). *React. Kinet. Catal. Lett.*, **39**, 101.
Balcar, H., Dosedlová, A., Petrusová, L. (1992a). *J. Mol. Catal.*, **77**, 289.
Balcar, H., Dosedlová, A. (1992b). *Makromol. Chem.*, **193**, 95.
Balcar, H., Petrusová, L. (1994a). *J. Mol. Catal.*, **90**, 135.
Balcar, H., Sedlácek, J. (1994b). *Macromol. Rapid Commun.*, **15**, 771.
Balcar, H., Sedlácek, J., Pacovská, M., Blechta, V. (1995). *Coll. Czech. Chem. Commun.*, **60**, 489.

Banasiak, D.S. (1985). *J. Mol. Catal.*, **28**, 107.
Banks, R.L. Bailey, G.C. (1964). *Ind. Eng. Chem., Prod. Res. Dev.*, **3**, 170.
Banks, R.L. (1969). *Chem. Abstr.*, **71**, 38315.
Banks, R.L., Regier, R.B. (1971). *Ind. Eng. Chem., Prod. Res. Dev.*, **10**, 46.
Banks, R.L. (1979). *Prepr. Am. Chem. Soc., Div. Petr. Chem.*, **24**, 399.
Banks, R.L. (1980). *J. Mol. Catal.*, **8**, 269.
Banks, R.L., Banasiak, D.S., Hudson, P.S., Norell, J.R. (1982). *J. Mol. Catal.*, **15**, 21.
Banks, R.L. (1984a). In '*Applied Industrial Catalysis*' (ed B.E. Leach), **3**, p.215. Academic Press, New York.
Banks, R.L. (1984b). *US Pat.* 4,454,368; *Chem. Abstr.*, **101**, 90389.
Banks, R.L., Kukes, S.G. (1985). *J. Mol. Catal.*, **28**, 117.
Banks, R.L. (1986a). *CHEMTECH*, **16**, 112.
Banks, R.L. (1986b). *US Pat.* 4,605,810; *Chem. Abstr.*, **105**, 155076.
Barnes, D.L., Eilerts, N.W., Heppert, J.A., Huang, W.H., Morton, M.D. (1994). *Polyhedron*, **13**, 1267.
Bartlett, B., Soto, C., Wu, R., Tysoe, W.T. (1993). *Catal. Lett.*, **21**, 1.
Bashkirov, A.N., Fridman, R.A., Nosakova, S.M., Liberov, L.G., Bol'shakov, A.M. (1974). *Kinet. Katal.*, **15**, 1607.
Basrur, A.G., Patwardhan, S.R., Vyas, S.N. (1991). *J. Catal.*, **127**, 86.
Basset, J., Mutin, R., Descotes, G., Sinou, D. (1975). *Compt. Rend. Acad. Sci.*, **C280**, 1181.
Basset, J.M., Leconte, M. (1979). *Fund. Res. Homog. Catal.*, **3**, 285.
Basset, J.M., Leconte, M., Ollivier, J., Quignard, F. (1988). *Chem. Abstr.*, **109**, 129831.
Bauch, C.G., Wagener, K.B., Boncella, J.M. (1991). *Makromol. Chem., Rapid Commun.*, **12**, 413.
Bazan, G., Schrock, R.R., Khosravi, E., Feast, W.J., Gibson, V.C. (1989). *Polymer Commun.*, **30**, 258.
Bazan, G.C., Khosravi, E., Schrock, R.R., Feast, W.J., Gibson, V.C., O'Regan, M.B., Thomas, J.K., Davis, W.M. (1990). *J. Am. Chem. Soc.*, **112**, 8378.
Bazan, G.C., Schrock, R.R. (1991a). *Macromolecules*, **24**, 817.
Bazan, G.C., Schrock, R.R., Cho, H-N., Gibson, V.C. (1991b). *Macromolecules*, **24**, 4495.
Bazan, G.C., Oskam, J.H., Cho, H-N., Park, L.Y., Schrock, R.R. (1991c). *J. Am. Chem. Soc.*, **113**, 6899.
Bazan, G.C., Miao, Y-J., Renak, M.L., Sun, B.J. (1996a). *J. Am. Chem. Soc.*, **118**, 2618.
Bazan, G.C., Renak, M.L., Sun, B.J. (1996b). *Macromolecules*, **29**, 1085.
Bechera, J.N., Bell, S.E.J., McGarvey, J.J., Rooney, J.J. (1986). *J. Chem. Soc., Chem. Commun.*, 1785.
Bell, A. (1992). *J. Mol. Catal.*, **76**, 165.
Bell, A. (1993). *US Pat.* 5,194,534.
Bell, A., Clegg, W., Dyer, P.W., Elsegood, M.R.J., Gibson, V.C., Marshall, E.L. (1994a). *J. Chem. Soc., Chem. Commun.*, 2247, 2547.
Bell, A. (1994b). *Am. Chem. Soc., Polymer Prepr.*, **35** (1), 694.
Bell, A.J. (1976). *Chem. Abstr.*, **85**, 7042.
Bell, B., Hamilton, J.G., Mackey, O.N.D., Rooney, J.J. (1992). *J. Mol. Catal.*, **77**, 61.
Bell, B., Hamilton, J.G., Law, E.E., Rooney, J.J. (1994). *Macromol. Rapid Commun.*, **15**, 543.
Bencheick, A., Petit, M., Mortreux, A., Petit, F. (1982). *J. Mol. Catal.*, **15**, 93.
Bencze, L., Ivin, K.J., Rooney, J.J. (1980). *J. Chem. Soc., Chem. Commun.*, 834.
Bencze, L., Kraut-Vass, A. (1984). *J. Organomet. Chem.*, **270**, 211.
Bencze, L., Kraut-Vass, A. (1985a). *J. Organomet. Chem.*, **280**, C14.
Bencze, L., Kraut-Vass, A. (1985b). *J. Mol. Catal.*, **28**. 369.
Bencze, L., Kraut-Vass, A., Prókai, L. (1985c). *J. Chem. Soc., Chem. Commun.*, 911.

Bencze, L., Halász, L., Szeifert, F., Ivin, K.J., Rooney, J.J. (1991). *J. Mol. Catal.*, **65**, 29.
Bencze, L., Szilagyi, R. (1992). *J. Mol. Catal.*, **76**, 145.
Bencze, L., Szilagyi, R. (1994a). *J. Mol. Catal.*, **90**, 157.
Bencze, L., Szilagyi, R. (1994b). *J. Organomet. Chem.*, **465**, 211.
Bencze, L., Szilagyi, R. (1994c). *J. Organomet. Chem.*, **475**, 183.
Bencze, L., Szilágyi, R. (1995). *J. Organomet. Chem.*, **505**, 81
Bencze, L., Szalai, G. (1996). *J. Mol. Catal.*, in press (ISOM 11 vol.).
Benedicto, A.D., Novak, B.M., Grubbs, R.H. (1992). *Macromolecules*, **25**, 5893.
Benedicto, A.D., Claverie, J.P., Grubbs, R.H. (1995). *Macromolecules*, **28**, 500.
Ben-Efraim, D.A., Batich, C., Wasserman, E. (1970). *J. Am. Chem. Soc.*, **92**, 2133.
Berezin, M.Y., Ignatov, V.M., Belov, P.S., Elev, I.V., Shelimov, B.N., Kazansky, V.B. (1991). *Kinet. Katal.*, **32**, 379.
Berglund, M., Andersson, C., Larsson R. (1985). *J. Organomet. Chem.*, **292**, C15.
Berglund, M., Andersson, C. (1986a). *J. Mol. Catal.*, **36**, 375.
Berglund, M., Andersson, C., Larsson R. (1986b). *J. Organomet. Chem.*, **314**, 61.
Bespalova, N.B., Babich, E.D., Vdovin, V.M., Nametkin N.S. (1975). *Dokl. Akad. Nauk SSSR*, **225**, 1071.
Bespalova, N.B., Bovina, M.A., Vdovin, V.M. (1988). *Izv. Akad. Nauk SSSR, Ser. Khim.*, **37**, 920.
Bespalova, N.B., Bovina, M.A. (1990). *Izv. Akad. Nauk SSSR, Ser. Khim.*, **39**, 172.
Bespalova, N.B., Bovina, M.A. (1992). *J. Mol. Catal.*, **76**, 181.
Bespalova, N.B., Bovina, M.A., Sergeeva, M.B., Oppengeim, V.D., Zaikin, V.G. (1994a). *J. Mol. Catal.*, **90**, 21.
Bespalova, N.B., Bovina, M.A., Sergeeva, M.B., Zaikin, V.G. (1994b). *Russ. Chem. Bull.*, **43**, 1425.
Betterton, K., Ebert, M., Haeussling, L., Lux, M.G., Twieg, R.J., Willson, C.G., Yoon, D., Burns, E.G., Grubbs, R.H. (1992). *Polymer Mater. Sci. Eng.*, **66**, 312.
Biagini, S.C.G., Bush, S.M., Gibson, V.C., Mazzariol, L., North, M., Teasdale, W.G., Williams, C.M., Zagotto, G., Zamuner, D. (1995). *Tetrahedron*, **51**, 7247.
Bilhou, J.L., Basset, J.M. (1977a). *J. Organomet. Chem.*, **132**, 395.
Bilhou, J.L., Basset, J.M., Mutin, R., Graydon, W.F. (1977b). *J. Am. Chem. Soc.*, **99**, 4083.
Bilhou, J.L., Mutin, R., Leconte, M., Basset, J.M. (1977c). *Rec. Trav. Chim. Pays-Bas*, **96**, M5.
Blackmore, P.M., Feast, W.J. (1986a). *Polymer*, **27**, 1296.
Blackmore, P.M., Feast, W.J. (1986b). *J. Mol. Catal.*, **36**, 145.
Blackmore, P.M., Feast, W.J., Taylor, P.C. (1987). *Brit. Polymer J.*, **19**, 205.
Blackmore, P.M., Feast, W.J. (1988). *J. Fluorine Chem.*, **40**, 331.
Blosch, L.L., Abboud, K., Boncella, J.M. (1991). *J. Am. Chem. Soc.*, **113**, 7066.
Blosch, L.L., Gamble, A.S., Abboud, K., Boncella, J.M. (1992a). *Organometallics*, **11**, 2342.
Blosch, L.L., Gamble, A.S., Boncella, J.M. (1992b). *J. Mol. Catal.*, **76**, 229.
Bobyshev, A.A., Kazansky, V.B., Kibardina, I.R., Shelimov, B.N. (1992). *Kinet. Katal.*, **33**, 363.
Boelhouwer, C., Mol, J.C. (1985). *Prog. Lipid. Res.*, **24**, 243.
Bogolepova, E.I., Fridman, R.A., Bashkirov, A.N. (1978). *Izv. Akad. Nauk SSSR, Ser. Khim.*, **27**, 2429.
Bogolepova, E.I., Fridman, R.A., Bashkirov, A.N. (1979). *Izv. Akad. Nauk SSSR, Ser. Khim.*, **28**, 1623.
Bogolepova, E.I., Verbovetskaya, S.B. (1982). *Neftekhimiya*, **22**, 207.
Bogolepova, E.I., Verbovetskaya, S.B., Kliger, G.A. (1984). *Neftekhimiya*, **24**, 668.

Bogolepova, E.I., Verbovetskaya, S.B., Vygodskaya, I.V., Kliger, G.A., Loktev, S.M. (1986). *Neftekhimiya*, **26**, 185.
Bogolepova, E.I., Vygodskaya, I.V., Bulanova, A.V., Kliger, G.A., Loktev., S.M. (1987). *Neftekhimiya*, **27**, 106.
Bogolepova, E.I., Vygodskaya, I.V., Bulanova, A.V., Kliger, G.A., Loktev, S.M. (1989). *Neftekhimiya*, **29**, 234.
Bogolepova, E.I., Vygodskaya, I.V., Kliger, G.A., Loktev., S.M. (1990). *Neftekhimiya*, **30**, 507.
Bogolepova, E.I., Vygodskaya, I.V., Kliger, G.A., Loktev, S.M. (1991). *Petrol. Chem.*, **31**, 809.
Bogolepova, E.I., Vygodskaya, I.V., Kliger, G.A., Loktev, S.M. (1992). *Petrol. Chem.*, **32**, 461.
Borer, B.C., Deerenberg, S., Bieräugel, H., Pandit, U.K. (1994). *Tetrahedron Lett.*, **35**, 3191.
Borowczak, D., Szymańska-Buzar, T., Ziólkowski, J.J. (1982). *React. Kinet. Catal. Lett.*, **20**, 389.
Borowczak, D., Szymańska-Buzar, T., Ziólkowski, J.J. (1984). *J. Mol. Catal.*, **27**, 355.
Bosma, R.H.A., Kouwenhoven, A.P., Mol, J.C. (1981a). *J. Chem. Soc., Chem. Commun.*, 1081.
Bosma, R.H.A., van den Aardweg, G.C.N., Mol, J.C. (1981b). *J. Chem. Soc., Chem. Commun.*, 1132.
Bosma, R.H.A., Xu, X., Mol, J.C. (1982). *J. Mol. Catal.*, **15**, 187.
Bosma, R.H.A., van den Aardweg, G.C.N., Mol, J.C. (1983). *J. Organomet. Chem.*, **255**, 159.
Bosma, R.H.A., van den Aardweg, G.C.N., Mol, J.C. (1985). *J. Organomet. Chem.*, **280**, 115.
Boutarfa, D., Quignard, F., Leconte, M., Basset, J.M., Hamilton, J.G., Ivin, K.J., Rooney, J.J. (1988). In '*Transition Metal Catalyzed Polymerizations*' (ed R.P. Quirk), p.695. Cambridge University, Cambridge.
Boutarfa, D., Paillet, C., Leconte, M., Basset, J.M. (1991). *J. Mol. Catal.*, **69**, 157.
Bozkurt, C., Imamoglu, Y., *Doga: Turk. Khim. Derg.*, (1989). **13**, 199.
Bradshaw, C.P.C., Howman, E.J., Turner, L. (1967). *J. Catal.*, **15**, 187.
Bradshaw, C.P.C., Turner, L. (1968). *Chem. Abstr.*, **69**, 76595.
Bray, A., Mortreux, A., Petit, F., Petit, M., Szymańska-Buzar, T. (1993). *J. Chem. Soc., Chem. Commun.*, 197.
Brégeault, J-M., El Ali, B., Martin, J., Martin, C., Derdar, F., Bugli, G., Delamar, M. (1988). *J. Mol. Catal.*, **46**, 37.
Brenner, A., Burwell, R.L. (1978). *J. Catal.*, **52**, 364.
Brenner, A., Hucul, D.A., Hardwick, S.J. (1979). *Inorg. Chem.*, **18**, 1478.
Breslow, D.S. (1990). *CHEMTECH*, **20**, 540.
Breslow, D.S. (1993). *Prog. Polymer Sci.*, **18**, 1141.
Breunig, S., Heroguez, V., Gnanou, Y., Fontanille, M. (1995). *Macromol. Symp.*, **95**, 151.
Brigodiot, M., Marechal, E. (1981). *Polymer Bull.*, **4**, 45.
Broeders, J., Feast, W.J., Gibson, V.C., Khosravi, E. (1996). *Chem. Commun.*, 343.
Brown, T.L. (1981). *J. Mol. Catal.*, **12**, 41.
Brunthaler, J.K., Stelzer, F., Leising, G. (1985). *J. Mol. Catal.*, **28**, 393.
Brzezinska, K., Wolfe, P.S., Watson, M.D., Wagener, K.B. (1996). *Macromol. Chem. Phys.*, **197**, 2065.
Buchacher, P., Fischer, W., Aichholzer, K.D., Stelzer, F. (1996). *J. Mol Catal.*, in press (ISOM 11 vol.)
Buchmeiser, M., Schrock, R.R. (1995). *Macromolecules*, **28**, 6642.

Buffon, R., Auroux, A., Lefebvre, F., Leconte, M., Choplin, A., Basset, J-M., Herrmann, W.A. (1992a). *J. Mol. Catal.*, **76**, 287.
Buffon, R., Choplin, A., Leconte, M., Basset, J.M., Touroude, R., Herrmann, W.A. (1992b). *J. Mol. Catal.*, **72**, L7.
Buffon, R., Leconte, M., Choplin, A., Basset, J-M. (1993). *J. Chem. Soc., Chem. Commun.*, 361.
Buffon, R., Leconte, M., Choplin, A., Basset, J-M. (1994). *J. Chem. Soc., Dalton Trans.*, 1723.
Buffon, R., Schuchardt, U., Abras, A. (1995). *J. Chem. Soc., Faraday Trans.*, **91**, 3511.
Burnett, R.L., Hughes, T.R. (1973). *J. Catal.*, **31**, 55.
Burwell, R.L., Brenner, A. (1975/76). *J. Mol. Catal.*, **1**, 77.
Burwell, R.L. (1984). *J. Catal.*, **86**, 301.
Bykov, V.I., Butenko, T.A., Finkel'shtein, E.S., Petrovskii, P.V., Vdovin, V.M. (1988). *Izv. Akad. Nauk SSSR, Ser. Khim.*, **37**, 1580.
Bykov, V.I., Butenko, T.A., Finkel'shtein, E.S. (1990). *Izv. Akad. Nauk SSSR, Ser. Khim.*, **39**, 168.
Bykov, V.I., Butenko, T.A., Finkel'shtein, E.S. (1991). *Dokl. Akad. Nauk SSSR*, **320**, 623.
Bykov, V.I., Butenko, T.A., Finkel'shtein, E.S., Henderson, P.T. (1994). *J. Mol. Catal.*, **90**, 111.
Calderon, N. *et al.* (1967a). *Chem. Eng. News*, **45**, 51.
Calderon, N., Chen, H.Y., Scott, K.W. (1967b). *Tetrahedron Lett.*, 3327.
Calderon, N., Ofstead, E.A., Judy, W.A. (1967c). *J. Polymer Sci.*, A-1, **5**, 2209.
Calderon, N., Ofstead, E.A., Ward, J.P., Judy, W.A., Scott, K.W. (1968). *J. Am. Chem. Soc.*, **90**, 4133.
Calderon, N., Hinrichs, R.L. (1974). *CHEMTECH*, **4**, 627.
Calderon, N., Ofstead, E.A., Judy, W.A. (1976). *Angew. Chem. Int. Ed. Engl.*, **15**, 401.
Calderon, N., Scott, K.W. (1981). *Chem. Abstr.*, **95**, 8034.
Cameron, R. (1983). *Europ. Pat. Appl.* 84,537.
Campistron, I., Reyx, D. (1986). *Makromol. Chem.*, **186**, 2801.
Candlin, J.P., Thomas, H. (1974). *Am. Chem. Soc., Adv. Chem. Ser.*, **132**, 212.
Canji, E., Perner, H. (1977). *Rec. Trav. Chim. Pays-Bas*, **96**, M70.
Canji, E., Perner, H. (1978). *Makromol. Chem.*, **179**, 567.
Cannizzo, L.F., Grubbs, R.H. (1987). *Macromolecules*, **20**, 1488.
Cannizzo, L.F., Grubbs, R.H. (1988). *Macromolecules*, **21**, 1961.
Cardin, D.J., Cetinkaya, B., Lappert, M.F. (1972a). *Chem. Rev.*, **72**, 545.
Cardin, D.J., Cetinkaya, B., Lappert, M.F. (1972b). *J. Chem. Soc., Chem. Commun.*, 927.
Casey, C.P., Burkhardt, T.J. (1974). *J. Am. Chem. Soc.*, **96**, 7808.
Casey, C.P., Tuinstra, H.E., Saeman, M.C. (1976). *J. Am. Chem. Soc.*, **98**, 608.
Casey, C.P., Polichnowski, S.W. (1977). *J. Am. Chem. Soc.*, **99**, 6097.
Casey, C.P., Tuinstra, H.E. (1978). *J. Am. Chem. Soc.*, **100**, 2270.
Casey, C.P., Shusterman, A.J. (1985). *Organometallics*, **4**, 736.
Casey, C.P., Hornung, N.L., Kosar, W.P. (1987). *J. Am. Chem. Soc.*, **109**, 4908.
Castner, K.F. (1977a). *Am. Chem. Soc., Polymer Prepr.*, **18**, 116.
Castner, K.F. (1977b). *Chem. Abstr.*, **87**, 152726.
Castner, K.F., Calderon, N. (1982). *J. Mol. Catal.*, **15**, 47.
Ceausescu, E. *et al.* (1985). *J. Mol. Catal.*, **28**, 351.
Ceausescu, E. *et al.* (1986). *J. Mol. Catal.*, **36**, 163.
Ceausescu, E. *et al.* (1988a). *J. Mol. Catal.*, **46**, 405.
Ceausescu, E. *et al.* (1988b). *J. Mol. Catal.*, **46**, 423.
Chan, Y.N.C., Schrock, R.R. (1992a). *Chem. Mater.*, **4**, 24.
Chan, Y.N.C., Craig, G.S.W., Schrock, R.R., Cohen, R.E. (1992b). *Chem. Mater.*, **4**, 885.
Chan, Y.N.C., Schrock, R.R. (1993). *Chem. Mater.*, **5**, 566.

Chasmawala, M., Chung, T.C. (1995). *Macromolecules*, **28**, 1333.
Chatani, N., Morimoto, T., Muto, T., Murai, S. (1994). *J. Am. Chem. Soc.*, **116**, 6049.
Chatt, J., Haines, R.J., Leigh, G.J. (1972). *J. Chem. Soc., Chem. Commun.*, 1202.
Chaumont, P., John, C.S. (1988). *J. Mol. Catal.*, **46**, 317.
Chauvin, Y., Commereuc, D., Cruypelinck, D. (1976). *Makromol. Chem.*, **177**, 2637.
Chauvin, Y., Commereuc, D., Zaborowski, G. (1977). *Rec. Trav. Chim. Pays-Bas*, **96**, M131.
Chauvin, Y., Commereuc, D., Zaborowski, G. (1978). *Makromol. Chem.*, **179**, 1285.
Chauvin, Y., Commereuc, D. (1992). *J. Chem. Soc., Chem. Commun.*, 462.
Chauvin, Y., Saussine, L. (1996). *Macromolecules*, **29**, 1163.
Chen, W., Pang, D., Zhang, Y., Zheng, Y., Yang, C., Shen, Z. (1981). *Chem. Abstr.*, **94** 84608.
Chen, Z-R., Claverie, J.P., Grubbs, R.H., Kornfield, J.A. (1995). *Macromolecules*, **28**, 2147.
Cherednichenko, V.M. (1979). *Polymer Sci. USSR*, **20**, 1225.
Chevalier, P., Sinou, D., Descotes, G. (1975). *Bull. Soc. Chim. France*, 2254.
Chevalier, P., Sinou, D., Descotes, G., Mutin, R., Basset, J.M. (1976). *J. Organomet. Chem.*, **113**, 1.
Chevalier-Seite, B., Commereuc, D., Chauvin, Y. (1983). *Fr. Demande* 2,521,872; *Chem. Abstr.*, **100**, 34157.
Chisholm, M.H., Folting, K., Hoffman, D.M., Huffman, J.C. (1984a). *J. Am. Chem. Soc.*, **106**, 6794.
Chisholm, M.H., Hoffman, D.M., Huffman, J.C. (1984b). *J. Am. Chem. Soc.*, **106**, 6806.
Chisholm, M.H., Eichhorn, B.W., Folting, K., Huffman, J.C. (1989). *Organometallics*, **8**, 49.
Cho, H-N., Choi, S-K. (1985). *J. Polymer Sci., Polymer Chem.*, **23**, 1469.
Cho, I., Bae, S.S. (1991). *Polymer J.*, **23**, 1283.
Cho, I., Hwang, K.M. (1993a). *J. Polymer Sci.*, A, *Polymer Chem.*, **31**, 1079.
Cho, I., Hwang, K.M. (1993b). *Polymer J.*, **25**, 111.
Cho, I., Moon, G.S. (1995). *J. Polymer Sci.*, A, *Polymer Chem.*, **33**, 1823.
Cho, O-K., Kim, Y-H., Choi, K.-Y., Choi, S-K. (1990). *Macromolecules*, **23**, 12.
Choi, S-J., Jin, S-H., Park, J-W., Cho, H-N., Choi, S-K. (1994a). *Macromolecules*, **27**, 309.
Choi, S-J., Kim, S-H., Ahn, W., Cho, H-N., Choi, S-K. (1994b). *Macromolecules*, **27**, 4871.
Choi, S-J., Cho, H-N., Choi, S-K. (1994c). *Polymer Bull.*, **32**, 11.
Choi, S-J., Cho, H-N., Choi, S-K. (1994d). *Polymer Bull.*, **32**, 179.
Choi, S-K. (1990). *Makromol. Chem., Macromol. Symp.*, **33**, 145.
Choung, S.J., Weller, S.W. (1983). *Ind. Eng. Chem., Proc. Des. Dev.*, **22**, 662.
Chung, T.C. (1991a). *J. Inorg. Organomet. Polymers*, **1**, 37.
Chung, T.C. (1991b). *Polymer*, **32**, 1336.
Chung, T.C. (1991c). *Macromolecules*, **24**, 2675.
Chung, T.C., (1991d). *Makromol. Chem., Macromol. Symp.*, **42/43**, 293.
Chung, T.C., Chasmawala, M. (1991e). *Macromolecules*, **24**, 3718.
Chung, T.C., Chasmawala, M. (1992a). *Macromolecules*, **25**, 5137.
Chung, T.C. (1992b). *J. Mol. Catal.*, **76**, 15.
Chung, T.C. (1992c). *Chem. Ind. (London)*, 255.
Churchill, M.R., Ziller, J.W., Freudenberger, J.H., Schrock, R.R. (1984). *Organometallics*, **3**, 1554.
Clark, A., Cook, C. (1969). *J. Catal.*, **15**, 420.
Clark, M., Farona, M.F. (1982). *Polymer Bull.*, **7**, 445.
Clark, T.D., Ghadiri, M.R. (1995). *J. Am. Chem. Soc.*, **117**, 12364.

Clawson, L., Soto, J., Buchwald, S.L., Steigerwald, M.L., Grubbs, R.H. (1985). *J. Am. Chem. Soc.*, **107**, 3377.

Coates, G.W., Grubbs, R.H. (1996). *J. Am. Chem. Soc.*, **118**, 229.

Coca, S. *et al.* (1994). *J. Mol. Catal.*, **90**, 101.

Coleman, C.G., McCarthy, T.J. (1988). *Am. Chem. Soc., Polymer Prepr.*, **29**(1), 283.

Coles, M.P., Gibson, V.C., Mazzariol, L., North, M., Teasdale, W.G., Williams, C.M., Zamuner, D. (1994). *J. Chem. Soc., Chem. Commun.*, 2505.

Commereuc, D. (1995). *J. Chem. Soc., Chem. Commun.*, 791.

Conticello, V.P., Gin, D.G., Grubbs, R.H. (1992). *J. Am. Chem. Soc.*, **114**, 9708.

Cooper, N.J., Green, M.L.H. (1974a). *J. Chem. Soc., Chem. Commun.*, 209.

Cooper, N.J., Green, M.L.H. (1974b). *J. Chem. Soc., Chem. Commun.*, 761.

Cooper, N.J., Green, M.L.H. (1979). *J. Chem. Soc., Dalton Trans.*, 1121.

Couturier, J-L., Paillet, C., Leconte, M., Basset, J-M., Weiss, K. (1992). *Angew. Chem. Int. Ed. Engl.*, **31**, 628.

Couturier, J-L., Leconte, M., Basset, J-M., Ollivier, J. (1993a). *J. Organomet. Chem.*, **451**, C7.

Couturier, J-L., Tanaka, K., Leconte, M., Basset, J-M., Ollivier, J. (1993b). *Angew. Chem. Int. Ed. Engl.*, **32**, 112.

Coverdale, A.K., Dearing, P.F., Ellison, A. (1983). *J. Chem. Soc., Chem. Commun.*, 567.

Craig, G.S.W., Cohen, R.E., Schrock, R.R., Silbey, R.J., Puccetti, G., Ledoux, I., Zyss, J. (1993). *J. Am. Chem. Soc.*, **115**, 860.

Craig, G.S.W., Cohen, R.E., Schrock, R.R., Dhenaut, C., Ledoux, I., Zyss, J. (1994). *Macromolecules*, **27**, 1875.

Crain, D.L. (1969). *J. Catal.*, **13**, 110.

Crain, D.L., Reusser, R.E. (1972). *Am. Chem. Soc., Petrol. Chem. Prepr.*, **17**, E80.

Cramail, H., Fontanille, M., Soum, A. (1990). *Am. Chem. Soc., Polymer Prepr.*, **31**(1), 51.

Cramail, H., Fontanille, M., Soum, A. (1991a). *J. Mol. Catal.*, **65**, 193.

Cramail, H., Fontanille, M., Soum, A. (1991b). *Makromol. Chem., Macromol. Symp.*, **42/43**, 281.

Crease, A.E., Egglestone, H., Taylor, N. (1982). *J. Organomet. Chem.*, **238**, C5.

Crisp, G.T., Collis, M.P. (1988). *Aust. J. Chem.*, **41**, 935.

Crowe, W.E., Mitchell, J.P., Gibson, V.C., Schrock, R.R. (1990). *Macromolecules*, **23**, 3534.

Crowe, W.E., Zhang, Z.J. (1993). *J. Am. Chem. Soc.*, **115**, 10998.

Crowe, W.E., Goldberg, D.R. (1995). *J. Am. Chem. Soc.*, **117**, 5162.

Crowe, W.E., Goldberg, D.R., Zhang, Z.J. (1996). *Tetrahedron Lett.*, **37**, 2117.

Cummings, S.K., Smith, D.W., Wagener, K.B. (1995). *Macromol. Rapid Commun.*, **16**, 347.

Cummins, C.C., Beachy, M.D., Schrock, R.R., Vale, M.G., Sankaran, V., Cohen, R.E. (1991). *Chem. Mater.*, **3**, 1153.

Cummins, C.C., Schrock, R.R., Cohen, R.E. (1992). *Chem. Mater.*, **4**, 27.

Cundari, T.R., Gordon, M.S. (1992). *Organometallics*, **11**, 55.

Dalcanale, E., Casagrande, F., Martinengo, T., Montanari, F. (1985). *J. Chem. Research (S)*, 294.

Dall'Asta, G., Mazzanti, G., Natta, G., Porri, L. (1962). *Makromol. Chem.*, **56**, 224.

Dall'Asta, G. (1968a). *J. Polymer Sci.*, A-1, **6**, 2397.

Dall'Asta, G., Manetti, R. (1968b). *Europ. Polymer J.*, **4**, 145.

Dall'Asta, G., Motroni, G. (1968c). *Chim. Ind. (Milan)*, 972.

Dall'Asta, G., Motroni, G. (1968d). *J. Polymer Sci.*, A-1, **6**, 2405.

Dall'Asta, G., Motroni, G., Manetti, R., Tosi, C. (1969). *Makromol. Chem.*, **130**, 153.

Dall'Asta, G., Motroni, G. (1971a). *Angew. Makromol. Chem.*, **16/17**, 51.

Dall'Asta, G., Motroni, G. (1971b). *Europ. Polymer J.*, **7**, 707.

Dall'Asta, G. (1972a). *Makromol. Chem.*, **154**, 1.
Dall'Asta, G., Motroni, G., Motta, L. (1972b). *J. Polymer Sci.*, A-1, **10**, 1601.
Dall'Asta, G., Stigliani, G., Greco, A., Motta, L. (1973). *Chim. Ind.* (*Milan*), **55**, 142.
Dall'Asta, G. (1973/74). *Pure Appl. Chem.*, **1**, 133.
Daly, D.G., McKervey, M.A. (1982). *Tetrahedron Lett.*, **23**, 2997.
Danilyuk, A.F., Kuznetsov, V.L., Yermakov, Y.I. (1983). *Kinet. Katal.*, **24**, 926.
Danilyuk, A.F., Kuznetsov, V.L., Shmachkov, V.A., Kochubey, D.I., Chuvilin, A.L., Yermakov, Y.I. (1988). *J. Mol. Catal.*, **46**, 209.
Davidson, J.L., Vasapollo, G., Nobile, C.F., Sacco, A. (1989). *J. Organomet. Chem.*, **371**, 297.
Davidson, T.A., Wagener, K.B., Priddy, D.B. (1996). *Macromolecules*, **29**, 786.
Davie, E.S., Whan, D.A., Kemball, C. (1971). *J. Chem. Soc., Chem. Commun.*, 1202.
Davie, E.S., Whan, D.A., Kemball, C. (1972a). *Proc. 5th Int. Congr. Catal.*, 1205.
Davie, E.S., Whan, D.A., Kemball, C. (1972b). *J. Catal.*, **24**, 272.
Davies, G.R. *et al.* (1993). *Makromol. Chem., Macromol. Symp.*, **66**, 289.
Davies, G.R., Hubbard, H.V.St.A., Ward, I.M., Feast, W.J., Gibson, V.C., Khosravi, E., Marshall, E.L. (1995). *Polymer*, **36**, 235.
Davies, G.R., Almond, P.J., Hubbard, V. St.A., Ward, I.M., Feast, W.J., Gibson, V.C., Khosravi, E., Marshall, E.L. (1996). *Macromol. Symp.*, **102**, 73.
Dawoodi, Z., Green, M.L.H., Mtetwa, V.S.B., Prout, K. (1982). *J. Chem. Soc., Chem. Commun.*, 1410.
Dedieu, A., Eisenstein, O. (1982). *Nouveau J. Chim.*, **6**, 337.
De Figueiredo, C.M.C., Gomes, A. de S. (1979). *J. Polymer Sci., Polymer Chem.*, **17**, 2845.
Dekker, F.H.M., Spronk, R., Mol, J.C. (1993). *Fluid Phase Equilibria*, **84**, 321.
Dekking, H.G.G. (1961). *J. Polymer Sci.*, **55**, 525.
de la Mata, F.J., Grubbs, R.H. (1996). *Organometallics*, **15**, 577.
Demel, H., Hummel. K.(1977). *Makromol. Chem.*, **178**, 1699.
Demonceau, A., Noels, A.F., Saive, E., Hubert, A.J. (1992). *J. Mol. Catal.*, **76**, 123.
Denisova, T.T., Syatkovskii, A.I., Babitskii, B.D. (1983a). *Polymer Sci. USSR*, **25**, 1408.
Denisova, T.T., Syatkovskii, A.I., Skuratova, T.B., Babitskii, B.D. (1983b). *Polymer Sci. USSR*, **25**, 798.
Descotes, G., Ramza, J., Basset, J-M., Pagano, S. (1994). *Tetrahedron Lett.*, **35**, 7379.
Descotes, G., Ramza, J., Basset, J-M., Pagano, S., Gentil, E., Banoub, J. (1996). *Tetrahedron*, **52**, 10903.
Devarajan, S., Walton, D.R.M., Leigh, G.J. (1979). *J. Organomet. Chem.*, **181**, 99.
Devine, G.I., Ho, H.T., Ivin, K.J., Mohamed, M.-A., Rooney, J.J. (1982). *J. Chem. Soc., Chem. Commun.*, 1229.
Devlin, P.A., Lutz, E.F. (1970). *US Pat.* 3,627,739.
de Vries, J.L.K.F., Pott, G.T. (1977). *Rec. Trav. Chim. Pays-Bas*, **96**, M115.
Diedrich, K.M. (1989). *Kautsch. Gummi, Kunstst.*, **42**, 1130.
Diefenbach, S.P. (1988). *Chem. Abstr.*, **108**, 40092.
Dietz, S.D., Eilerts, N.W., Heppert, J.A., Morton, M.D. (1993). *Inorg. Chem.*, **32**, 1698.
Dillon, K.B., Gibson, V.C., Sequeira, L.J. (1995). *J. Chem. Soc., Chem. Commun.*, 2429.
Dimonie, M., Coca, S., Dragutan,V. (1992). *J. Mol. Catal.*, **76**, 79.
Dodd, H.T., Rutt, K.J. (1982). *J. Mol. Catal.*, **15**, 103.
Dodd, H.T., Rutt, K.J. (1985). *J. Mol. Catal.*, **28**, 33.
Dodd, H.T., Rutt, K.J. (1988). *J. Mol. Catal.*, **47**, 67.
Doherty, M., Siove, A., Parlier, A., Rudler, H., Fontanille, M. (1986). *Makromol. Chem., Macromol. Symp.*, **6**, 33.
Dolgoplosk, B.A., Makovetsky, K.L., Golenko, T.G., Korshak, Y.V., Tinyakova, E.I. (1974). *Europ. Polymer J.*, **10**, 901.

Dolgoplosk, B.A., Makovetsky, K.L., Korshak, Y.V., Oreshkin, I.A., Tinyakova, E.I., Yakovlev, V.A. (1977a). *Rec. Trav. Chim. Pays-Bas*, **96**, M35.
Dolgoplosk, B.A., Tinyakova, E.I., Yakovlev, V.A. (1977b). *Dokl. Akad. Nauk SSSR*, **232**, 1075.
Dötz, K.H., Fischer, E.O. (1972). *Chem. Ber.*, **105**, 1356.
Dötz, K.H. (1984). *Angew. Chem. Int. Ed. Engl.*, **23**, 587.
Dounis, P., Feast, W.J., Kenwright, A.M. (1995). *Polymer*, **36**, 2787.
Dounis, P., Feast, W.J. (1996a). *Polymer*, **37**, 2547.
Dounis, P., Feast, W.J., Widawski, G. (1996b). *J. Mol. Catal.*, in press (ISOM 11 vol.).
Doxsee, K.M., Juliette, J.J.J., Mouser, J.K.M., Zientara, K. (1993). *Organometallics*, **12**, 4742.
Doyle, G. (1973). *J. Catal.*, **30**, 118.
Dragutan, V., Dimonie, M., Coca, S. (1994). *Am. Chem. Soc., Polymer Prepr.*, **35**(1), 698.
Drapeau, A., Leonard, J. (1985). *Macromolecules*, **18**, 144.
Dräxler, A. (1983a). *Elastomerics*, February, 16.
Dräxler, A. (1983b). *Kautsch. Gummi, Kunstst.*, **36**, 1037.
Dräxler, A. (1988). In '*Handbook of Elastomers*' (eds A.K. Bhowmick, H.L.Stephens), p.661, Dekker, New York.
Drent, E., Breed, A.J.M. (1991). *Chem. Abstr.*, **115**, 233216.
du Plessis, J.A.K., Heenop, P.J. (1979). *S. Afr. J. Chem.*, **32**, 1.
du Plessis, J.A.K., Heenop, P.J. (1980). *S. Afr. J. Chem.*, **33**, 51.
du Plessis, J.A.K., Heenop, P.J., Pienaar, J.J. (1982). *S. Afr. J.Chem.*, **35**, 42.
du Plessis, J.A.K., Pienaar, J.J., Vosloo, H.C.M. (1988). *J. Mol. Catal.*, **46**, 287.
du Plessis, J.A.K., Vosloo, H.C.M. (1991). *J. Mol. Catal.*, **65**, 51.
du Plessis, J.A.K., van Sittert, C.G.C.E., Vosloo, H.C.M. (1994). *J. Mol. Catal.*, **90**, 11.
Duquette, L.G., Cielinsky, R.C., Jung, C.W., Garrou, P.E. (1984). *J. Catal.*, **90**, 362.
Dürr, A., Hummel, K. (1971). *Kolloid Z., Z. Polymere*, **243**, 155.
Eberle, H.J., Kreuzer, F.H., Zeitler, N. (1986). *Chem. Abstr.*, **105**, 174798.
Eberle, H.J., Schweiger, C., Zeitler, N., Meixner, G. (1990). *Chem. Abstr.*, **113**, 5775.
Edreva-Kardjieva, R.M., Andreev, A.A. (1976). *React. Kinet. Catal. Lett.*, **5**, 465.
Edreva-Kardjieva, R.M., Andreev, A.A. (1977). *Rec. Trav. Chim. Pays-Bas*, **96**, M141.
Edreva-Kardjieva, R.M., Andreev, A.A. (1985). *J. Catal.*, **94**, 97.
Edreva-Kardjieva, R.M., Andreev, A.A. (1986). *J. Catal.*, **97**, 321.
Edreva-Kardjieva, R.M., Vuurman, M.A., Mol, J.C. (1992). *J. Mol. Catal.*, **76**, 297.
Edwards, J.H., Feast, W.J., Bott, D.C. (1984). *Polymer*, **25**, 395.
Edwige, C., Lattes, A., Laval, J.P., Mutin, R., Basset, J.M., Nouguier, R. (1980). *J. Mol. Catal.*, **8**, 297.
Efimov, V.A. (1994). *Polymer Sci.*, **A36**, 884.
Efimov, V.A. (1995). *Polymer Sci.*, **A37**, 574.
Eilerts, N.W., Heppert, J.A., Morton, M.D. (1992). *J. Mol. Catal.*, **76**, 157.
Eilerts, N.W., Heppert, J.A. (1995). *Polyhedron*, **14**, 3255.
Eleuterio, H.S. (1957). *US Pat.* 3,074,918; *Chem. Abstr.*, **55**, 16005 (1961).
Eleuterio, H.S. (1991). *J. Mol. Catal.*, **65**, 55.
Elev, I.V., Shelimov, B.N., Kazansky, V.B. (1987). *Kinet. Katal.*, **28**, 415.
Elev, I.V., Shelimov, B.N., Kazansky, V.B., Grinev, V.E., Krilov, O.V. (1988). *Izv. Akad. Nauk SSSR, Ser. Khim.*, **37**, 743.
Elev, I.V., Shelimov, B.N., Kazansky, V.B. (1989a). *Kinet. Katal.*, **30**, 895.
Elev, I.V., Shelimov, B.N., Kazansky, V.B., Berezin, M.Y., Usacheva, G.A., Belov, P.S. (1989b). *Kinet. Katal.*, **30**, 1101.
Ellison, A., Coverdale, A.K., Dearing, P.F. (1983). *Appl. Catal.*, **8**, 109.
Ellison, A., Coverdale, A.K., Dearing, P.F. (1985). *J. Mol. Catal.*, **28**, 141.
Ellison, A., Bickerstaffe, A., Diakun, G., Worthington, P. (1986). *J. Mol. Catal.*, **36**, 67.

Ellison, A., Worthington, P., Diakun, G. (1988). *J. Mol. Catal.*, **46**, 131.
Ellison, A. (1990). *NATO ASI Ser.*, **C326**, 335.
Ellsworth, M.W., Novak, B.M. (1991). *J. Am. Chem. Soc.*, **113**, 2756.
El-Saafin, I.F.A.F., Feast, W.J. (1982). *J. Mol. Catal.*, **15**, 61.
El-Sawi, M., Iannibello, A., Morelli, F., Catalano, G., Intrieri, F., Giordano, G. (1981). *J. Chem. Tech. Biotechnol.*, **31**, 388.
El-Sawi, M., Iannibello, A., Giordano, G., Fedele, U., Ricca, G. (1982). *J. Chem. Tech. Biotechnol.*, **32**, 1049.
Engelhardt, J., Goldwasser, J., Hall, W.K. (1981). *J. Catal.*, **70**, 364.
Engelhardt, J. (1982a). *Acta Chim. Hung.*, **111**, 465.
Engelhardt, J. (1982b). *React. Kinet. Catal. Lett.*, **21**, 7.
Engelhardt, J., Goldwasser, J., Hall, W.K. (1982c). *J. Catal.*, **76**, 48.
Engelhardt, J., Zsinka, I. (1982d). *React. Kinet. Catal. Lett.*, **21**, 529.
Engelhardt, J., Goldwasser, J., Hall, W.K. (1982e). *J. Mol. Catal.*, **15**, 173.
Engelhardt, J., Kalló, D. (1985a). *Acta Chim. Hung.*, **119**, 249.
Engelhardt, J., Zsinka, I. (1985b). *J. Mol. Catal.*, **28**, 169.
Ephritikhine, M., Green, M.L.H. (1976). *J. Chem. Soc., Chem. Commun.*, 926.
Ephritikhine, M., Francis, B., Green, M.L.H., Mackenzie, R.E., Smith, M.J. (1977). *J. Chem. Soc., Dalton Trans.*, 1131.
Espy, H.H., Matlack, S. (1988). *US Pat.* 4,751,337.
Evans, J., Gaunlett, J.T., Mosselmans, J.F.W. (1990). *Faraday Discuss. Chem. Soc.*, **89**, 107.
Farona, M.F., Lofgren, P.A., Woon, P.S. (1974). *J. Chem. Soc., Chem. Commun.*, 246.
Farona, M.F., Greenlee, W.S. (1975). *J. Chem. Soc., Chem. Commun.*, 759.
Farona, M.F., Motz, V.W. (1976). *J. Chem. Soc., Chem. Commun.*, 930.
Farona, M.F., Tucker, R.L. (1980). *J. Mol. Catal.*, **8**, 85.
Farren, T.R., Amass, A.J., Beevers, M.S., Stowell, J.A. (1987). *Makromol. Chem.*, **188**, 2535.
Farren, T.R., Amass, A.J., Beevers, M.S., Stowell, J.A. (1989). *Polymer*, **30**, 1008.
Fathikalajahi, J., Wills, G.B. (1980). *J. Mol. Catal.*, **8**, 127.
Feast, W.J., Wilson, B. (1979). *Polymer*, **20**, 1182.
Feast, W.J., Wilson, B. (1980). *J. Mol. Catal.*, **8**, 277.
Feast, W.J., Harper, K. (1985a). *J. Mol. Catal.*, **28**, 293.
Feast, W.J., Millichamp, I.S. (1985b). *J. Mol. Catal.*, **28**, 331.
Feast, W.J., Parker, D., Winter, J.N., Bott, D.C., Walker, N.S. (1985c). *Springer Ser. Solid State Sci.*, **63**, 45.
Feast, W.J., Shahada, L.A.H. (1986a). *Polymer*, **27**, 1289.
Feast, W.J., Harper, K. (1986b). *Brit. Polymer J.*, **18**, 161.
Feast, W.J., Taylor, M.J., Winter, J.N. (1987). *Polymer*, **28**, 593.
Feast, W.J., Gibson, V.C. (1989a). In '*The Chemistry of the Metal–Carbon Bond*' (ed F.R. Hartley), **5**, p.199. Wiley, New York.
Feast, W.J. (1989b). In '*Comprehensive Polymer Science*' (eds G.C. Eastmond, A. Ledwith, S. Russo, P. Sigwalt), **4**, p.135. Pergamon, Oxford.
Feast, W.J., Gibson, V.C., Khosravi, E., Schrock, R.R. (1990). *Am. Chem. Soc., Polymer Prepr.*, **31**(1), 310.
Feast, W.J., Harrison, D.B. (1991a). *J. Mol. Catal.*, **65**, 63.
Feast, W.J., Harrison, D.B. (1991b). *Polymer Bull.*, **25**, 343.
Feast, W.J., Shahada, L.A. (1991c). *Europ. Polymer J.*, **27**, 27.
Feast, W.J., Harrison, D.B. (1991d). *Polymer*, **32**, 558.
Feast, W.J., Gibson, V.C., Marshall, E.L. (1992a). *J. Chem. Soc., Chem. Commun.*, 1157.
Feast, W.J., Gibson, V.C., Khosravi, E., Marshall, E.L., Mitchell, J.P. (1992b). *Polymer*, **33**, 872.

Feast, W.J., Gibson, V.C., Ivin, K.J., Khosravi, E., Kenwright, A.M., Marshall, E.L., Mitchell, J.P. (1992c). *Makromol. Chem.*, **193**, 2103.
Feast, W.J., Gibson, V.C., Johnson, A.F., Khosravi, E., Mohsin, M.A. (1994a). *Polymer*, **35**, 3542.
Feast, W.J., Gibson, V.C., Khosravi, E., Marshall, E.L. (1994b). *J. Chem. Soc., Chem. Commun.*, 9.
Feast, W.J., Gibson, V.C., Ivin, K.J., Kenwright, A.M., Khosravi, E. (1994c). *J. Chem. Soc., Chem. Commun.*, 1399.
Feast, W.J., Gibson, V.C., Ivin, K.J., Kenwright, A.M., Khosravi, E., (1994d). *J. Mol. Catal.*, **90**, 87.
Feast, W.J., Gibson, V.C., Khosravi, E., Marshall, E.L., Wilson, B. (1995). In '*Ziegler Catalysts*' (eds G. Fink, R. Mülhaupt, H.H. Brintzinger,), p.470. Springer, Berlin.
Feast, W.J., Kenwright, A.M., Sugawara, K. (1996). *J. Mol. Catal.*, in press (ISOM 11 vol.).
Feher, F.J., Tajima, T.L. (1994). *J. Am. Chem. Soc.*, **116**, 2145.
Feldman, J., DePue, R.T., Schaverien, C.J., Davis, W.M., Schrock, R.R. (1989a). *NATO ASI Ser.* **C269**, 323.
Feldman, J., Davis, W.M., Schrock, R.R. (1989b). *Organometallics*, **8**, 2266.
Feldman, J., Schrock, R.R. (1991). *Progress in Inorg. Chem.*, **39**, 1.
Fellmann, J.D., Schrock, R.R., Traficante, D.D. (1982). *Organometallics*, **1**, 481.
Finkel'shtein, E.S., Strel'chik, B.S., Portnykh, E.B., Vdovin, V.M., Nametkin, N.S. (1977). *Dokl. Akad. Nauk SSSR*, **232**, 1322.
Finkel'shtein, E.S., Portnykh, E.B., Ushakov, N.V., Vdovin, B.M. (1979). *Izv. Akad. Nauk SSSR, Ser. Khim.*, **28**, 474.
Finkel'shtein, E.S., Portnykh, E.B., Ushakov, N.V., Vdovin, V.M. (1981). *Izv. Akad. Nauk SSSR, Ser. Khim.*, **30**, 641.
Finkel'shtein, E.S., Portnykh, E.B., Antipova, I.V., Vdovin, V.M. (1989). *Izv. Akad. Nauk SSSR, Ser. Khim.*, **38**, 1358.
Finkel'shtein, E.S. *et al.* (1991). *Makromol. Chem.*, **192**, 1.
Finkel'shtein, E.S., Bykov,V.I., Portnykh, E.B. (1992a). *J. Mol. Catal.*, **76**, 33.
Finkel'shtein, E.S., Ushakov, N.V., Portnykh, E.B. (1992b). *J. Mol. Catal.*, **76**, 133.
Finkel'shtein, E.S., Portnykh, E.B., Makovetskii, K.L., Ostrovskaya, I.Y., Gringol'ts, M.L., Dzhelimev, U.M., Gol'dberg, A.I. (1993). *Russ. Chem. Bull.*, **42**, 124.
Finkel'shtein, E.S., Portnykh, E.B., Ushakov, N.V., Greengolts, M.L., Fedorova, G.K., Platé, N.L. (1994). *Macromol. Rapid Commun.*, **15**, 155.
Fischer, E.O., Maasböl, A. (1964). *Angew. Chem. Int. Ed. Engl.*, **3**, 580.
Fischer, E.O., Dötz, K.H. (1972). *Chem. Ber.*, **105**, 3966.
Fischer, E.O., Wagner, W.R. (1976). *J. Organomet. Chem.*, **116**, C21.
Fischer, H., Dötz, K.H. (1980). *Chem. Ber.*, **113**, 193.
Fischer, H., Schmid, J. (1988). *J. Mol. Catal.*, **46**, 277.
Fischer, W., Stelzer, F., Heller, C., Leising, G. (1993). *Synthetic Metals*, **55–57**, 815.
Fisher, R.A., Grubbs, R.H. (1992). *Makromol. Chem., Macromol. Symp.*, **63**, 271.
Fitton, P., Whitesides, T. (1974). *US Pat.* 3,855,338.
Flatt, B.T., Grubbs, R.H., Blanski, R.L., Calabrese, J.C., Feldman, J. (1994). *Organometallics*, **13**, 2728.
Foley, H.C., Strubinger, L.M., Targos, T.S., Geoffroy, G.L. (1983). *J. Am. Chem. Soc.*, **105**, 3064.
Folga, E., Ziegler, T. (1993). *Organometallics*, **12**, 325.
Foltynowicz, Z., Marciniec, B. (1989). *J. Orgamometal. Chem.*, **376**, 15.
Foltynowicz, Z., Marciniec, B., Pietraszuk, C. (1991). *J. Mol. Catal.*, **65**, 113.
Foltynowicz, Z., Pietraszuk, C., Marciniec, B. (1993a). *Appl. Organomet. Chem.*, **7**, 539.
Foltynowicz, Z. (1993b). *Polish J. Chem.*, **67**, 1361.

Fontanille, M., Lucas, C., Soum, A. (1992). *Makromol. Chem.*, **193**, 411.
Forbes, M.D.E., Patton, J.T., Myers, T.L., Maynard, H.D., Smith, D.W., Schulz, G.R., Wagener, K.B. (1992). *J. Am. Chem. Soc.*, **114**, 10978.
Fox, H.H., Schrock, R.R. (1992). *Organometallics*, **11**, 2763.
Fox, H.H., Lee, J-K., Park, L.Y., Schrock, R.R. (1993). *Organometallics*, **12**, 759.
Fox, H.H., Schrock, R.R., O'Dell, R. (1994a). *Organometallics*, **13**, 635.
Fox, H.H., Wolf, M.O., O'Dell, R., Lin, B.L., Schrock, R.R., Wrighton, M.S. (1994b). *J. Am. Chem. Soc.*, **116**, 2827.
Fox, H.H., Schofield, M.H., Schrock, R.R. (1994c). *Organometallics*, **13**, 2804.
France, M.B., Paciello, R.A., Grubbs, R.H. (1993a). *Macromolecules*, **26**, 4739.
France, M.B., Grubbs, R.H., McGrath, D.V., Paciello, R.A. (1993b). *Macromolecules*, **26**, 4742.
France, M.B., Feldman, J., Grubbs, R.H. (1994). *J. Chem Soc., Chem. Commun.*, 1307.
Fraser, C., Grubbs, R.H. (1995a). *Macromolecules*, **28**, 7248.
Fraser, C., Hillmeyer, M.A., Gutierrez, E., Grubbs, R.H. (1995b). *Macromolecules*, **28**, 7256.
Freitas, E.R., Gum, C.R. (1979). *Chem. Eng. Progr.*, **75** (1), 73.
Freudenberger, J.H., Schrock, R.R., Churchill, R.R., Rheingold, A.L., Ziller, J.W. (1984). *Organometallics*, **3**, 1563.
Freudenberger, J.H., Schrock, R.R. (1986). *Organometallics*, **5**, 398.
Fridman, R.A., Nosakova, S.M., Kryukov, Y.B., Bashkirov, A.N., Nametkin, N.S., Vdovin, V.M. (1971). *Izv. Akad. Nauk SSSR, Ser. Khim.*, **20**, 2100.
Fridman, R.A., Bashkirov, A.N., Liberov, L.G., Nosakova, S.M., Smirnova, R.M., Verbovetskaya, S.B. (1977a). *Dokl. Akad. Nauk SSSR*, **234**, 1354.
Fridman, R.A., Nosakova, S.M., Liberov, L.G., Bashkirov, A.N. (1977b). *Izv. Akad. Nauk SSSR, Ser. Khim.*, **26**, 678.
Fridman, R.A., Liberov, L.G., Nosakova, S.M., Bashkirov, A.N. (1978). *Izv. Akad. Nauk SSSR, Ser. Khim.*, **27**, 1225.
Fridman, R.A., Liberov, L.G., Nosakova, S.M., Smirnova, R.M., Bashkirov, A.N. (1979). *Izv. Akad. Nauk SSSR, Ser. Khim.*, **28**, 2816.
Fu, G.C., Grubbs, R.H. (1992a). *J. Am. Chem. Soc.*, **114**, 5426.
Fu, G.C., Grubbs, R.H. (1992b). *J. Am. Chem. Soc.*, **114**, 7324.
Fu, G.C., Grubbs, R.H. (1993a). *J. Am. Chem. Soc.*, **115**, 3800.
Fu, G.C., Nguyen, S.T., Grubbs, R.H. (1993b). *J. Am. Chem. Soc.*, **115**, 9856.
Fujimori, J., Masuda, T., Higashimura, T. (1988). *Polymer Bull.*, **20**, 1.
Fujimura, O., Fu, G.C., Grubbs, R.H. (1994). *J. Org. Chem.*, **59**, 4029.
Fujimura, O., Fu, G.C., Rothemund, P.W.K., Grubbs, R.H. (1995). *J. Am. Chem. Soc.*, **117**, 2355.
Fujimura, O., Grubbs, R.H. (1996a). *J. Am. Chem. Soc.*, **118**, 2499.
Fujimura, O., de la Mata, F.J., Grubbs, R.H. (1996b). *Organometallics*, **15**, 1865.
Fürstner, A., Langemann, K. (1996). *J. Org. Chem.*, **61**, 3942.
Furukawa, J., Mizoe,Y. (1973). *J. Polymer Sci., Polymer Lett.*, **11**, 263.
Furukawa, J., Kobayashi, E., Jinda, K., Hayashi, O. (1978). *Chem. Abstr.*, **88**, 191783.
Gagné, M.R., Grubbs, R.H., Feldman, J., Ziller, J.W. (1992). *Organometallics*, **11**, 3933.
Gajda, G.J., Grubbs, R.H., Weinberg, W.H. (1987). *J. Am. Chem. Soc.*, **109**, 66.
Gal, Y.S., Cho, H.N., Choi, S.K. (1985). *Polymer (Korea)*, **9**, 361.
Gal, Y.S., Cho, H.N., Choi, S.K. (1986). *J. Polymer Sci.*, A, *Polymer Chem.*, **24**, 2021.
Gal, Y.S., Choi, S.K. (1987). *Polymer (Korea)*, **11**, 563.
Gal, Y-S., Choi, S-K. (1988). *J. Polymer Sci.*, C, *Polymer Lett.*, **26**, 115.
Gal, Y.S., Choi, S.K., Kim, C.Y. (1989). *J. Polymer Sci.*, A, *Polymer Chem.*, **27**, 31.
Gal, Y-S., Jung, B., Cho, H-N., Lee, W-C., Choi, S-K. (1990). *J. Polymer Sci.*, C, *Polymer Lett.*, **28**, 259.

Gal, Y-S., Jung, B., Cho, H-N., Lee, W-C., Choi, S-K. (1992a). *Bull. Korean Chem. Soc.*, **13**, 4.
Gal, Y-S., Jung, B., Lee, W-C., Choi, S-K. (1992b). *J. Polymer Sci.*, A, *Polymer Chem.*, **30**, 2657.
Gal, Y-S., Choi, S-K. (1993). *J. Polymer Sci.*, A, *Polymer Chem.*, **31**, 345.
Gal, Y-S., Jung, B., Lee, W-C., Choi, S-K. (1994a). *Macromol. Reports*, **A31** (Suppl. 3/4), 271.
Gal, Y-S. (1994b). *J. Macromol. Sci.-Pure Appl. Chem.*, **A31**, 703.
Gal, Y-S., Jung, B., Kim, J-H., Lee, W-C., Choi, S-K. (1994c). *J. Macromol. Sci.-Pure Appl. Chem.*, **A31**, 1177.
Gal, Y-S. (1994d). *J. Chem. Soc., Chem. Commun.*, 327.
Gal, Y-S., Jung, B., Kim, J-H., Lee, W-C., Choi, S-K. (1995a). *J. Polymer Sci.*, A, *Polymer Chem.*, **33**, 307.
Gal, Y-S. (1995b). *J. Macromol. Sci.-Pure Appl. Chem.*, **A32**, 61.
Gal, Y-S., Jung, B., Lee, W-C., Lee, H-J., Choi, S-K. (1995c). *Macromolecules*, **28**, 2086.
Gal, Y-S. (1995d). *Macromol. Reports*, **A32** (Suppl. 1/2), 55.
Gamble, A.S., Patton, J.T., Boncella, J.M. (1992). *Makromol. Chem., Rapid Commun.*, **13**, 109.
Gamble, A.S., Boncella, J.M. (1993). *Organometallics*, **12**, 2814.
Ganesamoorthy, S., Sundararajan, G. (1992). *J. Mol. Catal.*, **76**, 335.
Gangwal, S.K., Fathi-kalajahi, J., Wills, G.B. (1977). *Ind. Eng. Chem., Prod. Res. Dev.*, **16**, 237.
Gangwal, S.K., Wills, G.B. (1978). *J. Catal.*, **52**, 539.
Garnier, F., Krausz, P., Dubois, J.E. (1979). *J. Organomet. Chem.*, **170**, 195.
Garnier, F., Krausz, P. (1980a). *J. Mol. Catal.*, **8**, 91.
Garnier, F., Krausz, P., Rudler, H. (1980b). *J. Organomet. Chem.*, **186**, 77.
Garro-Hélion, F., Guibé, F. (1996). *Chem. Commun.*, 641.
Gassman, P.G., Johnson, T.H. (1976a). *J. Am. Chem. Soc.*, **98**, 6057.
Gassman, P.G., Johnson, T.H. (1976b). *J. Am. Chem. Soc.*, **98**, 6058.
Gassman, P.G., Johnson, T.H. (1977). *J. Am. Chem. Soc.*, **99**, 622.
Gassman, P.G., Macomber, D.W., Willging, S.M. (1985). *J. Am. Chem. Soc.*, **107**, 2380.
Gianotti, G., Dall'Asta, G., Valvassori, A., Zamboni, V. (1971). *Makromol. Chem.*, **149**, 117.
Gibson, T., Tulich, L. (1981). *J. Org. Chem.*, **46**, 1821.
Gibson, V.C. (1994). *Advanced Materials*, **6**, 37.
Gilbert, M., Herbert, I.R. (1993). *Polymer Bull.*, **30**, 83.
Gilet, M., Mortreux, A., Nicole, J., Petit, F. (1979). *J. Chem. Soc., Chem. Commun.*, 521.
Gilet, M., Mortreux, A., Folest, J-C., Petit, F. (1983). *J. Am. Chem. Soc.*, **105**, 3876.
Gillan, E.M.D., Hamilton, J.G., Mackey, O.N.D., Rooney, J.J. (1988). *J. Mol. Catal.*, **46**, 359.
Gillan, E.M.D., Hamilton, J.G., Rooney, J.J. (1989). *J. Mol. Catal.*, **50**, L23.
Gilliom, L.R., Grubbs, R.H. (1986a). *Organometallics*, **5**, 721.
Gilliom, L.R., Grubbs, R.H. (1986b). *J. Am. Chem. Soc.*, **108**, 733.
Gilliom, L.R., Grubbs, R.H. (1988). *J. Mol. Catal.*, **46**, 255.
Ginsburg, E.J., Gorman, C.B., Marder, S.R., Grubbs, R.H. (1989). *J. Am. Chem. Soc.*, **111**, 7621.
Gita, B., Sundararajan, G. (1993). *Tetrahedron Lett.*, **34**, 6123.
Goller, R., Schubert, U., Weiss, K. (1993). *J. Organomet. Chem.*, **459**, 229.
Goodall, B.L., Kroenke, W.J., Minchak, R.J., Rhodes, L.F. (1993). *J. Appl. Polymer Sci.*, **47**, 607.
Goodall, B.L., McIntosh, L.H., Rhodes, L.F. (1995). *Macromol. Symp.*, **89**, 421.
Goodyear Tire and Rubber Co. (1977). *Chem. Abstr.*, **86**, 5983.

Gorman, C.B, Ginsburg, E.J., Marder, S.R., Grubbs, R.H. (1989). *Angew. Chem. Int. Ed. Engl. Adv. Mater.*, **28**, 1571.
Gorman, C., Ginsburg, E., Marder, S., Grubbs, R. (1990). *Am. Chem. Soc., Polymer Prepr.*, **31**(1), 386.
Gorman, C.B., Ginsburg, E.J., Moore, J.S., Grubbs, R.H. (1991a). *Am. Chem. Soc., Polymer Prepr.*, **32**(3), 714.
Gorman, C.B., Ginsburg, E.J., Sailor M.J., Moore, J.S., Jozefiak, T.H., Lewis, N.S., Grubbs, R.H., Marder, S.R., Perry, J.W. (1991b). *Synthetic Metals*, **41–43**, 1033.
Gorman, C.B., Ginsburg, E.J., Grubbs, R.H. (1993). *J. Am. Chem. Soc.*, **115**, 1397.
Goto, K., Komiya, T., Yamahara, N., Iio, A. (1990a). *Chem. Abstr.*, **112**, 140102.
Goto, K., Iio, A., Komiya, T., Komya, T. (1990b). *Chem. Abstr.*, **113**, 7060.
Graham, J.R., Slaugh, L.H. (1971). *Tetrahedron Lett.*, 787.
Grahlert, W., Milowski, K., Langbein, U., Taeger, E. (1975). *Plaste, Kautschuk*, **22**, 229.
Graimann, C., Hönig, H., Hummel, K., Stelzer, F. (1985). *J. Computer Chem.*, **6**, 302.
Graulich, W., Swodenk, W., Theisen, D. (1972). *Hydrocarbon Process.*, 71.
Greene, R.M.E., Hamilton, J.G., Ivin, K.J., Rooney, J.J. (1986). *Makromol. Chem.*, **187**, 619.
Greene, R.M.E., Ivin, K.J., McCann, G.M., Rooney, J.J. (1987a). *Makromol. Chem.*, **188**, 1933.
Greene, R.M.E., Ivin, K.J., Milligan, B.D., Rooney, J.J. (1987b). *Brit. Polymer J.*, **19**, 339.
Greene, R.M.E., Ivin, K.J., Rooney, J.J., Kress, J., Osborn, J.A. (1988). *Makromol. Chem.*, **189**, 2797.
Greene, R.M.E., Ivin, K.J., Kress, J., Osborn, J.A., Rooney, J.J. (1989). *Brit. Polymer J.*, **21**, 237.
Greenlee, W.S., Farona, M.F. (1976). *Inorg. Chem.*, **15**, 2129.
Gregory, A.R., Mintz, E.A. (1985). *J. Am. Chem. Soc.*, **107**, 2179.
Gresham, W.F., Merckling, N.G. (1959). *Chem. Abstr.*, **53**, 7668, 23098.
Grubbs, R.H., Carr, D.D., Burk, P.L., (1975). In *Proc. 1st Jap. Am. Seminar on Organotransition Metal Chemistry* (eds Y. Ishii, M. Tsutsui), p.135. Plenum, New York.
Grubbs, R.H., Carr, D.D., Hoppin, C., Burk, P.L. (1976). *J. Am. Chem. Soc.*, **98**, 3478.
Grubbs, R.H., Swetnick, S., Su, S.C.H. (1977/78). *J. Mol. Catal.*, **3**, 11.
Grubbs, R.H. (1978). *Prog. Inorg. Chem.*, **24** 1.
Grubbs, R.H., Hoppin, C.R. (1979). *J. Am. Chem. Soc.*, **101**, 1499.
Grubbs, R.H., Swetnick, S. (1980). *J. Mol. Catal.*, **8**, 25.
Grubbs, R.H., Swager, T.M. (1987a). *Chem. Abstr.*, **106**, 177118.
Grubbs, R.H., Gilliom, L. (1987b). *NATO ASI* Ser., **C215**, 343.
Grubbs, R.H., Tumas, W. (1989). *Science*, **243**, 907.
Grubbs, R.H., Novak, B.M. (1990). *Chem. Abstr.*, **113**, 7050.
Grubbs, R.H. (1994a). *J. Macromol. Sci.-Pure Appl. Chem.*, **A31**, 1829.
Grubbs, R.H., Hillmyer, M., Benedicto, A., Wu, Z. (1994b). *Am. Chem. Soc., Polymer Prepr.*, **35**(1), 688.
Grubbs, R.H., Miller, S.J., Fu, G.C. (1995). *Acc. Chem. Res.*, **28**, 446.
Grünert, W., Shpiro, E.S., Feldhaus, R., Anders, K., Antoshin, G.V., Minachev, K.M. (1987). *J. Catal.*, **107**, 522.
Grünert, W., Feldhaus, R., Anders, K. (1988). *React. Kinet. Catal. Lett.*, **36**, 195.
Grünert, W., Mörke, W., Feldhaus, R., Anders, K. (1989a). *J. Catal.*, **117**, 485.
Grünert, W., Feldhaus, R., Anders, K., Shpiro, E.S., Minachev, K.M. (1989b). *J. Catal.*, **120**, 444.
Grünert, W., Stakheev, A.Y., Feldhaus, R., Anders, K., Shpiro, E.S., Minachev, K.M. (1992a). *J. Catal.*, **135**, 287.
Grünert, W. (1992b). *Indian J. Technol.*, **30**, 113.

Grupp, T., Kohl, V., Schäfer, H., Hofmann, H. (1991). *Appl. Catal.*, **76**, 61.
Gruter, G-J.M., Akkerman, O.S., Bickelhaupt, F. (1996). *Tetrahedron*, **52**, 2565.
Grutke, S., Hurley, J.H., Risse, W. (1994). *Macromol. Chem. Phys.*, **195**, 2875.
Guay, C., Léonard, J. (1995). *Polymer J.*, **27**, 1260.
Gudkov, B.S., Khrust, T.B., Ustinov, N.S. (1981). *Kinet. Katal.*, **22**, 525.
Guggenberger, L.J., Schrock, R.R. (1975). *J. Am. Chem. Soc.*, **97**, 6578.
Günther, P., Haas, F., Marwede, G., Nützel, K., Oberkirch, W., Pampus, G., Schön, N., Witte, J. (1970). *Angew. Makromol. Chem.*, **14**, 87.
Guo, X., Xu, Y., Shi, Y., Zhang, Y. (1988). *J. Mol. Catal.*, **46**, 119.
Haas, F., Nützel, K., Pampus, G., Theisen, D. (1970). *Rubber Chem. Technol.*, **43**, 1116.
Hamano, T., Masuda, T., Higashimura, T. (1988). *J. Polymer Sci.*, A, *Polymer Chem.*, **26**, 2603.
Hamilton, J.G., Ivin, K.J., Rooney, J.J., Waring, L.C. (1983). *J. Chem. Soc., Chem. Commun.*, 159.
Hamilton, J.G., Ivin, K.J., Rooney, J.J. (1984a). *Brit. Polymer J.*, **16**, 21.
Hamilton, J.G., Rooney, J.J. (1984b). *J. Chem. Soc., Faraday Trans.,1*, **80**, 129.
Hamilton, J.G., Ivin, K.J., McCann, G.M., Rooney, J.J. (1985a). *Makromol. Chem.*, **186**, 1477.
Hamilton, J.G., Ivin, K.J., Rooney, J.J. (1985b). *J. Mol. Catal.*, **28**, 255.
Hamilton, J.G., Ivin, K.J., Rooney, J.J. (1985c). *Brit. Polymer J.*, **17**, 41.
Hamilton, J.G., Ivin, K.J., Rooney, J.J. (1985d). *Brit. Polymer J.*, **17**, 11.
Hamilton, J.G., Ivin, K.J., Rooney, J.J. (1986). *J. Mol. Catal.*, **36**, 115.
Hamilton, J.G., Ivin, K.J., Rooney, J.J. (1988). *Brit. Polymer J.*, **20**, 91.
Hamilton, J.G., Mackey, O.N.D., Rooney, J.J., Gilheany, D.G. (1990a). *J. Chem. Soc., Chem. Commun.*, 1600.
Hamilton, J.G., Marquess, D.G., O'Neill, T.J., Rooney, J.J. (1990b). *J. Chem. Soc., Chem. Commun.*, 119.
Hamilton, J.G., Rooney, J.J. (1992). *J. Chem. Soc., Chem. Commun.*, 370.
Hamilton, J.G., Rooney, J.J., Snowden, D.G. (1993). *Makromol. Chem.*, **194**, 2907.
Hamilton, J.G., Rooney, J.J., Snowden, D.G. (1994). *J. Polymer Sci.*, A, *Polymer Chem.*, **32**, 993.
Hamilton, J.G., Rooney, J.J., Snowden, D.G. (1995). *Macromol. Chem. Phys.*, **196**, 1031.
Han, C-C., Katz, T.J. (1985). *Organometallics*, **4**, 2186.
Han, S.H., Kim, U.Y., Kang, Y.S., Choi, S.K. (1991). *Macromolecules*, **24**, 973.
Hara, S., Endo, Z., Mera, H. (1989a). *Chem. Abstr.*, **110**, 76287.
Hara, S., Endo, Z., Mera, H. (1989b). *Chem. Abstr.*, **110**, 95997.
Hara, S., Endo, Z. (1989c). *Chem. Abstr.*, **110**, 96705.
Hara, S., Endo, Z. (1989d). *Chem. Abstr.*, **110**, 115520.
Hara, S., Endo, Z. (1989e). *Chem. Abstr.*, **110**, 193649.
Hara, S., Endo, Z., Mera, H. (1989f). *Chem. Abstr.*, **110**, 213621.
Hara, S., Endo, Z. (1990). *Chem. Abstr.*, **113**, 7055.
Hara, S., Endo, Z. (1991). *Chem. Abstr.*, **114**, 44190.
Hardcastle, F.D., Wachs, I.E., Horsley, J.A., Via, G.H. (1988). *J. Mol. Catal.*, **46**, 15.
Hardee, J.R. (1979). *Diss. Abstr., Int. B.*, **40**, 1186.
Hardee, J.R., Hightower, J.W. (1983). *J. Catal.*, **83**, 182.
Hardiman, C.J. (1992). *Chem. Abstr.*, **116**, 129877.
Hasegawa, K., Masuda, T., Higashimura, T. (1975) *Macromolecules*, **8**, 255.
Hasegawa, K. (1977a). *Europ. Polymer J.*, **13**, 47.
Hasegawa, K. (1977b). *Europ Polymer J.*, **13**, 315.
Hasegawa, S., Tanaka, T., Kudo, M., Mamada, H., Hattori, H., Yoshida, S. (1992). *Catal. Lett.*, **12**, 255.
Hattikudur, U.R., Thodos, G. (1974). *Am. Chem. Soc., Adv. Chem. Ser.*, **133**, 80.

Hawkins, J.M., Grubbs, R.H. (1988). *J. Am. Chem. Soc.*, **110**, 2821.
Heckelsberg, L.F., Banks, R.L., Bailey, G.C. (1968). *Ind. Eng. Chem., Prod. Res. Dev.*, **7**, 29.
Heckelsberg, L.F., Banks, R.L., Bailey, G.C. (1969a). *Ind. Eng. Chem., Prod. Res. Dev.*, **8**, 259.
Heckelsberg, L.F., Banks, R.L., Bailey, G.C. (1969b). *J. Catal.*, **13**, 99.
Heckelsberg, L.F. (1983). *Anal. Chem.*, **55**, 398.
Heenop, P.J., du Plessis, J.A.K. (1978). *S. Afr. J. Chem.*, **31**, 23.
Heiling, P., Wewerka, D., Hummel, K. (1976). *Kautsch. Gummi, Kunstst.*, **29**, 238.
Hein, P.R. (1973). *J. Polymer Sci., Polymer Chem.*, **11**, 163.
Henrici-Olivé, G., Olivé, S. (1973). *Angew. Chem.*, **85**, 148.
Hepworth, P. (1973). *Chem. Abstr.*, **78**, 98287.
Hérisson, J-L., Chauvin, Y. (1971). *Makromol. Chem.*, **141**, 161 (usually misquoted as 1970 because of an error in the running header in the journal).
Heroguez, V., Soum, A., Fontanille, M. (1992). *Polymer*, **33**, 3302.
Heroguez, V., Fontanille, M. (1994). *J. Polymer Sci.*, A, *Polymer Chem.*, **32**, 1755.
Heroguez, V., Gnanou, Y., Fontanille, M. (1996a). *Macromol. Rapid Commun.*, **17**, 137.
Heroguez, V., Breunig, S., Gnanou, Y., Fontanille, M. (1996b). *Macromolecules*, **29**, 4459.
Herrmann, W.A., Kuchler, J.G., Felixberger, J.K., Herdtweck, E., Wagner, W. (1988). *Angew. Chem. Int. Ed. Engl.*, **27**, 394.
Herrmann, W.A., Wagner, W., Flessner, U.N., Volkhardt, U., Komber, H. (1991). *Angew. Chem. Int. Ed. Engl.*, **30**, 1636.
Herrmann, W.A., Schattenmann, W.C., Nuyken, O., Glander, S.C. (1996). *Angew. Chem. Int. Ed. Engl.*, **35**, 1087.
Hessen, B., Buijink, J-K.F., Meetsma, A., Teuben, J.H., Helgesson, G., Hakansson, M., Jagner, S., Spek, A.L. (1993). *Organometallics*, **12**, 2268.
Heveling, J. (1987). *J. Chem. Soc., Chem. Commun.*, 1152.
Heveling, J. (1990). *J. Mol. Catal.*, **58**, 1.
Hezhong, L., Qunxing, S., Decai, L., Huaxian, Y., Minquan, Z. (1992). *J. Mol. Catal.*, **76**, 327.
Hietala, J., Root, A., Knuuttila, P. (1994). *J. Catal.*, **150**, 46.
Higashimura, T., Deng, Y-X., Masuda, T. (1982). *Macromolecules*, **15**, 234.
Higashimura, T., Masuda, T., Okada, M. (1983). *Polymer Bull.*, **10**, 114.
Hill, A.F. (1991). *J. Mol. Catal.*, **65**, 85.
Hillmyer, M.A., Lepetit, C., McGrath, D.V., Grubbs, R.H. (1991). *Am. Chem. Soc., Polymer Prepr.*, **32**(1), 162.
Hillmyer, M.A., Lepetit, C., McGrath, D.V., Novak, B.M., Grubbs, R.H. (1992). *Macromolecules*, **25**, 3345.
Hillmyer, M.A., Grubbs, R.H. (1993). *Macromolecules*, **26**, 872.
Hillmyer, M.A., Benedicto, A.D., Nguyen, S.B., Wu, Z., Grubbs, R.H. (1995a). *Macromol. Symp.*, **89**, 411.
Hillmyer, M.A., Laredo, W.R., Grubbs, R.H. (1995b). *Macromolecules*, **28**, 6311.
Hillmyer, M.A., Grubbs, R.H. (1995c). *Macromolecules*, **28**, 8662.
Hiraki, K., Kuroiwa, A., Hirai, H. (1971). *J. Polymer Sci.*, A-1, **9**, 2323.
Hirano, M., Hirai, M., Ito, Y., Fukuoka, A., Komiya, S. (1994). *Chem. Lett.*, 165.
Hirata, M., Washiyama, J., Kawauchi, S., Nakamura, J. (1989). *Chem. Abstr.*, **111**, 59176.
Ho, H.T., Ivin, K.J., Rooney, J.J. (1982a). *J. Mol. Catal.*, **15**, 245.
Ho, H.T., Ivin, K.J., Rooney, J.J. (1982b). *Makromol. Chem.*, **183**, 1629.
Ho, H.T., Reddy, B.S.R., Rooney, J.J. (1982c). *J. Chem. Soc., Faraday Trans.,1*, **78**, 3307.
Ho, H.T., Ivin, K.J., Rooney, J.J. (1982d). *J. Chem. Soc., Faraday Trans.,1*, **78**, 2227.
Ho, H.T., Ivin, K.J., Reddy, B.S.R., Rooney, J.J. (1989). *Europ. Polymer J.*, **25**, 805.

Ho, T.H., Katz, T.J. (1985). *J. Mol. Catal.*, **28**, 359.
Höcker, H., Müsch, R. (1972). *Makromol. Chem.*, **157**, 201.
Höcker, H., Müsch, R. (1974). *Makromol. Chem.*, **175**, 1395.
Höcker, H., Reimann, W., Riebel, K., Szentivanyi, Z. (1976). *Makromol. Chem.*, **177**, 1707.
Höcker, H., Reif, L., Reimann, W., Riebel, K. (1977). *Rec. Trav. Chim. Pays-Bas*, **96**, M47.
Höcker, H., Reimann, W., Reif, L., Riebel, K. (1980). *J. Mol. Catal.*, **8**, 191.
Höcker, H., Reif, L., Thu, C.T. (1984). *Makromol. Chem., Suppl.*, **6**, 331.
Höcker, H. (1991). *J. Mol. Catal.*, **65**, 95.
Höcker, H. (1993). *J. Macromol. Sci., Pure Appl. Chem.*, **A30**, 595.
Hocks, L. (1975a). *Bull. Soc. Chim. France*, 1893.
Hocks, L., Berck, D., Hubert, A.J., Teyssié, P. (1975b). *J. Polymer Sci., Polymer Lett.*, **13**, 391.
Holder, S., Blechert, S. (1996). *SYNLETT*, 505.
Holtrup, W., Hammel, R. (1985a). *Chem.-Ztg.*, **109**, 267.
Holtrup, W., Streck, R., Zaar, W., Zerpner, D. (1985b). *Chem.-Ztg.* **109**, 435.
Holtrup, W., Streck, R., Zaar, W., Zerpner, D. (1986). *J. Mol. Catal.*, **36**, 127.
Houri, A.F., Xu, Z., Cogan, D.A., Hoveyda, A.H. (1995). *J. Am. Chem. Soc.*, **117**, 2943.
Howard, T.R., Lee, J.B., Grubbs, R.H. (1980a). *J. Am. Chem. Soc.*, **102**, 6876.
Howard, T.R. (1980b). *Diss. Abstr., Int. B*, **41**, 570.
Howe, R.F., Kemball, C. (1974). *J. Chem. Soc., Faraday Trans.,1*, **70**, 1153.
Hubmann, E., Loy, W.A., Pongratz, T., Hummel, K. (1987). *Makromol. Chem.*, **188**, 2481.
Hubmann, E., Hummel, K., Dösinger, M., Pongratz, T., Saf, R. (1990). *Makromol. Chem.*, **191**, 1799.
Hughes, W.B. (1969). *J. Chem. Soc., Chem. Commun.*, 431.
Hughes, W.B. (1970a). *J. Am. Chem. Soc.*, **92**, 532.
Hughes, W.B., Zuech, E.A. (1970b). *Chem. Abstr.*, **72**, 99985.
Hughes, W.B. (1974). *Am. Chem. Soc., Adv. Chem. Ser.*, **132**, 192.
Hughes, W.B. (1975). *CHEMTECH*, **5**, 486.
Hughes, W.B. (1977). *Ann. New York Acad. Sci.*, **295**, 271.
Hummel, K., Streck, R., Weber, H. (1970). *Naturwiss.*, **57**, 194.
Hummel, K., Imamoglu, Y. (1971). *Kautsch. Gummi, Kunstst.*, **24**, 383.
Hummel, K., Ast, W. (1973a). *Makromol. Chem.*, **166**, 39.
Hummel, K., Werwerka, D., Lorber, F., Zeplichal, G. (1973b). *Makromol. Chem.*, **166**, 45.
Hummel, K., Dürr, A. (1974). *Kolloid Z., Z. Polymere*, **252**, 928.
Hummel, K., Imamoglu, Y. (1975a). *Coll. Polymer Sci.*, **253**, 225.
Hummel, K., Ast, W. (1975b). *Coll. Polymer Sci.*, **253**, 474.
Hummel, K., Raithofer, G. (1976). *Angew. Makromol. Chem.*, **50**, 183.
Hummel, K., Demel, H., Werwerka, D. (1977a). *Makromol. Chem.*, **178**, 19.
Hummel, K., Kathan, W., Kovar, I., Wedam, O.A. (1977b). *Kautsch. Gummi, Kunstst.*, **30**, 7.
Hummel, K., Stelzer, F., Kathan, W., Demel, H., Wedam, O.A., Karoulis, C. (1977c). *Rec. Trav. Chim. Pays-Bas*, **96**, M75.
Hummel, K., Wedam, O., Kathan, W., Demel, H. (1978). *Makromol. Chem.*, **179**, 1159.
Hummel, K., Heiling, P., Karoulis, C., Kathan, W., Stelzer, F. (1980a). *Makromol. Chem.*, **181**, 1847.
Hummel, K., Stelzer, F., Heiling, P., Wedam, O.A., Griesser, H. (1980b). *J. Mol. Catal.*, **8**, 253.
Hummel, K., Wedam, O.A. (1981). *Makromol. Chem.*, **182**, 3041.
Hummel, K. (1982a). *Pure Appl. Chem.*, **54**, 351.

Hummel, K., Groyer, S., Lechner, H. (1982b). *Kautsch. Gummi, Kunstst.*, **35**, 731.
Hummel, K. Chemelli, R., Griesser, H., Kumar, V.N.G., Stelzer, F. (1983a). *Makromol. Chem., Rapid Commun.*, **4**, 429.
Hummel, K., Kathan, W. (1983b). *Makromol. Chem.*, **184**, 1561.
Hummel, K., Martl, M.G., Chemelli, R., Griesser, H., Wukovnig, S., Zekoll, H. (1984). *Makromol. Chem.*, **185**, 2489.
Hummel, K. (1985). *J. Mol. Catal.*, **28**, 381.
Hummel, K., Lechner, H., Pongratz, T. (1986). *J. Mol. Catal.*, **36**, 177.
Hummel, K., Lechner, H., Zekoll, H., Chemelli, R. (1987a). *Makromol. Chem.*, **188**, 1075.
Hummel, K., Stelzer, F., Hobisch, G., Hartmann, B. (1987b). *Angew. Makromol. Chem.*, **155**, 143.
Hummel, K., Hubmann, E., Pongratz, T. (1988). *Europ. Polymer J.*, **24**, 141.
Hummel, K. (1990a). *NATO ASI Ser.*, **C326**, 209.
Hummel, K., Hubmann, E. (1990b). *Makromol. Chem., Macromol. Symp.*, **40**, 139.
Hummel, K., Paschen, A. (1990c). *Polymer Bull.*, **24**, 391.
Hummel, K., Hubmann, E., Dösinger, M., Pongratz, T. (1991). *Europ. Polymer J.*, **27**, 303.
Hummel, K., Kiattanavith, N., Martl, K., Martl, C. (1992). *J. Mol. Catal.*, **76**, 239.
Hummel, K., Kiattanavith, N., Bernard, E. (1993a). *Angew. Makromol. Chem.*, **207**, 137.
Hummel, K. (1993b). *J. Macromol. Sci.-Pure Appl. Chem.*, **A30**, 621.
Huwe, C.M., Blechert, S. (1995). *Tetrahedron Lett.*, **36**, 1621.
Huwe, C.M., Kiehl, O.C., Blechert, S. (1996). *SYNLETT*, 65.
Ichikawa, K., Fukuzumi, K. (1976a). *J. Org. Chem.*, **41**, 2633.
Ichikawa, K., Fukuzumi, K. (1976b). *Yukagaku*, **25**, 779.
Ichikawa, K., Takagi, T., Fukuzumi, K. (1976c). *Transition Met. Chem. (Weinheim)*, **1**, 54.
Ichikawa, K., Takagi, T., Fukuzumi, K. (1976d). *Bull. Chem. Soc. Jpn.*, **49**, 750.
Ichikawa, K., Watanabe, O., Takagi, T., Fukuzumi, K. (1976e). *J. Catal.*, **44**, 416.
Ichikawa, K., Watanabe, O., Fukuzumi, K. (1976f). *Transition Met. Chem. (Weinheim)*, **1**, 183.
Ichikawa, M., Zhang, Q., Li, G.-J., Tanaka, K., Fujimoto, T., Fukuoka, A. (1993). In '*New Frontiers in Catalysis*' (eds L. Guczi, F. Solymosi, P. Tétényi), p.529. Elsevier Science, London.
Ichinose, H., Iwasawa, Y., Ogasawa, S. (1978). *Chem. Abstr.*, **89**, 169664.
Ignatov, V.M., Belov, P.S., Usacheva, G.A., Bochko, R.A. (1986). *Petrol. Chem. USSR*, **26**, 28.
Iio, A., Ohira, Y., Kayakawa, Y., Oka, H. (1991). *Chem. Abstr.*, **115**, 280872.
Ikariya, T., Ho, S.C.H., Grubbs, R.H. (1985). *Organometallics*, **4**, 199.
Ikeda, H., Matsumoto, S., Enyo, H. (1977a). *Chem. Abstr.*, **86**, 56195.
Ikeda, H., Matsumoto, S., Enyo, H. (1977b). *Chem. Abstr.*, **87**, 153199.
Imaizumi, F., Suzuki, K., Kawakami, M. (1977). *Chem. Abstr.*, **87**, 23950.
Imaizumi, F., Nakata, K., Ikeda, H. (1979). *Chem. Abstr.*, **90**, 152837.
Imamoglu, Y., Zümreoglu, B., Amass, A.J. (1986). *J. Mol. Catal.*, **36**, 107.
Imamoglu, Y., Zümreoglu-Karan, B., Amass, A.J. (1990). *NATO ASI Ser.*, **C326**, 1.
Imamura, H., Koda, K., Miura, H., Tsuchiya, S. (1989). *Chem. Lett.*, 171.
Indovina, V., Cimino, A., Cordischi, S., Della Bella, S., De Rossi, S., Ferraris, G., Gazzoli, D., Occhiuzzi, M., Valigi, M. (1993). In '*New Frontiers in Catalysis*' (eds L. Guczi, F. Solymosi, P. Tétényi), p.875. Elsevier Science, London.
Innes, R.A., Sabourin, E.T., Swift, H.E. (1979). *Am. Chem. Soc., Petrol. Chem. Prepr.*, **24**(4), 1065.
Innes, R.A., Swift, H.A. (1981). *CHEMTECH*, **11**, 244.
Isagulyants, G.V., Klimov, A.P., Kovalenko, L.I., Kogan, V.M. (1986). *Izv. Akad. Nauk SSSR, Ser. Khim.*, **35**, 2805.

Ismayel-Milanovic, A., Basset, J.M., Praliaud, H., Dufaux, M., de Mourgues, L. (1973). *J. Catal.*, **31**, 408.
Isobe, E., Masuda, T., Higashimura, T., Yamamoto, A. (1986). *J. Polymer Sci.*, A, *Polymer Chem.*, **24**, 1839.
Ivin, K.J. (1974). In '*Reactivity, Mechanism and Structure in Polymer Chemistry*' (eds A.D. Jenkins, A.D. Ledwith), Ch. 16. Wiley/Interscience, London.
Ivin, K.J., Laverty, D.T., Rooney, J.J. (1977a). *Makromol. Chem.*, **178**, 1545.
Ivin, K.J., Laverty, D.T., Rooney, J.J. (1977b). *Rec. Trav. Chim. Pays-Bas*, **96**, M54.
Ivin, K.J., Laverty, D.T., Rooney, J.J. (1978a). *Makromol. Chem.*, **179**, 253.
Ivin, K.J., Lillie, S., Rooney, J.J. (1978b). *Makromol. Chem.*, **179**, 2787.
Ivin, K.J., Rooney, J.J., Stewart, C.D. (1978c). *J. Chem. Soc., Chem. Commun.*, 604.
Ivin, K.J. (1979a). In '*Proc. 5th Europ. Symp. on Polymer Spectroscopy*' (ed D.O. Hummel), p.259. Verlag Chemie, Weinheim.
Ivin, K.J., Lapienis, G., Rooney, J.J. (1979b). *J. Chem. Soc., Chem. Commun.*, 1068.
Ivin, K.J., Laverty, D.T., O'Donnell, J.H., Rooney, J.J., Stewart, C.D. (1979c). *Makromol. Chem.*, **180**, 1989.
Ivin, K.J., O'Donnell, J.H., Rooney, J.J., Stewart, C.D. (1979d). *Makromol. Chem.*, **180**, 1975.
Ivin, K.J., Laverty, D.T., Reddy, B.S.R., Rooney, J.J. (1980a). *Makromol. Chem., Rapid Commun.*, **1**, 467.
Ivin, K.J., Lapienis, G., Rooney, J.J. (1980b). *Polymer*, **21**, 436.
Ivin, K.J., Lam, L.M., Rooney, J.J. (1981a). *Makromol. Chem.*, **182**, 1847.
Ivin, K.J., Reddy, B.S.R., Rooney, J.J. (1981b). *J. Chem. Soc., Chem. Commun.*, 1062.
Ivin, K.J., Rooney, J.J., Bencze, L., Hamilton, J.G., Lam, L.M., Lapienis, G., Reddy, B.S.R., Ho, H.T. (1982a). *Pure Appl. Chem.*, **54**, 447.
Ivin, K.J., Lapienis, G., Rooney, J.J. (1982b). *Makromol. Chem.*, **183**, 9.
Ivin, K.J. (1983). '*Olefin Metathesis*', Academic, London.
Ivin, K.J., Milligan, B.D. (1987). *Makromol. Chem., Rapid Commun.*, **8**, 269.
Ivin, K.J., Kress, J., Osborn, J.A. (1988). *J. Mol. Catal.*, **46**, 351.
Ivin, K.J., (1990). *NATO ASI Ser.*, **C326**, 187.
Ivin, K.J. (1991). *Makromol. Chem., Macromol. Symp.*, **42/43**, 1.
Ivin, K.J., Kress, J., Osborn, J.A. (1992). *Makromol. Chem.*, **193**, 1695.
Ivin, K.J., Lam, L-M., Rooney, J.J. (1993a). *Makromol. Chem.*, **194**, 3203.
Ivin, K.J., Lam, L-M., Rooney, J.J. (1993b). *Makromol. Chem.*, **194**, 3493.
Ivin, K.J., Lam, L-M., Rooney, J.J. (1994a). *Makromol. Chem. Phys.*, **195**, 1189.
Ivin, K.J., Lam, L-M., Rooney, J.J. (1994b). *Makromol. Chem. Phys.*, **195**, 3245.
Ivin, K.J. (1995). ISOM 11 Abstract L13.
Iwasawa, Y., Ogasawara, S., Soma, M. (1978). *Chem. Lett.*, 1039.
Iwasawa, Y., Kubo, H., Yamagishi, M. (1980). *Chem. Lett.*, 1165.
Iwasawa, Y., Ichinose, H., Ogasawara, S., Soma, M. (1981). *J. Chem. Soc., Faraday Trans.,1*, **77**, 1763.
Iwasawa, Y., Hamamura, H. (1983). *J. Chem. Soc., Chem. Commun.*, 130.
Iwasawa, Y., Kubo, H., Hamamura, H. (1985). *J. Mol. Catal.*, **28**, 191.
Jacobson, H., Stockmayer, W.H. (1950). *J. Chem. Phys.*, **18**, 1600.
Janda, D.J. (1990). *Chem. Abstr.*, **112**, 159970.
Jang, M-S., Kwon, S-K., Choi, S-K. (1990). *Macromolecules*, **23**, 4135.
Jin, S-H., Kim, S-H., Cho, H-N., Choi, S-K. (1991). *Macromolecules*, **24**, 6050.
Jin, S-H., Cho, H-N., Choi, S-K. (1993a). *J. Polymer Sci.*, A, *Polymer Chem.*, **31**, 69.
Jin, S-H., Choi, S-J. Ahn, W., Cho, H-N., Choi, S-K. (1993b). *Macromolecules*, **26**, 1487.
Johns, J.R., Howe, R.F. (1988). *Study. Surf. Sci. Catal.*, **36**, 615.
Johnson, L.K., Virgil, S.C., Grubbs, R.H., Ziller, J.W. (1990). *J. Am. Chem. Soc.*, **112**, 5384.

Johnson, L.K., Frey, M., Ulibarri, T.A., Virgil, S.C., Grubbs, R.H., Ziller, J.W. (1993). *J. Am. Chem. Soc.*, **115**, 8167.
Johnston, J.A., Tokles, M., Hatvany, G.S., Rinaldi, P.L., Farona, M.F. (1991a). *Macromolecules*, **24**, 5532.
Johnston, J.A., Farona, M.F. (1991b). *Polymer Bull.*, **25**, 625.
Johnston, J.A., Rinaldi, P.L., Farona M.F. (1992). *J. Mol. Catal.*, **76**, 209.
Jones, C.A., Lawrence, R.A., Martens, J., Friend, R.H. (1991). *Polymer*, **32**, 1200.
Jossifov, C., Schopov, I. (1991a). *Makromol. Chem.*, **192**, 857.
Jossifov, C., Schopov, I. (1991b). *Makromol. Chem.*, **192**, 863.
Jossifov, C. (1993a). *Europ. Polymer J.*, **29**, 9.
Jossifov, C., Schopov, I. (1993b). *Europ. Polymer J.*, **29**, 621.
Jozefiak, T.H., Ginsburg, E.J., Gorman, C.B., Sailor, M.J., Grubbs, R.H., Lewis, N.S. (1991). *49th Ann. Tech. Conf. - Soc. Plast. Eng.*, 872.
Junga, H., Blechert, S. (1993). *Tetrahedron Lett.*, **34**, 3731.
Kadushin, A.A., Aliev, R.K., Krylov, O.V., Andreev, A.A., Edreva-Kardjieva, R.M., Shopov, D.M. (1982a). *Kinet. Katal.*, **23**, 276.
Kadushin, A.A., Matyshak, V.A., Kutyreva, N.A., Sklyarov, A.V., Vlasenko, A.G. (1982b). *Kinet. Katal.*, **23**, 1167.
Kanaoka, S., Grubbs, R.H. (1995). *Macromolecules*, **28**, 4707.
Kaneshiro, H., Masuda, T., Higashimura, T. (1995). *Polymer Bull.*, **35**, 17.
Kaneta, N., Hirai, T., Mori, M. (1995a). *Chem. Lett.*, 627.
Kaneta, N., Hikichi, K., Asaka, S., Uemura, M., Mori, M. (1995b). *Chem. Lett.*, 1055.
Kang, K-L., Kim, S-H., Cho, H-N., Choi, K-Y., Choi, S-K. (1993). *Macromolecules*, **26**, 4539.
Kanischka, G., Höcker, H. (1988). *Makromol. Chem.*, **189**, 1447.
Kapellen, K.K., Stadler, R. (1994). *Polymer Bull.*, **32**, 3.
Kapteijn, F., Bredt, L.H.G., Mol, J.C. (1977). *Rec. Trav. Chim. Pays-Bas*, **96**, M139.
Kapteijn, F., Bredt, L.H.G., Homburg. E., Mol, J.C. (1981). *Ind. Eng. Chem., Prod. Res. Dev.*, **20**, 457.
Kapteijn, F., Mol, J.C. (1982). *J. Chem. Soc., Faraday Trans.,1*, **78**, 2583.
Kapteijn, F., van der Steen, A.J., Mol, J.C. (1983a). *J. Chem. Thermodyn.*, **15**, 137.
Kapteijn, F., Homburg, E., Mol, J.C. (1983b). *J. Chem. Thermodyn.*, **15**, 147.
Karan, B. (Zümreoglu), Imamoglu, Y. (1989). *NATO ASI Ser.*, **C269**, 347.
Karlen, T., Ludi, A. (1994). *J. Am. Chem. Soc.*, **116**, 11375.
Karlen, T., Ludi, A., Mühlebach, A., Bernhard, P., Pharisa, C. (1995). *J. Polymer Sci.*, A, *Polymer Chem.*, **33**, 1665.
Kathan, W., Wedam, O.A., Hummel, K. (1977). *Makromol. Chem.*, **178**, 1693.
Katz, T.J., McGinnis, J. (1975). *J. Am. Chem. Soc.*, **97**, 1592.
Katz, T.J., McGinnis, J., Altus, C. (1976a). *J. Am. Chem. Soc.*, **98**, 606.
Katz, T.J., Rothchild, R. (1976b). *J. Am. Chem. Soc.*, **98**, 2519.
Katz, T.J., Lee, S.J., Acton, N. (1976c). *Tetrahedron Lett.*, 4247.
Katz, T.J., Acton, N. (1976d). *Tetrahedron Lett.*, 4251.
Katz, T.J., McGinnis, J. (1977a). *J. Am. Chem. Soc.*, **99**, 1903.
Katz, T.J., Hersh,W.H. (1977b). *Tetrahedron Lett.*, 585.
Katz, T.J., Lee, S.J. (1980a). *J. Am. Chem. Soc.*, **102**, 422.
Katz, T.J., Lee, S.J., Shippey, M.A. (1980b). *J. Mol. Catal.*, **8**, 219.
Katz, T.J., Lee, S.J., Nair, M., Savage, E.B. (1980c). *J. Am. Chem. Soc.*, **102**, 7940.
Katz, T.J., Savage, E.B., Lee, S.J., Nair, M. (1980d). *J. Am. Chem. Soc.*, **102**, 7942.
Katz, T.J., Han, C-C. (1982). *Organometallics*, **1**, 1093.
Katz, T.J., Shippey, M.A. (1983). *Chem. Abstr.*, **97**, 183067.
Katz, T.J., Ho, T.H., Shih, N-Y., Ying, Y-C., Stuart, V.I.W. (1984). *J. Am. Chem. Soc.*, **106**, 2659.

Katz, T.J., Sivavec, T.M. (1985a). *J. Am. Chem. Soc.*, **107**, 737.
Katz, T.J., Hacker, S.M., Kendrick, R.D., Yannoni, C.S. (1985b). *J. Am. Chem. Soc.*, **107**, 2182.
Katz, T.J. (1989). *NATO ASI Ser.*, **C269**, 293.
Kauffmann, T., Abeln, R., Welke, S., Wingbermühle, D. (1986). *Angew. Chem. Int. Ed. Engl.*, **25**, 909.
Kawai, T., Yamazaki, Y., Tokumura, A. (1983). *J. Jpn. Petrol. Inst.*, **26**, 332.
Kawai, T., Yamazaki, Y., Taoka, T., (1984a). *J. Catal.*, **89**, 452.
Kawai, T., Yamazaki, Y., Nishikawa, M. (1984b). *J. Jpn. Petrol. Inst.*, **27**, 378.
Kawai, T., Goto, H., Yamazaki, Y. (1986). *J. Jpn. Petrol. Inst.*, **29**, 212.
Kawai, T., Goto, H., Ishikawa, T., Yamazaki, Y. (1987). *J. Mol. Catal.*, **39**, 369.
Kawai, T., Goto, H., Yamazaki, Y., Ishikawa, T. (1988). *J. Mol. Catal.*, **46**, 157.
Kawai, T., Maruoka, N., Goke, M., Ishikawa, T. (1989). *J. Mol. Catal.*, **49**, 261.
Kawai, T., Maruoka, N., Ishikawa, T. (1990). *J. Mol. Catal.*, **60**, 209.
Kawai, T., Okada, T., Ishikawa, T. (1992). *J. Mol. Catal.*, **76**, 249.
Kawai, T., Furuki, M., Ishikawa, T. (1994). *J. Mol. Catal.*, **90**, 1.
Kazansky, V.B., Shelimov, B.N., Vikulov, K.A. (1991a). *Catal. Sci. Techn., Proc. Tokyo Conf., 1st*, 159.
Kazansky, V.B., Shelimov, B.N. (1991b). *Res. Chem. Intermed.*, **15**, 1.
Kazansky, V.B., Shelimov, B.N., Vikulov, K.A. (1993). In '*New Frontiers in Catalysis*' (eds L. Guczi, F. Solymosi, P. Tétényi), p.515. Elsevier Science, London.
Kazuta, M., Tanaka, K. (1987). *J. Chem. Soc., Chem. Commun.*, 616.
Kazuta, M., Tanaka, K. (1988). *Catal. Lett.*, **1**, 7.
Kazuta, M., Tanaka, K. (1990). *J. Catal.*, **123**, 164.
Keller, A. (1987a). *Inorg. Chim. Acta*, **133**, 207.
Keller, A. (1987b). *Trans. Met. Chem.*, **12**, 320.
Keller, A. (1989a). *J. Mol. Catal.*, **53**, L9.
Keller, A., Szterenberg, L. (1989b). *J. Mol. Catal.*, **57**, 207.
Keller, A. (1990a). *J. Mol. Catal.*, **59**, 75.
Keller, A. (1990b). *J. Organomet. Chem.*, **385**, 285.
Keller, A. (1990c). *J. Organomet. Chem.*, **393**, 389.
Keller, A. (1991a). *J. Mol. Catal.*, **64**, 171.
Keller, A. (1991b). *J. Mol. Catal.*, **70**, 143.
Keller, A. (1991c). *J. Organomet. Chem.*, **407**, 237.
Keller, A., Szterenberg, L. (1992a). *Z. Naturforsch.*, **476**, 1469.
Keller, A. (1992b). *J. Organomet. Chem.*, **436**, 199.
Kelly, W.J., Calderon, N. (1975). *J. Macromol. Sci., Chem.*, **A9**, 911.
Kerkhof, F.P.J.M., Thomas, R., Moulijn, J.A. (1977). *Rec. Trav. Chim. Pays-Bas*, **96**, M121.
Kerkhof, F.P.J.M., Moulijn, J.A., Thomas, R. (1979). *J. Catal.*, **56**, 279.
Kershenbaum, I.L., Grebenyak, L.N., Oreshkin, I.A., Dolgopolosk, B.A., Tinyakova, E.I. (1978). *Dokl. Akad. Nauk SSSR*, **238**, 359.
Khachaturov, A.S., Abramenko, Y.L., Syatkowskii, A.I. (1982). *Polymer Sci. USSR*, **24**, 775.
Khasat, N.P., Patnaik, B.K. (1989). *US Pat.* 4,835,230.
Khodzhemirov, V.A., Evdokimova, V.A., Cherednichenko, V.M. (1976a). *Vysokomol. Soedin.*, **A18**, 509.
Khodzhemirov, V.A., Moisa, V.A. (1976b). *Vysokomol. Soedin.*, **B18**, 397.
Kim, S-H., Lee, H-J., Jin, S-H., Cho, H-N., Choi, S-K. (1993). *Macromolecules*, **26**, 846.
Kim, S-H., Choi, S-J., Park, J-W., Cho, H-N., Choi, S-K. (1994a). *Macromolecules*, **27**, 2339.
Kim, S-H., Bowden, N., Grubbs, R.H. (1994b). *J. Am. Chem. Soc.*, **116**, 10801.

Kim, S-H., Zuercher, W.J., Bowden, N.B., Grubbs, R.H. (1996). *J. Org. Chem.*, **61**, 1073.
Kim, Y-H., Gal, Y-S., Kim, U-Y., Choi, S-K. (1988). *Macromolecules*, **21**, 1991.
Kim, Y-H., Choi, K-Y., Choi, S-K. (1989). *J. Polymer Sci.*, C, *Polymer Lett.*, **27**, 443.
Kim, Y-H., Kwon, S-K., Choi, S-K. (1992). *Bull. Korean Chem. Soc.*, **13**, 459.
Kinoshita, A., Mori, M. (1994). *SYNLETT*, 1020.
Kirkland, C. (1990). *Plastics World*, 50.
Kirlin, P.S., Gates, B.C. (1985a). *J. Chem. Soc., Chem. Commun.*, 277.
Kirlin, P.S., Gates, B.C. (1985b). *Inorg. Chem.*, **24**, 3914.
Kittleman, E.T., Zuech, E.A. (1970). *Chem. Abstr.*, **72**, 31193.
Klavetter, F.L., Grubbs, R.H. (1988). *J. Am. Chem. Soc.*, **110**, 7807.
Klavetter, F.L., Grubbs, R.H. (1989). *Synthetic Metals*, **28**, D105.
Klimov, O.V., Krivoshchekova, E.A., Startsev, A.N. (1990a). *React. Kinet. Catal. Lett.*, **42**, 95.
Klimov, O.V., Startsev, A.N. (1990b). *React. Kinet. Catal. Lett.*, **41**, 135.
Klimov, O.V., Krivoshchekova, E.A., Startsev, A.N. (1991). *React. Kinet. Catal. Lett.*, **43**, 19.
Klimov, O.V., Khomyakova, E.A., Startsev, A.N. (1993). *Catal. Lett.*, **19**, 189.
Klimov, O.V., Fedotov, M.A., Kochubei, D.I., Degtyarev, S.P., Kalinkin, A.V., Startsev, A.N. (1995a). *Kinet. Katal.*, **36**, 330.
Klimov, O.V., Alekseev, O.S., Startsev, A.N. (1995b). *React. Kinet. Catal. Lett.*, **56**, 143.
Klosiewicz, D.W. (1983). *US Pat.* 4,400,340.
Klosiewicz, D.W. (1984). *US Pat.* 4,469,809.
Klosiewicz, D.W. (1985). *US Pat.* 4,520,181.
Klosiewicz, D.W. (1987). *US Pat.* 4,657,981.
Kmínek, I., Cimrová, V., Nespurek, S. (1989). *Makromol. Chem.*, **190**, 1025.
Knoche, H. (1971). *Chem. Abstr.*, **74**, 44118.
Knoll, K., Krouse, S.A., Schrock, R.R. (1988). *J. Am. Chem. Soc.*, **110**, 4424.
Knoll, K., Schrock, R.R. (1989). *J. Am. Chem. Soc.*, **111**, 7989.
Kobayashi, M., Kamijima, T., Kobayashi, S. (1978). *Chem. Abstr.*, **88**, 38328.
Kobayashi, S., Saegusa, T., Furukawa, K. (1967). *Kogyo Kagaku Zasshi*, **70**, 372.
Kobayashi, Y., Ueshima, T., Tanaka, Y., Kurosawa, S., Kobayashi, S. (1977). *Chem. Abstr.*, **86**, 156245.
Kobayashi, Y., Jio, R., Ueshima, T. (1979). *Chem. Abstr.*, **91**, 175962.
Kobylinski, T.P., Swift, H.E. (1972). *J. Catal.*, **26**, 416.
Kodemura, J., Natsuume, T. (1995). *Polymer J.*, **27**, 1167.
Kohashi, H., Foglia, T. (1985). *J. Am. Oil Chem. Soc.*, **62**, 549.
Kolesnikov, S.P., Povarova, N.I. (1979). *Izv. Akad. Nauk SSSR, Ser. Khim.*, **28**, 2398.
Kolesnikov, S.P., Mitenina, T.L., Radzinskii, S.A., Sheinker, A.P., Abkin, A.D., Nefedov, O.M. (1981). *Izv. Akad. Nauk SSSR, Ser. Khim.*, **30**, 2303.
Kolesnikov, S.P., Mitenina, T.L., Shteinschneider, A.Y., Nefedov, O.M. (1982). *Izv. Akad. Nauk SSSR, Ser. Khim.*, **31**, 942.
Komatsu, K., Matsumoto, S., Aotani, S. (1977). *Chem. Abstr.*, **86**, 56201.
Komatsu, T., Namba, S., Yashima, S. (1985). *Acta Phys. Chem.*, **31**, 251.
Komatsu, T., Yashima, T. (1987). *J. Mol. Catal.*, **40**, 83.
Komiya, T., Hisatomi, M., Iio, A., Oka, H. (1991). *Chem. Abstr.*, **115**, 72508.
Komiya, T., Iio, A., Oka, H. (1992a). *Chem. Abstr.*, **117**, 91085.
Komiya, T., Hisatomi, M., Iio, A., Oka, H. (1992b). *Chem. Abstr.*, **117**, 112260.
Komiya, Z., Pugh, C., Schrock, R.R. (1992a). *Macromolecules*, **25**, 3609.
Komiya, Z., Pugh, C., Schrock, R.R. (1992b). *Macromolecules*, **25**, 6586.
Komiya, Z., Schrock, R.R. (1993a). *Macromolecules*, **26**, 1387.
Komiya, Z., Schrock, R.R. (1993b). *Macromolecules*, **26**, 1393.
Konzelman, J., Wagener, K.B. (1995). *Macromolecules*, **28**, 4686.

Koo, K-M., Han, S-H., Kang, Y-S., Kim, U-Y., Choi, S-K. (1993). *Macromolecules*, **26**, 2485.
Korda, A., Giezynski, R., Krycinski, S. (1980). *J. Mol. Catal.*, **9**, 51.
Korda, A., Giezynski, R. (1982). *Polish J. Chem.*, **56**, 849.
Kormer, V.A., Poletayeva, I.A., Yufa, T.L. (1972). *J. Polymer Sci.*, A-1, **10**, 251.
Kormer, V.A., Dolinskaya, E.R., Khatchaturov, A.S. (1980). *Makromol. Chem., Rapid Commun.*, **1**, 531.
Korshak, Y.V., Tlenkopachev, M.A., Timofeeva, G.I., Pavlova, S.A., Dolgoplosk, B.A. (1976). *Dokl. Akad. Nauk SSSR*, **226**, 1344.
Korshak, Y.V., Dolgoplosk, B.A., Tlenkopachev, M.A. (1977). *Rec. Trav. Chim. Pays-Bas*, **96**, M64.
Korshak, Y.V., Berlin, A.A., Badamshina, E.R., Timofeeva, G.I., Pavlova, S.A. (1979). *Dokl. Akad. Nauk SSSR*, **248**, 372.
Korshak, Y.V., Turov, B.S., Vardanyan, L.M., Yefimov, V.A., Tlenkopachev, M.A., Koshevnik, A.Y., Dolgoplosk, B.A. (1980). *Vysokomol. Soedin.*, **A22**, 781.
Korshak, Y.V., Tlenkopatchev, M.A., Dolgoplosk, B.A., Avdeikina, E.G., Kutepov, D.F. (1982). *J. Mol. Catal.*, **15**, 207.
Korshak, Y.V., Korshak, V.V., Kanischka, G., Höcker, H. (1985). *Makromol. Chem., Rapid Commun.*, **6**, 685.
Kosaka, N., Sakai, Y., Tsuda, N. (1986). *J. Catal.*, **98**, 95.
Koutsantonis, G.A., Selegue, J.P. (1991). *J. Am. Chem. Soc.*, **113**, 2316.
Kouzai, H., Masuda, T., Higashimura, T. (1993). *Makromol. Chem.*, **194**, 3195.
Kouzai, H., Masuda, T., Higashimura, T. (1994a). *J. Polymer Sci.*, A, *Polymer Chem.*, **32**, 2523.
Kouzai, H., Masuda, T., Higashimura, T. (1994b). *Polymer*, **35**, 4920.
Kouzai, H., Masuda, T., Higashimura, T. (1994c). *Polymer Bull.*, **33**, 301.
Kouzai, H., Masuda, T., Higashimura, T. (1995). *Bull. Chem. Soc. Jpn.*, **68**, 398.
Kranz, D., Beck, M. (1972). *Angew. Makromol. Chem.*, **27**, 29.
Krausz, P., Garnier, F., Dubois, J.E. (1978). *J. Organomet. Chem.*, **161**, 97.
Kress, J., Russell, M.J.M., Wesolek, M.G., Osborn, J.A. (1980). *J. Chem. Soc., Chem. Commun.*, 431.
Kress, J., Wesolek, M., Le Ny, J.P., Osborn, J.A. (1981). *J. Chem. Soc., Chem. Commun.*, 1039.
Kress, J., Wesolek, M., Osborn, J.A. (1982). *J. Chem. Soc., Chem. Commun.*, 514.
Kress, J., Osborn, J.A. (1983). *J. Am. Chem. Soc.*, **105**, 6346.
Kress, J., Osborn, J.A., Greene, R.M.E., Ivin, K.J., Rooney, J.J. (1985). *J. Chem. Soc., Chem. Commun.*, 874.
Kress, J., Osborn, J.A. (1987a). *J. Am. Chem. Soc.*, **109**, 3953.
Kress, J., Osborn, J.A., Greene, R.M.E., Ivin, K.J., Rooney, J.J. (1987b). *J. Am. Chem. Soc.*, **109**, 899.
Kress, J., Osborn, J.A., Ivin, K.J., Rooney, J.J. (1987c). *NATO ASI Ser.*, **C215**, 363.
Kress, J., Osborn, J.A., Amir-Ebrahimi, V., Ivin, K.J., Rooney, J.J. (1988). *J. Chem. Soc., Chem. Commun.*, 1164.
Kress, J., Osborn, J.A., Ivin, K.J. (1989). *J. Chem. Soc., Chem. Commun.*, 1234.
Kress, J., Ivin, K.J., Amir-Ebrahimi, V., Weber, P. (1990). *Makromol. Chem.*, **191**, 2237.
Kress, J., Osborn, J.A. (1992). *Angew. Chem. Int. Ed. Engl.*, **31**, 1585.
Kress, J. (1995). *J. Mol. Catal.*, A, *Chem.*, **102**, 7.
Kroll, W.R., Doyle, G. (1971). *J. Chem. Soc., Chem. Commun.*, 839.
Kroll, W.R., Doyle, G. (1972). *J. Catal.*, **24**, 356.
Kropacheva, E.N., Sterenzat, D.E., Patrushin, Y.A., Dolgoplosk, B.A. (1972). *Dokl. Akad. Nauk SSSR*, **206**, 878.
Krouse, S.A., Schrock, R.R., Cohen, R.E. (1987). *Macromolecules*, **20**, 903.

Krouse, S.A., Schrock, R.R. (1988). *Macromolecules*, **21**, 1885.
Krouse, S.A., Schrock, R.R. (1989). *Macromolecules*, **22**, 2569.
Kubicek, D.H., Zuech, E.A. (1972). *Chem. Abstr.*, **77**, 100716.
Kumar, A., Eichinger, B.E. (1992). *Makromol. Chem., Rapid Commun.*, **13**, 311.
Kumar, V.N.G., Hummel, K., Hönig, H. (1981). *Angew. Makromol. Chem.*, **96**, 93.
Kumar, V.N.G., Hummel, K. (1982). *Angew. Makromol. Chem.*, **102**, 167.
Kumar, V.G., Hummel, K. (1983). *J. Polymer Sci., Polymer Chem.*, **21**, 1183.
Kumobayashi, H., Ogura, T., Akutagawa, S., Saito, K., Yamaguchi, T., Tanabe, K. (1976). *Chem. Lett.*, 317.
Kunzler, J., Percec, V. (1987a). *Am. Chem. Soc., Polymer Prepr.*, **28**(2), 252.
Kunzler, J., Percec, V. (1987b). *Polymer Bull.*, **18**, 303.
Kunzler, J., Percec, V. (1988a). *Am. Chem. Soc., Polymer Prepr.*, **29**(1), 80.
Kunzler, J., Percec, V. (1988b). *Am. Chem. Soc., Polymer Prepr.*, **29**(1), 219.
Kunzler, J., Percec, V. (1990a). *J. Polymer Sci.*, A, *Polymer Chem.*, **28**, 1043.
Kunzler, J., Percec, V. (1990b). *J. Polymer Sci.*, A, *Polymer Chem.*, **28**, 1221.
Küpper, F.W., Streck, R. (1973). *J. Organomet. Chem.*, **55**, C75.
Küpper, F.W., Streck, R. (1974). *Makromol. Chem.*, **175**, 2055.
Küpper, F.W., Streck, R. (1975). *Chem. Ztg.*, **99**, 464.
Küpper, F.W., Streck, R. (1976). *Z. Naturforsch.*, **B31**, 1256.
Küpper, F.W. (1979). *Angew. Makromol. Chem.*, **80**, 207.
Kurosawa, S., Ueshima, T., Tanaka, Y., Kobayashi, S. (1976a). *Chem. Abstr.*, **84**, 122565.
Kurosawa, S., Arioka, M., Ueshima, T., Takashi, S. (1976b). *Chem. Abstr.*, **85**, 78679.
Kuteinikov, V.M., Korshak, Y.V., Dolgoplosk, B.A. (1975). *Trans. Mosk. Khim. Tekhnol. Inst.*, **86**, 117.
Kuteinikov, V.M., Korshak, Y.V., Dolgoplosk, B.A. (1976). *Vysokomol. Soedin.*, **B18**, 363.
Kuznetsov, B.N., Startsev, A.N., Yermakov, Y.I. (1975). *React. Kinet. Catal. Lett.*, **2**, 29.
Kuznetsov, B.N., Startsev, A.N., Yermakov, Y.I. (1980). *J. Mol. Catal.*, **8**, 135.
Lahouste, J., Lemattre, M., Muller, J.C., Stern, C. (1976). *Chem. Abstr.*, **84**, 122568.
Lal, J., Smith, R.R., O'Connor, J.M. (1972). *Am. Chem. Soc., Polymer Prepr.*, **13**, 914.
Lal, J., Smith, R.R. (1975). *J. Org. Chem.*, **40**, 775.
Lambert, R.M., Ormerod, R.M., Tysoe, W.T. (1994). *Langmuir*, **10**, 730.
Lammens, H., Sartori, G., Siffert, J., Sprecher, N. (1971). *J. Polymer Sci., Polymer Lett.*, **9**, 341.
Landon, S.J., Shulman, P.M., Geoffroy, G.L. (1985). *J. Am. Chem. Soc.*, **107**, 6739.
LaPointe, A.M., Schrock, R.R. (1995). *Organometallics*, **14**, 1875.
Larroche, C., Laval, J.P., Lattes, A., Leconte, M., Quignard, F., Basset, J.M. (1982). *J. Org. Chem.*, **47**, 2019.
Larroche, C., Laval, J.P., Lattes, A., Leconte, M., Quignard, F., Basset, J.M. (1983). *J. Chem. Soc., Chem. Commun.*, 220.
Larroche, C., Laval, J.P., Lattes, A., Basset, J.M. (1984). *J. Org. Chem.*, **49**, 1886.
Laschewsky, A., Schulz-Hanke, W. (1993). *Makromol. Chem., Rapid Commun.*, **14**, 683.
Latesky, S.L., Selegue, J.P. (1987). *J. Am. Chem. Soc.*, **109**, 4731.
Latham, I.A., Sita, L.R., Schrock, R.R. (1986). *Organometallics*, **5**, 1508.
Lautens, M., Abd-el-Aziz, A.S., Reibel, J. (1989). *Macromolecules*, **22**, 4132.
Lautens, M., Abd-el-Aziz, A.S., Schmidt, G. (1990). *Macromolecules*, **23**, 2819.
Lautens, M., Crudden, C.M., Abd-el-Aziz, A.S. (1991). *Macromolecules*, **24**, 1425.
Lautens, M., Abd-el-Aziz, A.S., Crudden, C.M. (1993). *J. Polymer Sci.*, A, *Polymer Chem.*, **31**, 569.
Laval, J.P., Lattes, A., Mutin, R., Basset, J.M. (1977). *J. Chem. Soc., Chem. Commun.*, 502.

Laverty, D.T., McKervey, M.A., Rooney, J.J., Stewart, A. (1976a). *J. Chem. Soc., Chem. Commun.*, 193.

Laverty, D.T., Rooney, J.J., Stewart, A. (1976b). *J. Catal.*, **45**, 110.

Laverty, D.T., Rooney, J.J. (1983). *J. Chem. Soc., Faraday Trans.,1*, **79**, 869.

Lawrenson, M.J. (1984). *UK Pat. Appl.* 2,131,429.

Leach, D.R. (1984). *US Pat.* 4,458,037.

Leach, D.R. (1986). *US Pat.* 4,598,102.

Lebedev, B.V., Lityagov, V.Y. (1977a). *Vysokomol. Soedin.*, **B19**, 558.

Lebedev, B.V., Lityagov, V.Y. (1977b). *Vysokomol. Soedin.*, **B19**, 824.

Lebedev, B., Smirnova, N., Kulagina, T.G. (1988). *Makromol. Chem., Rapid Commun.*, **9**, 781.

Lebedev, B., Smirnova, N., Kiparisova, Y., Makovetsky, K. (1992). *Makromol. Chem.*, **193**, 1399.

Lebedev, B., Smirnova, N. (1994). *Macromol. Chem. Phys.*, **195**, 35.

Leconte, M., Bilhou, J.L., Reimann, W., Basset, J.M. (1978). *J. Chem. Soc., Chem. Commun.*, 341.

Leconte, M., Basset, J.M. (1979a). *Nouv. J. Chim.*, **3**, 429.

Leconte, M., Basset, J.M. (1979b). *J. Am. Chem. Soc.*, **101**, 7296.

Leconte, M., Ben Taarit, Y., Bilhou, J.L., Basset, J.M. (1980). *J. Mol. Catal.*, **8**, 263.

Leconte, M., Jourdan, I., Pagano, S., Lefebvre, F., Basset, J-M. (1995a). *J. Chem. Soc., Chem. Commun.*, 857.

Leconte, M., Pagano, S., Mutch, A., Lefebvre, F., Basset, J.M. (1995b). *Bull. Soc. Chim. Fr.*, **132**, 1069.

Lee, H.J., Shim, S.C. (1993). *J. Chem. Soc., Chem. Commun.*, 1420.

Lee, H.J., Shim, S.C. (1994a). *J. Polymer Sci.*, A, *Polymer Chem.*, **32**, 2437.

Lee, H-J., Oh, J-M., Choi, S-J., Kim, H-K., Choi, S-K. (1994b). *Polymer Bull.*, **32**, 433.

Lee, H-J., Gal, Y-S., Lee, W-C., Oh, J-M., Jin, S-H., Choi, S-K. (1995a). *Macromolecules*, **28**, 1208.

Lee, H-J., Choi, S-K., Gal, Y-S. (1995b). *J. Macromol. Sci.-Pure Appl. Chem.*, **A32**, 1863.

Lee, J.B., Gajda, G.J., Schaefer, W.P., Howard, T.R. (1981). *J. Am. Chem. Soc.*, **103**, 7358.

Lee, J.B., Ott, K.C., Grubbs, R.H. (1982). *J. Am. Chem. Soc.*, **104**, 7491.

Lee, J-H., Park, J-W., Ko, J.M., Chang, Y-H., Choi, S-K. (1993). *Polymer Bull.*, **31**, 339.

Lee, J.H., Park, J.W., Oh, J.M., Choi, S.K. (1995). *Macromolecules*, **28**, 377.

Lee, S.J., McGinnis, J., Katz, T.J. (1976). *J. Am. Chem. Soc.*, **98**, 7818.

Lee, W.C., Sohn, J.E., Gal, Y.S., Choi, S.W. (1988). *Bull. Korean Chem. Soc.*, **9**, 328.

Lefebvre, F., Leconte, M., Pagano, S., Mutch, A., Basset, J-M. (1995). *Polyhedron*, **14**, 3209.

Lehnert, G., Maertens, D., Pampus, G., Zimmermann, M. (1974). *Makromol. Chem.*, **175**, 2617.

Leising, G. (1984). *Polymer Bull.*, **11**, 401.

Lequan, M., Lequan, R-M., Villemin, D. (1984). *Chem. Ind.* (*London*), 379.

Levisalles, J., Rudler, H., Villemin, D. (1979). *J. Organomet. Chem.*, **164**, 251.

Levisalles, J., Rudler, H., Villemin, D. (1980a). *J. Organomet. Chem.*, **193**, 235.

Levisalles, J., Villemin, D. (1980b). *Tetrahedron*, **36**, 3181.

Levisalles, J., Rose-Munch, F., Rudler, H., Daran, J-C., Dromzée, Y., Jeannin, Y. (1981a). *J. Chem. Soc., Chem. Commun.*, 152.

Levisalles, J., Rose-Munch, F., Rudler, H., Daran, J-C., Dromzée, Y., Jeannin, Y., Ades, D. (1981b). *J. Chem. Soc., Chem. Commun.*, 1055.

Levisalles, J., Rudler, H., Cuzin, D., Rull, T., (1984). *J. Mol. Catal.*, **26**, 231.

Lewandos, G., Pettit, R. (1971a). *Tetrahedron Lett.*, 789.

Lewandos, G.S., Pettit, R. (1971b). *J. Am. Chem. Soc.*, **93**, 7087.

Lewis, M.J., Wills, G.B. (1969). *J. Catal.*, **15**, 140.
Lewis, M.J., Wills, G.B. (1971). *J. Catal.*, **20**, 182.
Leymet, I., Siove, A., Parlier, A., Rudler, H., Fontanille, M. (1989). *Makromol. Chem.*, **190**, 2397.
Liao, S., Guo, Y., Yu, S., Yu, D., Guo, H., He, B., Sun, J., Li, H. (1989). *Chem. Abstr.*, **110**, 57061.
Liaw, D.J., Soum, A., Fontanille, M., Parlier, A., Rudler, H. (1985). *Makromol. Chem., Rapid Commun.*, **6**, 309.
Liaw, D-J., Leu, S-D., Lin, C-L., Lin, C.F. (1992). *Polymer J.*, **24**, 889.
Liaw, D-J., Lin, C-L. (1993). *J. Polymer Sci.*, A, *Polymer Chem.*, **31**, 3151.
Liaw, D-J., Lin, C-L. (1995). *Polymer Int.*, **36**, 29.
Liaw, D-J., Chiang, H-H., Jin, B-H., Kang, E-T. (1996). *Europ. Polymer J.*, **32**, 215.
Lin, C.J., Aldag, A.W., Clark, A. (1976). *J. Catal.*, **45**, 287.
Lin, Z., Hall, M.B. (1994). *Organometallics*, **13**, 2878.
Lindmark-Hamberg, M., Wagener, K.B. (1987). *Macromolecules*, **20**, 2949.
Listemann, M.L., Schrock, R.R. (1985). *Organometallics*, **4**, 74.
Liu, D., Li, L., Min, L., Liu, D. (1986). *Chem. Abstr.*, **106**, 102736.
Logan, R.S., Banks, R.L. (1968a). *Hydrocarbon Process.*, **61**, 135.
Logan, R.S., Banks, R.L. (1968b). *Oil Gas J.*, **66**, 131.
Lombardo, E.A., LoJacono, M., Hall, W.K. (1978). *J. Catal.*, **51**, 243.
Lombardo, E.A., LoJacono, M., Hall, W.K. (1980). *J. Catal.*, **64**, 150.
Lorber, F., Hummel, K. (1973). *Makromol. Chem.*, **171**, 257.
Lorenz, B., Rommel, H., Wahren, M. (1982). *Z. Chem.*, **22**, 224.
Lu, S-Y., Quayle, P., Heatley, F., Booth, C., Yeates, S.G., Padget, J.C. (1992). *Macromolecules*, **25**, 2692.
Lu, S-Y., Quayle, P., Heatley, F., Booth, C., Yeates, S.G., Padget, J.C. (1993a). *Europ. Polymer J.*, **29**, 269.
Lu, S-Y., Quayle, P., Booth, C., Yeates, S.G., Padget, J.C. (1993b). *Polymer Int.*, **32**, 1.
Lu, S-Y., Amass, J.M., Majid, N., Glennon, D., Byerley, A., Heatley, F., Quayle, P., Booth, C., (1994). *Macromol. Chem. Phys.*, **195**, 1273.
Lucas, C., Soum, A., Fontanille, M. (1989). *Makromol. Chem.*, **190**, 377.
Lucas, C., Soum, A., Fontanille, M. (1993). *Makromol. Chem., Macromol. Symp.*, **66**, 297.
Luckner, R.C., McConchie, G.E., Wills, G.B. (1973a). *J. Catal.*, **28**, 63.
Luckner, R.C., Wills, G.B. (1973b). *J. Catal.*, **28**, 83.
Luyt, A.S., Vosloo, H.C.M. (1995). *J. Thermal Anal.*, **44**, 1277.
Lynn, D.M., Kanaoka, S., Grubbs, R.H. (1996). *J. Am. Chem. Soc.*, **118**, 784.
Lyons, D., MacAlpine, D.K. (1988). *Europ. Pat. Appl.* 282,313.
Maatman, R., Friesema, C. (1979). *Proc. Iowa Acad. Sci.*, **86**, 26.
Maier, M.E., Langenbacher, D., Rebien, F. (1995). *Liebigs Ann.*, 1843.
Makio, H., Masuda, T., Higashimura, T. (1993a). *Polymer*, **34**, 1490.
Makio, H., Masuda, T., Higashimura, T. (1993b). *Polymer*, **34**, 2218.
Makovetskii, K.L., Red'kina, L.I. (1976). *Dokl. Akad. Nauk SSSR*, **231**, 143.
Makovetskii, K.L., Red'kina, L.I., Oreshkin, I.A. (1981). *Izv. Akad. Nauk SSSR, Ser. Khim.*, **30**, 1928.
Makovetskii, K.L., Red'kina, L.I., Dolgoplosk, B.A. (1985). *Dokl. Akad. Nauk SSSR*, **284**, 170.
Makovetskii, K.L., Gorbacheva, L.I., Dolidze, A.V., Chediya, R.V. (1987). *Izv. Akad. Nauk SSSR, Ser. Khim.*, **36**, 1432.
Makovetsky, K.L., Gorbacheva, L.I., Ostrovskaya, I.Y., Golberg, A.I., Mikaya, A.I., Zakharian, A.A., Filatova, M.P. (1992a). *J. Mol. Catal.*, **76**, 65.
Makovetsky, K.L. *et al.* (1992b). *J. Mol. Catal.*, **76**, 107.
Makovetsky, K.L. (1994). *Polymer Sci.*, **A36**, 1433.

Maksimov, Y.V., Kushnerev, M.Y., Dumesic, J.A., Nechitailo, A.E., Fridman, R.A. (1976). *J. Catal.*, **45**, 114.
Maksimowski, P., Skupinski, W. (1991). *J. Mol. Catal.*, **65**, 187.
Marbach, A., Hupp, R. (1989). *Rubber World*, June, p.30.
March, J. (1977). In '*Advanced Organic Chemistry*' 2nd edn. p.690. McGraw-Hill, New York.
Marciniec, B., Gulinski, J. (1984). *J. Organomet. Chem.*, **266**, C19.
Marciniec, B., Rzejak, L., Gulinski, J., Foltynowicz, Z., Urbaniak, W. (1988). *J. Mol. Catal.*, **46**, 329.
Marciniec, B., Maciejewski, H., Gulinski, J., Rzejak, L. (1989). *J. Organomet. Chem.*, **362**, 273.
Marciniec, B., Pietraszuk, C. (1991). *J. Organomet. Chem.*, **412**, C1.
Marciniec, B., Pietraszuk, C., Foltynowicz, Z. (1992). *J. Mol. Catal.*, **76**, 307.
Marciniec, B., Pietraszuk, C. (1993). *J. Organomet. Chem.*, **447**, 163.
Marciniec, B., Foltynowicz, Z., Lewandowski, M. (1994a). *J. Mol. Catal.*, **90**, 125.
Marciniec, B., Foltynowicz, Z., Pietraszuk, C., Gulinski, J., Maciejewski, H. (1994b). *J. Mol. Catal.*, **90**, 213.
Marciniec, B., Pietraszuk, C., Foltynowicz, Z. (1994c). *J. Organomet. Chem.*, **474**, 83.
Marciniec, B., Lewandowski, M. (1995a). *J. Inorg. Org. Polymers*, **5**, 115.
Marciniec, B., Pietraszuk, C. (1995b). *J. Chem. Soc., Chem. Commun.*, 2003.
Marmo, J.C., Wagener, K.B. (1993). *Macromolecules*, **26**, 2137.
Marmo, J.C., Wagener, K.B. (1994). *Am. Chem. Soc., Polymer Prepr.*, **35**(1), 817.
Marmo, J.C., Wagener, K.B. (1995). *Macromolecules*, **28**, 2602.
Marshall, P.R., Ridgewell, B.J. (1969). *Europ. Polymer J.*, **5**, 29.
Martin, S.F., Liao, Y., Wong, Y., Rein, T. (1994a). *Tetrahedron Lett.*, **35**, 691.
Martin, S.F., Liao, Y., Chen, H-J., Pätzel, M., Ramser, M.N. (1994b). *Tetrahedron Lett.*, **35**, 6005.
Martin, S.F., Wagman, A.S. (1995). *Tetrahedron Lett.*, **36**, 1169.
Martin, S.F., Chen, H-J., Courtney, A.K., Liao, Y., Pätzel, M., Ramser, M.N., Wagman, A.S. (1996). *Tetrahedron*, **52**, 7251.
Martl, M.G., Zuegg, J., Hummel, K. (1991). *J. Mol. Catal.*, **65**, 127.
Martl, M.G., Hummel, K., Stelzer, F. (1998). *J. Mol. Catal.*, **46**, 373.
Maruyama, K., Terada, K., Yamamoto, Y. (1980). *J. Org. Chem.*, **45**, 737.
Marynick, D.S., Kirkpatrick, C.M. (1985). *J. Am. Chem. Soc.*, **107**, 1993.
Mashima, K., Tanaka, Y., Kaidzu, M., Nakamura, A. (1996). *Organometallics*, **15**, 2431.
Masuda, T., Hasegawa, K., Higashimura, T. (1974). *Macromolecules*, **7**, 728.
Masuda, T., Thieu, K.Q., Sasaki, N., Higashimura, T. (1976). *Macromolecules*, **9**, 661.
Masuda, T., Ohtori, T., Higashimura, T. (1979). *Am. Chem. Soc., Polymer Prepr.*, **20**, 731.
Masuda, T., Kuwane, Y., Yamamoto, K., Higashimura, T. (1980a). *Polymer Bull.*, **2**, 823.
Masuda, T., Okano, Y., Kuwane, Y., Higashimura, T. (1980b). *Polymer J.*, **12**, 907.
Masuda, T., Mouri, T., Higashimura, T. (1980c). *Bull. Chem. Soc. Jpn.*, **53**, 1152.
Masuda, T., Kuwane, Y., Higashimura, T. (1981). *Polymer J.*, **13**, 301.
Masuda, T., Kawasaki, M., Okano, Y., Higashimura, T. (1982a). *Polymer J.*, **14**, 371.
Masuda, T., Takahashi, T., Higashimura, T. (1982b). *J. Chem. Soc., Chem. Commun.*, 1297.
Masuda, T., Takahashi, T., Yamamoto, K., Higashimura T. (1982c). *J. Polymer Sci., Polymer Chem.*, **20**, 2603
Masuda, T., Kawai, M., Higashimura, T. (1982d). *Polymer*, **23**, 744.
Masuda, T., Kuwane, Y., Higashimura T. (1982e). *J. Polymer Sci., Polymer Chem.*, **20**, 1043.
Masuda, T., Deng, Y-X., Higashimura, T. (1983a). *Bull. Chem. Soc. Jpn.*, **56**, 2798.
Masuda, T., Isobe, E., Higashimura, T. (1983b). *J. Am. Chem. Soc.*, **105**, 7473.

Masuda, T., Yoshizawa, T., Okano, Y., Higashimura, T. (1984a). *Polymer*, **25**, 503.
Masuda, T., Yamagata, M., Higashimura, T. (1984b). *Macromolecules*, **17**, 126.
Masuda, T., Isobe, E., Higashimura, T. (1985a). *Macromolecules*, **18**, 841.
Masuda, T., Tamura, K., Higashimura, T. (1985b). *J. Chem. Soc., Chem. Commun.*, 1615.
Masuda, T., Takahashi, T., Higashimura, T. (1985c). *Macromolecules*, **18**, 311.
Masuda,T., Niki, A., Isobe, E., Higashimura, T. (1985d). *Macromolecules*, **18**, 2109.
Masuda, T., Okano, Y., Tamura, K., Higashimura, T. (1985e). *Polymer*, **26**, 793.
Masuda, T., Takahashi, T, Niki, A., Higashimura, T. (1986). *J. Polym Sci.*, A, *Polym. Chem.*, **24**, 809.
Masuda, T., Tajima,T., Yoshimura, T., Higashimura, T. (1987a). *Macromolecules*, **20**, 1467.
Masuda, T., Yoshimura, T., Tamura, K., Higashimura, T. (1987b). *Macromolecules*, **20**, 1734.
Masuda, T., Yoshimura, T., Fujimori, J., Higashimura, T. (1987c). *J. Chem. Soc., Chem. Commun.*, 1805.
Masuda, T., Higashimura, T. (1987d). *Adv. Polymer Sci.*, **81**, 121.
Masuda, T., Hamano, T., Higashimura, T., Ueda, T., Muramatsu, H. (1988). *Macromolecules*, **21**, 281.
Masuda, T., Tsuchihara, K., Ohmameuda, K., Higashimura, T. (1989a). *Macromolecules*, **22**, 1036.
Masuda, T., Yoshimura, T., Higashimura, T. (1989b). *Macromolecules*, **22**, 3804.
Masuda, T., Hamano, T., Tsuchihara, K., Higashimura, T. (1990a). *Macromolecules*, **23**, 1374.
Masuda, T., Matsumoto, T., Yoshimura, T., Higashimura, T. (1990b). *Macromolecules*, **23**, 4902.
Masuda, T., Yoshida, T., Makio, H., Rahman, M.Z.A., Higashimura, T. (1991a). *J. Chem. Soc., Chem. Commun.*, 503.
Masuda, T., Kouzai, H., Higashimura, T. (1991b). *J. Chem. Soc., Chem. Commun.*, 252.
Masuda, T., Makio, H., Higashimura, T. (1992a). *Polymer Mater. Sci. Eng.*, **67**, 37.
Masuda, T., Mishima, K., Fujimori, J-I., Nishida, M., Muramatsu, H., Higashimura, T. (1992b). *Macromolecules*, **25**, 1401.
Masuda, T., Katahira, S., Tsuchihara, K., Higashimura, T. (1992c). *Polymer J.*, **24**, 491.
Masuda, T., Fujimori, J-I., Rahman, M.Z.A., Higashimura, T. (1993). *Polymer J.*, **25**, 535.
Masuda, T., Abe, Y., Kouzai, H., Higashimura, T. (1994a). *Polymer J.*, **26**, 393.
Masuda, T., Tachimori, H. (1994b). *J. Macromol. Sci.-Pure Appl. Chem.*, **A31**, 1675.
Masuda, T., Mishima, K., Seki, H., Nishida, M., Higashimura, T. (1994c). *Polymer Bull.*, **32**, 19.
Masuyama, Y., Tomatsu, Y., Ishida, K., Kurusu, Y., Segawa, K. (1988). *J. Catal.*, **114**, 347.
Matejka, L., Houtman, C., Macosko, C.W. (1985). *J. Appl. Polym. Sci.*, **30**, 2787.
Matlack, A.S. (1987). *Chem. Abstr.*, **108**, 22479.
Matlack, A.S., Paisner, M.J. (1990). *Chem. Abstr.*, **112**, 21758.
Matsumoto, S., Nakamura, R., Suzuki, K., Komatsu, K. (1976). *Chem. Abstr.*, **85**, 33724.
Matsumoto, S., Komatsu, K., Igarishi, K. (1977a). *Am. Chem. Soc., Polymer Prepr.*, **18**, 110.
Matsumoto, S., Komatsu, K., Igarishi, K. (1977b). *Am. Chem. Soc. Symp. Ser.*, **59**, 303.
Matsumoto, T., Masuda, T., Higashimura, T. (1991). *J. Polymer Sci.*, A, *Polymer Chem.*, **29**, 295.
Matulewicz, E.R.A., Scheffer, B., Mol, J.C. (1983). *Thermochim. Acta*, **67**, 361.
Matyjaszewski, K., Zielinski, M., Kubisa, P., Slomkowski, S., Chojnowski, J., Penczek, S. (1980). *Makromol. Chem.*, **181**, 1469.

Matyshak, V.A., Kadushin, A.A., Kutyreva, N.A., Edreva-Kardjieva, R.M. (1983). *Kinet. Katal.*, **24**, 1255.
Matyska, B., Dosedlová, A., Petrusová, L., Balcar, H. (1989). *Coll. Czech. Chem. Commun.*, **54**, 455.
Maughon, B.R., Grubbs, R.H. (1995). *Am. Chem. Soc., Polymer Prepr.*, **36**(1), 471.
McArdle, C.M., Hamilton, J.G., Law, E.E., Rooney, J.J. (1995). *Macromol. Rapid Commun.*, **16**, 703.
McCann, M., Guinan, P. (1991). *Polyhedron*, **10**, 2283.
McCann, M., McDonnell, D. (1993). *J. Chem. Soc., Chem. Commun.*, 1718.
McCann, M., Coda, E.M.G., Maddock, K. (1994). *J. Chem. Soc., Dalton Trans.*, 1489.
McCann, M., McDonnell, D., Goodall, B.L. (1995). *J. Mol. Catal.*, A, *Chem.*, **96**, 31.
McCann, M., Beaumont, A.J. (1996). *J. Mol. Catal.*, A, *Chem.*, **108**, 23.
McConville, D.H., Wolf, J.R., Schrock, R.R. (1993). *J. Am. Chem. Soc.*, **115**, 4413.
McCoy, J.R., Farona, M.F. (1991). *J. Mol. Catal.*, **66**, 51.
McCullough, L.G., Listemann, M.L., Schrock, R.R., Churchill, M.R., Ziller, J.W. (1983). *J. Am. Chem. Soc.*, **105**, 6729.
McCullough, L.G., Schrock, R.R. (1984). *J. Am. Chem. Soc.*, **106**, 4067.
McCullough, L.G., Schrock, R.R., Dewan, J.C., Murdzek, J.C. (1985). *J. Am. Chem. Soc.*, **107**, 5987.
McGeary, M.J., Tonker, T.L., Templeton, J.L. (1985). *Organometallics*, **4**, 2102.
McGinnis, J., Katz, T.J., Hurwitz, S. (1976). *J. Am. Chem. Soc.*, **98**, 605.
McGrath, D.V., Grubbs, R.H., Ziller, J.W. (1991). *J. Am. Chem. Soc.*, **113**, 3611.
McKervey, M.A., Pitarch, M. (1996). *Chem. Commun.*, 1689.
McKinney, R.J., Stone, F.G.A. (1980). *Inorg. Chim. Acta*, **44**, L227.
McLain, S.J., Wood, C.D., Schrock, R.R. (1977). *J. Am. Chem. Soc.*, **99**, 3519.
Meisel, I., Hertel, G., Weiss, K. (1986). *J. Mol. Catal.*, **36**, 159.
Menapace, H.R., Maly, N.A., Wang, J.L., Wideman, L.G. (1975). *J. Org. Chem.*, **40**, 2983.
Meyer, K.E., Walsh, P.J., Bergman, R.G. (1994). *J. Am. Chem. Soc.*, **116**, 2669.
Meziane, D., Soum, A., Fontanille, M., Rudler, H. (1985). *Makromol. Chem.*, **186**, 367.
Meziane, D., Soum, A., Fontanille, M. (1988). *Makromol. Chem.*, **189**, 1407.
Miao, Y-J., Bazan, G.C. (1994a). *Macromolecules*, **27**, 1063.
Miao, Y-J., Bazan, G.C. (1994b). *J. Am. Chem. Soc.*, **116**, 9379.
Miao, Y-J., Sun, B.J., Bazan, G.C. (1995). *Macromol. Symp.*, **95**, 185.
Michajlov, L., Harwood, H.J. (1970). *Am. Chem. Soc., Polymer Prepr.*, **11**, 1197.
Michajlov, L., Harwood, H.J. (1971). In '*Polymer Characterization, Interdisciplinary Approaches*' (ed C.D. Craver) p.221. Plenum, New York.
Michelotti, F.W., Carter, J.H. (1965a). *Am. Chem. Soc., Polymer Prepr.*, **6**, 224.
Michelotti, F.W., Keaveney, W.P. (1965b). *J. Polymer Sci.*, **A3**, 895.
Miller, S.J., Kim, S-H., Chen, Z-R., Grubbs, R.H. (1995a). *J. Am. Chem. Soc.*, **117**, 2108.
Miller, S.J., Grubbs, R.H. (1995b). *J. Am. Chem. Soc.*, **117**, 5855.
Minchak, R.J., Tucker, H. (1972). *Am. Chem. Soc., Polymer Prepr.*, **13**, 885.
Minchak, R.J. (1977a). *US Pat.* 4,002,815.
Minchak, R.J. (1977b). *US Pat.* 4,025,708.
Minchak, R.J. (1978). *US Pat.* 4,069,376.
Minchak, R.J. (1979). *US Pat.* 4,138,448; *Chem. Abstr.*, **90**, 205547.
Minchak, R.J. (1981). *Chem. Abstr.*, **95**, 44525.
Minchak, R.J., Tucker, H. (1982). *Am. Chem. Soc. Symp. Ser.*, **193**, 155.
Minchak, R.J., Tucker, H., Macey, J.H. (1983a). In '*Transition Metal Catalyzed Polymerizations*' (ed R.P. Quirk), **4B**, p.863. Harwood, London.
Minchak, R.J., Kettering T.J., Kroenke, W.J. (1983b). *US Pat.* 4,380,617.
Minchak, R.J. (1984). *US Pat.* 4,426,502.

Minchak, R.J., Lane, P.C. (1987). *US Pat.* 4,701,510.
Mistele, C.D., Thorp, H.H., DeSimone, J.M. (1995). *Am. Chem. Soc., Polymer Prepr.*, **36**(1), 507.
Mistele, C.D., Thorp, H.H., De Simone, J.M. (1996). *J. Macromol. Sci.-Pure Appl. Chem.*, **A33**, 953.
Mitchell, J.P., Gibson, V.C., Schrock, R.R. (1991). *Macromolecules*, **24**, 1220.
Mizumoto, T., Masuda, T., Higashimura, T. (1993). *J. Polymer Sci.*, A, *Polymer Chem.*, **31**, 2555.
Mizumoto, T., Masuda, T., Higashimura, T. (1995). *Macromol. Chem. Phys.*, **196**, 1769.
Mocella, M.T., Busch, M.A., Muetterties, E.L. (1976a). *J. Am. Chem. Soc.*, **98**, 1283.
Mocella, M.T., Rovner, R., Muetterties, E.L. (1976b). *J. Am. Chem. Soc.*, **98**, 4689.
Moffat, A.J., Clark, A. (1970a). *J. Catal.*, **17**, 264.
Moffat, A.J., Johnson, M.M., Clark, A. (1970b). *J. Catal.*, **18**, 345.
Moffat, A.J., Clark, A., Johnson, M.M. (1971). *J. Catal.*, **22**, 379.
Mol, J.C., Moulijn, J.A., Boelhouwer, C. (1968). *J. Chem. Soc., Chem. Commun.*, 633.
Mol, J.C., Visser, F.R., Boelhouwer, C. (1970). *J. Catal.*, **17**, 114.
Mol, J.C., Moulijn, J.A. (1975). *Adv. Catal.*, **24**, 131.
Mol, J.C., Woerlee, E.F.G. (1979). *J. Chem. Soc., Chem. Commun.*, 330.
Mol, J.C. (1982). *J. Mol. Catal.*, **15**, 35.
Mol, J.C. (1983). *CHEMTECH*, **13**, 250.
Mol, J.C., Xu, X., Van den Aardweg, G.C.N. (1986). *Neth. Pat. Appl.* 8,403,050; *Chem. Abstr.*, **106**, 140086.
Mol, J.C., Moulijn, J.A. (1987). In 'Catalysis: Science and Technology' (eds J.R. Anderson, M. Boudart), **8**, p.69. Springer, Berlin.
Mol, J.C., Andreini, A. (1988). *J. Mol. Catal.*, **46**, 151.
Mol, J.C. (1990a). *NATO ASI Ser.*, **C326**, 115.
Mol, J.C. (1990b). *NATO ASI Ser.*, **C326**, 247.
Mol, J.C. (1991). *J. Mol. Catal.*, **65**, 145.
Mol, J.C. (1994a). *Catal. Lett.*, **23**, 113.
Mol, J.C. (1994b). *J. Mol. Catal.*, **90**, 185.
Moloy, K.G. (1994). *J. Mol. Catal.*, **91**, 291.
Monakov, Y.B., Khairullina, R.M., Svirskii, S.E., Sultanova, V.S., Karaseva, A.V. (1992). *Dokl. Akad. Nauk*, **323**, 691.
Monakov, Y.B., Dorogovets, T.E., Mudarisova, R.K., Khairullina, R.M., Svirskii, S.E., Farkhieva, I.T. (1995). *Dokl. Akad. Nauk*, **340**, 638.
Montecatini Edison SpA. (1968). *Chem. Abstr.*, **68**, 50817.
Montgomery, P.D., Moore, R.N., Knox, W.R. (1976). *Chem. Abstr.*, **85**, 123540.
Moore, J.S., Gorman, C.B., Grubbs, R.H. (1991). *J. Am. Chem. Soc.*, **113**, 1704.
Mori, T., Kosaka, N., Yamaguchi, S., Katsumi, T., Tsuda, N., Murata, Y., Tochihara, H., Kubota, M., Akahane, T. (1986). *J. Mol. Catal.*, **36**, 91.
Mori, M., Watanuki, S. (1992). *J. Chem. Soc., Chem. Commun.*, 1082.
Morken, J.P., Didiuk, M.T., Visser, M.S., Hoveyda, A.H. (1994). *J. Am. Chem. Soc.*, **116**, 3123.
Mortell, K.H., Gingras, M., Kiessling, L.L. (1994). *J. Am. Chem. Soc.*, **116**, 12053.
Mortell, K.H., Weatherman, R.V., Kiessling, L.L. (1996). *J. Am. Chem. Soc.*, **118**, 2297.
Mortreux, A., Blanchard, M. (1972). *Bull. Soc. Chim. France*, 1641.
Mortreux, A., Blanchard, M. (1974). *J. Chem. Soc., Chem. Commun.*, 786.
Mortreux, A.D.N., Dy, N., Blanchard, M. (1975/76). *J. Mol. Catal.*, **1**, 101.
Mortreux, A., Delgrange, J.C., Blanchard, M., Lubochinsky, B. (1977). *J. Mol. Catal.*, **2**, 73.
Mortreux, A., Petit, F., Blanchard, M. (1978). *Tetrahedron Lett.*, 4967.
Mortreux, A., Petit, F., Blanchard, M. (1980). *J. Mol. Catal.*, **8**, 97.

Mortreux, A., Petit, F., Petit, M., Szymańska-Buzar, T. (1995). *J. Mol. Catal.*, **96**, 95.
Motz, V.W., Farona, M.F. (1977). *Inorg. Chem.*, **16**, 2545.
Moulijn, J.A., Boelhouwer, C. (1971). *J. Chem. Soc., Chem. Commun.*, 1170.
Moulijn, J.A., Reitsma, H.J., Boelhouwer, C. (1972). *J. Catal.*, **25**, 434.
Moulijn, J.A., Mol, J.C. (1988). *J. Mol. Catal.*, **46**, 1.
Mowat, W., Smith, J., Whan, D.A. (1974). *J. Chem. Soc., Chem. Commun.*, 34.
Mühlebach, A., Bernhard, P., Bühler, N., Karlen, T., Ludi, A. (1994). *J. Mol. Catal.*, **90**, 143.
Muramatsu, H., Ueda, T., Ito, K. (1985). *Macromolecules*, **18**, 1634.
Murdzek, J.S., Schrock, R.R. (1987a). *Organometallics*, **6**, 1373.
Murdzek, J.S., Schrock, R.R. (1987b). *Macromolecules*, **20**, 2640.
Muthukumaru Pillai, S., Tembe, G.L., Ravindranathan, M. (1992). *Appl. Catal.*, **81**, 273.
Nagasawa, M., Kikukawa, K.,Takagi, M., Masuda, T. (1978). *Bull. Chem. Soc. Jpn.*, **51**, 1291.
Nahm, S.H. (1987). *US Pat.* 4,689,380.
Nakamura, J., Konuma, H., Kokuryo, S., Ueshima, T., Tsuge, C. (1979). *Chem. Abstr.*, **90**, 55730.
Nakamura, R., Echigoya, E. (1972a). *Chem. Lett.*, 273.
Nakamura, R., Echigoya, E. (1972b). *Bull. Jpn. Petrol. Inst.*, **14**, 187.
Nakamura, R., Fukuhara, S., Matsumoto, S., Komatsu, K. (1976a). *Chem. Lett.*, 253.
Nakamura, R., Matsumoto, S. (1976b). *Chem. Lett.*, 1019.
Nakamura, R., Iida, H., Echigoya, E. (1976c). *Nippon Kagaku Kaishi*, **2**, 221.
Nakamura, R., Echigoya, E. (1977a). *Chem. Lett.*, 1227.
Nakamura, R., Echigoya, E. (1977b). *Rec. Trav. Chim. Pays-Bas*, **96**, M31.
Nakamura, R., Ichikawa, K., Echigoya, E. (1978a). *Chem. Lett.*, 813.
Nakamura, R., Ichikawa, K., Echigoya, E. (1978b). *Nippon Kagaku Kaishi*, 1602.
Nakamura, R., Abe, F., Echigoya, E. (1981). *Chem. Lett.*, 51.
Nakamura, R., Echigoya, E. (1982). *J. Mol. Catal.*, **15**, 147.
Nakamura, S., Dedieu, A. (1982). *Nouveau J. Chim.*, **6**, 23.
Nakamura, T., Yamada, M., Yamaguchi, T. (1992). *Appl. Catal.*, **A87**, 69.
Nakano, M., Masuda, T., Higashimura, T. (1994). *Macromolecules*, **27**, 1344.
Nakano, M., Masuda, T., Higashimura, T. (1995). *Polymer Bull.*, **34**, 191.
Nakayama, Y., Mashima, K., Nakamura, A. (1992). *J. Chem. Soc., Chem. Commun.*, 1496.
Nakayama, Y., Mashima, K., Nakamura, A. (1993). *Macromolecules*, **26**, 6267.
Nametkin, N.S., Vdovin, V.M., Babich, E.D., Karelskii, V.N., Kacharmin, B.V. (1973). *Dokl. Akad. Nauk SSSR*, **213**, 356.
Nametkin, N.S., Vdovin, V.M., Babich, E.D., Kacharmin, B.V., Bespalova, N.B., Karelskii, V.N., (1975). *Dokl. Akad. Nauk SSSR*, **225**, 577.
Natta, G., Dall'Asta, G., Mazzanti, G., Pasquon, I. (1962). *Makromol. Chem.*, **54**, 95.
Natta, G., Dall'Asta, G., Mazzanti, G., Motroni, G. (1963). *Makromol. Chem.*, **69**, 163.
Natta, G., Dall'Asta, G., Mazzanti, G. (1964a). *Angew. Chem. Int. Ed. Engl.*, **3**, 723.
Natta, G., Dall'Asta, G., Motroni, G. (1964b). *J. Polymer Sci., Polymer Lett.*, **B2**, 349.
Natta, G., Dall'Asta, G., Porri, L. (1965a). *Makromol. Chem.*, **81**, 253.
Natta, G., Dall'Asta, G., Mazzanti, G., Pasquon, I., Valvassori, A., Zambelli, A. (1965b). *Chem. Abstr.*, **63**, 13540.
Natta, G., Dall'Asta, G., Bassi, I.W., Carella, G. (1966a). *Makromol. Chem.*, **91**, 87.
Natta, G., Dall'Asta, G., Mazzanti, G. (1966b). *Chem. Abstr.*, **65**, 3996.
Navarro, F.R., Farona, M.F. (1976). *J. Polymer Sci., Polymer Lett.*, **14**, 2335.
Nekrasov, N.V., Ustinova, N.S., Usov, Y.N., Kiperman, S.L. (1982). *Kinet. Katal.*, **23**, 649.

Nel, J.G., Wagener, K.B., Boncella, J.M. (1989). *Am. Chem. Soc., Polymer Prepr.*, **30**(2), 130.
Nelson, L.L. (1988). *US Pat.* 4,727,125.
Nelson, L.L. (1989). *US Pat.* 4,826,942.
Nelson, W.T., Heckelsberg, F. (1983). *Ind. Eng. Chem., Prod. Res. Dev.*, **22**, 178.
Newburg, N.R. (1984). *US Pat.* 4,481,344.
Nguyen, S.T., Johnson, L.K., Grubbs, R.H., Ziller, J.W. (1992). *J. Am. Chem. Soc.*, **114**, 3974.
Nguyen, S.T., Grubbs, R.H., Ziller, J.W. (1993). *J. Am. Chem. Soc.*, **115**, 9858.
Nguyen, S.T., Grubbs, R.H. (1995). *J. Organomet. Chem.*, **497**, 195.
Nguyen, S.T., Grubbs, R.H., Feldman, J. (1996). To be published.
Nicolaides, C.P., Gates, B.C. (1986). *J. Mol. Catal.*, **35**, 391.
Nicolaou, K.C., Postema, M.H.D., Claiborne, C.F. (1996). *J. Am. Chem. Soc.*, **118**, 1565.
Niki, A., Masuda, T., Higashimura, T. (1987). *J. Polymer Sci.*, A, *Polymer Chem.*, **25**, 1553.
Nishi, G., Oshima, M., Obara,T., Matsui, T., Natsume, Y. (1989). *Chem. Abstr.*, **111**, 24150.
Nishi, Y., Oshima, M., Kohara, T., Matsui, T., Natsuume, T. (1990a). *Chem. Abstr.*, **112**, 8081.
Nishi, Y., Oshima, M., Kohara, T., Natsuume, T. (1990b). *Chem. Abstr.*, **112**, 36838.
Nishi, Y., Ohshima, M., Kohara, T., Natsuume, T. (1990c). *Chem. Abstr.*, **112**, 37611.
Nishiguchi, T., Goto, S., Sugisaki, K., Fukuzumi, K. (1980). *J. Jap. Oil Chem. Soc.*, **29**, 15.
Nishiguchi, T., Fukuzumi, K., Sugisaki, K. (1981). *J. Catal.*, **70**, 24.
Nishimura, M., Asakura, K., Iwasawa, Y. (1986). *Chem. Lett.*, 1457.
Noels, A.F., Demonceau, A., Carlier, E., Hubert, A.J., Márquez-Silva, R-L., Sanchez-Delgado, R.A. (1988). *J. Chem. Soc., Chem. Commun.*, 783.
Nomura, K., Schrock, R.R. (1996). *Macromolecules*, **29**, 540.
Norton, R.L., McCarthy, T.J. (1989). *Macromolecules*, **22**, 1022.
Nouguier, R., Mutin, R., Laval, J.P., Chapelet, G., Basset, J.M., Lattes, A. (1977). *Rec. Trav. Chim. Pays-Bas*, **96**, M91.
Novak, B.M., Grubbs, R.H. (1987). *Polymer Mater. Sci. Eng.*, **57**, 651.
Novak, B.M., Grubbs, R.H. (1988a). *J. Am. Chem. Soc.*, **110**, 960.
Novak, B.M., Grubbs, R.H. (1988b). *J. Am. Chem. Soc.*, **110**, 7542.
Novak, B.M., Risse, W., Grubbs, R.H. (1992). *Adv. Polymer Sci.*, **102**, 47.
Nubel, P.O., Lutman, C.A., Yokelson, H.B. (1994). *Macromolecules*, **27**, 7000.
Nubel, P.O., Yokelson, H.B., Lutman, C.A., Bouslog, W.G., Behrends, R.T., Runge, K.D. (1996). *J. Mol. Catal.*, in press (ISOM 11 vol.).
Nugent, W.A., Feldman, J., Calabrese, J.C. (1995). *J. Am. Chem. Soc.*, **117**, 8992.
O'Dell, R., McConville, D.H., Hofmeister, G.E., Schrock, R.R. (1994). *J. Am. Chem. Soc.*, **116**, 3414.
O'Donoghue, M.B., Schrock, R.R., LaPointe, A.M., Davis, W.M. (1996). *Organometallics*, **15**, 1334.
Ofstead, E.A. (1969). *4th Int. Synth. Rubber Symp.*, **4**(2), 42.
Ofstead, E.A., Calderon, N. (1972). *Makromol. Chem.*, **154**, 21.
Ofstead, E.A. (1977). *Macromol. Synth.*, **6**, 69.
Ofstead, E.A., Senyek, M.L. (1979). *Chem. Abstr.*, **90**, 188325.
Ofstead, E.A., Lawrence, J.P., Senyek, M.L., Calderon, N. (1980). *J. Mol. Catal.*, **8**, 227.
O'Gara, J.E., Portmess, J.D., Wagener, K.B. (1993a). *Macromolecules*, **26**, 2837.
O'Gara, J.E., Wagener, K.B., Hahn, S.F. (1993b). *Macromol. Chem., Rapid Commun.*, **14**, 657.
Ogata, E., Kamiya, Y. (1973). *Chem. Lett.*, 603.

Ogata, E., Kamiya, Y. (1974). *Ind. Eng. Chem., Prod. Res. Dev.*, **13**, 226.
Ogata, E., Sodewasa, T., Kamiya, Y. (1976). *Bull. Chem. Soc. Japan*, **49**, 1317.
Oh, S-Y., Akagi, K., Shirakawa, H., Araya, K. (1993a). *Macromolecules*, **26**, 6203.
Oh, S-Y., Ezaki, R., Akagi, K., Shirakawa, H. (1993b). *J. Polymer Sci.*, A, *Polymer Chem.*, **31**, 2977.
Oh, S-Y., Oguri, F., Akagi, K., Shirakawa, H. (1993c). *J. Polymer Sci.*, A, *Polymer Chem.* **31**, 781.
Ohgane, T., Masuda, T., Higashimura, T. (1994). *Polymer Bull.*, **32**, 517.
Okamoto, Y., Kane, H., Imanaka, T. (1988). *Chem. Express*, **3**, 711.
Okano, T., Ito, K., Kodaira, K., Hosokawa, K., Nishida, M., Ueda, T., Muramatsu, H. (1988). *J. Fluorine Chem.*, **38**, 139.
Okano, Y., Masuda, T., Higashimura, T. (1982). *Polymer J.*, **14**, 477.
Okano, Y., Masuda, T., Higashimura, T. (1984). *J. Polymer Sci., Polymer Chem.*, **22**, 1603.
Okuhara, T., Tanaka, K. (1976). *J. Catal.*, **42**, 474.
Olsthoorn, A.A., Boelhouwer, C. (1976a). *J. Catal.*, **44**, 197.
Olsthoorn, A.A., Boelhouwer, C. (1976b). *J. Catal.*, **44**, 207.
Olsthoorn, A.A., Moulijn, J.A. (1980). *J. Mol. Catal.*, **8**, 147.
O'Neill, P.P., Rooney, J.J. (1972). *J. Chem. Soc., Chem. Commun.*, 104.
Ono, T., Anpo, M., Kubokawa,Y. (1986a). *J. Phys. Chem.*, **90**, 4780.
Ono, T., Anpo, M., Kubokawa,Y. (1986b). *Chem. Express*, **1**, 181.
Oreshkin, I.A., Red'kina, L.I., Makovetskii, K.L., Tinyakova, E.I., Dolgoplosk, B.A. (1971). *Izv. Akad. Nauk SSSR, Ser. Khim.*, **20**, 1123.
Oreshkin, I.A., Red'kina, L.I., Kershenbaum, I.L., Chernenko, G.M., Makovetskii, K.L., Tinyakova, E.I., Dolgoplosk, B.A. (1977). *Europ. Polymer J.*, **13**, 447.
Oshika, T., Tabuchi, H. (1968). *Bull. Chem. Soc. Jpn.*, **41**, 211.
Oskam, J.H., Schrock, R.R. (1992). *J. Am. Chem. Soc.*, **114**, 7588.
Oskam, J.H., Schrock, R.R. (1993a). *J. Am. Chem. Soc.*, **115**, 11831.
Oskam, J.H., Fox, H.H., Yap, K.B., McConville, D.H., O'Dell, R., Lichtenstein, B.J., Schrock, R.R. (1993b). *J. Organomet. Chem.*, **459**, 185.
Ott, K.C., Lee, J.B., Grubbs, R.H. (1982). *J. Am. Chem. Soc.*, **104**, 2942.
Otten, M.M., Lamb, H.H. (1992). *J. Mol. Catal.*, **74**, 305.
Otton, J. (1979). *Chem. Abstr.*, **90**, 121028.
Otton, J., Colleuille, Y., Varagnat, J. (1980). *J. Mol. Catal.*, **8**, 313.
Overkleeft, H.S., Pandit, U.K. (1996). *Tetrahedron Lett.*, **37**, 547.
Ozaki, M., Ikeda, Y., Arakawa, T. (1986). *Chem. Abstr.*, **105**, 134519.
Pacreau, A., Fontanille, M. (1987). *Makromol. Chem.*, **188**, 2585.
Pakuro, N.I., Agapova, E.I., Izyumnikov, A.L., Gantmakher, A.R., Dolgoplosk, B.A. (1976). *Dokl. Akad. Nauk SSSR*, **230**, 897.
Pakuro, N.I., Makovetskii, K.L., Gantmakher, A.R., Dolgoplosk, B.A. (1982). *Izv. Akad. Nauk SSSR, Ser. Khim.*, **31**, 509.
Pampus, G., Witte, J., Hoffmann, M. (1970). *Rev. Gen. Caout. Plast.*, **47**, 1343.
Pampus, G., Lehnert, G., Maertens, D. (1972). *Am. Chem. Soc., Polymer Prepr.*, **13**, 880.
Pampus, G., Lehnert, G. (1974). *Makromol. Chem.*, **175**, 2605.
Pandit, U.K., Borer, B.C., Bieräugel, H. (1996). *Pure Appl. Chem.*, **68**, 659.
Pantoja, A., Sarrin, J., Gonzales, L., Noguera, O., Perez Zurita, M.J., Galavis, J.L., Goldwasser, J. (1993). *J. Catal.*, **142**, 110.
Park, J.S., Serwon, S., Langner, A., Ehrlich, P. (1989). *J. Polymer Sci.*, A, *Polymer Chem.*, **27**, 4281.
Park, J-W., Lee, J-H., Cho, H-N., Choi, S-K. (1993). *Macromolecules*, **26**, 1191.
Park, J-W., Lee, J-H., Ko, J-M., Cho, H-M., Choi, S-K. (1994). *J. Polymer Sci.*, A, *Polymer Chem.*, **32**, 2789.

Park, L.Y., Schrock, R.R., Stieglitz, S.G., Crowe, W.E. (1991). *Macromolecules*, **24**, 3489.
Park, L.Y., Ofer, D., Gardner, T.J., Schrock, R.R., Wrighton, M.S. (1992). *Chem. Mater.*, **4**, 1388.
Parlier, A., Rudler, H., Platzer, N., Fontanille, M., Soum, A. (1987). *J. Chem. Soc., Dalton Trans.*, 1041.
Parshall, G.W. (1980). '*Homogeneous Catalysis*'. Wiley, New York.
Patton, J.T., Boncella, J.M., Wagener, K.B. (1992). *Macromolecules*, **25**, 3862.
Patton, P.A., Lillya, C.P., McCarthy, T.J. (1986). *Macromolecules*, **19**, 1266.
Patton, P.A., McCarthy, T.J. (1987). *Macromolecules*, **20**, 778.
Pedersen, S.F., Schrock, R.R., Churchill, M.R., Wasserman, H.J. (1982). *J. Am. Chem. Soc.*, **104**, 6808.
Pennella, F., Banks, R.L., Bailey, J.C. (1968). *J. Chem. Soc., Chem. Commun.*, 1548.
Pennella, F., Banks, R.L. (1973). *J. Catal.*, **34**, 52.
Pennella, F., Regier, R.B., Banks, R.L. (1974). *J. Catal.*, **31**, 304.
Penninger, J., Biermann, M., Krause, H-J. (1983). *Fette, Seifen, Anstrichm.*, **85**, 239.
Perez, E., Laval, J.P., Bon, M., Rico, I., Lattes, A. (1988). *J. Fluorine Chem.*, **39**, 173.
Perrott, M.G., Novak, B.M. (1995). *Macromolecules*, **28**, 3492.
Perrott, M.G., Novak, B.M. (1996). *Macromolecules*, **29**, 1817.
Pershin, A.N., Shelimov, B.N., Kazansky, V.B. (1980). *J. Catal.*, **64**, 426.
Petasis, N.A., Fu, D-K. (1993). *J. Am. Chem. Soc.*, **115**, 7208.
Petasis, N.A., Fu, D-K. (1996). *SYNLETT*, 155.
Peters, E.F., Evering, B.L. (1960). *US Pat.* 2,963,447; *Chem. Abstr.*, **55**, 5810.
Petit, M., Mortreux, A., Petit, F. (1982). *J. Chem. Soc., Chem. Commun.*, 1385.
Pienaar, J.J., du Plessis, J.A.K. (1985). *S. Afr. J. Chem.*, **38**, 95.
Pinazzi, C.P., Cattiaux, J., Brosse, J.C. (1973a). *Makromol. Chem.*, **169**, 45.
Pinazzi, C.P., Reyx, D. (1973b). *Compt. Rend. Acad. Sci.*, **C276**, 1077.
Pinazzi, C.P., Cattiaux, J., Pleurdeau, A., Brosse, J.C. (1974). *Makromol. Chem.*, **175**, 1795.
Pinazzi, C.P., Guilmet, I., Reyx, D. (1976). *Tetrahedron Lett.*, 989.
Pinazzi, C.P., Campistron, I., Reyx, D. (1977a). *Bull. Soc. Chim. France*, 896.
Pinazzi, C.P., Campistron, I., Reyx, D. (1977b). *Rec. Trav. Chim. Pays-Bas*, **96**, M59.
Pinazzi, C.P., Campistron, I., Croissandeau, M.C. (1980). *J. Mol. Catal.*, **8**, 325.
Plugge, M.F.C., Mol, J.C. (1991). *SYNLETT*, 507.
Popov, A.M., Fridman, R.A., Finkel'shtein, E.S., Nametkin, N.S., Vdovin, V.M., Bashkirov, A.N., Kryukov, Y.B., Liberov, L.G. (1973). *Izv. Akad. Nauk SSSR, Ser. Khim.*, **22**, 1429.
Porri, L., Rossi, R., Diversi, P., Lucherini, A. (1972). *Am. Chem. Soc., Polymer Prepr.*, **13**, 897.
Porri, L., Rossi, R., Diversi, P., Lucherini, A. (1974). *Makromol. Chem.*, **175**, 3097.
Porri, L., Diversi, P., Lucherini, A., Rossi, R. (1975). *Makromol. Chem.*, **176**, 3121.
Portmess, J.D., Wagener, K.B. (1995). *Am. Chem. Soc., Polymer Prepr.*, **36**(1), 614.
Portmess, J.D., Wagener, K.B. (1996). *J. Polymer Sci.*, A, *Polymer Chem.*, **34**, 1353.
Proulx, G., Bergman, R.G. (1993). *J. Am. Chem. Soc.*, **115**, 9802.
Pu, L., Wagaman, W., Grubbs, R.H. (1996). *Macromolecules*, **29**, 1138.
Pugh, C., Schrock, R.R. (1992). *Macromolecules*, **25**, 6593.
Pugh, C. (1994). *Macromol. Symp.*, **77**, 325.
Qiao, J., Baker, G.L. (1995). *Polymer Mater. Sci. Eng.*, **73**, 515.
Quignard, F., Leconte, M., Basset, J.M. (1985a). *J. Mol. Catal.*, **28**, 27.
Quignard, F., Leconte, M., Basset, J.M. (1985b). *J. Chem. Soc., Chem. Commun.*, 1816.
Quignard, F., Leconte, M., Basset, J.M. (1986). *J. Mol. Catal.*, **36**, 13.
Ramakrishnan, S., Chung, T.C. (1989a). *Macromolecules*, **22**, 3181.

Ramakrishnan, S., Hirsch, J.A., Martinez, J.R., Chien, J.C.W. (1989b). *J. Polymer Sci.*, A, *Polymer Chem.*, **27**, 3937.
Ramakrishnan, S., Chung, T.C. (1990). *Macromolecules*, **23**, 4519.
Ramza, J., Descotes, G., Basset, J-M., Mutch, A. (1996). *J. Carbohydrate Chem.*, **15**, 125.
Randall, M.L., Tallarico, J.A., Snapper, M.L. (1995). *J. Am. Chem. Soc.*, **117**, 9610.
Rappé, A.K., Goddard, W.A. (1982a). *J. Am. Chem. Soc.*, **104**, 297.
Rappé, A.K., Goddard, W.A. (1982b). *J. Am. Chem. Soc.*, **104**, 448.
Rappé, A.K., Upton, T.H. (1984). *Organometallics*, **3**, 1440.
Ray, G.C., Crain, D.L. (1966). See *Chem. Abstr.*, (1969) **70**, 114580.
Reddy, B.S.R. (1983). *Indian J. Chem.*, **A22**, 51.
Red'kina, L.I., Makovetskii, K.L., Kershenbaum, I.L., Dzemilev, U.M., Tolstikov, G.A. (1981). *Izv. Akad. Nauk SSSR, Ser. Khim.*, **30**, 133.
Reif, L., Höcker, H. (1981a). *Makromol. Chem., Rapid Commun.*, **2**, 183.
Reif, L., Höcker, H. (1981b). *Makromol. Chem., Rapid Commun.*, **2**, 745.
Reif, L., Höcker, H. (1983). *Makromol. Chem., Rapid Commun.*, **4**, 693.
Reif, L., Höcker, H. (1984). *Macromolecules*, **17**, 952.
Reyx, D., Campistron, I., Heiling, P. (1982a). *Makromol. Chem.*, **183**, 173.
Reyx, D., Croissandeau, M.C. (1982b). *Makromol. Chem.*, **183**, 1371.
Reyx, D., Campistron, I., Hamza, M. (1986). *J. Mol. Catal.*, **36**, 101.
Reyx, D., Hamza, M., Campistron, I. (1987). *J. Mol. Catal.*, **42**, 289.
Reyx, D., Campistron I. (1990). *Makromol. Chem.*, **191**, 251.
Rinehart, R.E., Smith, H.P. (1965). *J. Polymer Sci., Polymer Lett.*, **B3**, 1049.
Rinehart, R.E. (1968). *Chem. Abstr.*, **68**, 87750.
Rinehart, R.E. (1969). *J. Polymer Sci.*, **C27**, 7.
Risse, W., Grubbs, R.H. (1989a). *Macromolecules*, **22**, 1558.
Risse, W., Grubbs, R.H. (1989b). *Macromolecules*, **22**, 4462.
Risse, W., Wheeler, D.R., Cannizzo, L.F., Grubbs, R.H. (1989c). *Macromolecules*, **22**, 3205.
Risse, W., Grubbs, R.H. (1989d). *Makromol. Chem., Rapid Commun.*, **10**, 73.
Risse, W., Grubbs, R.H. (1991). *J. Mol. Catal.*, **65**, 211.
Rocklage, S.M., Fellmann, J.D., Rupprecht, G.A., Messerle, L.W., Schrock, R.R. (1981). *J. Am. Chem. Soc.*, **103**, 1440.
Rodríguez-Ramos, I., Guerrero-Ruiz, A., Homs, N., Ramírez de la Piscina, P., Fierro, J.L.G. (1995). *J. Mol. Catal., A, Chem.*, **95**, 147.
Rooney, J.J., Stewart, A. (1977). *Chem. Soc. Specialist Periodical Reports, Catalysis*, **1**, 277.
Roper, W.R., Waters, J.M., Wright, A.H. (1984). *J. Organomet. Chem.*, **275**, C13.
Rossi, R., Diversi, P., Lucherini, A., Porri, L. (1974). *Tetrahedron Lett.*, 879.
Rossi, R. (1975). *Chim. Ind.* (*Milan*), **57**, 242.
Rossi, R. (1976a). *Chim. Ind.* (*Milan*), **58**, 517.
Rossi, R., Giorgi, R. (1976b). *Chim. Ind.* (*Milan*), **58**, 518.
Rossi, R., Giorgi, R. (1976c). *Tetrahedron Lett.*, 1721.
Rossi, R. (1976d). *Gazz. Chim. Ital.*, **106**, 1103.
Rudler, H. (1980). *J. Mol. Catal.*, **8**, 53.
Ruhle, H.W. (1971). *Chem. Abstr.*, **75**, 151341.
Rybak, W.K., Ziolkowski, J.J. (1987). *J. Mol. Catal.*, **42**, 347.
Ryoo, M-S., Lee, W-C., Choi, S-K. (1990). *Macromolecules*, **23**, 3029.
Saegusa, T., Tsujino, T., Furukawa, J. (1964). *Makromol. Chem.*, **78**, 231.
Sailor, M.J., Ginsburg, E.J., Gorman, C.B., Kumar, A., Grubbs, R.H., Lewis, N.S. (1990). *Science*, **249**, 1146.
Saito, K., Yamaguchi, T., Tanabe, K., Ogura, T., Yagi, M. (1979). *Bull. Chem. Soc. Jpn.*, **52**, 3192.

Sancho, J., Schrock, R.R. (1982). *J. Mol. Catal.*, **15**, 75.
Sankaran, V., Cummins, C.C., Schrock, R.R., Cohen, R.E., Silbey, R.J. (1990). *J. Am. Chem. Soc.*, **112**, 6858.
Sartori, G., Ciampelli, F., Cameli, N. (1963). *Chim. Ind. (Milan)*, **45**, 1478.
Sasaki, M., Tanaka, K., Tanaka, K., Toyoshima, I. (1986). *J. Chem. Soc., Chem. Commun.*, 764.
Sasaki, N., Masuda, T., Higashimura, T. (1976). *Macromolecules*, **9**, 664.
Sato, H., Okimoto, K., Tanaka, Y. (1977a). *J. Macromol. Sci.*, **A11**, 767.
Sato, H., Tanaka, Y., Taketomi, T. (1977b). *Makromol. Chem.*, **178**, 1993.
Saunders, R.S., Cohen, R.E., Schrock, R.R. (1991). *Macromolecules*, **24**, 5599.
Saunders, R.S., Cohen, R.E., Wong, S.J., Schrock, R.R. (1992). *Macromolecules*, **25**, 2055.
Saunders, R.S. (1995). *Macromolecules*, **28**, 4347.
Savard, J., Alper, H. (1988). *Can. J. Chem.*, **66**, 2483.
Schaper, H., John, C.S. (1990). *Speciality Chemicals*, **10**, 224.
Schaverien, C.J., Dewan, J.C., Schrock, R.R. (1986). *J. Am. Chem. Soc.*, **108**, 2771.
Schilder, P.G.M., Stufkens, D.J., Oskam, A., Mol, J.C. (1992). *J. Organomet. Chem.*, **426**, 351.
Schimetta, M., Stelzer, F. (1994a). *Macromolecules*, **27**, 3769.
Schimetta, M., Stelzer, F. (1994b). *Makromol. Chem. Phys.*, **195**, 2699.
Schimetta, M., Leising, G., Stelzer, F. (1995). *Synthetic Metals*, **74**, 99.
Schlund, R., Schrock, R.R., Crowe, W.E. (1989). *J. Am. Chem. Soc.*, **111**, 8004.
Schmalz, H-G. (1995). *Angew. Chem. Int. Ed. Engl.*, **34**, 1833.
Schneider, M.F., Junga, H., Blechert, S. (1995). *Tetrahedron*, **51**, 13003.
Schneider, M.F., Blechert, S. (1996). *Angew. Chem. Int. Ed. Engl.*, **35**, 411.
Schneider, W.A., Müller, M.F. (1988). *J. Mol. Catal.*, **46**, 395.
Schoettel, G., Kress, J., Osborn, J.A. (1989). *J. Chem. Soc., Chem. Commun.*, 1062.
Schopov, I., Jossifov, C. (1983). *Makromol. Chem., Rapid Commun.*, **4**, 659.
Schopov, I., Mladenova, L. (1985). *Makromol. Chem., Rapid Commun.*, **6**, 659.
Schopov, I., Mladenova, L. (1989). *Makromol. Chem.*, **190**, 1483.
Schrock, R.R., Messerle, L.W., Wood, C.D., Guggenberger, L.J. (1978). *J. Am. Chem. Soc.*, **100**, 3793.
Schrock, R., Rocklage, S., Wengrovius, J., Rupprecht, G., Fellmann, J. (1980). *J. Mol. Catal.*, **8**, 73.
Schrock, R.R., Listemann, M.L., Sturgeoff, L.G. (1982). *J. Am. Chem. Soc.*, **104**, 4291.
Schrock, R.R., Freudenberger, J.H., Listemann, M.L., McCullough, L.G. (1985). *J. Mol. Catal.*, **28**, 1.
Schrock, R.R. (1986a). *J. Organomet. Chem.*, **300**, 249.
Schrock, R.R. (1986b). *Acc. Chem. Res.*, **19**, 342.
Schrock, R.R. (1987a). *Chem. Abstr.*, **107**, 79942.
Schrock, R.R., Feldman, J., Cannizzo, L.F., Grubbs, R.H. (1987b). *Macromolecules*, **20**, 1169.
Schrock, R.R. (1988a). *Polymer Mater. Sci. Eng.*, **58**, 92.
Schrock, R.R., DePue, R.T., Feldman, J., Schaverien, C. J., Dewan, J.C., Liu, A.H. (1988b). *J. Am. Chem. Soc.*, **110**, 1423.
Schrock, R.R., Krouse, S.A., Knoll, K., Feldman, J., Murdzek, J.S., Yang, D.C. (1988c). *J. Mol. Catal.*, **46**, 243.
Schrock, R.R., Weinstock, I.A., Horton, A.D., Liu, A.H., Schofield, M.H. (1988d). *J. Am. Chem. Soc.*, **110**, 2686.
Schrock, R.R., Yap, K.B., Yang, D.C., Sitzmann, H., Sita, L.R., Bazan, G.C. (1989). *Macromolecules*, **22**, 3191.
Schrock, R.R. (1990a). *Acc. Chem. Res.*, **23**, 158.

Schrock, R.R., Murdzek, J.S., Bazan, G.C., Robbins, J., DiMare, M., O'Regan, M. (1990b). *J. Am. Chem. Soc.*, **112**, 3875.
Schrock, R.R. (1993). In '*Ring-Opening Polymerization*', (ed D.J. Brunelle), p.129. Hauser, Munich.
Schrock, R.R.(1994a). *Pure Appl. Chem.*, **66**, 1447.
Schrock, R.R., Luo, S., Zanetti, N.C., Fox, H.H. (1994b). *Organometallics*, **13**, 3396.
Schrock, R.R. (1995a). *Polyhedron*, **14**, 3177.
Schrock, R.R., Lee, J-K., O'Dell, R., Oskam, J.H. (1995b). *Macromolecules*, **28**, 5933.
Schrock, R.R., Luo, S., Lee, J.C., Zanetti, N.C., Davis, W.M. (1996). *J. Am. Chem. Soc.*, **118**, 3883.
Schubert, U. (ed). (1989). '*Advances in Metal Carbene Chemistry*', *NATO ASI Ser.*, **C269**. Kluwer, Dordrecht.
Schultz, A.J., Brown, R.K., Williams, J.M., Schrock, R.R. (1981). *J. Am. Chem. Soc.*, **103**, 169.
Schwab, P., France, M.B., Ziller, J.W., Grubbs, R.H. (1995). *Angew. Chem. Int. Ed. Engl.*, **34**, 2039.
Schwab, P., Grubbs, R.H., Ziller, J.W. (1996). *J. Am. Chem. Soc.*, **118**, 100.
Scott, K.W., Calderon, N., Ofstead, E.A., Judy, W.A., Ward, J.P. (1969). *Am. Chem. Soc.*, *Adv. Chem. Ser.*, **91**, 399.
Scott, K.W., Calderon, N., Ofstead, E.A., Judy, W.A., Ward, J.P. (1971). *Rubber Chem. Technol.*, **44**, 1341.
Seehof, N., Mehler, C., Breunig, S., Risse, W. (1992). *J. Mol. Catal.*, **76**, 53.
Seehof, N., Risse, W. (1993a). *Macromolecules*, **26**, 5971.
Seehof, N., Grutke, S., Risse, W. (1993b). *Macromolecules*, **26**, 695.
Segawa, K., Kim, D.S., Kurusu, Y., Wachs, I.E. (1988). *Proc. 9th Int. Congr. Catalysis*, (eds M.J. Phillips, M. Ternan). p.1960. Chem. Inst. Canada, Ottawa.
Segawa, K., Soeya, T., Kim. D.S. (1991). In '*Catalytic Science and Technology*', **1** (eds S. Yoshida, N. Takezawa, M. Ono), p.153. Kodansha, Tokyo.
Seki, H., Masuda, T., Higashimura, T. (1995a). *J. Polymer Sci.*, A, *Polymer Chem.*, **33**, 117.
Seki, H., Masuda, T. (1995b). *J. Polymer Sci.*, A, *Polymer Chem.*, **33**, 1907.
Seki, Y., Takeshita, K., Kawamoto, K. (1989). *J. Organomet. Chem.*, **369**, 117.
Sentürk, B., Hummel, K., (1979). *Chim. Acta Turc.*, **7**, 15.
Severina, Y.N., Mavrenkova, G.V., Geiderikh, M.A., Davydov, B.E. (1987). *Polymer Sci. USSR*, **29**, 818.
Seyferth, K., Taube, R., Dahlke, M. (1981). *Chem. Abstr.*, **95**, 43943.
Seyferth, K., Taube, R. (1982a). *J. Organomet. Chem.*, **229**, 275.
Seyferth, K., Taube, R. (1982b). *J. Organomet. Chem.*, **229**, C19.
Seyferth, K., Rosenthal, K., Kühn, G., Taube, R. (1984a). *Z. Anorg. Allg. Chem.*, **513**, 57.
Seyferth, K., Taube, R. (1984b). *J. Organomet. Chem.*, **262**, 179.
Seyferth, K., Taube, R. (1984c). *J. Organomet. Chem.*, **262**, 191.
Seyferth, K., Taube, R. (1984d). *J. Organomet. Chem.*, **268**, 155.
Seyferth, K., Taube, R. (1985). *J. Mol. Catal.*, **28**, 53.
Shahada, L.A., Feast, W.J. (1990). *Europ. Polymer J.*, **26**, 25.
Shahada, L.A., Feast, W.J. (1992). *Polymer Int.*, **28**, 285.
Shahada, L.A., Alarab, H. (1994). *J. Appl. Polymer Sci.*, *Appl. Polymer Symp.*, **55**, 173.
Shapiro, Y.Y., Yefimov, V.A., Turov, B.S. (1983). *Polymer Sci. USSR*, **25**, 589.
Sheinker, A.P., Yakovleva, M.K., Bessonova, N.P., Il'icheva, Z.F., Godovskii, Y.K., Abkin, A.D. (1979). *Dokl. Akad. Nauk SSSR*, **247**, 160.
Shelimov, B.N., Elev, I.V., Kazansky, V.B. (1986). *J. Catal.*, **98**, 70.
Shelimov, B.N., Elev, I.V., Kazansky, V.B. (1988). *J. Mol. Catal.*, **46**, 187.
Sherman, A.M. (1976). *Diss. Abstr.*, *Int. B*, **36**, 5051.

Sherwood, M. (1982). *Chem. Ind. (London)*, 994.
Sheu, F.-C., Hong, C.-T., Hwang, W.-L., Shih, C.-J., Wu, J.-C., Yeh, C.-T. (1992). *Catal. Lett.*, **14**, 297.
Shiono, T., Naga, N., Soga, K. (1993). *Makromol. Chem.*, **14**, 323.
Shivasubramaniam, V., Sundararajan, G. (1991). *J. Mol. Catal.*, **65**, 205.
Shmidt, F.K., Grechkina, E.A., Lipovich, V.G. (1973). *Kinet. Katal.*, **14**, 1080.
Shmidt, F.K. Grechkina, Y.A., Levkovskii, Y.S., Shmidt, O.I., Lavrent'yeva, V.B. (1979). *Petrol. Chem. USSR*, **18**, 1.
Short, H.C. (1987). *Chem. Week*, 3 June, 30.
Sibeijn, M., Mol, J.C. (1991a). *Appl. Catal.*, **67**, 279.
Sibeijn, M., Spronk, R., van Veen, J.A.R., Mol, J.C. (1991b). *Catal. Lett.*, **8**, 201.
Sibeijn, M., Mol, J.C. (1992). *J. Mol. Catal.*, **76**, 345.
Sibeijn, M., van Veen, J.A.R., Bliek, A., Moulijn, J.A. (1994). *J. Catal.*, **145**, 416.
Singleton, D.M. (1970). *Chem. Abstr.*, **72**, 89734.
Siskos, M.G., Kyriakakou, G., Zarkadis, A.K. (1994). *Polymer Bull.*, **33**, 295.
Sita, L.R., Lyon, S.R. (1993). *J. Am. Chem. Soc.*, **115**, 10374.
Sita, L.R. (1995). *Macromolecules*, **28**, 656.
Sjardijn, W., Kramer, A.H. (1986). *US Pat.* 4,729,976.
Sjardijn, W., Kramer, A.H. (1987). *US Pat.* 4,810,762.
Sjardijn, W. (1990a). *Chem. Abstr.*, **113**, 212900.
Sjardijn, W. (1990b). *Chem. Abstr.*, **113**, 212901.
Slaugh, L.H. (1991). *US Pat.* 5,008,480.
Slomkowski, S. (1985). *Makromol. Chem.*, **186**, 2581.
Smith, D.W., Wagener, K.B. (1993a). *Macromolecules*, **26**, 1633.
Smith, D.W., Wagener, K.B. (1993b). *Macromolecules*, **26**, 3533.
Smith, J., Howe, R.F., Whan, D.A. (1974a). *J. Catal.*, **34**, 191.
Smith, J., Mowat, W., Whan, D.A., Ebsworth, E.A.V. (1974b). *J. Chem. Soc., Dalton Trans.*, 1742.
Sodewasa, T., Ogata, E., Kamiya, Y. (1977). *Bull. Chem. Soc. Jpn.*, **50**, 998.
Sodewasa, T., Ogata, E., Kamiya, Y. (1979). *Bull. Chem. Soc. Jpn.*, **52**, 1661.
Sodupe, M., Lluch, J.M., Oliva, A., Bertrán, J. (1991). *New J. Chem.*, **15**, 321.
Sohn, B.H., Gratt, J.A., Lee, J.K., Cohen, R.E. (1995). *J. Appl. Polymer Sci.*, **58**, 1041.
Soto, J., Steigerwald, M.L., Grubbs, R.H. (1982). *J. Am. Chem. Soc.*, **104**, 4479.
Soufflet, J.P., Commereuc, D., Chauvin, Y. (1973). *Compt. Rend. Acad. Sci.*, **C276**, 169.
Soum, A., Fontanille, M., Rudler, H., Gouarderes, R. (1986). *Makromol. Chem., Rapid Commun.*, **7**, 525.
Spinicci, R., Tofanari, A. (1983). *J. Thermal Anal.*, **27**, 391.
Spinicci, R., Tofanari, A. (1985a). *Appl. Catal.*, **14**, 261.
Spinicci, R., Tofanari, A. (1985b). *Mater. Chem. Phys.*, **12**, 321.
Spinicci, R. (1989). *Mater. Chem. Phys.*, **23**, 299.
Spronk, R., Andreini, A., Mol, J.C. (1991a). *J. Mol. Catal.*, **65**, 219.
Spronk, R., Mol, J.C. (1991b). *Appl. Catal.*, **70**, 295.
Spronk, R., Mol, J.C. (1991c). *Appl. Catal.*, **76**, 143.
Spronk, R., Dekker, F.H.M., Mol, J.C. (1992). *Appl. Catal.*, **83**, 213.
Spronk, R., Mol, J.C. (1993a). In '*New Frontiers in Catalysis*' (eds L. Guczi, F. Solymosi, P. Tétény), p.2071. Elsevier Science, London.
Spronk, R., van Veen, J.A.R., Mol, J.C. (1993b). *J. Catal.*, **144**, 472.
Srinivasan, R., Farona, M.F. (1988). *Polymer Bull.*, **20**, 359.
Stanton, C.E., Lee, T.R., Grubbs, R.H., Lewis, N.S., Pudelski, J.K., Callstrom, M.R., Erickson, M.S., McLaughlin, M.L. (1995). *Macromolecules*, **28**, 8713.
Startsev, A.N., Kuznetsov, B.N., Yermakov, Y.I. (1975). *React. Kinet. Catal. Lett.*, **3**, 321.

Startsev, A.N., Bogdanovic, B., Bönnemann, H., Rodin, V.N., Yermakov, Y.I. (1986). *J. Chem. Soc., Chem. Commun.*, 381.
Startsev, A.N., Klimov, O.V., Shkuropat, S.A. (1990). *React. Kinet. Catal. Lett.*, **41**, 121.
Startsev, A.N., Klimov, O.V., Khomyakova, E.A. (1993). *J. Catal.*, **139**, 134.
Startseva, G.P., Matkovskii, P.E., Brikenshtein, K-M.A. (1987). *Izv. Akad. Nauk SSSR, Ser. Khim.*, **36**, 499.
Stein, J., Lewis, L.N., Lettko, K.X., Sumpter, C.A. (1994). *J. Inorg. Organomet. Polym.*, **4**, 367.
Steinhäusler, T., Stelzer, F. (1994a). *J. Mol. Catal.*, **90**, 53.
Steinhäusler, T., Stelzer, F., Zenkl, E. (1994b). *Polymer*, **35**, 616.
Stelzer, F., Thummer, R., Wewerka, D., Hummel, K. (1973). *Kolloid Z., Z. Polymere*, **251**, 772.
Stelzer, F., Thummer, R., Hummel, K. (1977). *Coll. Polymer Sci.*, **255**, 664.
Stelzer, F., Graiman, C., Hummel, K. (1982). *Coll. Polymer Sci.*, **260**, 829.
Stelzer, F., Hummel, K., Sommer, F., Baumegger, A.E., Lesiak, M.C. (1987a). *Rubber Chem. Technol.*, **60**, 600.
Stelzer, F., Hummel, K., Graimann, C., Hobisch-Gosch, J., Martl, M.G. (1987b). *Makromol. Chem.*, **188**, 1795.
Stelzer, F., Hobisch, G., Pongratz, T., Hummel, K. (1988). *J. Mol. Catal.*, **46**, 433.
Stelzer, F., Leitner, O., Pressl, K., Leising, G., Grubbs, R.H. (1991). *Synthetic Metals*, **41–43**, 991.
Stelzer, F., Fischer, W., Leising, G., Heller, C. (1992). In '*Electronic Properties of Polymers*', (eds H. Kuzmany, M. Mehring, S. Roth), p.231. Springer, Berlin.
Stevenson, S., Sermon, P.A. (1991). *J. Mol. Catal.*, **65**, 181.
Stille, J.R., Grubbs, R.H. (1986). *J. Am. Chem. Soc.*, **108**, 855.
Stille, J.R., Santarsiero, B.D., Grubbs, R.H. (1990). *J. Org. Chem.*, **55**, 843.
Stockel, E. (1978). *Diss. Abstr., Int. B*, **38**, 5948.
Stonich, D.A., Weber, W.P. (1991a). *Polymer Bull.*, **26**, 493.
Stonich, D.A., Weber, W.P. (1991b). *Polymer Bull.*, **26**, 629.
Stonich, D.A., Weber, W.P. (1991c). *Polymer Bull.*, **27**, 243.
Stowell, J.A., Amass, A.J., Beevers, M.S., Farren, T.R. (1987). *Makromol. Chem.*, **188**, 1635.
Straus, D.A., Grubbs, R.H. (1985). *J. Mol. Catal.*, **28**, 9.
Streck, R., Weber, H. (1971). *Chem. Abstr.*, **74**, 142595.
Streck, R., Weber, H. (1972a). *Chem. Abstr.*, **76**, 113891.
Streck, R., Weber, H. (1972b). *Chem. Abstr.*, **76**, 113903.
Streck, R., Weber, H. (1972c). *Chem. Abstr.*, **76**, 127817.
Streck, R., Weber, H. (1972d). *Chem. Abstr.*, **77**, 140749.
Streck, R., Weber, H. (1973). *Chem. Abstr.*, **78**, 112522.
Streck, R., Weber, H. (1976). *Chem. Abstr.*, **85**, 124699.
Streck, R. (1982). *J. Mol. Catal.*, **15**, 3.
Streck, R. (1990a). *NATO ASI Ser.*, **C326**, 457.
Streck, R. (1990b). *NATO ASI Ser.*, **C326**, 489.
Streck, R. (1992). *J. Mol. Catal.*, **76**, 359.
Strel'chik, B.S., Fridman, R.A., Finkel'shtein, E.S., Vdovin, V.M. (1976). *Izv. Akad. Nauk SSSR, Ser. Khim.*, **25**, 579.
Strutz, H., Schrock, R.R. (1984). *Organometallics*, **3**, 1600.
Strutz, H., Dewan, J.C., Schrock, R.R. (1985). *J. Am. Chem. Soc.*, **107**, 5999.
Stumpf, A.W., Saive, E., Demonceau, A., Noels, A.F. (1995). *J. Chem. Soc., Chem. Commun.*, 1127.
Sumitomo Chem. Co. (1980). *Chem. Abstr.*, **93**, 150902.

Sunaga, T., Ivin, K.J., Hofmeister, G.E., Oskam, J.H., Schrock, R.R. (1994). *Macromolecules*, **27**, 4043.
Suter, U.W., Höcker, H. (1988). *Makromol. Chem.*, **189**, 1603.
Suzuki, M., Yamaguchi, S., Tsuda, M. (1987). *J. Phys. Soc. Jpn.*, **56**, 2982.
Suzuki, T., Tanaka, K., Toyoshima, I. (1989a). *React. Kinet. Catal. Lett.*, **38**, 209.
Suzuki, T., Hayashi, S., Hirai, T., Tanaka, K., Toyoshima, I. (1989b). *J. Mol. Catal.*, **49**, L43.
Suzuki, T., Hirai, T., Hayashi, S., Tanaka, K., Toyoshima, I. (1989c). *Chem. Express*, **4**, 417.
Suzuki, T., Tanaka, K., Toyoshima, I., Gotoh, H. (1989d). *Appl. Catal.*, **50**, 15.
Suzuki, T., Ishimaru, K., Tsunokai, Y. (1996). *Chem. Abstr.*, **125**, 59446.
Swager, T.M., Grubbs, R.H. (1987). *J. Am. Chem. Soc.*, **109**, 894.
Swager, T.M., Dougherty, D.A., Grubbs, R.H. (1988). *J. Am. Chem. Soc.*, **110**, 2973.
Swager, T.M., Grubbs, R.H. (1989). *J. Am. Chem. Soc.*, **111**, 4413.
Syatkovskii, A.I., Denisova, T.T., Gavrilova, L.V., Babitskii, B.D. (1981). *Chem. Abstr.*, **94**, 122054.
Syatkowsky, A.I., Denisova, T.T., Ikonitsky, I.V., Babitsky, B.D. (1979). *J. Polymer Sci., Polymer Chem.*, **17**, 3939.
Syatkowsky, A.I., Denisova, T.T., Buzina, N.A., Babitsky, B.D. (1980). *Polymer*, **21**, 1112.
Syatkowsky, A.I., Denisova, T.T., Abramenko, E.L., Khatchaturov, A.S., Babitsky, B.D. (1981). *Polymer*, **22**, 1554.
Szymańska-Buzar, T., Ziółkowski, J.J. (1987). *J. Mol. Catal.*, **43**, 161.
Szymańska-Buzar, T. (1991). *J. Mol. Catal.*, **68**, 177.
Szymańska-Buzar, T. (1994). *J. Mol. Catal.*, **93**, 137.
Tachimori, H., Masuda, T., Kouzai, H., Higashimura, T. (1994). *Polymer Bull.*, **32**, 133.
Tachimori, H., Masuda, T. (1995). *J. Polymer Sci.*, A, *Polymer Chem.*, **33**, 2079.
Taghizadeh, N., Quignard, F., Leconte, M., Basset, J.M., Larroche, C., Laval, J.P., Lattes, A. (1982). *J. Mol. Catal.*, **15**, 219.
Tajima, H., Masuda, T., Higashimura, T. (1987). *J. Polymer Sci.*, A, *Polymer Chem.*, **25**, 2033.
Takagi, T., Hamaguchi, T., Fukuzumi, K., Aoyamo, M. (1972). *J. Chem. Soc., Chem. Commun.*, 838.
Takagi, T., Ichikawa, K., Fukuzumi, K., Hamaguchi, T. (1975). *Yukagu*, **24**, 518; *Chem. Abstr.*, **83**, 163542.
Takahashi, H., Inoue, N., Ono, M., Ichihara, Y., Suzuki, T. (1987). *Chem. Abstr.*, **107**, 49752.
Tamagaki, S., Card, R.J., Neckers, D.C. (1978). *J. Am. Chem. Soc.*, **100**, 6635.
Tamura, K., Masuda, T., Higashimura, T. (1993). *Polymer Bull.*, **30**, 537.
Tamura, K., Masuda, T., Higashimura, T. (1994). *Polymer Bull.*, **32**, 289.
Tamura, K., Misumi, Y., Masuda, T. (1996). *Chem. Commun.*, 373.
Tanaka, K., Tanaka, K., Miyahara, K. (1979). *J. Chem. Soc., Chem. Commun.*, 314.
Tanaka, K., Miyakara, K., Tanaka, K. (1980a). *Chem. Lett.*, 623.
Tanaka, K., Tanaka, K., Miyahara, K. (1980b). *J. Chem. Soc., Chem. Commun.*, 666.
Tanaka, K., Miyakara, K., Tanaka, K. (1981). *Stud. Surf. Sci. Catal.*, **7B**, 1318.
Tanaka, K., Miyahara, K., Tanaka, K. (1982a). *J. Mol. Catal.*, **15**, 133.
Tanaka, K., Zhai, Y., Aomura, K. (1982b). In '*Chemistry and Uses of Molybdenum*', *Proc. 4th Int. Conf.*, p.278. Climax Molybdenum Co., Ann Arbor, Michigan.
Tanaka, K., Tanaka, K. (1984). *J. Chem. Soc., Chem. Commun.*, 748.
Tanaka, K., Tanaka, K. (1986a). *Proc. 5th Int. Symp. Relat. Homog. Heterog. Catal.*, (eds Y. Yermakov, V. Likholobov) p.245. VNU Science Press, Utrecht.

Tanaka, K, Tanaka, K., Takeo, H., Matsumura, C. (1986b). *J. Chem. Soc., Chem. Commun.*, 33.
Tanaka, K., Tanaka, K. (1987a). *Chem. Abstr.*, **106**, 220973.
Tanaka, K., Tanaka, K., Takeo, H., Matsumura, C. (1987b). *J. Am. Chem. Soc.*, **109**, 2422.
Tanaka, K. (1988a). *J. Mol. Catal.*, **46**, 87.
Tanaka, K., Sasaki, M., Toyoshima, I. (1988b). *J. Phys. Chem.*, **92**, 4730.
Tanaka, K., Tanaka, K. (1988c). *J. Chem. Soc., Faraday Trans.,1*, **84**, 601.
Tanaka, K. (1990a). *NATO ASI Ser.*, **C326**, 271.
Tanaka, K. (1990b). *NATO ASI Ser.*, **C326**, 303.
Tanaka, S., Okuhara, K., Kaeriyama, K. (1986). *Makromol. Chem.*, **187**, 2793.
Tanaka, Y., Ueshima, T., Kobayashi, S. (1976). *Chem. Abstr.*, **84**, 165412.
Tanaka, Y., Sato, H., Hatada, K., Terawaki, Y. (1977). *Makromol. Chem.*, **178**, 1823.
Tang, B-Z., Kotera, N. (1989). *Macromolecules*, **22**, 4388.
Taniélian, C., Kieffer, R., Harfouch, A. (1981). *J. Mol. Catal.*, **10**, 269.
Tao, D., Wagener, K.B. (1994). *Macromolecules*, **27**, 1281.
Tarasov, A.L., Shelimov, B.N., Kazansky, V.B., Mol, J.C. (1996). *J. Mol. Catal.*, in press (ISOM 11 vol.).
Tassoni, R., Schrock, R.R. (1994). *Chem. Mater.*, **6**, 744.
Taube, R., Seyferth, K. (1977). *Z. Anorg. Allg. Chem.*, **437**, 213.
Taube, R., Seyferth, K. (1983). *J. Organomet. Chem.*, **249**, 365.
Taylor, T.E., Hall, M.B. (1984). *J. Am. Chem. Soc.*, **106**, 1576.
Tebbe, F.N., Parshall, G.W., Reddy, G. (1978). *J. Am. Chem. Soc.*, **100**, 3611.
Tebbe, F.N., Parshall, G.W., Ovenall, D.W. (1979). *J. Am. Chem. Soc.*, **101**, 5074.
Tebbe, F.N., Harlow, R.L. (1980). *J. Am. Chem. Soc.*, **102**, 6149.
Tenney, L.P. (1979a). *US Pat.* 4,136,248; *Chem. Abstr.*, **90**, 169556.
Tenney, L.P., Lane, P.C. (1979b). *US Pat.* 4,136,247; *Chem. Abstr.*, **90**, 187884.
Teplyakov, V.V., Paul, D.R., Bespalova, N.B., Finkelshtein, E.S. (1992). *Macromolecules*, **25**, 4218.
Theophilou, N., Munardi, A., Aznar, R., Sledz, J., Schué, F., Naarmann, H. (1987). *Europ. Polymer J.*, **23**, 15.
Thoi, H.H., Ivin, K.J., Rooney, J.J. (1982). *J. Mol. Catal.*, **15**, 245.
Thomas, R., Moulijn, J.A., De Beer, V.H.J., Medema, J. (1980). *J. Mol. Catal.*, **8**, 161.
Thomas, R., Moulijn, J.A. (1982). *J. Mol. Catal.*, **15**, 157.
Thorn-Csányi, E., Perner, H. (1979). *Makromol. Chem.*, **180**, 919.
Thorn-Csányi, E., Abendroth, H., Perner, H. (1980). *Makromol. Chem.*, **181**, 2081.
Thorn-Csányi, E. (1981). *Angew. Makromol. Chem.*, **94**, 181.
Thorn-Csányi, E., Hennemann-Perner, C., Perner, H. (1982). *Makromol. Chem., Rapid Commun.*, **3**, 329.
Thorn-Csányi, E. (1985). *J. Mol. Catal.*, **28**, 49.
Thorn-Csányi, E., Kessler, M. (1986a). *J. Mol. Catal.*, **36**, 31.
Thorn-Csányi, E., Perner, H. (1986b). *J. Mol. Catal.*, **36**, 187.
Thorn-Csányi, E., Harder, C. (1991a). *Angew. Makromol. Chem.*, **185**, 283.
Thorn-Csányi, E., Kessler, M. (1991b). *J. Mol. Catal.*, **65**, 253.
Thorn-Csányi, E., Harder, C., Dahlher, B. (1992a). *J. Mol. Catal.*, **76**, 93.
Thorn-Csányi, E., Höhnk, H-D. (1992b). *J. Mol. Catal.*, **76**, 101.
Thorn-Csányi, E., Pflug, K.P. (1993a). *Makromol. Chem.*, **194**, 2287.
Thorn-Csányi, E., Pflug, K.P. (1993b). *Makromol. Chem., Rapid Commun.*, **14**, 619.
Thorn-Csányi, E., Höhnk, H-D., Pflug, K.P. (1993c). *J. Mol. Catal.*, **84**, 253.
Thorn-Csányi, E., Pflug, K.P. (1994a). *J. Mol. Catal.*, **90**, 29.
Thorn-Csányi, E., Pflug, K.P. (1994b). *J. Mol. Catal.*, **90**, 69.
Thorn-Csányi, E., Kraxner, P., Hammer, J. (1994c). *J. Mol. Catal.*, **90**, 15.

Thorn-Csányi, E. (1994d). *Rubber Chem. Technol.*, **67**, 786.
Thorn-Csányi, E., Hammer, J., Pflug, K.P., Zilles, J.U. (1995a). *Macromol. Chem. Phys.*, **196**, 1043.
Thorn-Csányi, E., Kraxner, P. (1995b). *Makromol. Chem., Rapid Commun.*, **16**, 147.
Thu, C.T., Bastelberger, T., Höcker, H. (1981a). *Makromol. Chem., Rapid Commun.*, **2**, 7.
Thu, C.T., Bastelberger, T., Höcker, H. (1981b). *Makromol. Chem., Rapid Commun.*, **2**, 383.
Thu, C.T., Bastelberger, T., Höcker, H. (1985). *J. Mol. Catal.*, **28**, 279.
Thummer, R., Stelzer, F., Hummel, K. (1975). *Makromol. Chem.*, **176**, 1703.
Tinland, B., Quignard, F., Leconte, M., Basset, J.M. (1983). *J. Am. Chem. Soc.*, **105**, 2924.
Tlenkopachev, M.A., Kop'eva, I.A., Bychkova, N.A., Korshak, Y.V., Timofeeva, G.I., Tinyakova, E.I., Dolgoplosk, B.A. (1976). *Dokl. Akad. Nauk SSSR*, **227**, 889.
Tlenkopachev, M.A., Avdeikina, E.G., Korshak, Y.V., Bondarenko, G.N., Dolgoplosk, B.A., Kutepov, D.F. (1983). *Dokl. Akad. Nauk SSSR*, **268**, 133.
Tlenkopachev, M.A., Korshak, Y.V., Orlov, A.V., Korshak, V.V. (1986). *Dokl. Akad. Nauk SSSR*, **291**, 409.
Tlenkopachev, M.A., Korshak, Y.V., Segizova, N.T., Bondarenko, G.N., Nechitailo, N.A., Dzyubina, M.A. (1989). *Polymer Sci. USSR*, **31**, 1995.
Tlenkopatchev, M.A., Korshak, Y.V., Miranda, E., Ogawa, T. (1995a). *Polymer Bull.*, **34**, 405.
Tlenkopatchev, M.A., Miranda, E., Canseco, M.A., Gaviño, R., Ogawa, T. (1995b). *Polymer Bull.*, **34**, 385.
Tlenkopatchiev, M.A., Miranda, E., Gabiño, R., Ogawa, T. (1995c). *Polymer Bull.*, **35**, 547.
Toledano, C.A., Rudler, H., Daran, J-C., Jeannin, Y. (1984a). *J. Chem. Soc., Chem. Commun.*, 574.
Toledano, C.A., Parlier, A., Rudler, H., Daran, J-C., Jeannin, Y. (1984b). *J. Chem. Soc., Chem. Commun.*, 576.
Tom, G.M. (1985). *US Pat.* 4,507,453.
Tom, G.M. (1987). *US Pat.* 4,661,575.
Toreki, R., Schrock, R.R. (1990). *J. Am. Chem. Soc.*, **112**, 2448.
Toreki, R., Vaughan, G.A., Schrock, R.R., Davis, W.M. (1993). *J. Am. Chem. Soc.*, **115**, 127.
Tritto, I., Sacchi, M.C., Grubbs, R.H. (1993). *J. Mol. Catal.*, **82**, 103.
Truett, W.L., Johnson, D.R., Robinson, I.M., Montague, B.A. (1960). *J. Am. Chem. Soc.*, **82**, 2337.
Tsonis, C., Farona, M.F. (1979). *J. Polymer Sci., Polymer Chem.*, **17**, 185.
Tsonis, C.P. (1992). *React. Kinet. Catal. Lett.*, **46**, 359.
Tsuchihara, K., Masuda, T., Higashimura, T. (1988). *Polymer Bull.*, **20**, 343.
Tsuchihara, K., Masuda, T., Higashimura, T., Nishida, M., Muramatsu, H. (1990). *Polymer Bull.*, **23**, 505.
Tsuchihara, K., Masuda, T., Higashimura, T. (1991). *J. Am. Chem. Soc.*, **113**, 8548.
Tsuchihara, K., Masuda, T., Higashimura, T. (1992). *Macromolecules*, **25**, 5816.
Tsuda, N., Fujimori, A. (1981). *J. Catal.*, **69**, 410.
Tsuda, N., Mori, T., Kosaka, N., Sakai, Y. (1985). *J. Mol. Catal.*, **28**, 183.
Tsuji, J., Hashiguchi, S. (1980). *Tetrahedron Lett.*, 2955.
Tsuji, J., Hashiguchi, S. (1981). *J. Organomet. Chem.*, **218**, 69.
Tsujino, T., Saegusa, T., Kobayashi, S., Furukawa, J. (1964). *J. Chem. Soc. Jpn.*, **67**, 1961.
Tsujino, T., Saegusa, T., Furukawa, J. (1965). *Makromol. Chem.*, **85**, 71.
Tsunokai, Y., Mizuno, H., Natsume, Y. (1995). *Chem. Abstr.*, **123**, 229392.
Turner, H.W., Schrock, R.R., Fellmann, J.D., Holmes, S.J. (1983). *J. Am. Chem. Soc.*, **105**, 4942.

Turner, L., Bradshaw, C.P.C. (1968). *Brit. Pat.* 1,105,565. *Chem. Abstr.*, **68**, 87758.
Turov, B.S., Yefimov, V.A., Shapiro, Y.Y., Musabekov, Y.Y., Sumerkin, A.N. (1983). *Polymer Sci., USSR*, **25**, 2370.
Uchida, A., Hamano, Y., Mukai, Y., Matsuda, S. (1971). *Ind. Eng. Chem., Prod. Res. Dev.*, **10**, 372.
Uchida, A., Ishikawa, T., Takagi, M. (1977). *Rec. Trav. Chim. Pays-Bas*, **96**, M13.
Uchida, A., Hinenoya, M., Yamamoto, T. (1981). *J. Chem. Soc., Dalton Trans.*, 1089.
Uchida, A., Ko-ori, M. (1985). *J. Mol. Catal.*, **28**, 209.
Uchida, Y., Hidai, M., Tatsumi, T. (1972). *Bull. Chem. Soc. Jpn.*, **45**, 1158.
Ueshima, T., Kobayashi, S. (1974). *Jpn. Plast.* (Engl.), **8**, 11.
Ueshima, T., Tanaka, Y., Kobayashi, S. (1975). *Chem. Abstr.*, **83**, 194035.
Ungerank, M., Winkler, B., Eder, E., Stelzer, F. (1995). *Macromol. Chem. Phys.*, **196**, 3623.
Upton, T.H., Rappé, A.K. (1985). *J. Am. Chem. Soc.*, **107**, 1206.
Ushakov, N.V., Finkel'shtein, E.S., Portnykh, E.B., Vdovin, V.M. (1981). *Izv. Akad. Nauk. SSSR, Ser. Khim.*, **30**, 2835.
Ushakov, N.V., Portnykh, E.B., Pritula, N.A., Finkel'shtein, E.S. (1989). *Izv. Akad. Nauk SSSR, Ser. Khim.*, **38**, 2797.
Ushakov, N.V., Portnykh, E.B., Finkel'shtein, E.S. (1991). *Izv. Akad. Nauk SSSR, Ser. Khim.*, **40**, 1207.
Ushio, J., Nakatsuji, H., Yonezawa, T. (1984). *J. Am. Chem. Soc.*, **106**, 5892.
Usov, Y. N., Pletneva, E.V., Skvortsova, E.V. (1983). *Kinet. Katal.*, **24**, 1017.
Vaghi, A., Castellan, A., Bart, J.C.J., Giordano, N. (1976). *J. Catal.*, **42**, 381.
van Dam, P.B., Mittelmeijer, M.C., Boelhouwer, C. (1972). *J. Chem. Soc., Chem. Commun.*, 1221.
van Dam, P.B., Mittelmeijer, M.C., Boelhouwer, C. (1974a). *React. Kinet. Catal. Lett.*, **1**, 481.
van Dam, P.B., Mittelmeijer, M.C., Boelhouwer, C. (1974b). *J. Am. Oil Chem. Soc.*, **51**, 389.
van Dam, P.B., Mittelmeijer, M.C., Boelhouwer, C. (1974c). *Fette, Seifen, Anstrichm.*, **76**, 264.
van den Aardweg, G.C.N., Bosma, R.H.A., Mol, J.C. (1983). *J. Chem. Soc., Chem. Commun.*, 262.
VanderLende, D.D., Abboud, K.A., Boncella, J.M. (1994). *Organometallics*, **13**, 3378.
van der Schaaf, P.A., Smeets, W.J.J., Spek, A.L., van Koten, G. (1992). *J. Chem. Soc., Chem. Commun.*, 717.
van der Schaaf, P.A., Grove, D.M., Smeets, W.J.J., Spek, A.L., van Koten, G. (1993). *Organometallics*, **12**, 3955.
van der Schaaf, P.A., Abbenhuis, R.A.T.M., van der Noort, W.P.A., de Graaf, R., Grove, D.M., Smeets, W.J.J., Spek, A.L., van Koten, G. (1994). *Organometallics*, **13**, 1433.
Vanderyse, L.M., Verdonck, L., Bossuyt, A.R., Van Wijnsberghe, F.M.G., van der Kelen, G.P. (1988a). *Bull. Soc. Chim. Belg.*, **97**, 317.
Vanderyse, L.M., Haemers, T., Bossuyt, A.R., Verdonck, L., van der Kelen, G.P. (1988b). *Bull. Soc. Chim. Belg.*, **97**, 723.
van Deursen, J.H., Sjardijn, W. (1989). *Chem. Mag.*, 669.
van Rijn, F.H.M., Mol, J.C. (1977). *Rec. Trav. Chim. Pays-Bas*, **96**, M96.
van Roosmalen, A.J., Mol, J.C. (1980a). *J. Chem. Soc., Chem. Commun.*, 704.
van Roosmalen, A.J., Polder, K., Mol, J.C. (1980b). *J. Mol. Catal.*, **8**, 185.
van Roosmalen, A.J., Mol, J.C. (1982). *J. Catal.*, **78**, 17.
van Ruyskensvelde, S., Bossuyt, A., Haemers, T., Verdonck, L. (1995). *Bull. Soc. Chim. Belg.*, **104**, 401.

VanWijnsberghe, F.M.G., Vanderyse, L.M., Bossuyt, A.R., Verdonck, L., van der Kelen, G.P. (1987). *Bull. Soc. Chim. Belg.*, **96**, 731.
Vardanyan, L.M., Korshak, Y.V., Teterina, M.P., Dolgoplosk, B.A. (1972). *Dokl. Akad. Nauk SSSR*, **207**, 345.
Vardanyan, L.M., Korshak, Y.V., Pavlova, S.A., Timofeeva, G.I., Dolgoplosk, B.A. (1973). *Dokl. Akad. Nauk SSSR*, **208**, 1349.
Vaughan, W.M., Abboud, K.A., Boncella, J.M. (1995). *Organometallics*, **14**, 1567.
Vdovin, V.M., Bespalova, N.B., Bovina, M.A., Kalinin, V.N., Zakharhin, L.I. (1984a). *Izv. Akad. Nauk SSSR, Ser. Khim.*, **33**, 474.
Vdovin, V.M., Ushakov, N.V., Portnykh, E.B., Finkel'shtein, E.S., Abashkina, N.P. (1984b). *Izv. Akad. Nauk SSSR, Ser. Khim.*, **33**, 1892.
Vdovin, V.M., Ushakov, N.V., Portnykh, E.B., Finkel'shtein, E.S., Abashkina, N.P. (1984c). *Izv. Akad. Nauk SSSR, Ser. Khim.*, **33**, 2121.
Vdovin, V.M., Bespalova, N.B., Bovina, M.A. (1986). *Izv. Akad. Nauk SSSR, Ser. Khim.*, **35**, 2618.
Vdovin, V.M., Kotov, S.V., Portnykh, E.B., Strel'chik, B.S., Finkel'shtein, E.S., Oppengeim, V.D. (1988). *Petrol. Chem. USSR*, **28**, 17.
Verdonck, L., Bossuyt, A.R., Praet, A., Verpoort, F., Geeraert, J., van de Vondel, D., van der Kelen, G.P. (1992). *J. Mol. Catal.*, **76**, 319.
Verkuijlen, E., Boelhouwer, C. (1974). *J. Chem. Soc., Chem. Commun.*, 793.
Verkuijlen, E., Boelhouwer, C. (1976). *Fette, Seifen, Anstrichm.*, **78**, 444.
Verkuijlen, E., Dirks, R.J., Boelhouwer, C. (1977a). *Rec. Trav. Chim. Pays-Bas*, **96**, M86.
Verkuijlen, E., Kapteijn, F., Mol, J.C., Boelhouwer, C. (1977b). *J. Chem. Soc., Chem. Commun.*, 198.
Verkuijlen, E. (1980). *J. Mol. Catal.*, **8**, 107.
Verpoort, F., Bossuyt, A.R., Verdonck, L. (1995). *J. Mol. Catal.*, A, *Chem.*, **95**, 75.
Verpoort, F., Bossuyt, A.R., Verdonck, L. (1996a). *J. Chem. Soc., Chem. Commun.*, 417.
Verpoort, F., Bossuyt, A.R., Verdonck, L., Coussens, B. (1996b). *J. Mol. Catal.*, in press (ISOM 11 vol.).
Vialle, J., Basset, J.M. (1975). *React. Kinet. Catal. Lett.*, **2**, 397.
Vijayaraj, T.A., Sundararajan, G. (1995). *J. Mol. Catal.*, A, *Chem.*, **99**, 47.
Vikulov, K.A., Elev, I.V., Shelimov, B.N., Kazansky, V.B. (1989a). *Catal. Lett.*, **2**, 121.
Vikulov, K.A., Elev, I.V., Shelimov, B.N., Kazansky, V.B. (1989b). *J. Mol. Catal.*, **55**, 126.
Vikulov, K.A., Shelimov, B.N., Kazansky, V.B. (1991). *J. Mol. Catal.*, **65**, 393.
Vikulov, K.A., Shelimov, B.N., Kazansky, V.B. (1992a). *J. Mol. Catal.*, **72**, 1.
Vikulov, K.A., Shelimov, B.N., Kazansky, V.B. (1992b). *J. Mol. Catal.*, **72**, 117.
Vikulov, K.A., Shelimov, B.N., Kazansky, V.B., Mol, J.C. (1994). *J. Mol. Catal.*, **90**, 61.
Villemin, D. (1980). *Tetrahedron Lett.*, **21**, 1715.
Villemin, D., Cadiot, P. (1982). *Tetrahedron Lett.*, **23**, 5139.
Villemin, D. (1983). *Tetrahedron Lett.*, **24**, 2855.
Viswanathan, T., Jethmalani, J., Toland, A. (1993a). *J. Appl. Polymer Sci.*, **47**, 1477.
Viswanathan, T., Jethmalani, J. (1993b). *J. Appl. Polymer Sci.*, **48**, 1289.
Viswanathan, T., Gomez, F., Wagener, K.B. (1994). *J. Polymer Sci.*, A, *Polymer Chem.*, **32**, 2469.
Vosloo, H.C.M., du Plessis, J.A.K. (1991). *J. Appl. Polymer Sci., Appl. Polymer Symp.*, **48**, 561.
Vosloo, H.C.M., du Plessis, J.A.K. (1993a). *J. Mol. Catal.*, **79**, 7.
Vosloo, H.C.M., du Plessis, J.A.K. (1993b). *Polymer Bull.*, **30**, 273.
Vosloo, H.C.M., Luyt, A.S. (1995). *J. Thermal Anal.*, **44**, 1261.
Vuurman, M.A., Stufkens, D.J., Oskam, A., Wachs, I.E. (1992). *J. Mol. Catal.*, **76**, 263.

Wache, S., Herrmann, W.A., Artus, G., Nuyken, O., Wolf, D. (1995a). *J. Organomet. Chem.*, **491**, 181.
Wache, S. (1995b). *J. Organomet. Chem.*, **494**, 235.
Wada, Y., Nakaoka, C., Morikawa, A. (1988). *Chem. Lett.*, 25.
Wagener, K.B., Nel, J.G., Konzelman, J., Boncella, J.M. (1990). *Macromolecules*, **23**, 5155.
Wagener, K.B., Puts, R.D., Smith, D.W. (1991a). *Makromol. Chem., Rapid Commun.*, **12**, 419.
Wagener, K.B., Puts, R.D. (1991b). *Am. Chem. Soc., Polymer Prepr.*, **32**(1), 379.
Wagener, K.B., Boncella, J.M., Nel, J.G. (1991c). *Macromolecules*, **24**, 2649.
Wagener, K.B., Konzelman, J. (1991d). *Am. Chem. Soc., Polymer Prepr.*, **32**(1), 375.
Wagener, K.B., Brzezinska, K. (1991e). *Macromolecules*, **24**, 5273.
Wagener, K.B., Smith, D.W. (1991f). *Macromolecules*, **24**, 6073.
Wagener, K.B., Brzezinska, K., Bauch, C.G. (1992). *Makromol. Chem., Rapid Commun.*, **13**, 75.
Wagener, K.B., Patton, J.T. (1993a). *Macromolecules*, **26**, 249.
Wagener, K.B., Patton, J.T., Forbes, M.D.E., Myers, T.L., Maynard, H.D. (1993b). *Polymer Int.*, **32**, 411.
Wagner, G.W., Hanson, B.E. (1987). *Organometallics*, **6**, 2494.
Wakatsuki, Y., Yamazaki, H., Nakano, M., Yamamoto, Y. (1991). *J. Chem. Soc., Chem. Commun.*, 703.
Walba, D.M., Keller, P., Shao, R., Clark, N.A., Hillymer, M., Grubbs, R.H. (1996). *J. Am. Chem. Soc.*, **118**, 2740.
Wallace, K.C., Dewan, J.C., Schrock, R.R. (1986). *Organometallics*, **5**, 2162.
Wallace, K.C., Schrock, R.R. (1987). *Macromolecules*, **20**, 448.
Wallace, K.C., Liu, A.H., Dewan, J.C., Schrock, R.R. (1988). *J. Am. Chem. Soc.*, **110**, 4964.
Wallace, K.C., Liu, A.H., Davis, W.M., Schrock, R.R. (1989). *Organometallics*, **8**, 644.
Wang, J-L., Menapace, H.R. (1973). *J. Catal.*, **28**, 300.
Wang, L., Hall, W.K. (1983). *J. Catal.*, **82**, 177.
Wang, L., Ko, Y.H., Weber, W.P. (1992). *Macromolecules*, **25**, 2828.
Wang, L.P., Tysoe, W.T. (1990). *Catal. Lett.*, **6**, 111.
Wang, L.P., Soto, C., Tysoe, W.T. (1993). *J. Catal.*, **143**, 92.
Warwel, S., Laarz, W. (1975). *Chem. Ztg.*, **99**, 502.
Warwel, S., Aivasidis, A. (1977a). *Chem. Ztg.*, **101**, 361.
Warwel, S., Laarz, W. (1977b). *Z. Naturforsch.*, **B32**, 1145.
Warwel, S. (1978). *Angew. Chem.*, **90**, 131.
Warwel, S., Hachen, G., Kirchmeijer, H., Ridder, H., Winkelmüller, W. (1982a). *Tenside Deterg.*, **19**, 321.
Warwel, S., Janssen, E. (1982b). *Chem. Ztg.*, **106**, 266.
Warwel, S., Ridder, H., Hachen, S. (1983a). *Chem. Ztg.*, **107**, 115.
Warwel, S., Siekermann, V. (1983b). *Makromol. Chem., Rapid Commun.*, **4**, 423.
Warwel, S., Winkelmüller, W. (1985). *J. Mol. Catal.*, **28**, 247.
Warwel, S., Döring, N., Biermans, F.J. (1987a). *Fat Sci. Technol.*, **89**, 335.
Warwel, S., Kätker, H., Rauenbusch, C. (1987b). *Angew. Chem. Int. Ed. Engl.*, **99**, 702.
Warwel, S., Kätker, H. (1987c). *Synthesis*, 935.
Warwel, S., Döring, N., Deckers, A. (1988a). *Fat Sci. Techn.*, **90**, 125.
Warwel, S., Pütz, G. (1988b). *Chem. Ztg.*, **112**, 15.
Warwel, S., Bachem, H., Deckers, A., Döring, N., Kätker, H., Rose, E. (1989a). *Seifen, Öle, Fette, Wachse*, **115**, 538.
Warwel, S., Deckers, A. (1989b). *Tenside Deterg.*, **26**, 252.
Warwel, S., Pütz, G. (1989c). *J. Organomet. Chem.*, **364**, 323.

Warwel, S., Jägers, H-G., Ercklentz, B. (1991). *Europ. Pat. Appl.* 444,265; *Chem. Abstr.*, **116**, 40912.
Warwel, S., Jägers, H-G., Thomas, S. (1992a). *Fat Sci. Technol.*, **94**, 323.
Warwel, S., Harperscheid, M., Jägers, H-G. (1992b). *German Pat. Appl.* 4,107,056.
Washiyama, J., Kawauchi, S., Hirata, M., Nakamura, J. (1989). *Chem. Abstr.*, **111**, 59177.
Wasserman, E., Ben-Efraim, D.A., Wolovsky, R. (1968). *J. Am. Chem. Soc.*, **90**, 3286.
Watanuki, S., Mori, M. (1993). *Heterocycles*, **35**, 679.
Watanuki, S., Ochifuji, N., Mori, M. (1994). *Organometallics*, **13**, 4129.
Watanuki, S., Ochifuji, N., Mori, M. (1995). *Organometallics*, **14**, 5062.
Watase, T., Tachimori, H., Masuda, T. (1995). *Bull. Chem. Soc. Jpn.*, **68**, 393.
Watkins, N., Quigley, P., Orton, M. (1994). *Macromol. Chem. Phys.*, **195**, 1147.
Watson, M.D., Wagener, K.B. (1996). *Am. Chem. Soc., Polymer Prepr.*, **37**(1), 609.
Weck, M., Schwab, P., Grubbs, R.H. (1996a). *Macromolecules*, **29**, 1789.
Weck, M., Maughon, B.R., Mohr, B., Grubbs, R.H. (1996b). *Am. Chem. Soc., Polymer Prepr.*, **37**(1), 587.
Weddigen, G. (1983). *Chem. Abstr.*, **98**, 180076.
Wei, J., Leonard, J. (1994). *Europ. Polymer J.*, **30**, 999.
Weinstock, I.A., Schrock, R.R., Davis, W.M. (1991). *J. Am. Chem. Soc.*, **113**, 135.
Weiss, K. (1986a). *Angew. Chem. Int. Ed. Engl.*, **25**, 359.
Weiss, K., Goller, R. (1986b). *J. Mol. Catal.*, **36**, 39.
Weiss, K., Schubert, U., Schrock, R.R. (1986c). *Organometallics*, **5**, 397.
Weiss, K., Hoffmann, K. (1987). *Z. Naturforsch.*, **42b**, 769.
Weiss, K., Guthmann, W., Maisuls, S. (1988a). *Angew. Chem. Int. Ed. Engl.*, **27**, 275.
Weiss, K., Denzer, M. (1988b). *J. Organomet. Chem.*, **355**, 273.
Weiss, K., Goller, R., Lössel, G. (1988c). *J. Mol. Catal.*, **46**, 267.
Weiss, K., Guthmann, W., Denzer, M. (1988d). *J. Mol. Catal.*, **46**, 341.
Weiss, K., Hoffmann, K. (1989a). *NATO ASI Ser.*, **C269**, 351.
Weiss, K., Lössel, G. (1989b). *NATO ASI Ser.*, **C269**, 355.
Weiss, K., Lössel, G. (1989c). *Angew. Chem. Int. Ed. Engl.*, **28**, 62.
Weiss, K., Guthmann, W., Denzer, M., Maisuls, S. (1990a). *NATO ASI Ser.*, **C326**, 517.
Weiss, K., Lössel, G., Denzer, M. (1990b). *NATO ASI Ser.*, **C326**, 521.
Wengrovius, J.H., Schrock, R.R., Churchill, M.R., Missert, J.R., Youngs, W.J. (1980). *J. Am. Chem. Soc.*, **102**, 4515.
Wengrovius, J.H., Sancho, J., Schrock, R.R. (1981). *J. Am. Chem. Soc.*, **103**, 3932.
Westhoff, R., Moulijn, J.A. (1977). *J. Catal.*, **46**, 414.
Wewerka, D., Hummel, K. (1976). *Kolloid Z., Z. Polymere*, **254**, 116.
Widawski, G., Feast, W.J., Dounis, P. (1995). *J. Mater. Chem.*, **5**, 1847.
Wideman, L.G. (1968). *J. Org. Chem.*, **33**, 4541.
Wilker, C.N., Hoffmann, R. (1983). *J. Am. Chem. Soc.*, **105**, 5285.
Williams, K.P.J., Harrison, K. (1990). *J. Chem. Soc., Faraday Trans.*, **86**, 1603.
Wills, G.B., Fathikalajahi, J., Gangwal, S.K., Tang, S. (1977). *Rec. Trav. Chim. Pays-Bas*, **96**, M110.
Wilson, S.R., Schalk, D.E. (1976). *J. Org. Chem.*, **41**, 3928.
Winstein, C.Z. (1977). *Chem. Abstr.*, **86**, 122050.
Witte, J., Hoffmann, M. (1978). *Makromol. Chem.*, **179**, 641.
Woerlee, E.F.G., Bosma, R.H.A., van Eijl, J.J.M., Mol, J.C. (1984). *Appl. Catal.*, **10**, 219.
Wolf, A., Wagener, K.B. (1991). *Am. Chem. Soc., Polymer Prepr.*, **32**(1), 535.
Wolfe, P.S., Wagener, K.B. (1996). *Am. Chem. Soc., Polymer Prepr.*, **37**(1), 439.
Wolovsky, R. (1970). *J. Am. Chem. Soc.*, **92**, 2132.
Woo, T., Folga, E., Ziegler, T. (1993). *Organometallics*, **12**, 1289.
Woody, F.F., Lewis, M.J., Wills, G.B. (1969). *J. Catal.*, **14**, 389.
Woon, P.S., Farona, M.F. (1974). *J. Polymer Sci., Polymer Chem.*, **12**, 1749.

Wu, Z., Wheeler, D.R., Grubbs, R.H. (1992). *J. Am. Chem. Soc.*, **114**, 146.
Wu, Z., Benedicto, A.D., Grubbs, R.H. (1993). *Macromolecules*, **26**, 4975.
Wu, Z., Grubbs, R.H. (1994a). *J. Mol. Catal.*, **90**, 39.
Wu, Z., Grubbs, R.H. (1994b). *Macromolecules*, **27**, 6700.
Wu, Z., Grubbs, R.H. (1995a). *Macromolecules*, **28**, 3502.
Wu, Z., Nguyen, S.T., Grubbs, R.H., Ziller, J.W. (1995b). *J. Am. Chem. Soc.*, **117**, 5503.
Xu, X., Andreini, A., Mol, J.C. (1985a). *J. Mol. Catal.*, **28**, 133.
Xu, X., Imhoff, P., van den Aardweg, G.C.N., Mol, J.C. (1985b). *J. Chem. Soc., Chem. Commun.*, 273.
Xu, X., Mol, J.C. (1985c). *J. Chem. Soc., Chem. Commun.*, 631.
Xu, X., Benecke, J.I., Boelhouwer, C., Mol, J.C. (1986a). *Appl. Catal.*, **28**, 271.
Xu, X., Boelhouwer, C., Benecke, J.I., Vonk, D., Mol, J.C. (1986b). *J. Chem. Soc., Faraday Trans.,1*, **82**, 1945.
Xu, X., Boelhouwer, C., Vonk, D., Benecke, J.I., Mol, J.C. (1986c). *J. Mol. Catal.*, **36**, 47.
Xu, X., Mol, J.C., Boelhouwer, C. (1986d). *J. Chem. Soc., Faraday Trans.,1*, **82**, 2707.
Xu, X., Vonk, D., Mol, J.C. (1986e). *Thermochimica Acta*, **105**, 135.
Xu, Y., Wei, X., Shi, Y., Zhang, Y., Guo, X. (1986). *J. Mol. Catal.*, **36**, 79.
Xu, Y., Huang, J., Lin, Z., Guo, X. (1991). *J. Mol. Catal.*, **65**, 275.
Yakovleva, M.K., Sheinker, A.P., Abkin, A.D. (1977a). *Vysokomol. Soedin.*, **B19**, 98.
Yakovleva, M.K., Sheinker, A.P., Kotin, E.B., Abkin, A.D. (1977b). *Vysokomol. Soedin.*, **B19**, 604.
Yamagata, M., Masuda, T., Higashimura, T. (1984). *J. Polymer Sci., Polymer Chem.*, **22**, 2275.
Yamaguchi, M., Tsunoda, Y. (1974). *Chem. Abstr.*, **80**, 132758.
Yamaguchi, M., Tsukamoto, Y., Ikeura, C., Nakamura, S., Minami, T. (1991). *Chem. Lett.*, 1259.
Yamaguchi, M., Hirama, M., Nishihara, H. (1992). *Chem. Lett.*, 1667.
Yao, J., Li, H., Sun, J., He, B. (1994). *Chinese J. React. Polym.*, **3**, 8.
Yashima, T., Komatsu, T., Namba, S. (1986). *Chem. Express*, **1**, 701.
Yefimov, V.A., Turov, B.S., Shilov, A.D. (1988). *Polymer Sci. USSR*, **30**, 2049.
Yermakov, Y.I., Kuznetsov, B.N., Startsev, A.N. (1974). *Kinet. Katal.*, **15**, 539.
Yermakov, Y.I, Kuznetzov, B.N., Grabovski, Y.P., Startzev, A.N., Lazutkin, A.M., Zakharov, V.A., Lazutkina, A.I. (1975/76). *J. Mol. Catal.*, **1**, 93.
Yokota, K., Ohtubo, M., Hirabayashi, T., Ina, Y. (1993). *Polymer J.*, **25**, 1079.
Yoshida, H., Tanaka, T., Matsuo, S., Funabiki, T., Yoshida, S. (1995). *J. Chem. Soc., Chem. Commun.*, 761.
Yoshida, T., Abe, Y., Masuda, T., Higashimura, T. (1996). *J. Polymer Sci.*, A. *Polymer Chem.*, **34**, 2229.
Yoshimura, T., Masuda, T., Higashimura, T., Ishihara, T. (1986). *J. Polymer Sci.*, A, *Polymer Chem.*, **24**, 3569.
Yoshimura, T., Masuda, T., Higashimura, T. (1988). *Macromolecules*, **21**, 1899.
Yoshimura, T., Masuda, T., Higashimura, T., Okuhara, K., Ueda, T. (1991). *Macromolecules*, **24**, 6053.
Yoshimura, T., Asano, M. (1994). *Polymer J.*, **26**, 159.
Youinou, M.T., Kress, J., Fischer, Aguero, A., Osborn, J.A. (1988). *J. Am. Chem. Soc.*, **110**, 1488.
Zaliznaya, N.F., Davydov, B.E., Zemtsov, L.M., Khoroshilova, V.V., Rabkina, A.L. (1990). *Vysokomol. Soedin.*, **B32**, 235.
Zemtsov, L.M., Davydov, B.E., Karpacheva, G.P., Khoroshilova, V.V., Samedova, T.G. (1977). *Kinet. Katal.*, **18**, 638.
Zenkl, E., Stelzer, F. (1992a). *J. Mol. Catal.*, **76**, 1.
Zenkl, E., Schimetta, M., Stelzer, F. (1992b). *Polymer Mater. Sci. Eng.*, **66**, 281.

Zerpner, D., Holtrup, W., Streck, R. (1986). *J. Mol. Catal.*, **36**, 153.
Zerpner, D., Streck, R. (1987). *Chem. Abstr.*, **106**, 177085.
Zhang, B., Li, Y., Lin, Q., Jin, D. (1988). *J. Mol. Catal.*, **46**, 229.
Zhang, B., Liu, N., Lin, Q., Jin, D. (1991). *J. Mol. Catal.*, **65**, 15.
Zhang, N., Schricker, S.R., Wudl, F., Prato, M., Maggini, M., Scorrano, G. (1995). *Chem. Mater.*, **7**, 441.
Zhang, X-P., Bazan, G.C. (1994). *Macromolecules*, **27**, 4627.
Zhuang, Q., Tanaka, K., Ichikawa, M. (1990). *J. Chem. Soc., Chem. Commun.*, 1477.
Zhuang, Q., Fukuoka, A., Fujimoto, T., Tanaka, K., Ichikawa, M. (1991). *J. Chem. Soc., Chem. Commun.*, 745.
Zimmermann, M., Lehnert, G. (1976a). *Chem. Abstr.*, **85**, 78660.
Zimmermann, M., Pampus, G., Maertens, D. (1976b). *Chem. Abstr.*, **85**, 95539.
Zowade, T., Höcker, H. (1973). *Makromol. Chem.*, **165**, 31.
Zuech, E.A. (1968). *J. Chem. Soc., Chem. Commun.*, 1182.
Zuech, E.A., Hughes, W.B., Kubicek, D.H., Kittleman, E.T. (1970). *J. Am. Chem. Soc.*, **92**, 528.
Zuercher, W.J., Hashimoto, M., Grubbs, R.H. (1996). *J. Am. Chem. Soc.*, **118**, 6634.
Zümreoglu-Karan, B., Imamoglu, Y. (1991). *J. Mol. Catal.*, **65**, 329.
Zümreoglu-Karan, B., Bozkurt, Ç., Imamoglu, Y. (1992). *Polymer J.*, **24**, 25.

Index

Note. The organic compounds listed in this index are generally either the reactants or products of one of the metathesis reactions described in this book.

FSC
www.fsc.org
MIX
Papier | Fördert
gute Waldnutzung
FSC® C083411

Kolibri 360 GmbH
Ferdinand-Jühlke-Straße 7
99095 Erfurt, Deutschland
produktsicherheit@kolibri360.de